AF581629

Novel Biomarkers in Alzheimer's Disease

Novel Biomarkers in Alzheimer's Disease

Editor

Chiara Villa

MDPI • Basel • Beijing • Wuhan • Barcelona • Belgrade • Manchester • Tokyo • Cluj • Tianjin

Editor
Chiara Villa
University of Milano-Bicocca
Italy

Editorial Office
MDPI
St. Alban-Anlage 66
4052 Basel, Switzerland

This is a reprint of articles from the Special Issue published online in the open access journal *Journal of Personalized Medicine* (ISSN 2075-4426) (available at: https://www.mdpi.com/journal/jpm/special_issues/Biomarkers_Alzheimer).

For citation purposes, cite each article independently as indicated on the article page online and as indicated below:

LastName, A.A.; LastName, B.B.; LastName, C.C. Article Title. *Journal Name* **Year**, *Volume Number*, Page Range.

ISBN 978-3-03943-903-4 (Hbk)
ISBN 978-3-03943-904-1 (PDF)

Contents

About the Editor

Chiara Villa is currently an Assistant Professor in Pathology at the University of Milano-Bicocca. She started her research activity in 2006 in Department of Neurological Sciences, University of Milan, Fondazione Cà Granda, IRCCS, Ospedale Maggiore Policlinico, where she obtained her Ph.D. in Molecular Medicine in 2009. Her research activity has mainly focused on the study of novel biomarkers (e.g., microRNAs, pro-inflammatory cytokines) and genetic risk factors in key genes for the early detection and/or progression of two neurodegenerative diseases: Alzheimer's disease and frontotemporal lobar degeneration. In 2014, she moved to the University of Milano-Bicocca and joined a research group that studies the molecular bases of autism spectrum disorders and sleep disorders, including restless legs syndrome and nocturnal frontal lobe epilepsy. Her work is now carried out by applying her previous research experience in this field through a molecular approach. At the moment, she is co-author of 58 research articles published in peer-reviewed journals with a personal H-index of 20.

Editorial

Biomarkers for Alzheimer's Disease: Where Do We Stand and Where Are We Going?

Chiara Villa

School of Medicine and Surgery, University of Milano-Bicocca, 20900 Monza, Italy; chiara.villa@unimib.it

Received: 5 November 2020; Accepted: 17 November 2020; Published: 20 November 2020

Alzheimer's disease (AD) is an age-related neurodegenerative and progressive disorder representing the most common form of dementia in older adults. AD is clinically characterized by significant cognitive impairments, behavioral changes, sleep disorders, and loss of functional autonomy until the patient becomes completely dependent on the care of family members and healthcare workers [1]. As the population ages worldwide, the number of people suffering from AD is growing rapidly, making this disorder a major public health issue. Actually, the leading biomarkers in clinical practice are directed at the early identification of the two neuropathological hallmarks of AD, namely, amyloid-β (Aβ) plaques and neurofibrillary tangles (NFTs), constituted by hyper-phosphorylated paired helical filaments of the microtubule-associated protein tau. The diagnostic criteria rely on the measures of Aβ, phosphorylated (p-tau), and total tau (t-tau) protein levels in the cerebrospinal fluid (CSF) of patients aided by advanced neuroimaging methods such as magnetic resonance imaging (MRI) and positron emission tomography (PET) [2]. However, the pathological changes silently accumulate in the brain over years or even decades before the onset of symptoms. Therefore, the current challenge is the searching for novel biomarkers to optimize the early diagnosis of AD in the pre-symptomatic stages, essential to start treatments and to propose personalized therapeutic solutions to individual patients.

This Special Issue gathers six original research articles, thirteen literature reviews, one commentary, and one protocol on recent efforts toward the discovery of novel biomarker candidates exploited in different research areas, including biological fluids, genetic/epigenetic factors, pathogens, inflammation, metabolism, nutrition, obesity, or neuropsychological changes (Figure 1). It is not surprising that the most of papers are addressed to review the current knowledge about biomarkers detected in different biological fluids, which are mainly related to pathophysiological processes occurring in AD (e.g., vascular dysfunction, neuroinflammation, and synaptic and neuronal integrity). These reviews largely describe and discuss potential biomarkers detected in CSF or blood as well as in alternative non-invasive body fluids and their possible use in early diagnosis [3–7] or ongoing research protocols on AD [8]. Among them, an emerging role of flotillin as promising biomarker for AD has been proposed by some authors [9]. Moreover, to partially overcome the limitations of biological fluids, advanced brain imaging techniques provide an attractive alternative for the identification of AD-related structural and functional biomarkers [10]. Integrated datasets of multi-faceted AD biomarkers and data-driven analytical methodologies may be involved in the application of the "precision medicine", aimed to unravel many aspects of AD heterogeneity and to expand the current treatment strategies to help guide more effective diagnosis and clinical management of the disorder [11].

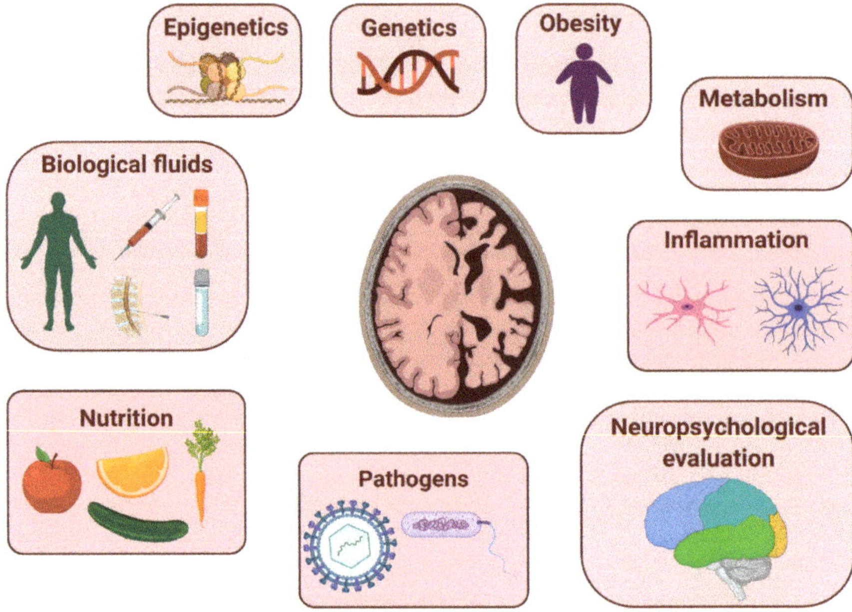

Figure 1. An overview of different research field for exploring potential biomarkers for Alzheimer's disease. This figure was created with the support of BioRender.com.

Given the central role of genetics in the development of AD, some authors reviewed emerging candidate genes for familial AD, as well as inherited risk factors, in order to improve the prognostic identification and management of the at-risk individuals. A better knowledge of these genes and their correlated molecular defects will further provide potential targets for the treatment of the disease [12]. One study has reported original results on the association between AD-related polymorphisms and cardiovascular risk factors, which influence the progression to the disorder. Therefore, understanding the molecular mechanisms of this interaction could allow the development of new personalized therapeutic approaches for treating AD [13]. Focusing on the occurrence of behavioral and psychological symptoms of dementia in AD (BPSD), other authors have found an interesting association between *APOE* and *MTHFR* genetic variants and BPSD, expanding the knowledge about the BPSD etiopathogenetic mechanisms, which in turn, leads not only improve the clinical/diagnostic assessment, but also to better definite suitable treatments [14].

As an early event in the pathogenesis of AD, some authors speculated that chronic inflammation should be considered as a potential biomarker in the treatment strategies for AD. Interestingly, inflammation is emerging as the central mechanistic link among diabetes, obesity, and cognitive decline in patients affected by AD. These authors discuss how diabetes and obesity could lead to both systemic and neuro-inflammation, hypothesizing an association with impaired mitochondrial health [15]. Indeed, AD has also been suggested as a metabolic disorder, owing to the fact that some genetic risk factors are key mediators in different metabolic pathways, including glucose, lipid, and energetic metabolism [16]. In this regard, Bell and collaborators demonstrate the strong correlation between fibroblast mitochondrial abnormalities and neuropsychological markers, suggesting the use of fibroblast metabolic assessment as an emergent biomarker of AD [17]. Similarly, another study reports that brain metabolism evaluated by 18F fluorodeoxyglucose (18F-FDG) uptake is moderately related to various neuropsychological tests [18]. Moreover, some authors conceived the "development of metabolic and functional markers of dementia in older people" (ODINO) protocol as an innovative multi-dimensional investigation in which clinical, functional, neuropsychological, and biological parameters are coupled with advanced statistical analyses in order to better identify possible biomarkers that can predict the conversion from mild cognitive impairment (MCI), the prodromal stage of dementia, to AD [19].

Among individuals with MCI, two additional papers reported original results. The randomized cognitive impairment study (CARES) clinical trial demonstrated that targeted nutritional intervention with ω-3FAs, carotenoids, and vitamin E significantly improves the cognitive performances [20]. Other authors showed that an experimental assessment of semantic priming in MCI seems to represent a good paradigm to evaluate subclinical impairment of the semantic system in the early stages of the AD pathology [21]. Finally, an outstanding review discussed how neurophysiological techniques, evaluating mechanisms of synaptic function and brain connectivity, may represent valid biomarkers for screening MCI individuals by the application of artificial intelligence (i.e., learning machine) [22].

Based on studies linking different pathogens with AD and age-related cognitive decline, Naughton and collaborators discuss an interesting role of pathogen-associated biomarkers as a novel tool for evaluating and decreasing AD risk across the population [23].

In conclusion, all articles appearing in this Special Issue cover attractive and current topics of a wide range of biomarkers in the basic research, clinical diagnosis, prognosis, and therapeutic strategies of AD, the most common form of neurodegenerative disorder and a major health challenge with significant social and economic consequences. Early diagnosis entailing the ability to detect AD in asymptomatic patients still remains a big challenge. Therefore, implementing a combination of the aforementioned biomarkers into a diagnostic setting may likely allow the identification of at-risk patients during pre-symptomatic stages necessary to start treatments and to suggest personalized therapeutic strategies.

Funding: The research received no external funding.

Acknowledgments: I am very grateful to all authors who have provided excellent contributions to this Special Issue. Moreover, I would like to thank *Journal of Personalized Medicine* for offering me the opportunity to make this Special Issue a reality, and in particular, Crystal Feng for her availability, professionalism, help, constant support, and kindness. Finally, I would like to acknowledge the excellent and efficient work of the expert reviewers who reviewed submissions in a timely, fair, and constructive manner.

Conflicts of Interest: The author declares no conflict of interest.

References

1. Villa, C.; Ferini-Strambi, L.; Combi, R. The Synergistic Relationship between Alzheimer's Disease and Sleep Disorders: An Update. *J. Alzheimers Dis.* **2015**, *46*, 571–580. [CrossRef] [PubMed]
2. Jack, C.R., Jr.; Bennett, D.A.; Blennow, K.; Carrillo, M.C.; Dunn, B.; Haeberlein, S.B.; Holtzman, D.M.; Jagust, W.; Jessen, F.; Karlawish, J.; et al. NIA-AA Research Framework: Toward a biological definition of Alzheimer's disease. *Alzheimers Dement.* **2018**, *14*, 535–562. [CrossRef] [PubMed]
3. Omar, S.H.; Preddy, J. Advantages and Pitfalls in Fluid Biomarkers for Diagnosis of Alzheimer's Disease. *J. Pers. Med.* **2020**, *10*, 63. [CrossRef]
4. Zou, K.; Abdullah, M.; Michikawa, M. Current biomarkers for Alzheimer's disease: From CSF to blood. *J. Pers. Med.* **2020**, *10*, 85. [CrossRef]
5. Ausó, E.; Gómez-Vicente, V.; Esquiva, G. Biomarkers for Alzheimer's disease early diagnosis. *J. Pers. Med.* **2020**, *10*, 114. [CrossRef]
6. D'Abramo, C.; D'Adamio, L.; Giliberto, L. Significance of blood and CSF biomarkers for Alzheimer's disease: Use and specificity. *J. Pers. Med.* **2020**, *10*, 116. [CrossRef]
7. Del Prete, E.; Beatino, M.F.; Campese, N.; Giampietri, L.; Siciliano, G.; Ceravolo, R.; Baldacci, F. Fluid candidate biomarkers for Alzheimer's Disease: A precision medicine approach. *J. Pers. Med.* **2020**, *10*, 221. [CrossRef]
8. Canevelli, M.; Remoli, G.; Bacigalupo, I.; Valletta, M.; Toccaceli Blasi, M.; Sciancalepore, F.; Bruno, G.; Cesari, M.; Vanacore, N. Use of Biomarkers in Ongoing Research Protocols on Alzheimer's Disease. *J. Pers. Med.* **2020**, *10*, 68. [CrossRef]
9. Angelopoulou, E.; Paudel, Y.N.; Shaikh, M.F.; Piperi, C. Flotillin: A Promising Biomarker for Alzheimer's Disease. *J. Pers. Med.* **2020**, *10*, 20. [CrossRef]

10. Villa, C.; Lavitrano, M.; Salvatore, E.; Combi, R. Molecular and Imaging Biomarkers in Alzheimer's Disease: A Focus on Recent Insights. *J. Pers. Med.* **2020**, *10*, 61. [CrossRef]
11. Lukiw, W.J.; Vergallo, A.; Lista, S.; Hampel, H.; Zhao, Y. Biomarkers for Alzheimer's disease (AD) and the application of Precision Medicine. *J. Pers. Med.* **2020**, *10*, 138. [CrossRef] [PubMed]
12. D'Argenio, V.; Sartanaro, D. New insights into the molecular bases of familial Alzheimer's disease. *J. Pers. Med.* **2020**, *10*, 26. [CrossRef] [PubMed]
13. Bessi, V.; Balestrini, J.; Bagnoli, S.; Mazzeo, S.; Giacomucci, G.; Padiglioni, S.; Piaceri, I.; Carraro, M.; Ferrari, C.; Bracco, L.; et al. Influence of ApoE Genotype and Clock T3111C Interaction with Cardiovascular Risk Factors on the Progression to Alzheimer's Disease in Subjective Cognitive Decline and Mild Cognitive Impairment Patients. *J. Pers. Med.* **2020**, *10*, 45. [CrossRef] [PubMed]
14. Scassellati, C.; Ciani, M.; Maj, C.; Geroldi, C.; Zanetti, O.; Gennarelli, M.; Bonvicini, C. Behavioural and Psychological Symptoms of Dementia (BPSD): Clinical characterization and genetic correlates in an Italian Alzheimer Disease cohort. *J. Pers. Med.* **2020**, *10*, 90. [CrossRef] [PubMed]
15. Khan, M.S.H.; Hegde, V. Obesity and Diabetes Mediated Chronic Inflammation: A Potential Biomarker in Alzheimer's Disease. *J. Pers. Med.* **2020**, *10*, 42. [CrossRef] [PubMed]
16. Argentati, C.; Tortorella, I.; Bazzucchi, M.; Emiliani, C.; Morena, M.; Martino, S. The other side of Alzheimer's Disease: Influence of metabolic disorder features for novel diagnostic biomarkers. *J. Pers. Med.* **2020**, *10*, 115. [CrossRef]
17. Bell, S.M.; De Marco, M.; Barnes, K.; Shaw, P.J.; Ferraiuolo, L.; Blackburn, D.J.; Mortiboys, H.; Venneri, A. Deficits in Mitochondrial Spare Respiratory Capacity Contribute to the Neuropsychological Changes of Alzheimer's Disease. *J. Pers. Med.* **2020**, *10*, 32. [CrossRef]
18. Chiaravalloti, A.; Ricci, M.; Di Biagio, D.; Filippi, L.; Martorana, A.; Schillaci, O. The Brain Metabolic Correlates of the Main Indices of Neuropsychological Assessment in Alzheimer's Disease. *J. Pers. Med.* **2020**, *10*, 25. [CrossRef]
19. Picca, A.; Ronconi, D.; Coelho-Junior, H.J.; Calvani, R.; Marini, F.; Biancolillo, A.; Gervasoni, J.; Primiano, A.; Pais, C.; Meloni, E.; et al. The "develOpment of metabolic and functional markers of Dementia IN Older people" (ODINO) Study: Rationale, Design and Methods. *J. Pers. Med.* **2020**, *10*, 22. [CrossRef]
20. Power, R.; Nolan, J.M.; Prado-Cabrero, A.; Coen, R.; Roche, W.; Power, T.; Howard, A.N.; Mulcahy, R. Targeted Nutritional Intervention for Patients with Mild Cognitive Impairment: The Cognitive impAiRmEnt Study (CARES) Trial 1. *J. Pers. Med.* **2020**, *10*, 43. [CrossRef]
21. Guglielmi, V.; Quaranta, D.; Mega, I.; Costantini, E.M.; Carrarini, C.; Innocenti, A.; Marra, C. Semantic Priming in Mild Cognitive Impairment and Healthy Subjects: Effect of Different Time of Presentation of Word-Pairs. *J. Pers. Med.* **2020**, *10*, 57. [CrossRef] [PubMed]
22. Rossini, P.M.; Miraglia, F.; Alù, F.; Cotelli, M.; Ferreri, F.; Iorio, R.D.; Iodice, F.; Vecchio, F. Neurophysiological Hallmarks of Neurodegenerative Cognitive Decline: The Study of Brain Connectivity as a Biomarker of Early Dementia. *J. Pers. Med.* **2020**, *10*, 34. [CrossRef] [PubMed]
23. Naughton, S.X.; Raval, U.; Pasinetti, G.M. The Viral Hypothesis in Alzheimer's Disease: Novel Insights and Pathogen-Based Biomarkers. *J. Pers. Med.* **2020**, *10*, 74. [CrossRef] [PubMed]

Review

Advantages and Pitfalls in Fluid Biomarkers for Diagnosis of Alzheimer's Disease

Syed Haris Omar [1,2,*] and John Preddy [1,2]

1 Rural Clinical School, Faculty of Medicine, University of New South Wales, Wagga Wagga, NSW 2650, Australia; j.preddy@unsw.edu.au
2 Murrumbidgee Local Health District, NSW Health, Wagga Wagga, NSW 2650, Australia
* Correspondence: haris.omar@unsw.edu.au or syedharisomar@gmail.com

Received: 8 June 2020; Accepted: 6 July 2020; Published: 17 July 2020

Abstract: Alzheimer's disease (AD) is a commonly occurring neurodegenerative disease in the advanced-age population, with a doubling of prevalence for each 5 years of age above 60 years. In the past two decades, there has been a sustained effort to find suitable biomarkers that may not only aide with the diagnosis of AD early in the disease process but also predict the onset of the disease in asymptomatic individuals. Current diagnostic evidence is supportive of some biomarker candidates isolated from cerebrospinal fluid (CSF), including amyloid beta peptide (Aβ), total tau (*t*-tau), and phosphorylated tau (p-tau) as being involved in the pathophysiology of AD. However, there are a few biomarkers that have been shown to be helpful, such as proteomic, inflammatory, oral, ocular and olfactory in the early detection of AD, especially in the individuals with mild cognitive impairment (MCI). To date, biomarkers are collected through invasive techniques, especially CSF from lumbar puncture; however, non-invasive (radio imaging) methods are used in practice to diagnose AD. In order to reduce invasive testing on the patients, present literature has highlighted the potential importance of biomarkers in blood to assist with diagnosing AD.

Keywords: Alzheimer's disease; cerebrospinal fluid; amyloid beta peptide; total tau; phosphorylated tau; diagnosis

---

1. Introduction

Alzheimer's disease (AD) is one of the most common neurodegenerative disease in an ageing population. AD is the most common cause of dementia and is characterized by cognitive impairment and the impedance of daily activities, including communication, decision making and behavioral changes [1]. It has been shown that the frequency of AD doubles for each 5 years of life above the age of sixty years. It is predicted that by 2050, 130 million globally will be symptomatic from AD [2]. In terms of risk factors, advanced age is the most important risk for the sporadic or late onset of AD as well as the presence of APOE e4 alleles. Inherited mutations in chromosome 11 amyloid precursor protein (APP), Presenilin-1 ($PSEN_1$), and Presenilin-2 ($PSEN_2$) are prevalent in the less common familial form of AD. In addition, women are more prone to AD compared with men, early menopause is also risk factor for AD. Cardiovascular disease and diabetes mellitus type-2 are associated with an increased risk of AD. The exact pathophysiology of AD is still under investigation; however, the deposition of senile plaques, neurofibrillary tangles (NFTs), and astrogliosis are cardinal features [3]. Moreover, studies have shown that pathological involvement of oxidative stress, neuron degeneration induced synaptic alteration, inflammation and microgliosis are important in the pathogenesis of AD [4]. Despite almost 3 decades of research into the exact molecular mechanism causing AD, unfortunately, none of the hypothesis completely answers the question. The still amyloid cascade hypothesis suggests a core pathological role of amyloid beta in AD [5]. The presence of Aβ peptides in cerebral and peripheral tissues mainly consists of amino acids and their sequences ranging from 1 to 43. $A\beta_{42}$ is very prone to aggregate and proceed to form the senile plaques found in hippocampus, neocortex and in the cerebrovasculature region [6]. Another highly aggregated peptide called tau (which undergoes extensive hyperphosphorylation) is responsible for the formation of neurofibrillary tangles inside neurons and ultimately results in extensive brain and nerve damage [7]. Currently, approved drugs only provided symptomatic relief for patients with AD without modifying the disease or slowing disease progression. However, for the treatment of mild cognitive impairment (MCI), there is no FDA-approved drugs available and suggested to consider off-label treatment, such as an acetylcholinesterase (AChE) inhibitor, which has provided a modest impact but is also associated with the risk of side effects. In order to reduce the side effects, research has been undertaken to modify the chemical moiety of drugs with compatible substitutes and also focused on natural products with the potential to act as disease modifying agents [8,9]. Several natural products including curcumin, ginkgolides, resveratrol, oleuropein etc. have been shown to be effective against AD pathology in vitro or in vivo models but have not shown success in randomized trails [10–13]. Lifestyle modification, including exercise and dietary modification, especially the Mediterranean diet (MedDi) and Mediterranean-Dietary Approaches to Stop Hypertension (DASH) diet Intervention for Neurological Delay (MIND) diet, have been associated with improved cognition among elderly subjects [14].

It has been shown that pathological changes of AD occur long before the appearance of clinical symptoms. Therefore, it is important to establish a diagnosis as early as possible especially for people above the age 60 years. Biomarkers offer essential tools for AD diagnosis, monitoring, early detection, therapeutic intervention, as well as prevention of inaccurate diagnoses. Body fluid biomarkers in cerebrospinal fluid (CSF) and blood have shown potential for AD diagnosis, individual prognosis and patient stratification. Despite the availability of numerous theoretical and clinical diagnostic tools, AD is still poor diagnosed, especially in the early stage of the disease. AD has a prolonged pre-symptomatic prodromal phase; however, the lack of specific biomarker, procedural and methodological inconsistencies, inconsistent cut-off values as well as a lack of assay standardization, have thwarted attempts to establish a diagnosis and treat AD during this early phase.

2. Search Methods

Potential studies were identified in electronic database PubMed, Embase, ScienceDirect, Cochrane Library, SpringerLink, Scopus and Google Scholar using combination of following keywords "Alzheimer's Disease", "biomarkers" and "Alzheimer's disease", "cerebrospinal fluid", "CSF", "invasive biomarkers", "non-invasive biomarkers", "plasma biomarkers", "blood biomarkers", "plasma amyloid", "plasma tau", "inflammatory biomarkers", "imaging biomarkers", "proteomic biomarkers", "salivary biomarkers", "olfactory biomarkers" and "ocular biomarkers". Selected studies published between 1990 and May 2020 were included to ensure that all randomized trial, pilot studies, and critical reviews or systematic reviews published evidence on potential Alzheimer's disease biomarkers for three decades were encompassed. The preclinical studies, in vitro studies, published media, as well as duplicate articles were excluded due to being outside the scope of the clinical study aim.

3. Biomarkers in Alzheimer's Disease

A biomarker is usually characterized by substances (synthetic molecules, specified cells, proteins, enzymes, hormones or genetic material) or imaging finding, which is used as a metric characteristic to indicate the presence of a specific physiological state and may assist with establishing a diagnosis well before a clinical diagnosis can be made. Furthermore, the use of biomarkers is increasingly for assisting with the prognosis and diagnosis of AD, reflected by a tremendous increase in research from 1980 to current time (Figure 1).

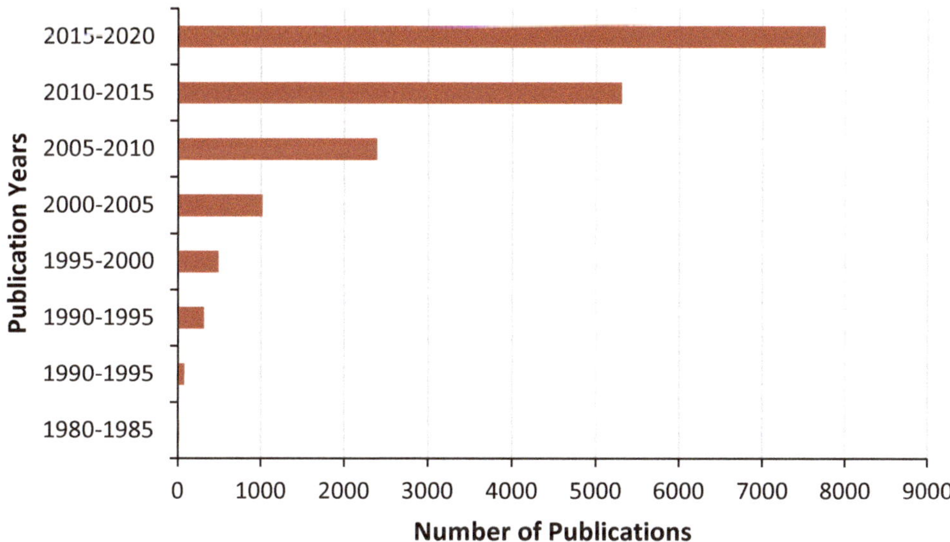

Figure 1. Publications statistics for "Biomarkers in Alzheimer's disease", source PubMed.

On the basis of AD pathogenesis and clinical condition, a set of diagnostic criteria were established in 1984, which was updated by the National Institute on Aging and Alzheimer's Association (NIA-AA) [15]. The updated NIA-AA guideline was mainly based upon the pathophysiological advancement in clinical, imaging, and research technologies in AD. Similarly, based upon clinical *probable*, *possible*, or *definite* symptoms, National Institute on Neurological and Communicative Disorder and Stroke and the Alzheimer's Disease and Related Disorders Association (NINCDS-ADRDA) have also published a diagnostic criteria for AD [16]. The clinical conditions of AD are considered to fall into three stages; however, some studies have expanded this to 1–5 or 1–7 stages. Of all the stages of AD, the prodromal period has the longest duration. This has resulted in a revision of NIA-AA diagnostic criteria, which are mainly based upon the identification of biomarkers, including CSF and imaging as valid diagnostic tools [17]. Based on modern diagnosis criteria, three sets of biomarkers are used as diagnostic tools, including Aβ peptide (A), tau peptide (T) and neurodegeneration (N), which are classified as A/T/N framework (Table 1) of AD diagnosis [17].

Table 1. Biomarkers based upon National Institute on Aging and Alzheimer's Association (NIA-AA) classification [17].

NIA-AA Classification	Alzheimer's Disease Biomarkers	Biomarkers Significance in AD
Amyloid (A) aggregates	CSF $A\beta_{42}$, $A\beta_{42}/A\beta_{40}$ ratio & Amyloid PET	↓ CSF $A\beta_{42}$ & $A\beta_{42}/A\beta_{40}$
Tau (T) aggregates	CSF phosphorylated tau & Tau PET	↑ CSF p-tau
Neurodegeneration (N)	CSF total tau & Anatomic MRI FDG PET	↑ t-tau

NIA-AA: National Institute on Aging and Alzheimer's Association; ↑: increase; ↓: decrease; CSF: cerebrospinal fluid; Aβ: β-amyloid; PET: positron emission tomography; FDG: fluorodeoxyglucose; MRI: magnetic resonance imaging.

4. Biomarkers Based upon Alzheimer's Disease Stages

Stage 1: assigned to the individuals who do not have functional impairment but might have cognitive impairment, which can only be detected through neuropsychological sensitive instruments. There is increasing evidence that certain biomarkers can predict the pathological changes at an early preclinical phase, namely the presence of amyloid imaging and a reduced CSF $A\beta_{42}$ concentration.

Early diagnosis based on biomarkers may assist with the approval of AD treatment, which could provide clinical benefits and improve outcomes [18]. Further trials are required to evaluate the reliability of clinical measurement and access the potential improvement with the drug-placebo conclusion.

Stage 2: Presence of biomarkers that predict pathophysiological changes of AD, a subtle cognitive effect, but no functional deficits in the patients, which can be detected with the use of sensitive instruments; however, they do not fulfil the criteria for dementia. According to the FDA guidance, sensitive neuropsychological testing should be considered alongside biomarker changes to diagnose AD stage 2 [18].

Stage 3: Pathophysiological biomarkers are present, and patients have started to have difficulty in doing some daily tasks which are measurable. This stage of the disease corresponds with mild cognitive impairment, whereas the first two stages are preclinical.

Stage 4, Stage 5 and Stage 6: Pathophysiological biomarkers are present with the consecutive stages of mild, moderate, and severe AD dementia with worsening cognitive impairment.

Assessment of Stages in AD

From the FDA classification of stages in Alzheimer's disease, stage 1 and 2 should be considered critical and monitored seriously. However, from stage 3 onwards, AD patients have similar pathophysiological biomarkers and ongoing cognitive decline. There are two basic questions that stem from the stages of AD. Which biomarkers may predict the presence of stage 1 and stage 2 AD in individuals? Secondly, from the treatment perspective, how can we establish the clinical effect of current FDA-approved drugs for patients with stage-1 AD, which is preclinical (based on the presence of amyloid and reduced CSF $A\beta_{42}$ concentration) without evidence of cognitive decline? There is a need to evaluate predictive biomarkers and establish whether changes in biomarkers is a predictor of treatment success.

5. Biomarkers through Invasive Diagnostic Methods

5.1. Cerebrospinal Fluid Biomarkers

Cerebrospinal fluid (CSF) is a clear liquid that is present in the subarachnoid space and ventricular system of the brain and spinal cord. The volume of CSF in the body varies between 125 and 150 mL. The composition of CSF can demonstrate minor biochemical change in the brain. Currently, CSF is considered an excellent biologic fluid that may contain potential biomarkers for AD, which may be able to identify without going through autopsy or biopsy. Furthermore, the presence and concentration of biomarkers may change in parallel to AD progression. The three most suggestive biomarkers in AD are $A\beta$, total tau (t-tau), and phosphorylated tau (p-tau) (Table 2). It has been suggested that CSF biomarkers did not vary with severity with stable levels noted in the follow-up patients with clinical AD [19].

Table 2. Established diagnostic biomarkers in the cerebrospinal fluid CSF of Alzheimer's disease (AD) [a] [20] and showed 85% sensitivity cutoff values for AD dementia diagnosis [21].

Biomarkers	Controls (pg/mL)	AD (pg/mL)	% Sensitivity (AD-Control)	% Sensitivity (MCI-Control)
$A\beta_{42}$	794 ± 20	<500 *	73 (≥75 years)	60 (≥75 years)
tau peptide	136 ± 89 (21–50 years)	[b]	74 (≤64 years)	65 (≤64 years)
	243 ± 127 (51–70 years)	>450	53 (65–74 years)	49 (65–74 years)
	341 ± 171 (>71 years)	>600 *	61 (≥75 years)	46 (≥75 years)
p-tau-181	23 ± 2	>60	37 (≥75 years)	30 (≥75 years)

[a] Data obtained using innogenetics single 96-well ELISA kits. [b] This is not relevant for sporadic AD, because it is only for patients >60 years of age. * $p < 0.001$.

5.1.1. CSF Aβ Biomarker

The amyloid isoforms $A\beta_{40}$ and $A\beta_{42}$ concentration in CSF are considered to be the most dependable biomarkers for the diagnosis of the AD disease. The production of both amyloid isoforms $A\beta_{40}$ and $A\beta_{42}$ was 24% higher for mutation carriers than noncarriers in the autosomal dominant AD patients. However, it was suggested that the fractional turnover rate of $A\beta_{42}$ was noted 65% higher in mutation carriers [22]. Interestingly, it was also reported that there is no change in CSF $A\beta_{40}$, while it is present in a 10-fold higher concentration than $A\beta_{42}$ in the CSF of AD patients. It was suggested that $A\beta_{42}$ be used as a proxy of total Aβ concentration. The amyloidogenic protein is found throughout the human body, and studies showed that $A\beta_{42}$ concentrations in CSF often correlate with Aβ levels in the patient's brain [23]. It was found that the $A\beta_{42}$ concentration was significantly reduced in CSF, which is a consequence of its presence in fibrils and plaques in the brains of patients with AD [24–26]. There are variations in quantification; however, it was found that $A\beta_{42}$ concentration declined by 50% in CSF of patients with AD as a result of its deposition in the brain parenchyma [27].

The underlying mechanisms of the reduction CSF $A\beta_{42}$ is not clear; however, few studies have suggested that it is due to the excessive hydrophobic aggregation of $A\beta_{42}$ sequestration in plaques, a reduction in its diffusion from interstitial fluid to CSF and/or decreased Aβ clearance as a consequence of an impaired blood–brain barrier [28,29]. It was reported that the other isoform of amyloid peptides, $A\beta_{38}$, was also found to have an increased concentration in CSF. Furthermore, the ratio of $A\beta_{42}/A\beta_{38}$ closely corresponds with imaging findings in patients with AD and thus results in a robust biomarkers for AD pathogenesis, which is more useful than the concentration of $A\beta_{42}$ alone in CSF [30,31]. In contrast, several studies have reported that the $A\beta_{40}$ concentration was unchanged in CSF from patients with AD and does not correlate with amyloid deposits in the brain [32]. In spite of the discrepancy in the diagnosis of CSF $A\beta_{40}$ levels, several studies supported a decrease CSF $A\beta_{42/40}$ ratio in the diagnosis of MCI patients compared to controls [33]. Because of the observed Aβ isoforms ratio and their positive relationship with AD pathogenesis, the NIA-AA has accepted $A\beta_{42}$ concentration as well as a comparative ratio of $A\beta_{42}/A\beta_{40}$ as important biomarkers in the diagnostic guideline for AD [17].

5.1.2. Assessment of CSF $A\beta_{42}$ Biomarker

CSF $A\beta_{42}$ biomarkers support a diagnosis of AD in its preclinical stage and are predictors of disease progression in cognitively unimpaired individuals and in those with MCI. One of the main limitations of CSF sampling is its invasive collection technique i.e., through lumbar puncture compared with blood sampling. Post lumbar puncturing, headache as most common adverse effect.

Most of the studies showed a significant decline in $A\beta_{42}$ levels as diagnostic biomarker and agreed that upto 40% reduction was observed in AD patients when compared with those of healthy individuals [34]. In contrast to the reduction of $A\beta_{42}$ concentration in CSF, some of the past studies have shown an unchanged $A\beta_{42}$ concentration [35] and an elevated level [36] in CSF $A\beta_{42}$ concentration compared to AD patients and healthy controls. CSF $A\beta_{42}$ was found to reach the plateau state early in the disease progression and produce a conflicting outcome, which demonstrates a process of preceding aggregation of Aβ mainly detected with amyloid PET analysis. There is a lack of standard protocol and universal agreement because of the varying biomarker concentrations and contradictory outcomes, which required further investigation in the age- and stage-matched individuals considering prodromal stage individuals with different ethnic groups. It has been demonstrated that CSF $A\beta_{42/40}$ ratio may predict abnormal cortical amyloid deposition (visualized with PET) compared with CSF $A\beta_{42}$. However, this diagnosis could result as false positive (low CSF $A\beta_{42}$) or false negative (high CSF $A\beta_{42}$) in fewer patients [25]. Further studies have been reported the presence of oligomers formation prior to the formation of Aβ fibrils in the pathogenesis of AD and suggested oligomers as potential early target in the prodromal stages, which required further confirmation in randomized trial as early diagnostic biomarker.

5.1.3. CSF Tau Biomarker

The Aβ and tau peptides have been suggested to interact mutually and prompt both aggregation and toxicity followed by proposed mechanisms including Aβ encourage tau pathology or tau induces Aβ toxicity, or synergistic toxicity exists between Aβ and tau [37,38]. Basically, tau is a microtubule associated protein have a pivotal action in intracellular transportation. Tau proteins classified as p-tau representing hyperphosphorylation, and t-tau representing several isomers of the tau protein. It was suggested that the hyperphosphorylation occur at threonine-231 (p-tau$_{231}$), threonine-181 (p-tau$_{181}$), and at serine-199 (p-tau$_{199}$). The involvement of p-tau in the assembly of neurofibrillary tangles represented as 'T' marker, and their presence in the CSF proving a sign of neuronal death as 'N' marker [39,40]. It was reported that CSF p-tau$_{231}$ was involved in the neurofibrillary neocortical pathology [40], and showed a significant increase in concentration correlated with a decline in cognitive performance and conversion to AD [41]. The concentration of t-tau in CSF was found to be highly age dependent and observed <300 pg/mL in 21–50 years, <450 pg/mL in 51–70 years, and <500 ng/L in 71–93 years age group of normal cognitive healthy individuals [42]. Several lines of studies have supported the results of significant rise in CSF tau peptide concentration in AD patients [39,43]. In particular, p-tau and t-tau were found to be increased by 200% 300% concentration in AD compared with nondemented elderly subjects [27,44]. Tau pathology cause an elevated level of CSF t-tau and p-tau and strongly associated with cognitive decline compared to the amyloid pathology.

Furthermore, the degree of neurodegeneration and neuronal/axonal damage in AD patients' brains marked by the presence of considerable CSF t-tau concentrations and constituted in the A,T,N Framework as a marker of 'N' [17,45,46]. A systematic review included 15 studies showed the presence and accuracy of CSF t-tau in seven studies, while six studies have showed the presence and accuracy of the CSF p-tau in mild cognitive impaired patients [47].

The intermediate filaments known as neurofilament light (NfL) were found to be present in the axons cytoplasm and may interfere with cytoplasmic function of axonal homeostasis as well as synaptic transmission. Studies have shown the presence of elevated NfL in AD patients and suggested end results of neuronal and axon damages [48]. More pronouncedly, the increase level of NfL could serve as a risk factor for MCI. However, elevated NfL levels were also recognized as a biomarker in other neurodegenerative diseases having marked axonal degeneration, white matter injury, or both, such as frontotemporal dementia, amyotrophic lateral sclerosis, Creutzfeldt–Jakob disease, multiple sclerosis and traumatic brain injury. A few post-mortem studies have reported significant elevated NfL levels in patients with amyotrophic lateral sclerosis and frontotemporal dementia more than the AD patients and suggested that NfL could be used for differentiation of two types of dementia [49]. Due to non-specificity in disease diagnosis, NfL is less popular as confirmatory biomarkers compared to Aβ and tau in AD.

5.1.4. Assessment of CSF Tau Biomarker

The increase in both CSF t-tau and p-tau concentrations are well settled in AD compared with controls, specifically the intensity of neuronal injury and neurodegeneration are indicated by CSF t-tau in AD. The concentration of CSF t-tau was reported two to three folds higher in patients with AD compared to the normal controls [34]. Still, it is not confirmed about the agreement in distinct tau phosphorylation sites for AD. Most studies have showed the significant rise in tau concentration with aging as well as in patients with AD; however, few studies have reported contradictory information and showed no significant change in the CSF tau level in normal healthy aged individuals [50]. It was postulated that the incidence of tau phosphorylation and the building of neurofibrillary tangles inside neuron is a results of cellular protective mechanism against oxidative stress and suggesting normal physiological pathway rather than a toxic pathway [51]. It is still unknown about the inconsistent results for tau analysis and required to develop a new technique for consistent outcome. It has been suggested that, due to the presence of heterogeneity in trial subjects, the consistency of the result varies and not being reproducible. Thus, there is a requirement to agree on one productive model for the diagnosis of CSF tau biomarkers and run the trial in a large cohort that can be reproducible or verify in a repetitive/confirmatory trial.

5.2. Blood Biomarkers in AD

Blood is a most commonly accessible biological sample than other body fluids such as CSF and offer inexpensive clinical diagnosis or screening methods and even convenient for getting reproducible results in clinical trials. In cardiovascular disease and cancer diagnosis and research, biofluid blood has been established as biomarkers; therefore, it may perform as a critical measure in the early diagnosis of AD [52]. Due to the presence of Aβ in the prodromal stage of AD and their ability to pass through the blood–brain barrier, for diagnostic purposes, Aβ received a considerable amount of attention as a potential blood biomarker.

5.2.1. Plasma Aβ Biomarker

In order to evaluate and understand amyloid clearance, studies reported a significant decline in amyloid clearance using the stable isotope labelling kinetic method in late-onset AD patient [53]. Both $A\beta_{40}$ and $A\beta_{42}$ production were found to be increased by 24% in individuals with mutation carriers than noncarriers autosomal dominant form of AD (ADAD), and, furthermore, the $A\beta_{42}$ fractional turnover rate was 65% faster deposition in mutation carriers individuals [22]. Several studies have reported the existence of Aβ in blood plasma and showed the rise of both $A\beta_{40}$ [54] and $A\beta_{42}$ [55] levels. In contrast, studies have also reported a fall in both $A\beta_{40}$ [56,57] and $A\beta_{42}$ [58] concentrations in individuals susceptible to AD. More recently, a comprehensive meta-analysis study showed the inconsistency in plasma for both $A\beta_{42}$ and $A\beta_{40}$ in AD [59]. In order to achieve consistency and accuracy in the biomarker analysis, trials based on immunoprecipitation-mass-spectrometry-based assays for evaluation showed a significant decline in both $A\beta_{40}$ and $A\beta_{42}$ concentration in plasma $A\beta_{42}/A\beta_{40}$ ratio (in the line with CSF test) with approximately 90% of diagnostic accuracy [60,61].

5.2.2. Assessment of Plasma Aβ Biomarkers in AD

Blood considered as highly complex fluid connective tissue containing cellular components and several compounds, such as proteinases nature compounds, genetic materials, and metabolites, appear in plasma. The primary barrier for inconsistency in the biomarkers analysis results was suggested due to the presence of low blood Aβ concentration as well as victims of matrix effects. In addition, a lack in assay sensitivity, specificity and methods selectivity are responsible for inconsistency in Aβ finding as biomarkers in blood. In general, biomarkers localized in the brain are not easily available in blood because of the restriction of movement through the blood–brain barrier, poor expression of AD pathology biomarkers in blood and the interference of blood containing endogenous antibodies with the assay reagents that finally resulted in a false rise and fall of measurement. There is a need to develop the analytical sensitive plasma-based assay, which can minimize the event of reaction of biomarkers such as Aβ with the reagent, and careful validation work. A few attempts have been made to analyze Aβ in blood through a new diagnostic technique; however, such attempts were unable to resolve the cerebral expression of Aβ including plasma protein and blood platelets [62] but still represent an important step forward.

5.2.3. Plasma Tau Biomarker

Studies have been reported that the elevated plasma tau levels but with overlapping ranges of results across diagnostic groups of AD patients compared with the normal control [63]. It has been suggested that the plasma tau is a late-stage marker of AD and did not show any change in the plasma of the MCI-stage individual followed by missing the interrelationship between tau levels in plasma and CSF due to the differential regulation of tau in both fluids [63]. Further study showed a positive associations between increased plasma tau level and AD hallmarks [64]. The elevated level of plasma total tau and $pTau_{181}$ were investigated in patients with dementia compared to the cognitively unimpaired individuals [65]. In general, total tau protein concentration was found to be approximately seven times higher than p-tau_{181} in human plasma. Based on immunomagnetic reduction technique, a study evaluated the concentration ratios of p-tau_{181} to t-tau in plasma are 14.4% for healthy controls, 13.6% for patients with MCI due to AD, and 19.5% for very mild AD, respectively, and suggested that p-tau_{181} in plasma can be used to differentiate memory disorder/cognitive decline in early-stage AD patients [66]. Further study based on the evaluation of plasma p-tau_{181} as a biomarker using ultrasensitive immunoassay methods showed a significantly higher plasma p-tau_{181} level in the AD group compared with the control group [67]. A recent study measured plasma-phosphorylated tau concentration and found a significantly higher concentration in the AD group compared with age-matched cognitively normal controls [68]. Tau hyperphosphorylation-induced neuronal damage was also investigated and suggested the presence of neurofilament as biomarker for neurodegeneration in AD [69]. Neurofilament was measured by using an ultrasensitive immunoassay method and showed increased serum neurofilament light (NfL) concentration in familial AD prior to symptomatic disease [70]. In the line of the previous report, a recent study outcome showed an early rise of serum neurofilament light in the presymptomatic phase of familial AD and continue to increase in neurofilament light level, respectively [71]. Increased plasma neurofilament light was found in MCI and AD dementia patients compared with controls and correlated with poor cognition [72].

NfL in plasma well determined, reported their elevated levels in the serum of familial AD patients usually appear a decade ahead to the onset of symptom and well correlated with whole-brain atrophy intensity through MRI and an assessment of cognition [73]. In addition, a high level of NfL was determined in plasma of MCI compared to AD patients and healthy controls, which can be used as a determinant assay to easily distinguish individual between MCI and AD [72].

5.2.4. Assessment of Plasma Tau Biomarkers in AD

In the past two decades, it is still in debate that tau pathology starts early in normal cognitive individuals upon ageing (>60 years) than Aβ pathology. However, research based on tau pathophysiology have showed high consistency in the increase level of tau protein (neurofilament light, p-tau_{181}, and total tau as biomarkers) and easily detected in serum. In contrast, Aβ ($A\beta_{40}$, $A\beta_{42}$ and $A\beta_{42}/A\beta_{40}$ as biomarkers) protein appearance in blood do not show those consistent results in early stage of AD progression regardless with the determination techniques. It was suggested that ultrasensitive immunoassays granted the accurate quantification of tau in blood. Further research is required to justify and validate tau protein as an early diagnostic biomarker in AD.

6. Biomarkers through Non-Invasive Diagnostic Methods

6.1. Cognitive Biomarkers

Currently, non-invasive diagnostic criteria for AD based on a group of assessments, including individual clinical history, cognitive and neuropsychological state and clinical rating score resulted from Mini-Mental State Examination (MMSE), Clinical Dementia Rating (CDR) and the Wechsler Memory Scale (WMS) Logical Memory (LM) test [74]. According to the MMSE score (0–30), if an individual has received a score between 20 and 24, then suggested mild dementia, followed by a score between 13 and 20 suggest moderate dementia and have a score less than 12 designated severe dementia [75]. A study evaluated the accuracy of the MMSE for diagnosing dementia subtypes in people aged 65 years who do not examine earlier for dementia and supported the diagnostic use of MMSE as part of the process for deciding whether or not someone has dementia [76]. The LM test consist of LM-I (immediate recall), LM-II (delayed recall), and LM Recognition (delayed recognition), used to investigate and measure verbal episodic memory in individuals [77]. The LM-I directed individual to immediately recall details of two short passages, while LM-II phase based on recall the passages after a 20 to 30-min delay. The LM recognition phase test provided questionnaire-based evaluation on an earlier provided passage in the form of yes or no.

6.2. Assessment of Cognitive Biomarkers

Several studies have indicated the psychometric limitations of MMSE analysis, including large ceiling and floor effects, and sensitivity to practice effects, limiting the clinical efficacy of MMSE in MCI and AD dementia investigation. Moreover, the scoring system in MMSE was found to lack accuracy in the investigation of individuals with MCI or mild AD dementia. The LM subtest is not only useful for distinguishing certain types of dementia such as AD dementia but is also known as a tool that can detect subtle memory changes in the individuals with MCI [78].

6.3. Imaging Biomarkers

The imaging of the brain in the diagnosis of AD has been used as a second line of diagnostic criteria, including magnetic resonance imaging (MRI), functional MRI (fMRI) and positron emission tomography (PET). In a clinical setting, current guidelines follow the structural imaging, i.e., magnetic resonance imaging (MRI) or computerized tomography (CT), mainly required for the evaluation of patients presenting with a cognitive/dementia syndrome [79]. In order to investigate the visualization of AD-linked cortical atrophy and changes in brain connectivity, MRI is used to provide a structural and functional imaging technique [80]. In addition, fMRI investigation provided the functional connectivity of the brain such as abnormality in the hippocampus [81]. Positron emission tomography (PET) is an advanced imaging technique using compounds labelled with short-lived positron-emitting radionuclides to detect Aβ associated metabolic activity and plaques deposition in AD [82]. Commonly used PET tracers, including Pittsburgh Compound-B (PiB) and Fluoro-2-deoxy-D-glucose (FDG), have a high sensitivity and specificity of diagnosis, particularly in the early stages, and are utilized in imaging biomarkers of amyloid plaque progression in individuals [83].

6.4. Assessment of Imaging Biomarkers

In a clinical setting for AD diagnosis, imaging techniques are often used as a second or third line of investigation due to the unavailability of imaging facility at every center, especially in rural areas. Imaging require rigorous measurement, expertise to interpret the findings and burden of high cost, preventing their frequent use in the routine clinical assessment of individual during their first visit. ^{18}F-FDG-PET analysis may differentiate dementia from normal aging and is used as an indicator biomarker for neurodegeneration; however, it is unable to track down pathology at an early stage of AD. In addition, the current practicing guidelines and expert opinions suggested that amyloid PET analysis was unable to anticipate the trajectory of AD disease progression for an individual patient [84,85]. Moreover, it is also uncertain how effective the PET analysis is at characterizing differences across the pathophysiological phase of AD.

7. Promising Biomarkers in AD

7.1. Proteomic or Enzymatic Biomarkers in AD

In order to investigate novel protein and their capacity to predict AD, an early study analyzed 120 plasma proteins and discovered 18 signaling proteins, which showed 90% of accuracy in diagnosis for AD patients and 91% for MCI patients [86]. Further studies have been reported a total of 1590 AD-related proteins, including 296 proteins encoded with 115 up-regulated and 181 down-regulated genes, and supposed to be blood-secretory proteins involved in the pathogenesis of AD [87]. It was suggested that around 35 AD-related proteins are consistent, including four key proteins (APP, apolipoprotein E, PSEN-1, and PSEN-2) involved in AD pathology [87]. Synaptic proteins such as synaptosomal-associated protein 25 (SNAP-25) and synaptotagmin-1 (SYT1) were found to be significantly increased in the CSF of AD dementia and prodromal AD patients; however, SNAP-25 and SYT1 were specified to decline in cortical areas [88,89]. To facilitate early diagnosis in AD, a protein profiling of blood samples in mild AD patients showed a downregulation of apolipoprotein A_1, α-2-HS-glycoprotein, and afamin, while, apolipoprotein A_4 and the fibrinogen gamma chain were identified upregulated [90].

The CSF BACE1 enzymatic activity and protein concentration has been elevated in AD patients [91], and represented as an indicator biomarker of MCI [92]. In support of elevated BACE1 activity, a study reported a significant rise in plasma BACE1 activity by 53.2% in subjects with MCI and by 68.9% in patients with AD and suggesting plasma BACE1 activity as a diagnostic biomarker [93]. The study focused on plasma-based biomarkers showed a significantly elevated level of BACE1, and soluble forms of APP were observed in AD patient [94]. Further study showed the presence of increased BACE1 enzymatic activity along with the increase in phospholipase-A_2 activity in platelets and brains of patient with AD [95]. One of the main interests for determining BACE1 biomarker is the development of specific BACE1 inhibitors, which may help in the reduction of amyloid production. It is unfortunate that the present plasma BACE1 investigation as biomarkers have not demonstrated a consistent result in terms of significant rise in BACE1 in AD patients compared to normal individuals. In search of other proteomic or enzymatic biomarker, the protein kinases such as glycogen synthase kinase 3β (GSK-3β) has been observed in tau protein hyperphosphorylation, and reported significantly elevated in plasma of MCI and AD patients compared with aged-matched controls [96] as well as their elevated level was also observed in white blood cells of AD and MCI patients compared with healthy individuals [97].

7.2. Assessment of Proteomic or Enzymatic Biomarkers

Several studies have shown the presence of proteins in plasma or serum, including albumin, lipoproteins, Aβ autoantibodies, fibrinogen, immunoglobulin, apolipoprotein-J, apolipoprotein-E, transthyretin, α-2-macroglobulin, serum amyloid p-component, plasminogen and amylin were found to be strongly interferes with the estimation of specific protein biomarkers. It was found that studies that have been conducted on proteomic or enzymatic biomarkers were only in cohorts with unmatched age patients and a target single protein or enzymatic marker; therefore, such studies were required to be focused on proteomics and enzymatic research in individuals with MCI and AD compared to aged-matched controls. Proteomic or enzymatic biomarkers could be an excellent and potential candidate for blood biomarkers as diagnostic tools, but due to the presence of low concentration and the incidence of protein-protein interactions in plasma result the inconsistency in outcomes.

Therefore, it is difficult to replicate due to the lack of specificity and accuracy involved in the current methods of investigation. Thus, the development of highly sensitive and reproducible novel methods that might have the capability to detect the plasma biomarkers in low concentration with accuracy is required.

7.3. Inflammatory Biomarkers in AD

Inflammation is one of the major cellular events considered at the initial pathogenic factor causing neurodegeneration in AD mainly through the activation of microglia. Out of the several investigated inflammatory biomarkers, the diminished C-reactive protein (CRP) level and increased level of triggering receptor expressed on myeloid cells-2 (TREM-2) were observed in the CSF of the patient with AD compared to the normal elderly controls [98]. Moreover, the reduced levels of plasma CRP was also observed in AD patients compared to MCI or normal cognition individuals [99]. The estimation of TREM-2 in CSF and blood is a marker of microglia response. Several studies have investigated the concentration of TREM-2 and reported their increased levels in CSF of AD patient, which was mainly associated with tau pathology compared to the controls [100,101].

Further studies have shown polymorphism in about 23 cytokines, and their 13 types were found to be observed in AD pathogenesis, including interleukins, TNF-α, TGF-β and IFN-γ [102]. Pro-inflammatory cytokines such as TNF-α level were significantly increased followed by decrease in the anti-inflammatory cytokine TGF-β level in CSF, which showed a positive correlation with a higher risk of disease progression from MCI to AD [103]. The concentration of cytokine I-309 level was found to be increased in CSF and suggested as a possible predictor of progression from MCI to AD patients [104]. It has been suggested that the presence of cytokines elevated steadily or reached the highest level upon progression from MCI to AD individuals, including IL-1β, IL-6, TNF-α, IL-18, monocyte chemotactic protein (MCP)-1 and IL-10 can be used as a predicting biomarker for early diagnosis [102].

Microglial activation involved in the stimulation of astrocytic expression of YKL-40 (also known as chitinase-3-like protein-1 (CHI3L1)), which was observed in significantly higher levels in the CSF of AD patients compared to the cognitively normal individuals [105]. Most of the studies have investigated the elevated effect of YKL-40 in CSF; however, less studies have been able to detect YKL-40 in blood, showing a similar finding of increased YKL-40 level in plasma [106].

7.4. Assessment of Inflammatory Biomarkers in AD

Despite the elevation of the inflammatory biomarker TREM-2 in AD, few studies have reported conflicting results with no significant change in CSF TREM-2 levels in MCI and AD patients compared to the controls [107]. In addition, to recognize TREM-2 as a potential inflammatory biomarker in AD diagnosis, their physiological function has been questioned whether it has a constructive or a harmful effect in human body. Due to non-specific diagnosis of TREM-2 in other neuroinflammatory diseases, the biomarkers have faced challenges and required further studies to answer these questions on the molecular basis. The other biomarker YKL-40 investigation in AD have shown a limited diagnostic value due to their non-specific findings in neurological disease other than AD. However, it has been suggested that YKL-40 has a potential role in astroglial activation and the assessment of neuroinflammation treatment, which required further investigation with a specified sensitive method and an inflammatory biomarker in AD.

The majority of studies investigated inflammatory cytokines or chemokines as biomarkers insufficiently sensitive in plasma and could not be reproduced. In addition, there was a lack in the conduction of study in a large cohort using imaging or CSF samples of individuals with MCI and AD for the detection of inflammatory biomarkers. Therefore, it is warranted to investigate and confirm inflammatory biomarkers targeting individuals for early intervention in future studies.

7.5. Oral, Ocular and Olfactory Fluid Biomarkers in AD

Body fluid other than CSF and blood, including oral, ocular and olfactory have received attention for detecting the biomarkers of AD because of their easily available, noninvasive nature and inexpensive sample collection methods. A decade ago, one pilot study reported a significant rise in saliva $A\beta_{42}$ levels in patients with mild AD compared to the control, while saliva $A\beta_{40}$ level was detected unchanged [108]. Further study reported a two-folds increase in salivary $A\beta_{42}$ concentration in AD patients compared to the controls, which have identical levels of salivary $A\beta_{42}$ regardless of sex or age [109]. Further pilot study using the 1H NMR metabolomics technique for detecting salivary $A\beta_{42}$ found a significant rise in salivary $A\beta_{42}$ concentration in MCI and AD patients compared to controls [110]. In the line of previous results showing the elevated level of salivary $A\beta_{42}$, a study based on enzyme-linked immunosorbent-type assays methods reported an increased $A\beta_{42}$ level in patients with AD compared to controls [111].

In order to attempt early diagnosis and the development of a non-invasive method for AD, studies have reported the presence of significant Aβ plaque concentration deposited in the retina [112]. A quantitative and histological study showed a two-fold increase in retinal Aβ deposition in the form of protofibrils and fibrils among AD patients versus controls [113]. A similar study has detected retinal Aβ plaque deposition two months prior to their presence in the hippocampus and cortices of AD murine models and suggested the appearance of Aβ in the retina is an early event in the pathogenesis of AD [113]. The emerging single visit, label-free, cost-effective eye scan along with the emergence of mobile imaging modalities for the early detection of Aβ plaques in retina through the polarization imaging of retinal Aβ and retinal fluorescence lifetime imaging ophthalmoscopy have been suggested as early diagnostic tools [114].

The olfactory dysfunction has been associated with tau pathology from early to advanced AD and is well recognized. A study based on the Braak stage of AD showed the presence of tau pathology 90% and Aβ 9% in stage 4 followed by 44.6% of tau pathology and Aβ 9% in stage 3 [115]. The olfactory tau pathology in stage 2 represented by 36.4% without Aβ deposition; however, Braak stages 0 and 1 do not show any positive association of tau pathology [115]. Further study showed significant elevation in t-tau protein and p-tau tau protein levels in the patients with AD suffering from loss of smell, compared to the healthy controls [116]. The results of a recent meta-analysis suggested that the identification of olfactory dysfunction was more profound in AD patients compared to MCI patients [117].

7.6. Assessment of Oral, Ocular and Olfactory Fluid Biomarkers in AD

The oral, ocular and olfactory biofluid as potential biomarkers in AD diagnosis are still in the preliminary stage of research, requiring further investigation to overcome the methodological heterogeneity and discrepancy in accuracy. Thus, there is a need to develop a reliable quantitative method among consecutive or random samples to determine these biomarkers as an effective tool for diagnosing AD.

8. Conclusions

AD is a multifactorial disease without a confirmatory biomarker; however, based on the NIA-AA guideline, the currently available biomarkers may provide a positive direction to identify individuals who are on risk of AD, and need of routine screening for early diagnosis. The most studied and practiced biomarkers for AD diagnosis are CSF $A\beta_{42}$, CSF $A\beta_{42}/A\beta_{40}$, CSF p-tau, amyloid PET, tau PET, structural MRI and fMRI. The accumulated evidences have been established that $A\beta_{42}$ and tau levels were found quite lower (~30 to 100 times) in plasma compared to CSF. Furthermore, because plasma and serum already enriched with immense level of several proteins (~50–70 g/L) which interferes in detection and face challenges in poor outcome of NIA-AA suggested biomarkers in blood compared to CSF. Therefore, there is no doubt that the investigation of biomarkers in blood face inconsistency in findings due to their low appearance (concentration) in blood (plasma or serum) than in the CSF. To date, there are weak evidences supporting the dominance of biomarker over the other (CSF/imaging) for diagnostic tools in AD. Thus, in terms of availability or diagnosis, CSF have less challenges in the A/T/N framework-based biomarkers compared to blood-based biomarkers (Figure 2). Despite of challenges in blood-based biomarkers, the early detection of biomarkers (plasma NfL) could be a promising diagnostic tool in body fluid blood which can be routinely investigate in the individual reaching the age of 60 years. In a nut-shell, blood-based biomarkers should pay more attention in order to support the patient's comfort, highly sensitive, least invasive and cost-effective inexpensive technologies for biomarker detection are required for detection and analysis in the range of 10^{-15} to 10^{-12} M in individuals before entering MCI stage.

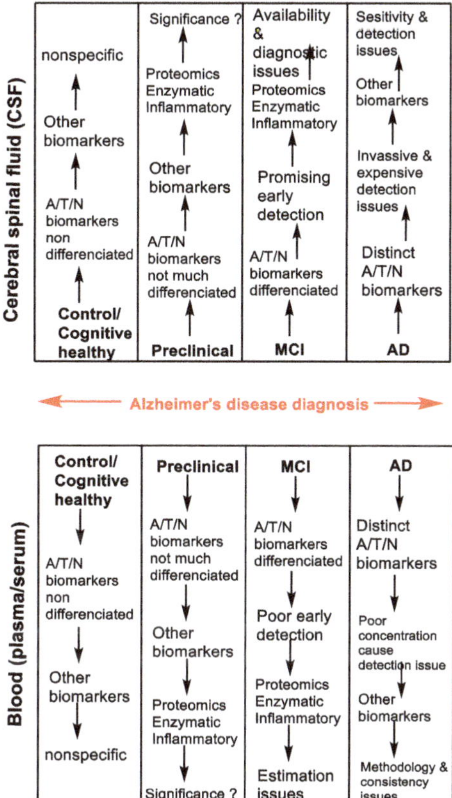

Figure 2. Overview of investigation and development CSF and blood-based biomarkers in Alzheimer's disease diagnosis.

Author Contributions: S.H.O. has created idea, designed and wrote the manuscript; J.P. has assisted in drafting the manuscript. All authors have read and agreed to the published version of the manuscript.

Funding: This research received no external funding.

Acknowledgments: Rural Clinical School, Faculty of Medicine, University of New South Wales, Wagga Wagga, NSW, Australia & *Murrumbidgee Local Health District*, NSW Health, Wagga Wagga, NSW 2650.

Conflicts of Interest: The authors declare no conflict of interest.

References

1. Scheltens, P.; Blennow, K.; Breteler, M.M.; de Strooper, B.; Frisoni, G.B.; Salloway, S.; Van der Flier, W.M. Alzheimer's disease. *Lancet* **2016**, *388*, 505–517. [CrossRef]
2. Alzheimer's, A. 2016 Alzheimer's disease facts and figures. *Alzheimers Dement.* **2016**, *12*, 459–509. [CrossRef] [PubMed]
3. Pozueta, J.; Lefort, R.; Shelanski, M.L. Synaptic changes in Alzheimer's disease and its models. *Neuroscience* **2013**, *251*, 51–65. [CrossRef]
4. Sharma, P.; Srivastava, P.; Seth, A.; Tripathi, P.N.; Banerjee, A.G.; Shrivastava, S.K. Comprehensive review of mechanisms of pathogenesis involved in Alzheimer's disease and potential therapeutic strategies. *Prog. Neurobiol.* **2019**, *174*, 53–89. [CrossRef]
5. Selkoe, D.J.; Hardy, J. The amyloid hypothesis of Alzheimer's disease at 25 years. *EMBO Mol. Med.* **2016**, *8*, 595–608. [CrossRef]

6. Sadigh-Eteghad, S.; Sabermarouf, B.; Majdi, A.; Talebi, M.; Farhoudi, M.; Mahmoudi, J. Amyloid-beta: A crucial factor in Alzheimer's disease. *Med. Princ. Pract.* **2015**, *24*, 1–10. [CrossRef]
7. Braak, H.; Del Tredici, K. Alzheimer's pathogenesis: Is there neuron-to-neuron propagation? *Acta Neuropathol.* **2011**, *121*, 589–595. [CrossRef] [PubMed]
8. Omar, S.H.; Scott, C.J.; Hamlin, A.S.; Obied, H.K. The protective role of plant biophenols in mechanisms of Alzheimer's disease. *J. Nutr. Biochem.* **2017**, *47*, 1–20. [CrossRef] [PubMed]
9. Omar, S.H.; Scott, C.J.; Hamlin, A.S.; Obied, H.K. Biophenols: Enzymes (beta-secretase, Cholinesterases, histone deacetylase and tyrosinase) inhibitors from olive (Olea europaea L.). *Fitoterapia* **2018**, *128*, 118–129. [CrossRef]
10. Omar, S.H.; Scott, C.J.; Hamlin, A.S.; Obied, H.K. Olive Biophenols Reduces Alzheimer's Pathology in SH-SY5Y Cells and APPswe Mice. *Int. J. Mol. Sci.* **2018**, *20*, 125. [CrossRef]
11. Omar, S.H. Biophenols pharmacology against the amyloidogenic activity in Alzheimer's disease. *Biomed. Pharm.* **2017**, *89*, 396–413. [CrossRef] [PubMed]
12. Omar, S.H. Oleuropein in olive and its pharmacological effects. *Sci. Pharm.* **2010**, *78*, 133–154. [CrossRef] [PubMed]
13. Omar, S.H. Ginkgolides and Neuroprotective Effects. In *Natural Products: Phytochemistry, Botany and Metabolism of Alkaloids, Phenolics and Terpenes*; Ramawat, K.G., Mérillon, J.-M., Eds.; Springer: Berlin/Heidelberg, Germany, 2013; pp. 3697–3741. [CrossRef]
14. Omar, S.H. Mediterranean and MIND Diets Containing Olive Biophenols Reduces the Prevalence of Alzheimer's Disease. *Int. J. Mol. Sci.* **2019**, *20*, 2797. [CrossRef] [PubMed]
15. McKhann, G.M.; Knopman, D.S.; Chertkow, H.; Hyman, B.T.; Jack, C.R., Jr.; Kawas, C.H.; Klunk, W.E.; Koroshetz, W.J.; Manly, J.J.; Mayeux, R.; et al. The diagnosis of dementia due to Alzheimer's disease: Recommendations from the National Institute on Aging-Alzheimer's Association workgroups on diagnostic guidelines for Alzheimer's disease. *Alzheimers Dement.* **2011**, *7*, 263–269. [CrossRef]
16. Reitz, C.; Brayne, C.; Mayeux, R. Epidemiology of Alzheimer disease. *Nat. Rev. Neurol.* **2011**, *7*, 137–152. [CrossRef]
17. Jack, C.R., Jr.; Bennett, D.A.; Blennow, K.; Carrillo, M.C.; Dunn, B.; Haeberlein, S.B.; Holtzman, D.M.; Jagust, W.; Jessen, F.; Karlawish, J.; et al. NIA-AA Research Framework: Toward a biological definition of Alzheimer's disease. *Alzheimers Dement.* **2018**, *14*, 535–562. [CrossRef]
18. Food, U.; Administration, D. Early Alzheimer's Disease: Developing Drugs for Treatment–Guidance for Industry. 2018. Available online: https://www.fda.gov/downloads/Drugs/GuidanceComplianceRegulatoryInformation/Guidances/UCM596728.pdf (accessed on 25 May 2020).
19. Hansson, O.; Zetterberg, H.; Buchhave, P.; Londos, E.; Blennow, K.; Minthon, L. Association between CSF biomarkers and incipient Alzheimer's disease in patients with mild cognitive impairment: A follow-up study. *Lancet Neurol.* **2006**, *5*, 228–234. [CrossRef]
20. Humpel, C. Identifying and validating biomarkers for Alzheimer's disease. *Trends Biotechnol.* **2011**, *29*, 26–32. [CrossRef]
21. Mattsson, N.; Rosen, E.; Hansson, O.; Andreasen, N.; Parnetti, L.; Jonsson, M.; Herukka, S.K.; van der Flier, W.M.; Blankenstein, M.A.; Ewers, M.; et al. Age and diagnostic performance of Alzheimer disease CSF biomarkers. *Neurology* **2012**, *78*, 468–476. [CrossRef]
22. Potter, R.; Patterson, B.W.; Elbert, D.L.; Ovod, V.; Kasten, T.; Sigurdson, W.; Mawuenyega, K.; Blazey, T.; Goate, A.; Chott, R.; et al. Increased in vivo amyloid-beta42 production, exchange, and loss in presenilin mutation carriers. *Sci. Transl. Med.* **2013**, *5*, 189ra177. [CrossRef]
23. Blennow, K. A Review of Fluid Biomarkers for Alzheimer's Disease: Moving from CSF to Blood. *Neurol. Ther.* **2017**, *6*, 15–24. [CrossRef] [PubMed]
24. Grimmer, T.; Riemenschneider, M.; Forstl, H.; Henriksen, G.; Klunk, W.E.; Mathis, C.A.; Shiga, T.; Wester, H.J.; Kurz, A.; Drzezga, A. Beta amyloid in Alzheimer's disease: Increased deposition in brain is reflected in reduced concentration in cerebrospinal fluid. *Biol. Psychiatry* **2009**, *65*, 927–934. [CrossRef] [PubMed]
25. Hansson, O.; Lehmann, S.; Otto, M.; Zetterberg, H.; Lewczuk, P. Advantages and disadvantages of the use of the CSF Amyloid beta (Abeta) 42/40 ratio in the diagnosis of Alzheimer's Disease. *Alzheimers Res. Ther.* **2019**, *11*, 34. [CrossRef] [PubMed]

26. Tapiola, T.; Pirttila, T.; Mikkonen, M.; Mehta, P.D.; Alafuzoff, I.; Koivisto, K.; Soininen, H. Three-year follow-up of cerebrospinal fluid tau, beta-amyloid 42 and 40 concentrations in Alzheimer's disease. *Neurosci. Lett.* **2000**, *280*, 119–122. [CrossRef]
27. Blennow, K.; Dubois, B.; Fagan, A.M.; Lewczuk, P.; de Leon, M.J.; Hampel, H. Clinical utility of cerebrospinal fluid biomarkers in the diagnosis of early Alzheimer's disease. *Alzheimers Dement.* **2015**, *11*, 58–69. [CrossRef]
28. Andreasen, N.; Hesse, C.; Davidsson, P.; Minthon, L.; Wallin, A.; Winblad, B.; Vanderstichele, H.; Vanmechelen, E.; Blennow, K. Cerebrospinal fluid beta-amyloid(1-42) in Alzheimer disease: Differences between early- and late-onset Alzheimer disease and stability during the course of disease. *Arch. Neurol.* **1999**, *56*, 673–680. [CrossRef] [PubMed]
29. Spies, P.E.; Verbeek, M.M.; van Groen, T.; Claassen, J.A. Reviewing reasons for the decreased CSF Abeta42 concentration in Alzheimer disease. *Front. Biosci.* **2012**, *17*, 2024–2034. [CrossRef]
30. Janelidze, S.; Zetterberg, H.; Mattsson, N.; Palmqvist, S.; Vanderstichele, H.; Lindberg, O.; van Westen, D.; Stomrud, E.; Minthon, L.; Blennow, K.; et al. CSF Abeta42/Abeta40 and Abeta42/Abeta38 ratios: Better diagnostic markers of Alzheimer disease. *Ann. Clin. Transl. Neurol.* **2016**, *3*, 154–165. [CrossRef]
31. Koychev, I.; Galna, B.; Zetterberg, H.; Lawson, J.; Zamboni, G.; Ridha, B.H.; Rowe, J.B.; Thomas, A.; Howard, R.; Malhotra, P.; et al. Abeta42/Abeta40 and Abeta42/Abeta38 Ratios Are Associated with Measures of Gait Variability and Activities of Daily Living in Mild Alzheimer's Disease: A Pilot Study. *J. Alzheimers Dis.* **2018**, *65*, 1377–1383. [CrossRef]
32. Mehta, P.D.; Pirttila, T.; Mehta, S.P.; Sersen, E.A.; Aisen, P.S.; Wisniewski, H.M. Plasma and cerebrospinal fluid levels of amyloid beta proteins 1-40 and 1-42 in Alzheimer disease. *Arch. Neurol.* **2000**, *57*, 100–105. [CrossRef]
33. Baldeiras, I.; Santana, I.; Leitao, M.J.; Gens, H.; Pascoal, R.; Tabuas-Pereira, M.; Beato-Coelho, J.; Duro, D.; Almeida, M.R.; Oliveira, C.R. Addition of the Abeta42/40 ratio to the cerebrospinal fluid biomarker profile increases the predictive value for underlying Alzheimer's disease dementia in mild cognitive impairment. *Alzheimers Res. Ther.* **2018**, *10*, 33. [CrossRef] [PubMed]
34. Shaw, L.M.; Korecka, M.; Clark, C.M.; Lee, V.M.; Trojanowski, J.Q. Biomarkers of neurodegeneration for diagnosis and monitoring therapeutics. *Nat. Rev. Drug Discov.* **2007**, *6*, 295–303. [CrossRef] [PubMed]
35. Southwick, P.C.; Yamagata, S.K.; Echols, C.L., Jr.; Higson, G.J.; Neynaber, S.A.; Parson, R.E.; Munroe, W.A. Assessment of amyloid beta protein in cerebrospinal fluid as an aid in the diagnosis of Alzheimer's disease. *J. Neurochem.* **1996**, *66*, 259–265. [CrossRef] [PubMed]
36. Bouwman, F.H.; van der Flier, W.M.; Schoonenboom, N.S.; van Elk, E.J.; Kok, A.; Rijmen, F.; Blankenstein, M.A.; Scheltens, P. Longitudinal changes of CSF biomarkers in memory clinic patients. *Neurology* **2007**, *69*, 1006–1011. [CrossRef]
37. Ittner, L.M.; Gotz, J. Amyloid-beta and tau–a toxic pas de deux in Alzheimer's disease. *Nat. Rev. Neurosci.* **2011**, *12*, 65–72. [CrossRef]
38. Bloom, G.S. Amyloid-beta and tau: The trigger and bullet in Alzheimer disease pathogenesis. *JAMA Neurol.* **2014**, *71*, 505–508. [CrossRef]
39. Tapiola, T.; Alafuzoff, I.; Herukka, S.K.; Parkkinen, L.; Hartikainen, P.; Soininen, H.; Pirttila, T. Cerebrospinal fluid {beta}-amyloid 42 and tau proteins as biomarkers of Alzheimer-type pathologic changes in the brain. *Arch. Neurol.* **2009**, *66*, 382–389. [CrossRef]
40. Buerger, K.; Ewers, M.; Pirttila, T.; Zinkowski, R.; Alafuzoff, I.; Teipel, S.J.; DeBernardis, J.; Kerkman, D.; McCulloch, C.; Soininen, H.; et al. CSF phosphorylated tau protein correlates with neocortical neurofibrillary pathology in Alzheimer's disease. *Brain* **2006**, *129*, 3035–3041. [CrossRef]
41. Buerger, K.; Teipel, S.J.; Zinkowski, R.; Blennow, K.; Arai, H.; Engel, R.; Hofmann-Kiefer, K.; McCulloch, C.; Ptok, U.; Heun, R.; et al. CSF tau protein phosphorylated at threonine 231 correlates with cognitive decline in MCI subjects. *Neurology* **2002**, *59*, 627–629. [CrossRef]
42. Sjogren, M.; Vanderstichele, H.; Agren, H.; Zachrisson, O.; Edsbagge, M.; Wikkelso, C.; Skoog, I.; Wallin, A.; Wahlund, L.O.; Marcusson, J.; et al. Tau and Abeta42 in cerebrospinal fluid from healthy adults 21-93 years of age: Establishment of reference values. *Clin. Chem.* **2001**, *47*, 1776–1781. [CrossRef]
43. Sunderland, T.; Linker, G.; Mirza, N.; Putnam, K.T.; Friedman, D.L.; Kimmel, L.H.; Bergeson, J.; Manetti, G.J.; Zimmermann, M.; Tang, B.; et al. Decreased beta-amyloid1-42 and increased tau levels in cerebrospinal fluid of patients with Alzheimer disease. *JAMA* **2003**, *289*, 2094–2103. [CrossRef]

44. Burger nee Buch, K.; Padberg, F.; Nolde, T.; Teipel, S.J.; Stubner, S.; Haslinger, A.; Schwarz, M.J.; Sunderland, T.; Arai, H.; Rapoport, S.I.; et al. Cerebrospinal fluid tau protein shows a better discrimination in young old (<70 years) than in old old patients with Alzheimer's disease compared with controls. *Neurosci. Lett.* **1999**, *277*, 21–24. [CrossRef]
45. Ost, M.; Nylen, K.; Csajbok, L.; Ohrfelt, A.O.; Tullberg, M.; Wikkelso, C.; Nellgard, P.; Rosengren, L.; Blennow, K.; Nellgard, B. Initial CSF total tau correlates with 1-year outcome in patients with traumatic brain injury. *Neurology* **2006**, *67*, 1600–1604. [CrossRef] [PubMed]
46. Skillback, T.; Zetterberg, H.; Blennow, K.; Mattsson, N. Cerebrospinal fluid biomarkers for Alzheimer disease and subcortical axonal damage in 5,542 clinical samples. *Alzheimers Res. Ther.* **2013**, *5*, 47. [CrossRef] [PubMed]
47. Ritchie, C.; Smailagic, N.; Noel-Storr, A.H.; Ukoumunne, O.; Ladds, E.C.; Martin, S. CSF tau and the CSF tau/ABeta ratio for the diagnosis of Alzheimer's disease dementia and other dementias in people with mild cognitive impairment (MCI). *Cochrane Database Syst. Rev.* **2017**, *3*, CD010803. [CrossRef] [PubMed]
48. Kern, S.; Syrjanen, J.A.; Blennow, K.; Zetterberg, H.; Skoog, I.; Waern, M.; Hagen, C.E.; van Harten, A.C.; Knopman, D.S.; Jack, C.R., Jr.; et al. Association of Cerebrospinal Fluid Neurofilament Light Protein With Risk of Mild Cognitive Impairment Among Individuals Without Cognitive Impairment. *JAMA Neurol.* **2019**, *76*, 187–193. [CrossRef] [PubMed]
49. Olsson, B.; Portelius, E.; Cullen, N.C.; Sandelius, A.; Zetterberg, H.; Andreasson, U.; Hoglund, K.; Irwin, D.J.; Grossman, M.; Weintraub, D.; et al. Association of Cerebrospinal Fluid Neurofilament Light Protein Levels With Cognition in Patients With Dementia, Motor Neuron Disease, and Movement Disorders. *JAMA Neurol.* **2019**, *76*, 318–325. [CrossRef]
50. Brys, M.; Pirraglia, E.; Rich, K.; Rolstad, S.; Mosconi, L.; Switalski, R.; Glodzik-Sobanska, L.; De Santi, S.; Zinkowski, R.; Mehta, P.; et al. Prediction and longitudinal study of CSF biomarkers in mild cognitive impairment. *Neurobiol. Aging* **2009**, *30*, 682–690. [CrossRef] [PubMed]
51. Castellani, R.J.; Nunomura, A.; Lee, H.G.; Perry, G.; Smith, M.A. Phosphorylated tau: Toxic, protective, or none of the above. *J. Alzheimers Dis.* **2008**, *14*, 377–383. [CrossRef]
52. O'Bryant, S.E.; Mielke, M.M.; Rissman, R.A.; Lista, S.; Vanderstichele, H.; Zetterberg, H.; Lewczuk, P.; Posner, H.; Hall, J.; Johnson, L.; et al. Blood-based biomarkers in Alzheimer disease: Current state of the science and a novel collaborative paradigm for advancing from discovery to clinic. *Alzheimers Dement.* **2017**, *13*, 45–58. [CrossRef]
53. Mawuenyega, K.G.; Sigurdson, W.; Ovod, V.; Munsell, L.; Kasten, T.; Morris, J.C.; Yarasheski, K.E.; Bateman, R.J. Decreased clearance of CNS beta-amyloid in Alzheimer's disease. *Science* **2010**, *330*, 1774. [CrossRef] [PubMed]
54. van Oijen, M.; Hofman, A.; Soares, H.D.; Koudstaal, P.J.; Breteler, M.M. Plasma Abeta(1-40) and Abeta(1-42) and the risk of dementia: A prospective case-cohort study. *Lancet Neurol.* **2006**, *5*, 655–660. [CrossRef]
55. Mayeux, R.; Honig, L.S.; Tang, M.X.; Manly, J.; Stern, Y.; Schupf, N.; Mehta, P.D. Plasma A[beta]40 and A[beta]42 and Alzheimer's disease: Relation to age, mortality, and risk. *Neurology* **2003**, *61*, 1185–1190. [CrossRef]
56. Sundelof, J.; Giedraitis, V.; Irizarry, M.C.; Sundstrom, J.; Ingelsson, E.; Ronnemaa, E.; Arnlov, J.; Gunnarsson, M.D.; Hyman, B.T.; Basun, H.; et al. Plasma beta amyloid and the risk of Alzheimer disease and dementia in elderly men: A prospective, population-based cohort study. *Arch. Neurol.* **2008**, *65*, 256–263. [CrossRef] [PubMed]
57. Pomara, N.; Willoughby, L.M.; Sidtis, J.J.; Mehta, P.D. Selective reductions in plasma Abeta 1-42 in healthy elderly subjects during longitudinal follow-up: A preliminary report. *Am. J. Geriatr. Psychiatry* **2005**, *13*, 914–917. [CrossRef]
58. Hansson, O.; Zetterberg, H.; Vanmechelen, E.; Vanderstichele, H.; Andreasson, U.; Londos, E.; Wallin, A.; Minthon, L.; Blennow, K. Evaluation of plasma Abeta(40) and Abeta(42) as predictors of conversion to Alzheimer's disease in patients with mild cognitive impairment. *Neurobiol. Aging* **2010**, *31*, 357–367. [CrossRef]
59. Olsson, B.; Lautner, R.; Andreasson, U.; Ohrfelt, A.; Portelius, E.; Bjerke, M.; Holtta, M.; Rosen, C.; Olsson, C.; Strobel, G.; et al. CSF and blood biomarkers for the diagnosis of Alzheimer's disease: A systematic review and meta-analysis. *Lancet Neurol.* **2016**, *15*, 673–684. [CrossRef]
60. Nakamura, A.; Kaneko, N.; Villemagne, V.L.; Kato, T.; Doecke, J.; Dore, V.; Fowler, C.; Li, Q.X.; Martins, R.; Rowe, C.; et al. High performance plasma amyloid-beta biomarkers for Alzheimer's disease. *Nature* **2018**, *554*, 249–254. [CrossRef]

61. Ovod, V.; Ramsey, K.N.; Mawuenyega, K.G.; Bollinger, J.G.; Hicks, T.; Schneider, T.; Sullivan, M.; Paumier, K.; Holtzman, D.M.; Morris, J.C.; et al. Amyloid beta concentrations and stable isotope labeling kinetics of human plasma specific to central nervous system amyloidosis. *Alzheimers Dement.* **2017**, *13*, 841–849. [CrossRef]
62. Li, Q.X.; Berndt, M.C.; Bush, A.I.; Rumble, B.; Mackenzie, I.; Friedhuber, A.; Beyreuther, K.; Masters, C.L. Membrane-associated forms of the beta A4 amyloid protein precursor of Alzheimer's disease in human platelet and brain: Surface expression on the activated human platelet. *Blood* **1994**, *84*, 133–142. [CrossRef]
63. Zetterberg, H.; Wilson, D.; Andreasson, U.; Minthon, L.; Blennow, K.; Randall, J.; Hansson, O. Plasma tau levels in Alzheimer's disease. *Alzheimers Res. Ther.* **2013**, *5*, 9. [CrossRef] [PubMed]
64. Mattsson, N.; Zetterberg, H.; Janelidze, S.; Insel, P.S.; Andreasson, U.; Stomrud, E.; Palmqvist, S.; Baker, D.; Tan Hehir, C.A.; Jeromin, A.; et al. Plasma tau in Alzheimer disease. *Neurology* **2016**, *87*, 1827–1835. [CrossRef] [PubMed]
65. Mielke, M.M.; Hagen, C.E.; Xu, J.; Chai, X.; Vemuri, P.; Lowe, V.J.; Airey, D.C.; Knopman, D.S.; Roberts, R.O.; Machulda, M.M.; et al. Plasma phospho-tau181 increases with Alzheimer's disease clinical severity and is associated with tau- and amyloid-positron emission tomography. *Alzheimers Dement.* **2018**, *14*, 989–997. [CrossRef]
66. Yang, C.C.; Chiu, M.J.; Chen, T.F.; Chang, H.L.; Liu, B.H.; Yang, S.Y. Assay of Plasma Phosphorylated Tau Protein (Threonine 181) and Total Tau Protein in Early-Stage Alzheimer's Disease. *J. Alzheimers Dis.* **2018**, *61*, 1323–1332. [CrossRef] [PubMed]
67. Tatebe, H.; Kasai, T.; Ohmichi, T.; Kishi, Y.; Kakeya, T.; Waragai, M.; Kondo, M.; Allsop, D.; Tokuda, T. Quantification of plasma phosphorylated tau to use as a biomarker for brain Alzheimer pathology: Pilot case-control studies including patients with Alzheimer's disease and down syndrome. *Mol. Neurodegener.* **2017**, *12*, 63. [CrossRef]
68. Fossati, S.; Ramos Cejudo, J.; Debure, L.; Pirraglia, E.; Sone, J.Y.; Li, Y.; Chen, J.; Butler, T.; Zetterberg, H.; Blennow, K.; et al. Plasma tau complements CSF tau and P-tau in the diagnosis of Alzheimer's disease. *Alzheimers Dement.* **2019**, *11*, 483–492. [CrossRef]
69. Khalil, M.; Teunissen, C.E.; Otto, M.; Piehl, F.; Sormani, M.P.; Gattringer, T.; Barro, C.; Kappos, L.; Comabella, M.; Fazekas, F.; et al. Neurofilaments as biomarkers in neurological disorders. *Nat. Rev. Neurol.* **2018**, *14*, 577–589. [CrossRef]
70. Weston, P.S.J.; Poole, T.; Ryan, N.S.; Nair, A.; Liang, Y.; Macpherson, K.; Druyeh, R.; Malone, I.B.; Ahsan, R.L.; Pemberton, H.; et al. Serum neurofilament light in familial Alzheimer disease: A marker of early neurodegeneration. *Neurology* **2017**, *89*, 2167–2175. [CrossRef]
71. Weston, P.S.J.; Poole, T.; O'Connor, A.; Heslegrave, A.; Ryan, N.S.; Liang, Y.; Druyeh, R.; Mead, S.; Blennow, K.; Schott, J.M.; et al. Longitudinal measurement of serum neurofilament light in presymptomatic familial Alzheimer's disease. *Alzheimers Res. Ther.* **2019**, *11*, 19. [CrossRef]
72. Mattsson, N.; Andreasson, U.; Zetterberg, H.; Blennow, K.; Alzheimer's Disease Neuroimaging, I. Association of Plasma Neurofilament Light With Neurodegeneration in Patients With Alzheimer Disease. *JAMA Neurol.* **2017**, *74*, 557–566. [CrossRef]
73. Preische, O.; Schultz, S.A.; Apel, A.; Kuhle, J.; Kaeser, S.A.; Barro, C.; Graber, S.; Kuder-Buletta, E.; LaFougere, C.; Laske, C.; et al. Serum neurofilament dynamics predicts neurodegeneration and clinical progression in presymptomatic Alzheimer's disease. *Nat. Med.* **2019**, *25*, 277–283. [CrossRef] [PubMed]
74. Lopez, O.L.; McDade, E.; Riverol, M.; Becker, J.T. Evolution of the diagnostic criteria for degenerative and cognitive disorders. *Curr. Opin. Neurol.* **2011**, *24*, 532–541. [CrossRef] [PubMed]
75. Folstein, M.F.; Folstein, S.E.; McHugh, P.R. "Mini-mental state". A practical method for grading the cognitive state of patients for the clinician. *J. Psychiatr. Res.* **1975**, *12*, 189–198. [CrossRef]
76. Creavin, S.T.; Wisniewski, S.; Noel-Storr, A.H.; Trevelyan, C.M.; Hampton, T.; Rayment, D.; Thom, V.M.; Nash, K.J.; Elhamoui, H.; Milligan, R.; et al. Mini-Mental State Examination (MMSE) for the detection of dementia in clinically unevaluated people aged 65 and over in community and primary care populations. *Cochrane Database Syst. Rev.* **2016**, CD011145. [CrossRef] [PubMed]
77. Chapman, K.R.; Bing-Canar, H.; Alosco, M.L.; Steinberg, E.G.; Martin, B.; Chaisson, C.; Kowall, N.; Tripodis, Y.; Stern, R.A. Mini Mental State Examination and Logical Memory scores for entry into Alzheimer's disease trials. *Alzheimers Res. Ther.* **2016**, *8*, 9. [CrossRef]
78. Li, M.; Ng, T.P.; Kua, E.H.; Ko, S.M. Brief informant screening test for mild cognitive impairment and early Alzheimer's disease. *Dement. Geriatr. Cogn. Disord.* **2006**, *21*, 392–402. [CrossRef]

79. Sheikh-Bahaei, N.; Sajjadi, S.A.; Manavaki, R.; Gillard, J.H. Imaging Biomarkers in Alzheimer's Disease: A Practical Guide for Clinicians. *J. Alzheimers Dis. Rep.* **2017**, *1*, 71–88. [CrossRef]
80. Scheltens, P.; Leys, D.; Barkhof, F.; Huglo, D.; Weinstein, H.C.; Vermersch, P.; Kuiper, M.; Steinling, M.; Wolters, E.C.; Valk, J. Atrophy of medial temporal lobes on MRI in "probable" Alzheimer's disease and normal ageing: Diagnostic value and neuropsychological correlates. *J. Neurol. Neurosurg. Psychiatry* **1992**, *55*, 967–972. [CrossRef]
81. Wang, L.; Zang, Y.; He, Y.; Liang, M.; Zhang, X.; Tian, L.; Wu, T.; Jiang, T.; Li, K. Changes in hippocampal connectivity in the early stages of Alzheimer's disease: Evidence from resting state fMRI. *Neuroimage* **2006**, *31*, 496–504. [CrossRef]
82. Kadir, A.; Almkvist, O.; Forsberg, A.; Wall, A.; Engler, H.; Langstrom, B.; Nordberg, A. Dynamic changes in PET amyloid and FDG imaging at different stages of Alzheimer's disease. *Neurobiol. Aging* **2012**, *33*, 198.e1–198.e4. [CrossRef]
83. Cohen, A.D.; Klunk, W.E. Early detection of Alzheimer's disease using PiB and FDG PET. *Neurobiol. Dis.* **2014**, *72 Pt A*, 117–122. [CrossRef]
84. Chiotis, K.; Saint-Aubert, L.; Boccardi, M.; Gietl, A.; Picco, A.; Varrone, A.; Garibotto, V.; Herholz, K.; Nobili, F.; Nordberg, A.; et al. Clinical validity of increased cortical uptake of amyloid ligands on PET as a biomarker for Alzheimer's disease in the context of a structured 5-phase development framework. *Neurobiol. Aging* **2017**, *52*, 214–227. [CrossRef] [PubMed]
85. Grill, J.D.; Apostolova, L.G.; Bullain, S.; Burns, J.M.; Cox, C.G.; Dick, M.; Hartley, D.; Kawas, C.; Kremen, S.; Lingler, J.; et al. Communicating mild cognitive impairment diagnoses with and without amyloid imaging. *Alzheimers Res. Ther.* **2017**, *9*, 35. [CrossRef] [PubMed]
86. Ray, S.; Britschgi, M.; Herbert, C.; Takeda-Uchimura, Y.; Boxer, A.; Blennow, K.; Friedman, L.F.; Galasko, D.R.; Jutel, M.; Karydas, A.; et al. Classification and prediction of clinical Alzheimer's diagnosis based on plasma signaling proteins. *Nat. Med.* **2007**, *13*, 1359–1362. [CrossRef] [PubMed]
87. Yao, F.; Zhang, K.; Zhang, Y.; Guo, Y.; Li, A.; Xiao, S.; Liu, Q.; Shen, L.; Ni, J. Identification of Blood Biomarkers for Alzheimer's Disease Through Computational Prediction and Experimental Validation. *Front. Neurol.* **2018**, *9*, 1158. [CrossRef]
88. Brinkmalm, A.; Brinkmalm, G.; Honer, W.G.; Frolich, L.; Hausner, L.; Minthon, L.; Hansson, O.; Wallin, A.; Zetterberg, H.; Blennow, K.; et al. SNAP-25 is a promising novel cerebrospinal fluid biomarker for synapse degeneration in Alzheimer's disease. *Mol. Neurodegener.* **2014**, *9*, 53. [CrossRef]
89. Ohrfelt, A.; Brinkmalm, A.; Dumurgier, J.; Brinkmalm, G.; Hansson, O.; Zetterberg, H.; Bouaziz-Amar, E.; Hugon, J.; Paquet, C.; Blennow, K. The pre-synaptic vesicle protein synaptotagmin is a novel biomarker for Alzheimer's disease. *Alzheimers Res. Ther.* **2016**, *8*, 41. [CrossRef] [PubMed]
90. Kitamura, Y.; Usami, R.; Ichihara, S.; Kida, H.; Satoh, M.; Tomimoto, H.; Murata, M.; Oikawa, S. Plasma protein profiling for potential biomarkers in the early diagnosis of Alzheimer's disease. *Neurol. Res.* **2017**, *39*, 231–238. [CrossRef]
91. Ewers, M.; Cheng, X.; Zhong, Z.; Nural, H.F.; Walsh, C.; Meindl, T.; Teipel, S.J.; Buerger, K.; He, P.; Shen, Y.; et al. Increased CSF-BACE1 activity associated with decreased hippocampus volume in Alzheimer's disease. *J. Alzheimers Dis.* **2011**, *25*, 373–381. [CrossRef]
92. Zhong, Z.; Ewers, M.; Teipel, S.; Burger, K.; Wallin, A.; Blennow, K.; He, P.; McAllister, C.; Hampel, H.; Shen, Y. Levels of beta-secretase (BACE1) in cerebrospinal fluid as a predictor of risk in mild cognitive impairment. *Arch. Gen. Psychiatry* **2007**, *64*, 718–726. [CrossRef]
93. Shen, Y.; Wang, H.; Sun, Q.; Yao, H.; Keegan, A.P.; Mullan, M.; Wilson, J.; Lista, S.; Leyhe, T.; Laske, C.; et al. Increased Plasma Beta-Secretase 1 May Predict Conversion to Alzheimer's Disease Dementia in Individuals With Mild Cognitive Impairment. *Biol. Psychiatry* **2018**, *83*, 447–455. [CrossRef] [PubMed]
94. Wu, G.; Sankaranarayanan, S.; Wong, J.; Tugusheva, K.; Michener, M.S.; Shi, X.; Cook, J.J.; Simon, A.J.; Savage, M.J. Characterization of plasma beta-secretase (BACE1) activity and soluble amyloid precursor proteins as potential biomarkers for Alzheimer's disease. *J. Neurosci. Res.* **2012**, *90*, 2247–2258. [CrossRef] [PubMed]
95. Veitinger, M.; Varga, B.; Guterres, S.B.; Zellner, M. Platelets, a reliable source for peripheral Alzheimer's disease biomarkers? *Acta Neuropathol. Commun.* **2014**, *2*, 65. [CrossRef] [PubMed]
96. Lista, S.; O'Bryant, S.E.; Blennow, K.; Dubois, B.; Hugon, J.; Zetterberg, H.; Hampel, H. Biomarkers in Sporadic and Familial Alzheimer's Disease. *J. Alzheimers Dis* **2015**, *47*, 291–317. [CrossRef] [PubMed]

97. Hye, A.; Kerr, F.; Archer, N.; Foy, C.; Poppe, M.; Brown, R.; Hamilton, G.; Powell, J.; Anderton, B.; Lovestone, S. Glycogen synthase kinase-3 is increased in white cells early in Alzheimer's disease. *Neurosci. Lett.* **2005**, *373*, 1–4. [CrossRef]
98. Brosseron, F.; Traschutz, A.; Widmann, C.N.; Kummer, M.P.; Tacik, P.; Santarelli, F.; Jessen, F.; Heneka, M.T. Characterization and clinical use of inflammatory cerebrospinal fluid protein markers in Alzheimer's disease. *Alzheimers Res. Ther.* **2018**, *10*, 25. [CrossRef]
99. Yarchoan, M.; Louneva, N.; Xie, S.X.; Swenson, F.J.; Hu, W.; Soares, H.; Trojanowski, J.Q.; Lee, V.M.; Kling, M.A.; Shaw, L.M.; et al. Association of plasma C-reactive protein levels with the diagnosis of Alzheimer's disease. *J. Neurol. Sci.* **2013**, *333*, 9–12. [CrossRef]
100. Rauchmann, B.S.; Schneider-Axmann, T.; Alexopoulos, P.; Perneczky, R.; Alzheimer's Disease Neuroimaging, I. CSF soluble TREM2 as a measure of immune response along the Alzheimer's disease continuum. *Neurobiol. Aging* **2019**, *74*, 182–190. [CrossRef]
101. Piccio, L.; Deming, Y.; Del-Aguila, J.L.; Ghezzi, L.; Holtzman, D.M.; Fagan, A.M.; Fenoglio, C.; Galimberti, D.; Borroni, B.; Cruchaga, C. Cerebrospinal fluid soluble TREM2 is higher in Alzheimer disease and associated with mutation status. *Acta Neuropathol.* **2016**, *131*, 925–933. [CrossRef]
102. Brosseron, F.; Krauthausen, M.; Kummer, M.; Heneka, M.T. Body fluid cytokine levels in mild cognitive impairment and Alzheimer's disease: A comparative overview. *Mol. Neurobiol.* **2014**, *50*, 534–544. [CrossRef]
103. Heneka, M.T.; Carson, M.J.; El Khoury, J.; Landreth, G.E.; Brosseron, F.; Feinstein, D.L.; Jacobs, A.H.; Wyss-Coray, T.; Vitorica, J.; Ransohoff, R.M.; et al. Neuroinflammation in Alzheimer's disease. *Lancet Neurol.* **2015**, *14*, 388–405. [CrossRef]
104. Hu, W.T.; Chen-Plotkin, A.; Arnold, S.E.; Grossman, M.; Clark, C.M.; Shaw, L.M.; Pickering, E.; Kuhn, M.; Chen, Y.; McCluskey, L.; et al. Novel CSF biomarkers for Alzheimer's disease and mild cognitive impairment. *Acta Neuropathol.* **2010**, *119*, 669–678. [CrossRef] [PubMed]
105. Muszynski, P.; Groblewska, M.; Kulczynska-Przybik, A.; Kulakowska, A.; Mroczko, B. YKL-40 as a Potential Biomarker and a Possible Target in Therapeutic Strategies of Alzheimer's Disease. *Curr. Neuropharmacol.* **2017**, *15*, 906–917. [CrossRef] [PubMed]
106. Craig-Schapiro, R.; Perrin, R.J.; Roe, C.M.; Xiong, C.; Carter, D.; Cairns, N.J.; Mintun, M.A.; Peskind, E.R.; Li, G.; Galasko, D.R.; et al. YKL-40: A novel prognostic fluid biomarker for preclinical Alzheimer's disease. *Biol. Psychiatry* **2010**, *68*, 903–912. [CrossRef] [PubMed]
107. Henjum, K.; Almdahl, I.S.; Arskog, V.; Minthon, L.; Hansson, O.; Fladby, T.; Nilsson, L.N. Cerebrospinal fluid soluble TREM2 in aging and Alzheimer's disease. *Alzheimers Res. Ther.* **2016**, *8*, 17. [CrossRef]
108. Bermejo-Pareja, F.; Antequera, D.; Vargas, T.; Molina, J.A.; Carro, E. Saliva levels of Abeta1-42 as potential biomarker of Alzheimer's disease: A pilot study. *BMC Neurol.* **2010**, *10*, 108. [CrossRef]
109. Lee, M.; Guo, J.P.; Kennedy, K.; McGeer, E.G.; McGeer, P.L. A Method for Diagnosing Alzheimer's Disease Based on Salivary Amyloid-beta Protein 42 Levels. *J. Alzheimers Dis.* **2017**, *55*, 1175–1182. [CrossRef]
110. Yilmaz, A.; Geddes, T.; Han, B.; Bahado-Singh, R.O.; Wilson, G.D.; Imam, K.; Maddens, M.; Graham, S.F. Diagnostic Biomarkers of Alzheimer's Disease as Identified in Saliva using 1H NMR-Based Metabolomics. *J. Alzheimers Dis.* **2017**, *58*, 355–359. [CrossRef]
111. Sabbagh, M.N.; Shi, J.; Lee, M.; Arnold, L.; Al-Hasan, Y.; Heim, J.; McGeer, P. Salivary beta amyloid protein levels are detectable and differentiate patients with Alzheimer's disease dementia from normal controls: Preliminary findings. *BMC Neurol.* **2018**, *18*, 155. [CrossRef]
112. Koronyo, Y.; Salumbides, B.C.; Black, K.L.; Koronyo-Hamaoui, M. Alzheimer's disease in the retina: Imaging retinal abeta plaques for early diagnosis and therapy assessment. *Neurodegener. Dis.* **2012**, *10*, 285–293. [CrossRef]
113. Koronyo, Y.; Biggs, D.; Barron, E.; Boyer, D.S.; Pearlman, J.A.; Au, W.J.; Kile, S.J.; Blanco, A.; Fuchs, D.T.; Ashfaq, A.; et al. Retinal amyloid pathology and proof-of-concept imaging trial in Alzheimer's disease. *JCI Insight* **2017**, *2*. [CrossRef] [PubMed]
114. Shah, T.M.; Gupta, S.M.; Chatterjee, P.; Campbell, M.; Martins, R.N. Beta-amyloid sequelae in the eye: A critical review on its diagnostic significance and clinical relevance in Alzheimer's disease. *Mol. Psychiatry* **2017**, *22*, 353–363. [CrossRef]
115. Attems, J.; Jellinger, K.A. Olfactory tau pathology in Alzheimer disease and mild cognitive impairment. *Clin. Neuropathol.* **2006**, *25*, 265–271. [PubMed]

116. Passali, G.C.; Politi, L.; Crisanti, A.; Loglisci, M.; Anzivino, R.; Passali, D. Tau Protein Detection in Anosmic Alzheimer's Disease Patient's Nasal Secretions. *Chemosens. Percept.* **2015**, *8*, 201–206. [CrossRef]
117. Jung, H.J.; Shin, I.S.; Lee, J.E. Olfactory function in mild cognitive impairment and Alzheimer's disease: A meta-analysis. *Laryngoscope* **2019**, *129*, 362–369. [CrossRef] [PubMed]

Journal of *Personalized Medicine*

Review

Current Biomarkers for Alzheimer's Disease: From CSF to Blood

Kun Zou *, Mohammad Abdullah and Makoto Michikawa

Department of Biochemistry, Nagoya City University Graduate School of Medical Sciences, Kawasumi 1, Mizuho-cyo, Mizuho-ku, Nagoya 467-8601, Aichi, Japan; mabdullahz49@gmail.com (M.A.); michi@med.nagoya-cu.ac.jp (M.M.)

* Correspondence: kunzou@med.nagoya-cu.ac.jp; Tel.: +81-52-853-8141; Fax: +81-52-841-3480

Received: 22 July 2020; Accepted: 10 August 2020; Published: 12 August 2020

Abstract: Alzheimer's disease (AD) is the most common cause of dementia and affects a large portion of the elderly population worldwide. Currently, a diagnosis of AD depends on the clinical symptoms of dementia, magnetic resonance imaging to determine brain volume, and positron emission tomography imaging to detect brain amyloid or tau deposition. The best characterized biological fluid markers for AD are decreased levels of amyloid β-protein (Aβ) 42 and increased levels of phosphorylated tau and total tau in cerebrospinal fluid (CSF). However, less invasive and easily detectable biomarkers for the diagnosis of AD, especially at the early stage, are still under development. Here, we provide an overview of various biomarkers identified in CSF and blood for the diagnostics of AD over the last 25 years. CSF biomarkers that reflect the three hallmarks of AD, amyloid deposition, neurofibrillary tangles, and neurodegeneration, are well established. Based on the need to start treatment in asymptomatic people with AD and to screen for AD risk in large numbers of young, healthy individuals, the development of biomarkers for AD is shifting from CSF to blood. Elements of the core pathogenesis of AD in blood, including Aβ42, tau proteins, plasma proteins, or lipids have shown their usefulness and capabilities in AD diagnosis. We also highlight some novel identified blood biomarkers (including Aβ42/Aβ43, p-tau 181, Aβ42/APP669-711, structure of Aβ in blood, and flotillin) for AD.

Keywords: Alzheimer's disease; biomarker; cerebrospinal fluid; blood

---

1. Introduction

Alzheimer disease (AD) is an age-dependent neurodegenerative disorder and the most prevalent form of dementia in the elderly population. AD is characterized by amyloid β-protein (Aβ) deposition in senile plaques in the brain parenchyma and by phosphorylated tau deposition in neurofibrillary tangles in cerebral neurons [1]. Until the 2000s, clinical diagnosis of AD depended on clinical symptoms, cognitive examination, and the exclusion of other etiologies of dementia. A definitive positive diagnosis of AD could only be made by post-mortem pathological confirmation of brain parenchymal Aβ deposition and neurofibrillary tangles [2]. Later, structural imaging of the hippocampus with magnetic resonance imaging became an integral part of the clinical assessment of patients with AD [3,4]. Recently, innovative imaging for brain Aβ deposition in patients using positron emission tomography (PET) technology was approved for clinical use [5].

The exploration of AD biomarkers in biological fluid has focused on the core molecules of AD pathogenesis, Aβ and tau proteins. Aβ in brain senile plaques contains 40–43 amino acids, and Aβ42 and Aβ40 are the major species generated by sequential proteolytic cleavage of amyloid precursor protein (APP) by β- and γ-secretases [6]. Most APP undergoes non-amyloidogenic processing by α-secretase, generating a non-amyloidogenic fragment of Aβ, called p3 [7]. The longer species, Aβ42 and Aβ43, are highly prone to aggregation, deposit early in the brain, and the oligomers are highly toxic to neurons [1,8,9], whereas Aβ40 may have antioxidant and anti-amyloidogenic effects [10–12]. Although numerous arguments remain regarding the causative molecule for AD, Aβ or phosphorylated tau (p-tau), extensive evidence suggests that Aβ42 deposition in the brain parenchyma appears earlier than p-tau deposition in neurofibrillary tangles and can be detected in the brain many years prior to the appearance of AD clinical symptoms [13]. In addition, the strongest evidence for Aβ42 as the causative molecule for AD is from genetic studies of familial AD (FAD) with mutations in *APP*, presenilin 1 (*PSEN1*, PS), or presenilin 2 (*PSEN2*, PS), which lead to the highest risk for AD among all AD risk genes identified so far [14]. PS1 and PS2 are the catalytic components of γ-secretase for Aβ generation. The involvement of either substrate (APP) or enzyme (PS) in FAD indicates a central role for Aβ42 in AD pathogenesis [15]. Thus, Aβ42 became the most important target in terms of both biomarkers and therapeutic strategy development for AD.

Current approved treatments for AD target its symptoms, but more and more clinical trials are testing potential disease-modifying drugs, which target the most upstream molecule of AD pathogenesis: Aβ42. However, over the past few decades, many anti-Aβ clinical trials have failed to treat symptomatic AD. Recently, Phase 3 clinical trials using a β-secretase inhibitor or a γ-secretase inhibitor to inhibit Aβ generation, or using Aβ antibodies to promote Aβ clearance in the early or mid-stage of AD, also failed to achieve their expected therapeutic effect [16–18]. One suggestion for these failures is that the anti-Aβ treatment could not rescue degenerative neurons or synapses that were already damaged by toxic Aβ42. Current clinical trials targeting Aβ have focused on preclinical patients without AD symptoms rather than symptomatic AD patients or patients with mild cognitive impairment (MCI). Thus, reliable biomarkers are required for rapid, early, and less-invasive detection in the predementia phase. Biomarkers identified from cerebrospinal fluid (CSF) and blood have shown high potential to diagnose AD at the early stage or to predict AD onset in the future.

In this mini review, we highlight biomarkers for AD diagnosis found in CSF and blood, including the core pathological proteins, Aβ42 and tau, and neurodegeneration- and metabolism-related biomarkers (Figure 1). In addition to the breakthrough finding of decreased Aβ42 and increased p-tau in CSF for AD diagnosis, additional, less-invasive, and easily accessible blood biomarkers are emerging.

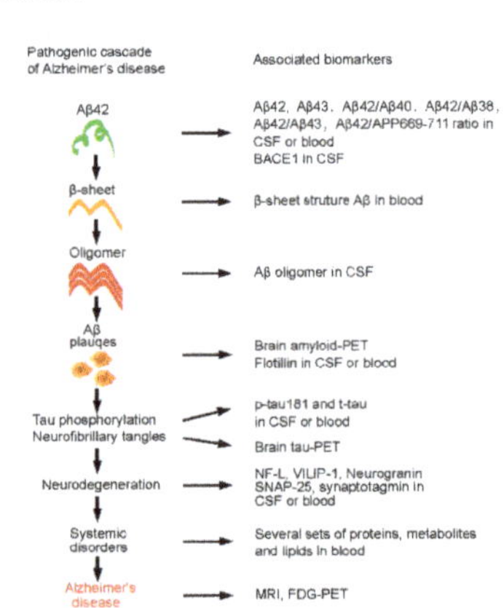

Figure 1. Pathogenic cascade and associated biomarkers of Alzheimer's disease (AD). Amyloid cascade hypothesis of AD is shown on the left side. Selected and associated biomarkers at each pathogenic stage of AD are shown on the right side. Aβ42, amyloid β-protein 42; Aβ43, amyloid β-protein 43; Aβ40, amyloid β-protein 40; Aβ38, amyloid β-protein 38; APP, amyloid precursor protein; BACE1, β-site APP-cleaving enzyme 1; CSF, cerebrospinal fluid; PET, positron emission tomography; p-tau, phosphorylated tau; t-tau, total tau; NF-L, neurofilament light; VILIP-1, visinin-like protein 1; SNAP-25, synaptosome-associated protein 25; MRI, magnetic resonance imaging; FDG, fluorodeoxyglucose.

2. CSF Markers of AD

2.1. AD Pathogenesis Molecule-Based Biomarkers in CSF

2.1.1. CSF Aβ Markers

CSF is in indirect contact with the extracellular space of the brain, and biochemical changes in the brain are therefore reflected in the CSF. Neurovascular and blood–brain barrier dysfunction may develop in neurodegenerative diseases. CSF is thus the optimal source for AD biomarkers [19]. The core pathogenic molecules for AD, Aβ42, total tau (t-tau), and p-tau in CSF, were studied beginning in the 1990s and have become reliable and sensitive biomarkers for AD diagnosis [20,21]. A marked decrease in CSF Aβ42 and a marked increases in CSF t-tau and p-tau can be used to identify symptomatic AD patients with a sensitivity and specificity above 80% [22]. The decrease in CSF Aβ42 is also useful for predicting later AD development. In a population-based study, Skoog et al. found that CSF Aβ42 is reduced before the onset of sporadic dementia [23]. Gustafson et al. reported that low levels of CSF Aβ42, but not high t-tau, may predict cognitive decline in older women [24]. Similarly, Stomrud et al. found that CSF Aβ42, but not t-tau or p-tau, can predict cognitive decline in healthy elderly people [25]. In contrast to CSF Aβ42, CSF Aβ40, the predominant species of Aβ, did not show a significant change in an AD group compared with a normal control group [20]. In another study by Nutu et al., the CSF Aβ40 level in AD was higher than that in Parkinson's disease dementia (PDD) and dementia with Lewy bodies (DLB), suggesting that the Aβ42/40 ratio may improve differentiation of AD from PDD and DLB [26]. Janelidze et al. also found that decreases in the Aβ42/40 and Aβ42/38 ratios may be better diagnostic markers of AD than CSF Aβ42 alone for discrimination of AD from non-AD conditions, especially from DLB, PDD, and vascular dementia (VaD) [27]. Similarly, Baldeiras et al. reported that addition of the Aβ42/40 ratio to the CSF biomarker profile increases the predictive value for underlying AD in MCI [28]. Recently, the usefulness of using CSF Aβ42 to predict preclinical AD was confirmed in cognitively normal individuals with inherited AD genes. An FAD cohort study from the Dominantly Inherited Alzheimer Network indicated that CSF Aβ42 levels may first increase and then start to decline 25 years before the onset of AD symptoms, whereas amyloid deposition measured with PET and Pittsburgh compound B and an increased concentration of t-tau in the CSF can be detected 15 years before expected symptom onset [29]. These findings raise the possibility that the decline in CSF Aβ42 may be the earliest marker for identifying preclinical AD, not only in FAD, but also in sporadic AD. However, obtaining CSF is invasive, risky, and unsuitable for screening healthy people.

Combined with hippocampal volumetry, fluorodeoxyglucose (FDG) PET, amyloid PET, decreased Aβ42, and increased tau or p-tau in CSF have been used in the National Institute for Aging-Alzheimer Association (NIA-AA) criteria to diagnose AD, to predict later onset of AD, and to differentiate AD from normal aging with MCI [1].

Oligomerization of Aβ42 has long been suggested as the central pathogenic event in AD [6,30]. The decrease in CSF Aβ42 was thought to be caused, at least in part, by deposition in amyloid plaques. Another interpretation is that oligomerization or aggregation of CSF Aβ42 reduces the detection of Aβ42 monomers with an enzyme-linked immunosorbent assay (ELISA). One ELISA method for examining Aβ oligomers was designed by Fumumoto et al., using the same antibody for capture and detection. Monomeric Aβ will not be detected because the epitope is already occupied by the captured antibody. They found that the level of Aβ oligomers in CSF was higher in AD or MCI patients compared with age-matched controls [31]. However, this study cannot identify the nature of the Aβ oligomers, e.g., dimers, trimers, or tetramers, and this finding needs to be confirmed in other larger, independent studies.

2.1.2. CSF t-tau and p-tau Markers

Tau proteins are located in neuronal axons and play roles in maintaining the stability of microtubules in neurons of the central nervous system (CNS). Tau proteins in the human brain are composed of six soluble isoforms and numerous phosphorylation sites [32]. Hyperphosphorylated tau proteins disassociate from microtubules and form insoluble aggregates in neurons, which are called neurofibrillary tangles [33,34]. CSF t-tau and p-tau are frequently studied in neurodegenerative diseases. CSF t-tau levels can serve as a neuronal injury marker and are increased in many neurodegenerative diseases, such as Creutzfeldt-Jakob disease [35], AD, DLB, and frontotemporal dementia (FTD), whereas CSF p-tau 181 or p-tau 231 (tau phosphorylated at threonine 181 or threonine 231) levels increase more specifically in AD than in other neurodegenerative diseases. Thus, p-tau may reflect the hyperphosphorylation of tau and the formation of neurofibrillary tangles in AD [2,36,37].

Blennow et al. demonstrated that marked elevations of t-tau and paired helical filaments (PHF)-tau (tau phosphorylated at serine 202 and threonine 205) are consistently found in the CSF of AD patients. However, moderate elevations of t-tau and PHF-tau are also found in other neurodegenerative diseases, such as VaD and frontal lobe dementia [38]. Later, Vanmechelen et al. reported a method using sandwich ELISA for detecting p-tau 181 and found that CSF p-tau 181 levels were increased in AD patients compared with age-matched controls, whereas levels were decreased in patients with FTD, suggesting that CSF p-tau 181 could be a more specific marker for AD [39]. Kohnken et al. developed a sandwich ELISA for detecting p-tau threonine 231 that shows 85% sensitivity and 97% specificity for discrimination of AD from non-AD controls [40]. These findings were confirmed in numerous subsequent studies. In a meta-analysis comprised of 164 cohorts with AD and 153 control cohorts representing 11,341 AD patients and 7086 controls, increased levels of CSF t-tau and p-tau were strongly associated with AD and MCI patients that developed AD [41]. Similar to CSF Aβ42, although CSF t-tau and p-tau are already included in the diagnostic criteria for symptomatic or prodromal AD, they are difficult to use in healthy people at the preclinical stage because of the limitation of obtaining CSF samples.

2.1.3. CSF β-Site APP-Cleaving Enzyme 1 (BACE1) Marker

BACE1 is the major β-secretase responsible for Aβ generation. Mutations in *BACE1* have not yet been reported in FAD. However, mutations in APP near the β-cleavage site may be responsible for early-onset FAD (Swedish mutation) or may be protective for preventing late-onset sporadic AD [42]. To study whether up-regulation of BACE1 is an early pathogenic event in AD, some human in vivo studies have reported good diagnostic performance of CSF BACE1 levels and activity for separating symptomatic AD patients and patients with MCI from cognitively healthy controls. Holsinger et al. found increased BACE1 activity in the CSF of AD patients, whereas Zhong et al. showed that increased CSF BACE1 levels can be a predictor of risk in patients with MCI [43,44]. Later, Ewers et al. reported that increased CSF BACE1 activity is not only associated with the *apoE4* genotype in MCI and AD patients, but is also associated with decreased hippocampal volume in AD [45,46]. However, the diagnostic value of CSF BACE1 activity requires further evaluation and confirmation in larger studies from different groups.

2.2. Neurodegeneration-Based Biomarkers in CSF

In addition to the three core CSF biomarkers, Aβ42, t-tau, and p-tau, biomarkers that reflect axonal neurodegeneration, synapse loss, and activation of glial cells have also been extensively explored [37,47].

Neurofilaments are intermediate filaments that serve as structural components of neuronal axons, particularly large myelinated axons. In the CNS, neurofilaments are heteropolymers and are composed of four subunits, neurofilament light (NF-L), neurofilament middle, neurofilament heavy, and α-internexin [48]. NF has been extensively examined in patients with neuronal injury and neurodegenerative diseases because it is released into CSF and systemic circulation when neurons are injured [49]. Recently, Sjogren et al. found that CSF NF-L levels are increased in patients with FTD and late onset AD compared with control subjects, and the increase in FTD patients is higher than in late onset AD [50]. In a meta-analysis, Olsson et al. found that NF-L has a large effect size for differentiating AD patients from control individuals [41]. However, high CSF levels of NF-L are also found in other neurodegenerative diseases, such VaD, normal-pressure hydrocephalus, multiple sclerosis, and amyotrophic lateral sclerosis [51–54]. Thus, CSF NF-L could be a representative marker of neurodegeneration, but not a specific marker for distinguishing AD from other neurological disorders. Nevertheless, Zetterberg et al. showed that higher CSF NF-L concentrations are associated with cognitive deterioration and brain atrophy over time in AD and MCI groups, indicating that CSF NF-L can be used as a marker for AD progression [49].

Visinin-like protein 1 (VILIP-1) is a neuronal calcium sensor protein that is exclusively expressed in neurons and can be used as a brain injury marker [55]. Lee et al. found that CSF VILIP-1 levels are significantly higher in AD patients compared with controls and that the diagnostic performance of VILIP-1 is similar to CSF Aβ42, t-tau, or p-tau [56]. Higher CSF VILIP-1 levels in AD patients compared with controls have also been reported in several other studies. Tarawneh et al. reported that CSF VILIP-1 and CSF VILIP-1/Aβ42 ratios are increased in early AD, suggesting the utility of them as diagnostic or prognostic markers of AD [57]. Later, they reported that CSF VILIP-1 levels can predict rates of whole-brain and regional atrophy, similar to tau and p-tau 181 [58]. CSF VILIP-1 levels have been consistently shown to be higher in AD patients compared with normal controls [59–61]. Luo et al. showed that CSF VILIP-1 levels are significantly increased in AD patients compared with both normal controls and DLB patients. However, a recent meta-analysis performed by Mavroudis et al. did not show a significant difference between AD and DLB [62]. Because the reports are limited, whether CSF VILIP-1 can be used as a specific marker for AD that discriminates AD from other neurodegenerative diseases needs to be further studied.

In addition to NF-L, which represents axonal damage, several pre- and postsynaptic proteins are increased in the CSF of AD patients, such as neurogranin, synaptosome-associated protein 25 (SNAP-25), and synaptotagmin. Neurogranin is a postsynaptic protein that is predominantly expressed in dendritic spines and plays a role in postsynaptic signaling pathways. Using immunoprecipitation enrichment of neurogranin, Thorsell et al. found a significant increase in neurogranin in the CSF of AD patients compared with a control group [63]. Several studies from different groups consistently confirmed higher levels of CSF neurogranin in AD patients compared to controls [64–66]. Keter et al. further showed that CSF levels of neurogranin are higher in patients with MCI who progressed to AD compared with those with stable MCI, indicating that neurogranin can be used as a predictive factor of progression from MCI to AD [67]. Tarawneh et al. proposed the diagnostic and prognostic utility of CSF neurogranin as a synaptic marker in early symptomatic AD [68]. However, the diagnostic value of CSF neurogranin in AD or MCI is still based on other diagnostic indexes of AD. Lista et al. have shown that CSF neurogranin concentrations are significantly higher in AD patients compared with FTD patients [66]. CSF neurogranin levels in other types of dementia or neurodegenerative diseases need to be studied further.

By using novel affinity mass spectrometry, Brinkmalm et al. found significantly higher levels of CSF SNAP-25 fragments in AD patients than controls [69]. In another longitudinal study, Sutphen et al. also revealed that CSF SNAP-25 levels are significantly higher in AD and MCI patients than controls, but decline over time in the AD group [61]. The increase in SNAP-25 fragments in CSF has the highest power among synaptic biomarkers to distinguish AD patients from non-AD patients [70].

Ohrfelt et al. reported that the presynaptic protein, synaptotagmin, is significantly increased in the CSF of patients with AD, or MCI patients that developed AD [71]. Recently, Tible et al. confirmed that all these synaptic biomarkers are significantly increased in patients with AD and MCI patients that developed AD. Given that the synaptic proteins are general markers of synaptic dysfunction, they likely can be used as supplementary diagnostic biomarkers for AD or MCI patients that have developed AD, but not as exclusive diagnostic markers for AD.

3. Blood Markers of AD

3.1. AD Pathogenesis Molecule-Based Biomarkers in Blood

3.1.1. Blood Aβ Markers

As with CSF Aβ levels, Aβ42 and Aβ40 are the most extensively studied blood markers for the diagnosis of symptomatic and prodromal AD. During the first decade of the 2000s, the findings regarding Aβ42 and Aβ40 levels in the plasma of AD patients were not consistent, and sometimes the results were contradictory. Mayeux et al. found increased plasma Aβ42 levels, but not plasma Aβ40, in AD patients at baseline and in those who developed AD within 3 years in a follow-up study. The risk of AD onset in individuals with high plasma Aβ42 was increased more than 2-fold compared to those with low plasma Aβ42 [72]. In later studies, van Oijen et al. reported that a high concentration of plasma Aβ40 is associated with an increased risk of dementia [73], whereas Yaffe et al. found that a lower plasma Aβ42/Aβ40 ratio is associated with greater cognitive decline among elderly persons without dementia over 9 years [74]. This discrepancy may come from the clinical stage of examination and/or the mix of other types of dementia.

Using magnetic resonance imaging for hippocampal volumetry and amyloid PET technology, patients with AD and MCI patients that developed AD can be specifically discriminated from patients with other types of dementia and MCI patients that did not develop AD in the last decade. Zou et al. showed that Aβ42 levels are lower, and Aβ43 levels are higher, in the serum of AD patients compared with age-matched normal controls, suggesting that a lower Aβ42/Aβ43 ratio can be used as a blood marker for AD diagnosis [9]. In two independent data sets, Nakamura et al. also revealed a significant decrease in plasma Aβ42 levels in brain amyloid-positive AD or MCI patients compared with cognitively normal individuals. They also found that the combination of decreased Aβ42/APP669–711 and the Aβ42/Aβ40 ratio showed the highest and most stable performance in predicting brain amyloid burden at an individual level [75]. Recently, a correlation between a lower plasma Aβ42/Aβ40 ratio and amyloid burden was consistently confirmed in other independent studies. Perez-Grijalba et al. showed that a decreased plasma Aβ42/Aβ40 ratio alone can accurately predict positivity and detect early stages of AD [76,77]. Using a multiplex sensor array, Kim et al. showed that a lower Aβ42/Aβ40 ratio and a higher plasma t-tau/Aβ42 and p-tau 181/Aβ42 ratios successfully discriminated AD patients from healthy controls [78]. In addition to plasma Aβ levels, Nabers et al. showed that a change in the secondary structure of Aβ in human blood plasma can be used as a blood amyloid indicator for prodromal AD. The change to an increased β-sheet structure of Aβ is correlated with CSF AD biomarkers and amyloid PET imaging [79]. Because the structure of Aβ is not stable and may change in several hours, this technology needs relatively high techniques and must be confirmed by other independent studies.

Consistent with the findings of lower CSF Aβ42 levels in AD patients, recent studies strongly suggest that plasma Aβ42 levels also decrease in AD patients or amyloid-positive MCI patients. Thus, the combined use of Aβ42/Aβ40, Aβ42/Aβ43, Aβ42/APP669-711, Aβ42/t-tau, or Aβ42/p-tau 181 may accurately diagnose or predict AD.

3.1.2. Blood p-tau Markers

Because of the invasiveness and high costs of examining CSF tau, plasma tau has also become a candidate blood marker for AD diagnosis, and many studies have focused on quantitation of tau in AD, MCI, and normal groups. Because the tau levels in plasma are much lower than in CSF, an ultra-sensitive assay was developed by Zetterberg et al. They found elevated t-tau levels in plasma from patients with AD compared with those from control or MCI patients, whereas no difference was found between MCI patients that developed AD and stable MCI patients [80]. In a later study, Mattsson et al. studied two large cohorts and reported that plasma t-tau may partly reflect AD pathology, but a large overlap was found between patients with AD and age-matched controls, suggesting that using plasma t-tau as an AD biomarker in individual people is difficult [81].

Recently, Tatebe et al. developed a novel ultrasensitive immunoassay for the quantitation of plasma p-tau 181. Although the number of participants was small, they clearly showed that plasma p-tau 181 is significantly increased in patients with AD, as well as in patients with Down's syndrome, compared with normal controls [82]. Karikari et al. further confirmed the increase in plasma p-tau 181 levels in patients with AD and MCI patients that developed AD and showed that plasma p-tau 181 can discriminate AD dementia from not only normal young and older adults, but also FTD, VaD, progressive supranuclear palsy, corticobasal syndrome, Parkinson's disease, and multiple system atrophy [83].

3.2. Other Biomarkers in Blood

Because neuronal or synaptic biomarkers indicate general neuronal injury in many neurodegenerative diseases, the blood levels of these proteins may not be specific markers for AD. Benussi et al. assessed the diagnostic and prognostic value of serum NF-L and serum p-tau 181. They found that serum NF-L levels are increased in both FTD and AD and cannot distinguish AD from FTD, whereas serum p-tau 181 levels are specifically increased in patients with AD [84]. Similarly, plasma NF-L is increased in both progressive supranuclear palsy and AD [85]. Regarding the use of other neurodegeneration-based biomarkers for AD diagnosis such as VILIP-1, neurogranin, SNAP-25, and synaptotagmin, very few studies were performed on their levels in plasma, and the results were inconsistent and very limited [37,47]. Therefore, these neuronal and synaptic biomarkers are currently considered to be representative of neurodegeneration, and may be parameters for assessing the progression or degree of AD and other types of dementia, but not useful for accurate diagnosis of AD.

In addition to the molecules related to AD core pathogenesis and neurodegeneration, other plasma proteins, lipids, and metabolites were also extensively studied in patients with AD. Ray et al. used ELISA and identified 18 signaling proteins in blood plasma. The change in the pattern of those proteins can distinguish AD and MCI that progressed to AD from control subjects with near 90% accuracy [86]. Using a multiplex assay, Doecke et al. identified another set of plasma proteins that distinguishes individuals with AD from healthy controls with high sensitivity and specificity [87], and Hye et al. identified 10 plasma proteins that are strongly associated with progression from MCI to AD [88]. The change in these plasma proteins seems to result from neurodegeneration or other systemic disorders in AD. Because the number of these plasma proteins is large, and examining all of the proteins is expensive, the patterns of change in these proteins in AD have still not been confirmed by other independent studies.

The systemic abnormalities in lipid metabolism in the blood of AD patients have also been studied by using quantitative and targeted metabolomics and mass spectrometry. Mapstone et al. identified 10 phospholipids from healthy elderly people that predicted conversion to either MCI or AD within 2–3 years with over 90% accuracy, suggesting their use in detection of early neurodegeneration in preclinical AD [89]. Recently, Varma et al. also identified four sphingolipids and found that their higher blood concentrations in cognitively normal individuals are associated with an increased risk of future conversion to incident AD [90]. The change in the levels of these phospholipids and sphingolipids in blood may reflect a disorder of lipid metabolism and/or neuronal degeneration in the CNS at the very early stage without cognitive symptoms. However, whether they distinguish AD from other types of dementia and neurodegenerative diseases needs to be further investigated. Given the high cost of quantifying a set of plasma proteins or lipids, single blood markers could be easier to use for screening for AD in large populations. In a recent study, Abdullah et al. identified flotillin, an abundant exosome protein, as a novel diagnostic marker for AD. Serum flotillin levels are significantly decreased in patients with AD and amyloid-positive MCI patients compared with age-matched patients with VaD and MCI patients without amyloid [91]. The decrease in flotillin levels may result from reduced exosome secretion caused by Aβ42 oligomers [92]. Thus, flotillin is likely to be a secondary responding molecule to pathogenetic Aβ42. The above CSF and blood biomarkers for AD were summarized in Table 1.

Table 1. Selected biomarkers of AD in cerebrospinal fluid (CSF) and blood.

Biomarker	Relevance in AD	Change in CSF/Blood of AD
Aβ42	Distinguishing AD, mild cognitive impairment (MCI) that developed AD and preclinical AD from normal controls and other neurodegenerative disease	Consistently decreased in CSF, also decreased in blood [20–29,74–78]
Aβ40	Inconsistent results for Aβ40 alone, Aβ42/Aβ40 ratio could be a better biomarker than Aβ42 alone	Aβ42/Aβ40 ratio consistently decreased in CSF, also decreased in blood [20–29,73]
Aβ38	Inconsistent results for Aβ38 alone, Aβ42/Aβ38 ratio could be a better biomarker than Aβ42 alone for discrimination of AD from other dementia	Aβ42/Aβ38 ratio decreased in CSF, very few studies [27]
Aβ43	Distinguishing AD from normal controls	Aβ43 increased and Aβ42/Aβ43 ratio decreased in blood, very few studies [9]
Aβ42/APP669-711	Distinguishing AD from normal controls and MCI that developed AD	Decreased in blood, one study [75]
BACE1	Distinguishing AD and MCI that developed AD from normal controls	Activity and levels increased in CSF, few studies [43–46]
β-sheet structure Aβ	Correlated with amyloid-PET and other established CSF AD biomarkers	Increased in blood, one study [79]
Aβ oligomer	Distinguishing AD and MCI that developed AD from normal controls	Increased in CSF, very few studies [31]
Flotillin	Distinguishing AD and MCI that developed AD from normal controls and vascular dementia (VaD); single blood marker	Decreased in CSF and blood, very few studies [91]
p-tau and t-tau	Distinguishing AD and MCI that developed AD from normal controls, p-tau 181 and p-tau 231 discriminates AD from other dementia	Consistently increased in CSF [38–41]; p-tau 181 increased in blood, several studies [80–83]
NF-L	Distinguishing AD from normal controls, but not other dementia; valuable for assessing neuronal injury	Increased in CSF and blood, several studies [41,49–54,84,85]
VILIP-1	Distinguishing early AD and AD from normal controls, but not other dementia	Increased in CSF, inconsistent and limited results in blood [37,47,55–62]
Synaptic proteins (neurogranin, SNAP-25, synaptotagmin)	Distinguishing AD and MCI developed to AD from normal controls, but not other dementia	Increased in CSF, inconsistent and limited results in blood [37,47,61,63–71]
18 Signaling proteins	Distinguishing AD and MCI developed to AD from normal controls	Pattern changed in blood, very few studies [86]
10 plasma proteins	Predicting progression from MCI to AD	Pattern changed in blood, very few studies [87,88]
10 phospholipids	Detecting preclinical AD from normal controls	Pattern changed in blood, very few studies [89]
4 sphingolipids	Detecting prodromal and preclinical AD from normal controls	Increased in blood, very few studies [90]

Aβ42, amyloid β-protein 42; MCI, mild cognitive impairment; Aβ40, amyloid β-protein 40; Aβ38, amyloid β-protein 38; Aβ43, amyloid β-protein 43; APP, amyloid precursor protein; CSF, cerebrospinal fluid; p-tau, phosphorylated tau; t-tau, total tau; VaD, vascular dementia; NF-L, neurofilament light; VILIP-1, visinin-like protein 1; SNAP-25, synaptosome-associated protein 25.

4. Advantages of Blood Biomarkers over CSF Biomarkers

CSF biomarkers for AD have been studied for more than 20 years, and many powerful markers have been identified for diagnosis, prognosis, or even prediction of the future onset of AD. The combined use of these CSF markers may largely improve the accuracy and sensitivity of AD diagnosis at the early stage. However, because obtaining CSF is invasive and may induce prognostic symptoms, physical examination using CSF samples to screen for the risk of AD in large populations of asymptomatic people is not practical.

Blood test indexes, such as blood cholesterol, triglyceride, high-density lipoprotein cholesterol, and low-density lipoprotein cholesterol, have been widely used for predicting the risks for arteriosclerosis and cerebrovascular and cardiovascular diseases in healthy and asymptomatic populations. However, a safe, less invasive, and readily accessible blood marker for AD diagnosis or for predicting the risk of AD is still not at clinical use stage. In annual physical examinations, blood samples were routinely collected in a large and healthy population from middle age to advanced age. The number and scale of blood samples will take great advantage over CSF samples of developing blood biomarkers for AD. In the past 15 years, many studies on blood markers for AD diagnosis, prognosis, and prediction have been performed, and some biomarkers have emerged as candidates for less invasive blood markers for AD (Table 1). For example, recent studies from different groups suggest that decreased blood Aβ42 and increased blood p-tau 181 may reflect brain amyloid deposition and neurofibrillary tangles, respectively, at the early stage of AD. The changes of some plasma proteins, lipids, Aβ43, and flotillin in the blood samples from AD patients are also needed to be confirmed by different groups. Nevertheless, using blood-borne biomarkers to make a clear AD diagnosis or prognosis will be available in the near future.

5. Conclusions

In this first quarter of the century, hundreds of biomarkers aiming for AD diagnosis and for the early detection of pathological changes in AD have been investigated and reported. Biomarkers reflecting the three hallmarks of AD, amyloid deposition, neurofibrillary tangles, and neurodegeneration, have shown a high accuracy in assisting with AD diagnosis. Of all the biomarkers, CSF biomarkers, including decreased Aβ42 and increased t-tau and p-tau, have been well-established for AD diagnosis and the prediction of future conversion to AD from MCI. These core pathogenesis markers of AD have been included in the diagnostic criteria of AD in NIA-AA; however, the invasiveness of obtaining CSF largely limits their utility in cognitively normal populations. Neurodegeneration-based markers in CSF, including NF-L, VILIP-1, neurogranin, and SNAP-25, also showed high positive correlations with neuronal damage in AD and MCI and can be used for evaluation and prediction of future cognitive decline in AD. Of note, these neurodegeneration markers change when neural damage occurs in various neurodegenerative diseases. Thus, they can be auxiliary markers, especially for evaluating the degree of neuronal damage in AD, but may not be suitable for differential diagnosis of AD dementia from other types of dementia.

In addition to CSF biomarkers, blood markers for AD diagnosis and prediction have been extensively studied. Although some contradictory results were reported regarding the blood levels of Aβ42 in AD patients, recent studies showed decreased blood Aβ42 levels in amyloid-positive AD and MCI patients. Furthermore, small but detectable amounts of p-tau and t-tau are increased in the plasma of AD and MCI patients. However, considerable overlap exists in the plasma Aβ42, t-tau, and p-tau levels between AD patients and age-matched controls. Further identification of other potential molecules and use of the ratios of these molecules to Aβ42 or tau proteins may significantly improve the accuracy and sensitivity for screening and discriminating prodromal or preclinical AD from the normal population. Some sets of plasma proteins and lipids may also have potential in AD diagnosis; however, more specific biomarkers are needed and the cost of examination needs to decrease.

Because effective drugs to stop the progression of AD are still not available, preventive therapies and disease-modifying treatments need to be started at the preclinical stage. Discovering new targets for early AD diagnosis and therapy is still necessary in the future direction of AD research. To screen for a risk of AD in healthy populations, the development of AD biomarkers has shifted to using less invasive (blood) or non-invasive (saliva or urine) samples. Given the extensive studies and convincing evidence provided for blood biomarkers, they are likely to be the next generation of biomarkers for AD diagnosis and risk screening.

Author Contributions: K.Z. wrote the paper. M.A. and M.M. provided information and modified the paper. All authors have read and agreed to the published version of the manuscript.

Funding: This research was funded by grants from the Ministry of Education, Culture, Sports, Science and Technology of Japan, a Grant-in-Aid for Scientific Research (C) (19K07846 to K.Z.), a Grant-in-aid for Early-career Scientists (to M.A.), and from the Daiko Foundation (to K.Z.) and Hirose International Scholarship Foundation (to K.Z.), and from AMED under Grant Number A-128 (to M.M.) and 20dk0207050h0001 (to M.M.).

Conflicts of Interest: The authors declare no conflict of interest.

References

1. Scheltens, P.; Blennow, K.; Breteler, M.M.; de Strooper, B.; Frisoni, G.B.; Salloway, S.; Van der Flier, W.M. Alzheimer's disease. *Lancet* **2016**, *388*, 505–517. [CrossRef]
2. Dubois, B.; Feldman, H.H.; Jacova, C.; Hampel, H.; Molinuevo, J.L.; Blennow, K.; DeKosky, S.T.; Gauthier, S.; Selkoe, D.; Bateman, R.; et al. Advancing research diagnostic criteria for Alzheimer's disease: The IWG-2 criteria. *Lancet Neurol.* **2014**, *13*, 614–629. [CrossRef]
3. Frisoni, G.B.; Jack, C.R.J.; Bocchetta, M.; Bauer, C.; Frederiksen, K.S.; Liu, Y.; Preboske, G.; Swihart, T.; Blair, M.; Cavedo, E.; et al. The EADC-ADNI Harmonized Protocol for manual hippocampal segmentation on magnetic resonance: Evidence of validity. *Alzheimers Dement.* **2015**, *11*, 111–125. [CrossRef]
4. Frisoni, G.B.; Fox, N.C.; Jack, C.R.J.; Scheltens, P.; Thompson, P.M. The clinical use of structural MRI in Alzheimer disease. *Nat. Rev. Neurol.* **2010**, *6*, 67–77. [CrossRef] [PubMed]
5. Herholz, K.; Ebmeier, K. Clinical amyloid imaging in Alzheimer's disease. *Lancet Neurol.* **2011**, *10*, 667–670. [CrossRef]
6. Selkoe, D.J.; Hardy, J. The amyloid hypothesis of Alzheimer's disease at 25 years. *EMBO Mol. Med.* **2016**, *8*, 595–608. [CrossRef]
7. Dewachter, I.; Van Leuven, F. Secretases as targets for the treatment of Alzheimer's disease: The prospects. *Lancet Neurol.* **2002**, *1*, 409–416. [CrossRef]
8. Saito, T.; Suemoto, T.; Brouwers, N.; Sleegers, K.; Funamoto, S.; Mihira, N.; Matsuba, Y.; Yamada, K.; Nilsson, P.; Takano, J.; et al. Potent amyloidogenicity and pathogenicity of Abeta43. *Nat. Neurosci.* **2011**, *14*, 1023–1032. [CrossRef]
9. Zou, K.; Liu, J.; Watanabe, A.; Hiraga, S.; Liu, S.; Tanabe, C.; Maeda, T.; Terayama, Y.; Takahashi, S.; Michikawa, M.; et al. Abeta43 is the earliest-depositing Abeta species in APP transgenic mouse brain and is converted to Abeta41 by two active domains of ACE. *Am. J. Pathol.* **2013**, *182*, 2322–2331. [CrossRef]
10. Zou, K.; Gong, J.S.; Yanagisawa, K.; Michikawa, M. A novel function of monomeric amyloid beta-protein serving as an antioxidant molecule against metal-induced oxidative damage. *J. Neurosci. Off. J. Soc. Neurosci.* **2002**, *22*, 4833–4841. [CrossRef]
11. Zou, K.; Kim, D.; Kakio, A.; Byun, K.; Gong, J.S.; Kim, J.; Kim, M.; Sawamura, N.; Nishimoto, S.; Matsuzaki, K.; et al. Amyloid beta-protein (Abeta)1-40 protects neurons from damage induced by Abeta1-42 in culture and in rat brain. *J. Neurochem.* **2003**, *87*, 609–619. [CrossRef] [PubMed]
12. Kim, J.; Onstead, L.; Randle, S.; Price, R.; Smithson, L.; Zwizinski, C.; Dickson, D.W.; Golde, T.; McGowan, E. Abeta40 inhibits amyloid deposition In Vivo. *J. Neurosci. Off. J. Soc. Neurosci.* **2007**, *27*, 627–633.
13. Blennow, K.; Zetterberg, H.; Fagan, A.M. Fluid biomarkers in Alzheimer disease. *Cold Spring Harb. Perspect. Med.* **2012**, *2*, a006221. [CrossRef]
14. Karch, C.M.; Goate, A.M. Alzheimer's disease risk genes and mechanisms of disease pathogenesis. *Biol. Psychiatry* **2015**, *77*, 43–51. [CrossRef] [PubMed]
15. Walsh, D.M.; Selkoe, D.J. Amyloid beta-protein and beyond: The path forward in Alzheimer's disease. *Curr. Opin. Neurobiol.* **2020**, *61*, 116–124. [CrossRef] [PubMed]
16. Honig, L.S.; Vellas, B.; Woodward, M.; Boada, M.; Bullock, R.; Borrie, M.; Hager, K.; Andreasen, N.; Scarpini, E.; Liu-Seifert, H.; et al. Trial of Solanezumab for Mild Dementia Due to Alzheimer's Disease. *N. Engl. J. Med.* **2018**, *378*, 321–330. [CrossRef]
17. Egan, M.F.; Kost, J.; Tariot, P.N.; Aisen, P.S.; Cummings, J.L.; Vellas, B.; Sur, C.; Mukai, Y.; Voss, T.; Furtek, C.; et al. Randomized Trial of Verubecestat for Mild-to-Moderate Alzheimer's Disease. *N. Engl. J. Med.* **2018**, *378*, 1691–1703. [CrossRef]
18. Doody, R.S.; Raman, R.; Farlow, M.; Iwatsubo, T.; Vellas, B.; Joffe, S.; Kieburtz, K.; He, F.; Sun, X.; Thomas, R.G.; et al. A phase 3 trial of semagacestat for treatment of Alzheimer's disease. *N. Engl. J. Med.* **2013**, *369*, 341–350. [CrossRef]
19. Zhao, Z.; Nelson, A.R.; Betsholtz, C.; Zlokovic, B.V. Establishment and Dysfunction of the Blood-Brain Barrier. *Cell* **2015**, *163*, 1064–1078. [CrossRef]

20. Shoji, M.; Matsubara, E.; Kanai, M.; Watanabe, M.; Nakamura, T.; Tomidokoro, Y.; Shizuka, M.; Wakabayashi, K.; Igeta, Y.; Ikeda, Y.; et al. Combination assay of CSF tau, A beta 1-40 and A beta 1-42(43) as a biochemical marker of Alzheimer's disease. *J. Neurol. Sci.* **1998**, *158*, 134–140. [CrossRef]
21. Blennow, K.; Hampel, H.; Weiner, M.; Zetterberg, H. Cerebrospinal fluid and plasma biomarkers in Alzheimer disease. *Nat. Rev. Neurol.* **2010**, *6*, 131–144. [CrossRef] [PubMed]
22. Blennow, K.; Hampel, H. CSF markers for incipient Alzheimer's disease. *Lancet Neurol.* **2003**, *2*, 605–613. [CrossRef]
23. Skoog, I.; Davidsson, P.; Aevarsson, O.; Vanderstichele, H.; Vanmechelen, E.; Blennow, K. Cerebrospinal fluid beta-amyloid 42 is reduced before the onset of sporadic dementia: A population-based study in 85-year-olds. *Dement. Geriatr. Cogn. Disord.* **2003**, *15*, 169–176. [CrossRef] [PubMed]
24. Gustafson, D.R.; Skoog, I.; Rosengren, L.; Zetterberg, H.; Blennow, K. Cerebrospinal fluid beta-amyloid 1-42 concentration may predict cognitive decline in older women. *J. Neurol. Neurosurg. Psychiatry* **2007**, *78*, 461–464. [CrossRef]
25. Stomrud, E.; Hansson, O.; Blennow, K.; Minthon, L.; Londos, E. Cerebrospinal fluid biomarkers predict decline in subjective cognitive function over 3 years in healthy elderly. *Dement. Geriatr. Cogn. Disord.* **2007**, *24*, 118–124. [CrossRef]
26. Nutu, M.; Zetterberg, H.; Londos, E.; Minthon, L.; Nagga, K.; Blennow, K.; Hansson, O.; Ohrfelt, A. Evaluation of the cerebrospinal fluid amyloid-beta1-42/amyloid-beta1-40 ratio measured by alpha-LISA to distinguish Alzheimer's disease from other dementia disorders. *Dement. Geriatr. Cogn. Disord.* **2013**, *36*, 99–110. [CrossRef]
27. Janelidze, S.; Zetterberg, H.; Mattsson, N.; Palmqvist, S.; Vanderstichele, H.; Lindberg, O.; van Westen, D.; Stomrud, E.; Minthon, L.; Blennow, K.; et al. CSF Abeta42/Abeta40 and Abeta42/Abeta38 ratios: Better diagnostic markers of Alzheimer disease. *Ann. Clin. Transl. Neurol.* **2016**, *3*, 154–165. [CrossRef]
28. Baldeiras, I.; Santana, I.; Leitao, M.J.; Gens, H.; Pascoal, R.; Tabuas-Pereira, M.; Beato-Coelho, J.; Duro, D.; Almeida, M.R.; Oliveira, C.R. Addition of the Abeta42/40 ratio to the cerebrospinal fluid biomarker profile increases the predictive value for underlying Alzheimer's disease dementia in mild cognitive impairment. *Alzheimers Res. Ther.* **2018**, *10*, 33. [CrossRef]
29. Bateman, R.J.; Xiong, C.; Benzinger, T.L.; Fagan, A.M.; Goate, A.; Fox, N.C.; Marcus, D.S.; Cairns, N.J.; Xie, X.; Blazey, T.M.; et al. Clinical and biomarker changes in dominantly inherited Alzheimer's disease. *N. Engl. J. Med.* **2012**, *367*, 795–804. [CrossRef]
30. Walsh, D.M.; Selkoe, D.J. A beta oligomers—A decade of discovery. *J. Neurochem.* **2007**, *101*, 1172–1184. [CrossRef]
31. Fukumoto, H.; Tokuda, T.; Kasai, T.; Ishigami, N.; Hidaka, H.; Kondo, M.; Allsop, D.; Nakagawa, M. High-molecular-weight beta-amyloid oligomers are elevated in cerebrospinal fluid of Alzheimer patients. *FASEB J.* **2010**, *24*, 2716–2726. [CrossRef] [PubMed]
32. Khan, S.S.; Bloom, G.S. Tau: The Center of a Signaling Nexus in Alzheimer's Disease. *Front. Neurosci.* **2016**, *10*, 31. [CrossRef] [PubMed]
33. Lee, V.M.; Brunden, K.R.; Hutton, M.; Trojanowski, J.Q. Developing therapeutic approaches to tau, selected kinases, and related neuronal protein targets. *Cold Spring Harb. Perspect. Med.* **2011**, *1*, a006437. [CrossRef] [PubMed]
34. Lee, V.M.; Balin, B.J.; Otvos, L.J.; Trojanowski, J.Q. A68: A major subunit of paired helical filaments and derivatized forms of normal Tau. *Science* **1991**, *251*, 675–678. [CrossRef] [PubMed]
35. Skillback, T.; Rosen, C.; Asztely, F.; Mattsson, N.; Blennow, K.; Zetterberg, H. Diagnostic performance of cerebrospinal fluid total tau and phosphorylated tau in Creutzfeldt-Jakob disease: Results from the Swedish Mortality Registry. *JAMA Neurol.* **2014**, *71*, 476–483. [CrossRef] [PubMed]
36. Van Harten, A.C.; Kester, M.I.; Visser, P.J.; Blankenstein, M.A.; Pijnenburg, Y.A.; van der Flier, W.M.; Scheltens, P. Tau and p-tau as CSF biomarkers in dementia: A meta-analysis. *Clin. Chem. Lab. Med.* **2011**, *49*, 353–366. [CrossRef] [PubMed]
37. Molinuevo, J.L.; Ayton, S.; Batrla, R.; Bednar, M.M.; Bittner, T.; Cummings, J.; Fagan, A.M.; Hampel, H.; Mielke, M.M.; Mikulskis, A.; et al. Current state of Alzheimer's fluid biomarkers. *Acta Neuropathol.* **2018**, *136*, 821–853. [CrossRef]

38. Blennow, K.; Wallin, A.; Agren, H.; Spenger, C.; Siegfried, J.; Vanmechelen, E. Tau protein in cerebrospinal fluid: A biochemical marker for axonal degeneration in Alzheimer disease? *Mol. Chem. Neuropathol.* **1995**, *26*, 231–245. [CrossRef]
39. Vanmechelen, E.; Vanderstichele, H.; Davidsson, P.; Van Kerschaver, E.; Van Der Perre, B.; Sjogren, M.; Andreasen, N.; Blennow, K. Quantification of tau phosphorylated at threonine 181 in human cerebrospinal fluid: A sandwich ELISA with a synthetic phosphopeptide for standardization. *Neurosci. Lett.* **2000**, *285*, 49–52. [CrossRef]
40. Kohnken, R.; Buerger, K.; Zinkowski, R.; Miller, C.; Kerkman, D.; DeBernardis, J.; Shen, J.; Moller, H.J.; Davies, P.; Hampel, H. Detection of tau phosphorylated at threonine 231 in cerebrospinal fluid of Alzheimer's disease patients. *Neurosci. Lett.* **2000**, *287*, 187–190. [CrossRef]
41. Olsson, B.; Lautner, R.; Andreasson, U.; Ohrfelt, A.; Portelius, E.; Bjerke, M.; Holtta, M.; Rosen, C.; Olsson, C.; Strobel, G.; et al. CSF and blood biomarkers for the diagnosis of Alzheimer's disease: A systematic review and meta-analysis. *Lancet Neurol.* **2016**, *15*, 673–684. [CrossRef]
42. Hampel, H.; Vassar, R.; De Strooper, B.; Hardy, J.; Willem, M.; Singh, N.; Zhou, J.; Yan, R.; Vanmechelen, E.; De Vos, A.; et al. The beta-Secretase BACE1 in Alzheimer's Disease. *Biol. Psychiatry* **2020**. [CrossRef] [PubMed]
43. Holsinger, R.M.; McLean, C.A.; Collins, S.J.; Masters, C.L.; Evin, G. Increased beta-Secretase activity in cerebrospinal fluid of Alzheimer's disease subjects. *Ann. Neurol.* **2004**, *55*, 898–899. [CrossRef]
44. Zhong, Z.; Ewers, M.; Teipel, S.; Burger, K.; Wallin, A.; Blennow, K.; He, P.; McAllister, C.; Hampel, H.; Shen, Y. Levels of beta-secretase (BACE1) in cerebrospinal fluid as a predictor of risk in mild cognitive impairment. *Arch. Gen. Psychiatry* **2007**, *64*, 718–726. [CrossRef] [PubMed]
45. Ewers, M.; Zhong, Z.; Burger, K.; Wallin, A.; Blennow, K.; Teipel, S.J.; Shen, Y.; Hampel, H. Increased CSF-BACE 1 activity is associated with ApoE-epsilon 4 genotype in subjects with mild cognitive impairment and Alzheimer's disease. *Brain* **2008**, *131*, 1252–1258. [CrossRef]
46. Ewers, M.; Cheng, X.; Zhong, Z.; Nural, H.F.; Walsh, C.; Meindl, T.; Teipel, S.J.; Buerger, K.; He, P.; Shen, Y.; et al. Increased CSF-BACE1 activity associated with decreased hippocampus volume in Alzheimer's disease. *J. Alzheimers Dis.* **2011**, *25*, 373–381. [CrossRef]
47. Park, S.A.; Han, S.M.; Kim, C.E. New fluid biomarkers tracking non-amyloid-beta and non-tau pathology in Alzheimer's disease. *Exp. Mol. Med.* **2020**, *52*, 556–568. [CrossRef]
48. Yuan, A.; Rao, M.V.; Nixon, R.A. Neurofilaments at a glance. *J. Cell Sci.* **2012**, *125*, 3257–3263. [CrossRef]
49. Zetterberg, H.; Skillback, T.; Mattsson, N.; Trojanowski, J.Q.; Portelius, E.; Shaw, L.M.; Weiner, M.W.; Blennow, K. Alzheimer's Disease Neuroimaging, I. Association of Cerebrospinal Fluid Neurofilament Light Concentration With Alzheimer Disease Progression. *JAMA Neurol.* **2016**, *73*, 60–67. [CrossRef]
50. Sjogren, M.; Rosengren, L.; Minthon, L.; Davidsson, P.; Blennow, K.; Wallin, A. Cytoskeleton proteins in CSF distinguish frontotemporal dementia from AD. *Neurology* **2000**, *54*, 1960–1964. [CrossRef]
51. Sjogren, M.; Blomberg, M.; Jonsson, M.; Wahlund, L.O.; Edman, A.; Lind, K.; Rosengren, L.; Blennow, K.; Wallin, A. Neurofilament protein in cerebrospinal fluid: A marker of white matter changes. *J. Neurosci. Res.* **2001**, *66*, 510–516. [CrossRef]
52. Agren-Wilsson, A.; Lekman, A.; Sjoberg, W.; Rosengren, L.; Blennow, K.; Bergenheim, A.T.; Malm, J. CSF biomarkers in the evaluation of idiopathic normal pressure hydrocephalus. *Acta Neurol. Scand.* **2007**, *116*, 333–339. [CrossRef] [PubMed]
53. Lycke, J.N.; Karlsson, J.E.; Andersen, O.; Rosengren, L.E. Neurofilament protein in cerebrospinal fluid: A potential marker of activity in multiple sclerosis. *J. Neurol. Neurosurg. Psychiatry* **1998**, *64*, 402–404. [CrossRef] [PubMed]
54. Forgrave, L.M.; Ma, M.; Best, J.R.; DeMarco, M.L. The diagnostic performance of neurofilament light chain in CSF and blood for Alzheimer's disease, frontotemporal dementia, and amyotrophic lateral sclerosis: A systematic review and meta-analysis. *Alzheimers Dement.* **2019**, *11*, 730–743. [CrossRef] [PubMed]
55. Laterza, O.F.; Modur, V.R.; Crimmins, D.L.; Olander, J.V.; Landt, Y.; Lee, J.M.; Ladenson, J.H. Identification of novel brain biomarkers. *Clin. Chem.* **2006**, *52*, 1713–1721. [CrossRef]
56. Lee, J.M.; Blennow, K.; Andreasen, N.; Laterza, O.; Modur, V.; Olander, J.; Gao, F.; Ohlendorf, M.; Ladenson, J.H. The brain injury biomarker VLP-1 is increased in the cerebrospinal fluid of Alzheimer disease patients. *Clin. Chem.* **2008**, *54*, 1617–1623. [CrossRef]

57. Tarawneh, R.; D'Angelo, G.; Macy, E.; Xiong, C.; Carter, D.; Cairns, N.J.; Fagan, A.M.; Head, D.; Mintun, M.A.; Ladenson, J.H.; et al. Visinin-like protein-1: Diagnostic and prognostic biomarker in Alzheimer disease. *Ann. Neurol.* **2011**, *70*, 274–285. [CrossRef]
58. Tarawneh, R.; Head, D.; Allison, S.; Buckles, V.; Fagan, A.M.; Ladenson, J.H.; Morris, J.C.; Holtzman, D.M. Cerebrospinal Fluid Markers of Neurodegeneration and Rates of Brain Atrophy in Early Alzheimer Disease. *JAMA Neurol.* **2015**, *72*, 656–665. [CrossRef]
59. Mroczko, B.; Groblewska, M.; Zboch, M.; Muszynski, P.; Zajkowska, A.; Borawska, R.; Szmitkowski, M.; Kornhuber, J.; Lewczuk, P. Evaluation of visinin-like protein 1 concentrations in the cerebrospinal fluid of patients with mild cognitive impairment as a dynamic biomarker of Alzheimer's disease. *J. Alzheimers Dis.* **2015**, *43*, 1031–1037. [CrossRef]
60. Luo, X.; Hou, L.; Shi, H.; Zhong, X.; Zhang, Y.; Zheng, D.; Tan, Y.; Hu, G.; Mu, N.; Chan, J.; et al. CSF levels of the neuronal injury biomarker visinin-like protein-1 in Alzheimer's disease and dementia with Lewy bodies. *J. Neurochem.* **2013**, *127*, 681–690. [CrossRef]
61. Sutphen, C.L.; McCue, L.; Herries, E.M.; Xiong, C.; Ladenson, J.H.; Holtzman, D.M.; Fagan, A.M.; Adni, O.B.O. Longitudinal decreases in multiple cerebrospinal fluid biomarkers of neuronal injury in symptomatic late onset Alzheimer's disease. *Alzheimers Dement.* **2018**, *14*, 869–879. [CrossRef] [PubMed]
62. Mavroudis, I.A.; Petridis, F.; Chatzikonstantinou, S.; Karantali, E.; Kazis, D. A meta-analysis on the levels of VILIP-1 in the CSF of Alzheimer's disease compared to normal controls and other neurodegenerative conditions. *Aging Clin. Exp. Res.* **2020**, 1–8. [CrossRef] [PubMed]
63. Thorsell, A.; Bjerke, M.; Gobom, J.; Brunhage, E.; Vanmechelen, E.; Andreasen, N.; Hansson, O.; Minthon, L.; Zetterberg, H.; Blennow, K. Neurogranin in cerebrospinal fluid as a marker of synaptic degeneration in Alzheimer's disease. *Brain Res.* **2010**, *1362*, 13–22. [CrossRef] [PubMed]
64. De Vos, A.; Jacobs, D.; Struyfs, H.; Fransen, E.; Andersson, K.; Portelius, E.; Andreasson, U.; De Surgeloose, D.; Hernalsteen, D.; Sleegers, K.; et al. C-terminal neurogranin is increased in cerebrospinal fluid but unchanged in plasma in Alzheimer's disease. *Alzheimers Dement.* **2015**, *11*, 1461–1469. [CrossRef] [PubMed]
65. Kvartsberg, H.; Portelius, E.; Andreasson, U.; Brinkmalm, G.; Hellwig, K.; Lelental, N.; Kornhuber, J.; Hansson, O.; Minthon, L.; Spitzer, P.; et al. Characterization of the postsynaptic protein neurogranin in paired cerebrospinal fluid and plasma samples from Alzheimer's disease patients and healthy controls. *Alzheimers Res. Ther.* **2015**, *7*, 40. [CrossRef]
66. Lista, S.; Toschi, N.; Baldacci, F.; Zetterberg, H.; Blennow, K.; Kilimann, I.; Teipel, S.J.; Cavedo, E.; Dos Santos, A.M.; Epelbaum, S.; et al. Cerebrospinal Fluid Neurogranin as a Biomarker of Neurodegenerative Diseases: A Cross-Sectional Study. *J. Alzheimers Dis.* **2017**, *59*, 1327–1334. [CrossRef]
67. Kester, M.I.; Teunissen, C.E.; Crimmins, D.L.; Herries, E.M.; Ladenson, J.H.; Scheltens, P.; van der Flier, W.M.; Morris, J.C.; Holtzman, D.M.; Fagan, A.M. Neurogranin as a Cerebrospinal Fluid Biomarker for Synaptic Loss in Symptomatic Alzheimer Disease. *JAMA Neurol.* **2015**, *72*, 1275–1280. [CrossRef]
68. Tarawneh, R.; D'Angelo, G.; Crimmins, D.; Herries, E.; Griest, T.; Fagan, A.M.; Zipfel, G.J.; Ladenson, J.H.; Morris, J.C.; Holtzman, D.M. Diagnostic and Prognostic Utility of the Synaptic Marker Neurogranin in Alzheimer Disease. *JAMA Neurol.* **2016**, *73*, 561–571. [CrossRef]
69. Brinkmalm, A.; Brinkmalm, G.; Honer, W.G.; Frolich, L.; Hausner, L.; Minthon, L.; Hansson, O.; Wallin, A.; Zetterberg, H.; Blennow, K.; et al. SNAP-25 is a promising novel cerebrospinal fluid biomarker for synapse degeneration in Alzheimer's disease. *Mol. Neurodegener.* **2014**, *9*, 53. [CrossRef]
70. Tible, M.; Sandelius, A.; Hoglund, K.; Brinkmalm, A.; Cognat, E.; Dumurgier, J.; Zetterberg, H.; Hugon, J.; Paquet, C.; Blennow, K. Dissection of synaptic pathways through the CSF biomarkers for predicting Alzheimer's disease. *Neurology* **2020**. [CrossRef]
71. Ohrfelt, A.; Brinkmalm, A.; Dumurgier, J.; Brinkmalm, G.; Hansson, O.; Zetterberg, H.; Bouaziz-Amar, E.; Hugon, J.; Paquet, C.; Blennow, K. The pre-synaptic vesicle protein synaptotagmin is a novel biomarker for Alzheimer's disease. *Alzheimers Res. Ther.* **2016**, *8*, 41. [CrossRef]
72. Mayeux, R.; Honig, L.S.; Tang, M.X.; Manly, J.; Stern, Y.; Schupf, N.; Mehta, P.D. Plasma A[beta]40 and A[beta]42 and Alzheimer's disease: Relation to age, mortality, and risk. *Neurology* **2003**, *61*, 1185–1190. [CrossRef] [PubMed]
73. Van Oijen, M.; Hofman, A.; Soares, H.D.; Koudstaal, P.J.; Breteler, M.M. Plasma Abeta(1-40) and Abeta(1-42) and the risk of dementia: A prospective case-cohort study. *Lancet Neurol.* **2006**, *5*, 655–660. [CrossRef]

74. Yaffe, K.; Weston, A.; Graff-Radford, N.R.; Satterfield, S.; Simonsick, E.M.; Younkin, S.G.; Younkin, L.H.; Kuller, L.; Ayonayon, H.N.; Ding, J.; et al. Association of plasma beta-amyloid level and cognitive reserve with subsequent cognitive decline. *JAMA* **2011**, *305*, 261–266. [CrossRef] [PubMed]
75. Nakamura, A.; Kaneko, N.; Villemagne, V.L.; Kato, T.; Doecke, J.; Dore, V.; Fowler, C.; Li, Q.X.; Martins, R.; Rowe, C.; et al. High performance plasma amyloid-beta biomarkers for Alzheimer's disease. *Nature* **2018**, *554*, 249–254. [CrossRef] [PubMed]
76. Perez-Grijalba, V.; Arbizu, J.; Romero, J.; Prieto, E.; Pesini, P.; Sarasa, L.; Guillen, F.; Monleon, I.; San-Jose, I.; Martinez-Lage, P.; et al. Plasma Abeta42/40 ratio alone or combined with FDG-PET can accurately predict amyloid-PET positivity: A cross-sectional analysis from the AB255 Study. *Alzheimers Res. Ther.* **2019**, *11*, 96. [CrossRef] [PubMed]
77. Perez-Grijalba, V.; Romero, J.; Pesini, P.; Sarasa, L.; Monleon, I.; San-Jose, I.; Arbizu, J.; Martinez-Lage, P.; Munuera, J.; Ruiz, A.; et al. Plasma Abeta42/40 Ratio Detects Early Stages of Alzheimer's Disease and Correlates with CSF and Neuroimaging Biomarkers in the AB255 Study. *J. Prev. Alzheimers Dis.* **2019**, *6*, 34–41.
78. Kim, K.; Kim, M.J.; Kim, D.W.; Kim, S.Y.; Park, S.; Park, C.B. Clinically accurate diagnosis of Alzheimer's disease via multiplexed sensing of core biomarkers in human plasma. *Nat. Commun.* **2020**, *11*, 119. [CrossRef]
79. Nabers, A.; Perna, L.; Lange, J.; Mons, U.; Schartner, J.; Guldenhaupt, J.; Saum, K.U.; Janelidze, S.; Holleczek, B.; Rujescu, D.; et al. Amyloid blood biomarker detects Alzheimer's disease. *EMBO Mol. Med.* **2018**, *10*, e8763. [CrossRef]
80. Zetterberg, H.; Wilson, D.; Andreasson, U.; Minthon, L.; Blennow, K.; Randall, J.; Hansson, O. Plasma tau levels in Alzheimer's disease. *Alzheimers Res. Ther.* **2013**, *5*, 9. [CrossRef]
81. Mattsson, N.; Zetterberg, H.; Janelidze, S.; Insel, P.S.; Andreasson, U.; Stomrud, E.; Palmqvist, S.; Baker, D.; Tan Hehir, C.A.; Jeromin, A.; et al. Plasma tau in Alzheimer disease. *Neurology* **2016**, *87*, 1827–1835. [CrossRef] [PubMed]
82. Tatebe, H.; Kasai, T.; Ohmichi, T.; Kishi, Y.; Kakeya, T.; Waragai, M.; Kondo, M.; Allsop, D.; Tokuda, T. Quantification of plasma phosphorylated tau to use as a biomarker for brain Alzheimer pathology: Pilot case-control studies including patients with Alzheimer's disease and down syndrome. *Mol. Neurodegener.* **2017**, *12*, 63. [CrossRef] [PubMed]
83. Karikari, T.K.; Pascoal, T.A.; Ashton, N.J.; Janelidze, S.; Benedet, A.L.; Rodriguez, J.L.; Chamoun, M.; Savard, M.; Kang, M.S.; Therriault, J.; et al. Blood phosphorylated tau 181 as a biomarker for Alzheimer's disease: A diagnostic performance and prediction modelling study using data from four prospective cohorts. *Lancet Neurol.* **2020**, *19*, 422–433. [CrossRef]
84. Benussi, A.; Karikari, T.K.; Ashton, N.; Gazzina, S.; Premi, E.; Benussi, L.; Ghidoni, R.; Rodriguez, J.L.; Emersic, A.; Simren, J.; et al. Diagnostic and prognostic value of serum NfL and p-Tau181 in frontotemporal lobar degeneration. *J. Neurol. Neurosurg. Psychiatry* **2020**. [CrossRef] [PubMed]
85. Rojas, J.C.; Karydas, A.; Bang, J.; Tsai, R.M.; Blennow, K.; Liman, V.; Kramer, J.H.; Rosen, H.; Miller, B.L.; Zetterberg, H.; et al. Plasma neurofilament light chain predicts progression in progressive supranuclear palsy. *Ann. Clin. Transl. Neurol.* **2016**, *3*, 216–225. [CrossRef] [PubMed]
86. Ray, S.; Britschgi, M.; Herbert, C.; Takeda-Uchimura, Y.; Boxer, A.; Blennow, K.; Friedman, L.F.; Galasko, D.R.; Jutel, M.; Karydas, A.; et al. Classification and prediction of clinical Alzheimer's diagnosis based on plasma signaling proteins. *Nat. Med.* **2007**, *13*, 1359–1362. [CrossRef]
87. Doecke, J.D.; Laws, S.M.; Faux, N.G.; Wilson, W.; Burnham, S.C.; Lam, C.P.; Mondal, A.; Bedo, J.; Bush, A.I.; Brown, B.; et al. Blood-based protein biomarkers for diagnosis of Alzheimer disease. *Arch. Neurol.* **2012**, *69*, 1318–1325. [CrossRef]
88. Hye, A.; Riddoch-Contreras, J.; Baird, A.L.; Ashton, N.J.; Bazenet, C.; Leung, R.; Westman, E.; Simmons, A.; Dobson, R.; Sattlecker, M.; et al. Plasma proteins predict conversion to dementia from prodromal disease. *Alzheimers Dement.* **2014**, *10*, 799–807. [CrossRef]
89. Mapstone, M.; Cheema, A.K.; Fiandaca, M.S.; Zhong, X.; Mhyre, T.R.; MacArthur, L.H.; Hall, W.J.; Fisher, S.G.; Peterson, D.R.; Haley, J.M.; et al. Plasma phospholipids identify antecedent memory impairment in older adults. *Nat. Med.* **2014**, *20*, 415–418. [CrossRef]
90. Varma, V.R.; Oommen, A.M.; Varma, S.; Casanova, R.; An, Y.; Andrews, R.M.; O'Brien, R.; Pletnikova, O.; Troncoso, J.C.; Toledo, J.; et al. Brain and blood metabolite signatures of pathology and progression in Alzheimer disease: A targeted metabolomics study. *PLoS ONE Med.* **2018**, *15*, e1002482. [CrossRef]

91. Abdullah, M.; Kimura, N.; Akatsu, H.; Hashizume, Y.; Ferdous, T.; Tachita, T.; Iida, S.; Zou, K.; Matsubara, E.; Michikawa, M. Flotillin is a Novel Diagnostic Blood Marker of Alzheimer's Disease. *J. Alzheimers Dis.* **2019**, *72*, 1165–1176. [CrossRef] [PubMed]
92. Abdullah, M.; Takase, H.; Nunome, M.; Enomoto, H.; Ito, J.; Gong, J.S.; Michikawa, M. Amyloid-beta Reduces Exosome Release from Astrocytes by Enhancing JNK Phosphorylation. *J. Alzheimers Dis.* **2016**, *53*, 1433–1441. [CrossRef] [PubMed]

Journal of *Personalized Medicine*

Review

Biomarkers for Alzheimer's Disease Early Diagnosis

Eva Ausó, Violeta Gómez-Vicente and Gema Esquiva *

Department of Optics, Pharmacology and Anatomy, University of Alicante, 03690 Alicante, Spain; eva.auso@ua.es (E.A.); vgvicente@ua.es (V.G.-V.)

* Correspondence: gema.esquiva@ua.es

Received: 8 July 2020; Accepted: 1 September 2020; Published: 4 September 2020

Abstract: Alzheimer's disease (AD) is the most common cause of dementia, affecting the central nervous system (CNS) through the accumulation of intraneuronal neurofibrillary tau tangles (NFTs) and β-amyloid plaques. By the time AD is clinically diagnosed, neuronal loss has already occurred in many brain and retinal regions. Therefore, the availability of early and reliable diagnosis markers of the disease would allow its detection and taking preventive measures to avoid neuronal loss. Current diagnostic tools in the brain, such as magnetic resonance imaging (MRI), positron emission tomography (PET) imaging, and cerebrospinal fluid (CSF) biomarkers (Aβ and tau) detection are invasive and expensive. Brain-secreted extracellular vesicles (BEVs) isolated from peripheral blood have emerged as novel strategies in the study of AD, with enormous potential as a diagnostic evaluation of therapeutics and treatment tools. In addition; similar mechanisms of neurodegeneration have been demonstrated in the brain and the eyes of AD patients. Since the eyes are more accessible than the brain, several eye tests that detect cellular and vascular changes in the retina have also been proposed as potential screening biomarkers. The aim of this study is to summarize and discuss several potential markers in the brain, eye, blood, and other accessible biofluids like saliva and urine, and correlate them with earlier diagnosis and prognosis to identify individuals with mild symptoms prior to dementia.

Keywords: Alzheimer's disease; biomarkers; early diagnosis; biofluids

---

1. Introduction

1.1. Pathophysiology of AD and Clinical Manifestations

The lesions of Alzheimer's disease (AD) include pathological changes in the brain such as the accumulation of proteins (amyloid-β (Aβ) peptide and Tau); the degeneration of neurons and synapses, most noticeably in the neocortex and the hippocampus, which leads to structural changes as well as to the loss of functional connectivity, and the alterations of reactive processes like neuroinflammation and plasticity, related to oxidative stress and mitochondrial dysfunction [1]. Some of these hallmarks can be detected in the prodromal stage of the disease, also referred to as mild cognitive impairment (MCI) due to AD, when the symptoms are not yet obvious.

Amyloid-β deposits are widely distributed in the brain and follow an anterograde sequence originating in five phases in which different brain regions are hierarchically involved [2–4]. The five phases go from phase 1, when the deposits are exclusively found in the isocortex, to phase 5, when the cerebellum and several brainstem nuclei, such as the pontine nuclei and the locus coeruleus, among others, are involved [2,4]. The progression of Tau pathology is also staggered from the transentorhinal and entorhinal cortex to the isocortex via the hippocampus, with a heterogeneous and area-specific neuronal loss [2–4]. It is well-established that the accumulation of Tau protein takes place specifically in neurons and occurs in their cell body as neurofibrillary tangles (NFTs), in their dendrites as neuropil threads (NT), and in their axons forming the senile plaque neuritic corona [3].

The Braak stages, based on phospho-Tau accumulation within connected brain regions, defines the progression of AD neuropathology. I–II refer to the entorhinal cortex, III–IV to the hippocampus/limbic system, and V–VI to the frontal and parietal lobes.

1.2. Diagnostic Tools

The progress in the diagnosis of AD has noticeably improved with the development in the last decades of noninvasive neuroimaging techniques that allow the visualization of structures in vivo. Some examples are novel magnetic resonance imaging (MRI), metabolic changes detected by positron emission tomography (PET), and amyloid imaging. These techniques permit the detection of pre-symptomatic diagnostic biomarkers in the brains of cognitively normal elderly individuals and also serve to monitor disease progression after the onset of symptoms [1]. Due to their reliability and high discriminative capacity in the pre-dementia state, volumetric approaches of the high-resolution subfield are useful, as well as diagnostic techniques in order to study the early changes in the most affected brain structures [2–4]. With all these tools, the typical lesions related to protein accumulation and the structural changes in certain brain areas are easier to detect and; therefore, constitute the basis of the diagnosis.

In addition, the advancement in the past few years of omics technologies (genomics, transcriptomics, proteomics, metabolomics, secretomics, etc.) has made possible the analysis of a wide range of AD hallmarks referring to both, sporadic and familial cases. These tools facilitate the analysis of human fluid samples of diverse nature such as blood, tears, urine, or saliva, whose collection in most cases does not require trained professionals and has the advantage of being noninvasive due to easy accessibility. The importance of identifying and developing reliable and sensitive tools for the early diagnosis of AD relies on the potential benefits for the patients, including timely access to medical treatments to slow down the progression of the disease and; therefore, preservation of longer cognitive capacity, or even the possibility to plan for the future.

2. Invasive Biomarkers

2.1. Changes in Specific Brain Areas as Early Biomarkers

The locus coeruleus (LC) is a neuromelanin-rich brainstem structure thought to modulate attention and memory and is the major source of noradrenaline in the brain. In the asymptomatic stage of AD, Tau NFTs are observed in the LC [5,6] prior to their presence in other cerebral areas such as the entorhinal cortex and the neocortex [7–10]. These Tau aggregates precede typical neuronal loss in the LC during AD progression [11]. Studies using unbiased stereology have revealed an average decrease in LC volume of 8.4% for each Braak stage increment, as well as neuronal loss mainly in the rostral/middle area of the LC, progressing from 30% in the prodromal stage to 55% when dementia is diagnosed [11]. Functionally, this neuronal loss has correlated with cognitive dysfunction [12] and reduced noradrenaline levels in the hippocampus and the cortex [13]. Additionally, a two-fold increase in Tau accumulation was also observed from Braak stage 0 to I [9]. Therefore, the detection using in vivo imaging of early structural tissue modifications such as the decrease in the LC volume or metabolic changes would support the diagnosis and could potentially slow down disease progression if the patient benefits from treatments in the appropriate time [14,15].

Although there is controversy regarding the accelerated rates of brain atrophy at the preclinical stage of the disease, it seems that the medial temporal lobe and, particularly, the hippocampus are brain structures early affected by NFTs and neurite loss. Studies using voxel-based morphometry and high-resolution MRI have revealed hippocampal atrophy in AD patients' brains at the preclinical stage, up to 10 years before the diagnosis of dementia [16], and even before MCI [17,18]. The magnitude of atrophy in the hippocampus and its subfields determines the progression to either MCI or AD [19–21]. Thus, studies using radial atrophy measurements have shown that CA1 and the subicular atrophy in cognitively healthy individuals is associated with an increased risk of developing MCI, while the gradual involvement of the CA1 and subiculum fields, along with atrophy spread to the rest of the hippocampus (CA2–3 subfields) in amnestic MCI, suggests the future diagnosis of AD [22,23]. Moreover, apolipoprotein E (ApoE) plays a significant role in AD pathogenesis by affecting amyloid and Tau pathology. The presence of the allele ε4 (APOEε4) [24] influences the reduction of the hippocampal volume and the accumulation of Aβ filaments in the brains of elderly people without cognitive impairment and normal levels of Aβ-peptides morphologically [25–27]. In this direction, MRI imaging has revealed a significant reduction in the hippocampal volume in amnestic MCI people carrying APOEε4, especially in those who progressed to AD [28]. Even cognitively normal APOEε4 carriers have shown hippocampal volume and cortical thickness reduction together with memory decline and accelerated brain atrophy rates before the onset of cognitive impairment [24,27].

Regarding neuronal connectivity dysfunction, a novel PET tracer that binds to synaptic vesicle glycoprotein 2A (SV2A) can be used to quantify synaptic density in vivo, predicting the stage of AD [29]. Several morphological studies have shown that synaptic loss appears early in the pathology [30,31], so the study of markers of neuronal death may derive in promising results for the early diagnosis of AD.

In conclusion, in vivo morphological studies of different brain areas (LC, hippocampus, etc.), along with genetic studies that detect alleles or mutations closely related to the pathology, point out their usefulness as biomarkers for the early detection of AD. Despite the high prognostic ability of these techniques in AD and MCI [32], sometimes, there are limitations that make it difficult to use them in the routine analysis [33]. For this reason, biomarkers in cerebrospinal fluid (CSF) are being extensively studied worldwide as potential candidates for the diagnosis of AD before the appearance of cognitive symptoms [34].

2.2. *Cerebrospinal Fluid*

There is no doubt the ideal fluid biomarker should have a series of characteristics—reliable, reproducible, and noninvasive in terms of collection, specific for a particular disease, simple, and inexpensive to measure, and easy to implement in large populations. In this regard, blood biomarkers meet several of these criteria and could be used in primary care to identify patients with risk of AD [35]. In contrast, CSF collection does not meet the criteria of being a noninvasive procedure, which certainly limits its use but given the close relationship between the brain and the CSF, this fluid could provide valuable information about the biochemical changes that occur in the brain at the preclinical stages of AD [36]. For instance, it is well established that decreased Aβ-42 and elevated total Tau and phospho-Tau in CSF are considered specific markers of AD [37,38], and that these biomarkers can predict cognitive decline over time [39]. The advantages and disadvantages of each category of fluid biomarkers (blood, CSF, and other matrices such as tears, saliva, and urine) are summarized in Table 1.

Table 1. Advantages and disadvantages for each category of biological fluids used to isolate Alzheimer's disease biomarkers.

	Advantages	Disadvantages
CSF	Close relationship with the brain High accuracy in the diagnostic process Ability to test a large number of candidate pathophysiological biomarkers High concentration of the biomarkers	Invasive Clinicians require training Positioned in later disease stages, after blood samples, as a confirmatory diagnostic modality Process less accepted by the population and at the risk of causing harm, anxiety, and fear to the patient

Table 1. *Cont.*

	Advantages	Disadvantages
Blood	Noninvasive, fast and convenient Inexpensive and reproducible Simple to measure (well-established as part of clinical routines globally) No prior training of the clinicians is required Can be performed in a large variety of settings (primary care, hospitals, patient's home . . .) Easy to implement in large populations Ability to test a large number of candidate pathophysiological biomarkers First-step of the multi-stage diagnostic process (identification of patients at the earliest stages of the disease)	Less accurate Presence of very low concentrations of the biomarkers once they have crossed the blood-brain barrier and decreased time window for testing Less consistent results (susceptibility to interference with other components)
Other matrices (tears, saliva, and urine)	Extremely noninvasive Repeatable collections Easy, no risk of infection, can be self-collected by the patient Cheap Stress-free	Remarkable lack of validated studies Lack of results replicated in larger, multicenter and longitudinal studies

Nowadays, novel molecular markers are being evaluated in CSF through omics technologies, which allow measuring a large number of analytes at a time (Figure 1). For example, a mass spectroscopy-based analysis revealed that similar levels of ApoE and its isoforms (ApoE2, ApoE3, and ApoE4) were found in the CSF of AD patients and non-AD individuals, independent of their APOE genotype (APOEε2, APOEε3, or APOEε4). However, CSF total ApoE concentrations were positively associated with CSF total Tau and phospho-Tau levels [40,41].

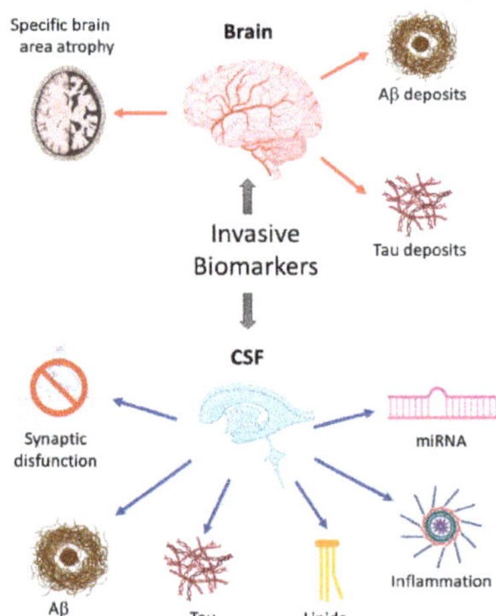

Figure 1. Schematic overview of invasive biomarkers. Different biomarkers have been used to detect early anatomical changes in the brains of people with mild cognitive impairment, including the atrophy of specific brain areas like the locus coeruleus or the hippocampus, and the presence of typical protein aggregates such as extracellular amyloid plaques or intracellular Tau-containing neurofibrillary tangles (upper panel). Additionally, biomarkers of Alzheimer's disease (AD)-related degenerative processes like synaptic dysfunction, neuroinflammation, oxidative stress, or neuronal loss can be measured in the cerebrospinal fluid of AD patients. The detection of miRNAs represents a novel and promising tool for the early AD diagnosis (lower panel).

Proteins involved in the pathological processing of the amyloid precursor protein (APP) could be biomarker candidates for early AD diagnosis and must be considered. Presenilin 1 (PSEN1) and β-secretase 1 (BACE1) are both enzymes involved in the cleavage of APP. In MCI patients, both PSEN1 and BACE1 levels and their activity were increased in CSF [42,43]. Moreover, elevated BACE1 expression has been associated with the APOEε4 genotype [43]. It is worth noting that BACE1 activity was only increased in MCI patients whose impairment was progressing to more advanced stages of dementia, and not occurring in stable MCI patients [44]. So, while BACE1 seems to be highlighted as a sensitive early biomarker to detect alterations in the amyloidogenic process in APOEε4 carriers [43], it does not seem to be a good candidate in APOEε4 non-carriers.

Other early aspects to highlight in AD are neuroinflammation and the synaptic dysfunction; thus, specific markers of these processes could also play a very important role and may correlate more directly with cognitive decline [45]. In this sense, many proteins involved in vesicular transport (secretogranin II (SCG2), chromogranin A (CHGA)), in synapses formation and stabilization (neurexins (NRXNs), neuronal pentraxin 1 (NPTX1), neurocan core protein (NCANP)), and in the immune system (lysozime C (LysC) and β_2-microglobulin (β_2M)) were significantly higher in the CSF of patients with MCI, especially in patients with MCI progressing to AD pathology than in AD and healthy control patients [46]. According to one study, higher levels of CHGA in the CSF of healthy elderly people predicted future decreases in Aβ-42 [47]. Other proteins that play a crucial role in inflammation are YKL-40 and visinin-like protein-1 (VILIP-1). Increased expression of these molecules has been seen in both MCI and AD patients, contrary to cognitively normal elderly subjects. While YKL-40 was increased from the prodromal stage until the severe stage of the disease, VILIP-1 was only increased in the prodromal stage [48]. Some studies have found an association between the upregulation of YKL-40 with an increased risk of progression from the normal conditions to MCI [49]. Another potential inflammatory marker is the interferon-γ-induced protein 10 (IP-10), whose level was increased in the CSF of asymptomatic elderly adults that also presented elevated levels of total-Tau and phospho-Tau [50]. Likewise, monocyte chemoattractant protein 1 (MCP-1), a low-molecular-weight cytokine involved in the inflammatory process, was found elevated in the CSF of MCI and AD patients [51]. The triggering receptor expressed on myeloid cells 2 (TREM2), expressed by microglial cells, among others, plays an important role in regulating immune responses in the brain and in the production of inflammatory cytokines [52]. Its haplodeficiency has been associated with increased axonal dystrophy and phospho-Tau accumulation around Aβ-plaques [53]. An increased level of CSF soluble TREM2 has been seen in carriers of an autosomal dominant AD mutation, at least five years before the onset of symptoms, although later to brain amyloidosis and Tau pathology [54]. All these findings reveal that a large number of proteins involved in the inflammatory response can be potential early biomarkers of AD.

One protein that plays an important role in memory enhancement is neurogranin. It is involved in post-synaptic signaling pathways, and its CSF levels differentiated patients with early symptomatic AD from controls with a comparable diagnostic utility to the other CSF biomarkers [55]. The potential of neurogranin as a biomarker of AD depended on the fragment measured [56].

Regarding neuronal damage, some proteins such as neurofilament light chain (NF-L), a protein involved in protecting neurites, and neuron-specific enolase (NSE), which plays a role in neuronal metabolism, have revealed increased CSF concentrations in MCI patients in comparison with cognitively elderly, and with patients at advanced AD stages [57,58]. In AD patients with advanced pathology, high CSF NF-L levels are associated with cognitive decline and morphological changes in the brain that indicate neuronal loss [57]. Schmidt et al. showed a correlation between high CSF NSE levels and Tau pathology [58]. These results agree with studies, where plasma protein levels were also studied and support the use of NF-L and NSE as early AD biomarkers [56].

Lipid alterations in CSF participate as well in the modulation of neuropathological events related to AD and can be an AD biomarker candidate. In patients with incipient dementia, a reduction of up to 40% of sulfatide levels was observed [59] The levels of some other lipids such as phosphocholine and sphingomyelin were increased in patients at the prodromal stage and correlated with amyloid and Tau pathology [60]. Another biomarker candidate is the fatty acid-binding protein (FABP3), which may play a role in neuronal synapse formation. In MCI and AD patients, the FABP3 level was higher than in cognitively healthy people [61], and it was related to early structural brain changes typical of AD patients (entorhinal cortex atrophy). Also, high FABP3 levels have been found in non-amnestic elderly APOEε4 carriers [62], showing their increase occurs at a very early stage of the disease [63].

Overall, all the findings mentioned above reflect changes in specific areas of the brain detected using novel imaging techniques (MRI, PET), the presence of two classical AD proteins (Aβ and Tau), and the progression of processes such as neuronal apoptosis, synaptic loss, and inflammation. Many of them are still under consideration as potential early biomarkers of AD, and larger longitudinal studies are required for validation of the results [56]. Unfortunately, most AD patients are asymptomatic during the preclinical stages, complicating the recruitment for these kinds of studies and emphasizing the importance of rapid diagnosis.

It is also relevant to investigate the presence of microRNAs (miRNAs), a big family of endogenous short non-coding RNAs that regulate the number of mature mRNAs at the post-transcriptional level [64]. About 70% of identified miRNAs are expressed in the brain, and some miRNAs species are present in exosomes, both good biomarker candidates in clinical diagnostics. Of the approximately 2000 human miRNAs identified to date, no more than 40 are abundantly expressed in the brain [65]. The core CSF biomarkers (Aβ-42, total-tau, and phospho-tau) are relatively stable in clinical AD, and although they are useful for diagnosis, they are not good enough as indicators of disease progression. Although CSF miRNAs are obtained in an invasive manner, which is far from ideal, they have the advantage of targeting important pathological AD genes. A single miRNA has the potential to interfere with the expression of a small family of genes. This is the case with miRNA-125b (upregulated in AD), which targets the synaptic protein synapsin-2 (SYN-2), the enzyme 15-lipoxygenase (15-LOX), and the cell cycle regulator CDKN2A. This opens up the door for the use of miRNAs as therapeutic agents in the future. Furthermore, the misregulation of specific miRNAs could contribute to AD etiopathogenesis [66] and partially explain the large number of brain mRNAs gradually and significantly downregulated in anatomical regions sensitive to AD progression (reviewed in Lukiv 2013)[65]. Other attractive points that favor the use of CSF miRNAs as diagnosis tools are their high stability in body fluids [67,68], the low concentration required for their detection by standard molecular biology techniques, such as quantitative RT-PCR, and the proven high predictive accuracy in the pathogenic process of AD. All of the above supports the huge potential miRNAs offer as diagnostic and prognostic biomarkers and, at the same time, as plausible therapeutic tools against AD [69].

In the hippocampus of AD patients, in comparison with healthy volunteers, upregulation of three out of the 13 brain-associated miRNAs studied was observed: miRNA-9, miRNA125b, and miRNA128 [66]. Very similar results were found in the CSF of AD patients. Using microarrays, qRT-PCR and novel highly sensitive LNA, EDC and DIG (LED)-Northern dot-blot (an improved northern blot-based protocol for small RNA detection that combines the use of digoxigenin (DIG)-labeled oligonucleotide probes containing locked nucleic acids (LNA) and 1-ethyl-3-(3-dimethylaminopropyl) carbodiimide (EDC) for cross-linking the RNA to the membrane), high amounts of proinflammatory miRNAs such as miRNA146a, miRNA-155, miRNA-9, and miRNA-125b have been detected in AD patients CSF compared with age-matched controls [65,70,71]. Briefly, miRNA-125b targets synaptic proteins, neurotrophic factors and cell regulator proteins, and miRNA146a targets immune system regulators and proteins involved in proinflammatory signaling, as well as in Aβ accumulation [65]. Both miRNA-125b and miRNA-146a can explain many of the pathogenic effects of AD, so they could be excellent candidates as AD biomarkers. In areas such as the frontal gyrus and the neocortex, the up- and downregulation of an elevated number of miRNAs has been seen, and even some of them showed different regulations according to the area studied [72]. miRNA-29a seems a promising biomarker because it targets BACE1, which promotes the formation of Aβ from APP. In the cortex of AD patients, decreased miRNA29a has been reported, while a two-fold increase in miR-29a levels was found in the

CSF of AD patients in comparison with cognitively healthy people [73,74]. The decrease in miR-29 brain expression in AD patients can be associated with an increase of BACE1, leading to the subsequent increase in Aβ levels [75]. Another miRNA that is upregulated in the neocortex, hippocampus, and CSF is miRNA-9 [65,70]. MiRNA-9 is mainly involved in neurogenesis and brain cell proliferation [76,77] and also targets BACE1, decreasing its expression [78]. In this sense, miRNA-29a and miRNA-9 could be indicators of pathology acting as biomarkers. We have only named a few microRNAs involved in AD, although it is worth noting that there are many more with promising results [79].

3. Noninvasive Biomarkers

3.1. Blood Biomarkers

Blood pressure has been pointed out as an early marker of AD. High blood pressure has been associated with senile plaques, neurofilament tangles, and hippocampal atrophy, and advanced age and hypertension have been linked to AD development [80]. In addition, selected low amounts of brain proteins/substances can cross the blood-brain barrier, reaching the bloodstream. Therefore, it is possible the detection in the blood of specific substances derived exclusively from the brain or systemic pathologies [81]. Despite blood is a more complex matrix for investigating neuronal processes, making the research of neurodegenerative biomarkers in blood challenging; its accessibility makes the study and validation interesting. To this effect, blood represents a noninvasive way of monitoring AD development and progression [82].

Compared with CSF, blood is easily collected and, therefore, represents the matrix of choice for the discovery of new accessible biomarkers. In addition, CSF and brain Aβ and Tau correlated with plasma Aβ and Tau in sporadic AD [83,84]. To this end, measurements in the bloodstream of proteins and peptide concentrations that originate in the brain are very promising. A decrease in the levels of Aβ-42, Aβ-40,and the Aβ-42/Aβ-40 ratio was found in the plasma of preclinical AD patients [85]. Other studies focused on Aβ-0 and showed higher levels in samples of AD patients [86]. Blood-based Tau levels have also been investigated in some studies and found to be elevated in the plasma of AD patients [87,88]. However, the relatively low levels of Aβ and Tau proteins in peripheral blood necessitate more sensitive detection techniques to consolidate as diagnostic biomarkers of AD.

Recent studies have revealed that serum neurofilament protein levels correlated with AD [89]. However, this fact was not specific of AD but was also reported in other neurodegenerative diseases [90]. Henceforth, Aβ, Tau, and neurofilaments are not strong enough biomarkers to predict sporadic AD [91], and it would be useful to account for additional molecules for a more accurate early diagnosis. It has been found that changes in the plasma concentration of brain-derived neurotrophic factor (BDNF) depend on the severity of AD [92]. Also, plasma clusterin levels are significantly increased in both, MCI and AD patients [93], and have been related to increased risk of progression from MCI to AD, but a slower cognitive decline in AD patients [94].

Extracellular RNA (exRNA) from human biofluids has been recently characterized. In neurological disorders, brain-derived exRNAs can reach the bloodstream in different ways. One possibility is the elimination of waste from the brain by the lymphatic system into the bloodstream [95]. A second one is that blood-brain barrier leakiness described in early AD facilitates the passage of all types of extracellular molecules [96]. Thus, the presence of exRNAs in the blood allows the study of gene expression in the central nervous system. In this context, it has been recently described that phosphoglycerate dehydrogenase (PHGDH) exhibits consistent upregulation in the AD brain transcriptome and is increased in presymptomatic AD plasma as compared to controls, suggesting the potential utility of plasma PHGDH exRNA as a presymptomatic indicator of AD [97].

Like in CSF, miRNAs are also considered to be one of the potential candidates for blood-based biomarkers. It has been reported that several miRNAs downregulate AD-related proteins, including BACE-1 and APP [98]. Four miRNAs (miR-31, miR-93, miR-143, and miR-146a) were significantly decreased in AD patients' serum, suggesting that these could be used as potential diagnostic and prognostic markers for dementia. Notably, miR-31, miR-146a, and miR-93 were related to inflammation, cell apoptosis, and fibrosis. Furthermore, miR-93 and miR-146a were significantly elevated in MCI compared to controls and miR-31, miR-93, and miR-146a can be used to discriminate AD from other types of dementia [99]. In addition, the level of miR-206, involved in cognitive decline and memory deficits, was increased in AD plasma, so it could also be a good AD biomarker candidate [100].

Circulating exRNAs are usually protected by exosomes and other extracellular carriers [101]. Extracellular vesicles (EVs), including exosomes, are small (50–150 nm) membrane microvesicles involved in cell-to-cell communication, which can go across a healthy blood-brain barrier. EVs contain not only exRNAs but also other biologically active cargo of molecules specific of their tissue of origin, such as metabolites and proteins, making EVs a good blood-based biomarker candidate, prognostic indicator, and therapeutic tool in AD. Isolation of brain-secreted EVs (BEVs) from the blood provides a minimally invasive way to sample components of brain tissue. Cerebrovascular-derived BEV studies are sparse in human AD patients, so more research would be needed in this field. Nonetheless, it has been reported that pathophysiological alterations in AD are, in fact, reflected in the number and composition of BEVs from neurons, neural precursor cells, and astrocytes [102,103].

As mentioned above, blood is a more complex matrix than CSF, and the high number of cells and soluble molecules contained in it can lead to interferences. Moreover, the low number of brain-derived biomarkers in blood requires highly sensitive techniques for their detection. Great variability in the results depending on the methodology used has been reported, and the fact that several studies show conflicting results represents a limitation for the use of blood biomarkers as an AD diagnostic tool [104]. For all these reasons, the use of blood biomarkers has not yet been validated [81], and the research of alternative fluids as urine, tears, and saliva, is challenging (Figure 2).

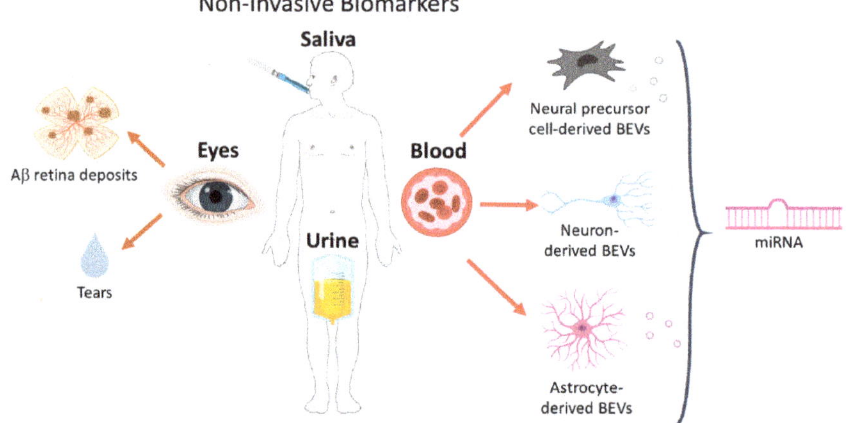

Figure 2. Schematic overview of noninvasive biomarkers: eyes, saliva, urine and blood. Besides fluid biomarkers (tears) that can be collected from the eyes, the promising advances in in vivo retinal imaging could provide an AD diagnosis tool in the near future. Blood-isolated brain secreted extracellular vesicles (BEVs) derive from three possible brain cell types: neural precursors, neurons and astrocytes. The content of this blood-isolated BEVs, mainly miRNAs, have been investigated as potential AD biomarker.

3.1.1. Neuron-Derived BEVs in Blood

Several studies reported significantly elevated Aβ-42 levels in plasma-isolated neuron-derived BEVs in AD dementia relative to cognitively normal individuals. EVs may mediate the transcellular spread of Aβ peptide by destabilizing calcium cell homeostasis and damaging mitochondrial function; thus making neurons more vulnerable to excitotoxicity [105]. Plasma Aβ-42 has the potential to be used both, as a novel earlier biomarkers of AD, as well as a marker of AD progression, having the same capacity as Aβ-42 in CSF for the diagnosis of AD [83]. Nonetheless, the EV concentration is higher in blood plasma than in CSF [105], making this a more sensitive biomarker.

Regarding Tau levels in the plasma, the association of neuron-derived BEVs with AD has yet to reach a consensus. Although three studies [83,106,107] found elevated phospho-Tau levels in AD dementia, these reached a plateau as early as 10 years before AD diagnosis, making it a worse marker of AD progression than Aβ-42 [107]. In addition, three other studies [108–110] showed no statistical difference in Tau fragments.

Protein cargo form plasma and serum isolated neuron-derived BEVs included synaptic proteins like neurogranin, synaptotagmin, synaptopodin, and synaptophysin, which was reduced in individuals with AD dementia [111]. However, the potential of being selective biomarkers of AD appears low because these synaptic proteins were also reduced in MCI and Parkinson's disease. Decreased levels of the growth-associated protein 43 (GAP43), synaptosomal-associated protein-25 (SNAP-25), and synapsin 1 were also observed in blood-isolated neuron-derived BEVs of AD patients [103]. Therefore, synaptic proteins cargo of neuron-derived BEVs demonstrates some biomarker potential in AD, although more studies are needed to confirm this.

Insulin pathway proteins are deregulated in AD as well. Specifically, higher phospho-Ser312-insulin receptor substrate-1 (IRS-1) and lower phospho-panTyr-IRS-1 levels were reported in blood isolated neuron-derived BEVS of AD patients [112,113]. So IRS-1 level, as well as being used to monitor insulin administration [114], could also be used as an AD biomarker.

Lysosomal proteins of neuron-derived BEVs from plasma were also found to discriminate AD dementia. Levels of cathepsin D, alysosome-associated membrane protein, and ubiquitinated proteins were significantly increased in AD patients, and levels of heat-shock protein-70 were diminished in preclinical and clinical AD, suggesting that neuronal lysosomal dysfunction is an early phenomenon in AD [115].

Finally, research from Winston et al. [106] demonstrated that the level of the repressor element 1-silencing transcription factor (REST) was significantly lower in AD patients and MCI compared to control subjects.

3.1.2. Neural Precursor Cell-Derived BEVs in Blood

Chondroitin sulfate proteoglycan (CSPG4) cells are a subtype of neuronal precursor cells that release neurotrophic factors implicated in neuronal growth and survival. Four assessed neurotrophic factors (hepatocyte growth factor, fibroblast growth factors 2 and 13, and type 1 insulin-like growth factor) were significantly lower in CSPG4 neuronal precursor cells-derived BEVs from preclinical AD patients, being able to use these neurotrophic factors as early biomarkers of AD. No significant further decrease was found during the course of the disease [116], though.

3.1.3. Astrocyte-Derived BEVs in Blood

Astrocyte-derived BEVs have also been reported to cargo Tau, Aβ-42, and APP pathway proteins. However, only levels of BACE1, sAPPβ, complement proteins, and glial-derived neurotrophic factor (GDNF) were significantly deregulated in AD. The levels of BACE1, sAPPβ, and complement proteins were higher [117], and the levels of GDNF were lower in AD patients compared to control individuals [118].

3.1.4. MicroRNA Cargo of Blood-Isolated EVs

The levels of miRNAs in peripheral blood can be affected by multiple factors and may also vary among different sample types. In this regard, exosomal miRNAs effectively avoid that problem because of their stable expression. Exosomes are a subtype of EV with a size of 40–100 nm that are released from most types of cells, including neurons [106]. Recent studies have shown that exosomes, in addition to functional proteins, carry mRNA and miRNAs [119], and abnormal expression of these exosomal miRNAs has been detected in AD [120].

More than 40 differentially expressed plasma- and serum-isolated EV-associated miRNAs have been described in AD and MCI relative to control individuals [121–127]. For example, exosomal miR-342-3p [123], miR-125a-5p, miR-125b-5p, and miR-451a, associated with fatty acid biosynthesis, hippo signaling, and protein processing in the endoplasmic reticulum were significantly lower in AD patients, and their level correlated with the extent of cognitive impairment [128]. Decreased levels of exosomal miRNA 23a-3p, ex-let-7i-5p, ex-miR-126-3p, and ex-miR-151a-3p, which target genes involved in cell death, among others, suggest that changes in the plasma level of AD individuals exhibit diagnostic value [129]. The exosomal miR-223, which regulates inflammation by interacting with different targets, was also significantly decreased in AD patients [125]. On the other hand, Barbagallo et al. found that exosomal miR-29a was significantly increased in AD patients [130] and Cheng et al. reported 14 significantly upregulated exosomal miRNAs [122]. It is also important to mention three of those exosomal miRNAs that have been reported in at least two different studies—the decrease of miR-193b and miR-342-3p [121,124], and both, the increase and decrease of miR3065-5p [122,123]. miR-193b is known to repress the expression of APP and PSEN1 mRNAs, so its reduction may promote amyloidosis, and miR 342-3p is suggested to affect Tau phosphorylation and aggregation.

These studies suggest that specific blood exosome miRNAs can be used as diagnostic biomarkers of AD and, additionally, are able to reflect the disease progression. It has also been reported that the combination of miR-135a, miR-193b, and miR-384, modulators of APP or BACE1 expression, are good for early AD diagnosis [124], demonstrating that a combined biomarker signature is better than a particular one for diagnosis. These studies have been carried out in already clinically diagnosed AD patients, so further studies will be necessary to evaluate the potential of these miRNAs as early biomarkers of AD. Table 2 summarizes all the information regarding AD-related miRNAs found in blood, as well as in CSF.

Table 2. AD-related main miRNA.

miRNAs	Regulation and Localization	References
miR-let-7d-5p, miR-let-7g-5p, miR-26b-5p, miR-191-5p	↓ Blood	[131]
miR-125a-5p	↓ Blood	[128]
miR-126-3p, miR-23a-3p, miR-151a-3p	↓ Blood	[129]
miR-135b	↓ Blood	[132]
miR-181a	↓ Blood	[133]
miR-194-5p	↓ Blood	[134]
miR-19b-3p, miR-29c-3p, miR-125b-3p	↓ Blood	[135]
miR-31, miR-93	↓ Blood	[99]
miR-3613-3p, miR-3916, miR-4772-3p, miR-185-5p, miR-20b-3p	↓ Blood	[123]
miR-501-3p	↓ Blood	[136]
miR-545-3p	↓ Blood	[137]
miR-181c	↓ Blood, ↓ Brain	[133,138]
miR-139-5p, miR-141-3p, miR-150-5p, miR-152-3p, miR-23b-3p, miR-24-3p, miR-338-3p, miR-342-3p, miR-125b-5p, miR-342-5p	↓ Blood, ↓ CSF	[123]
miR-1306-5p	↓ Blood, ↓ CSF	[122,139]
miR-143	↓ Blood, ↓ CSF	[99,133]
miR-15b	↓ Blood, ↓ CSF	[131,133]
miR-15b-3p	↓ Blood, ↓ CSF	[122,139]
miR-193b	↓ Blood, ↓ CSF	[121,124]
miR-223	↓ Blood, ↓ CSF	[125,140]
miR-451a	↓ Blood, ↓ CSF	[128,139]

Table 2. *Cont.*

miRNAs	Regulation and Localization	References
miR-106, miR-107, miR-181	↓ Brain	[69]
miR-106b	↓ Brain	[138]
miR-137, miR-139, miR-153, miR-183, miR-135, miR-124b	↓ Brain	[66]
miR-15a, miR-19b, miR-26b, miR-330	↓ Brain	[138]
miR-425	↓ Brain	[133]
miR-146b	↓ Brain, ↓ CSF	[133]
miR-210	↓ Brain, ↓ CSF	[133,141]
miR-10, miR-126, miR-127, miR-154, miR-194, miR-195, miR-199a, miR-214, miR-221, miR-338, miR-422b, miR-451, miR-455, miR-497, miR-99a, miR-27a-3p	↓ CSF	[133]
miR-16-2, miR-16-5p, miR-605-5p, mir-9-5p, miR-598, miR-136-3p	↓ CSF	[139]
miR-200b	↓ CSF	[142]
miR-214-3p, miR-299-5p	↓ CSF	[132,143]
miR-29b-3p	↓ CSF	[123]
miR-29c	↓ CSF	[134]
miR-29	↓ Blood, ↓ Brain, ↑Brain	[69,131,133]
miR-125b	↓ Blood, ↑ Brain, ↑ CSF	[65,66,123]
miR-146a	↓ Blood, ↑ Brain, ↑ CSF	[69,71,99]
miR-26a	↓ Brain (frontal cortex), ↑ Brain (hippocampus)	[133]
miR-3065-5p	↓ Blood, ↑ Brain	[122,123]
let-7i-5p	↓ Blood, ↑ CSF	[129,134]
miR-106a-5p, miR-20-5p, miR-425-5p, miR-18b-5p, miR-582-5p	↑ Blood	[122]
miR-106b-3p, miR-20b-5p, miR-146a-5p, miR-195-5p, niR-497-5p	↑ Blood	[135]
miR-455-3p, miR-4668-5p	↑ Blood	[144]
miR-5001-3p	↑ Blood	[123]
miR-519	↑ Blood	[140]
miR-548at-5p	↑ Blood	[123]
miR-590-5p	↑ Blood	[134]
miR-101-3p, miR-106b-5p, miR-143-3p, miR-335-5p, miR-361-5p,	↑ Blood, ↑ CSF	[122]
miR-138-5p	↑ Blood, ↑ CSF	[123]
miR-155	↑ Blood, ↑ CSF	[71,131]
miR-15a-5p	↑ Blood, ↑ CSF	[122,134]
miR-659-5p	↑ Blood, ↑ CSF	[123]
miR-100, miR-145, miR-148a, miR-27, miR-34a, miR-381, miR-422a, miR-423, miR-92	↑ Brain	[133]
miR-128	↑ Brain	[66]
miR-34	↑ Brain	[69]
miR-98	↑ Brain	[138]
miR-let-7b, miR-let7e	↑ CSF	[145]
miR-let-7f, miR-105, miR-138, miR-141, miR-151, miR-186, miR-191, miR-197, miR-204, miR-205, miR-216, miR-302b, miR-30a-3p, miR-30a-5p, miR-30b, miR-30d, miR-32, miR-345, miR-362, miR-371, miR-374, miR-375, miR-380-3p, miR-429, miR-448, miR-449, miR-494, miR-501, miR-517, miR-518, miR-520, miR-526	↑ CSF	[133]
miR-20a-5p	↑ CSF	[122]
miR-222	↑ CSF	[146]

Table 2. *Cont.*

miRNAs	Regulation and Localization	References
miR-331-5p, miR-485-5p, miR-132-5p	↑ CSF	[139]
miR-613	↑ CSF	[147]
miR-200b-5p	↑ Eyes	[148]
miR-93-5p	↑ ↓ Blood, ↑ CSF	[122,135]
miR-101	↑ Blood, ↓ Brain	[131,138]
miR-132, miR-212	↑ Blood, ↓ Brain	[126,133]
miR-200c	↑ Blood, ↓ Brain (frontal cortex), ↑ Brain (hippocampus)	[133,149]
miR-9	↑ Blood, ↓ Brain (frontal cortex, cortex), ↑ Brain (hippocampus), ↑ CSF	[66,71,131,133,138]
miR-30e-5p	↑ Blood, ↑ Brain, ↑ CSF,	[122,133]
miR-29a	↑ Blood, ↓ Brain, ↑ CSF	[73,74,130]
miR-206	↑ Blood, ↑ CSF, ↑ Eyes	[100,150]
miR-142-5p	↑ Blood, ↓ CSF	[133,134]
miR-384	↑ Blood, ↓ CSF	[124]
miR-135a	↑ Blood, ↓ CSF, ↑ CSF	[124,133,142]
miR-125a	↓ Brain, ↑ CSF	[66,133]
miR-29b	↓ Blood, ↓ Brain, ↑ CSF	[73,74,131]
miR-30c	↑ Brain (frontal cortex), ↓ Brain (hippocampus), ↑ CSF	[133]

3.2. Ocular Biomarkers

AD not only causes neurodegenerative changes in the brain but also produces structural and functional alterations in the retinal neural and non-neural ocular tissues [151]. Engagingly, specific biomarkers of AD have been reported as well in retinal degeneration and visual function impairment [152], sharing pathophysiological features with glaucoma and age-related macular degeneration (AMD) [153]. The similarity between ocular and cerebral tissues suggests that these ocular manifestations may be used as early biomarkers of AD.

Numerous studies have identified Aβ depositions in the lens and retina. Aβ-accumulates in the retina in an age-dependent manner in a mouse model of AD and correlates with amyloid plaques in the brain. Interestingly, the appearance of retinal amyloid plaques precedes that in the brain [154]. Elevated levels of Aβ1-42 and amyloid plaques were also reported in the retinas of confirmed AD patients [152]. This retinal Aβ can be detected noninvasively by using hyperspectral imaging microscopy or with modified endoscope applied to the corneal surface [155].

In the same way, Aβ has been identified in the lens of rodents, monkeys, and humans in several studies. The accumulation of the isoforms Aβ1-40 and Aβ1-42 has been demonstrated in the lenses of AD people *post mortem* at concentrations comparable with those in the brain [156]. In the same way, a clinical trial carried out on AD patients, and age-matched healthy volunteers confirmed the presence of Aβ in the lens of the first and its correlation in the brain using imaging techniques [157].

Moreover, changes such as thinning of the nerve cell layer, optic nerve atrophy, and the loss of retinal ganglion cells [156,158] were reported in AD, resulting in visual functional impairment and circadian disturbances [159]. Specifically, a study of melanopsin retinal ganglion cells (mRGCs), a photosensitive subtype of ganglion cells in charge of the circadian rhythms, has shown a significant decrease of this neuronal cell type in people with AD but not in healthy controls, with a prominent Aβ accumulation around mRGCs [160,161].

Other ocular tissues, such as the cornea, which is the outermost layer of the eye and, therefore, grants accessibility, could be used as potential biomarker for the diagnosis of AD. Dutescu et al. [162] found the cytoplasmic expression of APP in the epithelial cell layer of the cornea of transgenic mouse models of AD. More recently, Choi et al. corroborated the expression of APP, together with proteins involved in its processing such as BACE1, in human corneal fibroblasts, and the corneal epithelium [163].

For the development of a novel noninvasive screening and diagnostic tool, the ocular examination sector appears promising. In this context, tear fluid provides a viable source widely used for biomarker studies [164], including neurodegenerative diseases [165]. Tear samples are easy to collect and contain a lot of proteins, most notably lipocalin-1, lactotransferrin, and lysozyme C, involved in immune and inflammatory processes [166].

Against this backdrop, total protein concentration and composition modifications in tears and an abnormal flow rate and tear function have been described in AD patients [167], supporting the use of tears as a new noninvasive method to discriminate AD patients. Specifically, lipocalin-1, dermcidin, lysozyme C, and lacritin were shown to be potential biomarkers, with an 81% sensitivity and 77% specificity [168]. In addition, the elongation initiation factor 4E (eIF4E) was exclusively expressed in tear samples from AD patients. Total miRNA content was also higher in tears from AD individuals and miR-200b-5p was significantly elevated in AD tear fluid samples compared to controls [148]. Tears could be useful for first screening, and patients with a positive tear analysis test might be further evaluated to establish an early diagnosis. Assessments of pupillary responses and retinal vasculature have also been considered as biomarkers of AD, but are not yet conclusively validated for clinical diagnosis. Further research is needed in order to use ocular biomarkers as AD early diagnostic tools.

Regarding the use of ocular biomarkers, as well as other novel matrices, some limitations arise, like the amall volume of the samples or the standardization of the collection procedures.

3.3. Salivary Biomarkers

Due to the link between the decline of the salivary glands and AD [169,170], it seems likely that AD-specific proteins are expressed in the salivary glands. Salivary epithelial cells express APP and Aβ, and changes in the CSF may be reflected in the saliva [171,172]. Saliva is a novel matrix; therefore, there are still some conflicts between studies. One study has revealed no changes in Aβ-40 protein between AD patients and age-matched controls and high levels of Aβ-42 only in MCI patients (in severe AD stages, these levels returned to control values) [173]. A second study has shown that salivary levels of Aβ-40 and Aβ-42 peptides increase as the severity of AD grows, even as far as a three-fold increase in the case of Aβ-42 levels [174,175]. Moreover, studies where the genetic condition was considered, suggested that salivary Aβ-42 levels were associated with familial AD more than with sporadic AD [173,176]. Among the reasons for these different results may be the distinct Aβ peptide detection and saliva collection techniques, as well as the different disease stages of the patients. There is a clear need for more studies and larger sample size to conclude whether there is a relationship between salivary Aβ-42 and Aβ-40 levels and AD progression.

Another typical candidate protein for analysis in the saliva is Tau, which is expressed and secreted by the acinar epithelial cells of the salivary glands [177]. No changes have been seen in total Tau levels between healthy elderly, MCI, and AD patients [178]. In contrast, the phospho-Tau/total-Tau ratio looks promising in this regard. A high phospho-Tau/total-Tau ratio was found in AD patients compared with non-amnestic people [179]. Though no conclusive results were found that pointed in a particular direction, the few results available suggest that studies should be directed at investigating different sites of Tau protein phosphorylation as possible candidates for biomarkers of the pathology [180].

Interestingly, the study of one of the most important antimicrobial peptides in saliva, lactoferrin, seems to have a high accuracy in AD diagnosis. Lactoferrin participates in modulating the immune response and inflammation process due to its high defense action. It has been seen that low lactoferrin levels in saliva of healthy people mean a clear risk factor to develop amnestic MCI and AD dementia [181].

Another proposed candidate in the saliva is the acetylcholinesterase enzyme (AChE), which plays a role in removing the accumulation of acetylcholine (ACh) in Aβ-plaques and NFTs. Although there are still few studies, no evidence has been found to suggest that AChE activity is different between AD patients and control individuals of the same cognitive age not taking anticholinergics [182,183].

All these results support the need for longitudinal studies with a larger number of subjects to find conclusive results regarding the potential use of the expression of certain molecules in saliva as biomarkers in AD.

3.4. Urine Biomarkers

Oxidative stress plays an important role in AD, and the study of guide molecules of oxidative brain damage might be promising early hallmarks of AD in urine. Some metabolites such as the isoprostane 8,12-iso-iPF(2alpha)-VI, a free amino acid generated by lipid peroxidation whose levels in urine were higher at advanced AD stages, might predict the progression of the pathology from MCI to AD dementia [184,185].

Other metabolite candidates that reflect oxidative DNA damage can be oxidized nucleosides such as pseudouridine, 1-methyladenosine, 3-methyluridine, N2, N2-dimethylguanosine, 8-hydroxy-2′-deoxyguanosine, and 2-deoxyguanosine, whose levels were higher in AD patients than in healthy elderly [186].

Other promising markers are some proteins found in urine. Given that the AD-associated neuronal thread protein (AD7c-NTP) was isolated from brain tissue and was increased in the CSF of AD people, correlating with the severity of dementia [187], it has been suggested as a potential biomarker of AD. Recently, a high specificity of this protein has been demonstrated to predict Aβ-plaques in MCI patients when present in the urine [188]. These findings are supported by a meta-analysis proposing the use of urinary AD7c-NTP for the early diagnosis of probable AD [189]. In a longitudinal study carried out in 2018, high levels of albumin, a protein characteristic of chronic kidney disease, were detected in AD patients in comparison with age-matched healthy individuals [190]. According to Yao et al. [191], and in the same direction, the urine of patients with AD showed significantly decreased levels of osteopontin and increased levels of gelsolin and insulin-like growth factor-binding protein 7 in comparison with healthy elderly. All of them are proteins involved in several pathological processes of AD [192–194], and they may serve as potential novel urinary biomarkers.

Although it is evident the need for longitudinal studies with bigger samples to get conclusive results verifying the diagnostic value of these peripheral markers, the scientific community is hopeful that the biomarkers in these noninvasive new matrices will be able to demonstrate diagnostic value.

4. Concluding Remarks and Future Perspectives

In addition to the established Aβ-42, total-Tau, phospho-Tau, and CSF biomarkers, several candidate more accessible fluid biomarkers have shown potential for clinical use in AD to support diagnosis and prognosis.

Blood has been the most widely studied fluid biomarker, being neuron-derived BEVs the most investigated biomarker among blood-isolated EVs. The majority of the studies reported elevated Aβ-42 levels in blood-isolated neuron derived BEVs in AD and MCI patients early in the course of the disease, and also with disease progression. With regard to synaptic proteins assessed in neuron-derived BEVs, growth-associated protein 43 (GAP43) also showed some potential as a marker of AD progression. Henceforth, BEVs have emerged as a novel potential blood-based biomarker of AD. It would also be interesting to study BEVs from other components of the brain, such as the cerebrovasculature, that are affected in the early stages of AD and that would allow us to obtain more sensitive blood-based biomarkers.

Tear fluid also provides a viable source widely used for biomarker studies and lipocalin-1, dermcidin, lysozyme C, lacritin, eIF4E, and the microRNA-200b-5p were shown to be potential biomarkers. The relationship between salivary Aβ-42, Aβ-40, and lactoferrin levels and the progression of AD point out to also be biomarker candidates of AD and MCI in saliva. Lastly, several metabolites and proteins like AD7c-NTP, osteopontin, gelsolin, or insulin-like growth factor-binding protein 7 in urine are involved in several pathological processes in AD.

Inflammatory and oxidative stress markers are very common hallmarks of neurodegenerative diseases such as amyotrophic lateral sclerosis, Parkinson, Huntington, and AD. It has been suggested that these alterations reflect inflammatory mechanisms within the central nervous system that parallel the neurodegenerative process [56]. In this review, we have mentioned several neuroinflammation candidate biomarkers such as TREM2, MCP-1, and YKL-40, which have been extensively investigated in AD patients. Particularly, there is strong evidence regarding CSF YKL-40 levels, not only as a potential biomarker for AD diagnosis, but also as a predictor of disease progression from the asymptomatic stage to prodromal, and eventually dementia stages [51]. However, the idea that these inflammation-related proteins could differentiate AD from other dementias is controversial since neurodegeneration and neuroinflammation go hand in hand and inflammation on its own cannot be considered a marker of a specific pathology. There is a clear lack of reliable inflammatory biomarkers that can be used in the context of accurate diagnosis.

Despite recent studies strongly indicating the potential of fluid biomarkers as early diagnosis of AD, these biomarkers are not yet validated for clinical use and further research is needed before these can be regulatory qualified and applied clinically. Future work should establish normative ranges for the levels of these biomarkers to indicate a pathology that may find clinical applications. If we provided early diagnosis and treatment before the underlying pathology manifests clinically, AD patients' quality of life could notably improve and could be an approach to prevent its irreversible consequences.

Author Contributions: E.A. and G.E. carried out the literature review, conceptualized and prepared the initial draft. G.E. provided critical inputs. V.G.-V. edited and contributed in the final manuscript. All authors have read and agreed to the final version of the manuscript.

Funding: This research has not received any specific funding.

Acknowledgments: Figures created with BioRender.com.

Conflicts of Interest: The authors declare no conflict of interest.

References

1. Dubois, B.; Feldman, H.H.; Jacova, C.; DeKosky, S.T.; Barberger-Gateau, P.; Cummings, J.; Delacourte, A.; Galasko, D.; Gauthier, S.; Jicha, G.; et al. Research criteria for the diagnosis of Alzheimer's disease: Revising the NINCDS-ADRDA criteria. *Lancet Neurol.* **2007**, *6*, 734–746. [CrossRef]
2. Mueller, S.G.; Schuff, N.; Yaffe, K.; Madison, C.; Miller, B.; Weiner, M.W. Hippocampal atrophy patterns in mild cognitive impairment and alzheimer's disease. *Hum. Brain Mapp.* **2010**, *31*, 1339–1347. [CrossRef] [PubMed]
3. La Joie, R.; Perrotin, A.; De La Sayette, V.; Egret, S.; Doeuvre, L.; Belliard, S.; Eustache, F.; Desgranges, B.; Chételat, G. Hippocampal subfield volumetry in mild cognitive impairment, Alzheimer's disease and semantic dementia. *NeuroImage Clin.* **2013**, *3*, 155–162. [CrossRef] [PubMed]
4. Pluta, J.; Yushkevich, P.; Das, S.; Wolk, D. In vivo Analysis of Hippocampal Subfield Atrophy in Mild Cognitive Impairment via Semi-Automatic Segmentation of T2-Weighted MRI. *J. Alzheimer's Dis.* **2012**, *31*, 85–99. [CrossRef]
5. Arendt, T.; Brückner, M.K.; Morawski, M.; Jäger, C.; Gertz, H.J. Early neurone loss in Alzheimer's disease: Cortical or subcortical? *Acta Neuropathol. Commun.* **2015**, *3*, 10. [CrossRef]
6. Braak, H.; Del Tredici, K. Where, when, and in what form does sporadic Alzheimer's disease begin? *Curr. Opin. Neurol.* **2012**, *25*, 708–714. [CrossRef]
7. Andrés-Benito, P.; Fernández-Dueñas, V.; Carmona, M.; Escobar, L.A.; Torrejón-Escribano, B.; Aso, E.; Ciruela, F.; Ferrer, I. Locus coeruleus at asymptomatic early and middle Braak stages of neurofibrillary tangle pathology. *Neuropathol. Appl. Neurobiol.* **2017**, *43*, 373–392. [CrossRef]
8. Braak, H.; Thal, D.R.; Ghebremedhin, E.; Del Tredici, K. Stages of the pathologic process in alzheimer disease: Age categories from 1 to 100 years. *J. Neuropathol. Exp. Neurol.* **2011**, *70*, 960–969. [CrossRef]

9. Ehrenberg, A.J.; Nguy, A.K.; Theofilas, P.; Dunlop, S.; Suemoto, C.K.; Di Lorenzo Alho, A.T.; Leite, R.P.; Diehl Rodriguez, R.; Mejia, M.B.; Rüb, U.; et al. Quantifying the accretion of hyperphosphorylated tau in the locus coeruleus and dorsal raphe nucleus: The pathological building blocks of early Alzheimer's disease. *Neuropathol. Appl. Neurobiol.* **2017**, *43*, 393–408. [CrossRef]
10. Stratmann, K.; Heinsen, H.; Korf, H.-W.; Del Turco, D.; Ghebremedhin, E.; Seidel, K.; Bouzrou, M.; Grinberg, L.T.; Bohl, J.; Wharton, S.B.; et al. Precortical Phase of Alzheimer's Disease (AD)-Related Tau Cytoskeletal Pathology. *Brain Pathol.* **2016**, *26*, 371–386. [CrossRef]
11. Theofilas, P.; Ehrenberg, A.J.; Dunlop, S.; Di Lorenzo Alho, A.T.; Nguy, A.; Leite, R.E.P.; Rodriguez, R.D.; Mejia, M.B.; Suemoto, C.K.; Ferretti-Rebustini, R.E.D.L.; et al. Locus coeruleus volume and cell population changes during Alzheimer's disease progression: A stereological study in human postmortem brains with potential implication for early-stage biomarker discovery. *Alzheimer's Dement.* **2017**, *13*, 236–246. [CrossRef] [PubMed]
12. Kelly, S.C.; He, B.; Perez, S.E.; Ginsberg, S.D.; Mufson, E.J.; Counts, S.E. Locus coeruleus cellular and molecular pathology during the progression of Alzheimer's disease. *Acta Neuropathol. Commun.* **2017**, *5*, 8. [CrossRef] [PubMed]
13. Lyness, S.A.; Zarow, C.; Chui, H.C. Neuron loss in key cholinergic and aminergic nuclei in Alzheimer disease: A meta-analysis. *Neurobiol. Aging* **2003**, *24*, 1–23. [CrossRef]
14. Betts, M.J.; Cardenas-Blanco, A.; Kanowski, M.; Jessen, F.; Düzel, E. In vivo MRI assessment of the human locus coeruleus along its rostrocaudal extent in young and older adults. *Neuroimage* **2017**, *163*, 150–159. [CrossRef]
15. Priovoulos, N.; Jacobs, H.I.L.; Ivanov, D.; Uludağ, K.; Verhey, F.R.J.; Poser, B.A. High-resolution in vivo imaging of human locus coeruleus by magnetization transfer MRI at 3T and 7T. *Neuroimage* **2018**, *168*, 427–436. [CrossRef]
16. Tondelli, M.; Wilcock, G.K.; Nichelli, P.; de Jager, C.A.; Jenkinson, M.; Zamboni, G. Structural MRI changes detectable up to ten years before clinical Alzheimer's disease. *Neurobiol. Aging* **2012**, *33*, 825.e25–825.e36. [CrossRef]
17. Bernard, C.; Helmer, C.; Dilharreguy, B.; Amieva, H.; Auriacombe, S.; Dartigues, J.F.; Allard, M.; Catheline, G. Time course of brain volume changes in the preclinical phase of Alzheimer's disease. *Alzheimer's Dement.* **2014**, *10*, 143–151.e1. [CrossRef]
18. Smith, C.D.; Chebrolu, H.; Wekstein, D.R.; Schmitt, F.A.; Jicha, G.A.; Cooper, G.; Markesbery, W.R. Brain structural alterations before mild cognitive impairment. *Neurology* **2007**, *68*, 1268–1273. [CrossRef]
19. Apostolova, L.G.; Dutton, R.A.; Dinov, I.D.; Hayashi, K.M.; Toga, A.W.; Cummings, J.L.; Thompson, P.M. Conversion of mild cognitive impairment to alzheimer disease predicted by hippocampal atrophy maps. *Arch. Neurol.* **2006**, *63*, 693–699. [CrossRef]
20. Jack, C.R.; Petersen, R.C.; Xu, Y.C.; O'Brien, P.C.; Smith, G.E.; Ivnik, R.J.; Boeve, B.F.; Waring, S.C.; Tangalos, E.G.; Kokmen, E. Prediction of AD with MRI-based hippocampal volume in mild cognitive impairment. *Neurology* **1999**, *52*, 1397–1403. [CrossRef]
21. Korf, E.S.C.; Wahlund, L.O.; Visser, P.J.; Scheltens, P. Medial temporal lobe atrophy on MRI predicts dementia in patients with mild cognitive impairment. *Neurology* **2004**, *63*, 94–100. [CrossRef]
22. Apostolova, L.G.; Mosconi, L.; Thompson, P.M.; Green, A.E.; Hwang, K.S.; Ramirez, A.; Mistur, R.; Tsui, W.H.; de Leon, M.J. Subregional hippocampal atrophy predicts Alzheimer's dementia in the cognitively normal. *Neurobiol. Aging* **2010**, *31*, 1077–1088. [CrossRef] [PubMed]
23. Csernansky, J.G.; Wang, L.; Swank, J.; Miller, J.P.; Gado, M.; McKeel, D.; Miller, M.I.; Morris, J.C. Preclinical detection of Alzheimer's disease: Hippocampal shape and volume predict dementia onset in the elderly. *Neuroimage* **2005**, *25*, 783–792. [CrossRef] [PubMed]
24. Chen, K.; Reiman, E.M.; Alexander, G.E.; Caselli, R.J.; Gerkin, R.; Bandy, D.; Domb, A.; Osborne, D.; Fox, N.; Crum, W.R.; et al. Correlations between apolipoprotein E ε4 gene dose and whole brain atrophy rates. *Am. J. Psychiatry* **2007**, *164*, 916–921. [CrossRef] [PubMed]
25. Lim, Y.Y.; Mormino, E.C. APOE genotype and early β-amyloid accumulation in older adults without dementia. *Neurology* **2017**, *89*, 1028–1034. [CrossRef] [PubMed]
26. Castellano, J.M.; Kim, J.; Stewart, F.R.; Jiang, H.; DeMattos, R.B.; Patterson, B.W.; Fagan, A.M.; Morris, J.C.; Mawuenyega, K.G.; Cruchaga, C.; et al. Human apoE Isoforms Differentially Regulate Brain Amyloid-Peptide Clearance. *Sci. Transl. Med.* **2011**, *3*, 89ra57. [CrossRef]

27. Reiman, E.M.; Uecker, A.; Caselli, R.J.; Lewis, S.; Bandy, D.; De Leon, M.J.; De Santi, S.; Convit, A.; Osborne, D.; Weaver, A.; et al. Hippocampal volumes in cognitively normal persons at genetic risk for Alzheimer's disease. *Ann. Neurol.* **1998**, *44*, 288–291. [CrossRef]
28. Fang, Y.; Du, N.; Xing, L.; Duo, Y.; Zheng, L. Evaluation of hippocampal volume and serum brain-derived neurotrophic factor as potential diagnostic markers of conversion from amnestic mild cognitive impairment to Alzheimer disease A STROBE-compliant article. *Medicine (United States)* **2019**, *98*. [CrossRef]
29. Finnema, S.J.; Nabulsi, N.B.; Eid, T.; Detyniecki, K.; Lin, S.F.; Chen, M.K.; Dhaher, R.; Matuskey, D.; Baum, E.; Holden, D.; et al. Imaging synaptic density in the living human brain. *Sci. Transl. Med.* **2016**, *8*, 1–10. [CrossRef]
30. Terry, R.D.; Masliah, E.; Salmon, D.P.; Butters, N.; DeTeresa, R.; Hill, R.; Hansen, L.A.; Katzman, R. Physical basis of cognitive alterations in alzheimer's disease: Synapse loss is the major correlate of cognitive impairment. *Ann. Neurol.* **1991**, *30*, 572–580. [CrossRef]
31. Lista, S.; Hampel, H. Synaptic degeneration and neurogranin in the pathophysiology of Alzheimer's disease. *Expert Rev. Neurother.* **2017**, *17*, 47–57. [CrossRef] [PubMed]
32. Chandra, A.; Valkimadi, P.E.; Pagano, G.; Cousins, O.; Dervenoulas, G.; Politis, M. Applications of amyloid, tau, and neuroinflammation PET imaging to Alzheimer's disease and mild cognitive impairment. *Hum. Brain Mapp.* **2019**, *40*, 5424–5442. [CrossRef] [PubMed]
33. Johnson, K.A.; Fox, N.C.; Sperling, R.A.; Klunk, W.E. Brain Imaging in Alzheimer Disease. *Cold Spring Harb. Perspect. Med.* **2012**, *2*, a006213. [CrossRef] [PubMed]
34. Ferreira, D.; Perestelo-Pérez, L.; Westman, E.; Wahlund, L.O.; Sarrisa, A.; Serrano-Aguilar, P. Meta-review of CSF core biomarkers in Alzheimer's disease: The state-of-the-art after the new revised diagnostic criteria. *Front. Aging Neurosci.* **2014**, *6*, 1–24. [CrossRef]
35. Hampel, H.; O'Bryant, S.E.; Molinuevo, J.L.; Zetterberg, H.; Masters, C.L.; Lista, S.; Kiddle, S.J.; Batrla, R.; Blennow, K. Blood-based biomarkers for Alzheimer disease: Mapping the road to the clinic. *Nat. Rev. Neurol.* **2018**, *14*, 639–652. [CrossRef]
36. Fishman, R. Cerebrospinal fluid in diseases of the nervous system. *WB Saunders Co.* **1992**, 183–252.
37. Dubois, B.; Feldman, H.H.; Jacova, C.; Hampel, H.; Molinuevo, J.L.; Blennow, K.; Dekosky, S.T.; Gauthier, S.; Selkoe, D.; Bateman, R.; et al. Advancing research diagnostic criteria for Alzheimer's disease: The IWG-2 criteria. *Lancet Neurol.* **2014**, *13*, 614–629. [CrossRef]
38. McKhann, G.M.; Knopman, D.S.; Chertkow, H.; Hyman, B.T.; Jack, C.R.; Kawas, C.H.; Klunk, W.E.; Koroshetz, W.J.; Manly, J.J.; Mayeux, R.; et al. The diagnosis of dementia due to Alzheimer's disease: Recommendations from the National Institute on Aging-Alzheimer's Association workgroups on diagnostic guidelines for Alzheimer's disease. *Alzheimer's Dement.* **2011**, *7*, 263–269. [CrossRef]
39. Kester, M.I.; Van Der Vlies, A.E.; Blankenstein, M.A.; Pijnenburg, Y.A.L.; Van Elk, E.J.; Scheltens, P.; Van Der Flier, W.M. CSF biomarkers predict rate of cognitive decline in Alzheimer disease. *Neurology* **2009**, *73*, 1353–1358. [CrossRef]
40. Martínez-Morillo, E.; Hansson, O.; Atagi, Y.; Bu, G.; Minthon, L.; Diamandis, E.P.; Nielsen, H.M. Total apolipoprotein E levels and specific isoform composition in cerebrospinal fluid and plasma from Alzheimer's disease patients and controls. *Acta Neuropathol.* **2014**, *127*, 633–643. [CrossRef]
41. Strittmatter, W.J.; Saunders, A.M.; Goedert, M.; Weisgraber, K.H.; Dong, L.M.; Jakes, R.; Huang, D.Y.; Pericak-Vance, M.; Schmechel, D.; Roses, A.D. Isoform-specific interactions of apolipoprotein E with microtubule-associated protein tau: Implications for Alzheimer disease. *Proc. Natl. Acad. Sci. USA* **1994**, *91*, 11183–11186. [CrossRef] [PubMed]
42. García Ayllón, M.-S.; Campanari, M.-L.; Brinkmalm, G.; Rábano, A.; Alom, J.; Saura, C.A.; Andreasen, N.; Blennow, K.; Sáez-Valero, J. CSF Presenilin-1 complexes are increased in Alzheimer's disease. *Acta Neuropathol. Commun.* **2013**, *1*, 46. [CrossRef] [PubMed]
43. Ewers, M.; Zhong, Z.; Bürger, K.; Wallin, A.; Blennow, K.; Teipel, S.J.; Shen, Y.; Hampel, H. Increased CSF-BACE 1 activity is associated with ApoE-ε4 genotype in subjects with mild cognitive impairment and Alzheimer's disease. *Brain* **2008**, *131*, 1252–1258. [CrossRef] [PubMed]
44. Zetterberg, H.; Andreasson, U.; Hansson, O.; Wu, G.; Sankaranarayanan, S.; Andersson, M.E.; Buchhave, P.; Londos, E.; Umek, R.M.; Minthon, L.; et al. Elevated cerebrospinal fluid BACE1 activity in incipient alzheimer disease. *Arch. Neurol.* **2008**, *65*, 1102–1107. [CrossRef]

45. Calsolaro, V.; Edison, P. Neuroinflammation in Alzheimer's disease: Current evidence and future directions. *Alzheimer's Dement.* **2016**, *12*, 719–732. [CrossRef]
46. Duits, F.H.; Brinkmalm, G.; Teunissen, C.E.; Brinkmalm, A.; Scheltens, P.; Van Der Flier, W.M.; Zetterberg, H.; Blennow, K. Synaptic proteins in CSF as potential novel biomarkers for prognosis in prodromal Alzheimer's disease. *Alzheimer's Res. Ther.* **2018**, *10*, 1–9. [CrossRef]
47. Mattsson, N.; Insel, P.; Nosheny, R.; Zetterberg, H.; Trojanowski, J.Q.; Shaw, L.M.; Tosun, D.; Weiner, M. CSF protein biomarkers predicting longitudinal reduction of CSF β-amyloid42 in cognitively healthy elders. *Transl. Psychiatry* **2013**, *3*. [CrossRef]
48. Kester, M.I.; Teunissen, C.E.; Sutphen, C.; Herries, E.M.; Ladenson, J.H.; Xiong, C.; Scheltens, P.; Van Der Flier, W.M.; Morris, J.C.; Holtzman, D.M.; et al. Cerebrospinal fluid VILIP-1 and YKL-40, candidate biomarkers to diagnose, predict and monitor Alzheimer's disease in a memory clinic cohort. *Alzheimer's Res. Ther.* **2015**, *7*, 1–9. [CrossRef]
49. Craig-Schapiro, R.; Perrin, R.J.; Roe, C.M.; Xiong, C.; Carter, D.; Cairns, N.J.; Mintun, M.A.; Peskind, E.R.; Li, G.; Galasko, D.R.; et al. YKL-40: A Novel Prognostic Fluid Biomarker for Preclinical Alzheimer's Disease. *Biol. Psychiatry* **2010**, *68*, 903–912. [CrossRef]
50. Bettcher, B.M.; Johnson, S.C.; Fitch, R.; Casaletto, K.B.; Heffernan, K.S.; Asthana, S.; Zetterberg, H.; Blennow, K.; Carlsson, C.M.; Neuhaus, J.; et al. Cerebrospinal Fluid and Plasma Levels of Inflammation Differentially Relate to CNS Markers of Alzheimer's Disease Pathology and Neuronal Damage. *J. Alzheimer's Dis.* **2018**, *62*, 385–397. [CrossRef]
51. Dhiman, K.; Blennow, K.; Zetterberg, H.; Martins, R.N.; Gupta, V.B. *Cerebrospinal Fluid Biomarkers for Understanding Multiple Aspects of Alzheimer's Disease Pathogenesis*; Springer International Publishing: New York, NY, USA, 2019; Volume 76, ISBN 0001801903040.
52. Takahashi, K.; Rochford, C.D.P.; Neumann, H. Clearance of apoptotic neurons without inflammation by microglial triggering receptor expressed on myeloid cells-2. *J. Exp. Med.* **2005**, *201*, 647–657. [CrossRef] [PubMed]
53. Yuan, P.; Condello, C.; Keene, C.D.; Wang, Y.; Bird, T.D.; Paul, S.M.; Luo, W.; Colonna, M.; Baddeley, D.; Grutzendler, J. TREM2 Haplodeficiency in Mice and Humans Impairs the Microglia Barrier Function Leading to Decreased Amyloid Compaction and Severe Axonal Dystrophy. *Neuron* **2016**, *90*, 724–739. [CrossRef] [PubMed]
54. Suárez-Calvet, M.; Kleinberger, G.; Araque Caballero, M.Á.; Brendel, M.; Rominger, A.; Alcolea, D.; Fortea, J.; Lleó, A.; Blesa, R.; Gispert, J.D.; et al. sTREM 2 cerebrospinal fluid levels are a potential biomarker for microglia activity in early-stage Alzheimer's disease and associate with neuronal injury markers. *EMBO Mol. Med.* **2016**, *8*, 466–476. [CrossRef]
55. Tarawneh, R.; D'Angelo, G.; Crimmins, D.; Herries, E.; Griest, T.; Fagan, A.M.; Zipfel, G.J.; Ladenson, J.H.; Morris, J.C.; Holtzman, D.M. Diagnostic and Prognostic Utility of the Synaptic Marker Neurogranin in Alzheimer Disease. *JAMA Neurol.* **2016**, *73*, 561. [CrossRef]
56. Molinuevo, J.L.; Ayton, S.; Batrla, R.; Bednar, M.M.; Bittner, T.; Cummings, J.; Fagan, A.M.; Hampel, H.; Mielke, M.M.; Mikulskis, A.; et al. *Current State of Alzheimer's Fluid Biomarkers*; Springer: Berlin/Heidelberg, Germany, 2018; Volume 136, ISBN 0040101819.
57. Zetterberg, H.; Skillbäck, T.; Mattsson, N.; Trojanowski, J.Q.; Portelius, E.; Shaw, L.M.; Weiner, M.W.; Blennow, K. Association of Cerebrospinal Fluid Neurofilament Light Concentration With Alzheimer Disease Progression. *JAMA Neurol.* **2016**, *73*, 60. [CrossRef]
58. Schmidt, F.M.; Mergl, R.; Stach, B.; Jahn, I.; Gertz, H.J.; Schönknecht, P. Elevated levels of cerebrospinal fluid neuron-specific enolase (NSE) in Alzheimer's disease. *Neurosci. Lett.* **2014**, *570*, 81–85. [CrossRef]
59. Mulder, C.; Wahlund, L.O.; Teerlink, T.; Blomberg, M.; Veerhuis, R.; Van Kamp, G.J.; Scheltens, P.; Scheffer, P.G. Decreased lysophosphatidylcholine/phosphatidylcholine ratio in cerebrospinal fluid in Alzheimer's disease. *J. Neural Transm.* **2003**, *110*, 949–955. [CrossRef]
60. Mielke, M.M.; Haughey, N.J.; Bandaru, V.V.R.; Zetterberg, H.; Blennow, K.; Andreasson, U.; Johnson, S.C.; Gleason, C.E.; Blazel, H.M.; Puglielli, L.; et al. Cerebrospinal fluid sphingolipids, β-amyloid, and tau in adults at risk for Alzheimer's disease. *Neurobiol. Aging* **2014**, *35*, 2486–2494. [CrossRef]
61. Hu, W.T.; Chen-Plotkin, A.; Arnold, S.E.; Grossman, M.; Clark, C.M.; Shaw, L.M.; Pickering, E.; Kuhn, M.; Chen, Y.; Mccluskey, L.; et al. Novel CSF biomarkers for Alzheimer's disease and mild cognitive impairment. *Acta Neuropathol.* **2010**, *119*, 669–678. [CrossRef] [PubMed]

62. Desikan, R.S.; Thompson, W.K.; Holland, D.; Hess, C.P.; Brewer, J.B.; Zetterberg, H.; Blennow, K.; Andreassen, O.A.; McEvoy, L.K.; Hyman, B.T.; et al. Heart fatty acid binding protein and Aβ-associated Alzheimer's neurodegeneration. *Mol. Neurodegener.* **2013**, *8*, 1. [CrossRef] [PubMed]
63. Chiasserini, D.; Parnetti, L.; Andreasson, U.; Zetterberg, H.; Giannandrea, D.; Calabresi, P.; Blennow, K. CSF levels of heart fatty acid binding protein are altered during early phases of Alzheimer's disease. *J. Alzheimer's Dis.* **2010**, *22*, 1281–1288. [CrossRef] [PubMed]
64. Provost, P. Interpretation and applicability of microrna datato the context of Alzheimer's and age-related diseases. *Aging (Albany. NY)* **2010**, *2*, 166–169. [CrossRef] [PubMed]
65. Lukiw, W.J.; Andreeva, T.V.; Grigorenko, A.P.; Rogaev, E.I. Studying micro RNA function and dysfunction in Alzheimer's disease. *Front. Genet.* **2013**, *3*, 1–13. [CrossRef] [PubMed]
66. Lukiw, W.J. Micro-RNA speciation in fetal, adult and Alzheimer's disease hippocampus. *Neuroreport* **2007**, *18*, 297–300. [CrossRef]
67. Cortez, M.A.; Calin, G.A. MicroRNA identification in plasma and serum: A new tool to diagnose and monitor diseases. *Expert Opin. Biol. Ther.* **2009**, *9*, 703–711. [CrossRef]
68. Gallego, J.A.; Gordon, M.L.; Claycomb, K.; Bhatt, M.; Lencz, T.; Malhotra, A.K. In Vivo MicroRNA Detection and Quantitation in Cerebrospinal Fluid. *J. Mol. Neurosci.* **2012**, *47*, 243–248. [CrossRef]
69. Femminella, G.D.; Ferrara, N.; Rengo, G. The emerging role of microRNAs in Alzheimer's disease. *Front. Physiol.* **2015**, *6*, 1–5. [CrossRef]
70. Alexandrov, P.N.; Dua, P.; Hill, J.M.; Bhattacharjee, S.; Zhao, Y.; Lukiw, W.J. MicroRNA (miRNA) speciation in Alzheimer's disease (AD) cerebrospinal fluid (CSF) and extracellular fluid (ECF). *Int. J. Biochem. Mol. Biol.* **2012**, *3*, 365–373.
71. Lukiw, W.J.; Alexandrov, P.N.; Zhao, Y.; Hill, J.M.; Bhattacharjee, S. Spreading of Alzheimer's disease inflammatory signaling through soluble micro-RNA. *Neuroreport* **2012**, *23*, 621–626. [CrossRef] [PubMed]
72. Basavaraju, M.; De Lencastre, A. Alzheimer's disease: Presence and role of microRNAs. *Biomol. Concepts* **2016**, *7*, 241–252. [CrossRef] [PubMed]
73. Shioya, M.; Obayashi, S.; Tabunoki, H.; Arima, K.; Saito, Y.; Ishida, T.; Satoh, J. Aberrant microRNA expression in the brains of neurodegenerative diseases: MiR-29a decreased in Alzheimer disease brains targets neurone navigator 3. *Neuropathol. Appl. Neurobiol.* **2010**, *36*, 320–330. [CrossRef] [PubMed]
74. Müller, M.; Jäkel, L.; Bruinsma, I.B.; Claassen, J.A.; Kuiperij, H.B.; Verbeek, M.M. MicroRNA-29a Is a Candidate Biomarker for Alzheimer's Disease in Cell-Free Cerebrospinal Fluid. *Mol. Neurobiol.* **2016**, *53*, 2894–2899. [CrossRef]
75. Qiu, L.; Tan, E.K.; Zeng, L. microRNAs and Neurodegenerative Diseases. In *Advances in Experimental Medicine and Biology*; Academic Press: Cambridge, MA, USA, 2015; Chapter 6; pp. 85–105. ISBN 978-3-319-22671-2.
76. Zhang, F.; Gradinaru, V.; Adamantidis, A.R.; Durand, R.; Airan, R.D.; de Lecea, L.; Deisseroth, K. Optogenetic interrogation of neural circuits: Technology for probing mammalian brain structures. *Nat. Protoc.* **2010**, *5*, 439–456. [CrossRef]
77. Shibata, M.; Nakao, H.; Kiyonari, H.; Abe, T.; Aizawa, S. MicroRNA-9 regulates neurogenesis in mouse telencephalon by targeting multiple transcription factors. *J. Neurosci.* **2011**, *31*, 3407–3422. [CrossRef]
78. Vassar, R.; Kovacs, D.M.; Yan, R.; Wong, P.C. The β-secretase enzyme BACE in health and Alzheimer's disease: Regulation, cell biology, function, and therapeutic potential. *J. Neurosci.* **2009**, *29*, 12787–12794. [CrossRef]
79. Zhao, Y.; Zhang, Y.; Zhang, L.; Dong, Y.; Ji, H.; Shen, L. The potential markers of circulating micrornas and long non-coding RNAs in Alzheimer's disease. *Aging Dis.* **2019**, *10*, 1293–1301. [CrossRef]
80. Skoog, I.; Gustafson, D. Update on hypertension and Alzheimer's disease. *Neurol. Res.* **2006**, *28*, 605–611. [CrossRef]
81. Sancesario, G.; Bernardini, S. AD biomarker discovery in CSF and in alternative matrices. *Clin. Biochem.* **2019**, *72*, 52–57. [CrossRef]
82. Nakamura, A.; Kaneko, N.; Villemagne, V.L.; Kato, T.; Doecke, J.; Doré, V.; Fowler, C.; Li, Q.X.; Martins, R.; Rowe, C.; et al. High performance plasma amyloid-β biomarkers for Alzheimer's disease. *Nature* **2018**, *554*, 249–254. [CrossRef] [PubMed]

83. Jia, L.; Qiu, Q.; Zhang, H.; Chu, L.; Du, Y.; Zhang, J.; Zhou, C.; Liang, F.; Shi, S.; Wang, S.; et al. Concordance between the assessment of Aβ42, T-tau, and P-T181-tau in peripheral blood neuronal-derived exosomes and cerebrospinal fluid. *Alzheimer's Dement.* **2019**, *15*, 1071–1080. [CrossRef] [PubMed]
84. Lim, C.Z.J.; Zhang, Y.; Chen, Y.; Zhao, H.; Stephenson, M.C.; Ho, N.R.Y.; Chen, Y.; Chung, J.; Reilhac, A.; Loh, T.P.; et al. Subtyping of circulating exosome-bound amyloid β reflects brain plaque deposition. *Nat. Commun.* **2019**, *10*, 1–11. [CrossRef] [PubMed]
85. Janelidze, S.; Stomrud, E.; Palmqvist, S.; Zetterberg, H.; Van Westen, D.; Jeromin, A.; Song, L.; Hanlon, D.; Tan Hehir, C.A.; Baker, D.; et al. Plasma β-amyloid in Alzheimer's disease and vascular disease. *Sci. Rep.* **2016**, *6*, 1–11. [CrossRef]
86. Xia, W.; Yang, T.; Shankar, G.; Smith, I.M.; Shen, Y.; Walsh, D.M.; Selkoe, D.J. A specific enzyme-linked immunosorbent assay for measuring β-amyloid protein oligomers in human plasma and brain tissue of patients with Alzheimer Disease. *Arch. Neurol.* **2009**, *66*, 190–199. [CrossRef]
87. Mattsson, N.; Zetterberg, H.; Janelidze, S.; Insel, P.S.; Andreasson, U.; Stomrud, E.; Palmqvist, S.; Baker, D.; Tan Hehir, C.A.; Jeromin, A.; et al. Plasma tau in Alzheimer disease. *Neurology* **2016**, *87*, 1827–1835. [CrossRef]
88. Tatebe, H.; Kasai, T.; Ohmichi, T.; Kishi, Y.; Kakeya, T.; Waragai, M.; Kondo, M.; Allsop, D.; Tokuda, T. Quantification of plasma phosphorylated tau to use as a biomarker for brain Alzheimer pathology: Pilot case-control studies including patients with Alzheimer's disease and down syndrome. *Mol. Neurodegener.* **2017**, *12*, 1–11. [CrossRef]
89. Preische, O.; Schultz, S.A.; Apel, A.; Kuhle, J.; Kaeser, S.A.; Barro, C.; Gräber, S.; Kuder-Buletta, E.; LaFougere, C.; Laske, C.; et al. Serum neurofilament dynamics predicts neurodegeneration and clinical progression in presymptomatic Alzheimer's disease. *Nat. Med.* **2019**, *25*, 277–283. [CrossRef]
90. Hrubešová, K.; Fousková, M.; Habartová, L.; Fišar, Z.; Jirák, R.; Raboch, J.; Setnička, V. Search for biomarkers of Alzheimer's disease: Recent insights, current challenges and future prospects. *Clin. Biochem.* **2019**, *72*, 39–51. [CrossRef]
91. Kapogiannis, D.; Mustapic, M.; Shardell, M.D.; Berkowitz, S.T.; Diehl, T.C.; Spangler, R.D.; Tran, J.; Lazaropoulos, M.P.; Chawla, S.; Gulyani, S.; et al. Association of Extracellular Vesicle Biomarkers with Alzheimer Disease in the Baltimore Longitudinal Study of Aging. *JAMA Neurol.* **2019**, *76*, 1340–1351. [CrossRef]
92. Pláteník, J.; Fišar, Z.; Buchal, R.; Jirák, R.; Kitzlerová, E.; Zvěřová, M.; Raboch, J. GSK3β, CREB, and BDNF in peripheral blood of patients with alzheimer's disease and depression. *Prog. Neuro-Psychopharmacol. Biol. Psychiatry* **2014**, *50*, 83–93. [CrossRef] [PubMed]
93. Gupta, V.B.; Doecke, J.D.; Hone, E.; Pedrini, S.; Laws, S.M.; Thambisetty, M.; Bush, A.I.; Rowe, C.C.; Villemagne, V.L.; Ames, D.; et al. Plasma apolipoprotein J as a potential biomarker for Alzheimer's disease: Australian Imaging, Biomarkers and Lifestyle study of aging. *Alzheimer's Dement. Diagn. Assess. Dis. Monit.* **2016**, *3*, 18–26. [CrossRef] [PubMed]
94. Jongbloed, W.; Van Dijk, K.D.; Mulder, S.D.; Van De Berg, W.D.J.; Blankenstein, M.A.; Van Der Flier, W.; Veerhuis, R. Clusterin Levels in Plasma Predict Cognitive Decline and Progression to Alzheimer's Disease. *J. Alzheimer's Dis.* **2015**, *46*, 1103–1110. [CrossRef] [PubMed]
95. Da Mesquita, S.; Louveau, A.; Vaccari, A.; Smirnov, I.; Cornelison, R.C.; Kingsmore, K.M.; Contarino, C.; Onengut-Gumuscu, S.; Farber, E.; Raper, D.; et al. Functional aspects of meningeal lymphatics in ageing and Alzheimer's disease. *Nature* **2018**, *560*, 185–191. [CrossRef]
96. van de Haar, H.J.; Burgmans, S.; Jansen, J.F.A.; van Osch, M.J.P.; van Buchem, M.A.; Muller, M.; Hofman, P.A.M.; Verhey, F.R.J.; Backes, W.H. Blood-Brain Barrier Leakage in Patients with Early Alzheimer Disease. *Radiology* **2016**, *281*, 527–535. [CrossRef]
97. Yan, Z.; Zhou, Z.; Wu, Q.; Chen, Z.B.; Koo, E.H.; Zhong, S. Presymptomatic Increase of an Extracellular RNA in Blood Plasma Associates with the Development of Alzheimer's Disease. *Curr. Biol.* **2020**, *30*, 1–12. [CrossRef] [PubMed]
98. Chopra, N.; Wang, R.; Maloney, B.; Nho, K.; Beck, J.S.; Pourshafie, N.; Niculescu, A.; Saykin, A.J.; Rinaldi, C.; Counts, S.E.; et al. MicroRNA-298 reduces levels of human amyloid-β precursor protein (APP), β-site APP-converting enzyme 1 (BACE1) and specific tau protein moieties. *Mol. Psychiatry* **2020**, *1*. [CrossRef] [PubMed]

99. Dong, H.; Li, J.; Huang, L.; Chen, X.; Li, D.; Wang, T.; Hu, C.; Xu, J.; Zhang, C.; Zen, K.; et al. Serum MicroRNA profiles serve as novel biomarkers for the diagnosis of alzheimer's disease. *Dis. Markers* **2015**, *2015*. [CrossRef]
100. Kenny, A.; McArdle, H.; Calero, M.; Rabano, A.; Madden, S.F.; Adamson, K.; Forster, R.; Spain, E.; Prehn, J.H.M.; Henshall, D.C.; et al. Elevated plasma microRNA-206 levels predict cognitive decline and progression to dementia from mild cognitive impairment. *Biomolecules* **2019**, *9*, 734. [CrossRef]
101. Murillo, O.D.; Thistlethwaite, W.; Rozowsky, J.; Subramanian, S.L.; Lucero, R.; Shah, N.; Jackson, A.R.; Srinivasan, S.; Chung, A.; Laurent, C.D.; et al. exRNA Atlas Analysis Reveals Distinct Extracellular RNA Cargo Types and Their Carriers Present across Human Biofluids. *Cell* **2019**, *177*, 463–477.e15. [CrossRef]
102. Musunuri, S.; Khoonsari, P.E.; Mikus, M.; Wetterhall, M.; Häggmark-Mänberg, A.; Lannfelt, L.; Erlandsson, A.; Bergquist, J.; Ingelsson, M.; Shevchenko, G.; et al. Increased Levels of Extracellular Microvesicle Markers and Decreased Levels of Endocytic/Exocytic Proteins in the Alzheimer's Disease Brain. *J. Alzheimer's Dis.* **2016**, *54*, 1671–1686. [CrossRef] [PubMed]
103. Badhwar, A.; Haqqani, A.S. Biomarker potential of brain-secreted extracellular vesicles in blood in Alzheimer's disease. *Alzheimer's Dement. Diagn. Assess. Dis. Monit.* **2020**, *12*, 1–14. [CrossRef] [PubMed]
104. Lewczuk, P.; Ermann, N.; Andreasson, U.; Schultheis, C.; Podhorna, J.; Spitzer, P.; Maler, J.M.; Kornhuber, J.; Blennow, K.; Zetterberg, H. Plasma neurofilament light as a potential biomarker of neurodegeneration in Alzheimer's disease. *Alzheimer's Res. Ther.* **2018**, *10*, 1–10. [CrossRef] [PubMed]
105. Eitan, E.; Hutchison, E.R.; Marosi, K.; Comotto, J.; Mustapic, M.; Nigam, S.M.; Suire, C.; Maharana, C.; Jicha, G.A.; Liu, D.; et al. Extracellular vesicle-associated aβ mediates trans-neuronal bioenergetic and ca2+-handling deficits in alzheimer's disease models. *NPJ Aging Mech. Dis.* **2016**, *2*, 16019. [CrossRef] [PubMed]
106. Winston, C.N.; Goetzl, E.J.; Akers, J.C.; Carter, B.S.; Rockenstein, E.M.; Galasko, D.; Masliah, E.; Rissman, R.A. Prediction of conversion from mild cognitive impairment to dementia with neuronally derived blood exosome protein profile. *Alzheimer's Dement. Diagn. Assess. Dis. Monit.* **2016**, *3*, 63–72. [CrossRef] [PubMed]
107. Fiandaca, M.S.; Kapogiannis, D.; Mapstone, M.; Boxer, A.; Eitan, E.; Schwartz, J.B.; Abner, E.L.; Petersen, R.C.; Federoff, H.J.; Miller, B.L.; et al. Identification of preclinical Alzheimer's disease by a profile of pathogenic proteins in neurally derived blood exosomes: A case-control study. *Alzheimer's Dement.* **2015**, *11*, 600–607.e1. [CrossRef] [PubMed]
108. Shi, M.; Kovac, A.; Korff, A.; Cook, T.J.; Ginghina, C.; Bullock, K.M.; Yang, L.; Stewart, T.; Zheng, D.; Aro, P.; et al. CNS tau efflux via exosomes is likely increased in Parkinson's disease but not in Alzheimer's disease. *Alzheimer's Dement.* **2016**, *12*, 1125–1131. [CrossRef]
109. Winston, C.N.; Goetzl, E.J.; Baker, L.D.; Vitiello, M.V.; Rissman, R.A. Growth Hormone-Releasing Hormone Modulation of Neuronal Exosome Biomarkers in Mild Cognitive Impairment. *J. Alzheimer's Dis.* **2018**, *66*, 971–981. [CrossRef]
110. Guix, F.X.; Corbett, G.T.; Cha, D.J.; Mustapic, M.; Liu, W.; Mengel, D.; Chen, Z.; Aikawa, E.; Young-Pearse, T.; Kapogiannis, D.; et al. Detection of aggregation-competent tau in neuron-derived extracellular vesicles. *Int. J. Mol. Sci.* **2018**, *19*, 663. [CrossRef]
111. Goetzl, E.J.; Kapogiannis, D.; Schwartz, J.B.; Lobach, I.V.; Goetzl, L.; Abner, E.L.; Jicha, G.A.; Karydas, A.M.; Boxer, A.; Miller, B.L. Decreased synaptic proteins in neuronal exosomes of frontotemporal dementia and Alzheimer's disease. *FASEB J.* **2016**, *30*, 4141–4148. [CrossRef]
112. Mullins, R.J.; Mustapic, M.; Goetz, E.J.; Kapogiannis, D. Exosomal biomarkers of brain insulin resistance associated with regional atrophy in Alzheimer's disease. *Hum. Brain Mapp.* **2017**, *38*, 1933–1940. [CrossRef] [PubMed]
113. Kapogiannis, D.; Boxer, A.; Schwartz, J.B.; Abner, E.L.; Biragyn, A.; Masharani, U.; Frassetto, L.; Petersen, R.C.; Miller, B.L.; Goetzl, E.J. Dysfunctionally phosphorylated type 1 insulin receptor substrate in neural-derived blood exosomes of preclinical Alzheimer's disease. *FASEB J.* **2015**, *29*, 589–596. [CrossRef] [PubMed]
114. Mustapic, M.; Tran, J.; Craft, S.; Kapogiannis, D. Extracellular Vesicle Biomarkers Track Cognitive Changes Following Intranasal Insulin in Alzheimer's Disease. *J. Alzheimer's Dis.* **2019**, *69*, 489–498. [CrossRef] [PubMed]

115. Goetzl, E.J.; Boxer, A.; Schwartz, J.B.; Abner, E.L.; Petersen, R.C.; Miller, B.L.; Kapogiannis, D. Altered lysosomal proteins in neural-derived plasma exosomes in preclinical Alzheimer disease. *Neurology* **2015**, *85*, 40–47. [CrossRef]
116. Goetzl, E.J.; Nogueras-Ortiz, C.; Mustapic, M.; Mullins, R.J.; Abner, E.L.; Schwartz, J.B.; Kapogiannis, D. Deficient neurotrophic factors of CSPG4-type neural cell exosomes in Alzheimer disease. *FASEB J.* **2019**, *33*, 231–238. [CrossRef]
117. Goetzl, E.J.; Schwartz, J.B.; Abner, E.L.; Jicha, G.A.; Kapogiannis, D. High complement levels in astrocyte-derived exosomes of Alzheimer disease. *Ann. Neurol.* **2018**, *83*, 544–552. [CrossRef]
118. Goetzl, E.J.; Mustapic, M.; Kapogiannis, D.; Eitan, E.; Lobach, I.V.; Goetzl, L.; Schwartz, J.B.; Miller, B.L. Cargo proteins of plasma astrocyte-derived exosomes in Alzheimer's disease. *FASEB J.* **2016**, *30*, 3853–3859. [CrossRef]
119. Pant, S.; Hilton, H.; Burczynski, M.E. The multifaceted exosome: Biogenesis, role in normal and aberrant cellular function, and frontiers for pharmacological and biomarker opportunities. *Biochem. Pharmacol.* **2012**, *83*, 1484–1494. [CrossRef]
120. Burgos, K.; Malenica, I.; Metpally, R.; Courtright, A.; Rakela, B.; Beach, T.; Shill, H.; Adler, C.; Sabbagh, M.; Villa, S.; et al. Profiles of extracellular miRNA in cerebrospinal fluid and serum from patients with Alzheimer's and Parkinson's diseases correlate with disease status and features of pathology. *PLoS ONE* **2014**, *9*. [CrossRef]
121. Liu, C.G.; Song, J.; Zhang, Y.Q.; Wang, P.C. MicroRNA-193b is a regulator of amyloid precursor protein in the blood and cerebrospinal fluid derived exosomal microRNA-193b is a biomarker of Alzheimer's disease. *Mol. Med. Rep.* **2014**, *10*, 2395–2400. [CrossRef]
122. Cheng, L.; Doecke, J.D.; Sharples, R.A.; Villemagne, V.L.; Fowler, C.J.; Rembach, A.; Martins, R.N.; Rowe, C.C.; Macaulay, S.L.; Masters, C.L.; et al. Prognostic serum miRNA biomarkers associated with Alzheimer's disease shows concordance with neuropsychological and neuroimaging assessment. *Mol. Psychiatry* **2015**, *20*, 1188–1196. [CrossRef] [PubMed]
123. Lugli, G.; Cohen, A.M.; Bennett, D.A.; Shah, R.C.; Fields, C.J.; Hernandez, A.G.; Smalheiser, N.R. Plasma exosomal miRNAs in persons with and without Alzheimer disease: Altered expression and prospects for biomarkers. *PLoS ONE* **2015**, *10*, 1–18. [CrossRef]
124. Yang, T.T.; Liu, C.G.; Gao, S.C.; Zhang, Y.; Wang, P.C. The Serum Exosome Derived MicroRNA−135a, −193b, and −384 Were Potential Alzheimer's Disease Biomarkers. *Biomed. Environ. Sci.* **2018**, *31*, 87–96. [CrossRef] [PubMed]
125. Wei, H.; Xu, Y.; Xu, W.; Zhou, Q.; Chen, Q.; Yang, M.; Feng, F.; Liu, Y.; Zhu, X.; Yu, M.; et al. *Serum Exosomal miR-223 Serves as a Potential Diagnostic and Prognostic Biomarker for Dementia*; Elsevier: Amsterdam, The Netherlands, 2018; Volume 379, ISBN 8651188773631.
126. Cha, D.J.; Mengel, D.; Mustapic, M.; Liu, W.; Selkoe, D.J.; Kapogiannis, D.; Galasko, D.; Rissman, R.A.; Bennett, D.A.; Walsh, D.M. miR-212 and miR-132 Are Downregulated in Neurally Derived Plasma Exosomes of Alzheimer's Patients. *Front. Neurosci.* **2019**, *13*, 1–14. [CrossRef] [PubMed]
127. Galimberti, D.; Villa, C.; Fenoglio, C.; Serpente, M.; Ghezzi, L.; Cioffi, S.M.G.; Arighi, A.; Fumagalli, G.; Scarpini, E. Circulating miRNAs as potential biomarkers in alzheimer's disease. *J. Alzheimer's Dis.* **2014**, *42*, 1261–1267. [CrossRef]
128. Rani, A.; O'Shea, A.; Ianov, L.; Cohen, R.A.; Woods, A.J.; Foster, T.C. miRNA in circulating microvesicles as biomarkers for age-related cognitive decline. *Front. Aging Neurosci.* **2017**, *9*, 1–10. [CrossRef]
129. Gámez-Valero, A.; Campdelacreu, J.; Vilas, D.; Ispierto, L.; Reñé, R.; Álvarez, R.; Armengol, M.P.; Borràs, F.E.; Beyer, K. Exploratory study on microRNA profiles from plasma-derived extracellular vesicles in Alzheimer's disease and dementia with Lewy bodies. *Transl. Neurodegener.* **2019**, *8*, 1–17. [CrossRef]
130. Barbagallo, C.; Mostile, G.; Baglieri, G.; Giunta, F.; Luca, A.; Raciti, L.; Zappia, M.; Purrello, M.; Ragusa, M.; Nicoletti, A. Specific Signatures of Serum miRNAs as Potential Biomarkers to Discriminate Clinically Similar Neurodegenerative and Vascular-Related Diseases. *Cell. Mol. Neurobiol.* **2020**, *40*, 531–546. [CrossRef]
131. Chen, J.; Qi, Y.; Liu, C.-F.; Lu, J.-M.; Shi, J.; Shi, Y. MicroRNA expression data analysis to identify key miRNAs associated with Alzheimer's disease. *J. Gene Med.* **2018**, *20*, e3014. [CrossRef]

132. Zhang, Y.; Lv, X.; Liu, C.; Gao, S.; Ping, H.; Wang, J.; Wang, P. MiR-214-3p attenuates cognition defects via the inhibition of autophagy in SAMP8 mouse model of sporadic Alzheimer's disease. *Neurotoxicology* **2016**, *56*, 139–149. [CrossRef] [PubMed]
133. Cogswell, J.P.; Ward, J.; Taylor, I.A.; Waters, M.; Shi, Y.; Cannon, B.; Kelnar, K.; Kemppainen, J.; Brown, D.; Chen, C.; et al. Identification of miRNA Changes in Alzheimer's Disease Brain and CSF Yields Putative Biomarkers and Insights into Disease Pathways. *J. Alzheimer's Dis.* **2008**, *14*, 27–41. [CrossRef] [PubMed]
134. Sørensen, S.S.; Nygaard, A.B.; Christensen, T. miRNA expression profiles in cerebrospinal fluid and blood of patients with Alzheimer's disease and other types of dementia—An exploratory study. *Transl. Neurodegener.* **2016**, *5*, 1–12. [CrossRef] [PubMed]
135. Wu, Y.; Xu, J.; Xu, J.; Cheng, J.; Jiao, D.; Zhou, C.; Dai, Y.; Chen, Q. Lower serum levels of miR-29c-3p and miR-19b-3p as biomarkers for Alzheimer's disease. *Tohoku J. Exp. Med.* **2017**, *242*, 129–136. [CrossRef] [PubMed]
136. Hara, N.; Kikuchi, M.; Miyashita, A.; Hatsuta, H.; Saito, Y.; Kasuga, K.; Murayama, S.; Ikeuchi, T.; Kuwano, R. Serum microRNA miR-501-3p as a potential biomarker related to the progression of Alzheimer's disease. *Acta Neuropathol. Commun.* **2017**, *5*, 10. [CrossRef] [PubMed]
137. Cosín-Tomás, M.; Antonell, A.; Lladó, A.; Alcolea, D.; Fortea, J.; Ezquerra, M.; Lleó, A.; Martí, M.J.; Pallàs, M.; Sanchez-Valle, R.; et al. Plasma miR-34a-5p and miR-545-3p as Early Biomarkers of Alzheimer's Disease: Potential and Limitations. *Mol. Neurobiol.* **2017**, *54*, 5550–5562. [CrossRef] [PubMed]
138. Hébert, S.S.; Horré, K.; Nicolaï, L.; Papadopoulou, A.S.; Mandemakers, W.; Silahtaroglu, A.N.; Kauppinen, S.; Delacourte, A.; De Strooper, B. Loss of microRNA cluster miR-29a/b-1 in sporadic Alzheimer's disease correlates with increased BACE1/β-secretase expression. *Proc. Natl. Acad. Sci. USA* **2008**, *105*, 6415–6420. [CrossRef]
139. Wang, L.; Zhang, L. Circulating Exosomal miRNA as Diagnostic Biomarkers of Neurodegenerative Diseases. *Front. Mol. Neurosci.* **2020**, *13*, 1–11. [CrossRef]
140. Jia, L.-H.; Liu, Y.-N. Downregulated serum miR-223 servers as biomarker in Alzheimer's disease. *Cell Biochem. Funct.* **2016**, *34*, 233–237. [CrossRef]
141. Zhu, Y.; Li, C.; Sun, A.; Wang, Y.; Zhou, S. Quantification of microRNA-210 in the cerebrospinal fluid and serum: Implications for Alzheimer's disease. *Exp. Ther. Med.* **2015**, *9*, 1013–1017. [CrossRef]
142. Liu, C.G.; Wang, J.L.; Li, L.; Xue, L.X.; Zhang, Y.Q.; Wang, P.C. MicroRNA-135a and -200b, potential Biomarkers for Alzheimer's disease, regulate β secretase and amyloid precursor protein. *Brain Res.* **2014**, *1583*, 55–64. [CrossRef] [PubMed]
143. Zhang, Y.; Liu, C.; Wang, J.; Li, Q.; Ping, H.; Gao, S.; Wang, P. MIR-299-5p regulates apoptosis through autophagy in neurons and ameliorates cognitive capacity in APPswe/PS1dE9 mice. *Sci. Rep.* **2016**, *6*, 1–14. [CrossRef] [PubMed]
144. Kumar, S.; Reddy, P.H. MicroRNA-455-3p as a Potential Biomarker for Alzheimer's Disease: An Update. *Front. Aging Neurosci.* **2018**, *10*, 41. [CrossRef]
145. Derkow, K.; Rössling, R.; Schipke, C.; Krüger, C.; Bauer, J.; Fähling, M.; Stroux, A.; Schott, E.; Ruprecht, K.; Peters, O.; et al. Distinct expression of the neurotoxic microRNA family let-7 in the cerebrospinal fluid of patients with Alzheimer's disease. *PLoS ONE* **2018**, *13*, 1–18. [CrossRef] [PubMed]
146. Dangla-Valls, A.; Molinuevo, J.L.; Altirriba, J.; Sánchez-Valle, R.; Alcolea, D.; Fortea, J.; Rami, L.; Balasa, M.; Muñoz-García, C.; Ezquerra, M.; et al. CSF microRNA Profiling in Alzheimer's Disease: A Screening and Validation Study. *Mol. Neurobiol.* **2017**, *54*, 6647–6654. [CrossRef]
147. Li, W.; Li, X.; Xin, X.; Kan, P.C.; Yan, Y. MicroRNA-613 regulates the expression of brain-derived neurotrophic factor in Alzheimer's disease. *Biosci. Trends* **2016**, *10*, 372–377. [CrossRef]
148. Kenny, A.; Jiménez-Mateos, E.M.; Zea-Sevilla, M.A.; Rábano, A.; Gili-Manzanaro, P.; Prehn, J.H.M.; Henshall, D.C.; Ávila, J.; Engel, T.; Hernández, F. Proteins and microRNAs are differentially expressed in tear fluid from patients with Alzheimer's disease. *Sci. Rep.* **2019**, *9*, 1–14. [CrossRef]
149. Wu, Q.; Ye, X.; Xiong, Y.; Zhu, H.; Miao, J.; Zhang, W.; Wan, J. The Protective Role of microRNA-200c in Alzheimer's Disease Pathologies Is Induced by Beta Amyloid-Triggered Endoplasmic Reticulum Stress. *Front. Mol. Neurosci.* **2016**, *9*, 140. [CrossRef]
150. Tian, N.; Cao, Z.; Zhang, Y. MiR-206 decreases brain-derived neurotrophic factor levels in a transgenic mouse model of Alzheimer's disease. *Neurosci. Bull.* **2014**, *30*, 191–197. [CrossRef]

151. Crooke, A.; Huete-toral, F.; Martı, A.; Colligris, B.; Pintor, J. Ocular disorders and the utility of animal models in the discovery of melatoninergic drugs with therapeutic potential. *Expert Opin. Drug Discov.* **2012**, *7*, 989–1001. [CrossRef]
152. van Wijngaarden, P.; Hadoux, X.; Alwan, M.; Keel, S.; Dirani, M. Emerging ocular biomarkers of Alzheimer disease. *Clin. Exp. Ophthalmol.* **2017**, *45*, 54–61. [CrossRef] [PubMed]
153. Singh, A.K.; Verma, S. Use of ocular biomarkers as a potential tool for early diagnosis of Alzheimer's disease. *Indian J. Ophthalmol.* **2020**, *68*, 555–561. [CrossRef] [PubMed]
154. Ning, A.; Cui, J.; To, E.; Ashe, K.H.; Matsubara, J. Amyloid-β deposits lead to retinal degeneration in a mouse model of Alzheimer disease. *Investig. Ophthalmol. Vis. Sci.* **2008**, *49*, 5136–5143. [CrossRef] [PubMed]
155. More, S.S.; Vince, R. Hyperspectral imaging signatures detect amyloidopathy in alzheimers mouse retina well before onset of cognitive decline. *ACS Chem. Neurosci.* **2015**, *6*, 306–315. [CrossRef]
156. Goldstein, L.E.; Muffat, J.A.; Cherny, R.A.; Moir, R.D.; Ericsson, M.H.; Huang, X.; Mavros, C.; Coccia, J.A.; Faget, K.Y.; Fitch, K.A.; et al. Cytosolic β-amyloid deposition and supranuclear cataracts in lenses from people with Alzheimer's disease. *Lancet* **2003**, *361*, 1258–1265. [CrossRef]
157. Kerbage, C.; Sadowsky, C.H.; Tariot, P.N.; Agronin, M.; Alva, G.; Turner, F.D.; Nilan, D.; Cameron, A.; Cagle, G.D.; Hartung, P.D. Detection of Amyloid β Signature in the Lens and Its Correlation in the Brain to Aid in the Diagnosis of Alzheimer's Disease. *Am. J. Alzheimers Dis. Other Demen.* **2015**, *30*, 738–745. [CrossRef]
158. Koronyo-Hamaoui, M.; Koronyo, Y.; Ljubimov, A.V.; Miller, C.A.; Ko, M.K.; Black, K.L.; Schwartz, M.; Farkas, D.L. Identification of amyloid plaques in retinas from Alzheimer's patients and noninvasive in vivo optical imaging of retinal plaques in a mouse model. *Neuroimage* **2011**, *54*, S204–S217. [CrossRef]
159. Armstrong, R.A.; Syed, A.B. Alzheimer's disease and the eye. *Ophthalmic Physiol. Opt.* **2008**, *16*, S2–S8. [CrossRef]
160. La Morgia, C.; Ross-Cisneros, F.N.; Koronyo, Y.; Hannibal, J.; Gallassi, R.; Cantalupo, G.; Sambati, L.; Pan, B.X.; Tozer, K.R.; Barboni, P.; et al. Melanopsin retinal ganglion cell loss in Alzheimer disease. *Ann. Neurol.* **2016**, *79*, 90–109. [CrossRef]
161. Esquiva, G.; Hannibal, J. Melanopsin-expressing retinal ganglion cells in aging and disease. *Histol. Histopathol.* **2019**, *34*, 1299–1311.
162. Dutescu, R.M.; Li, Q.X.; Crowston, J.; Masters, C.L.; Baird, P.N.; Culvenor, J.G. Amyloid precursor protein processing and retinal pathology in mouse models of Alzheimer's disease. *Graefe's Arch. Clin. Exp. Ophthalmol.* **2009**, *247*, 1213–1221. [CrossRef] [PubMed]
163. Choi, S.I.; Lee, B.; Woo, J.H.; Jeong, J.B.; Jun, I.; Kim, E.K. APP processing and metabolism in corneal fibroblasts and epithelium as a potential biomarker for Alzheimer's disease. *Exp. Eye Res.* **2019**, *182*, 167–174. [CrossRef]
164. Csosz, É.; Boross, P.; Csutak, A.; Berta, A.; Tóth, F.; Póliska, S.; Török, Z.; Tozsér, J. Quantitative analysis of proteins in the tear fluid of patients with diabetic retinopathy. *J. Proteom.* **2012**, *75*, 2196–2204. [CrossRef] [PubMed]
165. Çomoğlu, S.S.; Güven, H.; Acar, M.; Öztürk, G.; Koçer, B. Tear levels of tumor necrosis factor-alpha in patients with Parkinson's disease. *Neurosci. Lett.* **2013**, *553*, 63–67. [CrossRef] [PubMed]
166. Zhou, L.; Zhao, S.Z.; Koh, S.K.; Chen, L.; Vaz, C.; Tanavde, V.; Li, X.R.; Beuerman, R.W. In-depth analysis of the human tear proteome. *J. Proteom.* **2012**, *75*, 3877–3885. [CrossRef] [PubMed]
167. Örnek, N.; Dag, E.; Örnek, K. Corneal sensitivity and tear function in neurodegenerative diseases. *Curr. Eye Res.* **2015**, *40*, 423–428. [CrossRef] [PubMed]
168. Kalló, G.; Emri, M.; Varga, Z.; Ujhelyi, B.; Tozsér, J.; Csutak, A.; Csosz, É. Changes in the chemical barrier composition of tears in Alzheimer's disease reveal potential tear diagnostic biomarkers. *PLoS ONE* **2016**, *11*, 1–14. [CrossRef] [PubMed]
169. Ship, J.A.; Puckett, S.A. Longitudinal Study on Oral Health in Subjects with Alzheimer's Disease. *J. Am. Geriatr. Soc.* **1994**, *42*, 57–63. [CrossRef]
170. Ship, J.A.; DeCarli, C.; Friedland, R.P.; Baum, B.J. Diminished submandibular salivary flow in dementia of the Alzheimer Type. *J. Gerontol.* **1990**, *45*, 61–66. [CrossRef]
171. Reuster, T.; Rilke, O.; Oehler, J. High correlation between salivary MHPG and CSF MHPG. *Psychopharmacology (Berl)* **2002**, *162*, 415–418. [CrossRef]

172. Formichi, P.; Battisti, C.; Radi, E.; Federico, A. Cerebrospinal fluid tau, Aß, and phosphorylated tau protein for the diagnosis of Alzheimer's disease. *J. Cell. Physiol.* **2006**, *208*, 39–46. [CrossRef] [PubMed]
173. Bermejo-Pareja, F.; Antequera, D.; Vargas, T.; Molina, J.A.; Carro, E. Saliva levels of Abeta1-42 as potential biomarker of Alzheimer's disease: A pilot study. *BMC Neurol.* **2010**, *10*, 108. [CrossRef] [PubMed]
174. Kim, C.-B.; Choi, Y.Y.; Song, W.K.; Song, K.-B. Antibody-based magnetic nanoparticle immunoassay for quantification of Alzheimer's disease pathogenic factor. *J. Biomed. Opt.* **2013**, *19*, 051205. [CrossRef] [PubMed]
175. Sabbagh, M.N.; Shi, J.; Lee, M.; Arnold, L.; Al-Hasan, Y.; Heim, J.; McGeer, P. Salivary beta amyloid protein levels are detectable and differentiate patients with Alzheimer's disease dementia from normal controls: Preliminary findings. *BMC Neurol.* **2018**, *18*, 155. [CrossRef] [PubMed]
176. Lee, M.; Guo, J.P.; Kennedy, K.; Mcgeer, E.G.; McGeer, P.L. A method for diagnosing Alzheimer's disease based on salivary amyloid-β protein 42 levels. *J. Alzheimer's Dis.* **2017**, *55*, 1175–1182. [CrossRef] [PubMed]
177. Conrad, C.; Vianna, C.; Freeman, M.; Davies, P. A polymorphic gene nested within an intron of the tau gene: Implications for Alzheimer's disease. *Proc. Natl. Acad. Sci. USA* **2002**, *99*, 7751–7756. [CrossRef]
178. Ashton, N.J.; Ide, M.; Schöll, M.; Blennow, K.; Lovestone, S.; Hye, A.; Zetterberg, H. No association of salivary total tau concentration with Alzheimer's disease. *Neurobiol. Aging* **2018**, *70*, 125–127. [CrossRef]
179. Shi, M.; Sui, Y.-T.; Peskind, E.R.; Li, G.; Hwang, H.; Devic, I.; Ginghina, C.; Edgar, J.S.; Pan, C.; Goodlett, D.R.; et al. Salivary Tau Species are Potential Biomarkers of Alzheimer's Disease. *J. Alzheimer's Dis.* **2011**, *27*, 299–305. [CrossRef]
180. Pekeles, H.; Qureshi, H.Y.; Paudel, H.K.; Schipper, H.M.; Gornistky, M.; Chertkow, H. Development and validation of a salivary tau biomarker in Alzheimer's disease. *Alzheimer's Dement. Diagn. Assess. Dis. Monit.* **2019**, *11*, 53–60. [CrossRef]
181. Carro, E.; Bartolomé, F.; Bermejo-Pareja, F.; Villarejo-Galende, A.; Molina, J.A.; Ortiz, P.; Calero, M.; Rabano, A.; Cantero, J.L.; Orive, G. Early diagnosis of mild cognitive impairment and Alzheimer's disease based on salivary lactoferrin. *Alzheimer's Dement. Diagn. Assess. Dis. Monit.* **2017**, *8*, 131–138. [CrossRef]
182. Boston, P.F.; Gopalkaje, K.; Manning, L.; Middleton, L.; Loxley, M. Developing a simple laboratory test for Alzheimer's disease: Measuring acetylcholinesterase in saliva—A pilot study. *Int. J. Geriatr. Psychiatry* **2008**, *23*, 439–440. [CrossRef] [PubMed]
183. Bakhtiari, S.; Moghadam, N.B.; Ehsani, M.; Mortazavi, H.; Sabour, S.; Bakhshi, M. Can salivary acetylcholinesterase be a diagnostic biomarker for Alzheimer? *J. Clin. Diagn. Res.* **2017**, *11*, ZC58–ZC60. [CrossRef] [PubMed]
184. García-Blanco, A.; Peña-Bautista, C.; Oger, C.; Vigor, C.; Galano, J.-M.; Durand, T.; Martín-Ibáñez, N.; Baquero, M.; Vento, M.; Cháfer-Pericás, C. Reliable determination of new lipid peroxidation compounds as potential early Alzheimer Disease biomarkers. *Talanta* **2018**, *184*, 193–201. [CrossRef] [PubMed]
185. Praticò, D.; Clark, C.M.; Liun, F.; Lee, V.Y.M.; Trojanowski, J.Q. Increase of brain oxidative stress in mild cognitive impairment: A possible predictor of Alzheimer disease. *Arch. Neurol.* **2002**, *59*, 972–976. [CrossRef] [PubMed]
186. Hee Lee, S.; Kim, I.; Chul Chung, B. Increased urinary level of oxidized nucleosides in patients with mild-to-moderate Alzheimer's disease. *Clin. Biochem.* **2007**, *40*, 936–938. [CrossRef]
187. De La Monte, S.M.; Ghanbari, K.; Frey, W.H.; Beheshti, I.; Averback, P.; Hauser, S.L.; Ghanbari, H.A.; Wands, J.R. Characterization of the AD7C-NTP cDNA expression in Alzheimer's disease and measurement of a 41-kD protein in cerebrospinal fluid. *J. Clin. Investig.* **1997**, *100*, 3093–3104. [CrossRef]
188. Zhang, N.; Zhang, L.; Li, Y.; Gordon, M.L.; Cai, L.; Wang, Y.; Xing, M.; Cheng, Y. Urine AD7c-NTP predicts amyloid deposition and symptom of agitation in patients with Alzheimer's disease and mild cognitive impairment. *J. Alzheimer's Dis.* **2017**, *60*, 87–95. [CrossRef]
189. Zhang, J.; Zhang, C.H.; Li, R.J.; Lin, X.L.; Chen, Y.D.; Gao, H.Q.; Shi, S.L. Accuracy of urinary AD7c-NTP for diagnosing alzheimer's disease: A systematic review and meta-analysis. *J. Alzheimer's Dis.* **2014**, *40*, 153–159. [CrossRef]
190. Takae, K.; Hata, J.; Ohara, T.; Yoshida, D.; Shibata, M.; Mukai, N.; Hirakawa, Y.; Kishimoto, H.; Tsuruya, K.; Kitazono, T.; et al. Albuminuria increases the risks for both Alzheimer disease and vascular dementia in community-dwelling Japanese elderly: The hisayama study. *J. Am. Heart Assoc.* **2018**, *7*. [CrossRef]

191. Yao, F.; Hong, X.; Li, S.; Zhang, Y.; Zhao, Q.; Du, W.; Wang, Y.; Ni, J. Urine-Based Biomarkers for Alzheimer's Disease Identified Through Coupling Computational and Experimental Methods. *J. Alzheimers Dis.* **2018**, *65*, 421–431. [CrossRef]
192. Agbemenyah, H.Y.; Agis-Balboa, R.C.; Burkhardt, S.; Delalle, I.; Fischer, A. Insulin growth factor binding protein 7 is a novel target to treat dementia. *Neurobiol. Dis.* **2014**, *62*, 135–143. [CrossRef] [PubMed]
193. Rentsendorj, A.; Sheyn, J.; Fuchs, D.; Daley, D.; Salumbides, B.C.; Schubloom, H.E.; Hart, N.J.; Li, S.; Hayden, E.Y.; Teplow, D.B.; et al. A novel role for osteopontin in macrophage-mediated amyloid-β clearance in Alzheimer's models. *Brain Behav. Immun.* **2018**, *67*, 163–180. [CrossRef] [PubMed]
194. Yang, W.; Chauhan, A.; Mehta, S.; Mehta, P.; Gu, F.; Chauhan, V. Trichostatin A increases the levels of plasma gelsolin and amyloid beta-protein in a transgenic mouse model of Alzheimer's disease. *Life Sci.* **2014**, *99*, 31–36. [CrossRef] [PubMed]

Review

Significance of Blood and Cerebrospinal Fluid Biomarkers for Alzheimer's Disease: Sensitivity, Specificity and Potential for Clinical Use

Cristina d'Abramo [1], Luciano D'Adamio [2] and Luca Giliberto [1,3,4,*]

1 The Litwin-Zucker Center for Alzheimer's Disease & Memory Disorders, Institute of Molecular Medicine, Feinstein Institutes for Medical Research, Northwell Health System, Manhasset, NY 11030, USA; cdabramo@northwell.edu
2 Rutgers Alzheimer's Disease & Dementia Research Center (RUADRC), Rutgers Brain Health Institute, Rutgers University, Piscataway, NJ 08854, USA; luciano.dadamio@rutgers.edu
3 Northwell Health Neuroscience Institute, Northwell Health System, Manhasset, NY 11030, USA
4 Donald and Barbara Zucker School of Medicine, Hofstra University, Hempstead, NY 11549, USA
* Correspondence: lgiliberto@northwell.edu; Tel.: +1-516-562-0423; Fax: +1-516-562-0401

Received: 1 August 2020; Accepted: 1 September 2020; Published: 8 September 2020

Abstract: Alzheimer's disease (AD) is the most common type of dementia, affecting more than 5 million Americans, with steadily increasing mortality and incredible socio-economic burden. Not only have therapeutic efforts so far failed to reach significant efficacy, but the real pathogenesis of the disease is still obscure. The current theories are based on pathological findings of amyloid plaques and tau neurofibrillary tangles that accumulate in the brain parenchyma of affected patients. These findings have defined, together with the extensive neurodegeneration, the diagnostic criteria of the disease. The ability to detect changes in the levels of amyloid and tau in cerebrospinal fluid (CSF) first, and more recently in blood, has allowed us to use these biomarkers for the specific in-vivo diagnosis of AD in humans. Furthermore, other pathological elements of AD, such as the loss of neurons, inflammation and metabolic derangement, have translated to the definition of other CSF and blood biomarkers, which are not specific of the disease but, when combined with amyloid and tau, correlate with the progression from mild cognitive impairment to AD dementia, or identify patients who will develop AD pathology. In this review, we discuss the role of current and hypothetical biomarkers of Alzheimer's disease, their specificity, and the caveats of current high-sensitivity platforms for their peripheral detection.

Keywords: Alzheimer's disease; biomarkers; cerebrospinal fluid; blood; amyloid; tau; soluble TREM2; NfL; Multiplex; SiMoA

1. Introduction

1.1. Clinicopathological Definition of Alzheimer's Disease

Alzheimer's disease (AD) is a neurodegenerative disorder, characterized by progressive cognitive impairment, affecting the domains of memory, speech, praxis, awareness and executive function, representing a decline from the prior functional state, with a relentless course [1]. It is the most common cause of dementia. Along with the destructive nature of the disease, which affects patients and their families, AD represents an enormous burden on public health. An estimated of 5+ million Americans have AD, and the prevalence worldwide reaches almost 60 million; the diseases hits the elderly population the most, with 11% of individuals 65 and older affected by the disease [2,3]. Two thirds of AD patients are women, implying hormonal, lifestyle and genetic factors as co-causative elements [2,4–15]. AD is the sixth leading cause of death in the US [16] and mortality due to AD has steadily increased in the last 20 years, in contrast to opposite trends for other chronic diseases, such as cardiovascular, cancer, HIV and more. The socio-economic burden of the disease, on patients, families and caregivers at large, and on society, is unparalleled; AD is the disease that contributes most to disability and poor health in the US, reducing life expectancy and productivity of the affected individuals and the wellbeing of caregivers, who often have to give up jobs and careers to care for their loved ones [17–19]; this is often disproportionally affecting women too [20–22]. Finally, AD has a tremendous economic impact: the sole health care costs, variably covered by insurances (including Medicare and Medicaid) or out of pocket, to care for this disease, is in excess of 230 billion [23]. Age-matched comparisons show that individuals affected by dementia had 3–4x the annual health care expenses than those not affected. It is estimated that these costs will reach, in 2050, 1 trillion dollars (2020 Alzheimer's Disease Facts and Figures. Available online: https://www.alz.org/media/Documents/alzheimers-facts-and-figures.pdf. Accessed on 29 July 2020), a figure comparable to the US budget deficit in 2020. Interestingly, these numbers should probably be corrected upward, as the disease is often under-recognized and under-reported. This is particularly relevant to our discussion, given current limitations in performing early and accurate diagnosis.

The pathological diagnosis of AD relies on the assessment of specific brain features [24–27], based on the presence, morphology and density of lesions and their topographic distribution, in the appropriate clinical setting. Alzheimer's brains are characterized by a spatiotemporally defined accumulation of amyloid beta (Aβ) plaques, neurofibrillary tangles (NFT), of hyperphosphorylated and aggregated tau and neuronal loss [25,28,29]. Aβ accumulates mostly as the Aβ peptide ending at residue 42, which is considered the most toxic and most prone to aggregation [30]; tau accumulates as a hyperphosphorylated and conformationally altered form, in both its 3- and 4-repeats variants [31]. The trajectory of protein accumulation is relevant. It is speculated that Aβ accumulates first; this is followed by the hyperphosphorylation, conformational changes and aggregation of tau; finally, neuronal loss ensues.

The link between these three features is, to some degree, still loose. Both Aβ [30,32] and tau [33,34] are per se neurotoxic; Aβ accumulation can trigger tau accumulation [35] but, in several models, tau can independently aggregate and cause neurodegeneration [36]. Furthermore, amyloid deposition, by pathological and imaging staging (amyloid positron emission tomography—PET), does not topographically and functionally correlate with the clinical cognitive decline seen in our patients, while tau does, and there are cases of aged individuals who have amyloid deposition in their brain, but no tau and no clinical dementia [25,37,38].

A significant degree of co-pathology, not strictly specific of AD, is common in these brains [39]. This includes the presence of activated microglia and reactive astrocytes, which appear initially around neuritic plaques, and increase in density and activation in proportion to the neuronal damage and the presence of tau pathology [40–42]; vascular alterations, in the form of amyloid deposition in the blood vessels (CAA: cerebral amyloid angiopathy); accumulation of other structural proteins in neurons, such as actin and actin-associated proteins in the form of Hirano bodies [43]; granulovacuolar degeneration (GVD), a feature often associated with the initial hyperphosphorylation and aggregation of tau in the hippocampus [44]. Argyrophilic grains of 4-repeat tau can also be found in neuronal dendrites, near oligodendrocytes inclusions and seems to be relevant to disease progression [45].

In some AD cases, specific neuronal populations are vulnerable to develop Lewy body pathology of α-synuclein [43,46], which is a signature of Parkinson's disease (PD), mostly at the axonal level, in cortico-limbic regions, olfactory bulbs and substantia nigra. Finally, TAR DNA-binding protein 43 (TDP-43) pathology, more commonly found in cases of frontotemporal lobe degeneration (FTD), can be present in AD, in the amygdala, hippocampus and eventually in the neocortex and more diffusely [47,48], being not specific of AD, but underlying the vast array of cellular dysfunction present in this disease, which can affect RNA-related metabolism as well [49]. The common path, which is likely consequent to all the above changes, is the dramatic loss of synapses, which precedes neuronal loss, and mostly follows the distribution of tau pathology [43,50–52].

The routine clinical diagnosis of possible or probable AD is currently based on the identification of symptoms by the clinician, mostly Neurologists, Geriatricians, Psychiatrists and Primary Care Physicians, followed by more in-depth assessment by Neuropsychology testing and using the tools available in clinical practice, e.g., Magnetic Resonance Imaging (MRI) of the brain and common blood works. Once AD is suspected, these tools allow to rule out secondary dementia and other, different pattern of primary dementia. On a more sophisticated level, the analysis of the cerebrospinal fluid (CSF), the use of PET imaging and genetics (in familial cases) may allow us to obtain a definitive diagnosis of AD in vivo: due to cost and perception of invasiveness, this is often limited to Academic centers or in the context of clinical trials.

Ideally, we should diagnose AD, and every dementia, as early as possible. Mild cognitive impairment (MCI) is a prodromal phase of dementia, where the patient shows symptoms of cognitive changes but maintains good functionality. This is a moment where the application of biomarkers is paramount and significant therapeutic intervention may still possible. Unfortunately, there is a significant underappreciation of the disease, which is missed in more than 20% of the cases. On the other hand, lack of familiarity with alternative diagnoses, leads to overdiagnosis of AD in more than 25% of the cases, where potentially curable conditions are instead the underlying cause of the cognitive and functional decline [53–56].

1.2. Pathogenesis

The majority of our understanding of the pathogenesis of AD derives from the identification of key genes involved in the etiology and pathology of the disease. In a small percentage of the cases, the disease is inherited as autosomal dominant (FAD: familial AD), with an early age of onset, as early as the 40s, and a faster progression [57]. FAD is caused by mutations of one of three genes, Amyloid Precursor Protein (APP), Presenilin 1 (PS1) or Presenilin 2 (PS2). APP is the precursor of the amyloid peptide, Aβ, which is generated from APP by two sequential cleavages: first, the beta-secretase activity of BACE1 (beta-site amyloid precursor protein cleaving enzyme 1) generates a peptide of ~99 amino acids (C99); subsequently, the activity of the gamma-secretase, a multi-peptide protease, progressively cleaves C99 to generate the Aβ peptide ending at residues 40 or 42. The gamma secretase complex is composed of PS1 or PS2 and three other proteins, nicastrin, Aph1 and Pen2 [58]. Mutations in APP mostly lead to increase affinity of APP itself for BACE1, with increased generation of C99. Mutations of PS1 and PS2, which result in an actual loss of function mutations of gamma-secretase [59–62], lead to increased generation of the Aβ peptide ending in residue 42. This is recognized as the most toxic of the Aβ peptides [63]. About 95% of AD cases are sporadic (SAD) and have a later onset (after 65 years of age): although SAD cases lack a Mendelian cause, variants of genes such as Apolipoprotein E (*APOE*) [64,65], and triggering receptors expressed on myeloid cells 2 (TREM2) [66] are known to increase the risk of developing SAD. It is not clear what the initial trigger to Aβ and tau accumulation is in the sporadic disease. At the subcellular and molecular level, the complexity of Aβ and tau accumulation and their specific contribution to pathology has been extensively studied, and is beyond the scope of the current publication. The general understanding is that both amyloid and tau oligomers have the ability to disrupt several of the essential functions of neurons, and possibly other brain cells, i.e., metabolic and mitochondrial activity [67,68], synaptic communication [69,70], membrane integrity [71], membrane pore formation with ions leakage [72,73], protein trafficking and degradation [74,75]. Both Aβ and tau possess the ability to trigger a cascade of functional changes in neurons, microglia, astrocytes and oligodendrocytes, which results in neuronal dysfunction, synaptic loss, innate immune system activation, inflammation, blood-brain barrier impairment and neurodegeneration [39,76]. Relevant to our discussion, the immune system is activated (early) in AD, and may contribute, by means of ineffective clearance of Aβ and tau by the microglia, to the diffusion of such pathology to neighboring or connected areas [77,78], and inflammation per se is a trigger to further production of Aβ and activation of kinases to hyper-phosphorylate tau [79], cascading more neuropathology.

1.3. The (Amyloid, Tau, Neurodegeneration) ATN System

Stemming from the above definitions, pathological correlates and pathogenic theories, AD can be viewed as the combination of specific brain pathology, Aβ and tau, and neurodegeneration, which eventually results in the clinical presentation. Recent advances in brain imaging and more sensitive and specific means to detect Aβ and tau in biological fluids, associated with our ability to detect loss of brain matter, metabolism or function, have allowed for development of a classification and staging system for AD, the ATN system [80], which is routinely revised [81], to include the everchanging horizon of new biomarkers available. Patients are classified by positivity to markers of amyloid deposition (A), tau load (T) and neurodegeneration (N). Amyloid markers include the positivity to Amyloid PET (several tracers are now available) or reduced levels of cerebrospinal fluid (CSF) Aβ42, Aβ40 or of the Aβ42/40 ratio (see below). T is defined by the high levels of total or phosphorylated tau in the CSF (see below), although the new available tau PET tracers may soon be included in the criteria. Finally, N is related to the neurodegenerative aspect, and can be defined by loss of volume/activity in specific areas of the brain by magnetic resonance imaging (MRI, e.g., NeuroQuant, [82,83]), fluorodeoxyglucose (FDG)-PET or by the presence of CSF biomarkers of synaptic or neuronal loss, e.g., Neurofilament Light chain (NfL).

This system has been a useful tool for many biomarkers studies, both in biological fluids and imaging, remaining the "benchmark" for AD diagnosis and prognosis determination. Importantly, the ATN system can be extremely useful to identify and correctly classify individuals who may have amyloid deposition but no neurodegeneration (A+T-N-) and vice versa (A-T-N+); this is extremely important when studying pathophysiological correlates and in longitudinal assessments. As hinted above, a good proportion of the ageing population will show deposition of amyloid in their brain, but no clinical dementia; we know that this amyloid may not be the same amyloid that accumulates in AD proper [84], hence not cause the disease, but also other factors may not be present in these individuals that predispose to tau accumulation and neurodegeneration. On the other hand, neurodegeneration and dementia may occur because of pathology other than Alzheimer's, e.g., vascular, TDP43, alpha-synuclein, etc. If rigorously determined and applied to our studies, the ATN system will be invaluable, as it prescinds from the subjective interpretation of the clinical status of the patients, and is purely descriptive of their biomarkers status.

It must be noted, though, that not all the pathophysiological events and biomarkers in AD can be inscribed in the ATN definition, hence we will be presenting a wider paradigm, that includes, for example, inflammation.

1.4. Treatment Options

No disease-modifying treatment is available as of today, despite decades of efforts aimed to curb the pathological changes of the disease [85–87]. The widest attempts to tackle the disease have been directed against the production/accumulation of Aβ and have overall been underwhelming [85]. Reducing its production via inhibition of BACE or gamma-secretase has not been effective, resulting in potential side effects, included hastening the cognitive decline [86]; active vaccination efforts against Aβ has also not been successful, mostly due to cases of dramatic adverse events, in the form of hemorrhagic encephalitis [88], despite some cognitive benefits in the survivors [89]. Alternative active vaccinations strategies, aimed at inducing a wide reactivity to Aβ without the excessive inflammatory response, have been tested [90,91] with scarce success. Passive immunization against Aβ was also tainted by questionable clinical efficacy and concerns of serious side effects [92–95], but efforts are well under way: at the moment that we write this review, a monoclonal antibody against amyloid has been submitted for U.S. Food and Drug Administration (FDA) expedited approval as a modifying drug against AD, with some skepticism among the scientific community in respect to the available efficacy data [95] (NCT02477800). Tau has been the other logical therapeutic target. Current strategies are directed to target total tau, which is thought to accumulate intra- and extracellularly, and may spread from cell to cell, aggregating and getting hyperphosphorylated, contributing to neurodegeneration.

Antibodies targeting extracellular tau have been in clinical trial now for some time, and we are all bracing for the upcoming results. Of note, this strategy has so far failed in progressive supranuclear palsy (PSP) cases with tau pathology (NCT03068468). An active immunization approach against tau is currently under way as well [96,97].

What clinicians are left with is the use of symptomatic treatments. Acetylcholine deficits are well recognized in AD [98], and are the basis for acetylcholine esterase inhibitors, which can ameliorate quality of life. Memantine, a noncompetitive N-methyl-D-aspartate-receptor antagonist, has also shown some benefit in maintaining cognitive function, alone and in association with acetylcholine esterase inhibitors. In all, unfortunately, the enthusiasm for these agents is very low, due to low efficacy and possible adverse events, such as behavioral changes, heart arrhythmias, loss of appetite and weight, which in the elderly population are a major concern.

The neuropathological changes described above are projected to begin decades before the presentation of symptoms. This makes any medical intervention, even in the prodromal stages, a "too little-too late" approach, with few chances of being meaningful. The challenges in AD are, in our opinion, 2-fold; first, defining the actual pathogenesis of the disease, by far not yet clear; second, be able to identify the risk to develop the disease early enough so that disease-modifying intervention can be effective.

1.5. Definition of AD Biomarkers and Diagnostic Challenges

What is a biomarker? What is an adequate and useful diagnostic biomarker for Alzheimer's disease? While it is fairly easy to respond to the first question, it is very difficult to answer the second. The reason for the latter is that, save for the genetically determined cases of AD (a minority of the cases), in the absence of a univocal and comprehensive pathogenic pathway, it is unlikely to be able to diagnose AD with certainty in the living adult in a time frame that would be useful for significant intervention.

A biomarker is a "defined characteristic that is measured as an indicator of normal biological processes, pathogenic processes, or responses to an exposure or intervention, including therapeutic interventions" (FDA-NIH Biomarker Working Group. BEST (Biomarkers, EndpointS, and other Tools). Available online: https://www.ncbi.nlm.nih.gov/books/NBK326791/. Accessed on 29 July 2020). This definition includes both a diagnostic and prognostic element. The current literature about AD biomarkers is extensive and rich of analyses and reviews. The popular website, Alzforum, has its own dedicated meta-analysis page (Alzforum. Available online: https://www.alzforum.org/alzbiomarker. Accessed on 29 July 2020).

AD biomarkers are quite reliable indicators of disease presence, topographical extension and stage. Levels of Aβ, total tau (t-tau), and phosphorylated tau (p-tau) are concurrent with AD pathology [99,100] and indicate its presence; at this stage amyloid [101] and tau PET imaging [102] define the topographical extension of such pathology, while MRI [103] and FDG-PET [104] define the topographical severity of the neurodegenerative process; finally, markers such as p-tau, NfL [105] and YKL40 [106] can be useful, and early, peripheral indicators of disease stage.

We can use a combination of fluid and imaging biomarkers to complement the clinical investigation in our patients, and fulfill the current diagnostic criteria [107–110] with ease. One caveat is that while these criteria can work well in research settings [111,112], the reality of clinical practice is tainted by the reluctance of patients to have invasive testing done (CSF) and by the lack or insurance support for the more "fancy" imaging [113]. Fortunately, as described below, the rapid evolution of detection techniques is allowing to reach remarkable results from blood biomarkers, which would improve access for patients.

Another element must be taken in consideration when defining a useful biomarker: time. The concept of early diagnosis has become crucial in neurodegeneration, since protein aggregation and accumulation might only reflect the last stage of the pathogenic cascade, a point of no return in the clinical manifestation and progression of the diseases, where therapeutic interventions have so far been ineffective. As mentioned above, the exact succession of pathogenic events in AD is still unclear;

Aβ and tau may be causative, but subtler prior changes in neuronal connectivity, metabolic patterns and inflammation could represent an early disruption of neuronal function, leading to secondary accumulation of Aβ and tau [36]. In this context, an early diagnosis should go beyond Aβ and tau. A "late" biomarker' profile may also be tainted by ongoing and long-standing pathology, and by other factors related to ageing, e.g., brain trauma, infections, microvascular disease/strokes, and co-morbidity with other frequent neurodegenerative conditions such as PD [114]. The clinical presentation of FAD and SAD is similar, except for the age of onset. For FAD mutations carriers, preclinical diagnosis is possible by genetic analysis; the more common SAD cases, instead, remain elusive until clinically evident. Hence, the challenge of procuring an early preclinical diagnosis of disease or susceptibility to it.

With these limitations in mind, the science of AD biomarkers has evolved tremendously in the last decade, and we are looking now at the possibility to have reliable peripheral non-invasive, scalable and affordable biomarkers of AD pathology and disease stage.

We will review below the current, most promising fluid (blood and CSF) biomarkers available, discuss the pathophysiologic correlate of each, their clinical significance, their general applicability in the clinical and research setting. We will then take tau as an example of biomarker detection complexity, with blessings and caveats, stemming from the need to accurately validate data emerging from new ultra-sensitive platforms, in light of their biochemical limitations.

Importantly, we have divided the below presentation in biomarkers that are specific of AD, where their appropriate combination in biological fluids defines the disease, from biomarkers that are representative of neurodegeneration or associated pathology, such as synaptic and neuronal loss, inflammation, vascular disease and others. The latter biomarkers may be present in neurodegenerative diseases different from AD, in head trauma, in stroke and infections, hence they are not specific for AD. They are, though, extremely useful to follow, and at time predict, AD progression and presentation. Their precise causative role in the neurodegenerative process is still unknown, hence we will present the current known hypotheses in descriptive fashion.

2. Biomarkers of AD Pathology

2.1. Aβ

Given its pathophysiological role, Aβ has revealed itself as the major biomarker of the disease, in addition to being one of main therapeutic targets so far. Over the last 30+ years, exponential efforts have demonstrated that CSF (and more recently blood) levels of Aβ (1) are reduced in AD compared to non-demented controls [115], (2) are decreased in mild cognitive impairment (MCI) patients compared to normal controls [116], (3) the trend of Aβ decrease is predictive of the clinical progression of MCI to AD and preclinical AD [117,118] to MCI. Similarly, and (4) the Aβ42/40 ratio is reduced in MCI and AD, with better clinical correlations to than total Aβ [119,120].

What is the pathogenic significance of Aβ reduction in biological fluids? Why is the Aβ42/40 ratio going down as the disease progresses? Is there ever a time, in the preclinical phases of the disease, when Aβ is instead increased in biological fluids, possibly marking a pivotal moment where a therapeutic intervention would still be effective?

In FAD, APP mutations lead to an increased overall production of Aβ, since the amyloidogenic cleavage of APP is favored. Mutations in PS1 or PS2 result instead in a preferential cleavage of C99 to form the Aβ42, while in reality the overall amount of Aβ may be reduced, since these are loss of function mutations [121]. In either case, there is an absolute or relative abundance of Aβ42 released in the brain parenchyma, which has a high tendency to aggregate and deposit as oligomers, fibrils and plaques in a process of nucleation [122]. Furthermore, BACE cleavage is seldom limited to residue #1 of Aβ; the amyloid peptide often starts at residue #3 or #11, a glutamate residue (E), which tends to be cyclized [123,124] as pyro-glutamate (pE). The resulting AβN3(pE) and AβN11(pE) peptides also have high tendency to aggregation, adding to the stickiness of Aβ42, and have been considered one of the first seeding species of Aβ [125]. In this context, it is easy to understand how the parenchymal sequestration of Aβ results in lower level of the soluble molecule available to diffuse to the interstitial fluid, CSF and blood, hence the low levels thereof. Further decrease of Aβ in the periphery may also be related to the progressive and massive loss of neuronal cells along the progression of the disease, as neurons are the highest manufacturers of the peptide [126]. Finally, other factors such as inflammation, infection or perturbation of Aβ clearance mechanism may affect its peripheral levels, though unlikely to have a chronic effect [127].

Other conditions show the opposite; in traumatic brain injury (TBI), for example, CSF and plasma levels of Aβ1-42 increase steadily after the acute event, and normalize with recovery in weeks [127–129]; in some cases, an increase of APP itself was noted when sampling intraventricular CSF [130]. These data tend to correlate with other markers of neuronal damage, i.e., the release of total tau, neuronal specific enolase (NSE) and APOE, but the actual mechanism of shedding is unclear. Notably, axonal injury after TBI is associated with redistribution of APP towards the axons and increased processing to generate Aβ, explaining part of the CSF levels [131,132].

Interestingly, whereas Aβ accumulation in the brain results in a reduction in the periphery, tau accumulation in the brain parallels its peripheral abundance [115] (discussed below). Tau accumulation follows, in time, Aβ accumulation. Can this time difference (maybe decades) explain the dichotomy? If so, is there a moment in time in AD pathogenesis when Aβ levels are actually increased or normal in the periphery? The current data on autosomal dominant AD show initial amyloid-PET abnormalities ca. 22 years prior the expected symptom onset (EYO-22) and CSF Aβ reduction (and tau increase) a few years later (EYO-19 to -14) [133]. This series of events signifies that (1) sensible Aβ deposition in the brain happens quite early in the pathogenesis, (2) current CSF (and blood) biomarkers follow brain changes by several years, and (3) measures of neuronal damage (peripheral tau, etc.) do not fall far behind peripheral Aβ changes. This leaves open the question of what happens even earlier to peripheral Aβ, is it ever elevated? Is that the ideal moment where anti-Aβ therapeutic strategies would be most effective? Only longitudinal studies with a very early CSF and blood sampling will be able to answer these questions.

A recent study in a Down syndrome (DS) cohort [134], with CSF sampling from age 30 to 61, showed Aβ42 reductions which was proportional to age and cognitive decline (from cognitive stable, to MCI to AD), with higher levels than those reported in corresponding older non-DS controls, MCI and AD [135], but did not have a non-DS age-matched control group. In another study [136], CSF levels of Aβ peptide were found to be elevated from 8 to 54 months of age in DS children, while tau was low, likely due to Aβ overproduction in the absence of significant parenchymal sequestration and neuronal loss at that age.

CSF. Sampling CSF by lumbar puncture (LP) is the most direct way to obtain biomarkers that reflect what happens in the brain. Being invasive, most patients do not feel comfortable undergoing a spinal tap, although the rate of complications, i.e., post-LP headache, bleeds or infections is very low [137]. CSF levels and variations of Aβ have long been the hallmark features of AD-type dementia. The literature on the matter is extremely extensive and will not be reviewed here. We will limit our discussion to the clinicopathological relevance of Aβ in CSF. The first attempts at measuring soluble APP-derived fragment in the CSF of AD patients were carried out by D. Selkoe and S. Younkin [138] and B. Frangione [139] labs, determining the reduction of soluble APP (sAPP) fragments in the CSF of AD patients. Soon after, scientists were able to map the Aβ region and retrieve the peptide in the CSF [140]. This led to the first studies looking at differences between AD and control subjects in soluble APP and Aβ peptides [141–145], with notable differences likely due to detection methods and populations [146,147].

CSF levels of Aβ 42 are generally lower in AD compared to control subjects, and have been seen to decline with disease progression [135,147–150]. Aβ1-42 levels can be reduced as much as 50% in patients with AD compared to control subjects [151]. There seems to be, though, significant overlap of total Aβ1-42 levels between AD and non-AD dementia, such as vascular dementia, Parkinson's disease dementia (PDD), Lewy body disease (LBD), frontotemporal lobe degeneration (FTD) and even amyotrophic lateral sclerosis (ALS), and this may be due to population selection, stage of disease or overlap of pathogenic events in advanced disease [152–158].

In the pre-dementia stages, both amnestic MCI (aMCI), the most frequent precursor of AD, and non-amnestic MCI, showed low levels of Aβ1-42 [154,159–162] compared to control subjects, also with significant overlap.

In all, the use of Aβ1-42 alone does not seem to be too promising for differentiating AD from non-AD dementia and less so at the MCI stage, due to heterogeneity and overlap. As mentioned above, the pathophysiology of APP processing in AD is characterized by an imbalance of the 1-42 over the 1-40 species of Aβ, with preferential deposition of the former in the brain. Additionally, the gamma-secretase complex cleaves APP in a sequential fashion, generating progressively shorter fragments, like the Aβ1-38. These fragments have been identified in CSF, and may be useful biomarkers as well [162]. Thus, measuring the CSF ratio of Aβ42/40 (and other 42/X peptides ratios, e.g., 42/38, etc.) has been attempted. Indeed, the CSF ratio of Aβ42/40 has proven to be a reliable marker of AD, both at the dementia and MCI stage, as well as of the risk of progression from MCI to AD [163–166]. Furthermore, this ratio was also useful to differentiate AD from other non-AD cognitive changes, such as subcortical deficits of vascular disease. Such correlations were validated by both amyloid PET and extensive clinical and conventional MRI approaches. CSF Aβ42/40 correlates almost completely with amyloid PET, and when there is discordance, longitudinal studies have shown that PET imaging turns positive within a few years [117,119,150]. Finally, it was shown that different assay platforms resulted in similar results [167].

Blood. Serum or plasma measurement of Aβ has become possible and reliable thanks to the increased sensitivity and specificity of the available diagnostic platforms. Initial testing was based on ELISA (enzyme-linked immunosorbent assay), but results have been equivocal at best [148,168]. Interference by plasma proteins was likely significant in these early attempts. ELISA have been progressively improved to reach higher sensitivity, and other platforms have been developed, including SiMoA (Single Molecule Array) [169–171], immunomagnetic reduction [172], multiplex assays (e.g., xMAP- [173]), MSD [174] and more. As discussed in more detail below, in order to overcome the complexity of the blood and the interference by blood proteins and heterophilic antibodies and low concentrations of the target biomarker [113], the more modern platforms work with very high dilution of the sample and amplification of the specific signal, allowing for detection of picomolar/mL (or less) concentrations of the target protein [175,176]. Thanks to these advances, we can now reliably measure the Aβ42/40 ratio in plasma.

Studies using SiMoA have, first of all, confirmed a significant correlation between CSF and plasma [170,177] Aβ42, Aβ40 and their ratio, and further confirmed the CSF data on AD and MCI vs. control patients; AD and MCI patients show reduced levels of Aβ42 and Aβ42/40 ratio compared to cognitively intact control subjects, confirmed by amyloid PET.

High-precision assays for plasma Aβ42/40, using immunoprecipitation followed by liquid chromatography-mass spectroscopy (IP–MS) [178], are strongly predictive of brain amyloidosis as validated by benchmark amyloid-β-PET, and reach 90% accuracy [175,179,180]. In one study, a novel fragment ratio to Aβ42 (APP669–711/Aβ42) was also used to predict amyloid-β-PET positivity [175].

Interestingly, in the case of Aβ, another factor that interferes with the accurate plasma measurement is its extra-cerebral production (e.g., platelets, [181]), which is difficult to control. Hence, while it is possible to detect a reduction of the Aβ42/40 ratio in plasma in AD vs. control subjects, plasma reduction only totals about 15%, while in CSF we can see as much as 50% reduction [115,176]. When analyzing the overall trajectory of Aβ variation in plasma vs. CSF, in parallel with Aβ deposition in the brain by PET, by a variety of modern assays, plasma Aβ does not perform as well as CSF [182], showing a narrower range, making it still a less sensitive tool.

2.2. Tau

In AD, tau undergoes significant conformational change and abnormal phosphorylation, leading to the formation of inclusions throughout the brain, that start in the entorhinal cortex (EC) and, as the disease evolves, spread progressively throughout the brain, leading to synapse loss and neuronal death, which correlate well with the clinical onset and progression of symptoms [25,183,184]. High levels of tau, both total (t-tau) and phosphorylated (p-tau), have consistently been detected in CSF of AD patients compared to healthy controls [115,185–192]. Studies in FAD and SAD suggest that the tau-PET signal associated with neurofibrillary tangle pathology increases at symptom onset and appears about one decade after CSF soluble tau increase [193–196]. Other studies suggest that, while neurodegeneration progresses, the rate of increase of p-tau and t-tau levels may actually be reduced [118,133,197], likely in relation to the massive neuronal loss. In this dynamic scenario, where the rate of p-tau accumulation changes, and Aβ may show fluctuations in time, as discussed previously, an ideal biomarker should be specific enough for an adequate pre-clinical detection of AD and to track the disease progression.

As previously reported by Augustinack et al. [183], distinct tau phosphorylation sites correlate with severity of neuronal cytopathology in AD. Pre-tangle state, intra- and extra- neuronal neurofibrillary tangles were used to characterize different stages of pathology. More specifically, p-T231, p-S262, and p-T153 (T: Threonine, S: Serine) were shown to stain non fibrillar tau and were identified as early phosphorylation markers; p-T175/181, p-S262/p-S356, p-S422, p-S46 and p-S214 co-localized with intra-neuronal fibrillar tau structures, and increase as disease progresses; p-S199/p-S202/p-T205, p-T212/p-S214 and p-S396/p-S404 identified instead extracellular tangles composed of substantial filamentous tau structures, seen in later stages of the disease.

In addition to focusing on crucial site-specific phosphorylation, it has been shown, by cryo-EM analysis in AD as well as in other tauopathies, that the most disease-specific change is a tridimensional folding of the tau filaments, which is a signature of the particular disease examined; these tau molecular conformers can in fact differentiate between AD, Pick's disease and corticobasal degeneration (CBD) [198–201]. Moreover, an AD-specific epitope formed by two discontinuous portions of tau, ${}_{7}$EFE${}_{9}$ and ${}_{313}$VDLSKVTSKC${}_{322}$ and recognized by the MC1 antibody [198,202,203], represents an early aberrant conformation of tau present both in a soluble form of the protein and in paired helical filaments (PHF) assemblies [204]. Interestingly, the level of MC1 reactivity correlates with the severity and progression of AD. To our knowledge, no assay has been developed to detect this early marker of pathology in biofluids, likely due to the possibility that tau truncated species [205–208] could prevent the identification of a conformational tau epitope using the currently available immunoassays, which are-based primarily on mid-region-directed antibodies. We believe that, although this pathological feature is not per se easily "translatable", it should be kept in consideration when designing and developing new tau biomarkers platforms.

In summary, the identification of precise phosphorylated tau species and conformers in the biological fluids will contribute not only to early preclinical diagnosis of AD, but will also allow for a "personalized" approach by staging the disease in its progression and possibly monitoring the effect of therapeutic interventions. Here below is a review of the classic and novel approaches to detect tau and its phosphorylated modifications in CSF and blood.

CSF. Increased phosphorylation and axonal degeneration are believed to induce passive release of tau from the microtubules into the extracellular space, possibly as a neuronal response to Aβ toxicity [209,210]. This process results in augmented levels of t-tau and p-tau in the interstitial fluid and consequently in the CSF. From a neuropathological perspective, analysis of postmortem brains has shown a certain degree of correlation between the neocortical tau burden and CSF p-tau [190,211]. Tau phosphorylated at Threonine 181 (p-T181), t-tau and Aβ42 have been extensively validated in the CSF as biomarkers of AD and are currently widely used as diagnostic benchmarks in clinical and research studies [54,109,212]. CSF p-T181 tau has also proven useful in differential diagnosis of dementia [213–215] and to predict cognitive decline in preclinical and prodromal disease stages [213,216,217], with specificity for AD.

An ongoing effort focuses on standardizing protocols and developing criteria for the appropriate use of CSF tau assessment in the diagnosis of AD, with the intent of integrating CSF biomarkers into clinical practice [218–220]. In this respect, the distinctive value of the CSF p-T181 tau level has still to be thoroughly demonstrated. In one comparison study, p-T181, tau phosphorylated at Serine 199 (p-S199), and tau phosphorylated at Threonine 231 (p-T231) showed similar diagnostic accuracies for AD [221]. On the other hand, Spiegel et al. [222] reported that CSF p-T231 tau levels are superior to p-T181 in clustering AD subjects from normal control, suggesting the potential for CSF p-T231 assay to be employed in early AD diagnosis. These data are in line with the AD neuropathological picture where p-tau231 shows greater sensitivity for neurofibrillary tangles than p-T181 [211,223], with propensity to accumulate in layer II of the entorhinal cortex [183], known to be the first limbic area in AD consistently targeted by neurofibrillary pathology [224].

Together with p-T181 and p-T231, other phospho-epitopes are under investigation with the hope to find even better AD diagnostics. Barthélemy et al. [225] reported that CSF tau phosphorylated at Threonine 217 (p-T217) is specifically associated with amyloid-β pathology, suggesting a strong interplay between AD amyloidosis and hyperphosphorylation of tau on p-T217; moreover, p-T217 outperforms p-T181 as a biomarker for detecting both the preclinical and advanced forms of Alzheimer's disease. In a second study, Barthelemy et al. [226] suggested a pattern of tau staging linking CSF site-specific phosphorylation to three different phases of the disease progression. By analyzing FAD, SAD and patients with unaffected cognition but at risk of disease (based on the presence of abnormal Aβ pathology); the authors identified an array of tau phosphorylation uniquely associated with structural, metabolic, neurodegenerative and clinical markers of disease. The study shows that (1) p-T217 and p-T181 rise in parallel to aggregated Aβ as early as 20 years before the development of neurofibrillary tangles, (2) p-T205 correlates with hypometabolism and atrophy closer to symptom onset, and (3) t-tau spikes when cognitive decline manifests. All of these data are in line with another study where the associations of p-T217 with tau PET, CSF and PET measures of amyloid pathology were tested in comparison to p-T181 [220]; the authors found that CSF p-T217 correlates better than p-T181 with PET measures of tau and amyloid pathologies in AD, and more accurately distinguishes AD from non-AD neurodegenerative disorders, suggesting the benefit of employing CSF p-T217 as a biomarker of AD in clinical practice.

Furthermore, investigating tau fragments in CSF, Chen et al. [227] developed a set of tau immunoassays able to detect different populations of tau fragments in CSF and blood. The authors were able to demonstrate that mid-region- and N-terminal-containing fragments increase with disease, while full-length tau is just a small fraction of the tau present in both normal and AD CSF. A recent study has also shown that the ratio between C-terminally truncated tau368 and t-tau is significantly decreased in patients with Alzheimer's disease, with a strong correlation with ^{18}F-GTP1 retention in brain [207] (a measure of tangle pathology in vivo); these data suggest that tau fragment may preferentially accumulate in tangles, and the CSF ratio tau368/t-tau reflects tangle pathology in brain and can be used as an additional tau biomarker to stage and improve the diagnosis of Alzheimer's disease.

Several studies indicate that TBI can be classified as an independent risk factor for developing Alzheimer's disease. As previously discussed for Aβ, the axonal injury resulting from the trauma drives the development of both Aβ plaques and NFT [228–232]. Interestingly, not only do tau levels increase in the CSF of the concussed patients undergoing repetitive trauma (i.e., sports players), but serum t-tau levels have been reported to increase in the acute stage following the concussion, while decreasing during rehabilitation [230,233,234]. Similarly, serum p-tau (p-T231 and p-S202) levels, though reduced, were shown to be still higher compared to the control levels even at six months post-injury [235].

When considering tauopathies different from AD with significant NFT pathology and neurodegeneration, CSF p-T181 and t-tau levels do not necessarily increase; this is the case for progressive supranuclear palsy (PSP) and some forms of FTD [221,236–238]. Why increased CSF t-tau and p-tau are specific to AD can be explained by considering that (1) the pattern of tau phosphorylation occurring in these disorders is disease-specific, (2) tangle pathology and neurodegeneration in AD may be consequent to Aβ accumulation, with disease-specific molecular cascades, such as kinases and proteases activation, (3) disease-specific tau truncation could prevent detection by currently available assays, and (4) specific pathological changes can be more severe in AD than in other tauopathies.

Blood. Until recently, AD biomarkers' quantification was limited to the CSF. However, while combining CSF biomarker analyses and PET scans have the potential to improve diagnostic accuracy, these methodologies show limitations, impeding their use as first-line diagnostics [113]. Developing an accurate and standardized blood test will be more cost effective than PET imaging [239,240], less invasive than CSF testing [241] and faster. Hence, blood-based biomarkers represent next-generation diagnostics for AD and other CNS diseases. One matter of concern is that blood tau has the potential to derive from other sources in addition to neuronal tau [242,243]. In this respect, novel ultrasensitive detection technologies have been developed to facilitate the detection of very low concentrations and variations of tau in the blood [186,244,245] using very high dilutions of the biological fluid: employing these technologies has pushed forward the potential for plasma tau to serve as a biomarker for neurodegeneration in AD, avoiding the interference of non-neuronal tau.

Longitudinal studies originally showed that higher levels of plasma t-tau are associated with greater cognitive decline and risk of MCI [246–248], with no correlation with elevated brain Aβ [247]. As discussed, plasma t-tau assays can identify neuronal injury in acute brain disorders, such as TBI, but work relatively poorly for AD diagnosis, and the correlation with CSF t-tau is weak [249]. In this context, t-tau could be intended as a non-specific prognostic marker for dementia and related endophenotypes [248]. Despite the lack of correlation between plasma and CSF t-tau, Fossati et al. [186] have recently demonstrated that by combining plasma t-tau with CSF t-tau or CSF p-T181 tau, the accuracy for the differentiation of AD versus cognitively normal controls increases, suggesting that plasma tau may be a useful biomarker to increase diagnostic power and determine therapeutic efficacy in clinical studies.

Analysis of phosphorylated tau in blood has so far been focused on the development and validation of assays for p-T181 [182,246,250–253], demonstrating that increased plasma p-T181 (1) correlates with amyloid and tau PET measures [250,254,255], allowing the identification of individuals with Aβ pathology but tau PET-negative measurements [253], (2) goes in parallel with CSF p-T181 [182,255], and (3) can help in the differential diagnosis of non-AD types of dementias [253–255].

While writing this review, two groups published strong data in support of the use of a blood-based assay detecting p-T217 tau. First, Barthelemy et al. [256] developed a protocol of purification and enrichment of tau from plasma, followed by liquid chromatography (LC-MS); this system (1) allows reliable detection of p-T217 showing high correlation with the CSF measurements, and (2) confirms the ability of this biomarker in clustering amyloid positive and negative patients. The limitation of this strategy is the high volume of blood needed to purify and extract tau and the elaborate methodology, not compatible with the idea of developing a fast and easy-to- use assay. In the second study, Palmqvist et al. [256] employed the Meso Scale Discovery (MSD)/Lilly platform (described below) and validated the value of p-T217 in plasma to discriminate Aβ+ MCI and AD patients from other neurodegenerative disorders and controls, including sensitive discrimination of autosomal-dominant pre-symptomatic Aβ+ patients. In all, given the high potential of measuring CSF p-T217 previously described, it will be of great interest to follow the development and validation of novel ultrasensitive platforms aimed at detecting p-T217 in blood, employing fully automated platforms to integrate this biomarker in clinical practice. Similarly, setting up sensitive and specific assays able to detect p-T231 (early phosphorylated epitope) and p-S396/404 (late phosphorylated epitope) in blood will also be relevant to stage the AD pathology and to monitor future therapeutic interventions.

Finally, an ultra-sensitive assay was developed in blood, to detect an N-terminal tau fragment (NT1) [227], showing that N-terminal fragments of tau mapping in the 6–198 sequence are increased in AD and MCI patients and may be useful as a screening blood-based test for AD. This approach is extremely significant due to the possible generation of disease-specific tau fragments (as for APP and Aβ), and will need further investigation.

Immunoassays and ultrasensitive platforms to detect tau in biological fluids. Several assays have been tested and extensively validated in CSF and plasma. In Table 1 we select and catalogue the most used assays in the tau biomarker field at the moment, with emphasis on their specificity, fluid samples' applicability and limit of detection. The list of assays includes (1) standard and second generation ELISAs (INNOTEST, EUROIMMUN, Elecsys), (2) MSD platforms (3) Single-Molecule enzyme-linked immunosorbent assay (SiMoA), (4) Mass spectrometry analysis, (5) superconducting-quantum-interference-device Immunomagnetic reduction (SQUID-IMR). The detection range spans from few pg/mL to fg/mL according to the sensitivity of the assay. EUROIMMUN and Elecsys are automated second-generation immune assays that have demonstrated good analytical performance with clinically relevant measuring ranges, supporting their use in clinical trials and practice [257,258]. MSD technology with electrochemiluminescence detection comes with several advantages compared to standard immunoassays, including high sensitivity, low background and wide dynamic range of detection [174]. Notably, Palmqvist et al. [182] recently ran a comparative analysis of Aβ and tau CSF assays from five different manufacturers: Elecsys (Aβ42, Aβ42/40, t-tau, p-tau181) EUROIMMUN (Aβ42, Aβ42/40, t-tau, p-tau181), INNOTEST (Aβ42, Aβ42/40, t-tau, p-tau181), MSD (Aβ42, Aβ42/40, t-tau) and Lilly (p-tau181, p-tau217): Aβ42, Aβ42/40, p-tau and t-tau had very similar trajectories between assays.

Moving to ultra-sensitive platforms, Rissn et al. previously developed [169,259] a Single-Molecule enzyme-linked immunosorbent assay (SiMoA) that detects serum proteins at sub-femtomolar concentrations, restricting the fluorophores produced by individual enzymes to considerably small volumes, hence capturing the fluorescence of each single molecule. This methodology has been extensively used to improve t-tau and p-tau detection in CSF and plasma (Quanterix).

SiMoA and INNOTEST ELISA have been also demonstrated to strongly correlate in CSF t-tau in a comparison study [186]. Moreover, in a recent study, Karikari et al. [253] validated a blood-based immunoassay measuring p-T181 using a sandwich immunoassay format on SiMoA technology.

To further quantify tau at high sensitivity, Barthelemy et al. has developed a protocol to detect tau and p-tau in CSF and plasma employing targeted high-resolution mass spectrometry (MS) [225,226,256,260,261]. Additionally, superconducting-quantum-interference-device Immunomagnetic reduction (SQUID-IMR) has proven efficient at quantifying t-tau and p-T181 in plasma in early stages of AD [172,252].

Table 1. Immunoassays and ultrasensitive platforms to detect tau in biological fluids.

Type of Assay	Platform/ Manufacturer	t-tau	p-tau	Fluid Sample	LOD pg/mL	References (PMID/link)
Standard ELISA	INNOTEST Fujirebio	Total tau Antibodies: Capture HT7/AT120 Detector BT2		CSF	34	8748926
	INNOTEST Fujirebio		p-T181 Antibodies: Capture HT7 Detector AT270	CSF	13	10788705
	Applied Neurosolutions (Dr. Peter Davies)		p-T231 Antibodies: Capture CP27 Detector CP9	CSF	9	26444757
Automated ELISA	Elecsys	Total tau Antibodies: Not available		CSF	63	31129184
	Elecsys		p-T181 Antibodies: N.A.	CSF	4	31129184
	EUROIMMUN	Total tau Antibodies: Capture ADx201 Detector ADx215		CSF	44.2	27447425
	EUROIMMUN		p-T181 Antibodies: N.A.	CSF	N.A.	27447425
MSD-ECL	MSD		p-T181 Antibodies: Capture biotinylated AT270 Detector SULFO-TAG-LRL (anti-tau, Lilly)	Plasma	N.A.	29626426
	MSD/Lilly		p-T181 Antibodies: Capture biotinylated AT270 Detector SULFO-TAG-LRL (anti-tau)	CSF	N.A.	31709776
	MSD/Lilly		p-T217 Antibodies: Capture biotinylated IBA413 Detector SULFO-TAG-LRL (anti-tau)	CSF	N.A.	31709776
	MSD/Lilly		p-T217 Antibodies: Capture Biotinylated-IBA493 Detector SULFO-TAG-4G10-E2 (anti-tau)	Plasma	N.A.	31709776

Table 1. *Cont.*

Type of Assay	Platform/ Manufacturer	t-tau	p-tau	Fluid Sample	LOD pg/mL	References (PMID/link)
SiMoA	Simoa™ Tau 2.0 kit HD-1 platform (Quanterix)	Total tau Antibodies: Capture sequence AA16–AA24 Detector sequence AA218–AA222		CSF	0.019	101444
	Janssen R&D Simoa technology HD-1 platform (Quanterix)	Total tau Antibodies: Capture HT7 Detector PT82		CSF	N.A.	32246036
	Modiefied version of Simoa™ Tau 2.0 kit HD-1 platform (Quanterix)		p-T181 Antibodies: Capture sequence AA16–AA24 Detector AT270	Plasma	0.0090	28866979
	Sandwich immunoassay format on Simoa technology (Quanterix)		p-T181 Antibodies: Capture AT270 Detector Tau12	Plasma Serum	N.A.	32333900
	Simoa™ p-181 kit HD-1 platform (Quanterix) NEW		p-T181 Antibodies: N/A	CSDF Plasma Serum		Alzheimer Association International Conference, 2020. Available online: https://alz.confex.com/alz/20amsterdam/meetingapp.cgi/Paper/41238. Accessed on 29 July 2020.
	Janssen R&D Simoa technology (Quanterix)		p-T217 Antibodies: Capture PT3 Detector PT82	CSF	N.A.	32246036
	Simoa® pTau-231 Advantage Kit HD-1 platform (Quanterix)		p-T231 Antibodies: Capture AT270 Detector Tau12	CSF	0.621	Quanterix.com. Available online: https://www.quanterix.com/sites/default/files/assays/Simoa_pTau-231_Data_Sheet_HD-1_HD-X_Rev02.pdf. Accessed on 29 July 2020.
	NT1 assay Simoa technology HD-1 platform (Quanterix)	N-terminal tau Antibodies: Capture Tau12 Detector BT2		CSF Plasma	0.2-0.7	30419228
Nano liquid chromatography-HRMS		Total tau	p-T181 p-S202 p-T205 p-T217	CSF	N.A.	26742856 32161412
(IP)-LC-MS			p-T217 p-T181	Plasma	p-T217:0.05 p-T181:0.2	32725127
SQUID-IMR		Total tau	p-T181	Plasma	N.A.	29376870

A detailed list of the original and the more updated assays used in the tau biomarker field is provided, with emphasis on total, phosphorylated or truncated tau. Type of assays and platforms: ELISA (Enzyme-linked immunosorbent assay); Meso Scale Discovery (MSD) technology with electrochemiluminescence; SiMoA: Single-Molecule enzyme-linked immunosorbent assay; Nano liquid chromatography–HRMMSD-ECLS: Nano liquid chromatography–high resolution mass spectrometry; (IP)-LC-MS: immunoprecipitation coupled with liquid chromatography-spectrometry; SQUID-IMR: superconducting-quantum-interference-device Immunomagnetic reduction. Fluid samples are specified (CSF, plasma, serum). Limit of detection (LOD) of the assay is provided where available, if not it is marked as N.A. References are listed as PMID or hyperlink.

3. Biomarkers of Synaptic and Neuronal Loss

Synaptic loss first and overt neuronal demise later are fundamental aspect of AD pathology and are, together with tau deposition, the real correlate of the manifestation of symptoms. Neurodegeneration is the "N" in the ATN system, and is a fundamental correlate of the functional disability of AD patients. When patients present to their doctors with memory loss or other cognitive changes, due to AD, it is estimated that a large number of neurons have already been lost to neurodegeneration, but recent models are looking at the effect of more subtle synaptic loss as an inciting event responsible for symptoms [70]. In this context, several markers of synaptic disruption have been proposed as indicators of disease onset and progression.

3.1. Neurofilament Light Chain (NfL)

NfL belongs to the type IV intermediate filaments, which make up the neuronal axoskeleton, maintain neuronal caliber and play a role in intracellular transport to axons and dendrites. With the neurodegenerative process, in AD as in other disease [262], NfL is released in the extracellular space. NfL has been detected in both CSF and blood, and appears to be extremely sensitive to AD onset and progression, predicting Aβ PET positivity as well; furthermore, longitudinal studies in familial AD, have shown accurate prediction of disease onset, as far as 10 years prior to symptoms [263–266]. The tight correlation between CSF and blood values [267] make NfL one of the most reliable blood biomarkers to predict onset and course of neurodegeneration.

3.2. Neurogranin

Neurogranin (Ng) is a neural-specific postsynaptic protein, concentrated on dendritic spines, in the hippocampus and basal forebrain. Given its localization at the epicenter of the pathogenic events in AD, it is not surprising that it would be released in the extracellular space and be detected in CSF. In AD, CSF levels of Ng grow quite rapidly prior to the presentation of cognitive deficits, before NfL does, in an Aβ-dependent way [268]. Plasma levels do not seem to be as reliable as CSF, possibly due to peripheral production [269]. Importantly, and contrarily to NfL, other neurodegenerative disease do not show an increase of Ng, raising the possibility of an AD-specific synaptic effect [185].

Other biomarkers in this category are being investigated, such as NSE and SNAP-25, and are the subject of extensive analysis, but are mostly used in complex and integrated predictive models [133,270] and will not be reviewed here.

4. Biomarkers of Inflammation and Microglia "Dysfunction"

Inflammation, microglia and astrocytes activation are well-recognized features of AD pathology, as for other neurodegenerative diseases. It is unclear if they are integral part of the original pathogenesis or are reactive features to the deposition of amyloid and tau. To really understand this aspect, we should investigate, with an unbiased approach, the very early features of patients with a predisposition to develop AD pathology, for example in large autosomal dominant cohorts, in prospective longitudinal studies, and some effort is under way in this context. For the time being, we will focus on the data available from the earliest detectable stages of AD pathology, the prodromal AD cases.

4.1. Soluble TREM2

The Triggering Receptor Expressed on Myeloid Cells 2 (TREM2) is a transmembrane receptor, belonging to the family of cell surface transmembrane glycoproteins [271] which mediates signaling and cell activation following ligand binding. It is expressed and functions in subsets of myeloid cells, e.g., granulocytes, dendritic cells, macrophages, and microglia in the brain [272–275]. Among its complex and protean roles, its function is to regulate immune cells' activation upon binding of a variety of ligands [66], including APOE and Aβ. In the brain, where it may be expressed preferentially in some regions, such as the hippocampus, and differentially in response to the pro/anti-inflammatory environment, it mediates microglia activation, proliferation, migration, apoptosis and expression of pro-inflammatory cytokines, in particular after binding with aggregated proteins such as Aβ, and facilitates the phagocytosis of apoptotic neurons [276]. TREM2 expression is increased in many neurodegenerative conditions, stroke, trauma and even ageing. Rare variants of TREM2, discovered by GWAS studies [277,278], have been associated with the risk of developing AD, in the range of the risk conferred by APOE4, of 2- to 4-fold (heterozygous). The most common AD-associated variant recognized in populations of Europeans descent, but not in Chinese and African-Americans, is the R47H [277,279]. This TREM2 mutant is associated with earlier onset of disease and higher CSF tau levels [280] and seems to demonstrate reduced activity in ligands-binding and hence signal transduction, with a net result of reduced microglia activation and, in AD models, less clearance of Aβ [274]. Given this premise, it is only natural to look at TREM2 as a possible biomarker for AD. The soluble form of TREM2 (sTREM2) is found in both CSF and blood [281,282]. In CSF, sTREM2 is elevated in AD patients vs. controls, with positive association with t-tau levels [212,283]. In familial AD, sTREM2 CSF levels increase prior to clinical onset, but after detectable changes of Aβ and tau in CSF [284]; this is a moment in time where the brain is already at an advanced stage of pathology, in particular tau, with significant neuronal loss and when patients are soon going to lose their cognitive reserve and show symptoms. From a clinical perspective, this is relatively late to be useful as early prediction of disease or point of intervention, but from a pathogenic point of view, it may signify a "breaking point" of the Aβ and tau pathology containment system, where microglia cannot keep up with plaques and tangles any more. This is extremely important to consider in the development of therapeutic strategies in the future. More work is needed to create more sensitive assays in order to detect sTREM2 blood changes, as early as possible. Additionally, since the degree of inflammation and microglia activation differs from patient to patient, and secondary to treatment (e.g., anti-amyloid strategies), a sensitive blood sTREM2 assay may be a very useful tool to monitor phenotypic variants and adverse events form upcoming treatments.

4.2. YKL-40

YKL-40 is a glycoprotein expressed in both astrocytes and microglia, and is the protein product of the CHI3L1 gene, a.k.a., Chitinase 3 Like 1. YKL-40 actually lacks chitinase activity and is secreted by activated macrophages, chondrocytes, neutrophils and synovial cells, and plays a role in inflammation and tissue remodeling [285–288]. The first indication that YKL-40 may be relevant for AD came from an aprioristic evaluation of potential novel AD biomarkers [289]. Since then, several studies have now looked at the both CSF and plasma levels of this protein. YKL-40 is elevated in AD vs. control subjects [106,289,290], although it was also significantly elevated in patients with FTD [291]. Similarly, in MCI, baseline higher levels of YKL-40 were associated with significant more risk to progress to AD [106,289] with a tendency to continue to raise with disease progression [182]. The CSF ratio of YKL-40/Aβ42 was shown to be predictive of the onset of cognitive symptoms [289]. Plasma levels of YKL-40 have shown similar trends, but further validation may be needed to confirm reliable use [292]. In all, YKL-40, with its trend toward increase with disease progression, can be an associated marker for progression and prognosis.

4.3. Other Biomarkers of Inflammation and Glial Activation

Several pro- and anti-inflammatory factors have been looked at in both CSF and blood, including IL1, IL6, TNFα, VILIP-1, sTNFR and IP-10 [293–295], with some correlation to AD and MCI. Their significant is still unclear, and some are now part of multiplex panes, used in association with specific AD markers, arguably as prognostic indicators of disease progression.

GFAP (glial fibrillary acidic protein) is one of the major intermediate filament proteins of mature astrocytes, and it is released by astrocytes during neurodegeneration [296]. It is not properly a marker of inflammation, but it is a sign of a reactive brain milieu, as astrocytes tend to proliferate in response to tissue injury, and eventually release large amounts of GFAP in the interstitial fluid. Recently, the astrocytic response to brain damage and the inflammatory microglia has been characterized [297–299], and is very relevant in the context of AD pathology. The currently available CSF GFAP data show a correlation between increased levels and Aβ and tau evidence of AD [115,300], and recent data on plasma GFAP showed increased concentrations in late and early onset AD patients compared with cognitively normal controls [301]. Given the non-specific nature of GFAP, its use may be in the future confined to monitoring disease onset and progression, and possible response to treatment.

In all, given both the convergent and divergent nature of the inflammatory response, which involves a myriad of molecules and pathways, leading to a fine balance between protective and deleterious inflammation, and the fact that inflammation is an unavoidable element of ageing, any single marker would not be useful in isolation. Rather, a comprehensive approach using combinations of such factors, associated with specific marker of AD and markers of synaptic/neuronal loss (see below), has the best chance to be useful in clinical practice and research.

5. Biomarkers of Other Associated Pathology

5.1. TAR DNA-Binding Protein 43 (TDP-43)

TDP-43 can be present as co-pathology in AD (up to 50% of cases), and is the main feature of other neurodegenerative diseases, such as FTD and ALS. Its role is that of a transcriptional repressor, which regulates genes' alternate splicing. An extensive study [302] was able to link the presence of TDP-43 in AD brains with the likelihood of clinical AD presentation; but to date, a reliable CSF or blood marker is not available, although notable attempts and data are present for ALS [303,304]. Part of the issue is the non-specificity of TDP-43 accumulation, as the concept of a spectrum of TDP-43 neurodegenerative presentation is becoming more evident [305]. In this context, TDP-43 may be useful as a biomarker in the future to predict course of disease or variants of AD presentation.

5.2. Alpha-Synuclein

Alpha-synuclein is the key component of Lewy bodies, and is classically found in Parkinson's pathology, LBD, multi-system atrophy (MSA) and AD (also 50% of cases) variants. It is involved in presynaptic signaling, inhibiting phospholipases and moderating vesicles transport at the terminals [306]. As for TDP-43, despite the feasible measurement of α-synuclein in CSF and blood, no significant correlation has been made to date, likely due to the non-specific nature. What is different here is the possibility to harness α-synuclein's specific tendency to aggregate in a "prion-like" fashion [307,308], for example by CSF RT-QuIC (real-time quaking-induced conversion), likely in virtue of its conformational changes [309]. Further studies will have to determine the sensitivity of the test for AD, as being able to detect synucleinopathy in AD cases would be highly significant for prognostic, therapeutic and research purposes.

6. Vascular Damage and the Blood Brain Barrier

Vascular damage is an increasingly recognized pathological component of AD, both as an independent consequence of the toxicity of Aβ and tau [310,311], and as a frequent co-morbidity [312,313] with microvascular disease. The diagnosis if "mixed dementia" is ever more frequent, as we scan more patients in our clinical practice. Although the real significance of the interplay between amyloid/tau pathology and the microvascular damage is still unclear, as the basic pathological mechanisms are quite complex, we have been able to detect vascular damage and blood-brain barrier (BBB) leak [314,315] in vivo. Modern magnetic resonance imaging (MRI) techniques, such as dynamic contrast-enhanced (DCE)-MRI, can show BBB permeability in specific regions, i.e., the hippocampus [315,316]. This abnormal permeability seems to be proportional to ageing [317], and is more frequent in MCI and AD patients compared to unimpaired control subjects [315,318,319]. An interesting correlate of MRI studies is the use of a marker of brain capillary damage, the soluble platelet-derived growth factor receptor-β (sPDGFRβ), which is highly expressed in brain capillary pericytes and vascular smooth muscle cells [315], and is shed in the CSF after brain injury, possibly in relation of the activation of proteases such as ADAM10 [320]. Emerging data show that ageing and cognitive decline, independently of AD pathology, are associated with increased CSF levels of sPDGFRβ, which correlate with MRI permeability [317]. Although the use of BBB permeability diagnostics is still in its infancy, and the specificity for AD pathology is unclear, it would certainly be a precious prognostic tool in the hands of clinicians. In this context, and somehow beyond the scope of the current manuscript, the use of gadolinium-free MRI techniques to assess BBB permeability (K-trans) would be extremely welcome.

7. Multi-Target Platforms

With so many biomarkers coming onto the scene, and in order to advance both the diagnostic and prognostic assessments, a new generation of assays is under development with the goal to simultaneously catch multiple targets within the same fluid sample. Using such types of assays will be extremely informative in order to identify AD co-pathologies, characterize disease progression and monitor treatment response. Table 2 contains a list of the more popular, commercially available, multi-target platforms.

Table 2. Multi-target platforms.

Type of Assay	Platform/ Manufacturer	Targets	Fluid Sample	LOD pg/mL	References (PMID/link)
SiMoA	Simoa® Neurology 3-Plex A Kit HD-1 platform (Quanterix)	t-tau Aβ40 Aβ42	CSF plasma	Tau: 0.019 Aβ40: 0.196 Aβ42: 0.045	Quanterix.com. Available online: https://www.quanterix.com/sites/default/files/assays/Simoa_N3PA_Data_Sheet_HD-1_HD-X_Rev04%20%281%29.pdf. Accessed on 29 July 2020
	Simoa® Neurology 4-Plex A Kit HD-1 platform (Quanterix)	t-tau NfL GFAP UCLH-1	CSF plasma	Tau: 0.024 NfL: 0.104 GFAP: 0.221 UCLH-1: 1.74	Quanterix.com. Available online: https://www.quanterix.com/sites/default/files/assays/Simoa_N4PA_Data_Sheet_HD-1_HD-X_DS-0074_rev7.pdf. Accessed on 29 July 2020
	Simoa® 4X-plex neurology Kit (Quanterix) NEW	Aβ40 Aβ42 GFAP NfL	CSF plasma	Aβ40: 0.084 Aβ42: 0.148 GFAP: 1.66 NfL: 1.157	Alzheimer Association International Conference, 2020. Available online: https://alz.confex.com/alz/20amsterdam/meetingapp.cgi/Paper/43506. Accessed on 29 July 2020.
Luminex®xMAP®	The MILLIPLEX® map 4X multiplex immunoassay kit (Millipore, Sigma)	t-tau p-T181 Aβ40 Aβ42	CSF	N.A.	Emdmillipore.com. Available online: https://www.emdmillipore.com/US/en/product/MILLIPLEX-MAP-Human-Amyloid-Beta-and-Tau-Magnetic-Bead-Panel-Multiplex-Assay,MM_NF-HNABTMAG-68K. Accessed on 29 July 2020.
NeuroToolKit	Roche's Elecsys electroluminescence immunoassay platform	t-tau, p-T181 Aβ40, Aβ42 α-synuclein S100b YKL-40 GFAP sTREM2 IL-6 NfL Neurogranin	CSF	N.A.	32573951

Listed are some of the assays available for the simultaneous detection of multiple biomarkers in CSF or plasma. Type of assays and platforms: SiMoA: Single-Molecule enzyme-linked immunosorbent assay; Luminex XMAP: bead-based multiplex immunoassay in a microplate format; NeuroToolKit: based on electroluminescence immunoassay platform. Targets and Fluid samples are specified. Limit of detection (LOD) of the assay is provided when available. References are listed as PMID or hyperlink.

The NeuroToolKit, in development with Roche's Elecsys electroluminescence immunoassay platform [300], includes (1) α-synuclein as a marker of synaptic dysfunction, (2) S100b, YKL-40, and glial fibrillary acidic protein (GFAP) as markers of astrocyte activation, (3) sTREM2 and IL-6 as markers of microglial activation and inflammation, and (4) NfL and neurogranin as markers of axonal injury and synaptic dysfunction.

Simultaneous measurement of Aβ42, t-tau and p-T181 in CSF has been previously developed by the xMAP technology [321]. The MILLIPLEX® map multiplex immunoassay kit is a 4Xplex panel enables the simultaneous measurement of Aβ1-40, Aβ1-42, t-tau and p-T181 in human CSF.

The NEUROLOGY 3-PLEX A (SiMoA, Quanterix) detects t-tau, Aβ1-40 and Aβ1-42 in both CSF and blood.

The NEUROLOGY 4-PLEX A (SiMoA, Quanterix) detects t-tau, NfL, GFAP and ubiquitin carboxyl-terminal hydrolase L1 (UCH-L1) both in CSF and blood.

A new 4-PLEX assay has been recently developed (SiMoA, Quanterix) with the ability to target Aβ1-40, Aβ1-4, GFAP and NfL in both CSF and blood (Alzheimer Association International Conference, 2020. Available online: https://alz.confex.com/alz/20amsterdam/meetingapp.cgi/Paper/43506. Accessed on 29 July 2020), and was validated against the parent assays. We await studies on larger number of patients, with diagnostic and prognostic implications.

Using a single platform with the ability to test multiple targets, such as in this case, would be invaluable in clinical practice, given how precious CSF and even blood samples are. Ideal kits should include markers specific of disease (e.g., Aβ42/40 ratio, p-T217 tau) and biomarkers of severity of disease, as NfL, sTREM2 etc.

8. Immunoassays, Ultrasensitive Platforms and Their Potential Limitations

One main concern in the development of an immunological assay is to exclude any factor that could potentially (1) interfere with the detection of the analyte and (2) alter the antibody binding. The first category includes pre-analytical factors (sample storage, anti-coagulants), anti-analyte antibodies, and hormone binding proteins that can change the measurable concentration of the analyte in the sample; the second group consists instead of antibodies such as heterophile antibodies, human anti-animal antibodies (HAAA), and human anti-mouse antibodies (HAMA) which are found in the biological fluids and are able to interfere with the antibody binding in the assay [322]. Unlike plasma (~7 g protein/100 mL) the CSF has only ~0.025 g protein/100 mL—mainly albumin [323]. Given the different composition of CSF and blood, it is relatively easy to anticipate the different susceptibility to assay interference of these two different biological fluids. While the CSF is a relatively "clean" fluid with respect to protein concentration, plasma contains heterophilic antibodies and other molecules that might interfere with the measurements of a biomarker of interest [245]. The presence of HAMA in human plasma has been described in 30% of the population, and consequent interference by human anti-globulin antibodies in immunoassay has been reported in several studies [324–328]. For these reasons, detecting biomarkers for CNS disease in blood not only calls for highly sensitive and specific assays, but require to efficiently remove all the potential confounders. Different methods for the reduction of heterophile antibody and anti-animal interferences in immunoassays have been previously described and include ways to remove or block the interfering antibody [322,329,330]. In this respect, our group has previously developed a protocol combining NaOAc (sodium acetate) and heat-treatment to 'purify' tau from the blood of transgenic animals undergone tau immunotherapy [328]. This protocol was validated using mouse and human serum. While the NaOAc/heat-treatment did not affect tau reactivity on CSF samples, human plasma samples displayed an impressive reduction of the apparent tau signals after undergoing the heat extraction protocol. The use of the ultrasensitive assays described here has overcome some of these technical issues. In addition, the use of blockers against heterophilic antibodies in the sample diluent is also helpful to achieve reliable quantification [245]. We still believe that the previous considerations need to be considered, especially when analyzing human samples derived from patients undergone immunotherapy, where the risk of assay interference could compromise the interpretations of the disease course and therapy.

Ageing is per se associated with a reduction of the integrity of the BBB, due to a multitude of causes, e.g., prior infections, general inflammatory state, trauma, tumors and more [315,316]. MCI, AD and cognitive changes in ageing have been associated with non-specific BBB permeability in certain brain regions [317]. The resulting permeability may result in falsification of CSF data for reasons analogous to the blood.

9. Novel Approaches to Unbiased Biomarker Discovery: The Case for Metabolomics

A successful preclinical diagnosis of AD may benefit from an unbiased metabolomic analysis of biological fluids. Given that CSF metabolites reflect brain changes better than blood, CSF biomarkers are widely used in research and clinical practice [331]. The strength of metabolomics is in its ability to measure a plethora of metabolites in an unbiased manner, providing a snapshot of an individual's biological status [331,332]. The ideal system would include a CSF metabolomic profile covering the whole lifespan of individuals belonging to both FAD and SAD cohorts, followed by plasma validation. While this analysis would require decades in human AD, or the longitudinal analysis of autosomal dominant families, running such studies on faithful animal models would make for an optimal surrogate, taking advantage of novel knock-in models, as opposed to transgenic ones, which reproduce the exact genetic defect found in humans [333–338].

Metabolomic studies have been performed in CSF, blood and post-mortem tissue from AD and MCI patients [331,339,340]. For example, elevated CSF glutamate and glutamine levels have been described in AD before [332,341,342], and other studies have shown blood and CSF metabolites changes in AD [343–345], some correlating with progression and risk of MCI to progress to AD [346]. Most identified pathways relate to energy metabolism, Krebs cycle, mitochondrial function, neurotransmitter and amino acids metabolism, and lipid biosynthesis. Longitudinal studies in FAD cases would be welcome in the near future to understand both the pathological relevance of the said pathways and their power to predict AD.

10. Discussion and Conclusions

Several studies have by now assessed the use of isolated biomarkers vs. their combinations to predict diagnosis of AD, or progression of MCI to AD. No single biomarker, in CSF, blood, imaging or cognitive alone is capable to predict onset and course reliably. Rather, the combination of multimodal biomarkers is required to reach reliable predictions [100,300,342,347,348]. CSF Aβ1-42 reduction (best proxied by the Aβ42/40 ratio) starts very early before symptoms onset, peaks a few years later when amyloid PET is at the zenith; CSF tau elevation follows Aβ, and represent clinical disease severity very well. NfL increase starts early, before tau, and is a very good indicator of neuronal damage early in the disease, though it is not specific for AD; still, its reliability in blood as well in CSF, makes it an excellent marker to track disease progression and possible response to therapies. Similarly, sTREM2 and YKL-40 are excellent markers of inflammation, and although their best use is in CSF, they could be extremely useful to monitor possible adverse events of therapeutic intervention. Aprioristic analysis of CSF and blood by metabolomic, proteomic, lipidomic and other large-scale approaches could be the key, if done early enough in the pathogenic course, to generate hypothesis on the primordial triggers of the disease, besides to identify AD patients and predict their trajectory. In this respect, autosomal dominant families and patients with Down's syndrome [134,349,350] may be the ideal subjects for dedicated longitudinal studies.

Given this premise, the biomarker field has to drive every resources toward the development of blood-based assays able to provide high specificity and sensitivity to excel in both diagnosis (Aβ, and tau) and prognosis (i.e., NfL) of AD. Implementing AD biomarkers in clinical practice will require blood-based assays (1) easy-to-use in terms of human specimen collection (few mL of blood) and methodology, (2) with high sensitivity (under the pg/mL threshold) and specificity, (3) free of interference produced by several confounders associated with the ageing process. The focus on pre-symptomatic Aβ+ populations is paramount (e.g., autosomal-dominant cohorts) if we really want to diagnose the disease in a phase where significant intervention is still possible.

Author Contributions: Conceptualization, L.D., L.G., C.d.; methodology, C.d., L.G.; writing—original draft preparation, C.d., L.G.; writing—review and editing, L.D., L.G.; visualization, C.d.; supervision, L.D. All authors have read and agreed to the published version of the manuscript.

Funding: This research received no external funding.

Acknowledgments: We would like to thank Peter Davies for his continued support and insight.

Conflicts of Interest: The authors declare no conflict of interest.

References

1. Apostolova, L.G. Alzheimer disease. *Continuum Minneap. Minn.* **2016**, *22*, 419–434. [CrossRef] [PubMed]
2. Hebert, L.E.; Weuve, J.; Scherr, P.A.; Evans, D.A. Alzheimer disease in the United States (2010–2050) estimated using the 2010 census. *Neurology* **2013**, *80*, 1778–1783. [CrossRef] [PubMed]
3. Hebert, L.E.; Bienias, J.L.; Aggarwal, N.T.; Wilson, R.S.; Bennett, D.A.; Shah, R.C.; Evans, D.A. Change in risk of Alzheimer disease over time. *Neurology* **2010**, *75*, 786–791. [CrossRef] [PubMed]
4. Fan, C.C.; Banks, S.J.; Thompson, W.K.; Chen, C.H.; McEvoy, L.K.; Tan, C.H.; Kukull, W.; Bennett, D.A.; Farrer, L.A.; Mayeux, R.; et al. Sex-dependent autosomal effects on clinical progression of alzheimer's disease. *Brain* **2020**, *143*, 2272–2280. [CrossRef]
5. Plassman, B.L.; Langa, K.M.; Fisher, G.G.; Heeringa, S.G.; Weir, D.R.; Ofstedal, M.B.; Burke, J.R.; Hurd, M.D.; Potter, G.G.; Rodgers, W.L.; et al. Prevalence of dementia in the united states: The aging, demographics, and memory study. *Neuroepidemiology* **2007**, *29*, 125–132. [CrossRef]
6. Seshadri, S.; Wolf, P.A.; Beiser, A.; Au, R.; McNulty, K.; White, R.; D'Agostino, R.B. Lifetime risk of dementia and alzheimer's disease. The impact of mortality on risk estimates in the framingham study. *Neurology* **1997**, *49*, 1498–1504. [CrossRef]
7. Hebert, L.E.; Scherr, P.A.; McCann, J.J.; Beckett, L.A.; Evans, D.A. Is the risk of developing alzheimer's disease greater for women than for men? *Am. J. Epidemiol.* **2001**, *153*, 132–136. [CrossRef]
8. Chêne, G.; Beiser, A.; Au, R.; Preis, S.R.; Wolf, P.A.; Dufouil, C.; Seshadri, S. Gender and incidence of dementia in the framingham heart study from mid-adult life. *Alzheimer's Dement.* **2015**, *11*, 310–320. [CrossRef]
9. Carter, C.L.; Resnick, E.M.; Mallampalli, M.; Kalbarczyk, A. Sex and gender differences in alzheimer's disease: Recommendations for future research. *J. Womens Health* **2012**, *21*, 1018–1023. [CrossRef]
10. Altmann, A.; Tian, L.; Henderson, V.W.; Greicius, M.D.; Investigators, A.s.D.N.I. Sex modifies the apoe-related risk of developing alzheimer disease. *Ann. Neurol.* **2014**, *75*, 563–573. [CrossRef]
11. Ungar, L.; Altmann, A.; Greicius, M.D. Apolipoprotein E, gender, and alzheimer's disease: An overlooked, but potent and promising interaction. *Brain Imaging Behav.* **2014**, *8*, 262–273. [CrossRef] [PubMed]
12. Yaffe, K.; Haan, M.; Byers, A.; Tangen, C.; Kuller, L. Estrogen use, apoe, and cognitive decline: Evidence of gene-environment interaction. *Neurology* **2000**, *54*, 1949–1954. [CrossRef] [PubMed]
13. Kang, J.H.; Grodstein, F. Postmenopausal hormone therapy, timing of initiation, apoe and cognitive decline. *Neurobiol. Aging* **2012**, *33*, 1129–1137. [CrossRef] [PubMed]
14. Barrett-Connor, E.; Laughlin, G.A. Endogenous and exogenous estrogen, cognitive function, and dementia in postmenopausal women: Evidence from epidemiologic studies and clinical trials. *Semin. Reprod. Med.* **2009**, *27*, 275–282. [CrossRef]
15. O'Brien, J.; Jackson, J.W.; Grodstein, F.; Blacker, D.; Weuve, J. Postmenopausal hormone therapy is not associated with risk of all-cause dementia and alzheimer's disease. *Epidemiol. Rev.* **2014**, *36*, 83–103. [CrossRef]
16. Xu, J.; Murphy, S.L.; Kochanek, K.D.; Bastian, B.A. Deaths: Final data for 2013. *Natl. Vital Stat. Rep.* **2016**, *64*, 1–119.
17. Murray, C.J.; Atkinson, C.; Bhalla, K.; Birbeck, G.; Burstein, R.; Chou, D.; Dellavalle, R.; Danaei, G.; Ezzati, M.; Fahimi, A.; et al. The state of us health, 1990–2010: Burden of diseases, injuries, and risk factors. *JAMA* **2013**, *310*, 591–608. [CrossRef]
18. Gaugler, J.E.; Kane, R.L.; Kane, R.A. Family care for older adults with disabilities: Toward more targeted and interpretable research. *Int. J. Aging Hum. Dev.* **2002**, *54*, 205–231. [CrossRef]
19. Schulz, R.; Czaja, S.J.; Lustig, A.; Zdaniuk, B.; Martire, L.M.; Perdomo, D. Improving the quality of life of caregivers of persons with spinal cord injury: A randomized controlled trial. *Rehabil. Psychol.* **2009**, *54*, 1. [CrossRef]
20. Kasper, J.D.; Freedman, V.A.; Spillman, B.C.; Wolff, J.L. The disproportionate impact of dementia on family and unpaid caregiving to older adults. *Health Aff. Millwood* **2015**, *34*, 1642–1649. [CrossRef]
21. Freedman, V.A.; Spillman, B.C. Disability and care needs among older americans. *Milbank Q.* **2014**, *92*, 509–541. [CrossRef] [PubMed]

22. Anderson, L.A.; Edwards, V.J.; Pearson, W.S.; Talley, R.C.; McGuire, L.C.; Andresen, E.M. Adult caregivers in the united states: Characteristics and differences in well-being, by caregiver age and caregiving status. *Prev. Chronic Dis.* **2013**, *10*, E135. [CrossRef] [PubMed]
23. Hurd, M.D.; Martorell, P.; Delavande, A.; Mullen, K.J.; Langa, K.M. Monetary costs of dementia in the united states. *N. Engl. J. Med.* **2013**, *368*, 1326–1334. [CrossRef] [PubMed]
24. Consensus recommendations for the postmortem diagnosis of alzheimer's disease. The national institute on aging, and reagan institute working group on diagnostic criteria for the neuropathological assessment of alzheimer's disease. *Neurobiol. Aging* **1997**, *18*, S1–S2.
25. Braak, H.; Braak, E. Neuropathological stageing of alzheimer-related changes. *Acta Neuropathol.* **1991**, *82*, 239–259. [CrossRef] [PubMed]
26. Hyman, B.T.; Phelps, C.H.; Beach, T.G.; Bigio, E.H.; Cairns, N.J.; Carrillo, M.C.; Dickson, D.W.; Duyckaerts, C.; Frosch, M.P.; Masliah, E.; et al. National institute on aging-alzheimer's association guidelines for the neuropathologic assessment of alzheimer's disease. *Alzheimer's Dement.* **2012**, *8*, 1–13. [CrossRef] [PubMed]
27. Montine, T.J.; Phelps, C.H.; Beach, T.G.; Bigio, E.H.; Cairns, N.J.; Dickson, D.W.; Duyckaerts, C.; Frosch, M.P.; Masliah, E.; Mirra, S.S.; et al. National institute on aging-alzheimer's association guidelines for the neuropathologic assessment of alzheimer's disease: A practical approach. *Acta Neuropathol.* **2012**, *123*, 1–13. [CrossRef]
28. Holtzman, D.M.; Mandelkow, E.; Selkoe, D.J. Alzheimer disease in 2020. *Cold Spring Harb. Perspect. Med.* **2012**, *2*, a011585. [CrossRef]
29. Braak, H.; Braak, E. Evolution of the neuropathology of alzheimer's disease. *Acta Neurol. Scand.* **1996**, *165*, 3–12. [CrossRef]
30. Sengupta, U.; Nilson, A.N.; Kayed, R. The role of amyloid-β oligomers in toxicity, propagation, and immunotherapy. *EBioMedicine* **2016**, *6*, 42–49. [CrossRef]
31. Goedert, M.; Spillantini, M.G. Ordered assembly of tau protein and neurodegeneration. *Adv. Exp. Med. Biol.* **2019**, *1184*, 3–21. [PubMed]
32. Walsh, D.M.; Klyubin, I.; Fadeeva, J.V.; Rowan, M.J.; Selkoe, D.J. Amyloid-beta oligomers: Their production, toxicity and therapeutic inhibition. *Biochem. Soc. Trans.* **2002**, *30*, 552–557. [CrossRef] [PubMed]
33. Goedert, M.; Eisenberg, D.S.; Crowther, R.A. Propagation of tau aggregates and neurodegeneration. *Annu. Rev. Neurosci.* **2017**, *40*, 189–210. [CrossRef] [PubMed]
34. Ittner, A.; Ittner, L.M. Dendritic tau in alzheimer's disease. *Neuron* **2018**, *99*, 13–27. [CrossRef] [PubMed]
35. Stancu, I.C.; Vasconcelos, B.; Terwel, D.; Dewachter, I. Models of β-amyloid induced tau-pathology: The long and "folded" road to understand the mechanism. *Mol. Neurodegener.* **2014**, *9*, 51. [CrossRef]
36. van der Kant, R.; Goldstein, L.S.B.; Ossenkoppele, R. Amyloid-β-independent regulators of tau pathology in alzheimer disease. *Nat. Rev. Neurosci.* **2020**, *21*, 21–35. [CrossRef]
37. La Joie, R.; Visani, A.V.; Baker, S.L.; Brown, J.A.; Bourakova, V.; Cha, J.; Chaudhary, K.; Edwards, L.; Iaccarino, L.; Janabi, M.; et al. Prospective longitudinal atrophy in alzheimer's disease correlates with the intensity and topography of baseline tau-pet. *Sci. Transl. Med.* **2020**, *12*, 5732–5745. [CrossRef]
38. Franzmeier, N.; Neitzel, J.; Rubinski, A.; Smith, R.; Strandberg, O.; Ossenkoppele, R.; Hansson, O.; Ewers, M.; (ADNI), A.s.D.N.I. Functional brain architecture is associated with the rate of tau accumulation in alzheimer's disease. *Nat. Commun.* **2020**, *11*, 347. [CrossRef]
39. DeTure, M.A.; Dickson, D.W. The neuropathological diagnosis of alzheimer's disease. *Mol. Neurodegener.* **2019**, *14*, 32. [CrossRef]
40. Serrano-Pozo, A.; Mielke, M.L.; Gómez-Isla, T.; Betensky, R.A.; Growdon, J.H.; Frosch, M.P.; Hyman, B.T. Reactive glia not only associates with plaques but also parallels tangles in alzheimer's disease. *Am. J. Pathol.* **2011**, *179*, 1373–1384. [CrossRef]
41. Wake, H.; Moorhouse, A.J.; Jinno, S.; Kohsaka, S.; Nabekura, J. Resting microglia directly monitor the functional state of synapses in vivo and determine the fate of ischemic terminals. *J. Neurosci.* **2009**, *29*, 3974–3980. [CrossRef] [PubMed]
42. Vehmas, A.K.; Kawas, C.H.; Stewart, W.F.; Troncoso, J.C. Immune reactive cells in senile plaques and cognitive decline in alzheimer's disease. *Neurobiol. Aging* **2003**, *24*, 321–331. [CrossRef]
43. Serrano-Pozo, A.; Frosch, M.P.; Masliah, E.; Hyman, B.T. Neuropathological alterations in alzheimer disease. *Cold Spring Harb. Perspect. Med.* **2011**, *1*, a006189. [CrossRef] [PubMed]
44. Perl, D.P. Neuropathology of alzheimer's disease. *Mt. Sinai J. Med.* **2010**, *77*, 32–42. [CrossRef] [PubMed]

45. Thal, D.R.; Schultz, C.; Botez, G.; Del Tredici, K.; Mrak, R.E.; Griffin, W.S.; Wiestler, O.D.; Braak, H.; Ghebremedhin, E. The impact of argyrophilic grain disease on the development of dementia and its relationship to concurrent alzheimer's disease-related pathology. *Neuropathol. Appl. Neurobiol.* **2005**, *31*, 270–279. [CrossRef] [PubMed]
46. Hansen, L.; Salmon, D.; Galasko, D.; Masliah, E.; Katzman, R.; DeTeresa, R.; Thal, L.; Pay, M.M.; Hofstetter, R.; Klauber, M. The lewy body variant of alzheimer's disease: A clinical and pathologic entity. *Neurology* **1990**, *40*, 1. [CrossRef]
47. Nelson, P.T.; Gal, Z.; Wang, W.X.; Niedowicz, D.M.; Artiushin, S.C.; Wycoff, S.; Wei, A.; Jicha, G.A.; Fardo, D.W. Tdp-43 proteinopathy in aging: Associations with risk-associated gene variants and with brain parenchymal thyroid hormone levels. *Neurobiol. Dis.* **2019**, *125*, 67–76. [CrossRef]
48. Josephs, K.A.; Murray, M.E.; Whitwell, J.L.; Parisi, J.E.; Petrucelli, L.; Jack, C.R.; Petersen, R.C.; Dickson, D.W. Staging tdp-43 pathology in alzheimer's disease. *Acta Neuropathol.* **2014**, *127*, 441–450. [CrossRef]
49. Prasad, A.; Bharathi, V.; Sivalingam, V.; Girdhar, A.; Patel, B.K. Molecular mechanisms of tdp-43 misfolding and pathology in amyotrophic lateral sclerosis. *Front. Mol. Neurosci.* **2019**, *12*, 25. [CrossRef]
50. Gómez-Isla, T.; Hollister, R.; West, H.; Mui, S.; Growdon, J.H.; Petersen, R.C.; Parisi, J.E.; Hyman, B.T. Neuronal loss correlates with but exceeds neurofibrillary tangles in alzheimer's disease. *Ann. Neurol.* **1997**, *41*, 17–24. [CrossRef]
51. Forner, S.; Baglietto-Vargas, D.; Martini, A.C.; Trujillo-Estrada, L.; LaFerla, F.M. Synaptic impairment in alzheimer's disease: A dysregulated symphony. *Trends Neurosci.* **2017**, *40*, 347–357. [CrossRef] [PubMed]
52. Overk, C.R.; Masliah, E. Pathogenesis of synaptic degeneration in alzheimer's disease and lewy body disease. *Biochem. Pharmacol.* **2014**, *88*, 508–516. [CrossRef] [PubMed]
53. Sabbagh, M.N.; Lue, L.F.; Fayard, D.; Shi, J. Increasing precision of clinical diagnosis of alzheimer's disease using a combined algorithm incorporating clinical and novel biomarker data. *Neurol. Ther.* **2017**, *6*, 83–95. [CrossRef]
54. Dubois, B.; Feldman, H.H.; Jacova, C.; Hampel, H.; Molinuevo, J.L.; Blennow, K.; DeKosky, S.T.; Gauthier, S.; Selkoe, D.; Bateman, R.; et al. Advancing research diagnostic criteria for alzheimer's disease: The iwg-2 criteria. *Lancet Neurol.* **2014**, *13*, 614–629. [CrossRef]
55. Boise, L.; Camicioli, R.; Morgan, D.L.; Rose, J.H.; Congleton, L. Diagnosing dementia: Perspectives of primary care physicians. *Gerontologist* **1999**, *39*, 457–464. [CrossRef] [PubMed]
56. Beach, T.G.; Monsell, S.E.; Phillips, L.E.; Kukull, W. Accuracy of the clinical diagnosis of alzheimer disease at national institute on aging alzheimer disease centers, 2005–2010. *J. Neuropathol. Exp. Neurol.* **2012**, *71*, 266–273. [CrossRef]
57. Cacace, R.; Sleegers, K.; Van Broeckhoven, C. Molecular genetics of early-onset alzheimer's disease revisited. *Alzheimer's Dement.* **2016**, *12*, 733–748. [CrossRef]
58. Kimberly, W.T.; LaVoie, M.J.; Ostaszewski, B.L.; Ye, W.; Wolfe, M.S.; Selkoe, D.J. Gamma-secretase is a membrane protein complex comprised of presenilin, nicastrin, aph-1, and pen-2. *Proc. Natl. Acad. Sci. USA* **2003**, *100*, 6382–6387. [CrossRef]
59. Van Broeckhoven, C. Presenilins and alzheimer disease. *Nat. Genet.* **1995**, *11*, 230–232. [CrossRef]
60. Rademakers, R.; Cruts, M.; Van Broeckhoven, C. Genetics of early-onset alzheimer dementia. *Sci. World J.* **2003**, *3*, 497–519. [CrossRef]
61. Sherrington, R.; Rogaev, E.I.; Liang, Y.; Rogaeva, E.A.; Levesque, G.; Ikeda, M.; Chi, H.; Lin, C.; Li, G.; Holman, K.; et al. Cloning of a gene bearing missense mutations in early-onset familial alzheimer's disease. *Nature* **1995**, *375*, 754–760. [CrossRef]
62. Levy-Lahad, E.; Wasco, W.; Poorkaj, P.; Romano, D.M.; Oshima, J.; Pettingell, W.H.; Yu, C.E.; Jondro, P.D.; Schmidt, S.D.; Wang, K. Candidate gene for the chromosome 1 familial alzheimer's disease locus. *Science* **1995**, *269*, 973–977. [CrossRef] [PubMed]
63. Welzel, A.T.; Maggio, J.E.; Shankar, G.M.; Walker, D.E.; Ostaszewski, B.L.; Li, S.; Klyubin, I.; Rowan, M.J.; Seubert, P.; Walsh, D.M.; et al. Secreted amyloid β-proteins in a cell culture model include n-terminally extended peptides that impair synaptic plasticity. *Biochemistry* **2014**, *53*, 3908–3921. [CrossRef] [PubMed]
64. Riedel, B.C.; Thompson, P.M.; Brinton, R.D. Age, apoe and sex: Triad of risk of alzheimer's disease. *J. Steroid Biochem. Mol. Biol.* **2016**, *160*, 134–147. [CrossRef] [PubMed]
65. Yamazaki, Y.; Zhao, N.; Caulfield, T.R.; Liu, C.C.; Bu, G. Apolipoprotein e and alzheimer disease: Pathobiology and targeting strategies. *Nat. Rev. Neurol.* **2019**, *15*, 501–518. [CrossRef] [PubMed]

66. Gratuze, M.; Leyns, C.E.G.; Holtzman, D.M. New insights into the role of trem2 in alzheimer's disease. *Mol. Neurodegener.* **2018**, *13*, 66. [CrossRef] [PubMed]
67. Cai, Q.; Tammineni, P. Alterations in mitochondrial quality control in alzheimer's disease. *Front. Cell. Neurosci.* **2016**, *10*, 24. [CrossRef]
68. Devi, L.; Prabhu, B.M.; Galati, D.F.; Avadhani, N.G.; Anandatheerthavarada, H.K. Accumulation of amyloid precursor protein in the mitochondrial import channels of human alzheimer's disease brain is associated with mitochondrial dysfunction. *J. Neurosci.* **2006**, *26*, 9057–9068. [CrossRef]
69. Lamprecht, R.; LeDoux, J. Structural plasticity and memory. *Nat. Rev. Neurosci.* **2004**, *5*, 45–54. [CrossRef]
70. Kashyap, G.; Bapat, D.; Das, D.; Gowaikar, R.; Amritkar, R.E.; Rangarajan, G.; Ravindranath, V.; Ambika, G. Synapse loss and progress of alzheimer's disease—A network model. *Sci. Rep.* **2019**, *9*, 6555. [CrossRef]
71. Fabiani, C.; Antollini, S.S. Alzheimer's disease as a membrane disorder: Spatial cross-talk among beta-amyloid peptides, nicotinic acetylcholine receptors and lipid rafts. *Front. Cell. Neurosci.* **2019**, *13*, 309. [CrossRef] [PubMed]
72. Di Scala, C.; Yahi, N.; Boutemeur, S.; Flores, A.; Rodriguez, L.; Chahinian, H.; Fantini, J. Common molecular mechanism of amyloid pore formation by alzheimer's β-amyloid peptide and α-synuclein. *Sci. Rep.* **2016**, *6*, 28781. [CrossRef] [PubMed]
73. Lasagna-Reeves, C.A.; Sengupta, U.; Castillo-Carranza, D.; Gerson, J.E.; Guerrero-Munoz, M.; Troncoso, J.C.; Jackson, G.R.; Kayed, R. The formation of tau pore-like structures is prevalent and cell specific: Possible implications for the disease phenotypes. *Acta Neuropathol. Commun.* **2014**, *2*, 56. [CrossRef] [PubMed]
74. Tseng, B.P.; Green, K.N.; Chan, J.L.; Blurton-Jones, M.; LaFerla, F.M. Abeta inhibits the proteasome and enhances amyloid and tau accumulation. *Neurobiol. Aging* **2008**, *29*, 1607–1618. [CrossRef] [PubMed]
75. Myeku, N.; Clelland, C.L.; Emrani, S.; Kukushkin, N.V.; Yu, W.H.; Goldberg, A.L.; Duff, K.E. Tau-driven 26s proteasome impairment and cognitive dysfunction can be prevented early in disease by activating camp-pka signaling. *Nat. Med.* **2016**, *22*, 46–53. [CrossRef]
76. Yamazaki, Y.; Kanekiyo, T. Blood-brain barrier dysfunction and the pathogenesis of alzheimer's disease. *Int. J. Mol. Sci.* **2017**, *18*, 1965. [CrossRef]
77. Venegas, C.; Kumar, S.; Franklin, B.S.; Dierkes, T.; Brinkschulte, R.; Tejera, D.; Vieira-Saecker, A.; Schwartz, S.; Santarelli, F.; Kummer, M.P.; et al. Microglia-derived asc specks cross-seed amyloid-β in alzheimer's disease. *Nature* **2017**, *552*, 355–361. [CrossRef]
78. Perea, J.R.; Llorens-Martín, M.; Ávila, J.; Bolós, M. The role of microglia in the spread of tau: Relevance for tauopathies. *Front. Cell. Neurosci.* **2018**, *12*, 172. [CrossRef]
79. Heneka, M.T.; Carson, M.J.; El Khoury, J.; Landreth, G.E.; Brosseron, F.; Feinstein, D.L.; Jacobs, A.H.; Wyss-Coray, T.; Vitorica, J.; Ransohoff, R.M.; et al. Neuroinflammation in alzheimer's disease. *Lancet Neurol.* **2015**, *14*, 388–405. [CrossRef]
80. Jack, C.R.; Bennett, D.A.; Blennow, K.; Carrillo, M.C.; Feldman, H.H.; Frisoni, G.B.; Hampel, H.; Jagust, W.J.; Johnson, K.A.; Knopman, D.S.; et al. A/t/n: An unbiased descriptive classification scheme for alzheimer disease biomarkers. *Neurology* **2016**, *87*, 539–547. [CrossRef]
81. Ekman, U.; Ferreira, D.; Westman, E. The a/t/n biomarker scheme and patterns of brain atrophy assessed in mild cognitive impairment. *Sci. Rep.* **2018**, *8*, 8431. [CrossRef] [PubMed]
82. Ross, D.E.; Ochs, A.L.; Seabaugh, J.; Henshaw, T. Neuroquant®revealed hippocampal atrophy in a patient with traumatic brain injury. *J. Neuropsychiatry Clin. Neurosci.* **2012**, *24*, E33. [CrossRef] [PubMed]
83. Brewer, J.B.; Magda, S.; Airriess, C.; Smith, M.E. Fully-automated quantification of regional brain volumes for improved detection of focal atrophy in alzheimer disease. *AJNR Am. J. Neuroradiol.* **2009**, *30*, 578–580. [CrossRef] [PubMed]
84. Piccini, A.; Russo, C.; Gliozzi, A.; Relini, A.; Vitali, A.; Borghi, R.; Giliberto, L.; Armirotti, A.; D'Arrigo, C.; Bachi, A.; et al. Beta-amyloid is different in normal aging and in alzheimer disease. *J. Biol. Chem.* **2005**, *280*, 34186–34192. [CrossRef]
85. Panza, F.; Lozupone, M.; Logroscino, G.; Imbimbo, B.P. A critical appraisal of amyloid-β-targeting therapies for alzheimer disease. *Nat. Rev. Neurol.* **2019**, *15*, 73–88. [CrossRef]
86. Selkoe, D.J. Resolving controversies on the path to alzheimer's therapeutics. *Nat. Med.* **2011**, *17*, 1060–1065. [CrossRef]
87. Hane, F.T.; Robinson, M.; Lee, B.Y.; Bai, O.; Leonenko, Z.; Albert, M.S. Recent progress in alzheimer's disease research, part 3: Diagnosis and treatment. *J. Alzheimer's Dis.* **2017**, *57*, 645–665. [CrossRef]

88. Orgogozo, J.M.; Gilman, S.; Dartigues, J.F.; Laurent, B.; Puel, M.; Kirby, L.C.; Jouanny, P.; Dubois, B.; Eisner, L.; Flitman, S.; et al. Subacute meningoencephalitis in a subset of patients with ad after abeta42 immunization. *Neurology* **2003**, *61*, 46–54. [CrossRef]
89. Vellas, B.; Black, R.; Thal, L.J.; Fox, N.C.; Daniels, M.; McLennan, G.; Tompkins, C.; Leibman, C.; Pomfret, M.; Grundman, M.; et al. Long-term follow-up of patients immunized with an1792: Reduced functional decline in antibody responders. *Curr. Alzheimer Res.* **2009**, *6*, 144–151. [CrossRef]
90. Lemere, C.A.; Masliah, E. Can alzheimer disease be prevented by amyloid-beta immunotherapy? *Nat. Rev. Neurol.* **2010**, *6*, 108–119. [CrossRef]
91. Wiessner, C.; Wiederhold, K.H.; Tissot, A.C.; Frey, P.; Danner, S.; Jacobson, L.H.; Jennings, G.T.; Lüönd, R.; Ortmann, R.; Reichwald, J.; et al. The second-generation active aβ immunotherapy cad106 reduces amyloid accumulation in app transgenic mice while minimizing potential side effects. *J. Neurosci.* **2011**, *31*, 9323–9331. [CrossRef] [PubMed]
92. Doody, R.S.; Thomas, R.G.; Farlow, M.; Iwatsubo, T.; Vellas, B.; Joffe, S.; Kieburtz, K.; Raman, R.; Sun, X.; Aisen, P.S.; et al. Phase 3 trials of solanezumab for mild-to-moderate alzheimer's disease. *N. Engl. J. Med.* **2014**, *370*, 311–321. [CrossRef] [PubMed]
93. Liu-Seifert, H.; Siemers, E.; Holdridge, K.C.; Andersen, S.W.; Lipkovich, I.; Carlson, C.; Sethuraman, G.; Hoog, S.; Hayduk, R.; Doody, R.; et al. Delayed-start analysis: Mild alzheimer's disease patients in solanezumab trials, 3.5 years. *Alzheimer's Dement. N. Y.* **2015**, *1*, 111–121. [CrossRef] [PubMed]
94. Siemers, E.R.; Sundell, K.L.; Carlson, C.; Case, M.; Sethuraman, G.; Liu-Seifert, H.; Dowsett, S.A.; Pontecorvo, M.J.; Dean, R.A.; Demattos, R. Phase 3 solanezumab trials: Secondary outcomes in mild alzheimer's disease patients. *Alzheimer's Dement.* **2016**, *12*, 110–120. [CrossRef]
95. Sevigny, J.; Chiao, P.; Bussière, T.; Weinreb, P.H.; Williams, L.; Maier, M.; Dunstan, R.; Salloway, S.; Chen, T.; Ling, Y.; et al. The antibody aducanumab reduces aβ plaques in alzheimer's disease. *Nature* **2016**, *537*, 50–56. [CrossRef]
96. Novak, P.; Zilka, N.; Zilkova, M.; Kovacech, B.; Skrabana, R.; Ondrus, M.; Fialova, L.; Kontsekova, E.; Otto, M.; Novak, M. Aadvac1, an active immunotherapy for alzheimer's disease and non alzheimer tauopathies: An overview of preclinical and clinical development. *J. Prev. Alzheimer's Dis.* **2019**, *6*, 63–69.
97. Theunis, C.; Crespo-Biel, N.; Gafner, V.; Pihlgren, M.; López-Deber, M.P.; Reis, P.; Hickman, D.T.; Adolfsson, O.; Chuard, N.; Ndao, D.M.; et al. Efficacy and safety of a liposome-based vaccine against protein tau, assessed in tau.P301l mice that model tauopathy. *PLoS ONE* **2013**, *8*, e72301. [CrossRef]
98. Davies, P.; Maloney, A.J. Selective loss of central cholinergic neurons in alzheimer's disease. *Lancet* **1976**, *2*, 1403. [CrossRef]
99. Jack, C.R.; Bennett, D.A.; Blennow, K.; Carrillo, M.C.; Dunn, B.; Haeberlein, S.B.; Holtzman, D.M.; Jagust, W.; Jessen, F.; Karlawish, J.; et al. Nia-aa research framework: Toward a biological definition of alzheimer's disease. *Alzheimer's Dement.* **2018**, *14*, 535–562. [CrossRef]
100. Pedrero-Prieto, C.M.; García-Carpintero, S.; Frontiñán-Rubio, J.; Llanos-González, E.; Aguilera García, C.; Alcaín, F.J.; Lindberg, I.; Durán-Prado, M.; Peinado, J.R.; Rabanal-Ruiz, Y. A comprehensive systematic review of csf proteins and peptides that define alzheimer's disease. *Clin. Proteom.* **2020**, *17*, 21. [CrossRef]
101. Villemagne, V.L.; Doré, V.; Burnham, S.C.; Masters, C.L.; Rowe, C.C. Imaging tau and amyloid-β proteinopathies in alzheimer disease and other conditions. *Nat. Rev. Neurol.* **2018**, *14*, 225–236. [CrossRef] [PubMed]
102. Mattsson-Carlgren, N.; Andersson, E.; Janelidze, S.; Ossenkoppele, R.; Insel, P.; Strandberg, O.; Zetterberg, H.; Rosen, H.J.; Rabinovici, G.; Chai, X.; et al. Aβ deposition is associated with increases in soluble and phosphorylated tau that precede a positive tau pet in alzheimer's disease. *Sci. Adv.* **2020**, *6*, eaaz2387. [CrossRef] [PubMed]
103. Lombardi, G.; Crescioli, G.; Cavedo, E.; Lucenteforte, E.; Casazza, G.; Bellatorre, A.G.; Lista, C.; Costantino, G.; Frisoni, G.; Virgili, G.; et al. Structural magnetic resonance imaging for the early diagnosis of dementia due to alzheimer's disease in people with mild cognitive impairment. *Cochrane Database Syst. Rev.* **2020**, *3*, CD009628. [CrossRef] [PubMed]
104. Khosravi, M.; Peter, J.; Wintering, N.A.; Serruya, M.; Shamchi, S.P.; Werner, T.J.; Alavi, A.; Newberg, A.B. 18f-fdg is a superior indicator of cognitive performance compared to 18f-florbetapir in alzheimer's disease and mild cognitive impairment evaluation: A global quantitative analysis. *J. Alzheimer's Dis.* **2019**, *70*, 1197–1207. [CrossRef]

105. Zetterberg, H.; Skillbäck, T.; Mattsson, N.; Trojanowski, J.Q.; Portelius, E.; Shaw, L.M.; Weiner, M.W.; Blennow, K.; Initiative, A.s.D.N. Association of cerebrospinal fluid neurofilament light concentration with alzheimer disease progression. *JAMA Neurol.* **2016**, *73*, 60–67. [CrossRef]
106. Janelidze, S.; Hertze, J.; Zetterberg, H.; Landqvist Waldö, M.; Santillo, A.; Blennow, K.; Hansson, O. Cerebrospinal fluid neurogranin and ykl-40 as biomarkers of alzheimer's disease. *Ann. Clin. Transl. Neurol.* **2016**, *3*, 12–20. [CrossRef]
107. Jack, C.R.; Albert, M.S.; Knopman, D.S.; McKhann, G.M.; Sperling, R.A.; Carrillo, M.C.; Thies, B.; Phelps, C.H. Introduction to the recommendations from the national institute on aging-alzheimer's association workgroups on diagnostic guidelines for alzheimer's disease. *Alzheimer's Dement.* **2011**, *7*, 257–262. [CrossRef]
108. McKhann, G.M.; Knopman, D.S.; Chertkow, H.; Hyman, B.T.; Jack, C.R.; Kawas, C.H.; Klunk, W.E.; Koroshetz, W.J.; Manly, J.J.; Mayeux, R.; et al. The diagnosis of dementia due to alzheimer's disease: Recommendations from the national institute on aging-alzheimer's association workgroups on diagnostic guidelines for alzheimer's disease. *Alzheimer's Dement.* **2011**, *7*, 263–269. [CrossRef]
109. Albert, M.S.; DeKosky, S.T.; Dickson, D.; Dubois, B.; Feldman, H.H.; Fox, N.C.; Gamst, A.; Holtzman, D.M.; Jagust, W.J.; Petersen, R.C.; et al. The diagnosis of mild cognitive impairment due to alzheimer's disease: Recommendations from the national institute on aging-alzheimer's association workgroups on diagnostic guidelines for alzheimer's disease. *Alzheimer's Dement.* **2011**, *7*, 270–279. [CrossRef]
110. Sperling, R.A.; Aisen, P.S.; Beckett, L.A.; Bennett, D.A.; Craft, S.; Fagan, A.M.; Iwatsubo, T.; Jack, C.R.; Kaye, J.; Montine, T.J.; et al. Toward defining the preclinical stages of alzheimer's disease: Recommendations from the national institute on aging-alzheimer's association workgroups on diagnostic guidelines for alzheimer's disease. *Alzheimer's Dement.* **2011**, *7*, 280–292. [CrossRef]
111. Dubois, B.; Feldman, H.H.; Jacova, C.; Dekosky, S.T.; Barberger-Gateau, P.; Cummings, J.; Delacourte, A.; Galasko, D.; Gauthier, S.; Jicha, G.; et al. Research criteria for the diagnosis of alzheimer's disease: Revising the nincds-adrda criteria. *Lancet Neurol.* **2007**, *6*, 734–746. [CrossRef]
112. Dubois, B.; Hampel, H.; Feldman, H.H.; Scheltens, P.; Aisen, P.; Andrieu, S.; Bakardjian, H.; Benali, H.; Bertram, L.; Blennow, K.; et al. Preclinical alzheimer's disease: Definition, natural history, and diagnostic criteria. *Alzheimer's Dement.* **2016**, *12*, 292–323. [CrossRef] [PubMed]
113. Hampel, H.; O'Bryant, S.E.; Molinuevo, J.L.; Zetterberg, H.; Masters, C.L.; Lista, S.; Kiddle, S.J.; Batrla, R.; Blennow, K. Blood-based biomarkers for alzheimer disease: Mapping the road to the clinic. *Nat. Rev. Neurol.* **2018**, *14*, 639–652. [CrossRef] [PubMed]
114. Twohig, D.; Rodriguez-Vieitez, E.; Sando, S.B.; Berge, G.; Lauridsen, C.; Møller, I.; Grøntvedt, G.R.; Bråthen, G.; Patra, K.; Bu, G.; et al. The relevance of cerebrospinal fluid α-synuclein levels to sporadic and familial alzheimer's disease. *Acta Neuropathol. Commun.* **2018**, *6*, 130. [CrossRef] [PubMed]
115. Olsson, B.; Lautner, R.; Andreasson, U.; Öhrfelt, A.; Portelius, E.; Bjerke, M.; Hölttä, M.; Rosén, C.; Olsson, C.; Strobel, G.; et al. Csf and blood biomarkers for the diagnosis of alzheimer's disease: A systematic review and meta-analysis. *Lancet Neurol.* **2016**, *15*, 673–684. [CrossRef]
116. Li, Q.X.; Villemagne, V.L.; Doecke, J.D.; Rembach, A.; Sarros, S.; Varghese, S.; McGlade, A.; Laughton, K.M.; Pertile, K.K.; Fowler, C.J.; et al. Alzheimer's disease normative cerebrospinal fluid biomarkers validated in pet amyloid-β characterized subjects from the australian imaging, biomarkers and lifestyle (aibl) study. *J. Alzheimer's Dis.* **2015**, *48*, 175–187. [CrossRef]
117. Mattsson, N.; Palmqvist, S.; Stomrud, E.; Vogel, J.; Hansson, O. Staging β-amyloid pathology with amyloid positron emission tomography. *JAMA Neurol.* **2019**, *76*, 1319–1329. [CrossRef]
118. Fagan, A.M.; Xiong, C.; Jasielec, M.S.; Bateman, R.J.; Goate, A.M.; Benzinger, T.L.; Ghetti, B.; Martins, R.N.; Masters, C.L.; Mayeux, R.; et al. Longitudinal change in csf biomarkers in autosomal-dominant alzheimer's disease. *Sci. Transl. Med.* **2014**, *6*, 226ra230. [CrossRef]
119. Lewczuk, P.; Matzen, A.; Blennow, K.; Parnetti, L.; Molinuevo, J.L.; Eusebi, P.; Kornhuber, J.; Morris, J.C.; Fagan, A.M. Cerebrospinal fluid aβ42/40 corresponds better than aβ42 to amyloid pet in alzheimer's disease. *J. Alzheimer's Dis.* **2017**, *55*, 813–822. [CrossRef]
120. Doecke, J.D.; Rembach, A.; Villemagne, V.L.; Varghese, S.; Rainey-Smith, S.; Sarros, S.; Evered, L.A.; Fowler, C.J.; Pertile, K.K.; Rumble, R.L.; et al. Concordance between cerebrospinal fluid biomarkers with alzheimer's disease pathology between three independent assay platforms. *J. Alzheimer's Dis.* **2018**, *61*, 169–183. [CrossRef]

121. Xia, D.; Watanabe, H.; Wu, B.; Lee, S.H.; Li, Y.; Tsvetkov, E.; Bolshakov, V.Y.; Shen, J.; Kelleher, R.J. Presenilin-1 knockin mice reveal loss-of-function mechanism for familial alzheimer's disease. *Neuron* **2015**, *85*, 967–981. [CrossRef] [PubMed]
122. Selkoe, D.J.; Hardy, J. The amyloid hypothesis of alzheimer's disease at 25 years. *EMBO Mol. Med.* **2016**, *8*, 595–608. [CrossRef] [PubMed]
123. Saido, T.C.; Iwatsubo, T.; Mann, D.M.; Shimada, H.; Ihara, Y.; Kawashima, S. Dominant and differential deposition of distinct beta-amyloid peptide species, a beta n3(pe), in senile plaques. *Neuron* **1995**, *14*, 457–466. [CrossRef]
124. Portelius, E.; Bogdanovic, N.; Gustavsson, M.K.; Volkmann, I.; Brinkmalm, G.; Zetterberg, H.; Winblad, B.; Blennow, K. Mass spectrometric characterization of brain amyloid beta isoform signatures in familial and sporadic alzheimer's disease. *Acta Neuropathol.* **2010**, *120*, 185–193. [CrossRef]
125. Nussbaum, J.M.; Schilling, S.; Cynis, H.; Silva, A.; Swanson, E.; Wangsanut, T.; Tayler, K.; Wiltgen, B.; Hatami, A.; Rönicke, R.; et al. Prion-like behaviour and tau-dependent cytotoxicity of pyroglutamylated amyloid-β. *Nature* **2012**, *485*, 651–655. [CrossRef]
126. Plant, L.D.; Boyle, J.P.; Smith, I.F.; Peers, C.; Pearson, H.A. The production of amyloid beta peptide is a critical requirement for the viability of central neurons. *J. Neurosci.* **2003**, *23*, 5531–5535. [CrossRef]
127. Sjögren, M.; Gisslén, M.; Vanmechelen, E.; Blennow, K. Low cerebrospinal fluid beta-amyloid 42 in patients with acute bacterial meningitis and normalization after treatment. *Neurosci. Lett.* **2001**, *314*, 33–36. [CrossRef]
128. Emmerling, M.R.; Morganti-Kossmann, M.C.; Kossmann, T.; Stahel, P.F.; Watson, M.D.; Evans, L.M.; Mehta, P.D.; Spiegel, K.; Kuo, Y.M.; Roher, A.E.; et al. Traumatic brain injury elevates the alzheimer's amyloid peptide a beta 42 in human csf. A possible role for nerve cell injury. *Ann. N. Y. Acad. Sci.* **2000**, *903*, 118–122. [CrossRef]
129. Mondello, S.; Buki, A.; Barzo, P.; Randall, J.; Provuncher, G.; Hanlon, D.; Wilson, D.; Kobeissy, F.; Jeromin, A. Csf and plasma amyloid-beta temporal profiles and relationships with neurological status and mortality after severe traumatic brain injury. *Sci. Rep.* **2014**, *4*, 6446. [CrossRef]
130. Olsson, A.; Csajbok, L.; Ost, M.; Höglund, K.; Nylén, K.; Rosengren, L.; Nellgård, B.; Blennow, K. Marked increase of beta-amyloid(1-42) and amyloid precursor protein in ventricular cerebrospinal fluid after severe traumatic brain injury. *J. Neurol.* **2004**, *251*, 870–876. [CrossRef]
131. Hortobágyi, T.; Wise, S.; Hunt, N.; Cary, N.; Djurovic, V.; Fegan-Earl, A.; Shorrock, K.; Rouse, D.; Al-Sarraj, S. Traumatic axonal damage in the brain can be detected using beta-app immunohistochemistry within 35 min after head injury to human adults. *Neuropathol. Appl. Neurobiol.* **2007**, *33*, 226–237. [CrossRef] [PubMed]
132. Tsitsopoulos, P.P.; Marklund, N. Amyloid-beta peptides and tau protein as biomarkers in cerebrospinal and interstitial fluid following traumatic brain injury: A review of experimental and clinical studies. *Front. Neurol.* **2013**, *4*, 79. [CrossRef] [PubMed]
133. Schindler, S.E.; Li, Y.; Todd, K.W.; Herries, E.M.; Henson, R.L.; Gray, J.D.; Wang, G.; Graham, D.L.; Shaw, L.M.; Trojanowski, J.Q.; et al. Emerging cerebrospinal fluid biomarkers in autosomal dominant alzheimer's disease. *Alzheimer's Dement.* **2019**, *15*, 655–665. [CrossRef] [PubMed]
134. Henson, R.L.; Doran, E.; Christian, B.T.; Handen, B.L.; Klunk, W.E.; Lai, F.; Lee, J.H.; Rosas, H.D.; Schupf, N.; Zaman, S.H.; et al. Cerebrospinal fluid biomarkers of alzheimer's disease in a cohort of adults with down syndrome. *Alzheimer's Dement.* **2020**, *12*, e12057. [CrossRef] [PubMed]
135. Mo, J.A.; Lim, J.H.; Sul, A.R.; Lee, M.; Youn, Y.C.; Kim, H.J. Cerebrospinal fluid β-amyloid1-42 levels in the differential diagnosis of alzheimer's disease–systematic review and meta-analysis. *PLoS ONE* **2015**, *10*, e0116802. [CrossRef]
136. Englund, H.; Anneren, G.; Gustafsson, J.; Wester, U.; Wiltfang, J.; Lannfelt, L.; Blennow, K.; Höglund, K. Increase in beta-amyloid levels in cerebrospinal fluid of children with down syndrome. *Dement. Geriatr. Cogn. Disord.* **2007**, *24*, 369–374. [CrossRef] [PubMed]
137. Peskind, E.R.; Riekse, R.; Quinn, J.F.; Kaye, J.; Clark, C.M.; Farlow, M.R.; Decarli, C.; Chabal, C.; Vavrek, D.; Raskind, M.A.; et al. Safety and acceptability of the research lumbar puncture. *Alzheimer Dis. Assoc. Disord.* **2005**, *19*, 220–225. [CrossRef]
138. Palmert, M.R.; Podlisny, M.B.; Witker, D.S.; Oltersdorf, T.; Younkin, L.H.; Selkoe, D.J.; Younkin, S.G. The beta-amyloid protein precursor of alzheimer disease has soluble derivatives found in human brain and cerebrospinal fluid. *Proc. Natl. Acad. Sci. USA* **1989**, *86*, 6338–6342. [CrossRef]

139. Ghiso, J.; Tagliavini, F.; Timmers, W.F.; Frangione, B. Alzheimer's disease amyloid precursor protein is present in senile plaques and cerebrospinal fluid: Immunohistochemical and biochemical characterization. *Biochem. Biophys. Res. Commun.* **1989**, *163*, 430–437. [CrossRef]
140. Palmert, M.R.; Siedlak, S.L.; Podlisny, M.B.; Greenberg, B.; Shelton, E.R.; Chan, H.W.; Usiak, M.; Selkoe, D.J.; Perry, G.; Younkin, S.G. Soluble derivatives of the beta amyloid protein precursor of alzheimer's disease are labeled by antisera to the beta amyloid protein. *Biochem. Biophys. Res. Commun.* **1989**, *165*, 182–188. [CrossRef]
141. Chong, J.K.; Miller, B.E.; Ghanbari, H.A. Detection of amyloid beta protein precursor immunoreactivity in normal and alzheimer's disease cerebrospinal fluid. *Life Sci.* **1990**, *47*, 1163–1171. [CrossRef]
142. Van Nostrand, W.E.; Wagner, S.L.; Shankle, W.R.; Farrow, J.S.; Dick, M.; Rozemuller, J.M.; Kuiper, M.A.; Wolters, E.C.; Zimmerman, J.; Cotman, C.W. Decreased levels of soluble amyloid beta-protein precursor in cerebrospinal fluid of live alzheimer disease patients. *Proc. Natl. Acad. Sci. USA* **1992**, *89*, 2551–2555. [CrossRef] [PubMed]
143. Urakami, K.; Takahashi, K.; Okada, A.; Oshima, T.; Adachi, Y.; Nakamura, S.; Kitaguchi, N.; Tokushima, Y.; Yamamoto, S.; Tanaka, S. Clinical course and csf amyloid beta protein precursor having the site of application of the protease inhibitor (appi) levels in patients with dementia of the alzheimer type. *Dementia* **1993**, *4*, 59–60.
144. Vigo-Pelfrey, C.; Lee, D.; Keim, P.; Lieberburg, I.; Schenk, D.B. Characterization of beta-amyloid peptide from human cerebrospinal fluid. *J. Neurochem.* **1993**, *61*, 1965–1968. [CrossRef]
145. van Gool, W.A.; Schenk, D.B.; Bolhuis, P.A. Concentrations of amyloid-beta protein in cerebrospinal fluid increase with age in patients free from neurodegenerative disease. *Neurosci. Lett.* **1994**, *172*, 122–124. [CrossRef]
146. Tabaton, M.; Nunzi, M.G.; Xue, R.; Usiak, M.; Autilio-Gambetti, L.; Gambetti, P. Soluble amyloid beta-protein is a marker of alzheimer amyloid in brain but not in cerebrospinal fluid. *Biochem. Biophys. Res. Commun.* **1994**, *200*, 1598–1603. [CrossRef]
147. Jensen, M.; Schröder, J.; Blomberg, M.; Engvall, B.; Pantel, J.; Ida, N.; Basun, H.; Wahlund, L.O.; Werle, E.; Jauss, M.; et al. Cerebrospinal fluid a beta42 is increased early in sporadic alzheimer's disease and declines with disease progression. *Ann. Neurol.* **1999**, *45*, 504–511. [CrossRef]
148. Fagan, A.M.; Roe, C.M.; Xiong, C.; Mintun, M.A.; Morris, J.C.; Holtzman, D.M. Cerebrospinal fluid tau/beta-amyloid(42) ratio as a prediction of cognitive decline in nondemented older adults. *Arch. Neurol.* **2007**, *64*, 343–349. [CrossRef]
149. Fagan, A.M.; Perrin, R.J. Upcoming candidate cerebrospinal fluid biomarkers of alzheimer's disease. *Biomark. Med.* **2012**, *6*, 455–476. [CrossRef]
150. Hansson, O.; Lehmann, S.; Otto, M.; Zetterberg, H.; Lewczuk, P. Advantages and disadvantages of the use of the csf amyloid β (aβ) 42/40 ratio in the diagnosis of alzheimer's disease. *Alzheimer's Res. Ther.* **2019**, *11*, 34. [CrossRef]
151. Holtzman, D.M. Csf biomarkers for alzheimer's disease: Current utility and potential future use. *Neurobiol. Aging* **2011**, *32* (Suppl. 1), S4–S9. [CrossRef]
152. Dumurgier, J.; Gabelle, A.; Vercruysse, O.; Bombois, S.; Laplanche, J.L.; Peoc'h, K.; Schraen, S.; Sablonnière, B.; Pasquier, F.; Touchon, J.; et al. Exacerbated csf abnormalities in younger patients with alzheimer's disease. *Neurobiol. Dis.* **2013**, *54*, 486–491. [CrossRef] [PubMed]
153. Reijn, T.S.; Rikkert, M.O.; van Geel, W.J.; de Jong, D.; Verbeek, M.M. Diagnostic accuracy of elisa and xmap technology for analysis of amyloid beta(42) and tau proteins. *Clin. Chem.* **2007**, *53*, 859–865. [CrossRef] [PubMed]
154. Mattsson, N.; Zetterberg, H.; Hansson, O.; Andreasen, N.; Parnetti, L.; Jonsson, M.; Herukka, S.K.; van der Flier, W.M.; Blankenstein, M.A.; Ewers, M.; et al. Csf biomarkers and incipient alzheimer disease in patients with mild cognitive impairment. *JAMA* **2009**, *302*, 385–393. [CrossRef] [PubMed]
155. Kapaki, E.N.; Paraskevas, G.P.; Tzerakis, N.G.; Sfagos, C.; Seretis, A.; Kararizou, E.; Vassilopoulos, D. Cerebrospinal fluid tau, phospho-tau181 and beta-amyloid1-42 in idiopathic normal pressure hydrocephalus: A discrimination from alzheimer's disease. *Eur. J. Neurol.* **2007**, *14*, 168–173. [CrossRef] [PubMed]
156. Kapaki, E.; Liappas, I.; Paraskevas, G.P.; Theotoka, I.; Rabavilas, A. The diagnostic value of tau protein, beta-amyloid (1-42) and their ratio for the discrimination of alcohol-related cognitive disorders from alzheimer's disease in the early stages. *Int. J. Geriatr. Psychiatry* **2005**, *20*, 722–729. [CrossRef]

157. Ewers, M.; Mattsson, N.; Minthon, L.; Molinuevo, J.L.; Antonell, A.; Popp, J.; Jessen, F.; Herukka, S.K.; Soininen, H.; Maetzler, W.; et al. Csf biomarkers for the differential diagnosis of alzheimer's disease: A large-scale international multicenter study. *Alzheimer's Dement.* **2015**, *11*, 1306–1315. [CrossRef]
158. Selnes, P.; Blennow, K.; Zetterberg, H.; Grambaite, R.; Rosengren, L.; Johnsen, L.; Stenset, V.; Fladby, T. Effects of cerebrovascular disease on amyloid precursor protein metabolites in cerebrospinal fluid. *Cerebrospinal Fluid Res.* **2010**, *7*, 10. [CrossRef]
159. Vos, S.J.; van Rossum, I.A.; Verhey, F.; Knol, D.L.; Soininen, H.; Wahlund, L.O.; Hampel, H.; Tsolaki, M.; Minthon, L.; Frisoni, G.B.; et al. Prediction of alzheimer disease in subjects with amnestic and nonamnestic mci. *Neurology* **2013**, *80*, 1124–1132. [CrossRef]
160. Hampel, H.; Teipel, S.J.; Fuchsberger, T.; Andreasen, N.; Wiltfang, J.; Otto, M.; Shen, Y.; Dodel, R.; Du, Y.; Farlow, M.; et al. Value of csf beta-amyloid1-42 and tau as predictors of alzheimer's disease in patients with mild cognitive impairment. *Mol. Psychiatry* **2004**, *9*, 705–710. [CrossRef]
161. Perneczky, R.; Tsolakidou, A.; Arnold, A.; Diehl-Schmid, J.; Grimmer, T.; Förstl, H.; Kurz, A.; Alexopoulos, P. Csf soluble amyloid precursor proteins in the diagnosis of incipient alzheimer disease. *Neurology* **2011**, *77*, 35–38. [CrossRef] [PubMed]
162. Lewczuk, P.; Kornhuber, J.; Vanderstichele, H.; Vanmechelen, E.; Esselmann, H.; Bibl, M.; Wolf, S.; Otto, M.; Reulbach, U.; Kölsch, H.; et al. Multiplexed quantification of dementia biomarkers in the csf of patients with early dementias and mci: A multicenter study. *Neurobiol. Aging* **2008**, *29*, 812–818. [CrossRef] [PubMed]
163. Bibl, M.; Gallus, M.; Welge, V.; Esselmann, H.; Wolf, S.; Rüther, E.; Wiltfang, J. Cerebrospinal fluid amyloid-β 2-42 is decreased in alzheimer's, but not in frontotemporal dementia. *J. Neural Transm.* **2012**, *119*, 805–813. [CrossRef] [PubMed]
164. Lewczuk, P.; Lelental, N.; Spitzer, P.; Maler, J.M.; Kornhuber, J. Amyloid-β 42/40 cerebrospinal fluid concentration ratio in the diagnostics of alzheimer's disease: Validation of two novel assays. *J. Alzheimer's Dis.* **2015**, *43*, 183–191. [CrossRef]
165. Struyfs, H.; Van Broeck, B.; Timmers, M.; Fransen, E.; Sleegers, K.; Van Broeckhoven, C.; De Deyn, P.P.; Streffer, J.R.; Mercken, M.; Engelborghs, S. Diagnostic accuracy of cerebrospinal fluid amyloid-β isoforms for early and differential dementia diagnosis. *J. Alzheimer's Dis.* **2015**, *45*, 813–822. [CrossRef]
166. Hansson, O.; Zetterberg, H.; Buchhave, P.; Andreasson, U.; Londos, E.; Minthon, L.; Blennow, K. Prediction of alzheimer's disease using the csf abeta42/abeta40 ratio in patients with mild cognitive impairment. *Dement. Geriatr. Cogn. Disord.* **2007**, *23*, 316–320. [CrossRef]
167. Janelidze, S.; Zetterberg, H.; Mattsson, N.; Palmqvist, S.; Vanderstichele, H.; Lindberg, O.; van Westen, D.; Stomrud, E.; Minthon, L.; Blennow, K.; et al. Csf aβ42/aβ40 and aβ42/aβ38 ratios: Better diagnostic markers of alzheimer disease. *Ann. Clin. Transl. Neurol.* **2016**, *3*, 154–165. [CrossRef]
168. Fukumoto, H.; Tennis, M.; Locascio, J.J.; Hyman, B.T.; Growdon, J.H.; Irizarry, M.C. Age but not diagnosis is the main predictor of plasma amyloid beta-protein levels. *Arch. Neurol.* **2003**, *60*, 958–964. [CrossRef]
169. Rissin, D.M.; Kan, C.W.; Campbell, T.G.; Howes, S.C.; Fournier, D.R.; Song, L.; Piech, T.; Patel, P.P.; Chang, L.; Rivnak, A.J.; et al. Single-molecule enzyme-linked immunosorbent assay detects serum proteins at subfemtomolar concentrations. *Nat. Biotechnol.* **2010**, *28*, 595–599. [CrossRef]
170. Janelidze, S.; Stomrud, E.; Palmqvist, S.; Zetterberg, H.; van Westen, D.; Jeromin, A.; Song, L.; Hanlon, D.; Tan Hehir, C.A.; Baker, D.; et al. Plasma β-amyloid in alzheimer's disease and vascular disease. *Sci. Rep.* **2016**, *6*, 26801. [CrossRef]
171. Zetterberg, H.; Mörtberg, E.; Song, L.; Chang, L.; Provuncher, G.K.; Patel, P.P.; Ferrell, E.; Fournier, D.R.; Kan, C.W.; Campbell, T.G.; et al. Hypoxia due to cardiac arrest induces a time-dependent increase in serum amyloid β levels in humans. *PLoS ONE* **2011**, *6*, e28263. [CrossRef] [PubMed]
172. Yang, S.Y.; Chiu, M.J.; Chen, T.F.; Horng, H.E. Detection of plasma biomarkers using immunomagnetic reduction: A promising method for the early diagnosis of alzheimer's disease. *Neurol. Ther.* **2017**, *6*, 37–56. [CrossRef] [PubMed]
173. Wang, L.S.; Leung, Y.Y.; Chang, S.K.; Leight, S.; Knapik-Czajka, M.; Baek, Y.; Shaw, L.M.; Lee, V.M.; Trojanowski, J.Q.; Clark, C.M. Comparison of xmap and elisa assays for detecting cerebrospinal fluid biomarkers of alzheimer's disease. *J. Alzheimer's Dis.* **2012**, *31*, 439–445. [CrossRef] [PubMed]
174. Pan, C.; Korff, A.; Galasko, D.; Ginghina, C.; Peskind, E.; Li, G.; Quinn, J.; Montine, T.J.; Cain, K.; Shi, M.; et al. Diagnostic values of cerebrospinal fluid t-tau and aβ_{42} using meso scale discovery assays for alzheimer's disease. *J. Alzheimer's Dis.* **2015**, *45*, 709–719. [CrossRef]

175. Nakamura, A.; Kaneko, N.; Villemagne, V.L.; Kato, T.; Doecke, J.; Doré, V.; Fowler, C.; Li, Q.X.; Martins, R.; Rowe, C.; et al. High performance plasma amyloid-β biomarkers for alzheimer's disease. *Nature* **2018**, *554*, 249–254. [CrossRef]
176. Schindler, S.E.; Bollinger, J.G.; Ovod, V.; Mawuenyega, K.G.; Li, Y.; Gordon, B.A.; Holtzman, D.M.; Morris, J.C.; Benzinger, T.L.S.; Xiong, C.; et al. High-precision plasma β-amyloid 42/40 predicts current and future brain amyloidosis. *Neurology* **2019**, *93*, e1647–e1659. [CrossRef]
177. Verberk, I.M.W.; Slot, R.E.; Verfaillie, S.C.J.; Heijst, H.; Prins, N.D.; van Berckel, B.N.M.; Scheltens, P.; Teunissen, C.E.; van der Flier, W.M. Plasma amyloid as prescreener for the earliest alzheimer pathological changes. *Ann. Neurol.* **2018**, *84*, 648–658. [CrossRef]
178. Kaneko, N.; Nakamura, A.; Washimi, Y.; Kato, T.; Sakurai, T.; Arahata, Y.; Bundo, M.; Takeda, A.; Niida, S.; Ito, K.; et al. Novel plasma biomarker surrogating cerebral amyloid deposition. *Proc. Jpn. Acad. Ser. B Phys. Biol. Sci.* **2014**, *90*, 353–364. [CrossRef]
179. Ovod, V.; Ramsey, K.N.; Mawuenyega, K.G.; Bollinger, J.G.; Hicks, T.; Schneider, T.; Sullivan, M.; Paumier, K.; Holtzman, D.M.; Morris, J.C.; et al. Amyloid β concentrations and stable isotope labeling kinetics of human plasma specific to central nervous system amyloidosis. *Alzheimer's Dement.* **2017**, *13*, 841–849. [CrossRef]
180. Pannee, J.; Törnqvist, U.; Westerlund, A.; Ingelsson, M.; Lannfelt, L.; Brinkmalm, G.; Persson, R.; Gobom, J.; Svensson, J.; Johansson, P.; et al. The amyloid-β degradation pattern in plasma–a possible tool for clinical trials in alzheimer's disease. *Neurosci. Lett.* **2014**, *573*, 7–12. [CrossRef]
181. Li, Q.X.; Berndt, M.C.; Bush, A.I.; Rumble, B.; Mackenzie, I.; Friedhuber, A.; Beyreuther, K.; Masters, C.L. Membrane-associated forms of the beta a4 amyloid protein precursor of alzheimer's disease in human platelet and brain: Surface expression on the activated human platelet. *Blood* **1994**, *84*, 133–142. [CrossRef] [PubMed]
182. Palmqvist, S.; Insel, P.S.; Stomrud, E.; Janelidze, S.; Zetterberg, H.; Brix, B.; Eichenlaub, U.; Dage, J.L.; Chai, X.; Blennow, K.; et al. Cerebrospinal fluid and plasma biomarker trajectories with increasing amyloid deposition in alzheimer's disease. *EMBO Mol. Med.* **2019**, *11*, e11170. [CrossRef]
183. Augustinack, J.C.; Schneider, A.; Mandelkow, E.M.; Hyman, B.T. Specific tau phosphorylation sites correlate with severity of neuronal cytopathology in alzheimer's disease. *Acta Neuropathol.* **2002**, *103*, 26–35. [CrossRef] [PubMed]
184. Haroutunian, V.; Davies, P.; Vianna, C.; Buxbaum, J.D.; Purohit, D.P. Tau protein abnormalities associated with the progression of alzheimer disease type dementia. *Neurobiol. Aging* **2007**, *28*, 1–7. [CrossRef] [PubMed]
185. Zetterberg, H.; Bendlin, B.B. Biomarkers for alzheimer's disease-preparing for a new era of disease-modifying therapies. *Mol. Psychiatry* **2020**, 1–13. [CrossRef] [PubMed]
186. Fossati, S.; Ramos Cejudo, J.; Debure, L.; Pirraglia, E.; Sone, J.Y.; Li, Y.; Chen, J.; Butler, T.; Zetterberg, H.; Blennow, K.; et al. Plasma tau complements csf tau and p-tau in the diagnosis of alzheimer's disease. *Alzheimer's Dement.* **2019**, *11*, 483–492. [CrossRef] [PubMed]
187. Vandermeeren, M.; Mercken, M.; Vanmechelen, E.; Six, J.; van de Voorde, A.; Martin, J.J.; Cras, P. Detection of tau proteins in normal and alzheimer's disease cerebrospinal fluid with a sensitive sandwich enzyme-linked immunosorbent assay. *J. Neurochem.* **1993**, *61*, 1828–1834. [CrossRef]
188. Kandimalla, R.J.; Prabhakar, S.; Bk, B.; Wani, W.Y.; Sharma, D.R.; Grover, V.K.; Bhardwaj, N.; Jain, K.; Gill, K.D. Cerebrospinal fluid profile of amyloid β42 (aβ42), htau and ubiquitin in north indian alzheimer's disease patients. *Neurosci. Lett.* **2011**, *487*, 134–138. [CrossRef]
189. Kandimalla, R.J.; Prabhakar, S.; Binukumar, B.K.; Wani, W.Y.; Gupta, N.; Sharma, D.R.; Sunkaria, A.; Grover, V.K.; Bhardwaj, N.; Jain, K.; et al. Apo-eε4 allele in conjunction with aβ42 and tau in csf: Biomarker for alzheimer's disease. *Curr. Alzheimer Res.* **2011**, *8*, 187–196. [CrossRef]
190. Tapiola, T.; Alafuzoff, I.; Herukka, S.K.; Parkkinen, L.; Hartikainen, P.; Soininen, H.; Pirttilä, T. Cerebrospinal fluid {beta}-amyloid 42 and tau proteins as biomarkers of alzheimer-type pathologic changes in the brain. *Arch Neurol.* **2009**, *66*, 382–389. [CrossRef]
191. Haense, C.; Buerger, K.; Kalbe, E.; Drzezga, A.; Teipel, S.J.; Markiewicz, P.; Herholz, K.; Heiss, W.D.; Hampel, H. Csf total and phosphorylated tau protein, regional glucose metabolism and dementia severity in alzheimer's disease. *Eur. J. Neurol.* **2008**, *15*, 1155–1162. [CrossRef] [PubMed]
192. Hampel, H.; Goernitz, A.; Buerger, K. Advances in the development of biomarkers for alzheimer's disease: From csf total tau and abeta(1-42) proteins to phosphorylated tau protein. *Brain Res. Bull.* **2003**, *61*, 243–253. [CrossRef]

193. Gordon, B.A.; Blazey, T.M.; Christensen, J.; Dincer, A.; Flores, S.; Keefe, S.; Chen, C.; Su, Y.; McDade, E.M.; Wang, G.; et al. Tau pet in autosomal dominant alzheimer's disease: Relationship with cognition, dementia and other biomarkers. *Brain* **2019**, *142*, 1063–1076. [CrossRef]
194. Quiroz, Y.T.; Sperling, R.A.; Norton, D.J.; Baena, A.; Arboleda-Velasquez, J.F.; Cosio, D.; Schultz, A.; Lapoint, M.; Guzman-Velez, E.; Miller, J.B.; et al. Association between amyloid and tau accumulation in young adults with autosomal dominant alzheimer disease. *JAMA Neurol.* **2018**, *75*, 548–556. [CrossRef] [PubMed]
195. Fleisher, A.S.; Chen, K.; Quiroz, Y.T.; Jakimovich, L.J.; Gutierrez Gomez, M.; Langois, C.M.; Langbaum, J.B.; Roontiva, A.; Thiyyagura, P.; Lee, W.; et al. Associations between biomarkers and age in the presenilin 1 e280a autosomal dominant alzheimer disease kindred: A cross-sectional study. *JAMA Neurol.* **2015**, *72*, 316–324. [CrossRef] [PubMed]
196. Toledo, J.B.; Xie, S.X.; Trojanowski, J.Q.; Shaw, L.M. Longitudinal change in csf tau and aβ biomarkers for up to 48 months in adni. *Acta Neuropathol.* **2013**, *126*, 659–670. [CrossRef]
197. McDade, E.; Wang, G.; Gordon, B.A.; Hassenstab, J.; Benzinger, T.L.S.; Buckles, V.; Fagan, A.M.; Holtzman, D.M.; Cairns, N.J.; Goate, A.M.; et al. Longitudinal cognitive and biomarker changes in dominantly inherited alzheimer disease. *Neurology* **2018**, *91*, e1295–e1306. [CrossRef]
198. Fitzpatrick, A.W.P.; Falcon, B.; He, S.; Murzin, A.G.; Murshudov, G.; Garringer, H.J.; Crowther, R.A.; Ghetti, B.; Goedert, M.; Scheres, S.H.W. Cryo-em structures of tau filaments from alzheimer's disease. *Nature* **2017**, *547*, 185–190. [CrossRef]
199. Falcon, B.; Zhang, W.; Schweighauser, M.; Murzin, A.G.; Vidal, R.; Garringer, H.J.; Ghetti, B.; Scheres, S.H.W.; Goedert, M. Tau filaments from multiple cases of sporadic and inherited alzheimer's disease adopt a common fold. *Acta Neuropathol.* **2018**, *136*, 699–708. [CrossRef]
200. Zhang, W.; Falcon, B.; Murzin, A.G.; Fan, J.; Crowther, R.A.; Goedert, M.; Scheres, S.H. Heparin-induced tau filaments are polymorphic and differ from those in alzheimer's and pick's diseases. *Elife* **2019**, *8*, e43584. [CrossRef]
201. Arakhamia, T.; Lee, C.E.; Carlomagno, Y.; Duong, D.M.; Kundinger, S.R.; Wang, K.; Williams, D.; DeTure, M.; Dickson, D.W.; Cook, C.N.; et al. Posttranslational modifications mediate the structural diversity of tauopathy strains. *Cell* **2020**, *180*, 633–644. [CrossRef] [PubMed]
202. Jicha, G.A.; Bowser, R.; Kazam, I.G.; Davies, P. Alz-50 and mc-1, a new monoclonal antibody raised to paired helical filaments, recognize conformational epitopes on recombinant tau. *J. Neurosci. Res.* **1997**, *48*, 128–132. [CrossRef]
203. Jicha, G.A.; Lane, E.; Vincent, I.; Otvos, L.; Hoffmann, R.; Davies, P. A conformation- and phosphorylation-dependent antibody recognizing the paired helical filaments of alzheimer's disease. *J. Neurochem.* **1997**, *69*, 2087–2095. [CrossRef] [PubMed]
204. Weaver, C.L.; Espinoza, M.; Kress, Y.; Davies, P. Conformational change as one of the earliest alterations of tau in alzheimer's disease. *Neurobiol. Aging* **2000**, *21*, 719–727. [CrossRef]
205. Cicognola, C.; Brinkmalm, G.; Wahlgren, J.; Portelius, E.; Gobom, J.; Cullen, N.C.; Hansson, O.; Parnetti, L.; Constantinescu, R.; Wildsmith, K.; et al. Novel tau fragments in cerebrospinal fluid: Relation to tangle pathology and cognitive decline in alzheimer's disease. *Acta Neuropathol.* **2019**, *137*, 279–296. [CrossRef]
206. Amadoro, G.; Corsetti, V.; Sancesario, G.M.; Lubrano, A.; Melchiorri, G.; Bernardini, S.; Calissano, P.; Sancesario, G. Cerebrospinal fluid levels of a 20-22 kda nh2 fragment of human tau provide a novel neuronal injury biomarker in alzheimer's disease and other dementias. *J. Alzheimer's Dis.* **2014**, *42*, 211–226. [CrossRef]
207. Blennow, K.; Chen, C.; Cicognola, C.; Wildsmith, K.R.; Manser, P.T.; Bohorquez, S.M.S.; Zhang, Z.; Xie, B.; Peng, J.; Hansson, O.; et al. Cerebrospinal fluid tau fragment correlates with tau pet: A candidate biomarker for tangle pathology. *Brain* **2020**, *143*, 650–660. [CrossRef]
208. Quinn, J.P.; Corbett, N.J.; Kellett, K.A.B.; Hooper, N.M. Tau proteolysis in the pathogenesis of tauopathies: Neurotoxic fragments and novel biomarkers. *J. Alzheimer's Dis.* **2018**, *63*, 13–33. [CrossRef]
209. Maia, L.F.; Kaeser, S.A.; Reichwald, J.; Hruscha, M.; Martus, P.; Staufenbiel, M.; Jucker, M. Changes in amyloid-β and tau in the cerebrospinal fluid of transgenic mice overexpressing amyloid precursor protein. *Sci. Transl. Med.* **2013**, *5*, 194re192. [CrossRef]
210. Sato, C.; Barthélemy, N.R.; Mawuenyega, K.G.; Patterson, B.W.; Gordon, B.A.; Jockel-Balsarotti, J.; Sullivan, M.; Crisp, M.J.; Kasten, T.; Kirmess, K.M.; et al. Tau kinetics in neurons and the human central nervous system. *Neuron* **2018**, *97*, 1284–1298. [CrossRef]

211. Buerger, K.; Ewers, M.; Pirttilä, T.; Zinkowski, R.; Alafuzoff, I.; Teipel, S.J.; DeBernardis, J.; Kerkman, D.; McCulloch, C.; Soininen, H.; et al. Csf phosphorylated tau protein correlates with neocortical neurofibrillary pathology in alzheimer's disease. *Brain* **2006**, *129*, 3035–3041. [CrossRef] [PubMed]
212. Molinuevo, J.L.; Ayton, S.; Batrla, R.; Bednar, M.M.; Bittner, T.; Cummings, J.; Fagan, A.M.; Hampel, H.; Mielke, M.M.; Mikulskis, A.; et al. Current state of alzheimer's fluid biomarkers. *Acta Neuropathol.* **2018**, *136*, 821–853. [CrossRef] [PubMed]
213. Blennow, K.; Hampel, H.; Weiner, M.; Zetterberg, H. Cerebrospinal fluid and plasma biomarkers in alzheimer disease. *Nat. Rev. Neurol.* **2010**, *6*, 131–144. [CrossRef] [PubMed]
214. Lleó, A.; Irwin, D.J.; Illán-Gala, I.; McMillan, C.T.; Wolk, D.A.; Lee, E.B.; Van Deerlin, V.M.; Shaw, L.M.; Trojanowski, J.Q.; Grossman, M. A 2-step cerebrospinal algorithm for the selection of frontotemporal lobar degeneration subtypes. *JAMA Neurol.* **2018**, *75*, 738–745. [CrossRef]
215. Schoonenboom, N.S.; Reesink, F.E.; Verwey, N.A.; Kester, M.I.; Teunissen, C.E.; van de Ven, P.M.; Pijnenburg, Y.A.; Blankenstein, M.A.; Rozemuller, A.J.; Scheltens, P.; et al. Cerebrospinal fluid markers for differential dementia diagnosis in a large memory clinic cohort. *Neurology* **2012**, *78*, 47–54. [CrossRef]
216. Buchhave, P.; Minthon, L.; Zetterberg, H.; Wallin, A.K.; Blennow, K.; Hansson, O. Cerebrospinal fluid levels of β-amyloid 1-42, but not of tau, are fully changed already 5 to 10 years before the onset of alzheimer dementia. *Arch. Gen. Psychiatry* **2012**, *69*, 98–106. [CrossRef]
217. Vos, S.J.; Xiong, C.; Visser, P.J.; Jasielec, M.S.; Hassenstab, J.; Grant, E.A.; Cairns, N.J.; Morris, J.C.; Holtzman, D.M.; Fagan, A.M. Preclinical alzheimer's disease and its outcome: A longitudinal cohort study. *Lancet Neurol.* **2013**, *12*, 957–965. [CrossRef]
218. Hansson, O.; Mikulskis, A.; Fagan, A.M.; Teunissen, C.; Zetterberg, H.; Vanderstichele, H.; Molinuevo, J.L.; Shaw, L.M.; Vandijck, M.; Verbeek, M.M.; et al. The impact of preanalytical variables on measuring cerebrospinal fluid biomarkers for alzheimer's disease diagnosis: A review. *Alzheimer's Dement.* **2018**, *14*, 1313–1333. [CrossRef]
219. Shaw, L.M.; Arias, J.; Blennow, K.; Galasko, D.; Molinuevo, J.L.; Salloway, S.; Schindler, S.; Carrillo, M.C.; Hendrix, J.A.; Ross, A.; et al. Appropriate use criteria for lumbar puncture and cerebrospinal fluid testing in the diagnosis of alzheimer's disease. *Alzheimer's Dement.* **2018**, *14*, 1505–1521. [CrossRef]
220. Janelidze, S.; Stomrud, E.; Smith, R.; Palmqvist, S.; Mattsson, N.; Airey, D.C.; Proctor, N.K.; Chai, X.; Shcherbinin, S.; Sims, J.R.; et al. Cerebrospinal fluid p-tau217 performs better than p-tau181 as a biomarker of alzheimer's disease. *Nat. Commun.* **2020**, *11*, 1683. [CrossRef]
221. Hampel, H.; Buerger, K.; Zinkowski, R.; Teipel, S.J.; Goernitz, A.; Andreasen, N.; Sjoegren, M.; DeBernardis, J.; Kerkman, D.; Ishiguro, K.; et al. Measurement of phosphorylated tau epitopes in the differential diagnosis of alzheimer disease: A comparative cerebrospinal fluid study. *Arch. Gen. Psychiatry* **2004**, *61*, 95–102. [CrossRef]
222. Spiegel, J.; Pirraglia, E.; Osorio, R.S.; Glodzik, L.; Li, Y.; Tsui, W.; Saint Louis, L.A.; Randall, C.; Butler, T.; Xu, J.; et al. Greater specificity for cerebrospinal fluid p-tau231 over p-tau181 in the differentiation of healthy controls from alzheimer's disease. *J. Alzheimer's Dis.* **2016**, *49*, 93–100. [CrossRef]
223. Buerger, K.; Alafuzoff, I.; Ewers, M.; Pirttilä, T.; Zinkowski, R.; Hampel, H. No correlation between csf tau protein phosphorylated at threonine 181 with neocortical neurofibrillary pathology in alzheimer's disease. *Brain* **2007**, *130*, e82. [CrossRef] [PubMed]
224. Braak, H.; Braak, E. On areas of transition between entorhinal allocortex and temporal isocortex in the human brain. Normal morphology and lamina-specific pathology in alzheimer's disease. *Acta Neuropathol.* **1985**, *68*, 325–332. [CrossRef] [PubMed]
225. Barthélemy, N.R.; Bateman, R.J.; Hirtz, C.; Marin, P.; Becher, F.; Sato, C.; Gabelle, A.; Lehmann, S. Cerebrospinal fluid phospho-tau t217 outperforms t181 as a biomarker for the differential diagnosis of alzheimer's disease and pet amyloid-positive patient identification. *Alzheimer's Res. Ther.* **2020**, *12*, 26. [CrossRef] [PubMed]
226. Barthélemy, N.R.; Li, Y.; Joseph-Mathurin, N.; Gordon, B.A.; Hassenstab, J.; Benzinger, T.L.S.; Buckles, V.; Fagan, A.M.; Perrin, R.J.; Goate, A.M.; et al. A soluble phosphorylated tau signature links tau, amyloid and the evolution of stages of dominantly inherited alzheimer's disease. *Nat. Med.* **2020**, *26*, 398–407. [CrossRef]
227. Chen, Z.; Mengel, D.; Keshavan, A.; Rissman, R.A.; Billinton, A.; Perkinton, M.; Percival-Alwyn, J.; Schultz, A.; Properzi, M.; Johnson, K.; et al. Learnings about the complexity of extracellular tau aid development of a blood-based screen for alzheimer's disease. *Alzheimer's Dement.* **2019**, *15*, 487–496. [CrossRef] [PubMed]

228. Ramos-Cejudo, J.; Wisniewski, T.; Marmar, C.; Zetterberg, H.; Blennow, K.; de Leon, M.J.; Fossati, S. Traumatic brain injury and alzheimer's disease: The cerebrovascular link. *EBioMedicine* **2018**, *28*, 21–30. [CrossRef] [PubMed]
229. Mendez, M.F. What is the relationship of traumatic brain injury to dementia? *J. Alzheimer's Dis.* **2017**, *57*, 667–681. [CrossRef]
230. Blennow, K.; Brody, D.L.; Kochanek, P.M.; Levin, H.; McKee, A.; Ribbers, G.M.; Yaffe, K.; Zetterberg, H. Traumatic brain injuries. *Nat. Rev. Dis. Primers* **2016**, *2*, 16084. [CrossRef]
231. Clinton, J.; Ambler, M.W.; Roberts, G.W. Post-traumatic alzheimer's disease: Preponderance of a single plaque type. *Neuropathol. Appl. Neurobiol.* **1991**, *17*, 69–74. [CrossRef] [PubMed]
232. Mortimer, J.A.; van Duijn, C.M.; Chandra, V.; Fratiglioni, L.; Graves, A.B.; Heyman, A.; Jorm, A.F.; Kokmen, E.; Kondo, K.; Rocca, W.A. Head trauma as a risk factor for alzheimer's disease: A collaborative re-analysis of case-control studies. Eurodem risk factors research group. *Int. J. Epidemiol.* **1991**, *20* (Suppl. 2), S28–S35. [CrossRef]
233. Shahim, P.; Blennow, K.; Zetterberg, H. Tau, s-100 calcium-binding protein b, and neuron-specific enolase as biomarkers of concussion-reply. *JAMA Neurol.* **2014**, *71*, 926–927. [CrossRef] [PubMed]
234. Shahim, P.; Tegner, Y.; Wilson, D.H.; Randall, J.; Skillback, T.; Pazooki, D.; Kallberg, B.; Blennow, K.; Zetterberg, H. Blood biomarkers for brain injury in concussed professional ice hockey players. *JAMA Neurol.* **2014**, *71*, 684–692. [CrossRef] [PubMed]
235. Rubenstein, R.; Chang, B.; Davies, P.; Wagner, A.K.; Robertson, C.S.; Wang, K.K. A novel, ultrasensitive assay for tau: Potential for assessing traumatic brain injury in tissues and biofluids. *J. Neurotrauma* **2015**, *32*, 342–352. [CrossRef]
236. Hu, W.T.; Watts, K.; Grossman, M.; Glass, J.; Lah, J.J.; Hales, C.; Shelnutt, M.; Van Deerlin, V.; Trojanowski, J.Q.; Levey, A.I. Reduced csf p-tau181 to tau ratio is a biomarker for ftld-tdp. *Neurology* **2013**, *81*, 1945–1952. [CrossRef]
237. Vanmechelen, E.; Vanderstichele, H.; Davidsson, P.; Van Kerschaver, E.; Van Der Perre, B.; Sjögren, M.; Andreasen, N.; Blennow, K. Quantification of tau phosphorylated at threonine 181 in human cerebrospinal fluid: A sandwich elisa with a synthetic phosphopeptide for standardization. *Neurosci. Lett.* **2000**, *285*, 49–52. [CrossRef]
238. Wagshal, D.; Sankaranarayanan, S.; Guss, V.; Hall, T.; Berisha, F.; Lobach, I.; Karydas, A.; Voltarelli, L.; Scherling, C.; Heuer, H.; et al. Divergent csf τ alterations in two common tauopathies: Alzheimer's disease and progressive supranuclear palsy. *J. Neurol. Neurosurg. Psychiatry* **2015**, *86*, 244–250. [CrossRef]
239. O'Brien, J.T.; Herholz, K. Amyloid imaging for dementia in clinical practice. *BMC Med.* **2015**, *13*, 163. [CrossRef]
240. Henriksen, K.; O'Bryant, S.E.; Hampel, H.; Trojanowski, J.Q.; Montine, T.J.; Jeromin, A.; Blennow, K.; Lonneborg, A.; Wyss-Coray, T.; Soares, H.; et al. The future of blood-based biomarkers for alzheimer's disease. *Alzheimer's Dement. J. Assoc.* **2014**, *10*, 115–131. [CrossRef]
241. de Almeida, S.M.; Shumaker, S.D.; LeBlanc, S.K.; Delaney, P.; Marquie-Beck, J.; Ueland, S.; Alexander, T.; Ellis, R.J. Incidence of post-dural puncture headache in research volunteers. *Headache* **2011**, *51*, 1503–1510. [CrossRef] [PubMed]
242. Mukaetova-Ladinska, E.B.; Abdell-All, Z.; Andrade, J.; da Silva, J.A.; Boksha, I.; Burbaeva, G.; Kalaria, R.N.; TO'Brien, J. Platelet tau protein as a potential peripheral biomarker in alzheimer's disease: An explorative study. *Curr. Alzheimer Res.* **2018**, *15*, 800–808. [CrossRef] [PubMed]
243. Slachevsky, A.; Guzmán-Martínez, L.; Delgado, C.; Reyes, P.; Farías, G.A.; Muñoz-Neira, C.; Bravo, E.; Farías, M.; Flores, P.; Garrido, C.; et al. Tau platelets correlate with regional brain atrophy in patients with alzheimer's disease. *J. Alzheimer's Dis.* **2017**, *55*, 1595–1603. [CrossRef] [PubMed]
244. Tzen, K.Y.; Yang, S.Y.; Chen, T.F.; Cheng, T.W.; Horng, H.E.; Wen, H.P.; Huang, Y.Y.; Shiue, C.Y.; Chiu, M.J. Plasma aβ but not tau is related to brain pib retention in early alzheimer's disease. *ACS Chem. Neurosci.* **2014**, *5*, 830–836. [CrossRef]
245. Zetterberg, H.; Blennow, K. Blood biomarkers: Democratizing alzheimer's diagnostics. *Neuron* **2020**, *106*, 881–883. [CrossRef]
246. Mattsson, N.; Zetterberg, H.; Janelidze, S.; Insel, P.S.; Andreasson, U.; Stomrud, E.; Palmqvist, S.; Baker, D.; Tan Hehir, C.A.; Jeromin, A.; et al. Plasma tau in alzheimer disease. *Neurology* **2016**, *87*, 1827–1835. [CrossRef]

247. Mielke, M.M.; Hagen, C.E.; Wennberg, A.M.V.; Airey, D.C.; Savica, R.; Knopman, D.S.; Machulda, M.M.; Roberts, R.O.; Jack, C.R.; Petersen, R.C.; et al. Association of plasma total tau level with cognitive decline and risk of mild cognitive impairment or dementia in the mayo clinic study on aging. *JAMA Neurol.* **2017**, *74*, 1073–1080. [CrossRef]
248. Dage, J.L.; Wennberg, A.M.V.; Airey, D.C.; Hagen, C.E.; Knopman, D.S.; Machulda, M.M.; Roberts, R.O.; Jack, C.R.; Petersen, R.C.; Mielke, M.M. Levels of tau protein in plasma are associated with neurodegeneration and cognitive function in a population-based elderly cohort. *Alzheimer's Dement.* **2016**, *12*, 1226–1234. [CrossRef]
249. Pereira, J.B.; Westman, E.; Hansson, O.; Initiative, A.s.D.N. Association between cerebrospinal fluid and plasma neurodegeneration biomarkers with brain atrophy in alzheimer's disease. *Neurobiol. Aging* **2017**, *58*, 14–29. [CrossRef]
250. Mielke, M.M.; Hagen, C.E.; Xu, J.; Chai, X.; Vemuri, P.; Lowe, V.J.; Airey, D.C.; Knopman, D.S.; Roberts, R.O.; Machulda, M.M.; et al. Plasma phospho-tau181 increases with alzheimer's disease clinical severity and is associated with tau- and amyloid-positron emission tomography. *Alzheimer's Dement.* **2018**, *14*, 989–997. [CrossRef]
251. Tatebe, H.; Kasai, T.; Ohmichi, T.; Kishi, Y.; Kakeya, T.; Waragai, M.; Kondo, M.; Allsop, D.; Tokuda, T. Quantification of plasma phosphorylated tau to use as a biomarker for brain alzheimer pathology: Pilot case-control studies including patients with alzheimer's disease and down syndrome. *Mol. Neurodegener.* **2017**, *12*, 63. [CrossRef] [PubMed]
252. Yang, C.C.; Chiu, M.J.; Chen, T.F.; Chang, H.L.; Liu, B.H.; Yang, S.Y. Assay of plasma phosphorylated tau protein (threonine 181) and total tau protein in early-stage alzheimer's disease. *J. Alzheimer's Dis.* **2018**, *61*, 1323–1332. [CrossRef] [PubMed]
253. Karikari, T.K.; Pascoal, T.A.; Ashton, N.J.; Janelidze, S.; Benedet, A.L.; Rodriguez, J.L.; Chamoun, M.; Savard, M.; Kang, M.S.; Therriault, J.; et al. Blood phosphorylated tau 181 as a biomarker for alzheimer's disease: A diagnostic performance and prediction modelling study using data from four prospective cohorts. *Lancet Neurol.* **2020**, *19*, 422–433. [CrossRef]
254. Thijssen, E.H.; La Joie, R.; Wolf, A.; Strom, A.; Wang, P.; Iaccarino, L.; Bourakova, V.; Cobigo, Y.; Heuer, H.; Spina, S.; et al. Diagnostic value of plasma phosphorylated tau181 in alzheimer's disease and frontotemporal lobar degeneration. *Nat. Med.* **2020**, *26*, 387–397. [CrossRef] [PubMed]
255. Janelidze, S.; Mattsson, N.; Palmqvist, S.; Smith, R.; Beach, T.G.; Serrano, G.E.; Chai, X.; Proctor, N.K.; Eichenlaub, U.; Zetterberg, H.; et al. Plasma p-tau181 in alzheimer's disease: Relationship to other biomarkers, differential diagnosis, neuropathology and longitudinal progression to Alzheimer's Dementia. *Nat. Med.* **2020**, *26*, 379–386. [CrossRef] [PubMed]
256. Barthélemy, N.R.; Horie, K.; Sato, C.; Bateman, R.J. Blood plasma phosphorylated-tau isoforms track cns change in alzheimer's disease. *J. Exp. Med.* **2020**, *217*, e20200861. [CrossRef]
257. Lifke, V.; Kollmorgen, G.; Manuilova, E.; Oelschlaegel, T.; Hillringhaus, L.; Widmann, M.; von Arnim, C.A.F.; Otto, M.; Christenson, R.H.; Powers, J.L.; et al. Elecsys. *Clin. Biochem.* **2019**, *72*, 30–38. [CrossRef]
258. Chiasserini, D.; Biscetti, L.; Farotti, L.; Eusebi, P.; Salvadori, N.; Lisetti, V.; Baschieri, F.; Chipi, E.; Frattini, G.; Stoops, E.; et al. Performance evaluation of an automated elisa system for alzheimer's disease detection in clinical routine. *J. Alzheimer's Dis.* **2016**, *54*, 55–67. [CrossRef]
259. Rissin, D.M.; Fournier, D.R.; Piech, T.; Kan, C.W.; Campbell, T.G.; Song, L.; Chang, L.; Rivnak, A.J.; Patel, P.P.; Provuncher, G.K.; et al. Simultaneous detection of single molecules and singulated ensembles of molecules enables immunoassays with broad dynamic range. *Anal Chem.* **2011**, *83*, 2279–2285. [CrossRef]
260. Barthélemy, N.R.; Fenaille, F.; Hirtz, C.; Sergeant, N.; Schraen-Maschke, S.; Vialaret, J.; Buée, L.; Gabelle, A.; Junot, C.; Lehmann, S.; et al. Tau protein quantification in human cerebrospinal fluid by targeted mass spectrometry at high sequence coverage provides insights into its primary structure heterogeneity. *J. Proteome Res.* **2016**, *15*, 667–676. [CrossRef]
261. Barthélemy, N.R.; Mallipeddi, N.; Moiseyev, P.; Sato, C.; Bateman, R.J. Tau phosphorylation rates measured by mass spectrometry differ in the intracellular brain vs. Extracellular cerebrospinal fluid compartments and are differentially affected by alzheimer's disease. *Front. Aging Neurosci.* **2019**, *11*, 121. [CrossRef] [PubMed]
262. Khalil, M.; Teunissen, C.E.; Otto, M.; Piehl, F.; Sormani, M.P.; Gattringer, T.; Barro, C.; Kappos, L.; Comabella, M.; Fazekas, F.; et al. Neurofilaments as biomarkers in neurological disorders. *Nat. Rev. Neurol.* **2018**, *14*, 577–589. [CrossRef] [PubMed]

263. Gaetani, L.; Blennow, K.; Calabresi, P.; Di Filippo, M.; Parnetti, L.; Zetterberg, H. Neurofilament light chain as a biomarker in neurological disorders. *J. Neurol. Neurosurg. Psychiatry* **2019**, *90*, 870–881. [CrossRef] [PubMed]
264. Bridel, C.; van Wieringen, W.N.; Zetterberg, H.; Tijms, B.M.; Teunissen, C.E.; Alvarez-Cermeño, J.C.; Andreasson, U.; Axelsson, M.; Bäckström, D.C.; Bartos, A.; et al. Diagnostic value of cerebrospinal fluid neurofilament light protein in neurology: A systematic review and meta-analysis. *JAMA Neurol.* **2019**, *76*, 1035–1048. [CrossRef] [PubMed]
265. Weston, P.S.J.; Poole, T.; O'Connor, A.; Heslegrave, A.; Ryan, N.S.; Liang, Y.; Druyeh, R.; Mead, S.; Blennow, K.; Schott, J.M.; et al. Longitudinal measurement of serum neurofilament light in presymptomatic familial alzheimer's disease. *Alzheimer's Res. Ther.* **2019**, *11*, 19. [CrossRef] [PubMed]
266. Schultz, S.A.; Strain, J.F.; Adedokun, A.; Wang, Q.; Preische, O.; Kuhle, J.; Flores, S.; Keefe, S.; Dincer, A.; Ances, B.M.; et al. Serum neurofilament light chain levels are associated with white matter integrity in autosomal dominant alzheimer's disease. *Neurobiol. Dis.* **2020**, *142*, 104960. [CrossRef]
267. Kuhle, J.; Barro, C.; Andreasson, U.; Derfuss, T.; Lindberg, R.; Sandelius, Å.; Liman, V.; Norgren, N.; Blennow, K.; Zetterberg, H. Comparison of three analytical platforms for quantification of the neurofilament light chain in blood samples: Elisa, electrochemiluminescence immunoassay and simoa. *Clin. Chem. Lab Med.* **2016**, *54*, 1655–1661. [CrossRef]
268. Kvartsberg, H.; Duits, F.H.; Ingelsson, M.; Andreasen, N.; Öhrfelt, A.; Andersson, K.; Brinkmalm, G.; Lannfelt, L.; Minthon, L.; Hansson, O.; et al. Cerebrospinal fluid levels of the synaptic protein neurogranin correlates with cognitive decline in prodromal alzheimer's disease. *Alzheimer's Dement.* **2015**, *11*, 1180–1190. [CrossRef]
269. De Vos, A.; Jacobs, D.; Struyfs, H.; Fransen, E.; Andersson, K.; Portelius, E.; Andreasson, U.; De Surgeloose, D.; Hernalsteen, D.; Sleegers, K.; et al. C-terminal neurogranin is increased in cerebrospinal fluid but unchanged in plasma in alzheimer's disease. *Alzheimer's Dement.* **2015**, *11*, 1461–1469. [CrossRef]
270. Ashton, N.J.; Nevado-Holgado, A.J.; Barber, I.S.; Lynham, S.; Gupta, V.; Chatterjee, P.; Goozee, K.; Hone, E.; Pedrini, S.; Blennow, K.; et al. A plasma protein classifier for predicting amyloid burden for preclinical alzheimer's disease. *Sci. Adv.* **2019**, *5*, eaau7220. [CrossRef]
271. Bouchon, A.; Dietrich, J.; Colonna, M. Cutting edge: Inflammatory responses can be triggered by trem-1, a novel receptor expressed on neutrophils and monocytes. *J. Immunol.* **2000**, *164*, 4991–4995. [CrossRef] [PubMed]
272. Ulland, T.K.; Song, W.M.; Huang, S.C.; Ulrich, J.D.; Sergushichev, A.; Beatty, W.L.; Loboda, A.A.; Zhou, Y.; Cairns, N.J.; Kambal, A.; et al. Trem2 maintains microglial metabolic fitness in alzheimer's disease. *Cell* **2017**, *170*, 649–663. [CrossRef] [PubMed]
273. Schlepckow, K.; Kleinberger, G.; Fukumori, A.; Feederle, R.; Lichtenthaler, S.F.; Steiner, H.; Haass, C. An alzheimer-associated trem2 variant occurs at the adam cleavage site and affects shedding and phagocytic function. *EMBO Mol. Med.* **2017**, *9*, 1356–1365. [CrossRef]
274. Yeh, F.L.; Wang, Y.; Tom, I.; Gonzalez, L.C.; Sheng, M. Trem2 binds to apolipoproteins, including apoe and clu/apoj, and thereby facilitates uptake of amyloid-beta by microglia. *Neuron* **2016**, *91*, 328–340. [CrossRef] [PubMed]
275. Filipello, F.; Morini, R.; Corradini, I.; Zerbi, V.; Canzi, A.; Michalski, B.; Erreni, M.; Markicevic, M.; Starvaggi-Cucuzza, C.; Otero, K.; et al. The microglial innate immune receptor trem2 is required for synapse elimination and normal brain connectivity. *Immunity* **2018**, *48*, 979–991. [CrossRef] [PubMed]
276. Kleinberger, G.; Yamanishi, Y.; Suárez-Calvet, M.; Czirr, E.; Lohmann, E.; Cuyvers, E.; Struyfs, H.; Pettkus, N.; Wenninger Weinzierl, A.; Mazaheri, F.; et al. Trem2 mutations implicated in neurodegeneration impair cell surface transport and phagocytosis. *Sci. Transl. Med.* **2014**, *6*, 243ra286. [CrossRef]
277. Guerreiro, R.; Wojtas, A.; Bras, J.; Carrasquillo, M.; Rogaeva, E.; Majounie, E.; Cruchaga, C.; Sassi, C.; Kauwe, J.S.; Younkin, S.; et al. Trem2 variants in alzheimer's disease. *N. Engl. J. Med.* **2013**, *368*, 117–127. [CrossRef]
278. Yaghmoor, F.; Noorsaeed, A.; Alsagaf, S.; Aljohani, W.; Scholtzova, H.; Boutajangout, A.; Wisniewski, T. The role of trem2 in alzheimer's disease and other neurological disorders. *J. Alzheimer's Dis. Parkinsonism* **2014**, *4*, 160. [CrossRef]

279. Jonsson, T.; Stefansson, H.; Steinberg, S.; Jonsdottir, I.; Jonsson, P.V.; Snaedal, J.; Bjornsson, S.; Huttenlocher, J.; Levey, A.I.; Lah, J.J.; et al. Variant of trem2 associated with the risk of alzheimer's disease. *N. Engl. J. Med.* **2013**, *368*, 107–116. [CrossRef]
280. Lill, C.M.; Rengmark, A.; Pihlstrøm, L.; Fogh, I.; Shatunov, A.; Sleiman, P.M.; Wang, L.S.; Liu, T.; Lassen, C.F.; Meissner, E.; et al. The role of trem2 r47h as a risk factor for alzheimer's disease, frontotemporal lobar degeneration, amyotrophic lateral sclerosis, and parkinson's disease. *Alzheimer's Dement.* **2015**, *11*, 1407–1416. [CrossRef]
281. Bekris, L.M.; Khrestian, M.; Dyne, E.; Shao, Y.; Pillai, J.A.; Rao, S.M.; Bemiller, S.M.; Lamb, B.; Fernandez, H.H.; Leverenz, J.B. Soluble trem2 and biomarkers of central and peripheral inflammation in neurodegenerative disease. *J. Neuroimmunol.* **2018**, *319*, 19–27. [CrossRef] [PubMed]
282. Ashton, N.J.; Suárez-Calvet, M.; Heslegrave, A.; Hye, A.; Razquin, C.; Pastor, P.; Sanchez-Valle, R.; Molinuevo, J.L.; Visser, P.J.; Blennow, K.; et al. Plasma levels of soluble trem2 and neurofilament light chain in trem2 rare variant carriers. *Alzheimer's Res. Ther.* **2019**, *11*, 94. [CrossRef] [PubMed]
283. Piccio, L.; Deming, Y.; Del-Águila, J.L.; Ghezzi, L.; Holtzman, D.M.; Fagan, A.M.; Fenoglio, C.; Galimberti, D.; Borroni, B.; Cruchaga, C. Cerebrospinal fluid soluble trem2 is higher in alzheimer disease and associated with mutation status. *Acta Neuropathol.* **2016**, *131*, 925–933. [CrossRef] [PubMed]
284. Suárez-Calvet, M.; Kleinberger, G.; Araque Caballero, M.; Brendel, M.; Rominger, A.; Alcolea, D.; Fortea, J.; Lleó, A.; Blesa, R.; Gispert, J.D.; et al. Strem2 cerebrospinal fluid levels are a potential biomarker for microglia activity in early-stage alzheimer's disease and associate with neuronal injury markers. *EMBO Mol. Med.* **2016**, *8*, 466–476. [CrossRef]
285. Hakala, B.E.; White, C.; Recklies, A.D. Human cartilage gp-39, a major secretory product of articular chondrocytes and synovial cells, is a mammalian member of a chitinase protein family. *J. Biol. Chem.* **1993**, *268*, 25803–25810.
286. Colton, C.A.; Mott, R.T.; Sharpe, H.; Xu, Q.; Van Nostrand, W.E.; Vitek, M.P. Expression profiles for macrophage alternative activation genes in ad and in mouse models of ad. *J. Neuroinflam.* **2006**, *3*, 27. [CrossRef]
287. Johansen, J.S. Studies on serum ykl-40 as a biomarker in diseases with inflammation, tissue remodelling, fibroses and cancer. *Dan Med. Bull.* **2006**, *53*, 172–209.
288. Johansen, J.S.; Schultz, N.A.; Jensen, B.V. Plasma ykl-40: A potential new cancer biomarker? *Future Oncol.* **2009**, *5*, 1065–1082. [CrossRef]
289. Craig-Schapiro, R.; Perrin, R.J.; Roe, C.M.; Xiong, C.; Carter, D.; Cairns, N.J.; Mintun, M.A.; Peskind, E.R.; Li, G.; Galasko, D.R.; et al. Ykl-40: A novel prognostic fluid biomarker for preclinical alzheimer's disease. *Biol. Psychiatry* **2010**, *68*, 903–912. [CrossRef]
290. Mattsson, N.; Tabatabaei, S.; Johansson, P.; Hansson, O.; Andreasson, U.; Månsson, J.E.; Johansson, J.O.; Olsson, B.; Wallin, A.; Svensson, J.; et al. Cerebrospinal fluid microglial markers in alzheimer's disease: Elevated chitotriosidase activity but lack of diagnostic utility. *Neuromol. Med.* **2011**, *13*, 151–159. [CrossRef]
291. Teunissen, C.E.; Elias, N.; Koel-Simmelink, M.J.; Durieux-Lu, S.; Malekzadeh, A.; Pham, T.V.; Piersma, S.R.; Beccari, T.; Meeter, L.H.; Dopper, E.G.; et al. Novel diagnostic cerebrospinal fluid biomarkers for pathologic subtypes of frontotemporal dementia identified by proteomics. *Alzheimer's Dement.* **2016**, *2*, 86–94. [CrossRef] [PubMed]
292. Choi, J.; Lee, H.W.; Suk, K. Plasma level of chitinase 3-like 1 protein increases in patients with early alzheimer's disease. *J. Neurol.* **2011**, *258*, 2181–2185. [CrossRef] [PubMed]
293. Shen, X.N.; Niu, L.D.; Wang, Y.J.; Cao, X.P.; Liu, Q.; Tan, L.; Zhang, C.; Yu, J.T. Inflammatory markers in alzheimer's disease and mild cognitive impairment: A meta-analysis and systematic review of 170 studies. *J. Neurol. Neurosurg. Psychiatry* **2019**, *90*, 590–598. [CrossRef] [PubMed]
294. Park, J.C.; Han, S.H.; Mook-Jung, I. Peripheral inflammatory biomarkers in alzheimer's disease: A brief review. *BMB Rep.* **2020**, *53*, 10–19. [CrossRef] [PubMed]
295. Zuena, A.R.; Casolini, P.; Lattanzi, R.; Maftei, D. Chemokines in alzheimer's disease: New insights into prokineticins, chemokine-like proteins. *Front. Pharmacol.* **2019**, *10*, 622. [CrossRef]
296. Colangelo, A.M.; Alberghina, L.; Papa, M. Astrogliosis as a therapeutic target for neurodegenerative diseases. *Neurosci. Lett.* **2014**, *565*, 59–64. [CrossRef]
297. Liddelow, S.; Barres, B. Snapshot: Astrocytes in health and disease. *Cell* **2015**, *162*, 1170. [CrossRef]

298. Liddelow, S.A.; Guttenplan, K.A.; Clarke, L.E.; Bennett, F.C.; Bohlen, C.J.; Schirmer, L.; Bennett, M.L.; Münch, A.E.; Chung, W.S.; Peterson, T.C.; et al. Neurotoxic reactive astrocytes are induced by activated microglia. *Nature* **2017**, *541*, 481–487. [CrossRef]
299. Liddelow, S.A. Modern approaches to investigating non-neuronal aspects of alzheimer's disease. *FASEB J.* **2019**, *33*, 1528–1535. [CrossRef]
300. Milà-Alomà, M.; Salvadó, G.; Gispert, J.D.; Vilor-Tejedor, N.; Grau-Rivera, O.; Sala-Vila, A.; Sánchez-Benavides, G.; Arenaza-Urquijo, E.M.; Crous-Bou, M.; González-de-Echávarri, J.M.; et al. Amyloid beta, tau, synaptic, neurodegeneration, and glial biomarkers in the preclinical stage of the alzheimer's continuum. *Alzheimer's Dement.* **2020**. [CrossRef]
301. Elahi, F.M.; Casaletto, K.B.; La Joie, R.; Walters, S.M.; Harvey, D.; Wolf, A.; Edwards, L.; Rivera-Contreras, W.; Karydas, A.; Cobigo, Y.; et al. Plasma biomarkers of astrocytic and neuronal dysfunction in early- and late-onset alzheimer's disease. *Alzheimer's Dement.* **2020**, *16*, 681–695. [CrossRef] [PubMed]
302. James, B.D.; Wilson, R.S.; Boyle, P.A.; Trojanowski, J.Q.; Bennett, D.A.; Schneider, J.A. Tdp-43 stage, mixed pathologies, and clinical alzheimer's-type dementia. *Brain* **2016**, *139*, 2983–2993. [CrossRef] [PubMed]
303. Majumder, V.; Gregory, J.M.; Barria, M.A.; Green, A.; Pal, S. Tdp-43 as a potential biomarker for amyotrophic lateral sclerosis: A systematic review and meta-analysis. *BMC Neurol.* **2018**, *18*, 90. [CrossRef] [PubMed]
304. Feneberg, E.; Gray, E.; Ansorge, O.; Talbot, K.; Turner, M.R. Towards a tdp-43-based biomarker for als and ftld. *Mol. Neurobiol.* **2018**, *55*, 7789–7801. [CrossRef] [PubMed]
305. Huang, W.; Zhou, Y.; Tu, L.; Ba, Z.; Huang, J.; Huang, N.; Luo, Y. Tdp-43: From alzheimer's disease to limbic-predominant age-related tdp-43 encephalopathy. *Front. Mol. Neurosci.* **2020**, *13*, 26. [CrossRef] [PubMed]
306. Meade, R.M.; Fairlie, D.P.; Mason, J.M. Alpha-synuclein structure and parkinson's disease—Lessons and emerging principles. *Mol. Neurodegener.* **2019**, *14*, 29. [CrossRef] [PubMed]
307. Paciotti, S.; Bellomo, G.; Gatticchi, L.; Parnetti, L. Are we ready for detecting α-synuclein prone to aggregation in patients? The case of "protein-misfolding cyclic amplification" and "real-time quaking-induced conversion" as diagnostic tools. *Front. Neurol.* **2018**, *9*, 415. [CrossRef]
308. Groveman, B.R.; Orrù, C.D.; Hughson, A.G.; Raymond, L.D.; Zanusso, G.; Ghetti, B.; Campbell, K.J.; Safar, J.; Galasko, D.; Caughey, B. Rapid and ultra-sensitive quantitation of disease-associated α-synuclein seeds in brain and cerebrospinal fluid by αsyn rt-quic. *Acta Neuropathol. Commun.* **2018**, *6*, 7. [CrossRef]
309. Schweighauser, M.; Shi, Y.; Tarutani, A.; Kametani, F.; Murzin, A.G.; Ghetti, B.; Matsubara, T.; Tomita, T.; Ando, T.; Hasegawa, K.; et al. Structures of α-synuclein filaments from multiple system atrophy. *Nature* **2020**, 1–6. [CrossRef]
310. Montagne, A.; Zhao, Z.; Zlokovic, B.V. Alzheimer's disease: A matter of blood-brain barrier dysfunction? *J. Exp. Med.* **2017**, *214*, 3151–3169. [CrossRef]
311. Bennett, R.E.; Robbins, A.B.; Hu, M.; Cao, X.; Betensky, R.A.; Clark, T.; Das, S.; Hyman, B.T. Tau induces blood vessel abnormalities and angiogenesis-related gene expression in p301l transgenic mice and human alzheimer's disease. *Proc. Natl. Acad. Sci. USA* **2018**, *115*, E1289–E1298. [CrossRef]
312. Snyder, H.M.; Corriveau, R.A.; Craft, S.; Faber, J.E.; Greenberg, S.M.; Knopman, D.; Lamb, B.T.; Montine, T.J.; Nedergaard, M.; Schaffer, C.B.; et al. Vascular contributions to cognitive impairment and dementia including alzheimer's disease. *Alzheimer's Dement.* **2015**, *11*, 710–717. [CrossRef] [PubMed]
313. Gottesman, R.F.; Schneider, A.L.; Zhou, Y.; Coresh, J.; Green, E.; Gupta, N.; Knopman, D.S.; Mintz, A.; Rahmim, A.; Sharrett, A.R.; et al. Association between midlife vascular risk factors and estimated brain amyloid deposition. *JAMA* **2017**, *317*, 1443–1450. [CrossRef] [PubMed]
314. Sweeney, M.D.; Sagare, A.P.; Zlokovic, B.V. Blood-brain barrier breakdown in alzheimer disease and other neurodegenerative disorders. *Nat. Rev. Neurol.* **2018**, *14*, 133–150. [CrossRef] [PubMed]
315. Montagne, A.; Barnes, S.R.; Sweeney, M.D.; Halliday, M.R.; Sagare, A.P.; Zhao, Z.; Toga, A.W.; Jacobs, R.E.; Liu, C.Y.; Amezcua, L.; et al. Blood-brain barrier breakdown in the aging human hippocampus. *Neuron* **2015**, *85*, 296–302. [CrossRef] [PubMed]
316. Verheggen, I.C.M.; de Jong, J.J.A.; van Boxtel, M.P.J.; Gronenschild, E.H.B.M.; Palm, W.M.; Postma, A.A.; Jansen, J.F.A.; Verhey, F.R.J.; Backes, W.H. Increase in blood-brain barrier leakage in healthy, older adults. *Geroscience* **2020**, *42*, 1183–1193. [CrossRef] [PubMed]

317. Nation, D.A.; Sweeney, M.D.; Montagne, A.; Sagare, A.P.; D'Orazio, L.M.; Pachicano, M.; Sepehrband, F.; Nelson, A.R.; Buennagel, D.P.; Harrington, M.G.; et al. Blood-brain barrier breakdown is an early biomarker of human cognitive dysfunction. *Nat. Med.* **2019**, *25*, 270–276. [CrossRef]
318. Wang, H.; Golob, E.J.; Su, M.Y. Vascular volume and blood-brain barrier permeability measured by dynamic contrast enhanced mri in hippocampus and cerebellum of patients with mci and normal controls. *J. Magn. Reson. Imaging* **2006**, *24*, 695–700. [CrossRef]
319. Starr, J.M.; Farrall, A.J.; Armitage, P.; McGurn, B.; Wardlaw, J. Blood-brain barrier permeability in alzheimer's disease: A case-control mri study. *Psychiatry Res.* **2009**, *171*, 232–241. [CrossRef]
320. Sagare, A.P.; Sweeney, M.D.; Makshanoff, J.; Zlokovic, B.V. Shedding of soluble platelet-derived growth factor receptor-β from human brain pericytes. *Neurosci. Lett.* **2015**, *607*, 97–101. [CrossRef]
321. Olsson, A.; Vanderstichele, H.; Andreasen, N.; De Meyer, G.; Wallin, A.; Holmberg, B.; Rosengren, L.; Vanmechelen, E.; Blennow, K. Simultaneous measurement of beta-amyloid(1-42), total tau, and phosphorylated tau (thr181) in cerebrospinal fluid by the xmap technology. *Clin. Chem.* **2005**, *51*, 336–345. [CrossRef] [PubMed]
322. Tate, J.; Ward, G. Interferences in immunoassay. *Clin. Biochem. Rev.* **2004**, *25*, 105–120. [PubMed]
323. Spector, R.; Robert Snodgrass, S.; Johanson, C.E. A balanced view of the cerebrospinal fluid composition and functions: Focus on adult humans. *Exp. Neurol.* **2015**, *273*, 57–68. [CrossRef] [PubMed]
324. Maiolini, R.; Bagrel, A.; Chavance, C.; Krebs, B.; Herbeth, B.; Masseyeff, R. Study of an enzyme immunoassay kit for carcinoembryonic antigen. *Clin. Chem.* **1980**, *26*, 1718–1722. [CrossRef]
325. Pimm, M.V.; Perkins, A.C.; Armitage, N.C.; Baldwin, R.W. The characteristics of blood-borne radiolabels and the effect of anti-mouse igg antibodies on localization of radiolabeled monoclonal antibody in cancer patients. *J. Nucl. Med. Off. Publ. Soc. Nucl. Med.* **1985**, *26*, 1011–1023.
326. Primus, F.J.; Kelley, E.A.; Hansen, H.J.; Goldenberg, D.M. "Sandwich"-type immunoassay of carcinoembryonic antigen in patients receiving murine monoclonal antibodies for diagnosis and therapy. *Clin. Chem.* **1988**, *34*, 261–264. [CrossRef]
327. Thompson, R.J.; Jackson, A.P.; Langlois, N. Circulating antibodies to mouse monoclonal immunoglobulins in normal subjects–incidence, species specificity, and effects on a two-site assay for creatine kinase-mb isoenzyme. *Clin. Chem.* **1986**, *32*, 476–481. [CrossRef]
328. d'Abramo, C.; Acker, C.M.; Schachter, J.B.; Terracina, G.; Wang, X.; Forest, S.K.; Davies, P. Detecting tau in serum of transgenic animal models after tau immunotherapy treatment. *Neurobiol. Aging* **2016**, *37*, 58–65. [CrossRef]
329. Kricka, L.J. Human anti-animal antibody interferences in immunological assays. *Clin. Chem.* **1999**, *45*, 942–956. [CrossRef]
330. Selby, C. Interference in immunoassay. *Ann Clin Biochem* **1999**, *36 Pt 6*, 704–721. [CrossRef]
331. Wilkins, J.M.; Trushina, E. Application of metabolomics in alzheimer's disease. *Front. Neurol.* **2017**, *8*, 719. [CrossRef] [PubMed]
332. Trushina, E.; Mielke, M.M. Recent advances in the application of metabolomics to alzheimer's disease. *Biochim. Biophys. Acta* **2014**, *1842*, 1232–1239. [CrossRef] [PubMed]
333. Tambini, M.D.; Yao, W.; D'Adamio, L. Facilitation of glutamate, but not gaba, release in familial alzheimer's app mutant knock-in rats with increased β-cleavage of app. *Aging Cell* **2019**, *18*, e13033. [CrossRef]
334. Tambini, M.D.; D'Adamio, L. Knock-in rats with homozygous. *J. Biol. Chem.* **2020**, *295*, 7442–7451. [CrossRef] [PubMed]
335. Tambini, M.D.; Norris, K.A.; D'Adamio, L. Opposite changes in app processing and human aβ levels in rats carrying either a protective or a pathogenic app mutation. *Elife* **2020**, *9*, e52612. [CrossRef]
336. Saito, T.; Matsuba, Y.; Mihira, N.; Takano, J.; Nilsson, P.; Itohara, S.; Iwata, N.; Saido, T.C. Single app knock-in mouse models of alzheimer's disease. *Nat. Neurosci.* **2014**, *17*, 661–663. [CrossRef]
337. Saito, T.; Mihira, N.; Matsuba, Y.; Sasaguri, H.; Hashimoto, S.; Narasimhan, S.; Zhang, B.; Murayama, S.; Higuchi, M.; Lee, V.M.Y.; et al. Humanization of the entire murine. *J. Biol. Chem.* **2019**, *294*, 12754–12765. [CrossRef]
338. Tambini, M.D.; D'Adamio, L. Trem2 splicing and expression are preserved in a human aβ-producing, rat knock-in model of trem2-r47h alzheimer's risk variant. *Sci. Rep.* **2020**, *10*, 4122. [CrossRef]
339. Niedzwiecki, M.M.; Walker, D.I.; Howell, J.C.; Watts, K.D.; Jones, D.P.; Miller, G.W.; Hu, W.T. High-resolution metabolomic profiling of alzheimer's disease in plasma. *Ann. Clin. Transl. Neurol.* **2020**, *7*, 36–45. [CrossRef]

340. Ibáñez, C.; Simó, C.; Barupal, D.K.; Fiehn, O.; Kivipelto, M.; Cedazo-Mínguez, A.; Cifuentes, A. A new metabolomic workflow for early detection of alzheimer's disease. *J. Chromatogr. A* **2013**, *1302*, 65–71. [CrossRef]
341. Pomara, N.; Singh, R.; Deptula, D.; Chou, J.C.; Schwartz, M.B.; LeWitt, P.A. Glutamate and other csf amino acids in alzheimer's disease. *Am. J. Psychiatry* **1992**, *149*, 251–254. [PubMed]
342. Madeira, C.; Vargas-Lopes, C.; Brandão, C.O.; Reis, T.; Laks, J.; Panizzutti, R.; Ferreira, S.T. Elevated glutamate and glutamine levels in the cerebrospinal fluid of patients with probable alzheimer's disease and depression. *Front. Psychiatry* **2018**, *9*, 561. [CrossRef] [PubMed]
343. Orešič, M.; Hyötyläinen, T.; Herukka, S.K.; Sysi-Aho, M.; Mattila, I.; Seppänan-Laakso, T.; Julkunen, V.; Gopalacharyulu, P.V.; Hallikainen, M.; Koikkalainen, J.; et al. Metabolome in progression to alzheimer's disease. *Transl. Psychiatry* **2011**, *1*, e57. [CrossRef] [PubMed]
344. Trushina, E.; Dutta, T.; Persson, X.M.; Mielke, M.M.; Petersen, R.C. Identification of altered metabolic pathways in plasma and csf in mild cognitive impairment and alzheimer's disease using metabolomics. *PLoS ONE* **2013**, *8*, e63644. [CrossRef] [PubMed]
345. Kaddurah-Daouk, R.; Rozen, S.; Matson, W.; Han, X.; Hulette, C.M.; Burke, J.R.; Doraiswamy, P.M.; Welsh-Bohmer, K.A. Metabolomic changes in autopsy-confirmed alzheimer's disease. *Alzheimer's Dement.* **2011**, *7*, 309–317. [CrossRef] [PubMed]
346. Mapstone, M.; Cheema, A.K.; Fiandaca, M.S.; Zhong, X.; Mhyre, T.R.; MacArthur, L.H.; Hall, W.J.; Fisher, S.G.; Peterson, D.R.; Haley, J.M.; et al. Plasma phospholipids identify antecedent memory impairment in older adults. *Nat. Med.* **2014**, *20*, 415–418. [CrossRef]
347. Frölich, L.; Peters, O.; Lewczuk, P.; Gruber, O.; Teipel, S.J.; Gertz, H.J.; Jahn, H.; Jessen, F.; Kurz, A.; Luckhaus, C.; et al. Incremental value of biomarker combinations to predict progression of mild cognitive impairment to Alzheimer's Dementia. *Alzheimer's Res. Ther.* **2017**, *9*, 84. [CrossRef]
348. Spies, P.E.; Claassen, J.A.; Peer, P.G.; Blankenstein, M.A.; Teunissen, C.E.; Scheltens, P.; van der Flier, W.M.; Olde Rikkert, M.G.; Verbeek, M.M. A prediction model to calculate probability of alzheimer's disease using cerebrospinal fluid biomarkers. *Alzheimer's Dement.* **2013**, *9*, 262–268. [CrossRef]
349. Antonaros, F.; Ghini, V.; Pulina, F.; Ramacieri, G.; Cicchini, E.; Mannini, E.; Martelli, A.; Feliciello, A.; Lanfranchi, S.; Onnivello, S.; et al. Plasma metabolome and cognitive skills in down syndrome. *Sci. Rep.* **2020**, *10*, 10491. [CrossRef]
350. Orozco, J.S.; Hertz-Picciotto, I.; Abbeduto, L.; Slupsky, C.M. Metabolomics analysis of children with autism, idiopathic-developmental delays, and down syndrome. *Transl. Psychiatry* **2019**, *9*, 243. [CrossRef]

Review

Fluid Candidate Biomarkers for Alzheimer's Disease: A Precision Medicine Approach

Eleonora Del Prete *, Maria Francesca Beatino, Nicole Campese, Linda Giampietri, Gabriele Siciliano, Roberto Ceravolo and Filippo Baldacci

Neurology Unit, Clinical and Experimental Medicine Department, University of Pisa, 56126 Pisa, Italy; mariaf.beatino@gmail.com (M.F.B.); campesenicole@gmail.com (N.C.); lindagiampietri@gmail.com (L.G.); gabriele.siciliano@unipi.it (G.S.); r.ceravolo@med.unipi.it (R.C.); filippo.baldacci@unipi.it (F.B.)

* Correspondence: ele.delprete86@gmail.com

Received: 11 September 2020; Accepted: 5 November 2020; Published: 11 November 2020

Abstract: A plethora of dynamic pathophysiological mechanisms underpins highly heterogeneous phenotypes in the field of dementia, particularly in Alzheimer's disease (AD). In such a faceted scenario, a biomarker-guided approach, through the implementation of specific fluid biomarkers individually reflecting distinct molecular pathways in the brain, may help establish a proper clinical diagnosis, even in its preclinical stages. Recently, ultrasensitive assays may detect different neurodegenerative mechanisms in blood earlier. ß-amyloid (Aß) peptides, phosphorylated-tau (p-tau), and neurofilament light chain (NFL) measured in blood are gaining momentum as candidate biomarkers for AD. P-tau is currently the more convincing plasma biomarker for the diagnostic workup of AD. The clinical role of plasma Aβ peptides should be better elucidated with further studies that also compare the accuracy of the different ultrasensitive techniques. Blood NFL is promising as a proxy of neurodegeneration process *tout court*. Protein misfolding amplification assays can accurately detect α-synuclein in cerebrospinal fluid (CSF), thus representing advancement in the pathologic stratification of AD. In CSF, neurogranin and YKL-40 are further candidate biomarkers tracking synaptic disruption and neuroinflammation, which are additional key pathophysiological pathways related to AD genesis. Advanced statistical analysis using clinical scores and biomarker data to bring together individuals with AD from large heterogeneous cohorts into consistent clusters may promote the discovery of pathophysiological causes and detection of tailored treatments.

Keywords: biomarkers; Alzheimer's disease; neurodegeneration; cerebrospinal fluid; mild cognitive impairment; synaptic biomarkers; neuroinflammation; neurofilament light chain

1. Introduction

Alzheimer's disease (AD) is the most common neurodegenerative disease (NDD), with 5.8 million Americans aged 65 years and older living with AD in 2020 [1]. Since Alois Alzheimer's first description of the typical histological alterations of neuritic plaques (NP) and neurofibrillary tangles (NFT) in 1906 [2], more than eighty years have passed before amyloid beta (Aβ) and phosphorylated-tau (p-tau) were identified as the main component of NP and NFT, respectively [3]. In 1984, the National Institute of Neurological and Communicative Disorders and Stroke-Alzheimer's Disease and Related Disorders Association (NINCDS-ADRDA) [4] set postmortem examination as the reference standard of AD diagnosis. Since then, the broad phenotypical variability of neurodegenerative diseases (NDDs) has pushed the efforts toward developing a classification based on the main misfolded protein deposition [5,6]. Nevertheless, the occurrence of these aggregates in multiple combinations is frequent, and NDDs are rather emerging as a spectrum of disorders characterized by the loss of proteostasis [7].

Due to the failure of numerous trials against amyloid pathology, the idea of "one drug fits all" treatment as an ultimate solution for an AD cure is fading [8]. In brief, the current framework on

AD is more complex than previously thought because AD is not a mere plaque and tangle disorder. The following pathophysiological pathways leading to neurodegeneration have been recognized as clearly implicated in AD pathogenesis: (1) accumulation of misfolded proteins in the brain (Aβ peptides, tau and p-tau proteins, other co-pathologies), (2) vascular dysfunction, (3) synaptic disruption, and (4) neuroinflammation. The discovery of biomarkers indicating the modification of these processes at different levels in space and time is gaining momentum, especially in design tailored disease-modifying trials (Figure 1).

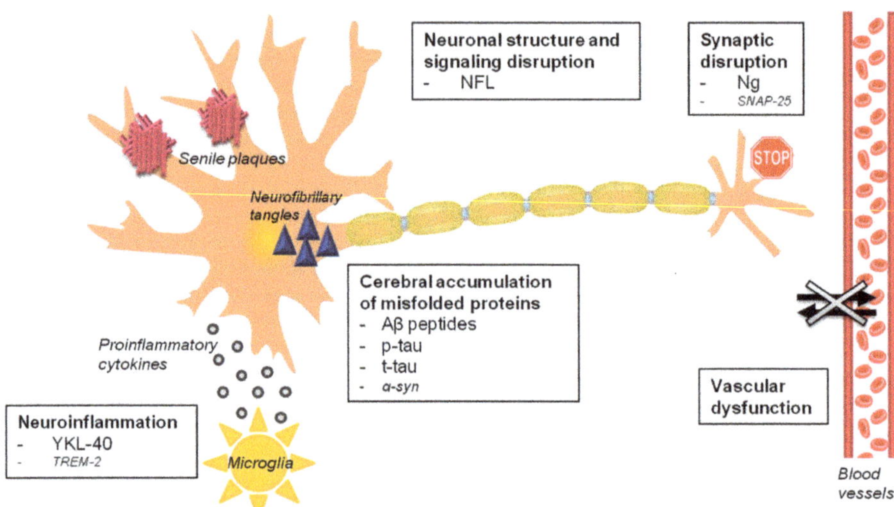

Figure 1. Alzheimer's disease fluid biomarkers. The major pathophysiological processes involved in Alzheimer's disease (in bold) with validated and proposed fluid biomarkers are schematically represented. Fluid biomarkers of vascular dysfunction, and of TAR DNA binding protein 43 (TDP-43) and α-syn pathologies are still missing. Abbreviations: *Aβ*, β-amyloid, *α-syn*, α-synuclein; *NFL*, neurofilament light chain; *Ng*, neurogranin; *p-tau*, phosphorylated tau protein; *t-tau*, total tau protein, synaptosomal-associated protein 25 (SNAP-25), and triggering receptor expressed on myeloid cells 2 (TREM2).

Our aim is to review the development of novel candidate fluid biomarkers tracking these key pathophysiological mechanisms in different matrices, especially cerebrospinal fluid (CSF) and blood. In relation to AD, we mainly focused on the diagnostic and prognostic value of these biomarkers, with particular attention to the novel ultrasensitive techniques.

2. Literature Search Methods

We performed a narrative review of literature focusing on novel candidate fluid biomarkers for AD. A systematic review of literature focused on plasma biomarkers detected by means of novel ultrasensitive techniques was performed in PubMed. We used the combination of the keywords "plasma", "serum", "amyloid-β", "NFL" (neurofilament light chain), "p-tau", "p-tau181", "phopsho-tau181", "phosphorylated tau181", "t-tau", "Simoa", "immunoassay", "immunomagnetic reduction", "fully automated", "immuno-infrared sensor", "mass spectrometry", and "multimer detection system". Only papers in English published between 2014 and July 2020 and focused on AD were included in the final analysis. Overall, we identified 21 studies that provided relevant diagnostic and/or prognostic information (Figure 2). Among them, 10 were focused on amyloid-β peptides, 7 were focused on p-tau or tau or both, and 4 were focused on NFL. For each paper, the study population, the study design (cross-sectional, perspective, retrospective), and the diagnostic and/or prognostic value of the investigated biomarker were analyzed. We classified the diagnostic value of each biomarker according to previously published recommendations as follows: "excellent" (area under ROC curve [AuROC] 0.90–1.00), "good" (AuROC 0.80–0.89), "fair or moderate" (AuROC 0.70–0.79), "poor" (AuROC 0.60–0.69), or "fail or insufficient" (i.e., no discriminatory capacity) (AuROC 0.50–0.59).

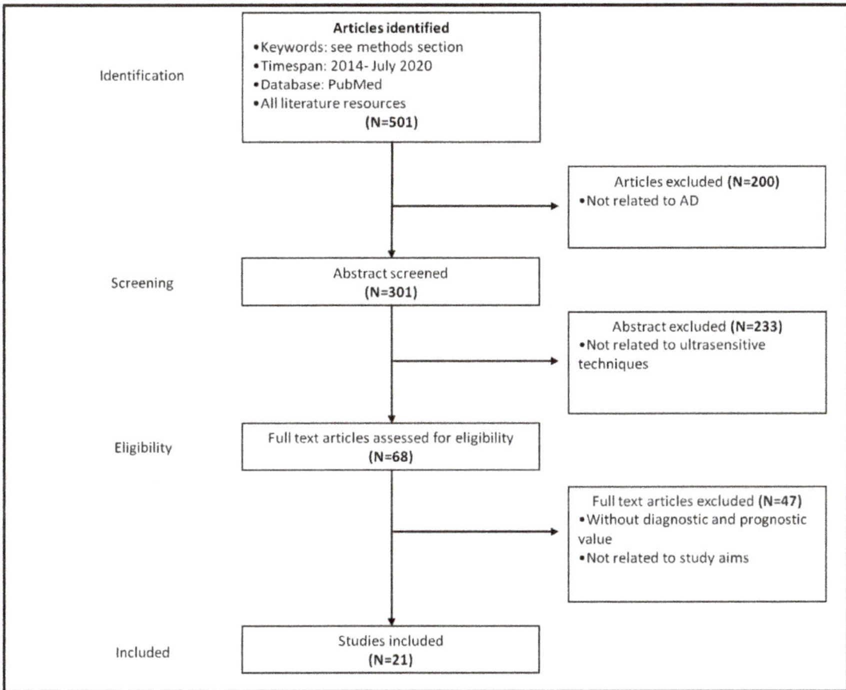

Figure 2. Flowchart displaying the article selection process.

3. Toward a Pathophysiological Definition of Alzheimer's Disease

With the 1984 NINCDS-ADRDA criteria, the accuracy for probable AD diagnosis was suboptimal, with sensitivity between 70.9% and 87.3% and specificity between 44.3% and 70.8% [9]. The definite diagnosis relied on postmortem examination, with obvious limitations, since it is not applicable in vivo. For this reason, the International Working Group (IWG) [10] and later the National Institute on Aging and Alzheimer's Association (NIA-AA) [11] published novel criteria for the diagnosis of AD incorporating in vivo biomarkers. According to the 2007 IWG criteria, AD can be identified in vivo by the presence of amnestic syndrome of the hippocampal type, which is characterized by low free recall that does not improve with cueing. Moreover, biomarkers must be consistent with AD pathology. These biomarkers are pathophysiological and topographical. The pathophysiological ones are low CSF Aβ1-42 peptide concentration, high CSF total tau (t-tau) or p-tau levels, and an increased cerebral uptake of amyloid tracers (e.g., Pittsburgh compound) with PET. The formers are hippocampal atrophy on volumetric Magnetic Resonance Imnaging (MRI) and cortical regional hypometabolism on fluorodeoxyglucose FDG-PET, involving bilateral temporal parietal regions and posterior cingulate. IWG criteria managed to move from the static and binary/dichotomic vision of AD as a clinicopathological entity to its current dynamic clinical-radio-biological description [10]. The subsequent 2010 revision of IWG criteria overtook the amnestic-centered concept of AD and broadened the spectrum, adding the rarer atypical forms of AD, such as primary progressive non-fluent aphasia, in particular logopenic aphasia, posterior cortical atrophy, and frontal variant AD. The-so-called "asymptomatic at risk of AD" condition without clinical symptoms but with positive biomarkers of AD pathology was stated out, as well as the concept of "mixed AD", implying the co-occurrence of clinical and biological features of other disease, such as parkinsonism (e.g., Lewy body pathology) or cerebrovascular disease [12]. Later on, these concepts were implemented in the IWG-2 criteria (2014) [13], where clinical diagnosis required specifying whether typical or atypical AD phenotypes occurred. Furthermore, the condition of a preclinical AD stage (for asymptomatic at risk and presymptomatic subjects) was defined in the presence of cognitive normal individuals with biomarkers indicative of AD pathophysiological process. Topographical biomarkers were used only for disease staging and monitoring. In parallel with IWG criteria, the NIA-AA diagnostic guidelines developed in 2011 [14–16] moved forward, defining the concept of mild cognitive impairment (MCI) due to AD (clinical MCI individuals with biomarkers indicating AD pathology). In fact, MCI due to AD had a high likelihood of developing AD over time. Subsequently, in 2016, the joint IWG-Alzheimer's Association (IWG-AA) formalized a purely biological definition of AD, based on the positivity of biomarkers of both amyloidosis and tauopathy [17]. In the same years, the "A/T/N" classification system for AD was published. In this classification, the validated AD biomarkers were reported into three binary categories (presence or absence) based on the nature of the pathophysiology. "A" refers to the ß-amyloid pathology (cerebral amyloid PET or CSF Aß42); "T," refers to taupathology (CSF p-tau, or cerebral tau PET); and "N" refers to neurodegeneration or neuronal injury *tout court* ([18F]-fluorodeoxyglucose-PET, structural MRI, or CSF total tau [18]). This unbiased biomarker-based scheme was recently incorporated in the current NIA-AA criteria published in 2018, with the addition of C for clinical change, to integrate the biomarkers condition with clinical cognitive status [19]. All A+ individuals are considered part of the "Alzheimer's continuum", while only A+ and T+ define AD. Non AD-specific parameters, namely neurodegenerative/neuronal injury biomarkers (N) and cognitive symptoms (C), define staging [19]. A- individuals fall either in the "normal AD biomarker" category with A-T-(N-), or "Suspected non-Alzheimer's pathophysiology" (SNAP) with A-T+(N)-, A-T-(N)+, or A-T+(N)+.

Among the mimics of typical AD-type dementia, Primary Age-Related Tauopathy (PART) should be mentioned [20,21]. PART identifies individuals with cerebral NFT indistinguishable from those of AD, in the absence of Aβ plaques; notably, NFT are restricted to the medial temporal lobe, basal forebrain, brainstem, and olfactory areas [21]. At a clinical level, associated manifestations range from normal cognition to amnesic cognitive impairment, but they are rarely a frank dementia. Similarly, a recently described entity is the limbic-predominant age-related TDP-43 encephalopathy (LATE) [22]. LATE is a common TDP-43 proteinopathy that generally affects older adults, and it is frequently associated with hippocampal sclerosis. Aβ plaques or tauopathy may also coexist. Generally, co-pathologies in AD subjects are common with approximately 30% of AD patients showing a cerebrovascular disease [23]. The concomitant deposition of Aβ and α-syn proteins is also described in postmortem examination in about 30% of AD individuals [24,25] but also in up to 40% of patients with Parkinson's disease (PD), Parkinson disease dementia (PDD), and dementia with Lewy bodies (DLB) clinical diagnoses [26–28].

Thus, in this scenario, it is likely that no single biomarker could reach a 100% diagnostic accuracy, being AD biologically multifaceted with a clinical picture reflecting pathology only in terms of probabilistic association. Despite these intrinsic limitations, the use of core biomarkers in the AD diagnostic workup improves accuracy (up to 90%) with a relevant impact on AD stratification and selection for disease-modifying trials tailored against Aß and tau pathologies [29].

To date, the neuropathological hallmarks of AD remain extracellular Aβ plaques and NFTs [30,31]. First proposed in 1992, the "amyloid cascade hypothesis" [32] has been later corroborated by genetic and biochemical data and currently represents the dominant pathogenetic model of AD. According to this hypothesis, the deposition of fibrillar Aβ plaques within the brain promotes the accumulation of NFTs, synaptic disintegration and neuronal death by inflammatory mechanisms, modification of ions homeostasis, kinase/phosphatase activity, and oxidative stress [33]. In particular, Aβ plaques create an unique environment that facilitates tau aggregation, initially as dystrophic neurites surrounding Aβ plaques, followed by the formation and spread throughout the brain in a prion-like manner of NFTs and neuropil threads [34]. NFTs are characteristic of AD and are composed of hyperphosphorylated tau [35–37]. The hyperphosphorylation of tau protein reduces its affinity for microtubules and promotes its capacity to aggregate and fibrillize [38]. Therefore, microtubules are destabilized, and axonal transport is impaired [39].

The hyperphosphorylated tau could also migrate in the somatodendridic compartments where it interacts with Aβ and enhances synaptotoxicity [40], finally causing cell death due to a toxic gain of function mechanism [41,42].

At the same time, much interest is growing around the role of inflammation in the pathogenesis of AD. The contribution of inflammation to the pathophysiology of AD has been already hypothesized more than 20 years ago [43–45]. The attention has been focused especially on microglia activation, which seems to occur decades before AD onset [46–48]. Furthermore, a correlation between neuroinflammation and amyloid or tau accumulation in the human brain has been reported in several investigations [46,49–51]. The microglial activation produces two different phenotypes. The microglial "pro-inflammatory" phenotype (M1) displays pro-inflammatory cytokines (IL-1β, IL-6, IL-12, tumor necrosis factor (TNF)-α, CCL2), nitric oxide, reactive oxygen and nitrogen species. The "anti-inflammatory" one (M2) sustains the production of IL-10 and TGF-β, and it increases the expression of neurotrophic factors (nerve-derived growth factor (NGF), brain-derived neurotrophic factor (BDNF), neurotrophins, glial cell–derived neurotrophic factor (GDNF)), and several other signals involved in downregulation, protection, or repair processes [52]. The chronic stimulus on microglia by Aβ peptides accumulation is likely to lead to a protracted inflammation, and, in turn, increase Aβ deposition, in a vicious circle [53]. The inflammatory state would promote the production and release of pro-inflammatory cytokines, which could themselves have a detrimental effect by inducing neuronal cell death.

Another relevant key pathophysiological mechanism that contributes to AD is vascular dysregulation [54]. Several pieces of evidence support the role of chronic cerebral hypoperfusion as the *primum movens* of AD pathology [55,56]. Hypoxia can activate β-secretase-1 and γ-secretase as well as increase Aβ peptides accumulation [57]. Furthermore, the reduced supply of oxygen and nutrients affects neurons per se, and, in turn, it promotes blood–brain barrier dysfunction, increasing oxidative stress and inflammation [58]. Since Aβ deposition derives essentially from an imbalance between production and removal, clearing system impairment is emerging as a further key pathophysiological mechanism leading to AD. In particular, this mechanism involves the alteration of astroglial-mediated interstitial fluid (ISF) bulk flow or glymphatic system [59,60]. This pathway is mainly modulated by the sleep–wake cycle and seems to be important for the sleep-driven clearance of Aβ [61]. Vascular pathology seems to be additive or even synergic to AD pathology as a cause of cognitive impairment [62,63]. This cross-talk is most evident for cerebral amyloid angiopathy (CAA), which shares Aβ deposition with AD typical neurotic plaques that are localized within leptomeningeal and intracortical arteries, arterioles, and capillaries. CAA is commonly found in AD brains: up to approximately 50% of subjects with severe NP load [64]. CAA can affect perivascular drainage impairing glymphatic circulation, thus reducing a major route of Aβ clearance from the brain [59]. Intracranial atherosclerosis was found to be an additional, although not strictly neurodegenerative, strong risk factor for AD dementia [65].

4. Fluid Biomarkers: Ultrasensitive Measurement Techniques

Due to several advantages over invasive (e.g., CSF Aβ peptides), expensive and scarcely available (e.g., cerebral amyloid-PET) diagnostic tools, technologies aiming at quantifying NDDs biomarkers in blood are gathering momentum.

However, the discovery of CNS-related biomarkers in blood presents challenging issues: (a) the concentration of a biomarker released in CNS is lower than in CSF, considering that it has to cross the blood–brain barrier and that the blood volume is consistently larger than the CSF one, (b) biomarkers could also be directly expressed peripherally, and the contribution of CNS might be difficult to quantify,

(c) proteolytic degradation of the analytes by plasma proteases and confounding blood proteins may interfere with biomarker measurement [66].

The traditional enzyme-linked immunosorbent assay (ELISA) was extensively used in the last few decades. It showed a substantial intrinsic variability in the quantification of plasma/serum biomarkers and provided overlapping results in the discrimination between NDDs and cognitively healthy subjects [67,68]. The large sample volume required in the analysis combined with a sensitivity limited to the picomolar range could be addressed as the main weakness of this method. Therefore, ultrasensitive techniques often representing ELISA-based evolutions have been developed for blood biomarker discovery.

The automated xMAP (multi-analyte profiling) Luminex technology, a flow cytometric method, allows the adaptation of several immunoassay formats to simultaneously detect multiple analytes on different sets of microspheres in a single well [69]. Through pre-made calibrators, it reduces measurement variability, partially overcoming some limitations of conventional ELISA methods (Table 1).

Another emerging technique is the single-molecule array (Simoa), which is essentially a digital ELISA. This fully automated method is based on capturing antibody-coated paramagnetic beads loaded in arrays of femtomolar-sized reaction chambers with a volume 2 billion times smaller than conventional ELISA. Ultimately, by acquiring fluorescence images, an increase in signal will reflect the presence of single enzyme-associated immunocomplexes [70] (Table 1). This method is a candidate prescreening tool but may be potentially useful throughout the whole AD spectrum [71]. Large-scale longitudinal multicenter studies are anyway needed for the standardization and harmonization of preanalytical and analytical variables [72].

Combining the unique advantages of highly specific immunoreactions and electrochemiluminescence (ECL) biosensors, ECL immunoassays (ECLIA) have been implemented in several automated platforms. A wide dynamic detection range, low background noise, and simple optical set-ups are the strengths of this technique [73,74] (Table 1).

For protein analysis, immunoprecipitation has also been coupled with mass-spectrometry (IP-MS) providing a robust quantitative tool to identify antigens based on their intrinsic chemicophysical properties [75]. A significant advantage of this method is the possibility to analyze complex mixtures of Aβ peptides at very low concentrations in a single assay. However, multi-step structured analysis strategies are required in order to reduce the influence of non-specific binders and improve the signal quality (Table 1) [76]. A tailored approach for the identification of oligomers has been adopted in the Multimer Detection System (MDS), which is a modified sandwich ELISA originally designed to detect prion proteins. As opposed to the conventional method, this strategy relies on two epitope-overlapping antibodies for capturing and detecting an epitope, so that only multimers will bind to both antibodies, allowing their selective detection over monomers, which conversely will only bind to one of them (Table 1) [77,78].

Table 1. Key points of ultrasensitive techniques for the detection of putative blood biomarkers for AD.

METHOD	PROS	CONS
xMAP	It is a flexible technology with a workflow ranging from semi- to fully automated options. It enables the concomitant evaluation of multiple analytes in a single sample representing a time-, cost-, and labor-saving method. It enables a shift from a hypothesis-based analysis of known targets to a data-driven approach [83,84].	The simultaneous measurement of multiple ligands may favor cross-reactivities ("matrix effect"). A rigorous adherence to the manufacturer's protocols is required to minimize any artifacts when using these kits [85].

Table 1. *Cont.*

METHOD	PROS	CONS
Simoa	It is a fully automated technology based on antibody-coated paramagnetic microbeads. It has a great sensitivity (×1000 greater compared to conventional immunoassays), being able to detect single proteins at subfemtomolar concentrations. It is capable of multiplexing with short turnaround times and a remarkable throughput (up to 66 samples/h). It represents the most established ultrasensitive technology for blood biomarkers of AD to date (kits to measure Aβ1-42, p-tau181, t-tau, and NFL are available). A higher sensitivity compared to both ELISA and ECLIA-based methods was shown for the detection of NFL in serum [86–88].	Wide longitudinal multicenter studies are warranted for the standardization of preanalytical and analytical protocols parameters [72,88].
ECLIA (MSD, Elecsys)	ECLIA-based methods are adopted in semi- to fully automated (MSD) and fully automated (Elecsys) platforms. MSD is a flexible multi-array technology enabling the detection of biomarkers in single and high throughput multiplex formats. It provides a high inter-laboratory reproducibility, low matrix effects, reliability and cost-effectiveness [73,74]. Aβ peptides measured with Elecsys showed among the best accuracies in predicting the Aβ status assessed by either amyloid-PET or CSF Aβ1-42/Aβ1-40 ratio when compared to other techniques [89]. MSD provides good to optimal accuracy regarding the discriminative role of plasma p-tau181 to detect AD [90,91].	The accuracy of the Aβ1-42 and Aβ1-40 Elecsys assays is still suboptimal and insufficient to enable the use of these techniques alone as clinical tests of Aβ positivity. Additional cross-evaluations are needed before these ECLIA-based methods can be recommended [89].
IP-MS	It is able to characterize and quantify peptides by introducing them into the mass spectrometer after isolation through antibody-driven immunoprecipitation. Using this technique, optimal discriminative accuracies in detecting AD were reached by the Aβ1-40/Aβ1-42 ratio measured in plasma [75].	Antibodies and solid matrices also isolate many non-specific "contaminants". To reduce the interferences with the signals and increase specificity in the detection of the antigens, targeted precautions are recommended (e.g., two rounds of repeated processing during the immunoprecipitation) [75,76]. Compared to automated ELISA-based techniques, IP-MS is a labor-intensive, low-throughput and time-consuming method not easily implementable on a wide scale [92].
MDS	It is an ELISA-based sandwich assay aiming at measuring oligomerization tendency in blood. It uses capture antibodies and epitope-overlapping detection antibodies to identify oligomers or multimers [93].	Its sensitivity in detecting Aβ oligomers failed to reach the cut-off of >80% that is needed for the validation of a biomarker [94].
Immuno-infrared sensor	It is an antibody-based method to extract all the Aβ peptides from blood samples, allowing the identification of β-sheet enriched conformations [79]. Compared to established ELISA-based tests, it does not measure the absolute biomarker concentration but the relative frequency shift in the infrared, reducing the influence of concentration fluctuations caused by biological variances [80]. Unique features of this assay are the absence of labels (enzymes, fluorescent or radioactive molecules) with potentially confounding effects, being the analytes detected based on their intrinsic physical properties, a simple and low-cost procedure and the low sample volume needed. It is able to identify the initial Aβ misfolding, occurring several years before clinical manifestation of AD [80].	Further tests in different clinical set-ups are needed to investigate the potential effects of sample handling and to evaluate their potential as screening-assays [79–81].
IMR	It measures the change in magnetic susceptibility caused by the association of antigens with antibody coated paramagnetic nanobeads [82]. In contrast to ultrasensitive digital ELISA methodologies, IMR is a single-antibody immunoassay. Less stereoscopical interferences and a better ability to detect Aβ1-42 molecules in different conformations (isolated, complex or oligomeric forms) are strengths of this technique [95].	In regard to Aβ peptides, it provides results that are not consistent with those of the ELISA- and MS-based methods. The unspecific detection of Aβ aggregates or Aβ binding proteins likely caused by the single-antibody nature of the technique may explain the increase of plasma Aβ1-42 levels in AD patients compared to healthy controls [96].

Abbreviations: Aβ: amyloid β; Aβ1-40: amyloid β-peptide 1-40; Aβ1-42: amyloid β-peptide 1-42; AD: Alzheimer's disease; CSF: cerebrospinal fluid; ECLIA: electrochemiluminescence immunoassay; ELISA: enzyme-linked immunosorbent assay; IMR: immunomagnetic reduction; IP-MS: immunoprecipitation coupled with mass spectrometry; MDS: multimer detection system; MSD: meso scale discovery; NFL: neurofilament light chain; p-tau: phosphorylated-tau; Simoa: single molecule array; t-tau: total tau; xMAP: multi-analyte profiling.

Among the recently developed biosensors in AD research, the immuno-infrared sensor represents a promising label-free technique not aiming at discriminating particular Aβ species, but rather aiming at identifying the secondary structure distribution of all misfolded peptides. Thus, it is potentially exploitable in a preclinical setting (Table 1) [79–81].

A further virtuous application of the immunoassay principles is part of the immunomagnetic reduction (IMR) technique, in which magnetic antibody-coated nanoparticles dispersed in aqueous solution oscillate under external multiple alternating current (AC) magnetic fields. The association of target molecules determines a reduction in the AC magnetic susceptibility of capturing nanoparticles that will be as high as the concentration of the analytes (Table 1) [82]. Compared to ELISA-based techniques (e.g., Simoa), this method does not make use of beads to purify or concentrate antigens, and it is virtually able to quantify smaller proteins in higher number. Whether this could represent an advantage to detect Aβ peptides in plasma is still to be elucidated (Table 1).

Huge efforts have been made to develop and refine these technologies. Nevertheless, there is an urgent need to promote unbiased cross-platform evaluations for an effective method standardization.

In view of a targeted-oriented approach to AD, the adoption of guidelines to systemize preanalytical and postanalytical procedures across laboratories, aiming at finding consensus on a high-performance scalable platform for the discovery and approval of blood biomarkers, would be strongly recommended.

5. Biomarkers Tracking Amyloid Pathology

Accumulating evidence from the clinical research consistently supports that CSF $A\beta_{1-42}$ peptide shows an inverse correlation with plaque load in the brain [97–101] and provides important diagnostic information throughout the continuum of the AD spectrum. Therefore, this biomarker is currently integrated in the diagnostic criteria of AD; it is used for subject selection in clinical trials and approved in medical practice as well [13,14,19]. On the contrary, CSF $A\beta_{1-40}$ peptide alone, albeit prevailing over the other Aβ species in both CNS and periphery, showed no relevant correlation with AD dementia [76,87–89]. Notably, the ratio of CSF $A\beta_{1-42}/A\beta_{1-40}$ has been found to predict cortical amyloid-PET positivity more accurately than CSF $A\beta_{1-42}$ alone [90–92], improving the discrimination of AD vs. non-AD demented patients (Table 2) [93,94].

Table 2. Overview on the possible context of use of fluid biomarkers in AD.

		Diagnostic Value			Prognostic Value	Monitoring Treatment
		Preclinical Phase	Prodromal Phase	Full-Blown Picture		
			Amyloid pathology			
Aβ peptides	Blood			+		
Aβ peptides	CSF		+	+	+	+
			Tau pathology			
p-tau	Blood	+	+	+	+	
			Neuroinflammation			
YKL-40	CSF			+	+	
			Synaptic dysfunction			
Ng	CSF		+	+	+	
			Neuronal structure and signaling disruption			
NFL	CSF		+	+	+	
	Blood		+	+	+	

Legend: plus sign (+): potential use, supportive data available. Abbreviations: Aβ: amyloid beta; t-tau: total tau; p-tau: phosphorylated-tau; YKL-40; Ng: neurogranin; NFL: neurofilaments; CSF: cerebrospinal fluid.

Some investigations suggest that also the CSF $A\beta_{1-42}/A\beta_{1-38}$ ratio turned out to improve cerebral amyloid deposition compared with CSF $A\beta_{1-42}$ alone [94–96]. In addition, in pharmacological trials, short Aβ peptides detection in CSF may help monitor patients receiving drugs that modulate γ-secretase (Table 2) [97,98]. Among Aβ species, Aβ oligomers that are likely to play a key role in AD pathogenesis could be potentially used as preclinical biomarkers in CSF. Unfortunately, the detection of Aβ species and Aβ oligomers is challenging due to their polymorphous and unstable nature. Moreover, their concentration in biofluid is low, and they compete with other proteins and Aβ monomers. For the aforementioned reasons, most of the existing techniques are not satisfactory and reporting conflicting results so far [99]. As it also emerged from the Olsson and colleagues meta-analysis in 2016, $A\beta 1_{1-42}$ and $A\beta_{1-40}$ peptides measured in blood with traditional ELISA methods did not discriminate AD from healthy controls [95].

Undoubtedly, the validation of relatively new technologies in the last few years—e.g., Simoa [100], immunoprecipitation-mass spectrometry (IP-MS) assays [74], stable labeling kinetics protocols [101], multimer detection system (MDS) [102], xMAP technology [103], immuno-infrared sensor [104], electrochemiluminescence immunoassays (ECLIA) [105–112]—led to a significant increased sensitivity in amyloid peptides detection in periphery when compared to the conventional ELISA technique, with drastically lower concentrations (up to the femtolitre) in blood than in CSF [65]. Particularly, in a 2017 study based on an IP-MS method, the plasma Aβ1-42/Aβ1-40 ratio was significantly lower in amyloid-PET positive compared with amyloid-PET negative participants. This ratio reported a good accuracy in distinguishing the two populations [101] (Table 3) [113–131].

In 2018, Nakamura and colleagues used the same technique to measure Aβ peptides in plasma of cognitive normal individuals, MCI and AD with dementia subjects, finding higher levels of plasma $A\beta_{1-40}/A\beta_{1-42}$ ratio in amyloid-PET positive compared with amyloid-PET negative individuals [75]. Regardless of the clinical diagnosis, the ratio of $A\beta_{1-40}$ and $A\beta_{1-42}$ peptides ($A\beta_{1-40}/A\beta_{1-42}$), and that of Aβ precursor protein fragment ($APP_{669-711}$) and $A\beta_{1-42}$ ($APP_{669-711}/A\beta_{1-42}$), had an excellent diagnostic accuracy in discriminating cerebral amyloid-PET positive and amyloid-PET negative subjects (Table 3). Similarly, in another study including a cohort of subjective memory complainers, a condition at risk for AD, the plasma $A\beta_{1-40}/A\beta_{1-42}$ ratio, turned out to be the best predictor of cerebral amyloidosis among a series of tested candidate biomarkers (e.g., β-site amyloid precursor protein cleaving enzyme 1 or BACE1, t-tau, NFL [72] (Table 3). Additional investigations performed using the Simoa technique reported a moderate accuracy of $A\beta_{1-42}/A\beta_{1-40}$ ratio in identifying the amyloid status. The $A\beta_{1-42}/A\beta_{1-40}$ ratio was lower in amyloid-PET positive compared with amyloid-PET negative participants [122,132] (Table 3). A further study in which Aβ peptides concentrations were assessed with a fully automated ECLIA technique confirmed the good discriminative role of $A\beta_{1-42}/A\beta_{1-40}$ ratio, in both validation and discovery cohorts [92] (Table 3). Similarly, Hanon and colleagues in a large investigation including 1040 MCI or AD participants reported that plasma $A\beta_{1-42}$ and $A\beta_{1-40}$ concentrations assessed by means of a kit based on a multiplex xMAP technique were lower in AD than in both MCI and non amnestic MCI (naMCI), suggesting a gradual decrease of these peripheral biomarkers with the course of the disease, in accordance to previous findings [109,133,134]. Conversely, in some studies where IMR was used, plasma $A\beta_{1-42}$ concentrations are higher in AD patients than in healthy subjects [95,121]. Lue and colleagues reported a moderate diagnostic accuracy of this plasma biomarker in one cohort and excellent in the other [121] (Table 3). Moreover, in a 2018 study carried out by Nabers and colleagues using an immuno-infrared sensor, when compared with controls, not only were the concentrations of β-sheet-enriched Aβ peptides higher in severe AD dementia patients, as previously demonstrated [80], but they were also higher in prodromal AD patients, reaching a good diagnostic accuracy in identifying the amyloid status assessed by PET (Table 3) [80]. A recently developed ELISA method detecting Aβ multimers from monomers (MDS) showed a good accuracy of plasma Aβ oligomers in distinguishing AD patients from healthy controls [118] (Table 3). In parallel, in a subsequent study in which traditional ELISA was applied, plasma BACE-1 increase has emerged as another surrogate hallmark of AD progression [135]. In an effort to assess the plasma concentrations of $A\beta_{1-38}$, $A\beta_{1-40}$ and $A\beta_{1-42}$ simultaneously, Shahpasand-Kroner and colleagues found that the $A\beta_{1-42}/A\beta_{1-40}$ and $A\beta_{1-42}/A\beta_{1-38}$ ratios are significantly lower in patients with AD dementia than in patients with dementia due to other causes and have a good accuracy in differentiating the two groups [120] (Table 3).

Table 3. Diagnostic and prognostic role of blood Aβ peptides, p-tau, t-tau, and NFL proteins measured with ultrasensitive techniques in AD.

Reference	Population	Study Design	Technique	Diagnostic Value	Prognostic Value
			Aβ peptides		
Ovod V. et al., 2017 [113]	N = 41 (CU, AD dementia)	Longitudinal	IP-MS and stable labeling kinetics protocols	$A\beta_{1-42}/A\beta_{1-40}$ in differentiating amyloid positive participants vs. negative: AuROC = 0.89 with amyloid-PET and CSF $A\beta_{1-42}$ as reference standards	NA
Wang M. et al., 2017 [114]	N = 61 (CU, AD dementia)	Cross-sectional	MDS	Aβ oligomers in differentiating AD patients vs. CU subjects: AuROC = 0.84 with clinical diagnosis (AD) as reference standard	NA
Lue L. et al., 2017 [115]	N = 124 (CU, AD dementia); U.S. cohort: N = 32; Taiwan cohort: N = 92	Cross-sectional	IMR	$A\beta_{1-42}$ in differentiating AD patients vs. CU subjects: AuROC = 0.69 (U.S. cohort); AuROC = 0.96 (Taiwan cohort) with clinical diagnosis (AD) as reference standard	NA
Nakamura A et al., 2018 [74]	N = 484 (CU, MCI, AD)	Cross-sectional (retrospective)	IP-MS	APP/$A\beta_{1-42}$ and $A\beta_{1-40}/A\beta_{1-42}$ in differentiating amyloid positive participants vs. negative: AuROC ≈0.90 compared with amyloid-PET as reference standard	NA
Nabers A. et al., 2018 [80]	N = 385 (CU, prodromal AD, AD); Sweden cohort: N = 73; Germany cohort: N = 312	Cross-sectional and nested case control	Immuno-infrared sensor	β-sheet-enriched Aβ peptides in differentiating: - amyloid positive participants vs. negative: AuROC = 0.78 (Sweden cohort) compared with amyloid-PET as reference standard; - AD vs. CU subjects: AuROC = 0.80 (Germany cohort)	NA
Shahpasand-Kroner H. et al., 2018 [120]	N = 40 (AD dementia, dementia due to other reasons)	Cross-sectional	ECLIA	$A\beta_{1-42}/A\beta_{1-40}$ in differentiating AD dementia vs. dementia due to other reasons: AuROC = 0.87 with clinical diagnosis as reference standard	NA
Verberk I. et al., 2018 [71]	N = 248 (SMC)	Longitudinal	Simoa	$A\beta_{1-42}/A\beta_{1-40}$ in differentiating amyloid positive SMC vs. negative: AuROC = 0.77 with CSF $A\beta_{1-42}$ and amyloid PET as reference standards	Low $A\beta_{1-40}/A\beta_{1-42}$ is associated to MCI or dementia conversion (HR = 2.0) also after correcting for age and sex (HR=1.67)
Palmqvist S. et al., 2019 [89]	N = 1079 (CU, MCI, AD) Sweden cohort: N = 842 Germany cohort: N = 237	Multicenter and longitudinal	ECLIA	$A\beta_{1-42}$ + $A\beta_{1-40}$ (used as separate predictors in a logistic regression) in differentiating amyloid negative participants vs. positive: AuROC = 0.80 (Sweden cohort) and AuROC = 0.86 (Germany cohort) compared with CSF $A\beta_{1-42}/A\beta_{1-40}$ ratio as reference standard	NA
Vergallo A. et al., 2019 [72]	N = 276 (SMC)	Longitudinal	Simoa	$A\beta_{1-40}/A\beta_{1-42}$ in differentiating amyloid positive SMC vs. negative: AuROC = 0.77 compared with amyloid-PET as reference standard	NA
Chatterjee P. et al., 2019 [122]	N = 95 (CU)	Cross-sectional	Simoa	$A\beta_{1-40}/A\beta_{1-42}$ along with age and APOE ε4 status in differentiating amyloid positive participants vs. negative: AuROC = 0.78 compared with amyloid-PET as reference standard	NA

Table 3. *Cont.*

Reference	Population	Study Design	Technique	Diagnostic Value	Prognostic Value
			p-tau and t-tau proteins		
Mielke MM. et al., 2017 [123]	N = 458 (CU, MCI)	Longitudinal	Simoa		Both the middle (HR = 2.43) and the highest (HR = 2.02) tertiles of plasma t-tau levels are associated with increased risk of MCI in CU participants
Mielke MM. et al., 2018 [124]	N = 269 (CU, MCI, AD)	Cross-sectional	Simoa	In the discrimination between amyloid negative participants vs. positive: - plasma p-tau181: AuROC = 0.80; - plasma t-tau: AuROC = 0.60 compared with amyloid-PET as reference standard	NA
Yang C. et al., 2018 [125]	N = 73 (CU, MCI, very mild AD)	Cross-sectional	IMR	Plasma p-tau181 discriminating: - CU vs. MCI due to AD: AuROC = 0.85; - MCI due to AD vs. mild AD: AuROC = 0.78 with clinical diagnosis as reference standard	NA
Park JC. et al., 2019 [126]	N = 76 (CU, MCI, AD)	Both cross-sectional and longitudinal designs	Simoa (tau protein)/xMAP($A\beta_{1-42}$)	In the discrimination between tau positive participants vs. negative: - plasma t-tau/$A\beta_{1-42}$ ratio: AuROC = 0.89; - plasma t-tau: AuROC = 0.80 with tau-PET as reference standard	NA
Janelidze S. et al., 2020 [90]	N = 589 (CU, MCI, AD dementia, non-AD dementia) cohort 1: N = 182 cohort 2: N = 344 cohort 3 (neuropathology cohort): N = 63	Both cross-sectional and longitudinal designs	ECLIA	Plasma p-tau181 in differentiating: - tau positive vs. negative participants: AuROC = 0.87–0.91 depending on different brain regions with tau-PET as reference standard (cohort 1); - AD dementia vs. non-AD dementia: AuROC = 0.94 with clinical diagnosis as reference standard (cohort 1); - Aβ positive vs. negative participants: AuROC ~ 0.80 (cohort 1 and cohort 2) with Aβ PET as reference standard; - AD dementia vs. non-AD dementia group: AuROC = 0.85 with neuropathology autopsy as reference standard (cohort 3)	High plasma p-tau levels are associated with future development of AD dementia in CU (HR = 2.48) and MCI (HR = 3.07) participants (cohort 2)
Thijssen E. et al., 2020 [91]	N = 404 (CU, MCI, AD, CBS, PSP, FTLD, nfvPPA, svPPA) 3 independent cohorts	Both cross-sectional (retrospective) and longitudinal designs	ECLIA	Plasma p-tau181 in differentiating: - AD (56) vs. FTLD (190) participants: AuROC = 0.89 with clinical diagnosis as reference standard; - Aβ-PET positive CU (11) vs. negative (29): AuROC = 0.86 with amyloid-PET as reference standard; - AD (15) vs. FTLD-tau participants (52): AuROC = 0.86 with neuropathology autopsy as reference standard	NA

Table 3. *Cont.*

Reference	Population	Study Design	Technique	Diagnostic Value	Prognostic Value
Karikari T. et al., 2020 [127]	kari	Longitudinal	Simoa	Plasma p-tau181 in differentiating AD participants vs: - amyloid β negative young adults: AuROC = 0.99; - CU older adults: AuROC = 0.90–0.98 across cohorts; - vascular dementia participants: AuROC = 0.92; - PSP or CBS participants: AuROC = 0.89; - PD or MSA participants: AuROC = 0.82 with clinical diagnosis as reference standard - tau-PET positive vs. tau-PET negative individuals AuROC = 0.83–0.93 across cohorts with tau-PET as reference standard	NA
			NFL protein		
Mattsson N. et al., 2017 [128]	*N* = 570 (CU, MCI, AD)	Case-control	Simoa	Plasma NFL in differentiating CU vs. AD participants: - AuROC = 0.87 with clinical diagnosis as reference standard	NA
Lewczuk P. et al., 2018 [129]	*N* = 99 (CU, MCI, AD)	Cross-sectional	Simoa	Plasma NFL in differentiating CU vs. diseased participants: - AuROC = 0.85 with clinical diagnosis as reference standard	NA
Steinacker P. et al., 2018 [130]	*N* = 132 (CU, MCI, AD, bvFTD)	Longitudinal	Simoa	Serum NFL in differentiating bvFTD vs: - AD: AuROC = 0.67 - MCI: AuROC = 0.90 - CU: AuROC = 0.85 with clinical diagnosis as reference standard Serum NFL in differentiating: - bvFTD vs. AD groups selected on the basis of CSF $A\beta_{1-42}$ levels: AuROC = 0.79 - bvFTD vs. AD groups selected on the basis of both CSF $A\beta_{1-42}$ and tau/p-tau levels: AuROC = 0.77	NA
Preische O. et al., 2019 [131]	*N* = 405 (controls - AD mutation non-carriers-, AD mutation carriers subdivided into presymptomatic mutation carriers, converters and symptomatic mutation carriers)	Longitudinal	Simoa	Rate of change of serum NFL in differentiating: - non-mutation carriers vs. presymptomatic mutation carriers: AuROC = 0.70 - non-mutation carriers vs. symptomatic non-mutation carriers: AuROC = 0.89 Baseline serum NFL levels in differentiating: - non-mutation carriers vs. presymptomatic mutation carriers: AuROC = 0.49 - non-mutation carriers vs. symptomatic non-mutation carriers: AuROC = 0.85	NA

Abbreviations: AD: Alzheimer's disease; AuROC: area under the receiver operating curve; Aβ: amyloid β; $A\beta_{1-40}$: amyloid β-peptide 1–40; $A\beta_{1-42}$: amyloid β-peptide 1–42; bvFTD: behavioral variant frontotemporal dementia; CBS: corticobasal syndrome; CSF: cerebrospinal fluid; CU: cognitively unimpaired; ECLIA: electrochemiluminescence immunoassay; FTD: frontotemporal dementia; FTLD: frontotemporal lobar degeneration; HR: hazard ratio; IMR: immunomagnetic reduction; IP: immunoprecipitation; IP MS: immunoprecipitation coupled to mass spectrometry; MCI: mild cognitive impairment; MDS: multimer detection system; MS: mass spectrometry; MSA: multiple system atrophy; NA: not assessed; NFL: neurofilament light chain; nfvPPA: non-fluent variant primary progressive aphasia; PD: Parkinson's disease; PPA: primary progressive aphasia; PSP: progressive supranuclear palsy; p-tau181: phospho-tau181; Simoa: single molecule array; SMC: subjective memory complainers; svPPA: semantic variant primary progressive aphasia; t-tau: total-tau; xMAP: multi-analyte profiling.

In summary, quite concordant results regarding low plasmatic concentrations of $A\beta_{1-42}$ and low $A\beta_{1-42}$ /$A\beta_{1-40}$ ratio in AD are reported in most of the studies in which an ultrasensitive technique has been applied. Overall, recent data suggest a low plasma $A\beta_{1-42}$ and $A\beta_{1-42}/A\beta_{1-40}$ ratio as being a specific feature of AD patients with a weak to moderate and a moderate to high concordance with amyloid-PET outcomes, respectively [136] (Table 2). Further investigations comparing the different ultrasensitive techniques in the same populations will provide more accurate information regarding advantages and drawbacks.

6. Biomarkers for Tau Pathology

Together with CSF $A\beta_{1-42}$, CSF t-tau and p-tau are considered as core biomarkers for AD diagnosis [11,137], and they are currently used for subject selection in clinical trials. Both CSF tau species are higher in patients compared to non-demented individuals. P-tau is more specific than t-tau for AD pathology and plays a main role in differential diagnosis being substantially normal in non-AD dementias [138]. Recent studies have shown that CSF tau can predict disease progression in both cognitively unimpaired and MCI subjects (Table 2) [139–141]. In 1999, Hulstaert and colleagues reported that the combined measurements of CSF $A\beta_{1-42}$ and tau had a better outcome than the individual biomarker in differentiating AD patients from controls and other dementias [142]. In an effort to establish the utility of both CSF t-tau/$A\beta_{1-42}$ and p-tau/$A\beta_{1-42}$ ratios, several authors ended up confirming those preliminary findings [98,143,144]. Moreover, the ability of both ratios to predict disease onset and progression was proven in other studies, including normal individuals and MCI subjects, respectively [140,145,146].

Since 2013, it has been reported that plasma t-tau levels, measured through an assay based on digital array technology, were higher in AD participants compared with both MCI and healthy subjects, but they did not show significant modifications in subsequent longitudinal evaluations (Table 2) [116,147]. Shortly thereafter, a large meta-analysis demonstrated the increase of plasma t-tau levels to be strongly associated with AD (Table 2) [115].

In the last four years, highly sensitive immunoassay techniques significantly implemented tau levels detection in peripheral blood. Indeed, Mattsson and colleagues to assess plasma t-tau concentration in two separate cohorts applied the Simoa technique. In AD patients compared with both MCI and healthy controls, an increase of plasma tau was demonstrated but with overlapping results. More interestingly, longitudinal evaluations revealed that high baseline levels of this biomarker were predictive of cognitive decline, higher atrophy rates at MRI, and hypometabolism at 18F-FDG-PET as well [148]. A longitudinal study carried out using the same technique found increased plasma tau levels associated with a higher risk of MCI and cognitive decline in MCI subjects, irrespective of the total Aβ-burden in the brain [123] (Table 3). In 2019, Park and colleagues highlighted that both plasma t-tau measured by the Simoa technique as well and the plasma t-tau/$A\beta_{42}$ ratio positively correlated with cerebral tau-PET uptake. Moreover, plasma tau-related biomarkers concentrations were significantly higher in Tau-PET+ subjects compared with Tau-PET- subjects and could differentiate the two groups with good accuracy (Table 3). It is noteworthy that the plasma t-tau and t-tau/$A\beta_{42}$ ratio could predict the cerebral accumulation of this misfolded protein after a 2-year follow-up [149]. A clear association between plasma and CSF levels of p-tau was also found in Aβ+ patients, including the presymptomatic stage (Aβ+ cognitively unimpaired), but not in Aβ− individuals [90]. Overall, these data suggest to some extent that plasma t-tau concentration is high in AD patients, but the substantial overlap with normal controls hinders its diagnostic utility (Table 2). Interestingly, an additional study found out that plasma p-tau concentrations improve diagnostic accuracy significantly compared to plasma t-tau by reaching a good capability in the discrimination of amyloid-positive and amyloid-negative subjects (Table 3) [124]. Moreover, plasma p-tau levels assessed by IMR have shown a good accuracy in differentiating unimpaired and MCI subjects [125] (Table 3).

The potential diagnostic role of plasma p-tau has been outlined in three recent investigations. In a first study performed with a novel ECLIA technique, plasma p-tau concentrations not only could discriminate AD and frontotemporal lobar degeneration (FTLD) participants with good accuracy, but they also identified amyloid-PET positive participants among elderly and MCI, and they predicted the rate of cognitive decline in AD and MCI over a 2-year follow-up (Table 3) [91]. The second study including three separate cohorts with 589 individuals (controls, MCI, AD, and non-AD NDDs) revealed that plasma p-tau levels likewise assessed by means of the MSD platform increase with disease progression (from preclinical to frank dementia phases encompassing prodromal/MCI stage) and can distinguish AD dementia from non-AD dementia with excellent accuracy (Table 3). Furthermore, plasma p-tau concentration was more elevated in Aβ+ cognitively unimpaired individuals than in Aβ− ones and in Aβ+ MCI who progressed to AD dementia compared to those who did not convert [90]. These results were confirmed in another study involving four independent cohorts (1131 total subjects) in different clinical contexts. Actually, plasma p-tau concentration measured by the Simoa technique discriminated AD dementia patients from both cognitively unimpaired subjects and other NDDs (including tauopathies such as Progressive Supranuclear Palsy and Corticobasal Syndrome) with optimal diagnostic accuracy (Table 3). Moreover, plasma p-tau predicted future cognitive decline over time. Interestingly, plasma p-tau concentration strongly correlates with cerebral amyloid-PET burden, even in amyloid-PET positive but tau-PET negative subjects, suggesting its crucial role in detecting the earliest disease phases. Thus, this biomarker might represent a screening tool implementable in different clinical settings and contexts of use [127].

7. Biomarkers for Neuroinflammation

The role of inflammation in AD pathogenesis was first suggested more than 20 years ago. Microglia, astrocytes, cytokines, and chemokines play a central role in disease pathogenesis since early phases [45,150]. Furthermore, amyloid and tau accumulation is linked to neuroinflammation [46,49,50,151], and Aβ accumulation evokes an exaggerated or heightened microglial response inducing and amplifying inflammatory reactions [152]. Therefore, biomarkers of neuroinflammation are gaining momentum in preclinical stages of AD and are useful to establish the eligibility of patients into new clinical trials [153–156].

A potential biomarker of neuroinflammation is the microglia/astrocyte-expressed protein YKL-40. YKL-40 is 45 a glycoprotein belonging to the family of 18 glycosyl hydrolases, and it is alternatively named human cartilage glycoprotein-39 (HC gp-39) or chitinase-3-like-1 protein (CHI3L1) [157]. CSF YKL-40 concentration is able to differentiate patients with typical AD dementia from cognitively normal controls with fair diagnostic accuracy [158,159]. Limited data regarding the ability of CSF YKL-40 to discriminate different NDDs are available so far. Actually, CSF YKL-40 differentiated AD from DLB, PD [160], FTLD [161], and non-AD MCI [162] with only a moderate diagnostic accuracy. Furthermore, CSF YKL-40 concentration is higher in AD versus Aβ-positive MCI subjects [163], and it significantly increases over time in the former (Table 2) [163]. CSF YKL-40 showed no ability in differentiating stable and progressing MCI [164,165], although it may predict progression to overt dementia in MCI [165] (Table 2). CSF YKL-40 levels negatively correlated with cortical thickness in specific AD-vulnerable areas, such as middle and inferior temporal areas in Aβ42-positive subjects [166] and grey matter volume in *APOE* ε4 carriers [167]. Interestingly, a positive association between CSF YKL-40 and t-tau has been reported in asymptomatic preclinical stages of AD and other NDDs [110,168], thus suggesting a link of YKL-40 with an underlying tau-driven neurodegenerative mechanisms [169].

YKL-40 has also been investigated in plasma, and elevated levels have been reported in patients with mild AD [170] and early AD [171] compared with controls. Unfortunately, plasma YKL-40 did not, so far, demonstrate utility as a diagnostic biomarker and for predicting cognitive decline (Table 2) [170,172]. To sum up, YKL-40 is an unspecific pathophysiological biomarker tracking immune/inflammatory response in NDDs, and it could be helpful as a monitoring biomarker for targeted anti-inflammatory therapies [169].

Another emerging biomarker of inflammation is "Triggering Receptor Expressed on Myeloid cells 2" (TREM2). TREM2 receptors play an important role in the pathogenesis of AD [173]. In the early stages of AD, TREM2 seems to be upregulated, probably in a protective intent [174]. However, due to the activation of inflammatory response, a detrimental role may prevail in later stages [175]. Some TREM2 genetic variants are related to AD possibly impairing microglia Aβ phagocytic ability and reducing, as a consequence, the cerebral Aß peptides clearance [167]. TREM2 has been proposed as AD biomarker in CSF, but with conflicting results so far. Some studies found higher CSF levels of TREM2 in AD [176–179] and MCI [176] compared to controls, and in subjects with MCI due to AD (or prodromal AD) compared with preclinical AD and AD dementia patients (Table 2) [179]. However, another study showed no difference between AD or MCI patients and cognitively normal controls [180]. A link between high CSF TREM2 value and neurodegeneration was proposed in MCI, considering that it positively correlated with gray matter volume and a negative correlation with mean diffusivity was detected [181]. Higher levels of TREM2 mRNA and TREM2 protein expressed in peripheral blood mononuclear cells were identified in AD patients compared to controls, with an inverse correlation with MMSE [182]. Moreover, TREM2 gene expression was found to be higher in MCI than AD patients [183]. Finally, a possible role of TREM2 as CSF and blood biomarker for AD has been suggested, but few data are currently available, and additional research is needed.

Another interesting candidate inflammatory biomarker is the monocyte chemoattractant protein-1 (MCP-1), which is a member of the C-C chemokine family and a potent chemotactic factor for monocytes [184]. Elevated CSF MCP-1 levels were found in AD patients compared to controls [185]. Noteworthy, also in blood, MCP-1 could be higher in MCI subjects than in controls [185].

Several other inflammatory biomarkers in CSF and blood have been investigated for their potential use as biomarkers for AD. A large metanalysis exploring inflammatory biomarkers in CSF demonstrated that AD patients could express higher levels of TGF-β compared with controls [350 Molievo]. TGF-ß1 is a neurotrophic, anti-inflammatory factor, and it enhances Aβ clearance by microglia activation. Since the early phases of disease, a reduced expression of TGF-ß has been described both in postmortem AD studies [186,187] and in animal models [188–190]. Two recently published meta-analyses investigating inflammatory biomarkers in blood reported an elevated tumor necrosis factor (TNF)-α, IL-12 [191], IL-1β, IL-2, IL-6, IL-18 [191,192], and reduced IL-1 receptor antagonist concentration in AD patients compared with controls. Mounting data about IL-6 as an AD biomarker are available. Blood IL-6 levels are associated with severity of cognitive decline in AD [193] and positively correlated with the cerebral ventricular volumes [194] and with matched CSF samples [195]. Blood IL-6 concentration was even higher in MCI individuals [135], suggesting that this biomarker is altered also in prodromal AD stages.

8. Biomarkers for Synaptic Dysfunction

Synaptic dysfunction is a core feature of AD, occurring early in the disease course. Synaptic density is strictly correlated with cognitive impairment and with Aβ and tau accumulation in AD, thus suggesting a central role in the underlying neurodegenerative process [196]. Based on these observations, several synaptic proteins have been investigated as potential diagnostic and prognostic biomarkers in AD. These include the quite well-characterized Neurogranin (Ng) and other emerging biomarkers such as Synaptosomal-Associated Protein 25 (SNAP-25), Synaptotagmin 1 and 2 (SYT-1 and SYT-2), Neuropentraxin 2 (NPTX-2), and Growth Associated Protein 43 (GAP-43) [197,198].

Ng is a post-synaptic protein largely expressed in the excitatory neurons of the hippocampus and cerebral cortex that acts as a calcium-sensitive modulator of post-synaptic signaling pathways and of long-term potentiation (LTP) [199]. Two recently published meta-analyses reported higher CSF Ng levels in AD compared to MCI and normal controls, thus supporting a role of CSF Ng as a useful diagnostic tool (Table 2) [200,201]. In particular, Ng reported good or even optimal diagnostic accuracy in differentiating AD patients with a full-blown clinical picture from cognitively normal subjects [202]. However, CSF Ng concentration discriminates between stable and converting MCI with poor diagnostic accuracy [200,201,203]. As regards its prognostic value, higher baseline CSF Ng levels are detected in controls and in MCI subjects who will convert to AD compared to non-converters, indicating a role of Ng in predicting progression to AD dementia in both cognitively normal [204] and MCI individuals (Table 2) [205,206]. CSF Ng could be also a reliable biomarker in the diagnostic workup of dementia being specifically more elevated in AD than non-AD dementias (FTD, DLB, but also VaD) [168,207–210]. Intriguingly, CSF Ng levels are high in AD patients with a typical amnesic phenotype, suggesting its role in the stratification and identification of AD subtypes, as a selective indicator of hippocampal degeneration [207]. In addition to Ng, other synaptic proteins have been explored as candidate biomarkers for AD. In particular, SNAP-25, a pre-synaptic protein involved in vesicle docking and neurotransmitter release, showed good accuracy in differentiating AD and MCI from normal controls (Table 2) [211–213]. Furthermore, high baseline CSF SNAP-25 levels predict future conversion to AD in MCI individuals [212]. Other pre-synaptic proteins such as SYT-1, SYT-2, NPTX-2, and GAP-43 are candidate as biomarkers to differentiate AD, MCI, and cognitive normal controls [197,213]. Importantly, an inverse correlation between CSF and neuron-derived plasma exosomes (NDE) levels of Ng, SYT-2, and GAP-43 has been observed [200,214]. Even if these results would deserve further supporting evidence, NDE may represent a window on the early synaptic dysfunction in AD and pave the way to a minimally invasive assessment (blood sample) of synaptic biomarkers in cognitively impaired and unimpaired subjects.

9. Biomarkers of Neuronal Injury

NFL is a subunit of neurofilaments that are neural cytoplasmic proteins designated to the structural stability of neurons; they are present in dendrites, soma, and also in axons. Axons physiologically release a low amount of NFL proteins that increase with aging [215]. The concentration of NFL significantly increases in both CSF and in blood as a result of axonal injury or neurodegeneration [216–219]. NFL in CSF is usually measured by sandwich ELISA technology. On the contrary, blood NFL concentration is 40-fold lower than in CSF, and it is below the sensitivity of ELISA or electrochemiluminescence assay technology [215]. Promising results came from recently developed ultrasensitive techniques capable of detecting even low concentrations of NFL in blood (Simoa) [220]. Despite being a sensitive biomarker of axonal injury, NFL is unspecific and did not discriminate between neurological diseases with a similar rate of axonal loss. However, growing data showed that CSF and, above all, blood NFL identifies neurodegeneration from early stages [215]. Indeed, NFL (CSF and blood) showed a good diagnostic accuracy in differentiating AD and FTD from healthy controls (Table 3) [128,129,131,208,221,222]. According to these results, a possible context-of-use of this biomarker is to rule out neurodegeneration in mimics such as psychiatric disturbances, or to early detect, within screening programs, the neurodegenerative process in a specific population at high risk (e.g., diabetes, elderly, genetic mutation carriers). Increased blood NFL concentration could also help clinicians to proceed or not with more invasive and expensive examinations in individuals with subjective memory complaints [215]. Moreover, CSF NFL but not t-tau, p-tau, and Ng might be a reliable risk biomarker being associated with a threefold higher risk to develop MCI over a median follow-up of 3.8 years in a population of cognitively healthy individuals [223]. CSF [224,225] and blood [128,226,227] NFL tightly correlated each other and with disease severity. In this regard, in a prospective case-control study including normal controls, MCI, and AD dementia patients, plasma NFL correlated with CSF NFL, poor cognition, cerebral atrophy, and brain hypometabolism [128].

Serum NFL concentration correlated with the estimated years to symptom onset and disease severity in autosomal dominant AD mutation carriers, suggesting its possible role as a risk biomarker in subjects with autosomal genetic mutations for AD (Table 3) [227]. Promising data concern the role of NFL in the differential diagnosis between FTD and AD. Actually, CSF NFL is higher in FTD patients compared to early onset AD, and the addition of NFL analysis improves the diagnostic accuracy of the traditional core biomarkers (p-tau181 and Aß42) up to a sensitivity of 86% and a specificity of 100% [228]. Similar findings were also reported in an autopsy-confirmed AD and FTD study (Table 3) [229]. Moreover, serum NFL could help in the differentiation of Primary Progressive Aphasia (PPA) variants. Indeed, serum NFL is increased in PPA compared to controls and discriminates between nfvPPA/svPPA (with a more likely FTD pathology) and lvPPA (where an AD pathology is expected in more than 50% of cases) with 81% and 67% of sensitivity and specificity, respectively [230].

Visinin-like protein 1 (VILIP-1) is emerging as a surrogate of signaling disruption and neuronal injury. VILIP-1 is a neuronal calcium sensor protein involved in signaling pathways related to synaptic plasticity [231]. A high intracellular concentration of Ca2+ induces the reversible translocation of VILIP-1 to the membrane components of the cell modulating signaling cascade in the neurons via the activation of specific membrane-bound targets [232]. The dysregulation of Ca2+ homeostasis is involved in AD neurodegeneration, bringing to a reduced intracellular expression of VILIP-1 and a quite selective damage of VILIP-1-containing neurons (cortical pyramidal cells, interneurons, septal, subthalamic, and hippocampal neurons) [232]. Therefore, this biomarker significantly increases in CSF [231]. Since VILIP-1 contributes to an altered Ca2+ homeostasis leading to neuronal loss [232], it is mainly considered a biomarker of neuronal injury. CSF VILIP-1 is higher in AD than in controls, but its diagnostic accuracy remains limited, especially in the prodromal stage (Table 2) [115,233,234]. Although VILIP-1 tightly correlated with p-tau and t-tau in CSF, conflicting results concern its relationship with Aß peptides [235,236]. CSF VILIP-1 and the VILIP-1/Aβ-42 ratio negatively correlate with MMSE and with the cerebral amyloid load, and they may predict a cognitive decline over time [233,234,236–240].

10. Toward Alternative Pathophysiological Pathways and Novel Matrices

The research on novel putative biomarkers in AD recently focused on two main directions: the exploration of new still under-characterized pathophysiological pathways, including mixed neuropathology models, and the identification of alternative easily accessible matrices.

TAR DNA-binding protein 43 (TDP43) is a DNA and RNA binding protein involved in transcription and splicing. TDP-43 contributes to neuroinflammation and may play a role in mitochondrial and neural dysfunction. In ALS and FTLD, its hyperphosphorylated and/or ubiquitinated cytoplasmic inclusions are detected [241,242] but also 20–50% of AD patients may show concomitant TDP-43 pathology [243–245]. Interestingly, TDP-43 pathology can be triggered by Aβ peptides [244]. In AD, increased plasma TDP-43 levels have been found compared to normal controls [246]; furthermore, plasma levels were increased also in the preclinical stage of subjects who subsequently progressed to AD dementia [247]. However, the evidence of a diagnostic and prognostic role of TDP-43 in AD is currently quite limited as well as its role in differentiating AD from other dementia mainly involving the hippocampus and memory (e.g., LATE) [22].

Lewy-related pathology (LRP), primarily consisting of α-synuclein (α-syn) aggregates, has been detected in more than half of autopsied AD brains, and higher levels of α-syn in the CSF of patients with MCI and AD have been associated with AD pathology and cognitive decline [248,249]. Moreover, CSF total α-syn (t-α-syn) and oligomeric α-synuclein (o-α-syn) levels were higher in AD [250] compared to PD, PD dementia and DLB individuals [251,252]. The use of standard ELISA methods to assess CSF α-syn levels does not ensure good diagnostic accuracy in discriminating AD from synucleinopathies [250]. Nevertheless, RT-QuIC [253] and protein misfolding cyclic amplification (PMCA) [254] are promising tools to identify AD individuals with α-syn co-pathology. Furthermore, growing interest toward the evaluation of α-syn heterocomplexes with $A\beta_{1-42}$ (α-syn/Aβ) or tau (α-syn/tau) measured in red blood cells (RBCs) as peripheral pathophysiological markers of NDDs has been displayed [254]. Despite both α-syn alone, α-syn/Aβ and α-syn/tau heteroaggregates being found lower in AD compared to cognitive normal controls when isolated from red blood cells (RBC), only RBC α-syn/Aβ and α-syn/tau heterodimers discriminated AD from controls with fair accuracy [254].

Exploring alternative easily accessible matrices as a source of putative biomarkers is another key point of the search for novel fluid biomarkers. In this frame, exosomes represent an innovative and promising non-invasive tool to track early neurodegenerative changes occurring within the central nervous system. Exosomes are vesicles containing potential biomarkers for NDDs released into the extracellular space (that can be isolated from several body fluids) [243,254]. Proteins reflecting key events of the neurodegenerative process have been isolated in exosomes extracted from CSF and blood by using proteomic analysis [244–246]; in particular, p-tau was isolated in CSF exosomes from patients with mild AD (Braak stage 3) [247], and increased levels of exosomes-associated tau and Aβ were found in AD patients compared to controls [248]. Finally, other easily accessible matrices such as the retina may represent an open window on early neurodegenerative events in AD [249]. Amyloid pathology was demonstrated in the retina, and high-resolution non-invasive retinal imaging [47–51] represents an in vivo approach for visualizing Aβ deposits [250–252]. Indeed, retinal Aβ accumulation positively correlated with cerebral amyloid plaques [8]. Furthermore, decreased flow velocities in the retinal central veins were found in both MCI and AD compared to controls, thus suggesting a strict correlation with the underlying early neurodegenerative changes [253]. However, this field of research remains in its pathfinding stages, and a consensus on retinal imaging modalities, methodologies, and measures is still missing [253].

11. Conclusions

Recent research efforts are expanding the array of biomarkers on detecting and stratifying NDDs. Since 2007, fluid biomarkers have been reported within the diagnostic criteria of AD. In particular, Aß42 peptide, p-tau, and t-tau proteins measured in CSF became essential for a "modern" AD definition. The conceptual shift from a phenotype to a biomarker-based (or a precision medicine) diagnostic approach allowed the inclusion of the atypical subtypes within the AD spectrum and the exclusion of AD-mimics. For instance, patients with early and predominant behavioral impairment but positive for core pathophysiological biomarkers are categorized as AD and not as FTD. By contrast, individuals showing cognitive impairment of the hippocampal type but negative for core biomarkers are not considered AD. Definitely, the identification of a specific pathophysiological process in vivo by one or more biomarkers prevails on clinical phenotype. Unfortunately, validated fluid biomarkers used for AD diagnosis are invasive, time-consuming, expensive, not easily repeatable and, most importantly, not applicable as screening tools in large asymptomatic populations. On the other hand, the preclinical or prodromal identification of AD is urgent for patient recruitment in future disease-modifying treatments. This is an "expert" opinion based on the current literature, reporting the diagnostic and prognostic value of fluid biomarkers in AD. Five candidate molecules—three in plasma measured using ultrasensitive techniques (Aβ peptides, p-tau, and NFL proteins) and two in CSF (Ng and YKL-40)—with different potential context-of-use (Table 2) may be proposed. These molecules may enrich the current array of fluid biomarkers—CSF Aβ42, t-tau, and p-tau—for a more precise management of AD, and, broadly speaking, NDDs. These biomarkers are useful to both classify patients in different diagnostic categories and to track the pathophysiological mechanisms underlying neurodegeneration. The blood biomarkers (Aβ peptides, p-tau, NFL) are probably not more accurate than the respective molecules measured in CSF. However, they may be easily repeated over time, proposed in screening programs, and monitor treatments in disease-modifying trials. On the other hand, CSF YKL-40 and Ng are proxies of additional pathophysiological mechanisms related to AD, namely neuroinflammation and synaptic disruption, that cannot be efficiently evaluated with peripheral blood biomarkers. Therefore, CSF YKL-40 and Ng may be used in a subsequent diagnostic step to better stratify patients with prodromal or definite AD.

Plasma p-tau is increased in AD patients compared to controls and MCI individuals, discriminating AD demented patients from both cognitively unimpaired subjects and other NDDs with optimal diagnostic accuracy [91]. In several recent studies from different research groups, its classificatory accuracy surprisingly overlaps with cerebral amyloid-PET [90,124]. Moreover, plasma p-tau predicts a future cognitive decline over time [91]. Therefore, plasma p-tau, being easily repeatable, could be proposed in screening, diagnostic, prognostic, and monitoring context-of-uses. Simoa is the ultrasensitive technique used with more successful results across the studies on plasma p-tau so far. Of course, additional studies are needed. In addition, plasma t-tau concentration correlated with future cognitive decline, increased atrophy rates measured by MRI, and cerebral hypometabolism in

FDG-PET images, but the results are less convincing and charted overlapping values among AD with dementia, prodromal, and preclinical AD groups [148].

Plasma Aβ peptides may represent a further significant improvement to facilitate the in vivo detection of amyloid pathology, substituting traditional core CSF biomarkers in next years. The combination of CSF traditional biomarkers (e.g., $A\beta_{1-42}/A\beta_{1-40}$ ratio, t-tau/$A\beta_{1-42}$, and p-tau/$A\beta_{1-42}$ ratios) can improve the diagnostic accuracy as well as the prediction of cognitive decline in AD patients. Similarly, the combination of plasma and serum biomarkers into ratios may increase the diagnostic power, although further evidence is needed [109,133]. Mounting data revealed that a low plasma $A\beta_{1-42}$ and $A\beta_{1-42}/A\beta_{1-40}$ ratio are quite specific of AD pathology, although the concordance with cerebral amyloid-PET examination is variable and should be carefully evaluated in future studies [75,92,120]. In brief, further investigations should clarify in larger prospective studies: (1) the more accurate method to detect Aβ peptides and related by-products in plasma, (2) the pathophysiological role of plasma Aβ peptides and Aβ oligomers, and (3) their diagnostic and prognostic value as biomarker of AD.

NFL is mainly a marker of axonal degeneration and considered an unspecific indicator of neurodegeneration. Importantly, CSF and plasma NFL strictly correlated in all studies, suggesting that plasma NFL would be a reliable peripheral biomarker consistently reflecting modifications within the CNS. Indeed, many studies on NDDs demonstrated that diagnostic and prognostic accuracies of plasma and CSF NFL overlap [254]. NFL is a good example as a versatile biomarker for multiple context-of-use. It is useful to differentiate NDDs from mimics such as psychiatric disturbances or to early detect neurodegenerative processes in particular populations at risk (e.g., diabetes, elderly, genetic mutation carriers). NFL values were associated with a threefold higher risk to develop MCI, demonstrating a potential prognostic value [128,226,227]. Finally, the negative predictive value of plasma NFL might be used as a first step in screening programs for neurodegeneration, involving individuals with subjective memory complaints and late-onset psychiatric disorders. Concerning AD, plasma NFL showed a promising role in differentiating AD from bvFTD patients. It is likely that bvFTD individuals with an underlying TDP-43 pathology (related to amyotrophic lateral sclerosis) reported significantly higher plasma NFL value than AD subjects. Plasma NFL could early discriminate AD from more aggressive neurodegenerative dementia such as CJD [244]. Simoa was the only ultrasensitive technique used to measure plasma NFL in AD studies.

An increasing number of studies are focused on the development of precise biomarkers tracking additional key pathophysiological pathways leading to neurodegeneration, such as synaptic disruption and neuroinflammation. The pre-synaptic protein Ng measured in CSF is the most promising indicator of a synaptic dysfunction and hippocampal damage [202]. It could help stratify patients suffering from NDDs involving the hippocampus, including AD but also hippocampal sclerosis, LATE, and PART. LATE and PART have been recently defined in postmortem examinations, but in vivo diagnostic biomarkers are needed. CSF Ng could be also used as predictive indicator of an anticholinesterasic treatment response in patients showing a prevalent hippocampal impairment (typical AD phenotype, etc.) [200,201,205,206]. CSF Ng demonstrated from good to optimal diagnostic accuracy in discriminating AD dementia patients from the control group and a reliable prognostic value for AD conversion in MCI individuals.

Growing data reported an abnormal neuroinflammatory response in AD. Currently, CSF YKL-40 is the most promising fluid biomarker of glia activation, and it has been extensively investigated in other NDDs as well. Neuroinflammation is a common pathway of several NDDs, and not surprisingly, YKL-40 is an unspecific biomarker. This biomarker could be helpful in monitoring tailored anti-inflammatory trials in AD. Studies exploring a possible correlation between CSF YKL-40 concentration and cerebral inflammatory tracer as the translocator protein (TSPO)-PET uptake could clarify the role of this fluid biomarker as an indicator of neuroinflammation. CSF YKL-40 showed a fair diagnostic accuracy to discriminate AD patients from the control group and other neurodegenerative dementias. YKL-40 also reported a certain predictive value for MCI-AD progression.

In summary, we found that reliable fluid biomarkers might track three out of four of the main pathophysiological pathways of AD (Figure 1). The concentration in different biofluids of Aβ peptides and p-tau proteins reflect the cerebral misfolded protein deposition in AD. Moreover, plasma NFL might help the early identification of a general neurodegenerative process independently from the specific pathology. The diffusion of ultrasensitive techniques in the last few years is radically revolutionizing the context-of-use of these biomarkers in AD. The possibility to measure biomarkers in blood opens a completely novel scenario for the detection of multiple neurodegenerative mechanisms with a low cost and minimally invasive examination. This should encourage the development of screening tools in selected populations and improve the monitoring of disease-modifying trials. Of note, validated surrogates of co-pathologies such as α-syn and TDP-43 protein accumulation are currently not available as well as of cerebrovascular impairment. Finally, YKL-40 and Ng measured in CSF are promising proxies of neuroinflammation and synaptic disruption, respectively.

The studies described have several shortcomings. Many investigations reported different inclusion criteria and sometimes, they were not biomarker-based. Moreover, there is frequently a lack of data about comorbidities, especially cerebrovascular burden, contemporary pharmacological treatments, or stratification for age, gender, and genetic profiles [53]. Hepatic and kidney dysfunctions may impact biomarker levels as well as modifications of blood cell counts and plasma protein composition [67]. Nonetheless, these variables were not systematically considered, thus constituting a possible methodological bias, since individuals with AD present relevant vascular comorbidities.

From the prospective of a precision medicine approach, increasing attention is paid to find biomarkers associated to pathogenic pathways leading to neurodegeneration. The contemporary use of multiple biomarkers can help dissect the pathological mechanisms dynamically acting in space and time providing an accurate stratification of AD population.

AD is more a spectrum of different pathological mechanisms that brings a loss of proteostasis, which is the accumulation of several misfolded and aggregated proteins in multiple combinations rather than a single entity. Advanced statistical analysis, including unsupervised clustering strategies, combining clinical, biomarkers, and genetic data to collect subjects from large diversified cohorts into consistent clusters might be an innovative representation of AD [6] and other NDDs [254] and significantly contribute to the discovery of causes and tailored treatments. Novel data-driven classifications based on quantitative measurements of biomarkers and clinical information (e.g., standardized clinical scores) could improve the identification of effective and personalized therapies [253,254]. In conclusion, we assume that the identification and inclusion of AD patients in disease-modifying trials will be soon changed, mainly based on the demonstration of specific pathophysiological mechanisms and minimally influenced by phenotypes.

Author Contributions: E.D.P., M.F.B., N.C. and L.G. wrote the paper, F.B. substantially contributed to the conception and design of the article and interpreting the relevant literature, R.C. and G.S. revised the article critically for important intellectual content. All authors have read and agreed to the published version of the manuscript.

Funding: This research received no external funding.

Conflicts of Interest: The authors declare no conflict of interest relevant to this work.

References

1. 2020 Alzheimer's disease facts and figures. *Alzheimer's Dement.* **2020**, *16*, 391–460. [CrossRef] [PubMed]
2. Schachter, A.S.; Davis, K.L. Alzheimer's disease. *Curr. Treat. Options Neurol.* **2000**, *2*, 51–60. [CrossRef] [PubMed]
3. Masters, C.L.; Simms, G.; Weinman, N.A.; Multhaup, G.; McDonald, B.L.; Beyreuther, K. Amyloid plaque core protein in Alzheimer disease and Down syndrome. *Proc. Natl. Acad. Sci. USA* **1985**, *82*, 4245–4249. [CrossRef] [PubMed]

4. McKhann, G.; Drachman, D.; Folstein, M.; Katzman, R.; Price, D.; Stadlan, E.M. Clinical diagnosis of Alzheimer's disease: Report of the NINCDS-ADRDA Work Group* under the auspices of Department of Health and Human Services Task Force on Alzheimer's Disease. *Neurology* **1984**, *34*, 939. [CrossRef] [PubMed]
5. Kovacs, G.G. Molecular Pathological Classification of Neurodegenerative Diseases: Turning towards Precision Medicine. *Int. J. Mol. Sci.* **2016**, *17*, 189. [CrossRef]
6. Baldacci, F.; Lista, S.; Garaci, F.; Bonuccelli, U.; Toschi, N.; Hampel, H. Biomarker-guided classification scheme of neurodegenerative diseases. *J. Sport Health Sci.* **2016**, *5*, 383–387. [CrossRef]
7. Jellinger, K.A. Basic mechanisms of neurodegeneration: A critical update. *J. Cell. Mol. Med.* **2010**, *14*, 457–487. [CrossRef]
8. Baldacci, F.; Mazzucchi, S.; Della Vecchia, A.; Giampietri, L.; Giannini, N.; Koronyo-Hamaoui, M.; Ceravolo, R.; Siciliano, G.; Bonuccelli, U.; Elahi, F.M.; et al. The path to biomarker-based diagnostic criteria for the spectrum of neurodegenerative diseases. *Expert Rev. Mol. Diagn.* **2020**, *20*, 421–441. [CrossRef]
9. Beach, T.G.; Monsell, S.E.; Phillips, L.E.; Kukull, W. Accuracy of the Clinical Diagnosis of Alzheimer Disease at National Institute on Aging Alzheimer Disease Centers, 2005–2010. *J. Neuropathol. Exp. Neurol.* **2012**, *71*, 266–273. [CrossRef]
10. Dubois, B.; Feldman, H.H.; Jacova, C.; DeKosky, S.T.; Barberger-Gateau, P.; Cummings, J.L.; Delacourte, A.; Galasko, D.; Gauthier, S.; Jicha, G.; et al. Research criteria for the diagnosis of Alzheimer's disease: Revising the NINCDS–ADRDA criteria. *Lancet Neurol.* **2007**, *6*, 734–746. [CrossRef]
11. Jack, C.R.; Albert, M.S.; Knopman, D.S.; McKhann, G.M.; Sperling, R.A.; Carrillo, M.C.; Thies, W.; Phelps, C.H. Introduction to the recommendations from the National Institute on Aging-Alzheimer's Association workgroups on diagnostic guidelines for Alzheimer's disease. *Alzheimer's Dement.* **2011**, *7*, 257–262. [CrossRef] [PubMed]
12. Dubois, B.; Feldman, H.H.; Jacova, C.; Cummings, J.L.; DeKosky, S.T.; Barberger-Gateau, P.; Delacourte, A.; Frisoni, G.; Fox, N.C.; Galasko, D.; et al. Revising the definition of Alzheimer's disease: A new lexicon. *Lancet Neurol.* **2010**, *9*, 1118–1127. [CrossRef]
13. Dubois, B.; Feldman, H.H.; Jacova, C.; Hampel, H.; Molinuevo, J.L.; Blennow, K.; DeKosky, S.T.; Gauthier, S.; Selkoe, D.; Bateman, R.; et al. Advancing research diagnostic criteria for Alzheimer's disease: The IWG-2 criteria. *Lancet Neurol.* **2014**, *13*, 614–629. [CrossRef]
14. McKhann, G.M.; Knopman, D.S.; Chertkow, H.; Hyman, B.T.; Jack, C.R.; Kawas, C.H.; Klunk, W.E.; Koroshetz, W.J.; Manly, J.J.; Mayeux, R.; et al. The diagnosis of dementia due to Alzheimer's disease: Recommendations from the National Institute on Aging-Alzheimer's Association workgroups on diagnostic guidelines for Alzheimer's disease. *Alzheimer's Dement.* **2011**, *7*, 263–269. [CrossRef] [PubMed]
15. Albert, M.S.; DeKosky, S.T.; Dickson, D.; Dubois, B.; Feldman, H.H.; Fox, N.C.; Gamst, A.; Holtzman, D.M.; Jagust, W.J.; Petersen, R.C.; et al. The diagnosis of mild cognitive impairment due to Alzheimer's disease: Recommendations from the National Institute on Aging-Alzheimer's Association workgroups on diagnostic guidelines for Alzheimer's disease. *Alzheimer's Dement.* **2011**, *7*, 270–279. [CrossRef]
16. Sperling, R.A.; Aisen, P.S.; Beckett, L.A.; Bennett, D.A.; Craft, S.; Fagan, A.M.; Iwatsubo, T.; Jack, C.R., Jr.; Kaye, J.; Montine, T.J.; et al. Toward defining the preclinical stages of Alzheimer's disease: Recommendations from the National Institute on Aging-Alzheimer's Association workgroups on diagnostic guidelines for Alzheimer's disease. *Alzheimer's Dement.* **2011**, *7*, 280–292. [CrossRef] [PubMed]
17. Dubois, B.; Hampel, H.; Feldman, H.H.; Scheltens, P.; Aisen, P.; Andrieu, S.; Bakardjian, H.; Benali, H.; Bertram, L.; Blennow, K.; et al. Preclinical Alzheimer's disease: Definition, natural history, and diagnostic criteria. *Alzheimer's Dement.* **2016**, *12*, 292–323. [CrossRef]
18. Jack, C.R.; Bennett, D.A.; Blennow, K.; Carrillo, M.C.; Feldman, H.H.; Frisoni, G.B.; Hampel, H.; Jagust, W.J.; Johnson, K.A.; Knopman, D.S.; et al. A/T/N: An unbiased descriptive classification scheme for Alzheimer disease biomarkers. *Neurology* **2016**, *87*, 539–547. [CrossRef]
19. Jack, C.R.; Bennett, D.A.; Blennow, K.; Carrillo, M.C.; Dunn, B.; Haeberlein, S.B.; Holtzman, D.M.; Jagust, W.; Jessen, F.; Karlawish, J.; et al. NIA-AA Research Framework: Toward a biological definition of Alzheimer's disease. *Alzheimer's Dement.* **2018**, *14*, 535–562. [CrossRef]
20. Crary, J.F.; Trojanowski, J.Q.; Schneider, J.A.; Abisambra, J.F.; Abner, E.L.; Alafuzoff, I.; Arnold, S.E.; Attems, J.; Beach, T.G.; Bigio, E.H.; et al. Primary age-related tauopathy (PART): A common pathology associated with human aging. *Acta Neuropathol.* **2014**, *128*, 755–766. [CrossRef]

21. Nelson, P.T.; Dickson, D.W.; Trojanowski, J.Q.; Jack, C.R.; Boyle, P.; Arfanakis, K.; Rademakers, R.; Alafuzoff, I.; Attems, J.; Brayne, C.; et al. Limbic-predominant age-related TDP-43 encephalopathy (LATE): Consensus working group report. *Brain* **2019**, *142*, 1503–1527. [CrossRef]
22. Toledo, J.B.; Arnold, S.E.; Raible, K.; Brettschneider, J.; Xie, S.X.; Grossman, M.; Monsell, S.E.; Kukull, W.A.; Trojanowski, J.Q. Contribution of cerebrovascular disease in autopsy confirmed neurodegenerative disease cases in the National Alzheimer's Coordinating Centre. *Brain* **2013**, *136*, 2697–2706. [CrossRef] [PubMed]
23. Hamilton, R.L. Lewy Bodies in Alzheimer's Disease: A Neuropathological Review of 145 Cases Using α-Synuclein Immunohistochemistry. *Brain Pathol.* **2006**, *10*, 378–384. [CrossRef] [PubMed]
24. Schneider, J.A.; Arvanitakis, Z.; Bang, W.; Bennett, D.A. Mixed brain pathologies account for most dementia cases in community-dwelling older persons. *Neorology* **2007**, *69*, 2197–2204. [CrossRef] [PubMed]
25. Colom-Cadena, M.; Grau-Rivera, O.; Planellas, L.; Cerquera, C.; Morenas, E.; Helgueta, S.; Muñoz, L.; Kulisevsky, J.; Martí, M.J.; Tolosa, E.; et al. Regional Overlap of Pathologies in Lewy Body Disorders. *J. Neuropathol. Exp. Neurol.* **2017**, *76*, 216–224. [CrossRef]
26. Irwin, D.J.; Xie, S.X.; Coughlin, D.; Nevler, N.; Akhtar, R.S.; McMillan, C.T.; Lee, E.B.; Wolk, D.A.; Weintraub, D.; Chen-Plotkin, A.; et al. CSF tau and β-amyloid predict cerebral synucleinopathy in autopsied Lewy body disorders. *Neorology* **2018**, *90*, e1038–e1046. [CrossRef]
27. Parnetti, L.; Paciotti, S.; Farotti, L.; Bellomo, G.; Sepe, F.N.; Eusebi, P. Parkinson's and Lewy body dementia CSF biomarkers. *Clin. Chim. Acta* **2019**, *495*, 318–325. [CrossRef]
28. Van Bulck, M.; Sierra-Magro, A.; Alarcon-Gil, J.; Perez-Castillo, A.; Morales-García, J.A. Novel Approaches for the Treatment of Alzheimer's and Parkinson's Disease. *Int. J. Mol. Sci.* **2019**, *20*, 719. [CrossRef]
29. Masters, C.L.; Bateman, R.; Blennow, K.; Rowe, C.C.; Sperling, R.A.; Cummings, J.L. Alzheimer's disease. *Nat. Rev. Dis. Prim.* **2015**, *1*, 1–18. [CrossRef]
30. Scheltens, P.; Blennow, K.; Breteler, M.M.B.; de Strooper, B.; Frisoni, G.B.; Salloway, S.; Van der Flier, W.M. Alzheimer's disease. *Lancet* **2016**, *388*, 505–517. [CrossRef]
31. Hardy, J.; Higgins, G.; Mayford, M.; Barzilai, A.; Keller, F.; Schacher, S.; Kandel, E. Alzheimer's disease: The amyloid cascade hypothesis. *Sci.* **1992**, *256*, 184–185. [CrossRef] [PubMed]
32. Selkoe, D.J.; Hardy, J. The amyloid hypothesis of Alzheimer's disease at 25 years. *EMBO Mol. Med.* **2016**, *8*, 595–608. [CrossRef] [PubMed]
33. Christopher, D.; Guo, J.L.; McBride, J.D.; Narasimhan, S.; Kim, H.; Changolkar, L.; Zhang, B.; Gathagan, R.J.; Yue, C.; Dengler, C.; et al. Amyloid-β plaques enhance Alzheimer's brain tau-seeded pathologies by facilitating neuritic plaque tau aggregation. *Nat. Med.* **2018**, *24*, 29–38. [CrossRef]
34. Bakota, L.; Brandt, R. Tau Biology and Tau-Directed Therapies for Alzheimer's Disease. *Drugs* **2016**, *76*, 301–313. [CrossRef] [PubMed]
35. Lee, V.M.; Balin, B.J.; Otvos, L.; Trojanowski, J.Q. A68: A major subunit of paired helical filaments and derivatized forms of normal Tau. *Science* **1991**, *251*, 675–678. [CrossRef]
36. Irwin, D.J. Tauopathies as clinicopathological entities. *Park. Relat. Disord.* **2016**, *22*, S29–S33. [CrossRef]
37. Sanabria-Castro, A.; Alvarado-Echeverría, I.; Monge-Bonilla, C. Molecular Pathogenesis of Alzheimer's Disease: An Update. *Ann. Neurosci.* **2017**, *24*, 46–54. [CrossRef]
38. Zhang, B.; Carroll, J.; Trojanowski, J.Q.; Yao, Y.; Iba, M.; Potuzak, J.S.; Hogan, A.-M.L.; Xie, S.X.; Ballatore, C.; Smith, A.B.; et al. The Microtubule-Stabilizing Agent, Epothilone D, Reduces Axonal Dysfunction, Neurotoxicity, Cognitive Deficits, and Alzheimer-Like Pathology in an Interventional Study with Aged Tau Transgenic Mice. *J. Neurosci.* **2012**, *32*, 3601–3611. [CrossRef]
39. Khan, S.S.; Bloom, G.S. Tau: The Center of a Signaling Nexus in Alzheimer's Disease. *Front. Neurosci.* **2016**, *10*, 31. [CrossRef]
40. Leschik, J.; Welzel, A.; Weissmann, C.; Eckert, A.; Brandt, R. Inverse and distinct modulation of tau-dependent neurodegeneration by presenilin 1 and amyloid-beta in cultured cortical neurons: Evidence that tau phosphorylation is the limiting factor in amyloid-?-induced cell death. *J. Neurochem.* **2007**, *101*, 1303–1315. [CrossRef]
41. Fath, T.; Eidenmüller, J.; Brandt, R. Tau-Mediated Cytotoxicity in a Pseudohyperphosphorylation Model of Alzheimer's Disease. *J. Neurosci.* **2002**, *22*, 9733–9741. [CrossRef] [PubMed]
42. Rogers, J.; Webster, S.; Lue, L.-F.; Brachova, L.; Civin, W.H.; Emmerling, M.; Shivers, B.; Walker, D.; McGeer, P. Inflammation and Alzheimer's disease pathogenesis. *Neurobiol. Aging* **1996**, *17*, 681–686. [CrossRef]

43. Hull, M.; Berger, M.; Bauer, J.; Strauss, S.; Volk, B. Inflammatory mechanisms in Alzheimer's disease. *Eur. Arch. Psychiatry Clin. Neurosci.* **1996**, *246*, 124–128. [CrossRef] [PubMed]
44. Eikelenboom, P. Neuroinflammation and Alzheimer's Disease. *Neurochemistry* **1997**, *14*, 15–19. [CrossRef]
45. Nordengen, K.; Kirsebom, B.-E.; Henjum, K.; Selnes, P.; Gísladóttir, B.; Wettergreen, M.; Torsetnes, S.B.; Grøntvedt, G.R.; Waterloo, K.K.; Aarsland, D.; et al. Glial activation and inflammation along the Alzheimer's disease continuum. *J. Neuroinflamm.* **2019**, *16*, 1–13. [CrossRef] [PubMed]
46. Morgan, A.R.; Touchard, S.; Leckey, C.; O'Hagan, C.; Nevado-Holgado, A.J.; Barkhof, F.; Bertram, L.; Blin, O.; Bos, I.; Dobricic, V.; et al. Inflammatory biomarkers in Alzheimer's disease plasma. *Alzheimer's Dement.* **2019**, *15*, 776–787. [CrossRef]
47. Brosseron, F.; Krauthausen, M.; Kummer, M.; Heneka, M.T. Body Fluid Cytokine Levels in Mild Cognitive Impairment and Alzheimer's Disease: A Comparative Overview. *Mol. Neurobiol.* **2014**, *50*, 534–544. [CrossRef]
48. Fan, Z.; Brooks, D.J.; Okello, A.; Edison, P. An early and late peak in microglial activation in Alzheimer's disease trajectory. *Brain* **2017**, *140*, 792–803. [CrossRef]
49. Edison, P.; Brooks, D.J. Role of Neuroinflammation in the Trajectory of Alzheimer's Disease and in vivo Quantification Using PET. *J. Alzheimer's Dis.* **2018**, *64*, S339–S351. [CrossRef]
50. Parbo, P.; Ismail, R.; Hansen, K.V.; Amidi, A.; Marup, F.; Gottrup, H.; Braendgaard, H.; Eriksson, B.; Eskildsen, S.; Lung, T.; et al. Brain inflammation accompanies amyloid in the majority of mild cognitive impairment cases due to Alzheimer's disease. *Brain* **2017**, *140*, 2002–2011. [CrossRef]
51. Sarlus, H.; Heneka, M.T. Microglia in Alzheimer's disease. *J. Clin. Investig.* **2017**, *127*, 3240–3249. [CrossRef] [PubMed]
52. Hampel, H.; Caraci, F.; Cuello, A.C.; Caruso, G.; Nisticò, R.; Corbo, M.; Baldacci, F.; Toschi, N.; Garaci, F.; Chiesa, P.A.; et al. A Path Toward Precision Medicine for Neuroinflammatory Mechanisms in Alzheimer's Disease. *Front. Immunol.* **2020**, *11*, 456. [CrossRef] [PubMed]
53. Snyder, H.M.; Corriveau, R.A.; Craft, S.; Faber, J.E.; Greenberg, S.M.; Knopman, D.; Lamb, B.T.; Montine, T.J.; Nedergaard, M.; Schaffer, C.B.; et al. Vascular contributions to cognitive impairment and dementia including Alzheimer's disease. *Alzheimer's Dement.* **2015**, *11*, 710–717. [CrossRef] [PubMed]
54. Zlokovic, B.V. Neurovascular pathways to neurodegeneration in Alzheimer's disease and other disorders. *Nat. Rev. Neurosci.* **2011**, *12*, 723–738. [CrossRef] [PubMed]
55. Kelleher, R.J.; Soiza, R.L. Evidence of endothelial dysfunction in the development of Alzheimer's disease: Is Alzheimer's a vascular disorder? *Am. J. Cardiovasc. Dis.* **2013**, *3*, 197–226. [PubMed]
56. Salminen, A.; Kauppinen, A.; Kaarniranta, K. Hypoxia/ischemia activate processing of Amyloid Precursor Protein: Impact of vascular dysfunction in the pathogenesis of Alzheimer's disease. *J. Neurochem.* **2017**, *140*, 536–549. [CrossRef] [PubMed]
57. Di Marco, L.Y.; Venneri, A.; Farkas, E.; Evans, P.C.; Marzo, A.; Frangi, A.F. Vascular dysfunction in the pathogenesis of Alzheimer's disease — A review of endothelium-mediated mechanisms and ensuing vicious circles. *Neurobiol. Dis.* **2015**, *82*, 593–606. [CrossRef]
58. Tarasoff-Conway, J.M.; Carare, R.O.; Osorio, R.S.; Glodzik, L.; Butler, T.; Fieremans, E.; Axel, L.; Rusinek, H.; Nicholson, C.; Zlokovic, B.V.; et al. Clearance systems in the brain—implications for Alzheimer disease. *Nat. Rev. Neurol.* **2015**, *11*, 457–470. [CrossRef]
59. Iliff, J.J.; Wang, M.; Liao, Y.; Plogg, B.A.; Peng, W.; Gundersen, G.A.; Benveniste, H.; Vates, G.E.; Deane, R.; Goldman, S.A.; et al. A Paravascular Pathway Facilitates CSF Flow Through the Brain Parenchyma and the Clearance of Interstitial Solutes, Including Amyloid. *Sci. Transl. Med.* **2012**, *4*, 147ra111. [CrossRef]
60. Xie, L.; Kang, H.; Xu, Q.; Chen, M.J.; Liao, Y.; Thiyagarajan, M.; O'Donnell, J.; Christensen, D.J.; Nicholson, C.; Iliff, J.J.; et al. Sleep Drives Metabolite Clearance from the Adult Brain. *Science* **2013**, *342*, 373–377. [CrossRef]
61. Attems, J.; Jellinger, K. The overlap between vascular disease and Alzheimer's disease - lessons from pathology. *BMC Med.* **2014**, *12*, 1–12. [CrossRef] [PubMed]
62. Snowdon, D.A.; Greiner, L.H.; Mortimer, J.A.; Riley, K.P.; Greiner, P.A.; Markesbery, W.R. Brain Infarction and the Clinical Expression of Alzheimer DiseaseThe Nun Study. *JAMA* **1997**, *277*, 813–817. [CrossRef] [PubMed]
63. Brenowitz, W.D.; Nelson, P.T.; Besser, L.M.; Heller, K.B.; Kukull, W.A. Cerebral amyloid angiopathy and its co-occurrence with Alzheimer's disease and other cerebrovascular neuropathologic changes. *Neurobiol. Aging* **2015**, *36*, 2702–2708. [CrossRef] [PubMed]

64. Dolan, H.; Crain, B.; Troncoso, J.; Resnick, S.M.; Zonderman, A.B.; O'Brien, R.J. Atherosclerosis, dementia, and alzheimer's disease in the BLSA cohort. *Ann. Neurol.* **2010**, *68*, 231–240. [CrossRef]
65. Andreasson, U.; Blennow, K.; Zetterberg, H. Update on ultrasensitive technologies to facilitate research on blood biomarkers for central nervous system disorders. *Alzheimer's Dement. Diagn. Assess. Dis. Monit.* **2016**, *3*, 98–102. [CrossRef]
66. Baldacci, F.; Lista, S.; Vergallo, A.; Palermo, G.; Giorgi, F.S.; Hampel, H. A frontline defense against neurodegenerative diseases:the development of early disease detection methods. *Expert Rev. Mol. Diagn.* **2019**, *19*, 559–563. [CrossRef]
67. Hampel, H.; O'Bryant, S.E.; Molinuevo, J.L.; Zetterberg, H.; Masters, C.L.; Lista, S.; Kiddle, S.J.; Batrla, R.; Blennow, K. Blood-based biomarkers for Alzheimer disease: Mapping the road to the clinic. *Nat. Rev. Neurol.* **2018**, *14*, 639–652. [CrossRef]
68. Kang, J.-H.; Vanderstichele, H.; Trojanowski, J.Q.; Shaw, L.M. Simultaneous analysis of cerebrospinal fluid biomarkers using microsphere-based xMAP multiplex technology for early detection of Alzheimer's disease. *Methods* **2012**, *56*, 484–493. [CrossRef]
69. Kuhle, J.; Barro, C.; Andreasson, U.; Derfuss, T.; Lindberg, R.; Sandelius, Å.; Liman, V.; Norgren, N.; Blennow, K.; Zetterberg, H. Comparison of three analytical platforms for quantification of the neurofilament light chain in blood samples: ELISA, electrochemiluminescence immunoassay and Simoa. *Clin. Chem. Lab. Med.* **2016**, *54*, 1655–1661. [CrossRef]
70. Verberk, I.M.W.; Slot, R.E.; Verfaillie, S.C.J.; Heijst, H.; Prins, N.D.; Van Berckel, B.N.M.; Scheltens, P.; Teunissen, C.E.; Van Der Flier, W.M. Plasma Amyloid as Prescreener for the Earliest A lzheimer Pathological Changes. *Ann. Neurol.* **2018**, *84*, 648–658. [CrossRef]
71. Vergallo, A.; Mégret, L.; Lista, S.; Cavedo, E.; Zetterberg, H.; Blennow, K.; Vanmechelen, E.; De Vos, A.; Habert, M.-O.; Potier, M.-C. Plasma amyloid β 40/42 ratio predicts cerebral amyloidosis in cognitively normal individuals at risk for Alzheimer's disease. *Alzheimer's Dement.* **2019**, *15*, 764–775. [CrossRef] [PubMed]
72. Gowan, S.M.; Hardcastle, A.; Hallsworth, A.E.; Valenti, M.R.; Hunter, L.-J.K.; Brandon, A.K.D.H.; Garrett, M.D.; Raynaud, F.I.; Workman, P.; Aherne, W.; et al. Application of Meso Scale Technology for the Measurement of Phosphoproteins in Human Tumor Xenografts. *ASSAY Drug Dev. Technol.* **2007**, *5*, 391–402. [CrossRef] [PubMed]
73. Fichorova, R.N.; Richardson-Harman, N.; Alfano, M.; Belec, L.; Carbonneil, C.; Chen, S.; Cosentino, L.; Curtis, K.; Dezzutti, C.S.; Donoval, B.; et al. Biological and Technical Variables Affecting Immunoassay Recovery of Cytokines from Human Serum and Simulated Vaginal Fluid: A Multicenter Study. *Anal. Chem.* **2008**, *80*, 4741–4751. [CrossRef] [PubMed]
74. Nakamura, A.; Kaneko, N.; Villemagne, V.L.; Kato, T.; Doecke, J.; Doré, V.; Fowler, C.; Li, Q.-X.; Martins, R.; Rowe, C.; et al. High performance plasma amyloid-β biomarkers for Alzheimer's disease. *Nat. Cell Biol.* **2018**, *554*, 249–254. [CrossRef] [PubMed]
75. Have, S.T.; Boulon, S.; Ahmad, Y.; Lamond, A.I. Mass spectrometry-based immuno-precipitation proteomics—The user's guide. *Proteomics* **2011**, *11*, 1153–1159. [CrossRef]
76. Seubert, P.; Vigo-Pelfrey, C.; Esch, F.; Lee, M.; Dovey, H.F.; Davis, D.; Sinha, S.; Schiossmacher, M.; Whaley, J.; Swindlehurst, C.; et al. Isolation and quantification of soluble Alzheimer's β-peptide from biological fluids. *Nat. Cell Biol.* **1992**, *359*, 325–327. [CrossRef]
77. Kim, S.H.; Kang, S.; Suh, J.W.; Park, Y.H.; Kang, M.J.; Pyun, J.-M.; Choi, S.H.; Jeong, J.H.; Park, K.W.; Lee, H.-W.; et al. Blood amyloid-β oligomerization associated with neurodegeneration of Alzheimer's disease. *Alzheimer's Res. Ther.* **2019**, *11*, 40. [CrossRef]
78. Nabers, A.; Ollesch, J.; Schartner, J.; Kötting, C.; Genius, J.; Hafermann, H.; Klafki, H.; Gerwert, K.; Wiltfang, J. Amyloid-β-Secondary Structure Distribution in Cerebrospinal Fluid and Blood Measured by an Immuno-Infrared-Sensor: A Biomarker Candidate for Alzheimer's Disease. *Anal. Chem.* **2016**, *88*, 2755–2762. [CrossRef]
79. Nabers, A.; Perna, L.; Lange, J.; Mons, U.; Schartner, J.; Güldenhaupt, J.; Saum, K.; Janelidze, S.; Holleczek, B.; Rujescu, D.; et al. Amyloid blood biomarker detects Alzheimer's disease. *EMBO Mol. Med.* **2018**, *10*, e8763. [CrossRef]
80. Nabers, A.; Hafermann, H.; Wiltfang, J.; Gerwert, K. Aβ and tau structure-based biomarkers for a blood- and CSF-based two-step recruitment strategy to identify patients with dementia due to Alzheimer's disease. *Alzheimer's Dement. Diagn. Assess. Dis. Monit.* **2019**, *11*, 257–263. [CrossRef]

81. Yang, S.-Y.; Chiu, M.-J.; Chen, T.-F.; Horng, H.-E. Detection of Plasma Biomarkers Using Immunomagnetic Reduction: A Promising Method for the Early Diagnosis of Alzheimer's Disease. *Neurol. Ther.* **2017**, *6*, 37–56. [CrossRef] [PubMed]
82. Reslova, N.; Michna, V.; Kasny, M.; Mikel, P.; Kralik, P. xMAP Technology: Applications in Detection of Pathogens. *Front. Microbiol.* **2017**, *8*, 55. [CrossRef] [PubMed]
83. Graham, H.; Chandler, D.J.; Dunbar, S. The genesis and evolution of bead-based multiplexing. *Methods* **2019**, *158*, 2–11. [CrossRef] [PubMed]
84. ElShal, M.F.; McCoy, J.P. Multiplex bead array assays: Performance evaluation and comparison of sensitivity to ELISA☆. *Methods* **2006**, *38*, 317–323. [CrossRef] [PubMed]
85. Wilson, D.H.; Rissin, D.M.; Kan, C.W.; Fournier, D.R.; Piech, T.; Campbell, T.G.; Meyer, R.E.; Fishburn, M.W.; Cabrera, C.; Patel, P.P.; et al. The Simoa HD-1 Analyzer. *J. Lab. Autom.* **2016**, *21*, 533–547. [CrossRef]
86. Kuhle, J.; Barro, C.; Disanto, G.; Mathias, A.; Soneson, C.; Bonnier, G.; Yaldizli, Ö.; Regeniter, A.; Derfuss, T.; Canales, M.; et al. Serum neurofilament light chain in early relapsing remitting MS is increased and correlates with CSF levels and with MRI measures of disease severity. *Mult. Scler. J.* **2016**, *22*, 1550–1559. [CrossRef]
87. Li, D.; Mielke, M.M. An Update on Blood-Based Markers of Alzheimer's Disease Using the SiMoA Platform. *Neurol. Ther.* **2019**, *8*, 73–82. [CrossRef]
88. Palmqvist, S.; Janelidze, S.; Stomrud, E.; Zetterberg, H.; Karl, J.; Zink, K.; Bittner, T.; Mattsson, N.; Eichenlaub, U.; Blennow, K.; et al. Performance of Fully Automated Plasma Assays as Screening Tests for Alzheimer Disease–Related β-Amyloid Status. *JAMA Neurol.* **2019**, *76*, 1060–1069. [CrossRef]
89. Janelidze, S.; Mattsson, N.; Palmqvist, S.; Smith, R.; Beach, T.G.; Serrano, G.E.; Chai, X.; Proctor, N.K.; Eichenlaub, U.; Zetterberg, H.; et al. Plasma P-tau181 in Alzheimer's disease: Relationship to other biomarkers, differential diagnosis, neuropathology and longitudinal progression to Alzheimer's dementia. *Nat. Med.* **2020**, *26*, 379–386. [CrossRef]
90. Thijssen, E.H.; Advancing Research and Treatment for Frontotemporal Lobar Degeneration (ARTFL) investigators; La Joie, R.; Wolf, A.; Strom, A.; Wang, P.; Iaccarino, L.; Bourakova, V.; Cobigo, Y.; Heuer, H.; et al. Diagnostic value of plasma phosphorylated tau181 in Alzheimer's disease and frontotemporal lobar degeneration. *Nat. Med.* **2020**, *26*, 387–397. [CrossRef]
91. Wang, M.J.; Yi, S.; Han, J.-Y.; Park, S.Y.; Jang, J.-W.; Chun, I.K.; Kim, S.E.; Lee, B.S.; Kim, G.J.; Yu, J.S.; et al. Oligomeric forms of amyloid-β protein in plasma as a potential blood-based biomarker for Alzheimer's disease. *Alzheimer's Res. Ther.* **2017**, *9*, 1–10. [CrossRef] [PubMed]
92. Teunissen, C.E.; Chiu, M.-J.; Yang, C.-C.; Yang, S.-Y.; Scheltens, P.; Zetterberg, H.; Blennow, K. Plasma Amyloid-β (Aβ42) Correlates with Cerebrospinal Fluid Aβ42 in Alzheimer's Disease. *J. Alzheimer's Dis.* **2018**, *62*, 1857–1863. [CrossRef] [PubMed]
93. Zetterberg, H. Blood-based biomarkers for Alzheimer's disease—An update. *J. Neurosci. Methods* **2019**, *319*, 2–6. [CrossRef] [PubMed]
94. Grimmer, T.; Riemenschneider, M.; Förstl, H.; Henriksen, G.; Klunk, W.E.; Mathis, C.A.; Shiga, T.; Wester, H.-J.; Kurz, A.; Drzezga, A. Beta Amyloid in Alzheimer's Disease: Increased Deposition in Brain Is Reflected in Reduced Concentration in Cerebrospinal Fluid. *Biol. Psychiatry* **2009**, *65*, 927–934. [CrossRef]
95. Fagan, A.M.; Shaw, L.M.; Xiong, C.; Vanderstichele, H.; Mintun, M.A.; Trojanowski, J.Q.; Coart, E.; Morris, J.C.; Holtzman, D.M. Comparison of Analytical Platforms for Cerebrospinal Fluid Measures of β-Amyloid 1-42, Total tau, and P-tau181 for Identifying Alzheimer Disease Amyloid Plaque Pathology. *Arch. Neurol.* **2011**, *68*, 1137–1144. [CrossRef]
96. Jagust, W.J.; Landau, S.M.; Shaw, L.M.; Trojanowski, J.Q.; Koeppe, R.A.; Reiman, E.M.; Foster, N.L.; Petersen, R.C.; Weiner, M.W.; Price, J.C.; et al. Relationships between biomarkers in aging and dementia. *Neurology* **2009**, *73*, 1193–1199. [CrossRef]
97. Strozyk, D.; Blennow, K.; White, L.R.; Launer, L.J. CSF Aβ 42 levels correlate with amyloid-neuropathology in a population-based autopsy study. *Neorology* **2003**, *60*, 652–656. [CrossRef]
98. Tapiola, T.; Alafuzoff, I.; Herukka, S.-K.; Parkkinen, L.; Hartikainen, P.; Soininen, H.; Pirttilä, T. Cerebrospinal Fluid β-Amyloid 42 and Tau Proteins as Biomarkers of Alzheimer-Type Pathologic Changes in the Brain. *Arch. Neurol.* **2009**, *66*, 382–389. [CrossRef]
99. Suzuki, N.; Iwatsubo, T.; Odaka, A.; Ishibashi, Y.; Kitada, C.; Ihara, Y. High tissue content of soluble β1-40 is linked to cerebral amyloid angiopathy. *Am. J. Pathol.* **1994**, *145*, 452–460.

100. Mehta, P.D.; Pirttilä, T.; Mehta, S.P.; Sersen, E.A.; Aisen, P.S.; Wisniewski, H.M. Plasma and cerebrospinal fluid levels of amyloid β proteins 1-40 and 1-42 in Alzheimer disease. *Arch. Neurol.* **2000**, *57*, 100–105. [CrossRef]
101. Slaets, S.; Le Bastard, N.; Martin, J.-J.; Sleegers, K.; Van Broeckhoven, C.; De Deyn, P.P.; Engelborghs, S. Cerebrospinal Fluid Aβ1-40 Improves Differential Dementia Diagnosis in Patients with Intermediate P-tau181P Levels. *J. Alzheimer's Dis.* **2013**, *36*, 759–767. [CrossRef] [PubMed]
102. Leuzy, A.; Chiotis, K.; Hasselbalch, B.J.; Rinne, J.O.; De Mendonça, A.; Otto, M.; Lleó, A.; Castelo-Branco, M.; Santana, I.; Johansson, J.; et al. Pittsburgh compound B imaging and cerebrospinal fluid amyloid-β in a multicentre European memory clinic study. *Brain* **2016**, *139*, 2540–2553. [CrossRef] [PubMed]
103. Lewczuk, P.; Matzen, A.; Blennow, K.; Parnetti, L.; Molinuevo, J.L.; Eusebi, P.; Kornhuber, J.; Morris, J.C.; Fagan, A.M. Cerebrospinal Fluid Aβ42/40 Corresponds Better than Aβ42 to Amyloid PET in Alzheimer's Disease. *J. Alzheimer's Dis.* **2017**, *55*, 813–822. [CrossRef] [PubMed]
104. Pannee, J.; Portelius, E.; Minthon, L.; Gobom, J.; Andreasson, U.; Zetterberg, H.; Hansson, O.; Blennow, K. Reference measurement procedure for CSF amyloid beta $(A\beta)_{1-42}$ and the CSF $A\beta_{1-42}/A\beta_{1-40}$ ratio—A cross-validation study against amyloid PET. *J. Neurochem.* **2016**, *139*, 651–658. [CrossRef] [PubMed]
105. Dorey, A.; Perret-Liaudet, A.; Tholance, Y.; Fourier, A.; Quadrio, I. Cerebrospinal Fluid Aβ40 Improves the Interpretation of Aβ42 Concentration for Diagnosing Alzheimer's Disease. *Front. Neurol.* **2015**, *6*, 247. [CrossRef]
106. Janelidze, S.; Stomrud, E.; Palmqvist, S.; Zetterberg, H.; Van Westen, D.; Jeromin, A.; Song, L.; Hanlon, D.; Hehir, C.A.T.; Baker, D.; et al. Plasma β-amyloid in Alzheimer's disease and vascular disease. *Sci. Rep.* **2016**, *6*, 26801. [CrossRef]
107. Olsson, B.; Lautner, R.; Andreasson, U.; Öhrfelt, A.; Portelius, E.; Bjerke, M.; Hölttä, M.; Rosén, C.; Olsson, C.; Strobel, G.; et al. CSF and blood biomarkers for the diagnosis of Alzheimer's disease: A systematic review and meta-analysis. *Lancet Neurol.* **2016**, *15*, 673–684. [CrossRef]
108. Mulugeta, E.; Londos, E.; Ballard, C.; Alves, G.; Zetterberg, H.; Blennow, K.; Skogseth, R.; Minthon, L.; Aarsland, D. CSF amyloid 38 as a novel diagnostic marker for dementia with Lewy bodies. *J. Neurol. Neurosurg. Psychiatry* **2010**, *82*, 160–164. [CrossRef]
109. Olsson, F.; Schmidt, S.; Althoff, V.; Munter, L.M.; Jin, S.; Rosqvist, S.; Lendahl, U.; Multhaup, G.; Lundkvist, J. Characterization of Intermediate Steps in Amyloid Beta (Aβ) Production under Near-native Conditions. *J. Biol. Chem.* **2013**, *289*, 1540–1550. [CrossRef]
110. Soares, H.D.; Gasior, M.; Toyn, J.H.; Wang, J.-S.; Hong, Q.; Berisha, F.; Furlong, M.T.; Raybon, J.; Lentz, K.A.; Sweeney, F.; et al. The γ-secretase modulator, BMS-932481, modulates Aβ peptides in the plasma and cerebrospinal fluid of healthy volunteerss. *J. Pharmacol. Exp. Ther.* **2016**, *358*, 138–150. [CrossRef]
111. Schuster, J.; Funke, S.A. Methods for the Specific Detection and Quantitation of Amyloid-β Oligomers in Cerebrospinal Fluid. *J. Alzheimer's Dis.* **2016**, *53*, 53–67. [CrossRef] [PubMed]
112. Zetterberg, H.; Wilson, D.; Andreasson, U.; Minthon, L.; Blennow, K.; Randall, J.; Hansson, O. Plasma tau levels in Alzheimer's disease. *Alzheimer's Res. Ther.* **2013**, *5*, 1–3. [CrossRef] [PubMed]
113. Ovod, V.; Ramsey, K.N.; Mawuenyega, K.G.; Bollinger, J.G.; Hicks, T.; Schneider, T.; Sullivan, M.; Paumier, K.; Holtzman, D.M.; Morris, J.C.; et al. Amyloid β concentrations and stable isotope labeling kinetics of human plasma specific to central nervous system amyloidosis. *Alzheimer's Dement.* **2017**, *13*, 841–849. [CrossRef] [PubMed]
114. Shahpasand-Kroner, H.; Klafki, H.-W.; Bauer, C.; Schuchhardt, J.; Hüttenrauch, M.; Stazi, M.; Bouter, C.; Wirths, O.; Vogelgsang, J.; Wiltfang, J. A two-step immunoassay for the simultaneous assessment of Aβ38, Aβ40 and Aβ42 in human blood plasma supports the Aβ42/Aβ40 ratio as a promising biomarker candidate of Alzheimer's disease. *Alzheimer's Res. Ther.* **2018**, *10*, 1–14. [CrossRef]
115. Lue, L.-F.; Sabbagh, M.N.; Chiu, M.-J.; Jing, N.; Snyder, N.L.; Schmitz, C.; Guerra, A.; Belden, C.M.; Chen, T.-F.; Yang, C.-C.; et al. Plasma Levels of Aβ42 and Tau Identified Probable Alzheimer's Dementia: Findings in Two Cohorts. *Front. Aging Neurosci.* **2017**, *9*, 226. [CrossRef]
116. Chatterjee, P.; Elmi, M.; Goozee, K.; Shah, T.; Sohrabi, H.R.; Dias, C.B.; Pedrini, S.; Shen, K.; Asih, P.R.; Dave, P.; et al. Ultrasensitive Detection of Plasma Amyloid-β as a Biomarker for Cognitively Normal Elderly Individuals at Risk of Alzheimer's Disease. *J. Alzheimer's Dis.* **2019**, *71*, 775–783. [CrossRef] [PubMed]

117. Mielke, M.M.; Hagen, C.E.; Wennberg, A.M.V.; Airey, D.C.; Savica, R.; Knopman, D.S.; Machulda, M.M.; Roberts, R.O.; Jack, C.R.; Petersen, R.C.; et al. Association of Plasma Total Tau Level With Cognitive Decline and Risk of Mild Cognitive Impairment or Dementia in the Mayo Clinic Study on Aging. *JAMA Neurol.* **2017**, *74*, 1073–1080. [CrossRef]
118. Mielke, M.M.; Hagen, C.E.; Xu, J.; Chai, X.; Vemuri, P.; Lowe, V.J.; Airey, D.C.; Knopman, D.S.; Roberts, R.O.; Machulda, M.M.; et al. Plasma phospho-tau181 increases with Alzheimer's disease clinical severity and is associated with tau- and amyloid-positron emission tomography. *Alzheimer's Dement.* **2018**, *14*, 989–997. [CrossRef]
119. Yang, C.-C.; Chiu, M.-J.; Chen, T.-F.; Chang, H.-L.; Liu, B.-H.; Yang, S.-Y. Assay of Plasma Phosphorylated Tau Protein (Threonine 181) and Total Tau Protein in Early-Stage Alzheimer's Disease. *J. Alzheimer's Dis.* **2018**, *61*, 1323–1332. [CrossRef]
120. Park, J.C.; Han, S.-H.; Yi, D.; Byun, M.S.; Lee, J.H.; Jang, S.; Ko, K.; Jeon, S.Y.; Lee, Y.-S.; Kim, Y.K.; et al. Plasma tau/amyloid-β1–42 ratio predicts brain tau deposition and neurodegeneration in Alzheimer's disease. *Brain* **2019**, *142*, 771–786. [CrossRef]
121. Karikari, T.K.; Pascoal, T.; Ashton, N.J.; Janelidze, S.; Benedet, A.L.; Rodriguez, J.L.; Chamoun, M.; Savard, M.; Kang, M.S.; Therriault, J.; et al. Blood phosphorylated tau 181 as a biomarker for Alzheimer's disease: A diagnostic performance and prediction modelling study using data from four prospective cohorts. *Lancet Neurol.* **2020**, *19*, 422–433. [CrossRef]
122. Mattsson, N.; Andreasson, U.; Zetterberg, H.; Blennow, K.; Alzheimer's Disease Neuroimaging Initiative. Association of Plasma Neurofilament Light with Neurodegeneration in Patients With Alzheimer Disease. *JAMA Neurol.* **2017**, *74*, 557–566. [CrossRef] [PubMed]
123. Lewczuk, P.; Ermann, N.; Andreasson, U.; Schultheis, C.; Podhorna, J.; Spitzer, P.; Maler, J.M.; Kornhuber, J.; Blennow, K.; Zetterberg, H. Plasma neurofilament light as a potential biomarker of neurodegeneration in Alzheimer's disease. *Alzheimer's Res. Ther.* **2018**, *10*, 1–10. [CrossRef] [PubMed]
124. Steinacker, P.; Anderl-Straub, S.; Diehl-Schmid, J.; Semler, E.; Uttner, I.; Von Arnim, C.A.; Barthel, H.; Danek, A.; Fassbender, K.; Fliessbach, K.; et al. Serum neurofilament light chain in behavioral variant frontotemporal dementia. *Neurology* **2018**, *91*, e1390–e1401. [CrossRef]
125. Hanon, O.; Vidal, J.-S.; Lehmann, S.; Bombois, S.; Allinquant, B.; Tréluyer, J.-M.; Gelé, P.; Delmaire, C.; Blanc, F.; Mangin, J.-F.; et al. Plasma amyloid levels within the Alzheimer's process and correlations with central biomarkers. *Alzheimer's Dement.* **2018**, *14*, 858–868. [CrossRef]
126. Shen, X.-N.; Niu, L.-D.; Wang, J.; Cao, X.-P.; Liu, Q.; Tan, L.; Zhang, C.; Yu, J.-T. Inflammatory markers in Alzheimer's disease and mild cognitive impairment: A meta-analysis and systematic review of 170 studies. *J. Neurol. Neurosurg. Psychiatry* **2019**, *90*, 590–598. [CrossRef] [PubMed]
127. Ashton, N.J.; Hye, A.; Rajkumar, A.P.; Leuzy, A.; Snowden, S.; Suárez-Calvet, M.; Karikari, T.K.; Schöll, M.; La Joie, R.; Rabinovici, G.D.; et al. An update on blood-based biomarkers for non-Alzheimer neurodegenerative disorders. *Nat. Rev. Neurol.* **2020**, *16*, 265–284. [CrossRef]
128. Blennow, K.; Hampel, H.; Weiner, M.W.; Zetterberg, H. Cerebrospinal fluid and plasma biomarkers in Alzheimer disease. *Nat. Rev. Neurol.* **2010**, *6*, 131–144. [CrossRef]
129. Roe, C.M.; Fagan, A.M.; Grant, E.A.; Hassenstab, J.; Moulder, K.L.; Dreyfus, D.M.; Sutphen, C.L.; Benzinger, T.L.; Mintun, M.A.; Holtzman, D.M.; et al. Amyloid imaging and CSF biomarkers in predicting cognitive impairment up to 7.5 years later. *Neurology* **2013**, *80*, 1784–1791. [CrossRef]
130. Ferreira, D.; Rivero-Santana, A.; Perestelo-Perez, L.; Westman, E.; Wahlund, L.O.; Sarría, A.; Serrano-Aguilar, P. Improving CSF Biomarkersâ€™ Performance for Predicting Progression from Mild Cognitive Impairment to Alzheimer's Disease by Considering Different Confounding Factors: A Meta-Analysis. *Front. Aging Neurosci.* **2014**, *6*, 287. [CrossRef]
131. Petersen, R.C.; Aisen, P.; Boeve, B.F.; Geda, Y.E.; Ivnik, R.J.; Knopman, D.S.; Mielke, M.; Pankratz, V.S.; Roberts, R.; Rocca, W.A.; et al. Mild cognitive impairment due to Alzheimer disease in the community. *Ann. Neurol.* **2013**, *74*, 199–208. [CrossRef] [PubMed]
132. Hulstaert, F.; Blennow, K.; Ivanoiu, A.; Schoonderwaldt, H.C.; Riemenschneider, M.; Deyn, P.P.D.; Bancher, C.; Cras, P.; Wiltfang, J.; Mehta, P.D.; et al. Improved discrimination of AD patients using -amyloid(1-42) and tau levels in CSF. *Neurology* **1999**, *52*, 1555. [CrossRef] [PubMed]

133. Rivero-Santana, A.; Ferreira, D.; Perestelo-Pérez, L.; Westman, E.; Wahlund, L.O.; Sarría, A.; Serrano-Aguilar, P. Cerebrospinal Fluid Biomarkers for the Differential Diagnosis between Alzheimer's Disease and Frontotemporal Lobar Degeneration: Systematic Review, HSROC Analysis, and Confounding Factors. *J. Alzheimer's Dis.* **2017**, *55*, 625–644. [CrossRef] [PubMed]
134. Seeburger, J.L.; Holder, D.J.; Combrinck, M.; Joachim, C.; Laterza, O.; Tanen, M.; Dallob, A.; Chappell, D.; Snyder, K.; Flynn, M.; et al. Cerebrospinal Fluid Biomarkers Distinguish Postmortem-Confirmed Alzheimer's Disease from Other Dementias and Healthy Controls in the OPTIMA Cohort. *J. Alzheimer's Dis.* **2015**, *44*, 525–539. [CrossRef] [PubMed]
135. Fagan, A.M.; Roe, C.M.; Xiong, C.; Mintun, M.A.; Morris, J.C.; Holtzman, D.M. Cerebrospinal Fluid tau/β-Amyloid42 Ratio as a Prediction of Cognitive Decline in Nondemented Older Adults. *Arch. Neurol.* **2007**, *64*, 343–349. [CrossRef]
136. Van Rossum, I.; Vos, S.; Handels, R.L.H.; Visser, P.J. Biomarkers as Predictors for Conversion from Mild Cognitive Impairment to Alzheimer-Type Dementia: Implications for Trial Design. *J. Alzheimer's Dis.* **2010**, *20*, 881–891. [CrossRef]
137. Mattsson, N.; Zetterberg, H.; Janelidze, S.; Insel, P.S.; Andreasson, U.; Stomrud, E.; Palmqvist, S.; Baker, D.; Hehir, C.A.T.; Jeromin, A.; et al. Plasma tau in Alzheimer disease. *Neurology* **2016**, *87*, 1827–1835. [CrossRef]
138. Businaro, R.; Corsi, M.; Asprino, R.; Di Lorenzo, C.; Laskin, D.; Corbo, R.; Ricci, S.; Pinto, A. Modulation of Inflammation as a Way of Delaying Alzheimer's Disease Progression: The Diet's Role. *Curr. Alzheimer Res.* **2018**, *15*, 363–380. [CrossRef]
139. Li, J.-W.; Zong, Y.; Cao, X.-P.; Tan, L.; Tan, L. Microglial priming in Alzheimer's disease. *Ann. Transl. Med.* **2018**, *6*, 176. [CrossRef]
140. Guzman-Martinez, L.; Maccioni, R.B.; Andrade, V.; Navarrete, L.P.; Pastor, M.G.; Ramos-Escobar, N. Neuroinflammation as a Common Feature of Neurodegenerative Disorders. *Front. Pharmacol.* **2019**, *10*, 1008. [CrossRef]
141. Hampel, H.; Vergallo, A.; Giorgi, F.S.; Kim, S.H.; Depypere, H.; Graziani, M.; Saidi, A.; Nisticò, R.; Lista, S. Precision medicine and drug development in Alzheimer's disease: The importance of sexual dimorphism and patient stratification. *Front. Neuroendocr.* **2018**, *50*, 31–51. [CrossRef] [PubMed]
142. Hampel, H.; Goetzl, E.J.; Kapogiannis, D.; Lista, S.; Vergallo, A. Biomarker-Drug and Liquid Biopsy Co-development for Disease Staging and Targeted Therapy: Cornerstones for Alzheimer's Precision Medicine and Pharmacology. *Front. Pharmacol.* **2019**, *10*. [CrossRef] [PubMed]
143. Dhiman, K.; Blennow, K.; Zetterberg, H.; Martins, R.N.; Gupta, V.B. Cerebrospinal fluid biomarkers for understanding multiple aspects of Alzheimer's disease pathogenesis. *Cell. Mol. Life Sci.* **2019**, *76*, 1833–1863. [CrossRef] [PubMed]
144. Baldacci, F.; Lista, S.; Palermo, G.; Giorgi, F.S.; Vergallo, A.; Hampel, H. The neuroinflammatory biomarker YKL-40 for neurodegenerative diseases: Advances in development. *Expert Rev. Proteom.* **2019**, *16*, 593–600. [CrossRef] [PubMed]
145. Llorens, F.; Thüne, K.; Tahir, W.; Kanata, E.; Diaz-Lucena, D.; Xanthopoulos, K.; Kovatsi, E.; Pleschka, C.; Garcia-Esparcia, P.; Schmitz, M.; et al. YKL-40 in the brain and cerebrospinal fluid of neurodegenerative dementias. *Mol. Neurodegener.* **2017**, *12*, 1–21. [CrossRef]
146. Wennström, M.; Surova, Y.; Hall, S.; Nilsson, C.; Minthon, L.; Hansson, O.; Nielsen, H.M. The Inflammatory Marker YKL-40 Is Elevated in Cerebrospinal Fluid from Patients with Alzheimer's but Not Parkinson's Disease or Dementia with Lewy Bodies. *PLoS ONE* **2015**, *10*, e0135458. [CrossRef]
147. Baldacci, F.; Toschi, N.; Lista, S.; Zetterberg, H.; Blennow, K.; Kilimann, I.; Teipel, S.J.; Cavedo, E.; Dos Santos, A.M.; Epelbaum, S.; et al. Two-level diagnostic classification using cerebrospinal fluid YKL-40 in Alzheimer's disease. *Alzheimer's Dement.* **2017**, *13*, 993–1003. [CrossRef]
148. Hellwig, K.; Kvartsberg, H.; Portelius, E.; Andreasson, U.; Oberstein, T.J.; Lewczuk, P.; Blennow, K.; Kornhuber, J.; Maler, J.M.; Zetterberg, H.; et al. Neurogranin and YKL-40: Independent markers of synaptic degeneration and neuroinflammation in Alzheimer's disease. *Alzheimer's Res. Ther.* **2015**, *7*, 1–8. [CrossRef]
149. Sutphen, C.L.; McCue, L.; Herries, E.M.; Xiong, C.; Ladenson, J.H.; Holtzman, D.M.; Fagan, A.M.; Adni, O.B.O. Longitudinal decreases in multiple cerebrospinal fluid biomarkers of neuronal injury in symptomatic late onset Alzheimer's disease. *Alzheimer's Dement.* **2018**, *14*, 869–879. [CrossRef]

150. Zhang, H.; Ng, K.P.; Therriault, J.; Kang, M.S.; Pascoal, T.A.; Rosa-Neto, P.; Gauthier, S. Cerebrospinal fluid phosphorylated tau, visinin-like protein-1, and chitinase-3-like protein 1 in mild cognitive impairment and Alzheimer's disease 11 Medical and Health Sciences 1109 Neurosciences. *Transl. Neurodegener.* **2018**, *7*. [CrossRef]
151. Janelidze, S.; Mattsson, N.; Stomrud, E.; Lindberg, O.; Palmqvist, S.; Zetterberg, H.; Blennow, K.; Hansson, O. CSF biomarkers of neuroinflammation and cerebrovascular dysfunction in early Alzheimer disease. *Neurology* **2018**, *91*, e867–e877. [CrossRef]
152. Alcolea, D.; Vilaplana, E.; Pegueroles, J.; Montal, V.; Sanchez-Juan, P.; González-Suárez, A.; Pozueta, A.; Rodríguez-Rodríguez, E.; Bartrés-Faz, D.; Vidal-Pineiro, D.; et al. Relationship between cortical thickness and cerebrospinal fluid YKL-40 in predementia stages of Alzheimer's disease. *Neurobiol. Aging* **2015**, *36*, 2018–2023. [CrossRef] [PubMed]
153. Gispert, J.D.; Monté, G.C.; Suárez-Calvet, M.; Falcon, C.; Tucholka, A.; Rojas, S.; Rami, L.; Sánchez-Valle, R.; Lladó, A.; Kleinberger, G.; et al. The APOE ε4 genotype modulates CSF YKL-40 levels and their structural brain correlates in the continuum of Alzheimer's disease but not those of sTREM2. *Alzheimer's Dement. Diagn. Assess. Dis. Monit.* **2017**, *6*, 50–59. [CrossRef] [PubMed]
154. Janelidze, S.; Hertze, J.; Zetterberg, H.; Waldö, M.L.; Santillo, A.; Blennow, K.; Hansson, O. Cerebrospinal fluid neurogranin and YKL -40 as biomarkers of Alzheimer's disease. *Ann. Clin. Transl. Neurol.* **2015**, *3*, 12–20. [CrossRef]
155. Baldacci, F.; Lista, S.; Cavedo, E.; Bonuccelli, U.; Hampel, H. Diagnostic function of the neuroinflammatory biomarker YKL-40 in Alzheimer's disease and other neurodegenerative diseases. *Expert Rev. Proteom.* **2017**, *14*, 285–299. [CrossRef] [PubMed]
156. Craig-Schapiro, R.; Perrin, R.J.; Roe, C.M.; Xiong, C.; Carter, D.; Cairns, N.J.; Mintun, M.A.; Peskind, E.R.; Li, G.; Galasko, D.R.; et al. YKL-40: A Novel Prognostic Fluid Biomarker for Preclinical Alzheimer's Disease. *Biol. Psychiatry* **2010**, *68*, 903–912. [CrossRef] [PubMed]
157. Choi, J.; Lee, H.-W.; Suk, K. Plasma level of chitinase 3-like 1 protein increases in patients with early Alzheimer's disease. *J. Neurol.* **2011**, *258*, 2181–2185. [CrossRef] [PubMed]
158. Vergallo, A.; Lista, S.; Lemercier, P.; Chiesa, P.A.; Zetterberg, H.; Blennow, K.; Potier, M.-C.; Habert, M.-O.; Baldacci, F.; Cavedo, E.; et al. Association of plasma YKL-40 with brain amyloid-β levels, memory performance, and sex in subjective memory complainers. *Neurobiol. Aging* **2020**, *96*, 22–32. [CrossRef]
159. Li, J.-T.; Zhang, Y. TREM2 regulates innate immunity in Alzheimer's disease. *J. Neuroinflamm.* **2018**, *15*, 1–7. [CrossRef]
160. Hickman, S.E.; El Khoury, J. TREM2 and the neuroimmunology of Alzheimer's disease. *Biochem. Pharmacol.* **2014**, *88*, 495–498. [CrossRef]
161. Jay, T.R.; Von Saucken, V.E.; Landreth, G.E. TREM2 in Neurodegenerative Diseases. *Mol. Neurodegener.* **2017**, *12*, 1–33. [CrossRef] [PubMed]
162. Brosseron, F.; Traschütz, A.; Widmann, C.N.; Kummer, M.P.; Tacik, P.; Santarelli, F.; Jessen, F.; Heneka, M.T. Characterization and clinical use of inflammatory cerebrospinal fluid protein markers in Alzheimer's disease. *Alzheimer's Res. Ther.* **2018**, *10*, 1–14. [CrossRef] [PubMed]
163. Heslegrave, A.J.; Heywood, W.; Paterson, R.W.; Magdalinou, N.K.; Svensson, J.; Johansson, P.; Öhrfelt, A.; Blennow, K.; Hardy, J.; Schott, J.M.; et al. Increased cerebrospinal fluid soluble TREM2 concentration in Alzheimer's disease. *Mol. Neurodegener.* **2016**, *11*, 1–7. [CrossRef] [PubMed]
164. Piccio, L.; Deming, Y.; Del-Águila, J.L.; Ghezzi, L.; Holtzman, D.M.; Fagan, A.M.; Fenoglio, C.; Galimberti, D.; Borroni, B.; Cruchaga, C. Cerebrospinal fluid soluble TREM2 is higher in Alzheimer disease and associated with mutation status. *Acta Neuropathol.* **2016**, *131*, 925–933. [CrossRef] [PubMed]
165. Suárez-Calvet, M.; Kleinberger, G.; Caballero, M.; Ángel, A.; Brendel, M.; Rominger, A.; Alcolea, D.; Fortea, J.; Lleó, A.; Blesa, R.; et al. sTREM 2 cerebrospinal fluid levels are a potential biomarker for microglia activity in early-stage Alzheimer's disease and associate with neuronal injury markers. *EMBO Mol. Med.* **2016**, *8*, 466–476. [CrossRef]
166. Henjum, K.; Almdahl, I.S.; Årskog, V.; Minthon, L.; Hansson, O.; Fladby, T.; Nilsson, L.N.G. Cerebrospinal fluid soluble TREM2 in aging and Alzheimer's disease. *Alzheimer's Res. Ther.* **2016**, *8*, 1–11. [CrossRef]

167. Gispert, J.D.; Suárez-Calvet, M.; Monté, G.C.; Tucholka, A.; Falcon, C.; Rojas, S.; Rami, L.; Sánchez-Valle, R.; Lladó, A.; Kleinberger, G.; et al. Cerebrospinal fluid sTREM2 levels are associated with gray matter volume increases and reduced diffusivity in early Alzheimer's disease. *Alzheimer's Dement.* **2016**, *12*, 1259–1272. [CrossRef]
168. Hu, N.; Tan, M.-S.; Yu, J.-T.; Sun, L.; Tan, L.; Wang, Y.-L.; Jiang, T.; Tan, L. Increased Expression of TREM2 in Peripheral Blood of Alzheimer's Disease Patients. *J. Alzheimer's Dis.* **2013**, *38*, 497–501. [CrossRef]
169. Casati, M.; Ferri, E.; Gussago, C.; Mazzola, P.; Abbate, C.; Bellelli, G.; Mari, D.; Cesari, M.; Arosio, B. Increased expression of TREM2 in peripheral cells from mild cognitive impairment patients who progress into Alzheimer's disease. *Eur. J. Neurol.* **2018**, *25*, 805–810. [CrossRef]
170. Deshmane, S.L.; Kremlev, S.; Amini, S.; Sawaya, B.E. Monocyte Chemoattractant Protein-1 (MCP-1): An Overview. *J. Interf. Cytokine Res.* **2009**, *29*, 313–326. [CrossRef]
171. Ueberham, U.; Ueberham, E.; Gruschka, H.; Arendt, T. Altered subcellular location of phosphorylated Smads in Alzheimer's disease. *Eur. J. Neurosci.* **2006**, *24*, 2327–2334. [CrossRef] [PubMed]
172. Tesseur, I.; Zou, K.; Esposito, L.; Bard, F.; Berber, E.; Van Can, J.; Lin, A.H.; Crews, L.; Tremblay, P.; Mathews, P.; et al. Deficiency in neuronal TGF-β signaling promotes neurodegeneration and Alzheimer's pathology. *J. Clin. Investig.* **2006**, *116*, 3060–3069. [CrossRef] [PubMed]
173. Torrisi, S.A.; Geraci, F.; Tropea, M.R.; Grasso, M.; Caruso, G.; Fidilio, A.; Musso, N.; Sanfilippo, G.; Tascedda, F.; Palmeri, A.; et al. Fluoxetine and Vortioxetine Reverse Depressive-Like Phenotype and Memory Deficits Induced by Aβ1-42 Oligomers in Mice: A Key Role of Transforming Growth Factor-β1. *Front. Pharmacol.* **2019**, *10*. [CrossRef] [PubMed]
174. Caraci, F.; Spampinato, S.; Sortino, M.A.; Bosco, P.; Battaglia, G.; Bruno, V.; Drago, F.; Nicoletti, F.; Copani, A. Dysfunction of TGF-β1 signaling in Alzheimer's disease: Perspectives for neuroprotection. *Cell Tissue Res.* **2011**, *347*, 291–301. [CrossRef] [PubMed]
175. Liu, C.; Liu, Q.; Song, L.; Gu, Y.; Jie, J.; Bai, X.; Yang, Y. Dab2 attenuates brain injury in APP/PS1 mice via targeting transforming growth factor-beta/SMAD signaling. *Neural Regen. Res.* **2014**, *9*, 41–50. [CrossRef]
176. Swardfager, W.; Lanctôt, K.; Rothenburg, L.; Wong, A.; Cappell, J.; Herrmann, N. A Meta-Analysis of Cytokines in Alzheimer's Disease. *Biol. Psychiatry* **2010**, *68*, 930–941. [CrossRef]
177. Junttila, A.; Kuvaja, M.; Hartikainen, P.; Siloaho, M.; Helisalmi, S.; Moilanen, V.; Kiviharju, A.; Jansson, L.; Tienari, P.J.; Remes, A.M.; et al. Cerebrospinal Fluid TDP-43 in Frontotemporal Lobar Degeneration and Amyotrophic Lateral Sclerosis Patients with and without the C9ORF72 Hexanucleotide Expansion. *Dement. Geriatr. Cogn. Dis. Extra* **2016**, *6*, 142–149. [CrossRef]
178. Lai, K.S.P.; Liu, C.S.; Rau, A.; Lanctôt, K.L.; Köhler, C.; Pakosh, M.; Carvalho, A.F.; Herrmann, N. Peripheral inflammatory markers in Alzheimer's disease: A systematic review and meta-analysis of 175 studies. *J. Neurol. Neurosurg. Psychiatry* **2017**, *88*, 876–882. [CrossRef]
179. Leung, R.; Proitsi, P.; Simmons, A.; Lunnon, K.; Güntert, A.; Kronenberg-Versteeg, D.; Pritchard, M.; Tsolaki, M.; Mecocci, P.; Kloszewska, I.; et al. Inflammatory Proteins in Plasma Are Associated with Severity of Alzheimer's Disease. *PLoS ONE* **2013**, *8*, e64971. [CrossRef]
180. Sun, Y.-X.; Minthon, L.; Wallmark, A.; Warkentin, S.; Blennow, K.; Janciauskiene, S. Inflammatory Markers in Matched Plasma and Cerebrospinal Fluid from Patients with Alzheimer's Disease. *Dement. Geriatr. Cogn. Disord.* **2003**, *16*, 136–144. [CrossRef]
181. Scheff, S.W.; Price, D.A.; Schmitt, F.A.; Mufson, E.J. Hippocampal synaptic loss in early Alzheimer's disease and mild cognitive impairment. *Neurobiol. Aging* **2006**, *27*, 1372–1384. [CrossRef] [PubMed]
182. Galasko, D.; Xiao, M.; Xu, D.; Smirnov, D.; Salmon, D.P.; Dewit, N.; Vanbrabant, J.; Jacobs, D.; Vanderstichele, H.; Vanmechelen, E.; et al. Synaptic biomarkers in CSF aid in diagnosis, correlate with cognition and predict progression in MCI and Alzheimer's disease. *Alzheimer's Dement. Transl. Res. Clin. Interv.* **2019**, *5*, 871–882. [CrossRef] [PubMed]
183. Molinuevo, J.L.; Ayton, S.; Batrla, R.; Bednar, M.M.; Bittner, T.; Cummings, J.; Fagan, A.M.; Hampel, H.; Mielke, M.M.; Mikulskis, A.; et al. Current state of Alzheimer's fluid biomarkers. *Acta Neuropathol.* **2018**, *136*, 821–853. [CrossRef]
184. Zhong, L.; Cherry, T.; E Bies, C.; Florence, M.; Gerges, N.Z. Neurogranin enhances synaptic strength through its interaction with calmodulin. *EMBO J.* **2009**, *28*, 3027–3039. [CrossRef] [PubMed]

185. Liu, W.; Lin, H.; He, X.; Chen, L.; Dai, Y.; Jia, W.; Xue, X.; Tao, J.; Chen, L. Neurogranin as a cognitive biomarker in cerebrospinal fluid and blood exosomes for Alzheimer's disease and mild cognitive impairment. *Transl. Psychiatry* **2020**, *10*, 1–9. [CrossRef] [PubMed]
186. Mavroudis, I.A.; Petridis, F.; Chatzikonstantinou, S.; Kazis, D. A meta-analysis on CSF neurogranin levels for the diagnosis of Alzheimer's disease and mild cognitive impairment. *Aging Clin. Exp. Res.* **2019**, *32*, 1639–1646. [CrossRef] [PubMed]
187. Mazzucchi, S.; Palermo, G.; Campese, N.; Galgani, A.; Della Vecchia, A.; Vergallo, A.; Siciliano, G.; Ceravolo, R.; Hampel, H.; Baldacci, F. The role of synaptic biomarkers in the spectrum of neurodegenerative diseases. *Expert Rev. Proteom.* **2020**. [CrossRef]
188. Mattsson, N.; Insel, P.S.; Palmqvist, S.; Portelius, E.; Zetterberg, H.; Weiner, M.; Blennow, K.; Hansson, O.; Initiative, T.A.D.N. Cerebrospinal fluid tau, neurogranin, and neurofilament light in Alzheimer's disease. *EMBO Mol. Med.* **2016**, *8*, 1184–1196. [CrossRef]
189. Tarawneh, R.; D'Angelo, G.; Crimmins, D.; Herries, E.; Griest, T.; Fagan, A.M.; Zipfel, G.J.; Ladenson, J.H.; Morris, J.C.; Holtzman, D.M. Diagnostic and Prognostic Utility of the Synaptic Marker Neurogranin in Alzheimer Disease. *JAMA Neurol.* **2016**, *73*, 561–571. [CrossRef]
190. Kester, M.I.; Teunissen, C.E.; Crimmins, D.L.; Herries, E.M.; Ladenson, J.H.; Scheltens, P.; Van Der Flier, W.M.; Morris, J.C.; Holtzman, D.M.; Fagan, A.M. Neurogranin as a Cerebrospinal Fluid Biomarker for Synaptic Loss in Symptomatic Alzheimer Disease. *JAMA Neurol.* **2015**, *72*, 1275–1280. [CrossRef]
191. Kvartsberg, H.; Duits, F.H.; Ingelsson, M.; Andreasen, N.; Öhrfelt, A.; Andersson, K.; Brinkmalm, G.; Lannfelt, L.; Minthon, L.; Hansson, O.; et al. Cerebrospinal fluid levels of the synaptic protein neurogranin correlates with cognitive decline in prodromal Alzheimer's disease. *Alzheimer's Dement.* **2014**, *11*, 1180–1190. [CrossRef] [PubMed]
192. Wellington, H.; Paterson, R.W.; Portelius, E.; Törnqvist, U.; Magdalinou, N.; Fox, N.C.; Blennow, K.; Schott, J.M.; Zetterberg, H. Increased CSF neurogranin concentration is specific to Alzheimer disease. *Neurology* **2016**, *86*, 829–835. [CrossRef] [PubMed]
193. Lista, S.; Toschi, N.; Baldacci, F.; Zetterberg, H.; Blennow, K.; Kilimann, I.; Teipel, S.J.; Cavedo, E.; Santos, A.M.D.; Epelbaum, S.; et al. Cerebrospinal fluid neurogranin as a biomarker of neurodegenerative diseases: A cross-sectional study. *J. Alzheimer's Dis.* **2017**, *59*, 1327–1334. [CrossRef] [PubMed]
194. Portelius, E.; Olsson, B.; Höglund, K.; Cullen, N.C.; Kvartsberg, H.; Andreasson, U.; Zetterberg, H.; Sandelius, Å.; Shaw, L.M.; Lee, V.M.Y.; et al. Cerebrospinal fluid neurogranin concentration in neurodegeneration: Relation to clinical phenotypes and neuropathology. *Acta Neuropathol.* **2018**, *136*, 363–376. [CrossRef] [PubMed]
195. Brinkmalm, A.; Brinkmalm, G.; Honer, W.G.; Frölich, L.; Hausner, L.; Minthon, L.; Hansson, O.; Wallin, A.; Zetterberg, H.; Blennow, K.; et al. SNAP-25 is a promising novel cerebrospinal fluid biomarker for synapse degeneration in Alzheimer's disease. *Mol. Neurodegener.* **2014**, *9*, 1–13. [CrossRef] [PubMed]
196. Zhang, H.; Initiative, T.A.D.N.; Therriault, J.; Kang, M.S.; Ng, K.P.; Pascoal, T.A.; Rosa-Neto, P.; Gauthier, S. Cerebrospinal fluid synaptosomal-associated protein 25 is a key player in synaptic degeneration in mild cognitive impairment and Alzheimer's disease. *Alzheimer's Res. Ther.* **2018**, *10*, 1–11. [CrossRef]
197. Öhrfelt, A.; Brinkmalm, A.; Dumurgier, J.; Zetterberg, H.; Bouaziz-Amar, E.; Hugon, J.; Paquet, C.; Blennow, K. A Novel ELISA for the Measurement of Cerebrospinal Fluid SNAP-25 in Patients with Alzheimer's Disease. *Neuroscience* **2019**, *420*, 136–144. [CrossRef]
198. Goetzl, E.J.; Kapogiannis, D.; Schwartz, J.B.; Lobach, I.V.; Goetzl, L.; Abner, E.L.; Jicha, G.A.; Karydas, A.M.; Boxer, A.; Miller, B.L. Decreased synaptic proteins in neuronal exosomes of frontotemporal dementia and Alzheimer's disease. *FASEB J.* **2016**, *30*, 4141–4148. [CrossRef]
199. Gaetani, L.; Blennow, K.; Calabresi, P.; Di Filippo, M.; Parnetti, L.; Zetterberg, H. Neurofilament light chain as a biomarker in neurological disorders. *J. Neurol. Neurosurg. Psychiatry* **2019**, *90*, 870–881. [CrossRef]
200. Brureau, A.; Blanchard-Bregeon, V.; Pech, C.; Hamon, S.; Chaillou, P.; Guillemot, J.-C.; Barneoud, P.; Bertrand, P.; Pradier, L.; Rooney, T.; et al. NF-L in cerebrospinal fluid and serum is a biomarker of neuronal damage in an inducible mouse model of neurodegeneration. *Neurobiol. Dis.* **2017**, *104*, 73–84. [CrossRef]
201. Bacioglu, M.; Maia, L.F.; Preische, O.; Schelle, J.; Apel, A.; Kaeser, S.A.; Schweighauser, M.; Eninger, T.; Lambert, M.; Pilotto, A.; et al. Neurofilament Light Chain in Blood and CSF as Marker of Disease Progression in Mouse Models and in Neurodegenerative Diseases. *Neuron* **2016**, *91*, 56–66. [CrossRef] [PubMed]

202. Barro, C.; Benkert, P.; Disanto, G.; Tsagkas, C.; Amann, M.; Naegelin, Y.; Leppert, D.; Gobbi, C.; Granziera, C.; Yaldizli, Ö.; et al. Serum neurofilament as a predictor of disease worsening and brain and spinal cord atrophy in multiple sclerosis. *Brain* **2018**, *141*, 2382–2391. [CrossRef] [PubMed]
203. Disanto, G.; Barro, C.; Benkert, P.; Naegelin, Y.; Schädelin, S.; Giardiello, A.; Zecca, C.; Blennow, K.; Zetterberg, H.; Leppert, D.; et al. Serum Neurofilament light: A biomarker of neuronal damage in multiple sclerosis. *Ann. Neurol.* **2017**, *81*, 857–870. [CrossRef] [PubMed]
204. Rohrer, J.D.; Woollacott, I.O.; Dick, K.M.; Brotherhood, E.; Gordon, E.; Fellows, A.; Toombs, J.; Druyeh, R.; Cardoso, M.J.; Ourselin, S.; et al. Serum neurofilament light chain protein is a measure of disease intensity in frontotemporal dementia. *Neurology* **2016**, *87*, 1329–1336. [CrossRef]
205. Alcolea, D.; Vilaplana, E.; Suárez-Calvet, M.; Illán-Gala, I.; Blesa, R.; Clarimón, J.; Lladó, A.; Sánchez-Valle, R.; Molinuevo, J.L.; García-Ribas, G.; et al. CSF sAPPβ, YKL-40, and neurofilament light in frontotemporal lobar degeneration. *Neurology* **2017**, *89*, 178–188. [CrossRef]
206. Kern, S.; Syrjanen, J.A.; Blennow, K.; Zetterberg, H.; Skoog, I.; Waern, M.; Hagen, C.E.; Van Harten, A.C.; Knopman, D.S.; Jack, C.R.; et al. Association of Cerebrospinal Fluid Neurofilament Light Protein With Risk of Mild Cognitive Impairment Among Individuals Without Cognitive Impairment. *JAMA Neurol.* **2018**, *76*, 187. [CrossRef]
207. Skillbäck, T.; Farahmand, B.; Bartlett, J.W.; Rosén, C.; Mattsson, N.; Nägga, K.; Kilander, L.; Religa, D.; Wimo, A.; Winblad, B.; et al. CSF neurofilament light differs in neurodegenerative diseases and predicts severity and survival. *Neurology* **2014**, *83*, 1945–1953. [CrossRef]
208. Zetterberg, H.; Skillbäck, T.; Mattsson, N.; Trojanowski, J.Q.; Portelius, E.; Shaw, L.M.; Weiner, M.W.; Blennow, K.; Alzheimer's Disease Neuroimaging Initiative. Association of Cerebrospinal Fluid Neurofilament Light Concentration With Alzheimer Disease Progression. *JAMA Neurol.* **2016**, *73*, 60–67. [CrossRef]
209. Sánchez-Valle, R.; Heslegrave, A.; Foiani, M.S.; Bosch, B.; Antonell, A.; Balasa, M.; Lladó, A.; Toombs, J.; Fox, N.C. Serum neurofilament light levels correlate with severity measures and neurodegeneration markers in autosomal dominant Alzheimer's disease. *Alzheimer's Res. Ther.* **2018**, *10*, 113. [CrossRef]
210. De Jong, D.; Jansen, R.W.M.M.; Pijnenburg, Y.A.L.; Van Geel, W.J.; Borm, G.F.; Kremer, H.P.H.; Verbeek, M.M. CSF neurofilament proteins in the differential diagnosis of dementia. *J. Neurol. Neurosurg. Psychiatry* **2007**, *78*, 936–938. [CrossRef]
211. Olsson, B.; Portelius, E.; Cullen, N.C.; Sandelius, Å.; Zetterberg, H.; Andreasson, U.; Höglund, K.; Irwin, D.J.; Grossman, M.; Weintraub, D.; et al. Association of Cerebrospinal Fluid Neurofilament Light Protein Levels With Cognition in Patients With Dementia, Motor Neuron Disease, and Movement Disorders. *JAMA Neurol.* **2019**, *76*, 318–325. [CrossRef] [PubMed]
212. Steinacker, P.; Semler, E.; Anderl-Straub, S.; Diehl-Schmid, J.; Schroeter, M.L.; Uttner, I.; Foerstl, H.; Landwehrmeyer, B.; Von Arnim, C.A.; Kassubek, J.; et al. Neurofilament as a blood marker for diagnosis and monitoring of primary progressive aphasias. *Neorology* **2017**, *88*, 961–969. [CrossRef] [PubMed]
213. Groblewska, M.; Muszyński, P.; Wojtulewska-Supron, A.; Kulczyńska-Przybik, A.; Mroczko, B. The Role of Visinin-Like Protein-1 in the Pathophysiology of Alzheimer's Disease. *J. Alzheimer's Dis.* **2015**, *47*, 17–32. [CrossRef] [PubMed]
214. Bernstein, H.-G.; Baumann, B.; Danos, P.; Diekmann, S.; Bogerts, B.; Gundelfinger, E.D.; Braunewell, K.-H. Regional and cellular distribution of neural visinin-like protein immunoreactivities (VILIP-1 and VILIP-3) in human brain. *J. Neurocytol.* **1999**, *28*, 655–662. [CrossRef]
215. Mroczko, B.; Groblewska, M.; Zboch, M.; Muszyński, P.; Zajkowska, A.; Borawska, R.; Szmitkowski, M.; Kornhuber, J.; Lewczuk, P. Evaluation of Visinin-Like Protein 1 Concentrations in the Cerebrospinal Fluid of Patients with Mild Cognitive Impairment as a Dynamic Biomarker of Alzheimer's Disease. *J. Alzheimer's Dis.* **2014**, *43*, 1031–1037. [CrossRef]
216. Leko, M.B.; Borovečki, F.; Dejanović, N.; Hof, P.R.; Šimić, G. Predictive Value of Cerebrospinal Fluid Visinin-Like Protein-1 Levels for Alzheimer's Disease Early Detection and Differential Diagnosis in Patients with Mild Cognitive Impairment. *J. Alzheimer's Dis.* **2016**, *50*, 765–778. [CrossRef]
217. Braunewell, K.H. The visinin-like proteins VILIP-1 and VILIP-3 in Alzheimer's disease—old wine in new bottles. *Front. Mol. Neurosci.* **2012**, *5*, 20. [CrossRef]
218. Lee, J.-M.; Blennow, K.; Andreasen, N.; Laterza, O.; Modur, V.; Olander, J.; Gao, F.; Ohlendorf, M.; Ladenson, J.H. The Brain Injury Biomarker VLP-1 Is Increased in the Cerebrospinal Fluid of Alzheimer Disease Patients. *Clin. Chem.* **2008**, *54*, 1617–1623. [CrossRef]

219. Luo, X.; Hou, L.; Shi, H.; Zhong, X.; Zhang, Y.; Zheng, D.; Tan, Y.; Hu, G.; Mu, N.; Chan, J.; et al. CSF levels of the neuronal injury biomarker visinin-like protein-1 in Alzheimer's disease and dementia with Lewy bodies. *J. Neurochem.* **2013**, *127*, 681–690. [CrossRef]
220. Tarawneh, R.M.; D'Angelo, G.; Ba, E.M.; Xiong, C.; Carter, D.; Cairns, N.J.; Fagan, A.M.; Head, D.; Mintun, M.A.; Ladenson, J.H.; et al. Visinin-like protein-1: Diagnostic and prognostic biomarker in Alzheimer disease. *Ann. Neurol.* **2011**, *70*, 274–285. [CrossRef]
221. Tarawneh, R.M.; Head, D.; Allison, S.; Buckles, V.; Fagan, A.M.; Ladenson, J.H.; Morris, J.C.; Holtzman, D.M. Cerebrospinal Fluid Markers of Neurodegeneration and Rates of Brain Atrophy in Early Alzheimer Disease. *JAMA Neurol.* **2015**, *72*, 656–665. [CrossRef] [PubMed]
222. Tarawneh, R.; Lee, J.-M.; Ladenson, J.H.; Morris, J.C.; Holtzman, D.M. CSF VILIP-1 predicts rates of cognitive decline in early Alzheimer disease. *Neurology* **2012**, *78*, 709–719. [CrossRef] [PubMed]
223. Chen-Plotkin, A.S.; Lee, V.M.-Y.; Trojanowski, J.Q. TAR DNA-binding protein 43 in neurodegenerative disease. *Nat. Rev. Neurol.* **2010**, *6*, 211–220. [CrossRef] [PubMed]
224. Neumann, M.; Sampathu, D.M.; Kwong, L.K.; Truax, A.C.; Micsenyi, M.C.; Chou, T.T.; Bruce, J.; Schuck, T.; Grossman, M.; Clark, C.M.; et al. Ubiquitinated TDP-43 in Frontotemporal Lobar Degeneration and Amyotrophic Lateral Sclerosis. *Science* **2006**, *314*, 130–133. [CrossRef] [PubMed]
225. Amador-Ortiz, C.; Lin, W.-L.; Ahmed, Z.; Ms, D.P.; Davies, P.; Duara, R.; Graff-Radford, N.R.; Hutton, M.L.; Dickson, D.W. TDP-43 immunoreactivity in hippocampal sclerosis and Alzheimer's disease. *Ann. Neurol.* **2007**, *61*, 435–445. [CrossRef]
226. Chang, X.-L.; Tan, M.-S.; Tan, L.; Yu, J.-T. The Role of TDP-43 in Alzheimer's Disease. *Mol. Neurobiol.* **2015**, *53*, 3349–3359. [CrossRef]
227. James, B.D.; Wilson, R.S.; Boyle, P.A.; Trojanowski, J.Q.; Bennett, D.A.; Schneider, J.A. TDP-43 stage, mixed pathologies, and clinical Alzheimer's-type dementia. *Brain* **2016**, *139*, 2983–2993. [CrossRef]
228. Foulds, P.; McAuley, E.; Gibbons, L.; Davidson, Y.; Pickering-Brown, S.; Neary, D.; Snowden, J.S.; Allsop, D.; Mann, D.M.A. TDP-43 protein in plasma may index TDP-43 brain pathology in Alzheimer's disease and frontotemporal lobar degeneration. *Acta Neuropathol.* **2008**, *116*, 141–146. [CrossRef]
229. Williams, S.M.; Schulz, P.; Rosenberry, T.L.; Caselli, R.J.; Sierks, M.R. Blood-Based Oligomeric and Other Protein Variant Biomarkers to Facilitate Pre-Symptomatic Diagnosis and Staging of Alzheimer's Disease. *J. Alzheimer's Dis.* **2017**, *58*, 23–35. [CrossRef]
230. Vergallo, A.; Bun, R.-S.; Toschi, N.; Baldacci, F.; Zetterberg, H.; Blennow, K.; Cavedo, E.; Lamari, F.; Habert, M.-O.; Dubois, B.; et al. Association of cerebrospinal fluid α-synuclein with total and phospho-tau181 protein concentrations and brain amyloid load in cognitively normal subjective memory complainers stratified by Alzheimer's disease biomarkers. *Alzheimer's Dement.* **2018**, *14*, 1623–1631. [CrossRef]
231. Twohig, D.; Dominantly Inherited Alzheimer Network (DIAN); Rodriguez-Vieitez, E.; Sando, S.B.; Berge, G.; Lauridsen, C.; Møller, I.; Grøntvedt, G.R.; Bråthen, G.; Patra, K.; et al. The relevance of cerebrospinal fluid α-synuclein levels to sporadic and familial Alzheimer's disease. *Acta Neuropathol. Commun.* **2018**, *6*, 1–19. [CrossRef] [PubMed]
232. Twohig, D.; Nielsen, H.M. α-synuclein in the pathophysiology of Alzheimer's disease. *Mol. Neurodegener.* **2019**, *14*, 1–19. [CrossRef] [PubMed]
233. Gao, L.; Tang, H.; Nie, K.; Wang, L.; Zhao, J.; Gan, R.; Huang, J.; Zhu, R.; Feng, S.; Duan, Z.; et al. Cerebrospinal fluid alpha-synuclein as a biomarker for Parkinson's disease diagnosis: A systematic review and meta-analysis. *Int. J. Neurosci.* **2014**, *125*, 645–654. [CrossRef] [PubMed]
234. Esako, W.; Murakami, N.; Izumi, Y.; Kaji, R. Reduced alpha-synuclein in cerebrospinal fluid in synucleinopathies: Evidence from a meta-analysis. *Mov. Disord.* **2014**, *29*, 1599–1605. [CrossRef]
235. Fairfoul, G.; McGuire, L.I.; Pal, S.; Ironside, J.W.; Neumann, J.; Christie, S.; Joachim, C.; Esiri, M.; Evetts, S.G.; Rolinski, M.; et al. Alpha-synuclein RT -Qu IC in the CSF of patients with alpha-synucleinopathies. *Ann. Clin. Transl. Neurol.* **2016**, *3*, 812–818. [CrossRef]
236. Shahnawaz, M.; Tokuda, T.; Waragai, M.; Mendez, N.; Ishii, R.; Trenkwalder, C.; Mollenhauer, B.; Soto, C. Development of a Biochemical Diagnosis of Parkinson Disease by Detection of α-Synuclein Misfolded Aggregates in Cerebrospinal Fluid. *JAMA Neurol.* **2017**, *74*, 163–172. [CrossRef]
237. Baldacci, F.; Daniele, S.; Piccarducci, R.; Giampietri, L.; Pietrobono, D.; Giorgi, F.S.; Nicoletti, V.; Frosini, D.; Libertini, P.; Gerfo, A.L.; et al. Potential Diagnostic Value of Red Blood Cells α-Synuclein Heteroaggregates in Alzheimer's Disease. *Mol. Neurobiol.* **2019**, *56*, 6451–6459. [CrossRef]

238. Johnstone, R.M.; Adam, M.; Hammond, J.R.; Orr, L.; Turbide, C. Vesicle formation during reticulocyte maturation. Association of plasma membrane activities with released vesicles (exosomes). *J. Biol. Chem.* **1987**, *262*, 9412–9420.
239. Coleman, B.M.; Hill, A.F. Extracellular vesicles—Their role in the packaging and spread of misfolded proteins associated with neurodegenerative diseases. *Semin. Cell Dev. Biol.* **2015**, *40*, 89–96. [CrossRef]
240. Vella, L.J.; Greenwood, D.L.; Cappai, R.; Scheerlinck, J.-P.Y.; Hill, A.F. Enrichment of prion protein in exosomes derived from ovine cerebral spinal fluid. *Veter- Immunol. Immunopathol.* **2008**, *124*, 385–393. [CrossRef]
241. Street, J.M.; E Barran, P.; Mackay, C.L.; Weidt, S.; Balmforth, C.; Walsh, T.S.; Chalmers, R.T.; Webb, D.J.; Dear, J.W. Identification and proteomic profiling of exosomes in human cerebrospinal fluid. *J. Transl. Med.* **2012**, *10*, 5. [CrossRef] [PubMed]
242. Chiasserini, D.; Van Weering, J.R.; Piersma, S.R.; Pham, T.V.; Malekzadeh, A.; Teunissen, C.E.; De Wit, H.; Jiménez, C.R. Proteomic analysis of cerebrospinal fluid extracellular vesicles: A comprehensive dataset. *J. Proteom.* **2014**, *106*, 191–204. [CrossRef] [PubMed]
243. Saman, S.; Kim, W.; Raya, M.; Visnick, Y.; Miro, S.; Saman, S.; Jackson, B.; McKee, A.C.; Alvarez, V.E.; Lee, N.C.Y.; et al. Exosome-associated Tau Is Secreted in Tauopathy Models and Is Selectively Phosphorylated in Cerebrospinal Fluid in Early Alzheimer Disease. *J. Biol. Chem.* **2012**, *287*, 3842–3849. [CrossRef] [PubMed]
244. Fiandaca, M.S.; Kapogiannis, D.; Mapstone, M.; Boxer, A.; Eitan, E.; Schwartz, J.B.; Abner, E.L.; Petersen, R.C.; Federoff, H.J.; Miller, B.L.; et al. Identification of preclinical Alzheimer's disease by a profile of pathogenic proteins in neurally derived blood exosomes: A case-control study. *Alzheimer's Dement.* **2015**, *11*, 600–607. [CrossRef]
245. Hart, N.J.; Koronyo, Y.; Black, K.L.; Koronyo-Hamaoui, M. Ocular indicators of Alzheimer's: Exploring disease in the retina. *Acta Neuropathol.* **2016**, *132*, 767–787. [CrossRef]
246. Koronyo-Hamaoui, M.; Koronyo, Y.; Ljubimov, A.V.; Miller, C.A.; Ko, M.K.; Black, K.L.; Schwartz, M.; Farkas, D.L. Identification of amyloid plaques in retinas from Alzheimer's patients and noninvasive in vivo optical imaging of retinal plaques in a mouse model. *NeuroImage* **2011**, *54*, S204–S217. [CrossRef]
247. Koronyo, Y.; Salumbides, B.C.; Black, K.L.; Koronyo-Hamaoui, M. Alzheimer's Disease in the Retina: Imaging Retinal Aß Plaques for Early Diagnosis and Therapy Assessment. *Neurodegener. Dis.* **2012**, *10*, 285–293. [CrossRef]
248. Koronyo, Y.; Biggs, D.; Barron, E.; Boyer, D.S.; Pearlman, J.A.; Au, W.J.; Kile, S.J.; Blanco, A.; Fuchs, D.-T.; Ashfaq, A.; et al. Retinal amyloid pathology and proof-of-concept imaging trial in Alzheimer's disease. *JCI Insight* **2017**, *2*. [CrossRef]
249. Feke, G.T.; Hyman, B.T.; Stern, R.A.; Pasquale, L.R. Retinal blood flow in mild cognitive impairment and Alzheimer's disease. *Alzheimer's Dement. Diagn. Assess. Dis. Monit.* **2015**, *1*, 144–151. [CrossRef]
250. Alber, J.; Goldfarb, D.; Thompson, L.I.; Arthur, E.; Hernandez, K.; Cheng, D.; DeBuc, D.C.; Cordeiro, F.; Provetti-Cunha, L.; Haan, J.D.; et al. Developing retinal biomarkers for the earliest stages of Alzheimer's disease: What we know, what we don't, and how to move forward. *Alzheimer's Dement.* **2020**, *16*, 229–243. [CrossRef]
251. Palermo, G.; Mazzucchi, S.; Della Vecchia, A.; Siciliano, G.; Bonuccelli, U.; Azuar, C.; Ceravolo, R.; Lista, S.; Hampel, H.; Baldacci, F. Different Clinical Contexts of Use of Blood Neurofilament Light Chain Protein in the Spectrum of Neurodegenerative Diseases. *Mol. Neurobiol.* **2020**, *57*, 4667–4691. [CrossRef] [PubMed]
252. Al-Chalabi, A.; Hardiman, O.; Kiernan, M.C.; Chiò, A.; Rix-Brooks, B.; Berg, L.H.V.D. Amyotrophic lateral sclerosis: Moving towards a new classification system. *Lancet Neurol.* **2016**, *15*, 1182–1194. [CrossRef]
253. Castrillo, J.I.; Lista, S.; Hampel, H.; Ritchie, C.W. Systems biology methods for Alzheimer's disease research toward molecular signatures, subtypes, and stages and precision medicine: Application in cohort studies and trials. In *Methods in Molecular Biology*; Humana Press Inc.: Totowa, NJ, USA, 2018; pp. 31–66.
254. Geerts, H.; Dacks, P.A.; Devanarayan, V.; Haas, M.; Khachaturian, Z.S.; Gordon, M.F.; Maudsley, S.; Romero, K.; Stephenson, D.; Brain Health Modeling Initiative (BHMI). Big data to smart data in Alzheimer's disease: The brain health modeling initiative to foster actionable knowledge. *Alzheimer's Dement.* **2016**, *12*, 1014–1021. [CrossRef] [PubMed]

Review

Use of Biomarkers in Ongoing Research Protocols on Alzheimer's Disease

Marco Canevelli [1,2,*], Giulia Remoli [1], Ilaria Bacigalupo [2], Martina Valletta [1], Marco Toccaceli Blasi [1], Francesco Sciancalepore [1], Giuseppe Bruno [1], Matteo Cesari [3,4] and Nicola Vanacore [2]

1 Department of Human Neuroscience, Sapienza University, 00185 Rome, Italy; remoligiulia92@gmail.com (G.R.); m.valletta@live.it (M.V.); toccaceliblasi.1529394@studenti.uniroma1.it (M.T.B.); sciancalepore.1679543@studenti.uniroma1.it (F.S.); giuseppe.bruno@uniroma1.it (G.B.)
2 National Center for Disease Prevention and Health Promotion, Italian National Institute of Health, 00161 Rome, Italy; ilaria.bacigalupo@iss.it (I.B.); nicola.vanacore@iss.it (N.V.)
3 Department of Clinical Sciences and Community Health, University of Milan, 20122 Milan, Italy; macesari@gmail.com
4 Geriatric Unit, Fondazione IRCCS Ca' Granda Ospedale Maggiore Policlinico, 20122 Milan, Italy
* Correspondence: marco.canevelli@gmail.com; Tel./Fax: +39-(06)-4991-4604

Received: 30 June 2020; Accepted: 22 July 2020; Published: 24 July 2020

Abstract: The present study aimed to describe and discuss the state of the art of biomarker use in ongoing Alzheimer's disease (AD) research. A review of 222 ongoing phase 1, 2, 3, and 4 protocols registered in the clinicaltrials.gov database was performed. All the trials (i) enrolling subjects with clinical disturbances and/or preclinical diagnoses falling within the AD continuum; and (ii) testing the efficacy and/or safety/tolerability of a therapeutic intervention, were analyzed. The use of biomarkers of amyloid deposition, tau pathology, and neurodegeneration among the eligibility criteria and/or study outcomes was assessed. Overall, 58.2% of ongoing interventional studies on AD adopt candidate biomarkers. They are mostly adopted by studies at the preliminary stages of the drug development process to explore the safety profile of novel therapies, and to provide evidence of target engagement and disease-modifying properties. The biologically supported selection of participants is mostly based on biomarkers of amyloid deposition, whereas the use of biomarkers as study outcomes mostly relies on markers of neurodegeneration. Biomarkers play an important role in the design and conduction of research protocols targeting AD. Nevertheless, their clinical validity, utility, and cost-effectiveness in the "real world" remain to be clarified.

Keywords: Alzheimer's disease; biomarkers; drug development; clinical trials; diagnostic research

1. Introduction

In the last few decades, the notion of Alzheimer's disease (AD) has substantially changed. The enhanced understanding of the underlying neuropathophysiological mechanisms has supported a gradual shift from a clinical-pathological conception of AD [1] to a more biology-oriented framework [2,3].

The progressive evolution of the AD construct has been mainly sustained by the identification of key biomarkers of the AD neuropathological process [4]. According to the latest research definition, AD can be diagnosed in vivo, independently of cognitive manifestations, by the means of biomarkers reflecting β-amyloid deposition, tau pathology, and neurodegeneration [3]. Although their use is currently recommended only for research purposes [3,5], biomarkers are increasingly used in specialist clinical settings [6,7]. Nevertheless, some concerns and methodological shortcomings still affect the use of candidate AD biomarkers in routine clinical care. First, their clinical validity and utility have not yet been fully proven, and inconclusive evidence exists about their analytical validity [8]. Moreover, the lifetime risk of dementia among individuals with positive AD biomarkers varies considerably. In this regard, it is estimated that most persons affected by preclinical AD based on abnormal biomarker status will not develop dementia in their lifetimes [9]. Furthermore, they are either expensive, invasive, or both.

Biomarkers often play an important role in the design and conduction of research protocols targeting AD [10]. They may improve the selection of participants and render more biologically homogeneous the sampled populations. Moreover, their use may help demonstrating target engagement by tested intervention, providing evidence on disease modification, informing analytic stratification, and monitoring adverse effects [10]. At the same time, the sustainability of biomarker's translation into care models and the generalizability (i.e., transferability) of interventions to the "real world" should always be taken into account [11].

In the present study, we describe and discuss the state of the art of biomarker use in AD research by reviewing the ongoing protocols registered in the clinicaltrials.gov database, updating and extending previous analyses on the topic [12]. In particular, we focus on the adoption of candidate biomarkers of amyloid deposition, tau pathology, and neurodegeneration as (i) eligibility criteria; and (ii) study outcomes. This analysis may inform how AD biomarkers are currently adopted in the drug development process, shedding a light on potential methodological, clinical, and ethical issues. Moreover, this approach may provide useful insights into a public health perspective considering that AD biomarkers and diagnostic procedures tested in research protocols will likely be increasingly translated in the daily practice to support clinical activities (e.g., risk prediction [13]) and regulatory decisions (e.g., drug accessibility and reimbursability [14]).

2. Materials and Methods

2.1. Data Source and Search Strategy

Clinicaltrials.gov was used as the reference source for the present study. Clinicaltrials.gov is an online database provided by the U.S. National Library of Medicine, which collects information from clinical studies that are conducted worldwide on a wide range of diseases and conditions.

The database was explored on 10 May 2020, by using the following terms and fields in the advanced search function: "Alzheimer" [CONDITION OR DISEASE] AND "interventional studies (clinical trials)" [STUDY TYPE] AND ("not yet recruiting" OR "recruiting" OR "enrolling by invitation" OR "active, not recruiting") [STATUS: RECRUITMENT] AND ("phase 1" OR "phase 2" OR "phase 3" OR "phase 4") [PHASE]. No restriction on age, sex, date, and location was applied.

Two authors (M.V. and G.R.) independently screened the resulting records and assessed their adherence to the following inclusion criteria:

i) Enrolling subjects with clinical disturbances and/or preclinical diagnoses falling within the AD continuum (i.e., preclinical AD, subjective cognitive decline, mild cognitive impairment, prodromal AD, and AD dementia) [3];
ii) Testing the efficacy and/or safety/tolerability of a therapeutic (both pharmacological and non-pharmacological) intervention.

Thus, trials enrolling participants with non-AD dementias or healthy volunteers, and primarily aiming at investigating diagnostic procedures (e.g., a novel neuroimaging technique) were excluded.

The flow chart presented in Figure 1 shows the process of protocols selection.

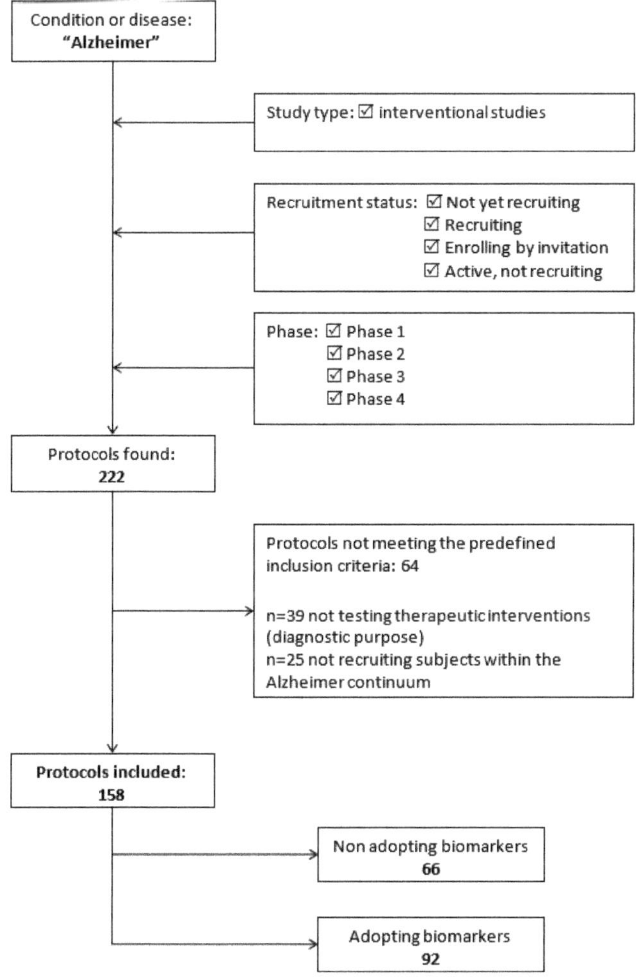

Figure 1. Flow-chart of protocols' selection.

2.2. Data Extraction

Two reviewers (M.V. and G.R.) independently extracted the following data from the selected protocols: NCT number; phase; status; study design; expected end date; sponsor; target condition; intervention; mechanism of action; primary and secondary outcome measure (s); planned number of participants. The use of the following AD biomarkers among the eligibility criteria and/or study outcomes was also assessed according to the AT (N) biomarker grouping [3]:

- A, amyloid deposition: (i) low cerebrospinal fluid (CSF) Aβ42 or Aβ42/Aβ40 ratio; and (ii) positive amyloid positron emission tomography (PET) scan;
- T, tau pathology: (i) elevated CSF phospho-tau (*p*-tau); and (ii) positive tau PET scan;
- N, neurodegeneration: (i) atrophy on anatomic magnetic resonance imaging (MRI); (ii) elevated CSF total tau (*t*-tau); and (iii) fluorodeoxyglucose (FDG) PET hypometabolism.

Disagreements in the selection process and/or extraction of data were solved by consensus or by involving two additional reviewers (M.C. and M.T.B.).

2.3. Data Analysis

Data were provided for two categories of protocols: (i) those using biomarkers in the selection of participants; and (ii) those using biomarkers as study outcomes. These categories were partially overlapping because some studies adopted biomarkers both as eligibility criteria and endpoints. Percentages and median values were calculated to summarize the abstracted categorical and continuous variables. Chi-square and median tests were used to compare the methodological characteristics of protocols adopting versus non-adopting biomarkers to ascertain eligibility and/or as endpoints. Statistical significance was set at $p < 0.05$.

3. Results

3.1. Search Results

A total of 222 protocols of phase 1, 2, 3, and 4 interventional studies were retrieved by the structured search on clinicaltrials.gov. Sixty-four of them were subsequently excluded because they did not fulfill the predefined set of inclusion criteria. Specifically, 39 studies were not testing therapeutic interventions, but were evaluated novel diagnostic procedures. Moreover, 25 trials were recruiting participants with clinical conditions not falling within the AD continuum (e.g., healthy volunteers, patients with other neurodegenerative dementias). Thus, 158 protocols were ultimately included in the analysis, as shown in Figure 1. A high agreement (>90%) was reported by the two reviewers involved in the selection process.

3.2. Characteristics of Protocols Adopting Biomarkers

Overall, 92 out of the 158 identified interventional studies (58.2%) adopted candidate AD biomarkers (Table 1). Compared with protocols not using biomarkers (n = 66), these studies were more frequently in the earlier phases of drug development (i.e., phase 1 and 2), more commonly aimed at evaluating the safety/tolerability of the tested interventions or their impact on AD underlying pathophysiological mechanisms (e.g., amyloid deposition, tau pathology, neuroinflammation), and almost exclusively focused on the assessment of pharmacological therapies (Table 1). On the contrary, studies not adopting biomarkers mostly adopted cognitive, functional, and neuropsychiatric measures as primary outcomes and were more commonly finalized at testing non-pharmacological interventions.

A total of 27,566 participants will tentatively be enrolled in the 92 protocols using biomarkers, with estimated sample sizes ranging between 12 and 2400 (median 120) subjects. These protocols are mostly conducted in the U.S. and funded by the biopharma industry. Their reported starting dates vary from November 2010 to January 2021, with completion dates indicated as ranging between December 2019 and August 2026. In the majority of studies, patients with overt AD dementia were indicated as eligible for participation, whereas nearly 20% of protocols restricted participation to subjects with preclinical or prodromal AD. Measures of clinical improvement were indicated as primary outcomes by 41.3% of studies with biomarkers.

Concerning how biomarkers are planned to be used, 62 studies adopt biomarkers to ascertain the eligibility of participants, 66 as study outcomes, whereas 36 both as inclusion criteria and endpoints.

The main characteristics of the ongoing protocols on AD using biomarkers are resumed in the Supplementary Material.

Table 1. Characteristics of the included protocols according to the adoption of biomarkers.

	Adopting Biomarkers (n = 92)	Non Adopting Biomarkers (n = 66)	p
Participants per study (n)	120 (43–382)	133 (43–267)	0.77
Phases			<0.01
Phase 1 and Phase 1–2	22 (23.9)	14 (21.2)	
Phase 2 and Phase 2–3	52 (56.5)	24 (36.4)	
Phase 3	15 (16.3)	18 (27.3)	
Phase 4	3 (3.3)	10 (15.2)	
Condition			0.21
Enrolling participants with AD dementia	74 (80.4)	58 (87.9)	
Not enrolling participants with AD dementia	18 (19.6)	8 (12.1)	
Main sponsor			0.12
Industry	56 (60.9)	32 (48.5)	
Other	36 (39.1)	34 (51.5)	
Primary outcome			<0.001
Safety	33 (35.9)	17 (25.8)	
Clinical improvement	38 (41.3)	47 (71.2)	
AD biological change	21 (14.6)	2 (3.0)	
Intervention			0.04
Pharmacological	87 (94.6)	56 (84.8)	
Non-pharmacological	5 (5.4)	10 (15.2)	

AD: Alzheimer's disease; MCI: Mild cognitive impairment; SCD: Subjective cognitive decline. Data are shown as median (IQR) or n (%).

3.3. Use of Biomarkers in the Selection of Participants

A total of 62 phase 1 (n = 7), phase 1–2 (n = 11), phase 2 (n = 26), phase 2–3 (n = 3), phase 3 (n = 14), and phase 4 (n = 1) studies are currently using AD biomarkers in the selection of participants. Twenty-one studies are testing anti-amyloid therapies, 7 anti-tau compounds, and 34 novel treatments with other mechanisms of action.

The biomarker-based criteria that is most commonly adopted to determine the eligibility of participants in the selected protocols are a positive amyloid PET scan (72.6% of the protocols) and a low CSF Aβ42 or Aβ42/Aβ40 ratio (48.4% of the protocols) (Figure 2). These two criteria are interchangeably used in 21 studies (33.9%). No study requires the documentation of amyloid positivity at both the CSF and PET assessment. CSF t-tau and p-tau levels, tau-PET, volumetric MRI, and FDG PET are instead less commonly used. The biologically supported selection of participants is mostly based on biomarkers of amyloid deposition (87.1% of the cases) rather than on those reflective of neurodegeneration (21.0%) or tau pathology (9.7%). Few studies are using combinations of biomarkers, often simultaneously measuring those indicative of amyloid deposition and neuronal injury (11.3% of the protocols) (Figure 2). No statistically significant difference was found when comparing the biomarkers adopted by phase 1 and 2 studies versus phase 3 and 4 studies (data not shown).

Only 6 out of the 36 protocols adopting CSF measurements in the ascertainment of participants' eligibility indicated the considered cut-points (Table 2). The majority of protocols performing amyloid PET in the screening phase does not provide sufficient information about the adopted radiotracer (e.g., florbetapir, florbetaben).

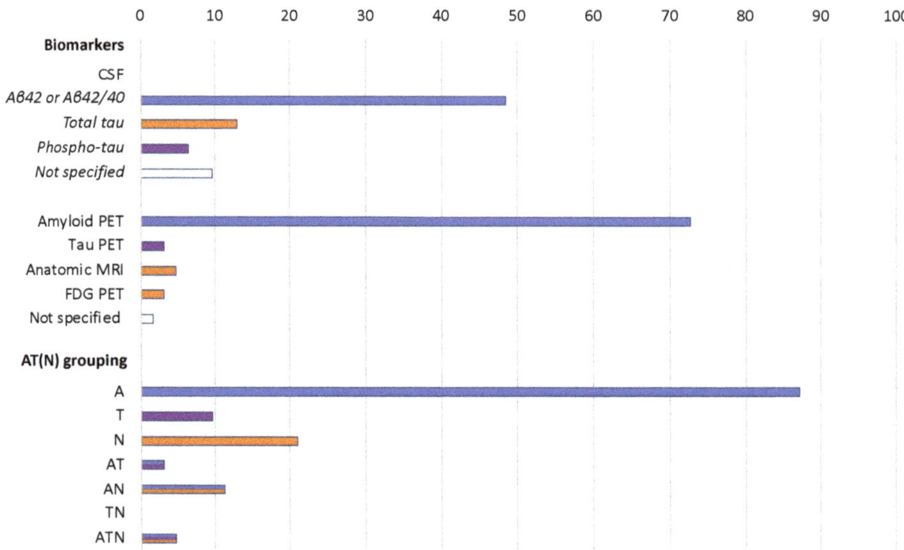

Figure 2. Biomarkers adopted to ascertain eligibility in the selected protocols (n = 62) according to the AT (N) grouping. Data are shown as %. A: aggregated Aβ or associated pathologic state; T: aggregated tau (neurofibrillary tangles) or associated pathologic state; N: neurodegeneration or neuronal injury. Colors refer to AT (N) biomarker grouping: blue: A; purple: T; orange: N.

Table 2. CSF cut-offs adopted in the six protocols with available information.

	CSF Cut-Offs
NCT02547818	Aβ42 ≥ 180 pg/mL and ≤ 690 pg/mL
NCT02947893	Aβ42 < 600 ng/mL
NCT03061474	Aβ42 ≤ 600 pg/mL or *t*-tau/Aβ42 ratio ≥ 0.39
NCT03069014	Aβ42 < 550 ng/L or Aβ40/42 ratio < 0.89
NCT04079803	*t*-tau/Aβ42 ratio ≥ 0.28
NCT04191486	Aβ ≤ 1000 pg/mL and *p*-tau 181 ≥ 19 pg/mL

Values and measurement units are reported as specified in the protocols.

3.4. Use of Biomarkers as Study Outcomes

Overall, 66 phase 1 (n = 8), phase 1–2 (n = 9), phase 2 (n = 31), phase 2–3 (n = 7), phase 3 (n = 9), and phase 4 (n = 2) studies are using AD biomarkers as study outcomes. Eighteen of them are testing anti-amyloid therapies, 3 anti-tau molecules, and 45 novel compounds with different biological properties. Eight protocols indicate biomarkers as primary outcomes, 46 as secondary outcomes, whereas 12 both as primary and secondary outcomes.

Changes of MRI or FDG PET findings are the most commonly adopted biomarker-based primary endpoints, being selected by 5 out of the 20 studies using biomarkers as primary outcome measures (Figure 3). Among protocols using biomarkers as secondary outcomes, the most frequently adopted measures are brain MRI (56.9% of the studies), CSF *t*-tau (36.2%), Aβ42 or Aβ42/Aβ40 (34.5%), *p*-tau (32.8%), and amyloid PET (31.0%). The use of biomarkers as study outcomes mostly relies on markers of neurodegeneration (48.1% of the cases) rather than on those of amyloid (21.6%) and tau (21.6%) pathology (Figure 3). In 21 studies, an ATN combination of biomarkers is adopted. No statistically significant difference was observed when comparing the biomarkers used as outcomes by phase 1 and 2 studies versus phase 3 and 4 studies (data not shown).

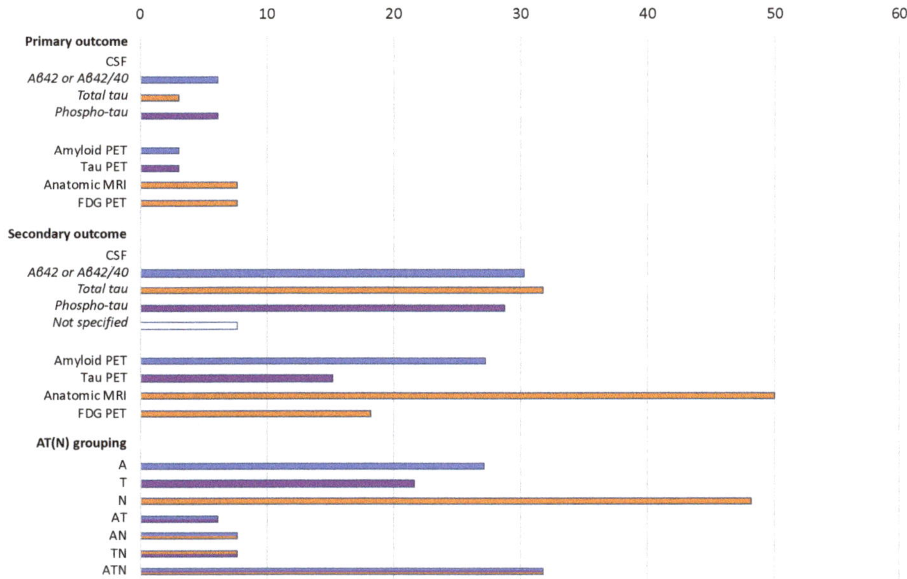

Figure 3. Biomarkers adopted as primary or secondary outcomes in the selected protocols (n = 66) according to the AT (N) grouping. Data are shown as %. A: aggregated Aβ or associated pathologic state; T: aggregated tau (neurofibrillary tangles) or associated pathologic state; N: neurodegeneration or neuronal injury. Colors refer to AT (N) biomarker grouping blue: A; purple: T; orange: N.

For the studies with biomarkers as primary outcomes, information on the sample size calculation has never been provided.

3.5. Use of Biomarkers Both as Eligibility Criteria and Study Outcomes

A total of 36 phase 1 (n = 5), phase 1–2 (n = 8), phase 2 (n = 13), phase 2–3 (n = 2), and phase 3 (n = 8) studies are using AD biomarkers both in the selection of participants and as study outcomes. These studies mostly include biomarkers of amyloid deposition among the eligibility criteria (91.7% of studies) and biomarkers of amyloid pathology (75.0%) and/or neuronal injury (75.0%) as endpoints.

4. Conclusions

In the present study, we provided a snapshot of the current use of candidate biomarkers in AD research protocols based on the data registered on the clinicaltrials.gov database. Overall, more than half of ongoing interventional studies targeting the AD continuum are adopting measures of amyloid deposition and/or tau pathology and/or neuronal injury. Biomarkers are used both for the selection of participants and to ascertain the efficacy of the tested interventions. They are mostly adopted by studies at the preliminary stages of the drug development process to explore the safety profile of novel therapies, and to provide evidence of target engagement (e.g., receptor occupancy) and disease-modifying properties. On the contrary, their use in trials primarily looking at improving the clinical manifestations of the disease is still limited. Indeed, only nearly 40% of studies with biomarkers adopted cognitive, functional, neuropsychiatric, and other clinical measures as their primary outcomes.

Based on our findings, a relevant number of clinical trials are incorporating measures of A to refine the clinical diagnosis of AD in participating subjects with dementia. It is well-established that a sizeable proportion of individuals clinically diagnosed with AD dementia have amyloid biomarkers incompatible with the diagnosis [15]. Patients with negative amyloid status tend to exhibit a slower clinical progression over time as compared with patients with amyloid positivity at the CSF and/or PET [15]. Introducing these procedures as enrichment tools can, therefore, allow to homogenize study samples, reduce the heterogeneity in clinical outcomes, and increase the probability of detecting drug-placebo differences. The same considerations can be extended to participants with mild cognitive impairment (MCI) in whom the demonstration of underlying amyloid pathology reduces the chance of observing alternative clinical trajectories (e.g., reversion to normal cognition) that may bias the study findings and their interpretation [16].

Different reflections may instead be applied to the preclinical stages of AD. In fact, among cognitively intact individuals, the detection of isolated amyloid positivity (or "Alzheimer's pathological change" [3]) is associated with a more considerable heterogeneity of clinical outcomes, thus complicating the assessment of clinical endpoints. Moreover, most of the subjects testing positive for A biomarkers will likely not exhibit a significant cognitive and functional worsening over time [9,17,18]. For instance, it has been calculated that the lifetime risks of developing AD dementia of a 60-year-old man and a 60-year-old woman with amyloidosis are only 23% and 31%, respectively [9]. Therefore, there is the risk of including in research protocols a disproportionate number of subjects who will never develop the target condition, thus only exposing them to the risk of serious adverse events with an unfavorable harm/benefit balance. For these reasons, it is fundamental to develop and implement appropriate and ethical strategies of risk disclosure to avoid the misinterpretation of testing results and the generation of adverse psychological reactions [19]. It is noteworthy that only a few studies combined biomarkers of A and T in the selection of participants. Including measures of tau pathology is instead required to determine if someone who is in the Alzheimer's disease continuum has indeed AD [3], consequently making more uniform the study populations and risk profiles of participants in clinical trials.

Other important considerations are raised concerning the use of biomarkers as screening tools in clinical trials [20]. In particular, the "real world" transferability of interventions emerging as effective in restricted samples of asymptomatic individuals with positive biomarkers will hardly be sustainable for our healthcare systems. Pathological AD markers are very common (i.e., 46%) in representative samples of older people [21], and it has been estimated that about 46.7 million Americans had preclinical AD in 2017 [22]. So, how can we translate these procedures on such a large scale to identify candidates for future disease-modifying treatments? Other controversies such as the standardization of CSF cut-points, the interchangeability of PET radiotracers and CSF assays, and the agreement between amyloid testing procedures remain to be addressed.

Candidate AD biomarkers are also frequently used as study outcomes. Specifically, measures of neurodegeneration are often adopted as downstream biomarkers to provide objective evidence that a drug ameliorates neurodegeneration. Indeed, the attenuation of the neurodegenerative process may be demonstrated (or at least suggested) by changes in the rates of glucose metabolism in cortical neurons (FDG PET), brain atrophy (volumetric MRI), and neuronal injury (CSF *t*-tau) [23]. N biomarkers may, therefore, support disease modification if a drug–placebo difference is detected and coupled with a similar clinical benefit [10,24,25]. It is noteworthy that, to date, no AD biomarker has yet qualified as a surrogate endpoint [26], which is a measurement that substitutes for clinical endpoints and is expected to predict clinical benefit [25]. Different to other areas of clinical neuroscience, such as multiple sclerosis, no AD biomarker has been consistently associated with the clinical trajectory of the disease, especially in its preclinical and prodromal stages [25]. Markers of amyloid and tau pathology as outcome measures are instead mostly useful to demonstrate drug engagement. Especially in the early phases of drug development, longitudinal CSF and PET measures of amyloid and tau pathology may confirm that a given compound affects the target brain protein [10]. For instance, several anti-Aβ monoclonal antibodies have shown a dose and time-dependent plaque reduction at the amyloid imaging, reflecting target engagement [27].

All these reflections on the use of biomarkers in AD research reinforce the need pursue their thorough clinical validation, as various methodological issues still limit their translation in the daily practice [8]. In particular, there is a lack of studies analyzing the distribution of biomarkers in healthy subjects and their interindividual variability according to major clinical and biological characteristics (e.g., age, sex, frailty) [28]. This information would be of great relevance in the promise of personalized medicine approaches. Moreover, no candidate AD biomarker has yet passed all the phases on which the architecture of diagnostic research is based [29].

The present study has some limitations to be mentioned and discussed. Our analysis cannot constitute an exhaustive overview on the topic because there are other registries for clinical trials on AD ongoing worldwide besides clinicaltrials.gov. Only a limited amount of information is available in the registered protocols, making it challenging to achieve an unbiased analysis of the methodological aspects of the studies. For example, in most of the selected protocols, brain MRI was simply mentioned among the participant selection procedures to exclude non-AD diagnoses. However, we cannot determine whether, in some cases, volumetric sequences were performed. This may have led to an underestimation of results concerning the use of MRI as a biomarker of neurodegeneration. Moreover, only the biomarkers incorporated in the AT (N) framework were considered. There is emerging evidence that other measures may represent promising biomarkers to be used in clinical trials [10,25].

In conclusion, biomarkers are largely used in ongoing clinical trials targeting the AD continuum. Their adoption may relevantly improve the selection of participants and the assessment of the efficacy and safety/tolerability of novel treatments. Nevertheless, their use in drug development must be accompanied by a demonstration of their clinical validity, utility, and cost-effectiveness in the "real world". Otherwise, there is a risk of an experimenting model of care and access to treatments that are unsustainable for our healthcare systems.

Supplementary Materials: The following are available online at http://www.mdpi.com/2075-4426/10/3/68/s1. Table S1: Characteristics of the selected protocols registered on the clinicltrials.gov database.

Author Contributions: M.C. (Marco Canevelli): Conception and design of the study, writing of the manuscript. G.R.: Data collection, drafting of the manuscript. I.B.: Data collection, drafting of the manuscript. M.V.: Data collection, drafting of the manuscript. M.T.B.: Data collection, drafting of the manuscript. F.S.: Data collection, drafting of the manuscript. G.B.: revising the manuscript critically for important intellectual content. M.C. (Matteo Cesari): Revising the manuscript critically for important intellectual content. N.V.: Conception of the study, writing of the manuscript. All authors have read and agreed to the published version of the manuscript.

Funding: Authors have no funding source to disclose for the present study.

Acknowledgments: Marco Canevelli is supported by a research grant of the Italian Ministry of Health (GR-2016-02364975) for the project "Dementia in immigrants and ethnic minorities living in Italy: clinical-epidemiological aspects and public health perspectives" (ImmiDem). Matteo Cesari has received honoraria for presentations at scientific meetings and/or research funding from Nestlé and Pfizer. He is involved in the coordination of an Innovative Medicines Initiative-funded project (including partners from the European Federation Pharmaceutical Industries and Associates [Sanofi, Novartis, Servier, GSK, Lilly]).

Conflicts of Interest: Authors have no conflicts of interest to disclose for the present study.

References

1. McKhann, G.; Drachman, D.; Folstein, M.; Katzman, R.; Price, D.; Stadlan, E.M. Clinical diagnosis of Alzheimer's disease: Report of the NINCDS-ADRDA Work Group under the auspices of Department of Health and Human Services Task Force on Alzheimer's Disease. *Neurology* **1984**, *34*, 939–944. [CrossRef] [PubMed]
2. Jack, C.R.; Albert, M.S.; Knopman, D.S.; McKhann, G.M.; Sperling, R.A.; Carrillo, M.C.; Thies, B.; Phelps, C.H. Introduction to the recommendations from the National Institute on Aging-Alzheimer's Association workgroups on diagnostic guidelines for Alzheimer's disease. *Alzheimers Dement.* **2011**, *7*, 257–262. [CrossRef] [PubMed]
3. Jack, C.R.; Bennett, D.A.; Blennow, K.; Carrillo, M.C.; Dunn, B.; Haeberlein, S.B.; Holtzman, D.M.; Jagust, W.; Jessen, F.; Karlawish, J.; et al. NIA-AA Research Framework: Toward a biological definition of Alzheimer's disease. *Alzheimers Dement.* **2018**, *14*, 535–562. [CrossRef] [PubMed]

4. Jack, C.R.; Bennett, D.A.; Blennow, K.; Carrillo, M.C.; Feldman, H.H.; Frisoni, G.B.; Hampel, H.; Jagust, W.J.; Johnson, K.A.; Knopman, D.S.; et al. A/T/N: An unbiased descriptive classification scheme for Alzheimer disease biomarkers. *Neurology* **2016**, *87*, 539–547. [CrossRef] [PubMed]
5. McKhann, G.M.; Knopman, D.S.; Chertkow, H.; Hyman, B.T.; Jack, C.R.; Kawas, C.H.; Klunk, W.E.; Koroshetz, W.J.; Manly, J.J.; Mayeux, R.; et al. The diagnosis of dementia due to Alzheimer's disease: Recommendations from the National Institute on Aging-Alzheimer's Association workgroups on diagnostic guidelines for Alzheimer's disease. *Alzheimers Dement.* **2011**, *7*, 263–269. [CrossRef]
6. Frisoni, G.B.; Boccardi, M.; Barkhof, F.; Blennow, K.; Cappa, S.; Chiotis, K.; Démonet, J.-F.; Garibotto, V.; Giannakopoulos, P.; Gietl, A.; et al. Strategic roadmap for an early diagnosis of Alzheimer's disease based on biomarkers. *Lancet Neurol.* **2017**, *16*, 661–676. [CrossRef]
7. Sancesario, G.M.; Toniolo, S.; Chiasserini, D.; Di Santo, S.G.; Zegeer, J.; Bernardi, G.; Musicco, M.; Caltagirone, C.; Parnetti, L.; Bernardini, S. The Clinical Use of Cerebrospinal Fluid Biomarkers for Alzheimer's Disease Diagnosis: The Italian Selfie. *J. Alzheimers Dis.* **2017**, *55*, 1659–1666. [CrossRef]
8. Canevelli, M.; Bacigalupo, I.; Gervasi, G.; Lacorte, E.; Massari, M.; Mayer, F.; Vanacore, N.; Cesari, M. Methodological Issues in the Clinical Validation of Biomarkers for Alzheimer's Disease: The Paradigmatic Example of CSF. *Front. Aging Neurosci.* **2019**, *11*. [CrossRef]
9. Brookmeyer, R.; Abdalla, N. Estimation of lifetime risks of Alzheimer's disease dementia using biomarkers for preclinical disease. *Alzheimers Dement.* **2018**, *14*, 981–988. [CrossRef]
10. Cummings, J. The Role of Biomarkers in Alzheimer's Disease Drug Development. *Adv. Exp. Med. Biol.* **2019**, *1118*, 29–61. [CrossRef]
11. Rothwell, P.M. External validity of randomised controlled trials: "To whom do the results of this trial apply?". *Lancet* **2005**, *365*, 82–93. [CrossRef]
12. Cummings, J.; Lee, G.; Ritter, A.; Sabbagh, M.; Zhong, K. Alzheimer's disease drug development pipeline: 2019. *Alzheimers Dement.* **2019**, *5*, 272–293. [CrossRef] [PubMed]
13. Van Maurik, I.S.; Slot, R.E.R.; Verfaillie, S.C.J.; Zwan, M.D.; Bouwman, F.H.; Prins, N.D.; Teunissen, C.E.; Scheltens, P.; Barkhof, F.; Wattjes, M.P.; et al. Personalized risk for clinical progression in cognitively normal subjects-the ABIDE project. *Alzheimers Res. Ther.* **2019**, *11*, 33. [CrossRef]
14. Rossini, P.M.; Cappa, S.F.; Lattanzio, F.; Perani, D.; Spadin, P.; Tagliavini, F.; Vanacore, N. The Italian INTERCEPTOR Project: From the Early Identification of Patients Eligible for Prescription of Antidementia Drugs to a Nationwide Organizational Model for Early Alzheimer's Disease Diagnosis. *J. Alzheimers Dis.* **2019**, *72*, 373–388. [CrossRef] [PubMed]
15. Landau, S.M.; Horng, A.; Fero, A.; Jagust, W.J. Alzheimer's Disease Neuroimaging Initiative Amyloid negativity in patients with clinically diagnosed Alzheimer disease and MCI. *Neurology* **2016**, *86*, 1377–1385. [CrossRef] [PubMed]
16. Canevelli, M.; Grande, G.; Lacorte, E.; Quarchioni, E.; Cesari, M.; Mariani, C.; Bruno, G.; Vanacore, N. Spontaneous Reversion of Mild Cognitive Impairment to Normal Cognition: A Systematic Review of Literature and Meta-Analysis. *J. Am. Med. Dir. Assoc.* **2016**, *17*, 943–948. [CrossRef]
17. Rosenberg, A.; Solomon, A.; Jelic, V.; Hagman, G.; Bogdanovic, N.; Kivipelto, M. Progression to dementia in memory clinic patients with mild cognitive impairment and normal β-amyloid. *Alzheimers Res. Ther.* **2019**, *11*, 99. [CrossRef]
18. Jansen, W.J.; Ossenkoppele, R.; Knol, D.L.; Tijms, B.M.; Scheltens, P.; Verhey, F.R.J.; Visser, P.J.; Amyloid Biomarker Study Group; Aalten, P.; Aarsland, D.; et al. Prevalence of cerebral amyloid pathology in persons without dementia: A meta-analysis. *JAMA* **2015**, *313*, 1924–1938. [CrossRef]
19. Roberts, J.S.; Dunn, L.B.; Rabinovici, G.D. Amyloid imaging, risk disclosure and Alzheimer's disease: Ethical and practical issues. *Neurodegener. Dis. Manag.* **2013**, *3*, 219–229. [CrossRef]
20. Mattsson, N.; Carrillo, M.C.; Dean, R.A.; Devous, M.D.; Nikolcheva, T.; Pesini, P.; Salter, H.; Potter, W.Z.; Sperling, R.S.; Bateman, R.J.; et al. Revolutionizing Alzheimer's disease and clinical trials through biomarkers. *Alzheimer's Dement. Diagn. Assess. Dis. Monit.* **2015**, *1*, 412–419. [CrossRef]
21. Kern, S.; Zetterberg, H.; Kern, J.; Zettergren, A.; Waern, M.; Höglund, K.; Andreasson, U.; Wetterberg, H.; Börjesson-Hanson, A.; Blennow, K.; et al. Prevalence of preclinical Alzheimer disease. *Neurology* **2018**, *90*, e1682–e1691. [CrossRef] [PubMed]
22. Brookmeyer, R.; Abdalla, N.; Kawas, C.H.; Corrada, M.M. Forecasting the prevalence of preclinical and clinical Alzheimer's disease in the United States. *Alzheimers Dement.* **2018**, *14*, 121–129. [CrossRef] [PubMed]

23. Blennow, K.; Hampel, H.; Zetterberg, H. Biomarkers in amyloid-β immunotherapy trials in Alzheimer's disease. *Neuropsychopharmacology* **2014**, *39*, 189–201. [CrossRef]
24. Cummings, J. Disease modification and Neuroprotection in neurodegenerative disorders. *Transl. Neurodegener.* **2017**, *6*, 25. [CrossRef] [PubMed]
25. Vellas, B.; Andrieu, S.; Sampaio, C.; Wilcock, G. Disease-modifying trials in Alzheimer's disease: A European task force consensus. *Lancet Neurol.* **2007**, *6*, 56–62. [CrossRef]
26. Barcikowska, M. European Medicines Agency Guideline on the clinical investigation of medicines for the treatment of Alzheimer's disease. *Lek. POZ* **2018**, *4*, 370–374.
27. Cummings, J.; Feldman, H.H.; Scheltens, P. The "rights" of precision drug development for Alzheimer's disease. *Alzheimer's Res. Ther.* **2019**, *11*, 76. [CrossRef]
28. Wallace, L.M.K.; Theou, O.; Godin, J.; Andrew, M.K.; Bennett, D.A.; Rockwood, K. Investigation of frailty as a moderator of the relationship between neuropathology and dementia in Alzheimer's disease: A cross-sectional analysis of data from the Rush Memory and Aging Project. *Lancet Neurol.* **2019**, *18*, 177–184. [CrossRef]
29. Sackett, D.L.; Haynes, R.B. The architecture of diagnostic research. *BMJ* **2002**, *324*, 539–541. [CrossRef]

Review

Flotillin: A Promising Biomarker for Alzheimer's Disease

Efthalia Angelopoulou [1], Yam Nath Paudel [2], Mohd. Farooq Shaikh [2,*] and Christina Piperi [1,*]

[1] Department of Biological Chemistry, Medical School, National and Kapodistrian University of Athens, 11527 Athens, Greece; angelthal@med.uoa.gr

[2] Neuropharmacology Research Strength, Jeffrey Cheah School of Medicine and Health Sciences, Monash University Malaysia, Bandar Sunway 47500, Selangor, Malaysia; yam.paudel@monash.edu

* Correspondence: farooq.shaikh@monash.edu (M.F.S.); cpiperi@med.uoa.gr (C.P.)

Received: 6 March 2020; Accepted: 20 March 2020; Published: 26 March 2020

Abstract: Alzheimer's disease (AD) is characterized by the accumulation of beta amyloid (Aβ) in extracellular senile plaques and intracellular neurofibrillary tangles (NFTs) mainly consisting of tau protein. Although the exact etiology of the disease remains elusive, accumulating evidence highlights the key role of lipid rafts, as well as the endocytic pathways in amyloidogenic amyloid precursor protein (APP) processing and AD pathogenesis. The combination of reduced Aβ42 levels and increased phosphorylated tau protein levels in the cerebrospinal fluid (CSF) is the most well established biomarker, along with Pittsburgh compound B and positron emission tomography (PiB-PET) for amyloid imaging. However, their invasive nature, the cost, and their availability often limit their use. In this context, an easily detectable marker for AD diagnosis even at preclinical stages is highly needed. Flotillins, being hydrophobic proteins located in lipid rafts of intra- and extracellular vesicles, are mainly involved in signal transduction and membrane–protein interactions. Accumulating evidence highlights the emerging implication of flotillins in AD pathogenesis, by affecting APP endocytosis and processing, Ca^{2+} homeostasis, mitochondrial dysfunction, neuronal apoptosis, Aβ-induced neurotoxicity, and prion-like spreading of Aβ. Importantly, there is also clinical evidence supporting their potential use as biomarker candidates for AD, due to reduced serum and CSF levels that correlate with amyloid burden in AD patients compared with controls. This review focuses on the emerging preclinical and clinical evidence on the role of flotillins in AD pathogenesis, further addressing their potential usage as disease biomarkers.

Keywords: flotillin; Alzheimer's disease; biomarker; exosomes; beta-amyloid; Tau

---

1. Introduction

Alzheimer's disease (AD) is pathologically characterized by the accumulation of beta amyloid (Aβ), a peptide of 40 or 42 amino acids, in extracellular senile plaques, as well as intracellular neurofibrillary tangles (NFTs) mainly consisting of tau protein [1,2]. Aβ is generated by the sequential cleavage of the transmembrane amyloid β precursor protein (APP) via β- and γ-secretases [1]. It has been shown that amyloidogenic APP processing mainly occurs in lipid rafts, which are caveolae-like membrane microdomains enriched in sphingolipids, glycolipids, and cholesterol, serving as dynamic platforms for signal transduction, protein trafficking, and clathrin-independent endocytosis [3,4]. Most AD cases are sporadic, whereas a small percentage (1%–5%) of them are caused by mutations in genes including *APP*, *presenilin-1*, and *presenilin-2* [5]. Although the exact etiology and pathophysiological mechanisms of AD remain elusive, growing preclinical and clinical evidence highlights the key role of the homeostasis of lipid rafts and both intra- and extracellular vesicles in AD pathogenesis, even at early stages of the disease [6].

Currently, there is no easy and objective method available for early AD diagnosis. The combined estimation of reduced Aβ_{42} levels and increased phosphorylated tau protein levels in the cerebrospinal fluid (CSF) is the most well established biomarker for the diagnosis of the disease, along with Pittsburgh compound B and positron emission tomography (PiB-PET) for amyloid imaging [7–9]. Although both methods can relatively reliably detect AD at prosymptomatic stages [8,9], the lumbar puncture required for CSF collection is an invasive procedure demanding specialized staff technical skills and cannot be carried out in primary care settings [10]. On the other hand, PiB-PET is quite expensive and not available worldwide. Specific patterns of various sets of plasma proteins or phospholipids have been proposed as promising AD biomarkers by several studies, usually involving specialized experimental procedures, such as mass spectrometry, or necessitating the measurement of multiple factors [7]. In this context, an easily accessible and detectable marker with high sensitivity and specificity for AD diagnosis, reflecting Aβ burden or underlying AD pathophysiological mechanisms even at preclinical stages, is highly needed.

Flotillins are highly conserved hydrophobic myristoylated and palmitoylated proteins that belong to the prohibitin (PHB) family [3]. They are located in the inner part of the plasma membrane and play an essential role in the formation of lipid rafts [3,4]. Moreover, intracellular and extracellular vesicles are enriched in flotillins, whose subcellular localization is constantly changing [3,4]. Flotillins were first detected in the lung plasma membrane of mice, described as "float like a flotilla of ships in the Triton-insoluble, buoyant fraction" [3], and since then they have been widely utilized as markers of lipid rafts and exosomes for many years. In metazoans, there are two flotillin paralogues, termed as flotillin-1/reggie-2 and flotillin-2/reggie-1. Within cells, flotillins form flotillin-1/flotillin-2 hetero-oligomers mediated by C-terminal interactions [11]. Their expression is particularly high in cell types lacking caveolin, including neurons. Functionally, flotillins are implicated in various cellular processes, including membrane–cytoskeletal interaction, membrane and vesicular trafficking, signal transduction, axonal regeneration, cell migration, and clathrin-independent endocytosis. It has been demonstrated that flotillins (mainly flotillin-1) are abundantly expressed in pyramidal neurons in the cortex, as well as in the astrocytes of the white matter of normal human brain tissue [12]. Notably, their levels have been shown to be higher in brain samples from non-demented patients with amyloid plaques, subjects with Down syndrome (who overexpress APP), and AD patients compared with non-demented individuals without amyloid plaques [12], suggesting a key role in AD pathogenesis. Since then, several preclinical and human studies have investigated the implication of flotillins in AD pathophysiology, also highlighting their potential as promising molecular biomarkers.

In this mini review, we summarize the emerging preclinical and clinical evidence on the role of flotillins in AD pathogenesis, and discuss their potential usage as biomarker candidates (Figure 1) for AD along with possible limitations.

2. The Role of Endocytic Pathway and Exosome Release in AD Pathogenesis

Abnormal APP processing, trafficking, and turnover are considered to play a major role in AD pathogenesis [1]. Several studies indicate that amyloidogenic APP processing is predominantly carried out in lipid rafts, since APP, Aβ, presenilin-1, β-secretase, and specific components of the γ-secretase complex have been located in the membrane microdomains of neurons [13–17]. On the contrary, non-amyloidogenic APP processing usually occurs in areas of the plasma membrane that are enriched in phospholipids [18]. However, there is also evidence suggesting that amyloidogenic APP cleavage may occur outside lipid rafts. In particular, membrane cholesterol reduction in hippocampal neuronal cells from AD patients has been associated with enhanced β-site APP-cleaving enzyme 1 (BACE1)–APP colocalization and increased Aβ load, implying that amyloidogenic APP processing may be carried out in more fluid membrane domains [19]. In addition, depletion of the cholesterol-synthesizing enzyme seladin-1 in mice has been shown to be related to decreased cholesterol levels, disrupted cholesterol-rich detergent-resistant membrane domains (DRMs), displacement of BACE1 from DRMs to membrane fractions containing APP, enhanced APP cleavage, and increased Aβ load [20]. It has been proposed that BACE1 in cholesterol-rich membrane domains could represent a relatively inactive pool of the enzyme, which may be transferred to APP-containing domains under specific conditions. Based on these contradictory findings, more studies are needed in order to clarify this issue. Aβ can be produced by proteolytic cleavage of APP in the endoplasmic reticulum (ER) and *trans*-Golgi network, resulting in the formation of secretory vesicles. Alternatively, after APP internalization it can be directed from the plasma membrane into the endosomal system and subsequently targeted to lysosomes [21,22]. Early endosomes are the first vesicles of endocytic pathway responsible for APP endocytosis [23]. Internalization of APP into endosomes has been required for its cleavage by BACE1 [24], which proteolyzes APP [25]. Aβ generated in early endosomes has been also shown to be transported into late endosomes in a retrograde manner, leading to their fusion with cellular membrane and subsequent extracellular secretion as exosomes [26]. Impaired APP endocytosis has been associated with reduction in the secretion of Aβ in vitro [27], while abnormally enlarged endosomes have been identified in early stages of sporadic AD [23]. Therefore, dysregulation of the endocytic pathway has been implicated in AD pathophysiology, mainly by affecting APP processing and trafficking.

Exosomes are small membrane vesicles (20–120 nm in diameter) derived from endosomes through the formation of multivesicular bodies, being able to transport their internalized content into recipient cells, thus mainly contributing to cell-to-cell communication [28]. Exosomes can be secreted by various cell types, including neurons, astrocytes, oligodendrocytes, and microglia in the extracellular space and subsequently CSF, containing several kinds of cargos, including RNAs, micro-RNAs, and proteins [29,30]. It has been suggested that exosomes may be involved in AD pathogenesis by affecting Aβ metabolism and aggregation, as well as through tau-related molecular mechanisms [31]. More specifically, in vitro evidence has indicated that neuron-derived exosomes can bind to and promote conformational alterations in extracellular Aβ, leading to the formation of nontoxic amyloid fibrils. They can also enhance Aβ clearance by microglia through the facilitation of Aβ transportation to lysosomes for degradation [31]. Exosome secretion by neurons has been shown to be regulated by enzymes involved in the metabolism of sphingolipids, such as sphingomyelin synthase 2 (SMS2) and neutral sphingomyelinase 2 (nSMase2) [31]. Upregulation of exosome release with SMS2 siRNA use has been associated with increased Aβ uptake by microglia and reduced levels of extracellular Aβ in vitro [31]. In vivo evidence has revealed that exogenous intracerebral injection of neuroblastoma-derived exosomes into APP transgenic mice led to decreased Aβ depositions and Aβ-induced synaptotoxicity in their hippocampus [32]. In this study, glycosphingolipids (GSLs) enriched with glycans, which are essential components of exosome membrane, have exerted a contributing role in Aβ binding and internalization by the exosomes [32]. Notably, apart from Aβ, exosomes also contain β- and γ-secretase complexes, full-length APP, as well as APP C-terminal fragments (CTFs) in APP transgenic mice, highlighting their potential contribution to Aβ metabolism in the brain [33]. Apart from their role in Aβ pathology, secretion of phosphorylated tau proteins has been shown to be at least partially exosome-mediated in tauopathy models in vitro [34]. Moreover, exosome-associated phosphorylated tau has been found released in the cerebrospinal fluid of AD patients [34], further confirming the significant implication of exosomes in tau-related pathophysiology in human AD. It has been also demonstrated that exosomes can attenuate the Aβ-mediated disruption of synaptic plasticity in vivo [35]. Another study has shown that $A\beta_{1-42}$ treatment inhibited exosome release from astrocytes, as demonstrated by decreased flotillin levels in vitro, by inducing c-Jun N-terminal kinase (JNK) signaling pathway [29]. Furthermore, exosomes, as Aβ carriers, have been proposed to contribute to the prion-like spreading of Aβ aggregates in AD [5]. Collectively, these findings reveal the key role of exosomes in AD pathogenesis by affecting Aβ pathology, tau-related mechanisms, and Aβ-mediated neurotoxicity.

3. The Role of Flotillin in AD Pathogenesis

3.1. Evidence from Human Studies

Several human studies support the potential implication of flotillin in AD pathogenesis. In this regard, it has been indicated that flotillin accumulates in the endosomes of neurons of AD patients' brains [36], possibly playing a key role in neuronal endosomal pathway. More specifically, flotillin-1 immunoreactivity has been shown to be higher in neurons bearing NFTs in the amygdala, hippocampus, and isocortex of AD patients compared with controls [36], and flotillin-1 was found to be co-localized with cathepsin-D, a lysosomal protease, indicating its contribution to lysosomal degradation [36]. Of note, some neurons bearing NFTs did not contain flotillin-1, implying a possible secondary reduction in synthesis of flotillin-1 in these cells [36]. Flotillin-1 has been also demonstrated to be highly enriched in extracellular Aβ plaques and neurons bearing NFTs in brain specimens of AD patients compared with PD patients and controls [4]. Furthermore, it has been reported that flotillin-positive extracellular vesicles (EVs) isolated from the CSF and plasma of sporadic, late-onset AD patients contained an increased amounts of Aβ42 in their external surface compared with age-matched neurologically healthy controls, which could be internalized by cortical neurons and acted in a neurotoxic manner, by impairing Ca^{2+} homeostasis, causing mitochondrial dysfunction, increasing the vulnerability of neurons to excitotoxicity, and triggering neuronal apoptosis [37]. On the contrary, the EVs derived from controls displayed no neurotoxic effects [37]. These effects were also observed in vivo in transgenic APP and presenilin-1 mice, as well *as* in vitro in neural cells expressing presenilin-1 mutations of familial AD [37]. It was also reported that impaired autophagy could enhance the release of EVs [37]. Of note, these flotillin-positive EV-mediated effects have been shown to be prevented via Aβ antibody treatment, highlighting the essential role of Aβ in the EV-induced neurotoxicity [37]. Furthermore, another study has demonstrated that flotillin-1-positive exosomes isolated from postmortem brain sections of AD patients were highly enriched in toxic Aβ oligomers compared with healthy controls, which could be transferred into recipient's cultured neurons [38]. Disruption of the production, release, or uptake of these exosomes was shown to attenuate the spreading of Aβ oligomers and Aβ-induced neurotoxicity [38]. Furthermore, a neuropathological study in brain tissues from AD patients has also reported that flotillins are also present in granulovacuolar degeneration (GVD) bodies [39], which are basophilic perinuclear vacuoles accumulating in neurons of AD patients [40]. Moreover, it has been indicated that levels of flotillin-2, along with other endocytic-associated proteins, were increased in brain sections of older non-demented humans [41]. Given the fact that age is a significant risk factor for AD development, these data further strengthen the potential contribution of flotillins in the pathophysiology of AD. These findings suggest that flotillin-positive exosomes may play a key role in AD pathogenesis and particularly, be implicated in the prion-like propagation of Aβ pathology in AD.

3.2. Evidence from In Vivo Models

In addition to human studies, a growing body of preclinical evidence demonstrates the key role of flotillin in the pathogenesis of AD, shedding more light into the underlying molecular mechanisms (Table 1).

Table 1. In vitro and in vivo evidence about the role of flotillin in Alzheimer's disease (AD).

S.N.	Type of Study	Main Findings	Reference
1	*In vitro*	• Flotillin-1 directly interacted with the intracellular domain of APP (AICD), and the residues 189-282 of flotillin-1 were essential for this interaction	[42]
2	*In vitro*	• FKBP12 overexpression enhanced the amyloidogenic APP processing, possibly by affecting the affinity of AICD to flotillin-1	[43]
3	*In vitro*	• Flotillin-1 could stabilize LGI3, and LGI3/Flo1 complex downregulation could directly affect APP trafficking via the disruption of exosome formation	[44]
4	*In vitro*	• si-RNA-mediated flotillin-2 knockdown impaired APP endocytosis in primary cultures of hippocampal neurons, and reduced Aβ production in tissue cell cultures • Flotillin-2 acted as a scaffold protein and enhanced APP clustering at the plasma membrane and subsequent APP endocytosis in a clathrin-dependent manner	[45]
5	*In vitro*	• Flotillin-1 bound to BACE1, whereas flotillin-1 overexpression was associated with increased recruitment into rafts and reduced activity of BACE1	[46]
6	*In vitro*	• Flotillin-1 could directly interact with the dileucine motif in the cytoplasmic domain of BACE1, thus affecting its endosomal sorting • Flotillin-2-BACE1 interaction was at least partially mediated by flotillin-1 • Depletion of flotillin-1 and -2 led to increased accumulation of overexpressed BACE1 into late endosomes in shRNA-mediated flotillin knockdown HeLa cells • Flotillin-1 knockdown resulted in overexpression of BACE1 • Flotillin-2 knockdown increased amyloidogenic processing of APP	[47]
7	*In vitro*	• Neuroblastoma cells transfected with mutant *DNM 2* gene displayed higher flotillin levels, and APP was mainly localized in the lipid rafts	[48]
8	*In vivo & in vitro*	• Copper inhibited the association of flotillin-2 with lipid rafts via its redistribution into non-raft fractions, resulting in the reduction of endocytosis and processing of APP in membrane microdomains both in vitro and in mutant APP transgenic mice	[49]
9	*In vivo*	• Endosomes of neurons in the amygdala, hippocampus, and isocortex of APP and presenilin-1 transgenic mice contained flotillin-1	[21]
10	*In vivo*	• Intracellular Aβ was accumulated in early and late endosomes positive for flotillin-1 in transgenic ArcAβ mice	[4]

Table 1. *Cont.*

S.N.	Type of Study	Main Findings	Reference
11	*In vivo*	• Knockout of *flotillin-1* gene (with or without additional knockout of flotillin-2 gene) in APP and presenilin-1 transgenic mice was correlated with reduced Aβ accumulation and plaque formation • No differences on APP clustering or endocytosis were indicated in mouse embryonic fibroblasts with flotillin-1 depletion	[50]
12	*In vivo*	• Treadmill exercise inhibited amyloidogenic APP cleavage and Aβ production in APP and presenilin-1 transgenic mice by inhibiting flotillin-1 levels and lipid raft formation	[51]
13	*In vivo*	• Treatment of senescence-accelerated mouse prone 8 mice with Yizhijiannao granule reduced flotillin-1 levels	[52]
14	*In vivo*	• Flotillin levels were not altered in mutant APP transgenic mice, compared with proteins associated with clathrin-dependent pathways	[53]

AD, Alzheimer's disease; APP, amyloid precursor protein; Aβ, amyloid β; LGI3, leucine-rich glioma inactivated 3; BACE1, beta-site APP cleaving enzyme 1; DNM, dynamin; FKBP12, FK506-binding protein; shRNAs, short hairpin RNAs; siRNAs, small interfering RNAs.

More specifically, flotillin-1 was present in endosomes of neurons in the amygdala, hippocampus, and isocortex of transgenic mice expressing human mutant APP and presenilin-1 [21]. Furthermore, intracellular Aβ has been shown to accumulate in early and late endosomes of the endocytic pathway that is positive for flotillin-1 in transgenic ArcAβ mice [4], overexpressing human APP 695 (and containing the Swedish and Arctic mutations in a single construct). Another study demonstrated that knockout of *flotillin-1* gene (with or without additional knockout of *flotillin-2* gene) in transgenic mice overexpressing APP and mutant presenilin-1 was associated with reduced Aβ accumulation and plaque deposition [50]. However, no significant differences in APP clustering or endocytosis were observed in mouse embryonic fibroblasts with no flotillin-1 expression [50]. Another study has indicated that treadmill exercise could inhibit amyloidogenic APP cleavage and subsequent Aβ production in the brain of APP and presenilin-1 transgenic mouse models, by suppressing lipid raft formation and flotillin-1 levels [51]. Furthermore, it has been reported that treatment of senescence-accelerated mouse prone 8 mice with *Yizhijiannao* granule, a Chinese medicinal compound, could downregulate flotillin-1 levels in the temporal lobes of the animals [52]. Interestingly, it has been demonstrated that copper could inhibit the association of flotillin-2 with lipid rafts by redistributing it into non-raft fractions, thus decreasing the endocytosis and processing of APP in membrane microdomains both in vitro and in mutant APP transgenic mice [49]. Previous works have shown that copper can attenuate Aβ production [54], and suppression of flotillin-2-mediated APP processing in rafts may represent one potential underlying mechanism. On the other hand, there is also evidence indicating that clathrin-independent endocytic pathways may not play a significant role in APP processing, since flotillin levels have been shown to remain unchanged in mutant APP transgenic mice, compared with proteins associated with clathrin-dependent pathways [53]. Nevertheless, these findings highlight the important implication of flotillins in the pathogenesis of AD, however, further studies are needed for deeper understanding of the underlying molecular mechanisms.

3.3. Evidence from In Vitro Studies

There is also a growing amount of in vitro evidence investigating the role of flotillins at a molecular level. Specifically, flotillin-1 was shown to directly interact with the intracellular domain of APP

(AICD), the C-terminus of APP consisting of 57–59 amino acids, while the residues 189–282 of flotillin-1 were found essential for this interaction. [42]. In this context, it has been demonstrated that the overexpression of FKBP12, a protein interacting with AICD, could trigger the amyloidogenic APP processing pathway in vitro, possibly by altering the affinity of AICD to flotillin-1 [43]. AICD has been demonstrated to play an important role in signal transduction via the interaction with specific PTB domain-containing proteins, including X11 and Fe65 [55]. Flotillin-1 has been also found to interact with leucine-rich glioma inactivated 3 (LGI3), a protein that is implicated in the internalization of APP in neurons [44]. More specifically, flotillin-1 was indicated to stabilize LGI3, and downregulation of the LGI3/Flo1 complex directly affected APP trafficking, by disrupting the formation of exosomes [44]. Another study has revealed that knockdown of flotillin-2 by small interfering RNA (siRNA) could impair APP endocytosis in primary cultures of hippocampal neuronal cells, resulting in reduced Aβ production in tissue cell cultures [45]. In addition, this study has demonstrated that flotillin-2 was able to act as a scaffold protein and enhance APP clustering at the surface of the plasma membrane, leading to APP endocytosis via a clathrin-dependent pathway [45]. Moreover, it has been indicated that flotillin-1 could bind to BACE1, whereas flotillin-1 overexpression was associated with increased recruitment of BACE1 into rafts and reduced activity of β-secretase in vitro [46]. Another study showed that flotillin-1 was able to directly interact with the dileucine motif in the cytoplasmic domain of BACE1, thus affecting its endosomal sorting. The flotillin-2–BACE1 interaction was shown to be indirect at least partially via flotillin-1 [47]. Depletion of both flotillins was also indicated to be associated with more prominent perinuclear localization and accumulation of overexpressed BACE1 into late endosomes in HeLa cells in which the expression of flotillin expression was knocked down via lentiviral shRNAs, suggesting that flotillins may affect its subcellular localization [47]. Additionally, flotillin-1 knockdown was demonstrated to result in overexpression of BACE1, and flotillin-2 knockdown led to increased amyloidogenic processing of endogenous APP in the same cellular system, possibly by affecting BACE1 trafficking [47]. Therefore, flotillins play a major role in BACE1 trafficking and expression, thus affecting APP cleavage. Moreover, neuroblastoma cells that were transfected with mutant *dynamin (DNM) 2* gene, in which a specific polymorphism has been associated with late-onset AD in non-carriers of the apolipoprotein E-ε4 allele [56], displayed higher levels of flotillin, and APP was predominantly localized in the lipid rafts [48].

Taken together, flotillins have been shown to affect APP endocytosis and processing, Ca^{2+} homeostasis, mitochondrial dysfunction, neuronal apoptosis, Aβ-induced neurotoxicity, and prion-like spreading of Aβ, thus playing an emerging role in AD pathogenesis.

4. Flotillin as a Novel Biomarker Candidate for AD: Clinical Evidence

CSF analysis is considered as the most reliable biofluid for detection of biomarkers of the central nervous system (CNS) disorders since it allows a more accurate elucidation of the underlying molecular processes in the brain. However, its acquisition requires lumbar puncture, which is an invasive procedure. Alternatively, blood biomarkers present a more optimal approach, and enormous research efforts are made towards this direction. However, the abundance of proteins in the plasma compared with the CSF limits the efficacy of this strategy for several molecules including $A\beta_{42}$, and the results of respective clinical studies are often inconsistent [57]. In this regard, plasma neuron-derived exosomes obtained from AD patients have been shown to contain higher levels of Aβ42 and phosphorylated tau protein compared with controls, highlighting their potential usage as AD biomarkers [57]. Several methodological concerns remain to be clarified for the estimation of exosomes per se in AD, since their detection in blood or CSF may be affected by conditions of sample storage, temperature, and the use of anticoagulants [57]. Although various techniques for the isolation of exosomes from biofluids have been developed, there is still no established validated reference method for their effective purification [57].

Flotillin, as one of the main components of exosomes, has been recently proposed as an alternative single molecule that could be used as an AD biomarker. Proteomic analyses have already indicated that flotillins are constituents of CSF-derived exosomes [58,59]. Importantly, flotillin levels can be relatively easily measured by ELISA method and/or immunoblot analysis [10], providing a significant advantage compared with other biomarker candidates that may require more expensive and specialized equipment.

Interestingly, a recent clinical study has indicated that flotillin levels were lower not only in the CSF, but also in the serum of AD patients, compared with subjects with mild cognitive impairment (MCI) or age-matched non-AD controls [10]. Additionally, CSF and serum flotillin levels have been shown to be reduced in patients with AD-related MCI, compared with non-AD-related MCI (determined by PiB-PET) [10]. Investigation of postmortem brain tissues further revealed that flotillin levels were also lower in the cerebroventricular fluid (CVF) samples obtained from AD patients compared with subjects with vascular dementia [10]. Furthermore, flotillin levels have been found to be negatively correlated with amyloid burden, as shown by the mean cortical PiB retention levels in PiB-PET [10]. However, flotillin-2 levels were shown elevated in brain sections of older non-demented humans compared with younger individuals [41], as well as levels of plasma neuron-derived exosomes were lower in aged HIV-infected individuals compared with controls [60], raising concerns regarding the efficacy of flotillin levels to effectively discriminate between AD and normal ageing. In this regard, flotillin levels were found to remain stable with advancing age in healthy controls in the abovementioned study [10], highlighting the possibility that their levels may not be affected by ageing process itself.

Although clinical evidence on the diagnostic utility of flotillins in AD is limited and the subject is still in its infancy, these findings strongly support the promising role of flotillin as a novel early CSF and serum diagnostic biomarker for AD (Figure 1). Future studies with larger sample size are therefore required for the validation of these results.

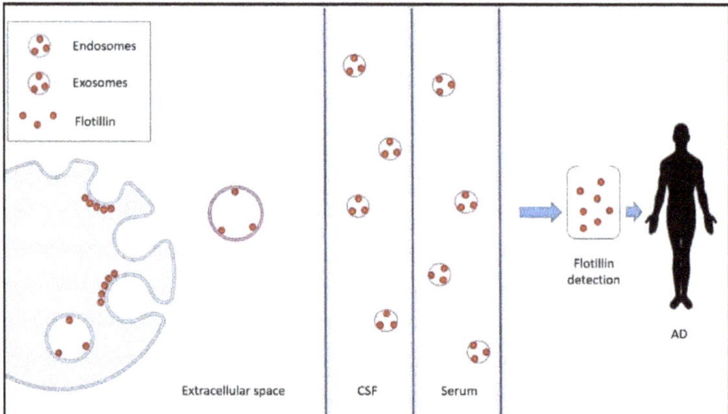

Flotillin can be detected in the CSF and serum of AD patients, and it has been demonstrated that its levels are lower in AD patients compared with controls, suggesting their use as potential AD biomarkers. AD, Alzheimer's disease; CSF: Cerebrospinal fluid.

Figure 1. Lipid rafts, endosomes, and exosomes are enriched in flotillin.

5. Discussion and Future Perspectives

Accumulating evidence highlights the emerging implication of flotillins in AD pathogenesis, through their implication in APP endocytosis and processing, Ca^{2+} homeostasis, mitochondrial dysfunction, neuronal apoptosis, Aβ-induced neurotoxicity, and prion-like spreading of Aβ. Importantly, there is also clinical evidence suggesting their potential use as biomarker candidates for AD, based on the reduced serum and CSF levels observed in AD patients compared with controls that correlated with amyloid burden.

On the other hand, there are specific challenges related to the development of flotillins as valid AD biomarkers. Firstly, it is important to note that flotillins are universally expressed by all cell types, and consist as a component of all subtypes of extracellular vesicles, independently to their density or size [61]. In addition, it has been indicated that the concentration of plasma neuron-derived exosomes was lower in neurocognitively impaired individuals (regardless of the cause) in comparison with controls [60]. Furthermore, flotillin-1 has been reported to be enriched in exosomes from CSF samples obtained from patients at early stage of severe traumatic brain injury, although no comparisons with control subjects were made in this study [62]. Moreover, levels of CSF exosomal α-synuclein have been shown to distinguish patients with Parkinson's disease, dementia with Lewy bodies, and controls with other neurological disorders, and correlate well with cognitive impairment [63]. However, flotillin-1 levels were shown to be higher in neurons bearing NFTs in brain specimens of AD compared with PD patients [4]. Nevertheless, these findings raise important concerns about the specificity of flotillin detection in AD, and highlight the need for future comparative studies in age-matched patients with other neurological disorders.

Notably, accumulating evidence reveals the potential role of tissue flotillin levels as prognostic biomarkers in various types of solid tumors, including breast cancer and lung carcinoma [64]. Although data about plasma flotillin levels in cancer patients are lacking, other comorbidities including tumors should be also considered for the evaluation of flotillin levels in AD patients. Furthermore, preclinical evidence has demonstrated that several types of statins may alter the expression of flotillins in brain plasma membrane [65], implying that treatment with statins may affect flotillin levels in AD patients. In addition, the potential effects of anti-cholinesterase inhibitors, currently prescribed for AD patients, on flotillin levels should be also considered.

Another issue that should be mentioned is the selective detection of flotillin-1 and -2 levels in the CSF. In this context, it has been indicated that flotillin-2 levels were positively correlated with flotillin-1 levels in the CSF obtained from patients after severe traumatic brain injury [62]. Nevertheless, further research is required regarding the differential role of both types of flotillins in the serum and CSF of AD patients.

Apart from flotillins, another mechanism of clathrin-independent endocytosis involves caveolins, which have been also implicated in Aβ production [66]. Caveolin-1 levels have been increased in the cortex and hippocampus of brain specimens from AD patients compared with age-matched controls [67], indicating another potential endocytosis-associated biomarker candidate for AD that could be further investigated.

Importantly, it has been shown that specific gene polymorphisms of the *neuronal sortilin-related receptor* (*SORL1*), which is highly implicated in the APP endosomal trafficking, may affect the susceptibility to AD development [68]. It has been shown that specific polymorphisms of *flotillin-2* gene may be associated with coronary artery disease in the Chinese population. Given the connection of hypercholesterolemia with AD development [18,69], further research is needed for the potential link between these *flotillin-2* polymorphisms and AD risk.

6. Conclusions

Collectively, flotillin may serve as a single CSF or blood biomarker, or be used supplementary to CSF Aβ42 and tau levels, as well as PET neuroimaging findings for more efficient and earlier AD diagnosis. However, additional larger comparative studies with age-matched controls and patients with other neurodegenerative disorders are needed for the validation of their usage as biomarkers for AD in clinical settings.

Author Contributions: E.A. carried out the literature review, conceptualized, and prepared the initial draft. Y.N.P. and M.F.S. edited and contributed in the final manuscript. C.P. provided critical inputs, edited and contributed to the final version of the manuscript. All authors have read and agreed to the published version of the manuscript.

Funding: This research has not received any specific grant from funding agencies in the public, Commercial, or not-for-profit sectors.

Acknowledgments: Y.N.P. would like to acknowledge Monash University Malaysia for supporting with HDR Scholarship.

Conflicts of Interest: The authors have no conflict of interest to declare.

Abbreviations

AD: Alzheimer's disease; APP, Amyloid precursor protein; Aβ, Amyloid β; NFTs, neurofibrillary tangles;LGI3, Leucine-rich glioma inactivated 3; BACE1, Beta-site APP cleaving enzyme 1; DNM, Dynamin; FKBP12, FK506-binding protein; shRNAs, Short hairpin RNAs; siRNAs, Small-interfering RNAs; PiB-PET; Pittsburgh compound B and positron emission tomography; PHB, Prohibitin; ER, Endoplasmic reticulum;SMS2, Sphingomyelin synthase 2;GSLs, Glycosphingolipids; CTFs, C-terminal fragments; EVs, Extracellular vesicles; GVD, Granulovacuolar degeneration; CNS, Central nervous system; CVF, Cerebroventricular fluid; SORL1, sortilin-related receptor 1; JNK, C-jun N-terminal kinases.

References

1. Tiwari, S.; Atluri, V.; Kaushik, A.; Yndart, A.; Nair, M. Alzheimer's disease: Pathogenesis, diagnostics, and therapeutics. *Int. J. Nanomed.* **2019**, *14*, 5541–5554. [CrossRef] [PubMed]
2. Paudel, Y.N.; Angelopoulou, E.; Piperi, C.; Othman, I.; Aamir, K.; Shaikh, M. Impact of HMGB1, RAGE, and TLR4 in Alzheimer's Disease (AD): From Risk Factors to Therapeutic Targeting. *Cells* **2020**, *9*, 383. [CrossRef] [PubMed]
3. Bickel, P.E.; Scherer, P.E.; Schnitzer, J.E.; Oh, P.; Lisanti, M.P.; Lodish, H.F. Flotillin and epidermal surface antigen define a new family of caveolae-associated integral membrane proteins. *J. Biol. Chem.* **1997**, *272*, 13793–13802. [CrossRef] [PubMed]
4. Rajendran, L.; Knobloch, M.; Geiger, K.D.; Dienel, S.; Nitsch, R.; Simons, K.; Konietzko, U. Increased Abeta production leads to intracellular accumulation of Abeta in flotillin-1-positive endosomes. *Neurodegener. Dis.* **2007**, *4*, 164–170. [CrossRef]
5. Xiao, T.; Zhang, W.; Jiao, B.; Pan, C.Z.; Liu, X.; Shen, L. The role of exosomes in the pathogenesis of Alzheimer' disease. *Transl. Neurodegener.* **2017**, *6*, 3. [CrossRef]
6. Tate, B.A.; Mathews, P.M. Targeting the role of the endosome in the pathophysiology of Alzheimer's disease: A strategy for treatment. *Sci. Aging Knowl. Environ.* **2006**, *2006*, re2. [CrossRef]
7. Irizarry, M.C. Biomarkers of Alzheimer disease in plasma. *NeuroRx J. Am. Soc. Exp. Neurother.* **2004**, *1*, 226–234. [CrossRef]
8. Racine, A.M.; Koscik, R.L.; Nicholas, C.R.; Clark, L.R.; Okonkwo, O.C.; Oh, J.M.; Hillmer, A.T.; Murali, D.; Barnhart, T.E.; Betthauser, T.J.; et al. Cerebrospinal fluid ratios with Abeta42 predict preclinical brain beta-amyloid accumulation. *Alzheimer's Dement.* **2016**, *2*, 27–38. [CrossRef]
9. Vlassenko, A.G.; Benzinger, T.L.; Morris, J.C. PET amyloid-beta imaging in preclinical Alzheimer's disease. *Biochim. Biophys. Acta* **2012**, *1822*, 370–379. [CrossRef]
10. Abdullah, M.; Kimura, N.; Akatsu, H.; Hashizume, Y.; Ferdous, T.; Tachita, T.; Iida, S.; Zou, K.; Matsubara, E.; Michikawa, M. Flotillin is a Novel Diagnostic Blood Marker of Alzheimer's Disease. *J. Alzheimer's Dis.* **2019**, *72*, 1165–1176. [CrossRef]
11. Neumann-Giesen, C.; Falkenbach, B.; Beicht, P.; Claasen, S.; Luers, G.; Stuermer, C.A.; Herzog, V.; Tikkanen, R. Membrane and raft association of reggie-1/flotillin-2: Role of myristoylation, palmitoylation and oligomerization and induction of filopodia by overexpression. *Biochem. J.* **2004**, *378*, 509–518. [CrossRef] [PubMed]
12. Kokubo, H.; Lemere, C.A.; Yamaguchi, H. Localization of flotillins in human brain and their accumulation with the progression of Alzheimer's disease pathology. *Neurosci. Lett.* **2000**, *290*, 93–96. [CrossRef]
13. Bouillot, C.; Prochiantz, A.; Rougon, G.; Allinquant, B. Axonal amyloid precursor protein expressed by neurons in vitro is present in a membrane fraction with caveolae-like properties. *J. Biol. Chem.* **1996**, *271*, 7640–7644. [CrossRef] [PubMed]
14. Lee, S.J.; Liyanage, U.; Bickel, P.E.; Xia, W.; Lansbury, P.T., Jr.; Kosik, K.S. A detergent-insoluble membrane compartment contains A beta in vivo. *Nat. Med.* **1998**, *4*, 730–734. [CrossRef] [PubMed]
15. Riddell, D.R.; Christie, G.; Hussain, I.; Dingwall, C. Compartmentalization of beta-secretase (Asp2) into low-buoyant density, noncaveolar lipid rafts. *Curr. Biol.* **2001**, *11*, 1288–1293. [CrossRef]
16. Vetrivel, K.S.; Cheng, H.; Lin, W.; Sakurai, T.; Li, T.; Nukina, N.; Wong, P.C.; Xu, H.; Thinakaran, G. Association of gamma-secretase with lipid rafts in post-Golgi and endosome membranes. *J. Biol. Chem.* **2004**, *279*, 44945–44954. [CrossRef]

17. Kokubo, H.; Kayed, R.; Glabe, C.G.; Yamaguchi, H. Soluble Abeta oligomers ultrastructurally localize to cell processes and might be related to synaptic dysfunction in Alzheimer's disease brain. *Brain Res.* **2005**, *1031*, 222–228. [CrossRef]
18. Vetrivel, K.S.; Thinakaran, G. Membrane rafts in Alzheimer's disease beta-amyloid production. *Biochim. Biophys. Acta* **2010**, *1801*, 860–867. [CrossRef]
19. Abad-Rodriguez, J.; Ledesma, M.D.; Craessaerts, K.; Perga, S.; Medina, M.; Delacourte, A.; Dingwall, C.; De Strooper, B.; Dotti, C.G. Neuronal membrane cholesterol loss enhances amyloid peptide generation. *J. Cell Biol.* **2004**, *167*, 953–960. [CrossRef]
20. Crameri, A.; Biondi, E.; Kuehnle, K.; Lutjohann, D.; Thelen, K.M.; Perga, S.; Dotti, C.G.; Nitsch, R.M.; Ledesma, M.D.; Mohajeri, M.H. The role of seladin-1/DHCR24 in cholesterol biosynthesis, APP processing and Abeta generation in vivo. *EMBO J.* **2006**, *25*, 432–443. [CrossRef]
21. Langui, D.; Girardot, N.; El Hachimi, K.H.; Allinquant, B.; Blanchard, V.; Pradier, L.; Duyckaerts, C. Subcellular topography of neuronal Abeta peptide in APPxPS1 transgenic mice. *Am. J. Pathol.* **2004**, *165*, 1465–1477. [CrossRef]
22. Yamazaki, T.; Koo, E.H.; Selkoe, D.J. Trafficking of cell-surface amyloid beta-protein precursor. II. Endocytosis, recycling and lysosomal targeting detected by immunolocalization. *J. Cell Sci.* **1996**, *109*, 999–1008. [PubMed]
23. Cataldo, A.M.; Barnett, J.L.; Pieroni, C.; Nixon, R.A. Increased neuronal endocytosis and protease delivery to early endosomes in sporadic Alzheimer's disease: Neuropathologic evidence for a mechanism of increased beta-amyloidogenesis. *J. Neurosci. Off. J. Soc. Neurosci.* **1997**, *17*, 6142–6151. [CrossRef]
24. Vetrivel, K.S.; Thinakaran, G. Amyloidogenic processing of beta-amyloid precursor protein in intracellular compartments. *Neurology* **2006**, *66*, S69–S73. [CrossRef]
25. Refolo, L.M.; Sambamurti, K.; Efthimiopoulos, S.; Pappolla, M.A.; Robakis, N.K. Evidence that secretase cleavage of cell surface Alzheimer amyloid precursor occurs after normal endocytic internalization. *J. Neurosci. Res.* **1995**, *40*, 694–706. [CrossRef]
26. Rajendran, L.; Honsho, M.; Zahn, T.R.; Keller, P.; Geiger, K.D.; Verkade, P.; Simons, K. Alzheimer's disease beta-amyloid peptides are released in association with exosomes. *Proc. Natl. Acad. Sci. USA* **2006**, *103*, 11172–11177. [CrossRef]
27. Perez, R.G.; Soriano, S.; Hayes, J.D.; Ostaszewski, B.; Xia, W.; Selkoe, D.J.; Chen, X.; Stokin, G.B.; Koo, E.H. Mutagenesis identifies new signals for beta-amyloid precursor protein endocytosis, turnover, and the generation of secreted fragments, including Abeta42. *J. Biol. Chem.* **1999**, *274*, 18851–18856. [CrossRef]
28. Elkin, S.R.; Lakoduk, A.M.; Schmid, S.L. Endocytic pathways and endosomal trafficking: A primer. *Wien. Med. Wochenschr.* **2016**, *166*, 196–204. [CrossRef]
29. Abdullah, M.; Takase, H.; Nunome, M.; Enomoto, H.; Ito, J.; Gong, J.S.; Michikawa, M. Amyloid-beta Reduces Exosome Release from Astrocytes by Enhancing JNK Phosphorylation. *J. Alzheimer's Dis.* **2016**, *53*, 1433–1441. [CrossRef]
30. Paolicelli, R.C.; Bergamini, G.; Rajendran, L. Cell-to-cell Communication by Extracellular Vesicles: Focus on Microglia. *Neuroscience* **2019**, *405*, 148–157. [CrossRef]
31. Yuyama, K.; Sun, H.; Mitsutake, S.; Igarashi, Y. Sphingolipid-modulated exosome secretion promotes clearance of amyloid-beta by microglia. *J. Biol. Chem.* **2012**, *287*, 10977–10989. [CrossRef] [PubMed]
32. Yuyama, K.; Sun, H.; Sakai, S.; Mitsutake, S.; Okada, M.; Tahara, H.; Furukawa, J.; Fujitani, N.; Shinohara, Y.; Igarashi, Y. Decreased amyloid-beta pathologies by intracerebral loading of glycosphingolipid-enriched exosomes in Alzheimer model mice. *J. Biol. Chem.* **2014**, *289*, 24488–24498. [CrossRef] [PubMed]
33. Perez-Gonzalez, R.; Gauthier, S.A.; Kumar, A.; Levy, E. The exosome secretory pathway transports amyloid precursor protein carboxyl-terminal fragments from the cell into the brain extracellular space. *J. Biol. Chem.* **2012**, *287*, 43108–43115. [CrossRef] [PubMed]
34. Saman, S.; Kim, W.; Raya, M.; Visnick, Y.; Miro, S.; Saman, S.; Jackson, B.; McKee, A.C.; Alvarez, V.E.; Lee, N.C.; et al. Exosome-associated tau is secreted in tauopathy models and is selectively phosphorylated in cerebrospinal fluid in early Alzheimer disease. *J. Biol. Chem.* **2012**, *287*, 3842–3849. [CrossRef] [PubMed]
35. An, K.; Klyubin, I.; Kim, Y.; Jung, J.H.; Mably, A.J.; O'Dowd, S.T.; Lynch, T.; Kanmert, D.; Lemere, C.A.; Finan, G.M.; et al. Exosomes neutralize synaptic-plasticity-disrupting activity of Abeta assemblies in vivo. *Mol. Brain* **2013**, *6*, 47. [CrossRef] [PubMed]

36. Girardot, N.; Allinquant, B.; Langui, D.; Laquerriere, A.; Dubois, B.; Hauw, J.J.; Duyckaerts, C. Accumulation of flotillin-1 in tangle-bearing neurones of Alzheimer's disease. *Neuropathol. Appl. Neurobiol.* **2003**, *29*, 451–461. [CrossRef]
37. Eitan, E.; Hutchison, E.R.; Marosi, K.; Comotto, J.; Mustapic, M.; Nigam, S.M.; Suire, C.; Maharana, C.; Jicha, G.A.; Liu, D.; et al. Extracellular Vesicle-Associated Abeta Mediates Trans-Neuronal Bioenergetic and Ca(2+)-Handling Deficits in Alzheimer's Disease Models. *NPJ Aging Mech. Dis.* **2016**, *2*. [CrossRef]
38. Sardar Sinha, M.; Ansell-Schultz, A.; Civitelli, L.; Hildesjo, C.; Larsson, M.; Lannfelt, L.; Ingelsson, M.; Hallbeck, M. Alzheimer's disease pathology propagation by exosomes containing toxic amyloid-beta oligomers. *Acta Neuropathol.* **2018**, *136*, 41–56. [CrossRef]
39. Nishikawa, T.; Takahashi, T.; Nakamori, M.; Yamazaki, Y.; Kurashige, T.; Nagano, Y.; Nishida, Y.; Izumi, Y.; Matsumoto, M. Phosphatidylinositol-4,5-bisphosphate is enriched in granulovacuolar degeneration bodies and neurofibrillary tangles. *Neuropathol. Appl. Neurobiol.* **2014**, *40*, 489–501. [CrossRef]
40. Woodard, J.S. Clinicopathologic significance of granulovacuolar degeneration in Alzheimer's disease. *J. Neuropathol. Exp. Neurol.* **1962**, *21*, 85–91. [CrossRef]
41. Alsaqati, M.; Thomas, R.S.; Kidd, E.J. Proteins Involved in Endocytosis Are Upregulated by Ageing in the Normal Human Brain: Implications for the Development of Alzheimer's Disease. *J. Gerontol. Ser. A Biol. Sci. Med Sci.* **2018**, *73*, 289–298. [CrossRef] [PubMed]
42. Chen, T.Y.; Liu, P.H.; Ruan, C.T.; Chiu, L.; Kung, F.L. The intracellular domain of amyloid precursor protein interacts with flotillin-1, a lipid raft protein. *Biochem. Biophys. Res. Commun.* **2006**, *342*, 266–272. [CrossRef] [PubMed]
43. Liu, F.L.; Liu, T.Y.; Kung, F.L. FKBP12 regulates the localization and processing of amyloid precursor protein in human cell lines. *J. Biosci.* **2014**, *39*, 85–95. [CrossRef] [PubMed]
44. Okabayashi, S.; Kimura, N. LGI3 interacts with flotillin-1 to mediate APP trafficking and exosome formation. *Neuroreport* **2010**, *21*, 606–610. [CrossRef] [PubMed]
45. Schneider, A.; Rajendran, L.; Honsho, M.; Gralle, M.; Donnert, G.; Wouters, F.; Hell, S.W.; Simons, M. Flotillin-dependent clustering of the amyloid precursor protein regulates its endocytosis and amyloidogenic processing in neurons. *J. Neurosci. Off. J. Soc. Neurosci.* **2008**, *28*, 2874–2882. [CrossRef] [PubMed]
46. Hattori, C.; Asai, M.; Onishi, H.; Sasagawa, N.; Hashimoto, Y.; Saido, T.C.; Maruyama, K.; Mizutani, S.; Ishiura, S. BACE1 interacts with lipid raft proteins. *J. Neurosci. Res.* **2006**, *84*, 912–917. [CrossRef]
47. John, B.A.; Meister, M.; Banning, A.; Tikkanen, R. Flotillins bind to the dileucine sorting motif of beta-site amyloid precursor protein-cleaving enzyme 1 and influence its endosomal sorting. *FEBS J.* **2014**, *281*, 2074–2087. [CrossRef]
48. Kamagata, E.; Kudo, T.; Kimura, R.; Tanimukai, H.; Morihara, T.; Sadik, M.G.; Kamino, K.; Takeda, M. Decrease of dynamin 2 levels in late-onset Alzheimer's disease alters Abeta metabolism. *Biochem. Biophys. Res. Commun.* **2009**, *379*, 691–695. [CrossRef]
49. Hung, Y.H.; Robb, E.L.; Volitakis, I.; Ho, M.; Evin, G.; Li, Q.X.; Culvenor, J.G.; Masters, C.L.; Cherny, R.A.; Bush, A.I. Paradoxical condensation of copper with elevated beta-amyloid in lipid rafts under cellular copper deficiency conditions: Implications for Alzheimer disease. *J. Biol. Chem.* **2009**, *284*, 21899–21907. [CrossRef]
50. Bitsikas, V.; Riento, K.; Howe, J.D.; Barry, N.P.; Nichols, B.J. The role of flotillins in regulating abeta production, investigated using flotillin 1-/-, flotillin 2-/- double knockout mice. *PLoS ONE* **2014**, *9*, e85217. [CrossRef]
51. Zhang, X.L.; Zhao, N.; Xu, B.; Chen, X.H.; Li, T.J. Treadmill exercise inhibits amyloid-beta generation in the hippocampus of APP/PS1 transgenic mice by reducing cholesterol-mediated lipid raft formation. *Neuroreport* **2019**, *30*, 498–503. [CrossRef] [PubMed]
52. Zhu, H.; Luo, L.; Hu, S.; Dong, K.; Li, G.; Zhang, T. Treating Alzheimer's disease with Yizhijiannao granules by regulating expression of multiple proteins in temporal lobe. *Neural Regen. Res.* **2014**, *9*, 1283–1287. [CrossRef] [PubMed]
53. Thomas, R.S.; Lelos, M.J.; Good, M.A.; Kidd, E.J. Clathrin-mediated endocytic proteins are upregulated in the cortex of the Tg2576 mouse model of Alzheimer's disease-like amyloid pathology. *Biochem. Biophys. Res. Commun.* **2011**, *415*, 656–661. [CrossRef] [PubMed]
54. Kong, G.K.; Miles, L.A.; Crespi, G.A.; Morton, C.J.; Ng, H.L.; Barnham, K.J.; McKinstry, W.J.; Cappai, R.; Parker, M.W. Copper binding to the Alzheimer's disease amyloid precursor protein. *Eur. Biophys. J.* **2008**, *37*, 269–279. [CrossRef]

55. Guenette, S.; Strecker, P.; Kins, S. APP Protein Family Signaling at the Synapse: Insights from Intracellular APP-Binding Proteins. *Front. Mol. Neurosci.* **2017**, *10*, 87. [CrossRef]
56. Aidaralieva, N.J.; Kamino, K.; Kimura, R.; Yamamoto, M.; Morihara, T.; Kazui, H.; Hashimoto, R.; Tanaka, T.; Kudo, T.; Kida, T.; et al. Dynamin 2 gene is a novel susceptibility gene for late-onset Alzheimer disease in non-APOE-epsilon4 carriers. *J. Hum. Genet.* **2008**, *53*, 296–302. [CrossRef]
57. Lee, S.; Mankhong, S.; Kang, J.H. Extracellular Vesicle as a Source of Alzheimer's Biomarkers: Opportunities and Challenges. *Int. J. Mol. Sci.* **2019**, *20*, 1728. [CrossRef]
58. Chiasserini, D.; van Weering, J.R.; Piersma, S.R.; Pham, T.V.; Malekzadeh, A.; Teunissen, C.E.; de Wit, H.; Jimenez, C.R. Proteomic analysis of cerebrospinal fluid extracellular vesicles: A comprehensive dataset. *J. Proteom.* **2014**, *106*, 191–204. [CrossRef]
59. Street, J.M.; Barran, P.E.; Mackay, C.L.; Weidt, S.; Balmforth, C.; Walsh, T.S.; Chalmers, R.T.; Webb, D.J.; Dear, J.W. Identification and proteomic profiling of exosomes in human cerebrospinal fluid. *J. Transl. Med.* **2012**, *10*, 5. [CrossRef]
60. Sun, B.; Dalvi, P.; Abadjian, L.; Tang, N.; Pulliam, L. Blood neuron-derived exosomes as biomarkers of cognitive impairment in HIV. *Aids* **2017**, *31*, F9–F17. [CrossRef]
61. Kowal, J.; Arras, G.; Colombo, M.; Jouve, M.; Morath, J.P.; Primdal-Bengtson, B.; Dingli, F.; Loew, D.; Tkach, M.; Thery, C. Proteomic comparison defines novel markers to characterize heterogeneous populations of extracellular vesicle subtypes. *Proc. Natl. Acad. Sci. USA* **2016**, *113*, E968–E977. [CrossRef] [PubMed]
62. Kuharic, J.; Grabusic, K.; Tokmadzic, V.S.; Stifter, S.; Tulic, K.; Shevchuk, O.; Lucin, P.; Sustic, A. Severe Traumatic Brain Injury Induces Early Changes in the Physical Properties and Protein Composition of Intracranial Extracellular Vesicles. *J. Neurotrauma* **2019**, *36*, 190–200. [CrossRef]
63. Stuendl, A.; Kunadt, M.; Kruse, N.; Bartels, C.; Moebius, W.; Danzer, K.M.; Mollenhauer, B.; Schneider, A. Induction of alpha-synuclein aggregate formation by CSF exosomes from patients with Parkinson's disease and dementia with Lewy bodies. *Brain J. Neurol.* **2016**, *139*, 481–494. [CrossRef] [PubMed]
64. Ou, Y.X.; Liu, F.T.; Chen, F.Y.; Zhu, Z.M. Prognostic value of Flotillin-1 expression in patients with solid tumors. *Oncotarget* **2017**, *8*, 52665–52677. [CrossRef] [PubMed]
65. Kirsch, C.; Eckert, G.P.; Mueller, W.E. Statin effects on cholesterol micro-domains in brain plasma membranes. *Biochem. Pharmacol.* **2003**, *65*, 843–856. [CrossRef]
66. Dominguez-Prieto, M.; Velasco, A.; Tabernero, A.; Medina, J.M. Endocytosis and Transcytosis of Amyloid-beta Peptides by Astrocytes: A Possible Mechanism for Amyloid-beta Clearance in Alzheimer's Disease. *J. Alzheimer's Dis.* **2018**, *65*, 1109–1124. [CrossRef]
67. Gaudreault, S.B.; Dea, D.; Poirier, J. Increased caveolin-1 expression in Alzheimer's disease brain. *Neurobiol. Aging* **2004**, *25*, 753–759. [CrossRef]
68. Cong, L.; Kong, X.; Wang, J.; Du, J.; Xu, Z.; Xu, Y.; Zhao, Q. Association between SORL1 polymorphisms and the risk of Alzheimer's disease. *J. Integr. Neurosci.* **2018**, *17*, 185–192. [CrossRef]
69. Reitz, C. Dyslipidemia and the risk of Alzheimer's disease. *Curr. Atheroscler. Rep.* **2013**, *15*, 307. [CrossRef]

Review

Molecular and Imaging Biomarkers in Alzheimer's Disease: A Focus on Recent Insights

Chiara Villa [1,*], Marialuisa Lavitrano [1,2], Elena Salvatore [3] and Romina Combi [1,*]

1 School of Medicine and Surgery, University of Milano-Bicocca, 20900 Monza, Italy
2 Institute for the Experimental Endocrinology and Oncology, National Research Council (IEOS-CNR), 80131 Naples, Italy; m.lavitrano@ieos.cnr.it
3 Department of Neurosciences and Reproductive and Odontostomatological Sciences, Federico II University, 80131 Naples, Italy; elena.salvatore@unina.it
* Correspondence: chiara.villa@unimib.it (C.V.); romina.combi@unimib.it (R.C.)

Received: 25 May 2020; Accepted: 7 July 2020; Published: 10 July 2020

Abstract: Alzheimer's disease (AD) is the most common neurodegenerative disease among the elderly, affecting millions of people worldwide and clinically characterized by a progressive and irreversible cognitive decline. The rapid increase in the incidence of AD highlights the need for an easy, efficient and accurate diagnosis of the disease in its initial stages in order to halt or delay the progression. The currently used diagnostic methods rely on measures of amyloid-β (Aβ), phosphorylated (p-tau) and total tau (t-tau) protein levels in the cerebrospinal fluid (CSF) aided by advanced neuroimaging techniques like positron emission tomography (PET) and magnetic resonance imaging (MRI). However, the invasiveness of these procedures and the high cost restrict their utilization. Hence, biomarkers from biological fluids obtained using non-invasive methods and novel neuroimaging approaches provide an attractive alternative for the early diagnosis of AD. Such biomarkers may also be helpful for better understanding of the molecular mechanisms underlying the disease, allowing differential diagnosis or at least prolonging the pre-symptomatic stage in patients suffering from AD. Herein, we discuss the advantages and limits of the conventional biomarkers as well as recent promising candidates from alternative body fluids and new imaging techniques.

Keywords: Alzheimer's disease; biomarker; amyloid beta; neuroimaging; cerebrospinal fluid

1. Introduction

Alzheimer's disease (AD) represents the most common form of dementia in the elderly population worldwide, accounting for up to 80% of all diagnoses [1]. AD is clinically characterized by irreversible and progressive neurodegeneration leading to memory deterioration, behavioral changes and cognitive dysfunction, resulting in autonomy loss, which ultimately requires full-time medical care [2]. The neuropathological hallmarks include the presence of extracellular senile plaques constituted by the amyloid-β (Aβ) peptides and intracellular neurofibrillary tangles (NFTs) consisting of hyper-phosphorylated paired helical filaments (PHFs) of the microtubule-associated protein tau (MAPT) [3]. Aβ plaques are composed of various Aβ peptides, including the 40 and 42 amino acid products ($A\beta_{40}$ and $A\beta_{42}$), generated as a result of the sequential proteolytic cleavage of Aβ precursor protein (APP) by β-site APP-cleaving enzyme 1 (BACE-1) and the γ-secretase complex [4]. During the early stage of the disease, extraneuronal Aβ deposits, intraneuronal NFTs and neuritic threads are found in the entorhinal cortex and in the hippocampus, which are the key regions of memory and learning functions. However, in addition to Aβ and tau pathology, other processes, such as synaptic dysfunctions and microglia-mediated inflammation also play an important role in AD pathogenesis and may correlate with cognitive decline (Figure 1) [5,6].

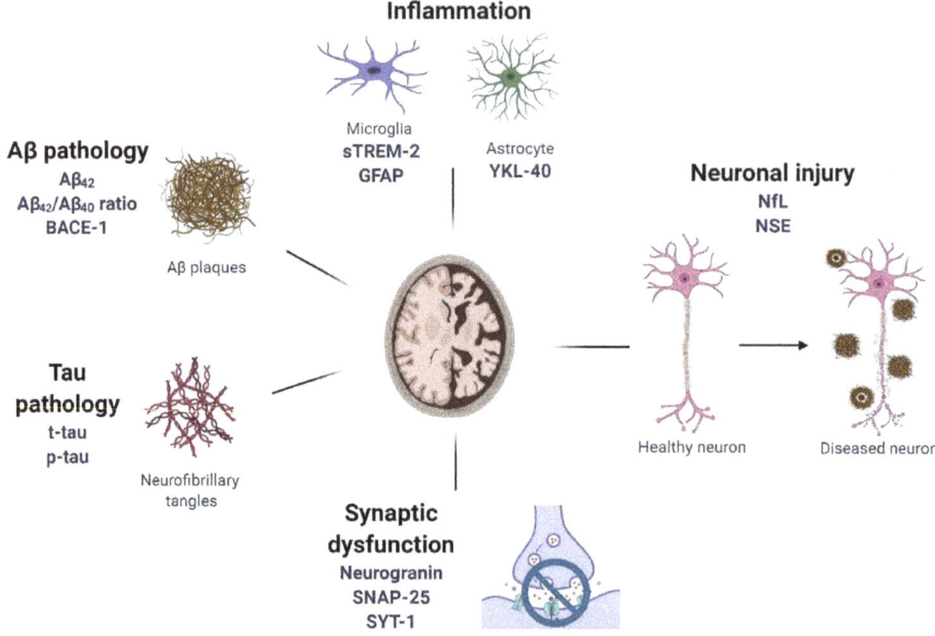

Figure 1. Main pathological mechanisms occurring in Alzheimer's disease and their associated fluid biomarkers.

Most AD cases are sporadic with a late onset, usually occurring in individuals aged 65 or older, and the main risk factors are aging and carrying the ε4 allele of Apolipoprotein E (ApoE) [7]. Conversely, the rare early-onset forms of AD affect individuals under 65 years of age and show an autosomal dominant pattern of inheritance, generally presenting a positive family history. These patients carry mutations in one of the known genes, namely *PSEN1*, *PSEN2* and *APP*, encoding the presenilin-1, presenilin-2 and APP proteins, respectively. All of them are involved in the maturation and processing of APP, leading to an increased production or aggregation of Aβ peptide [8].

Although behavioral symptoms can be alleviated by actual therapeutic strategies, drugs that prevent or halt the disease course are still not available [9]. The lack of success of disease-modifying therapy may be partially explained by the complex etiology in its pathophysiology and the limitations in past clinical trials designed on enrolled participants with mild-to-moderate AD or with no Aβ pathology [10]. In this regard, a biomarker holds promise for enabling more effective drug development in AD and establishing a more personalized medicine approach [11]. It would be a suitable indicator of the stage of disease progression, treatment monitoring and a valuable tool for epidemiological and therapeutic research. According to the latest guidelines of the National Institute on Aging and Alzheimer's Association (NIA-AA), the recommendations for the diagnosis of pre-clinical, mild cognitive impairment (MCI) and AD dementia have been updated, unifying biological markers and imaging into AT(N) groups. This novel classification summarizes biomarkers into three categories: Aβ deposition (A), pathological fibrillary tau (T) and neurodegeneration (N) [12,13]. Currently, group A includes low levels of $A\beta_{42}$ in the cerebrospinal fluid (CSF) or Aβ positron emission tomography (PET) ligand binding; group T includes elevated levels of CSF phosphorylated tau at threonine 181 (p-tau) and tau PET ligand binding, whereas group N includes elevated CSF total tau (t-tau), fluorodeoxyglucose (FDG)-PET hypometabolism and atrophy on magnetic resonance imaging (MRI) [13]. Increasing efforts have been made in recent years to detect biomarkers in more accessible biological matrices; therefore, in this review we discuss the clinical relevance of emerging candidate biomarkers in CSF and in other promising alternative non-invasive biological fluids as well as novel approaches in "dry" biomarkers like neuroimaging or neurophysiological techniques.

2. Emerging AD Biomarkers in Biological Fluids

2.1. Invasive CSF Biomarkers

Despite its invasiveness of collection, CSF still represents the most reliable biological fluid for biomarker detection of the central nervous system (CNS) disorders, allowing the most accurate elucidation about the molecular processes occurring during neurodegeneration. Compared with blood, CSF has the advantage of its proximity to the brain parenchyma and that it contains brain proteins which are directly secreted from the brain extracellular space.

In addition to the well-established core AD CSF biomarkers like Aβ and tau proteins, a number of candidate molecules have been investigated as potential AD biomarker, mainly related with pathological mechanisms or to other aspects of the disease pathophysiology, such as enzymatic deficits, degrading pathway, biochemical modifications or clearance. Currently, one of the most studied biomarkers is neurofilament light chain (NfL), a scaffold protein found in the neuronal cytoskeleton. After axonal injury, intracellular NfL is released in the extracellular space, leading to an increased concentration in the CSF. Therefore, it represents a non-specific marker for neuronal damage and has been largely studied in the context of neurodegenerative diseases, including multiple sclerosis (MS), Parkinson's disease (PD), frontotemporal dementia (FTD) and amyotrophic lateral sclerosis (ALS) [14]. Elevated levels of CSF NfL were found in patients with MCI and AD, associating with the severity of memory impairment as a marker of disease progression [15–17]. Altogether, studies reported a very good performance of CSF NfL to distinguish AD cases from cognitively healthy controls with no evidence of structural brain damage [18]. However, the currently available evidence does not support the ability of CSF NfL to differentiate AD from disease mimics or MCI [14].

Similarly, also the presence of neuron-specific enolase (NSE) in CSF represents a marker of neuronal damage. NSE is a glycolytic enzyme involved in neuronal energy metabolism, axoplasmatic transport and cell survival. It is physiologically not secreted in the extracellular space, so elevated CSF NSE levels are regarded as the result of an upregulation of neuronal metabolic activity that follows increased energy demand. Significantly higher protein levels were found in AD patients, and alone or in combination with t-tau and p-tau, NSE further showed both high specificity and sensitivity to distinguish AD cases from healthy controls, suggesting a clinical application of this potential biomarker [19]. However, NSE could not discriminate AD from other forms of dementia [20].

The post-synaptic protein neurogranin, which is exclusively expressed in the cortex and hippocampus by excitatory neurons, seems to be a promising biomarker candidate. It is known to play an important role in learning and memory by maintaining long-term potentiation and synaptic plasticity. Neurogranin expression is highest in cortical areas, but its levels are markedly low in the frontal cortex and the hippocampus, indicating that the measurement of neurogranin in CSF could serve as a biomarker for synaptic degeneration and dendritic instability [21]. Synapse loss is a downstream effect of amyloidosis, tauopathy, inflammation and other pathological mechanisms occurring in AD and strongly correlates with decline in cognitive performance. High CSF levels of neurogranin in AD and prodromal AD have been confirmed by several studies using immunoassay recognizing both the full-length protein and the fragment peptides [22–25]. Moreover, encouraging data showed that increased neurogranin fragments in CSF correlate with cognitive decline, hippocampal atrophy measured by MRI and reduced glucose metabolism on FDG-PET [22,26]. Interestingly, the increase in CSF neurogranin seems to be specific for AD and not found in other neurodegenerative disorders, including FTD, PD, Lewy body dementia (LBD), progressive supranuclear palsy or multiple system atrophy [27].

Additional synaptic proteins, including synaptosomal-associated protein 25 (SNAP-25) and synaptotagmin-1 (SYT-1), also showed promising results as CSF biomarkers for synaptic damage and loss. Whereas SNAP-25 is found at synaptic vesicles, SYT-1 is located in the pre-synaptic plasma membrane and is essential for synaptic vesicle exocytosis and therefore neurotransmitter release [23]. The levels of both SNAP-25 and SYT-1 are decreased in the cortical areas of AD brain, reflecting the synaptic loss and degeneration occurring in AD [28,29]. Interestingly, a marked increase in both SNAP-25 and STY-1 levels in CSF was found in patients with AD or MCI as compared with controls [28,30,31]. Although these results need validation in further studies, they may represent a valuable tool regarding the relevance of synaptic degeneration and loss in AD pathogenesis and also in the clinical evaluation of patients.

Recent research proposed markers of glial activation as potential biomarkers for AD. Among them, one of the most promising is the triggering receptor expressed on myeloid cells 2 (TREM-2), mostly because there is a strong genetic association between TREM-2 and AD. TREM-2 play several roles in microglia, including cytokine release, proliferation, APOE binding and shielding of Aβ plaques [32,33]. It is a transmembrane protein and its soluble domain (sTREM-2) is released into the extracellular space and can be measured in both CSF and blood. The majority of studies reported increased levels of sTREM-2 in AD vs. controls which dynamically change during the disease course, reaching the peak in the later asymptomatic stage and early symptomatic phase of late-onset AD or in the genetic forms of AD [34–37]. However, CSF sTREM-2 is closely associated with tau-related neurodegeneration but not with Aβ pathology [38], and it increases also during MS and other neuroinflammatory disorders, suggesting that the microglia response mediated by TREM-2 occurs whenever there is a neuronal injury, so not only in AD [39,40]. Another marker of glial activation is the glial fibrillary acidic protein (GFAP), one of the cytoskeletal filament proteins in astrocytes, which is activated and then released from these cells during neurodegeneration. CSF GFAP levels were reported to inversely correlate with the cognitive function, although an increase of this protein production was found not only in patients affected by AD, but also with FTD and LBD, suggesting its potential use in the prediction of dementia progression [41]. Regarding the microglial and astrocyte marker YKL-40, several studies observed higher levels in the CSF of AD patients as compared with controls [42], and these results have been also confirmed by a recent meta-analysis [43]. CSF YKL-40 increases with disease progression and is positively correlated with biomarkers of neurodegeneration [44]. Some studies have even reported that its levels can predict the progression from cognitively unimpaired to MCI and from MCI to AD dementia [42,45]. Also in the case of YKL-40, its increased levels in CSF are not specific for AD, albeit they are unchanged or even decreased in PD patients without dementia [46].

Finally, several studies also focused on BACE-1 as a possible AD biomarker, but with conflicting results. Most of them showed an increase in activity or protein levels of BACE-1 in AD individuals and also in subjects with MCI who developed AD later, being a good progression marker [47–49]. Conversely, other authors reported different results, including no differences in BACE-1 activity between controls, MCI and AD cases, or even decreased CSF levels in AD as compared with healthy controls [50,51]. A summary of levels of CSF biomarkers is reported in Table 1.

Table 1. Changes in levels of CSF biomarkers in Alzheimer's disease (AD) patients and controls.

Biomarker	AD	Controls	AD/Controls	Technique	Reference
NfL (pg/mL)	1574.04–2827.96	995.41–2016.59	↑	ELISA	[16]
NSE (ng/mL)	15.63–20.60	5.98–10.94	↑	ECLIA	[19]
neurogranin (pg/mL)	349–744	161–453	↑	ELISA	[26]
SNAP-25 (ng/mL)	22–38	18–20	↑	MS	[28]
SYT-1 pM	131.7–449.7	166.9–309.7	↑	ELISA	[30]
sTREM-2 (pg/mL)	172.5–305.4	131.0–240.7	↑	ELISA	[34]
GFAP (ng/mL)	1.77–4.26	1.31–3.21	↑	ELISA	[41]
YKL-40 (ng/mL)	400–422	254–293	↑	ELISA	[20,42]
BACE-1 activity (pM)	30–43	23–42	↑≅	ELISA	[47,50]

ECLIA, electrochemiluminescence immunoassay; ELISA, enzyme-linked immunosorbent assay; MS, mass spectrometry.Overall, the value of these molecules as AD biomarkers has to be validated [52,53]. Moreover, given the fact that CSF collection requires lumbar puncture, there is still a need to discover additional non-invasive, reproducible, reliable, inexpensive and simple to measure biomarkers in alternative biological fluids, e.g., blood, saliva, urine and tears (Figure 2).

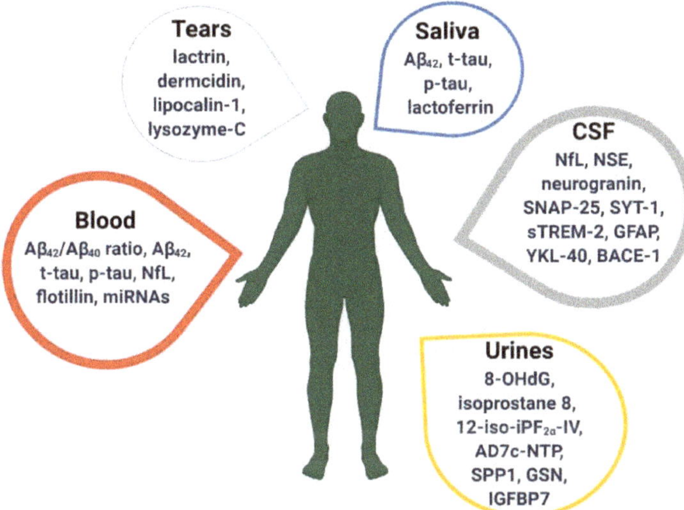

Figure 2. Emerging biomarkers in cerebrospinal fluid (CSF) and alternative biological fluids. For each of them, the thickness of the box is proportional to the number of articles in English searched in Pubmed database on 24 June 2020 using the keywords: "biomarker" AND "Alzheimer's disease" AND the type of fluid.

2.2. Non-Invasive Biomarkers

2.2.1. Blood

As blood is more accessible than CSF, potential biomarkers have largely been studied in this biological fluid [54,55]. Apart from the much less invasive procedure of blood collection as compared to the lumbar puncture, it allows repeated sampling and measurements to monitor the AD progression or to evaluate the efficacy of the newly developed drugs during clinical trials. However, developing blood biomarkers for AD and other brain disorders is still challenging and requires highly sensitive technologies for detection and careful validation work. The blood-brain barrier (BBB) represents a major issue in finding suitable blood-based biomarkers. Whilst CSF is continuous with the brain extracellular fluid, with a free passage of molecules from the brain to the CSF, the highly selective semipermeable membrane allows entrance only to selected brain-derived proteins, which are typically present at low concentrations [56]. Moreover, some biomarkers related to AD pathology are expressed in non-cerebral tissues and this could confound their measurements in the blood. Finally, the activity of various proteases or protein carriers in blood may participate in degrading or masking the epitopes of a potential biomarker, interfering with its detection and measurement [57].

Several studies on plasma biomarkers have indeed reported inconsistent results, even for the core AD CSF biomarkers like Aβ and tau proteins. While some authors reported high concordance between the levels of these proteins detected in CSF and blood [58], conversely, other studies on Aβ plasma levels demonstrated a lack of correlation between CSF and blood in both $A\beta_{40}$ and $A\beta_{42}$ in AD [43,59–61]. This discrepancy may probably be due to the low levels of Aβ peptide in blood or to the analytical sensitivities of enzyme-linked immunosorbent assay (ELISA) [62]. In order to overcome the limits in the detection using traditional ELISA methods, ultrasensitive technologies have been developed with promising results, including single molecule array (Simoa) [62,63], immunoprecipitation coupled with mass-spectrometry (IP-MS) analysis [46] and stable isotope labeling kinetics followed by IP-MS [64]. Using Simoa assay, Aβ_{42} levels and the ratio of Aβ_{42}/Aβ_{40} in plasma were shown to correlate with CSF levels and Aβ deposition measured using PET [62,63]. A decreased Aβ_{42}/Aβ_{40} ratio was even found in plasma of patients with MCI and preclinical AD [62]. Novel approaches based on MS technology have the advantage of allowing the investigation of various species of Aβ peptides at very low concentrations. Immunoprecipitation coupled with MS was useful to pull down different Aβ fragments, showing a decreased Aβ_{42}/Aβ_{40} ratio in plasma with around 90% diagnostic accuracy and a great ability to predict the presence of Aβ plaques in the brain of AD patients detected using Aβ PET imaging [64,65]. Although these assays overcome several comparison and standardization limits, they do not solve the confounding problem with non-cerebral expression of Aβ which is produced by platelets, the primary source of Aβ peptide accounting for 90% of total blood Aβ [66]. Similarly, ultrasensitive techniques have also been used to measure tau protein in blood samples. Several groups reported elevated t-tau levels in the plasma of AD patients as compared with controls, but the overlapping values between these two groups exclude its potential use as a diagnostic tool [67–69]. Additionally, plasma t-tau predicted cognitive decline and the risk of dementia [68–70]. On the other hand, plasma p-tau achieved promising results, showing higher concentration in AD patients than in control individuals and a strong correlation with CSF p-tau [71].

Advancements in ultrasensitive assays also enabled the accurate measurements of NfL levels, not only in CSF, but also in blood, revealing a tight correlation with its concentration in CSF. Therefore, blood NfL represents a well-replicated and reliable biomarker useful for screening neurodegenerative processes, monitoring disease progression or therapy [72–75]. Serum or plasma NfL levels were highly increased in AD and MCI cases when compared to controls and in other neurological disorders [16,72,74,75]. Interestingly, CSF and blood NfL levels are related with AD severity markers, including brain atrophy detected using MRI, glucose hypometabolism measured using FDG-PET and cognitive deterioration evaluated using MMSE, suggesting its use as a disease stage biomarker [72,76,77]. More importantly, blood NfL levels were increased in symptomatic familial AD cases but also in pre-symptomatic carriers of AD mutations and correlated with estimated years of symptom onset as well as both cognitive and MRI measures of AD stage [75,78]. Given all these findings, emerging agreements recommend the use of NfL instead of t-tau as an independent marker of neurodegeneration (N) in the AT(N) classification for AD [13].

Recent evidence has pointed out a role of flotillin as a novel AD biomarker [79,80]. This is a membrane-associated protein located in lipid rafts of intra- and extracellular vesicles; therefore, it plays important roles in signal transduction and membrane–protein interactions. Regarding AD pathogenesis, flotillin is involved in several pathological processes, such as APP processing and endocytosis, mitochondrial dysfunction, Aβ-induced neurotoxicity and neuronal apoptosis [80]. A clinical study reported that flotillin levels were decreased in both CSF and serum of AD patients compared with MCI individuals or age-matched non-AD controls. Moreover, flotillin levels in serum negatively correlated with Aβ burden detected using PET, whereas they remained stable with advancing age in healthy controls [79]. Despite the clinical evidence of a diagnostic utility of flotillin in AD being still in its infancy, emerging findings support that it may be used in support of CSF Aβ_{42} and tau levels as well as PET neuroimaging for more efficient and earlier diagnosis for AD [80]. A summary of levels of blood biomarkers is reported in Table 2.

Epigenetics is also of increasing interest in biomarker discovery, with gene regulation by micro(mi)RNAs representing one of the most investigated molecules [81–83]. Since miRNAs are dysregulated in the brain, CSF and blood, they may be used as diagnostic and prognostic biomarkers for AD. Several studies identified panels of miRNAs to discriminate AD patients from controls with a good specificity and sensitivity [84–87]. Most dysregulated miRNAs are associated with molecular mechanisms occurring during AD pathogenesis, such as inflammation, apoptosis, Aβ and tau signaling pathways [88], suggesting them as an alternative and more sensitive approach for detection and management of AD [89]. Although promising, the use of miRNAs in clinical practice still has several limitations, including variations in analytical platforms and different methods of data validation and normalization.

Table 2. Changes in levels of blood biomarkers in AD patients and controls.

Biomarker	AD	Controls	AD/Controls	Technique	Reference
$A\beta_{42}/A\beta_{40}$ ratio	0.04–0.08	0.05–0.1	↓	Simoa	[62]
$A\beta_{42}$ (pg/mL)	5.9–20.5	14.4–24.8	↓	Simoa	[62]
t-tau (pg/mL)	2.61–4.73	1.98–3.50	↑	Simoa	[67]
p-tau (pg/mL)	>0.0921		↑	ELISA	[71]
NfL (ng/L)	24.1–77.9	13.3–56.1	↑	ELISA	[72]
flotillin (% over controls)	−30		↓	WB	[79]

Simoa, single molecule array; ELISA, enzyme-linked immunosorbent assay; WB, Western blotting.

Finally, a number of additional emerging biomarkers have been investigated in various studies aiming to find a potential link with AD pathological processes, including glial activation, inflammation, neurodegeneration, Aβ pathology or degrading enzymes [43,90]. Although most of them displayed significant association with AD in the CSF, the corresponding levels in blood did not reflect such alteration [43].

2.2.2. Saliva

Saliva represents an attractive marker for monitoring diseases as its collection is non-invasive, convenient and cost-effective. Since salivary secretion decreases with aging, it is supposed that changes in biochemical composition may be related to the development of age-associated diseases. Interestingly, it has been reported that proteins from the CNS are excreted into the saliva [91]. Several studies have evaluated salivary Aβ levels but with conflicting results: increased or unaltered levels of $A\beta_{42}$ were found in AD patients as compared with controls [92]. Similarly to Aβ, tau protein was also investigated as a potential salivary biomarker for AD. Specifically, the p-tau/t-tau ratio was shown to be significantly increased in AD patients. Moreover, while data on salivary t-tau are consistently negative, p-tau species in saliva could have greater utility [93,94].

A recent study suggested the use of lactoferrin as a salivary biomarker with high sensitivity and specificity. Lactoferrin, one of the major antimicrobial peptides, is abundantly present in human saliva and shows Aβ-binding properties. Decreased levels of this peptide were detected in patients with both AD and amnestic mild cognitive impairment (aMCI) compared to healthy controls, resulting in 100% sensitivity and specificity. In addition, authors also reported a positive correlation with CSF $A\beta_{42}$, t-tau and the mini-mental state examination (MMSE) [95].

With contrasting results, several other salivary candidate biomarkers have been examined, including mucins, acetylcholinesterase and cortisol; therefore, it is necessary for data to be replicated in larger cohorts or longitudinal studies [96]. Although saliva seems to represent an interesting source of markers, its content may be influenced by the circadian rhythm, flow rate and time of sample collection. Moreover, the levels of biomarkers require normalization and are unstable because of the presence of degradative enzymes (Table 3).

2.2.3. Urine

In contrast with saliva, the use of urine as a diagnostic biomarker has the advantage of it being easily normalized on measured levels of creatinine, which is physiologically stable. Urine has so far represented a good marker source for the diagnosis and monitoring of renal dysfunctions and urinary tract and prostate cancers. As mentioned above, oxidative stress and oxidative DNA damage play an important role in the early stages of the disorder and are currently being explored for possible biomarkers in AD (Table 4). Specifically, ROS combines with mitochondrial and nuclear DNA to produce 8-hydroxy-2′-deoxyguanosine (8-OHdG), a marker used to monitor cellular dysfunction in urine. It has been reported that AD patients exhibit high urinary levels of 8-OHdG as compared with healthy elderly controls [97]. Another biomarker for oxidative injury is represented by isoprostane 8, 12-iso-iPF$_{2\alpha}$-IV, generated from arachidonic acid peroxidation by free radicals. Elevated levels of isoprostane 8, 12-iso-iPF$_{2\alpha}$-IV were found in all biofluids of AD patients, including CSF, urine and plasma [98].

Table 3. Changes in levels of salivary biomarkers in AD patients and controls.

Biomarker	AD	Controls	AD/Controls	Technique	Reference
Aβ$_{42}$ (pg/mL)	41–51.7	21.1–29	↑	ELISA	[92]
t-tau (pg/mL)	~11.5–14.5	~14–17	≅	ELISA	[93]
p-tau (pg/mL)	~90–140	~85–105	↑	ELISA	[93]
p-tau/t-tau ratio	~9–12	~6.5–7.5	↑	ELISA	[93]
lactoferrin (μg/mL)	3.67–5.89	8.28–12.20	↓	ELISA	[95]

ELISA, enzyme-linked immunosorbent assay.

Table 4. Changes in levels of urinary biomarkers in AD patients and controls.

Biomarker	AD	Controls	AD/Controls	Technique	Reference
8-OHdG (nmol/L)	99–159	16.5–33.1	↑	HPLC	[97]
isoprostane 8, 12-iso-iPF$_{2\alpha}$-IV (ng/mg creatinine)	4.51–5.35	1.60–1.94	↑	MS	[98]
AD7c-NTP (μg/mL)	>22		↑	ELISA	[99]
SPP1 (ng/mg total protein)	~8–10	~12–18	↑	ELISA	[100]
GSN (pg/mg total protein)	~1300–1800	~1000–1200	↑	ELISA	[100]
IGFBP7 (pg/mg total protein)	~6–8	~4.8–5.2	↑	ELISA	[100]

HPLC, high performance liquid chromatography; MS, mass spectrometry; ELISA, enzyme-linked immunosorbent assay.

Increasing evidence supports the use of urinary Alzheimer-associated neuronal thread protein (AD7c-NTP) as a potential candidate for AD early diagnosis. The transmembrane phosphoprotein AD7c-NTP co-localizes with NFTs and is positively associated with phosphorylated tau accumulation in CSF from AD patients. Several studies reported increased AD7c-NTP levels in CSF and urine in the early course of neurodegeneration in AD, which is positively associated with the disease severity. Moreover, a meta-analysis suggested a possible use of urinary AD7c-NTP in the early diagnosis of AD [99].

A recent study coupling computational and experimental methods revealed three differentially expressed proteins in the urine of AD patients: SPP1 (secreted phosphoprotein 1), GSN (gelsolin) and IGFBP7 (insulin-like growth factor-binding protein-7). Interestingly, all of them have already been reported to play an important role in AD pathogenesis. In AD models, SSP1 is involved in modulating the macrophage immunological profile towards a better capacity in mediating Aβ clearance. GSN binds to Aβ, solubilizing its existed fibrils or inhibiting Aβ fibrillization in the brain, whereas IGFBP7 contributes to brain cell homeostasis and is a critical mediator of memory function [100].

2.2.4. Tears

Tear samples have been already suggested as an excellent biomarker candidate, providing information not only on pathological conditions affecting the ocular system, but also on systemic pathophysiological processes [101,102]. Interestingly, Aβ plaques were found in the retina and lens of AD patients, as well as changes in the retinal morphology and vasculature, resulting in an impaired visual performance [103]. Thus, an alteration of the eye microenvironment may be reflected at the level of tear proteins. In a recent pilot study using quantitative proteomics techniques, the authors found a change of tear flow rate, total tear protein concentration and composition of the chemical barrier specific to AD (Table 5). Moreover, a very high accuracy in discriminating AD patients from healthy donor controls has been reached by the combination of a panel of proteins, including the extracellular glycoprotein lacritin, dermcidin, lipocalin-1 and lysozyme-C, which are mostly involved in the inflammatory processes and defense against pathogens [104].

Table 5. Changes in levels of biomarkers in tears from AD patients and controls.

Biomarker (Log2 Fold Change Over Controls)	AD	AD/Controls	Technique	Reference
extracellular glycoprotein lacritin	−2.04	↓	MS	[104]
dermcidin	0.85	↑	MS	[104]
lipocalin-1	−0.76	↓	MS	[104]
lysozyme-C	−1.11	↓	MS	[104]

MS, mass spectrometry.

3. Emerging AD "Dry" Biomarkers: Structural and Functional Techniques

Several brain-imaging and neurophysiological techniques could be used for studying morphological and functional changes occurring in AD [105]. Quantitative electroencephalography (EEG) and other neurophysiological biomarkers like event-related potentials and transcranial magnetic stimulation (TMS) have been used to predict MCI conversion to AD and for dementia differential diagnosis, but more research should examine their sensitivity and specificity for diagnostic purposes [106–108]. Morpho-functional imaging studies have been used according to the NIA-AA guidelines and have the advantage over fluid biomarkers to provide crucial disease-staging information, as imaging can distinguish the different phases of the disease both temporally and anatomically [12]. We could subdivide AD imaging techniques into two main categories: structural and functional imaging (Figure 3).

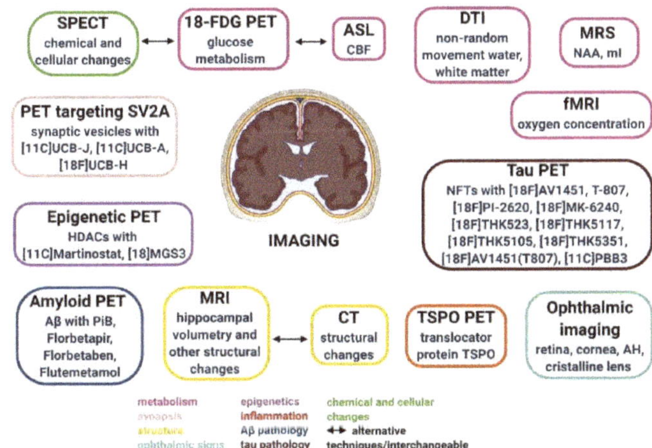

Figure 3. Neuroimaging techniques and their validated molecular targets for studying morphological and functional changes occurring in AD.

3.1. Structural Imaging

Structurally, AD patients show characteristic patterns of atrophy involving several structures of the medial temporal lobe (MTL) including the hippocampi, amygdala, the cingulate cortex, parahippocampal gyrus and the entorhinal cortex [109–112]. The two main techniques providing structural data on the brain of AD patients are computed tomography (CT) and MRI. Several guidelines suggest using one of them for evaluation of patients presenting with a cognitive/dementia syndrome (CDS) in the clinical setting [110]. Diffuse cerebral atrophy, with enlargement of the cortical sulci and increased size of ventricles are easily detectable in AD using both CT and MRI. However, despite the fact the CT imaging techniques are preferable to MRI in non-collaborative patients, where you need speed, for a minor cost, and also higher availability in developing nations, quantitative measurements are limited using CT. Moreover, CT is not sensitive for early changes, and so it is not considered as a standard biomarker for early diagnosis of AD.

3.1.1. MRI

The MRI technique is the most used in AD diagnosis due to its high spatial resolution, which allows the difference between two arbitrarily similar but not identical tissues and multiparametric acquisitions to be distinguished [113,114]. MRI does not involve X-rays or the use of ionizing radiation, it is non-invasive and has no significant adverse health effects. Moreover, this technique has revealed the ability to define a spectrum of changes related to vascular pathologies and white matter diseases in the brain and permits detection of microstructural and biochemically changes, thanks to diffusion tensor imaging (DTI) and MR spectroscopy (MRS) techniques, respectively, and functional studies as well.

Since 2008, Whitwell and colleagues, studying AD patients, demonstrated the existence of a correlation between MRI volumetric measurements of MTL and, in particular, the hippocampus and NFT accumulation in the lobes [115]. Hence, the measure of hippocampal volume has become a well-established biomarker for AD diagnosis and follow-up [116–118]. Some MRI studies have also shown a correlation between the extent of hippocampal and entorhinal volume decline and the increasing age predicted performance on memory tasks [119–122]. However, the underlying pathological mechanism is not clear. In particular, it could not be inferred that these changes are actually the result of cell loss with age, because it could also be plausible that they really are secondary to synaptic and dendritic loss.

Hippocampal volumetry has shown an important limit, which is the absence of great sensitivity during the prodromal stage of the disease. Therefore, new MR-derived biomarkers are needed with a higher sensitivity compared to whole hippocampal volumetry at an early disease stage. Moreover, clinical routine has shown that it is very challenging to quantify the degree of atrophy for each single case, while it is evident based on several reports in the literature that in large research studies based on group level analysis, an AD "signature" atrophy could be easily detected. As stated above, the evaluation of whole hippocampal volume was not a well-suited biomarker during the pre-clinical AD stage. Several papers reported contradictory findings, potentially due to the use of different approaches in performing the analyses [123]. This could be related to the fact that the manual volumetric technique is a very challenging and time-consuming approach also requiring an excellent working knowledge of neuroanatomy as well as good skill in delineating regions of interest (ROI). Several studies on hippocampus volumetry in middle-aged adults below the age of 55 years reported a contradictory pattern. In particular, while Okonkwo and colleagues found characteristic left posterior hippocampal atrophy [124] and O'Dwyer et al. reported reduced right hippocampal volume in young APOE carriers [125]; others were not able to find differences between the high- and low-risk groups [119]. To overcome difficulties in whole hippocampal volumetry analysis, more recent studies have used segmentation techniques to quantify volumes of each functionally specialized subfield of the hippocampus, which has discrete histological characteristics. This approach demonstrated a higher sensitivity in capturing subtle atrophy patterns compared to whole hippocampal volumetry [126]. A recent work suggested that surface deflections across all hippocampal subfields (CA1 lateral zone, dentate gyrus/CA2-4 superior zone and subiculum inferior medial zone) could be used as biomarkers to differentiate early AD patients from non-demented controls [127].

Currently, the most recurrent volumetry assessment tools and image analysis methods for AD include manual tracing for image processing analysis and visualization; automated Voxel-based morphometry (VBM) for investigation of focal differences in brain anatomy (it allows healthy controls and patients to be distinguished based on the volume of brain and ROI) [128,129]; individual brain atlases using statistical parametric mapping (IBASPM) to automatically create segments of cerebral structures and compute the volume of gross anatomical structures to distinguish an AD group from a normal control group; the insight segmentation and registration toolkit-SNake automatic partitioning (ITK-SNAP), which is a semi-automated 3D brain segmentation technique; and FreeSurfer, which automatically identifies and measures brain regions detecting hippocampal atrophy in patients and providing information about the shape, position or volume of brain tissue.

FreeSurfer software is one of the two main methods for hippocampal segmentation. The other one is automatic segmentation of hippocampal subfields (ASHS). Both support multispectral segmentation relying on atlases based on histological information and high-resolution ex vivo MRI scans. However, the use of the segmentation methods has not yet demonstrated its validity in the study of healthy adults at risk of AD as the results have been inconclusive [130]. Very recently, an MRI biomarker for in vivo AD diagnosis based on the use of FreeSurfer and a supervised machine learning approach was reported [131]. Based on an individual's pattern of brain atrophy, a continuous AD score is assigned which measures the similarity with brain atrophy patterns seen in clinical cases of AD [131].

Besides hippocampal volume, structural MRI also allows cortical thinning to be studied in the entorhinal cortex (EC), another biomarker identified as a highly sensitive measure of structural change both in MCI and AD [132]. The decline in EC thickness occurs earlier and could be used as a predictive marker of hippocampus volume decline.

Structural MRI methodological limitations are, in particular, susceptibility to movement, differences in spatial resolution across scans that make it difficult to perform a comparison and difficulties in determining the real neural source of volume or thickness loss (cell loss vs. dendritic and synaptic loss) in the absence of high-resolution scanning. This is a concern due to the fact that high resolution and post-processing studies are not feasible in many institutions.

3.1.2. Diffusion Tensor Imaging (DTI) and Biomarker of White Matter Integrity

DTI is an advanced MRI which measures non-random movement of water molecules and also tracks the fibers of tracts within the brain. It has been used to study the microstructural features of white matter [133]. In AD, these studies showed fiber alteration in the temporal and frontal lobes and also corpus callosum and posterior cingulate gyrus [134] with a posterior to anterior gradient [135]. A modest diagnostic power in discriminating AD from controls has been reported in a meta-analysis using DTI measurements of limbic regions [136].

DTI could be used in clinical trials to monitor response to disease modifying drugs or as an indicator of disease progression owing to the fact that functional modifications are detectable prior to structural abnormalities. However, DTI use in AD diagnosis is at the early stages of evaluation and requires further studies. DTI has a number of technical limits that reduce its utility in clinical settings: there is significant variability of DTI-based diffusion metrics between MRI scanners, and traditional approaches cannot resolve intra-voxel complexities such as fiber bending, crossing and twisting [137]. The latter limits may be overcome in the future using high angular resolution diffusion imaging (HARDI), which provides correct information to model diffusion with an orientation distribution function (ODF), a more versatile diffusion representation that captures multiple orientations in a voxel [138].

3.1.3. Proton Magnetic Resonance Spectroscopy (MRS) and Metabolic Markers of AD

MRS is an imaging method for which three decades of research indicate a potential role as a biochemical imaging marker in AD [139]. MRS aims to measure chemical concentration in the brain. The most common compounds analyzed in MRS are represented by N-acetylaspartate (NAA), choline (Chol), myoinositol (mI), glutamate plus glutamine (Glx) and creatine (Cr). The last one is generally considered as an internal reference due to it having no significant changes in different conditions, and MRS is then reported as a ratio of the tested metabolite over Cr [140].

Since 1992, a reduction in NAA neuronal metabolite has been demonstrated using MRS performed on autopsy brain samples of AD patients in respect to levels detected in healthy controls. The observed decrease of NAA was also demonstrated to correlate with the amount of tangle and plaque in the brain [141]. The decrease in NAA or NAA/Cr has been then considered as a marker of neuroaxonal density and viability [142]. AD patients show reduced NAA/Cr ratio in posterior cingulate voxels [143] and in the medial temporal lobe [144] and hippocampi [145] compared to controls. The same decrement is not detectable in MCI patients because these areas are not yet involved in the neurodegeneration [142].

Myoinositol, a glial marker, was also analyzed in AD patients through MRS and its increase was detected in several brain regions. This is thought to be an early event in the course of AD pathology preceding NAA reduction [142].

Choline has also been analyzed in AD as a possible biomarker in MRS studies. However, contrasting data have been reported [145–151]. Inconsistent changes were also reported using MRS testing GLX [143,152]. The advantages of MRS compared to other functional imaging techniques are that it is more widely available, much less expensive, has no radiation risk and can be added to the structural MRI sequences to extract useful information to help diagnosis. MRS can be used as a follow-up imaging tool in therapeutic trials correlating the level of metabolites and pathological changes, or it can be used as a tool to predict cognitive impairment in combination with other biomarkers.

However, MRS is still not routinely used clinically because further research is required to standardize the techniques, to compare the results of MRS with other functional biomarkers and to better understand the pathological substrates for metabolite abnormalities.

3.2. *Functional Imaging*

Functional imaging includes recently developed techniques not yet well applied in worldwide routine clinical settings. In particular, these techniques include PET with different tracers, single photon emission computed tomography (SPECT) and advanced MRI techniques such as functional MRI (fMRI) and arterial spin labeling (ASL).

3.2.1. Single Photon Emission Computed Tomography (SPECT) and Perfusion Imaging

Another molecular technique useful in discriminating individuals with AD is SPECT, which is able to detect both chemical and cellular changes linked to a disease through the use of highly targeted radiotracers [153]. SPECT evaluates cerebral perfusion and shows good correlation with metabolic changes. Perfusion hexamethylpropyleneamine oxime (HMPAO)-SPECT is seen as an alternative to 18F-FDG PET even though SPECT has lower sensitivity and specificity than FDG for the diagnosis of AD [154]. Abnormalities on perfusion SPECT in AD are represented by hypoperfusions commonly affecting temporoparietal areas in a bilateral distribution, with the posterior cingulate and medial temporal areas particularly affected in AD and the sensory motor cortices, including the cerebellum, largely spared [155]. Studies of the accuracy of SPECT for diagnosing AD report sensitivities of 65–85% and specificities (for other dementias) of 72–87% [156]. SPECT has the advantages of being more widely available, cheaper than FDG PET and better tolerated by the patient. (HMPAO)-SPECT was demonstrated to have less diagnostic accuracy than FDG PET; thus, it could be helpful for the dementia/no-dementia comparison but not in differentiating AD from DLB [154].

3.2.2. PET

PET scanning is a well-established molecular imaging technique resulting in 3D images of what is happening in a patient's brain at the molecular and cellular level [157]. It is very accurate at diagnosing AD and differentiating it from other dementias. Consensus agreements suggest the use of PET biomarkers in patients suspected to be affected by AD without meeting NIA-AA criteria for dementia and with a well-defined clinical and cognitive profile. In these cases, PET imaging might support or exclude the clinically suspected etiology avoiding the unnecessary use of costly and potentially harmful treatments and allowing a rapid and accurate diagnosis. This has important positive impacts on the patient, his family and society [158].

Several tracers with different specificity have been developed to study AD patterns in different stages of severity. In particular, ligands were designed to study Aβ deposition, synaptic dysfunction (mainly cortical hypometabolism) or tau fibrillary tangle deposition.

In the following sections we will briefly describe the most used PET ligands having diagnostic and prognostic significance in AD.

PET and Metabolic Biomarkers

18F-FDG PET measures the glucose metabolism in different regions of the brain and represents a metabolic marker in early diagnosis and preclinical detection of dementia, metabolic changes usually becoming impaired before the appearance of detectable structural changes in the brain. This technique, which also points to vascular deficits and impairment in the blood-brain barrier frequently found in AD, can be used when the clinical diagnosis still remains in doubt to identify the causes of dementia and to have a differential diagnosis. Hypometabolism in 18F-FDG PET is considered a biomarker of neuronal dysfunction, neurodegeneration and synaptic disease.

Cerebral glucose hypometabolism has been consistently demonstrated in the medial temporal lobe, posterior cingulate gyrus, parieto-temporal regions and/or frontal cortices of typical AD patients, while it has been detected to be moderate in the basal ganglia, cerebellum, thalamus and sensorimotor and visual cortices [159,160]. The observed metabolic changes associated with neocortical dysfunctions may predict successive atrophy and suggest that a conversion of MCI to AD could occur within several years [161,162]. Moreover, a correlation was reported between the performance of patients on a cognitive test and the hypometabolic patterns during disease progress [163] and clinical symptoms of cognitive impairment [164].

The use of FDG PET instead of structural MRI is suggested in rapidly converting early MCI individuals, while structural MRI may outperform FDG PET in late MCI subjects [165,166]. The FDG PET technique sensitivity as well as specificity are very high (>90%) in discriminating healthy elderly individuals from AD, while the specificity was reported to be reduced (78%) in differential diagnosis of AD and other dementia [167]. The use of FDG PET is limited by the fact that it is an expensive technique that also requires exposure to ionizing radiation.

Like in the case of MRI reported above, PET images can also be assessed in different way. One of them is visual inspection, but the efficacy of this largely depends on the experience of the reader, and it might be very challenging in the study of patients with mild disease due to a lack of clear cut-off between normal and abnormal values. In these cases, the use of standardized uptake value ratio (SUVr) quantification is usually suggested instead of visual inspection. SUVr utilizes static imaging and is evaluated in respect to a region without altered metabolism or with mildly affected metabolism in AD [168].

PET and Amyloid Imaging Biomarkers

In the past decades, the possibility of direct imaging of AD brain lesions and in particular of the presence of Aβ aggregates has been investigated and several tracers with different half-lives and affinities to plaques have been engineered to be used for this purpose [169].

A lot of studies on Aβ deposition in AD are now available linking this pathological finding with several other correlates such as aberrant entorhinal activity among cognitively normal older adults and cortical thinning in frontoparietal regions [170,171]. Moreover, it was reported that Aβ deposition may predict tau deposition during aging [172]. In fact, amyloid imaging studies highlighted that the deposition onset usually starts in the posterior cingulate, restrosplenial cortex and precuneus regions [111]. These brain regions are interconnected with the MTL, which, on the contrary, is a site for early tau aggregation, and this suggests the possibility of an influence of anatomical and functional connectivity between all these brain areas in the progression of the disease.

Amyloid imaging uses PET technology to acquire images of the brain in order to display foci of abnormal amyloid accumulation after an injection of a radiolabelled ligand specifically targeting amyloid aggregates. Among tracers, the first ligand developed was the Pittsburg compound B (PiB), a fluorescent derivative of the amyloid-binding histological dye, thioflavin-T [173]. Being the first reported, it is also the best characterized amyloid tracer. It was demonstrated that it can bind selectively with high affinity to fibrillar Aβ aggregates but not to amorphous Aβ deposits or NFTs [170,171]. In fact, studies reported in the literature demonstrate a high affinity of PiB to frontal, temporo-parietal and posterior cingulate cortices, while minimal binding was reported in the cerebellum and other regions with typical low density of Aβ plaques [174,175].

The use of the PiB tracer has a number of limitations including a very short life (around 20 min), the need of an on-site cyclotron and 11C radiochemistry expertise, the fact that it has also displayed retention in the brain of nondemented people [176] and high affinity for vascular deposits that could also be observed in non-AD conditions [177] and, finally, it has low ability to detect soluble oligomeric Aβ conformations considered to be highly pathogenic [178].

Due to these limitations in the use of PiB, newer fluorinated tracers with longer half-lives have been developed and are increasingly used. They include Florbetapir F 18-Florbetapir F 18 (18F-AV-45), Florbetaben (BAY 94-9172) and 18F Flutemetamol (Flute), which is structurally identical to PIB except for one fluorine atom in the place of a carbon atom [179].

Despite the common agreement in considering amyloid PET as the most specific and sensitive biomarker for AD, its usefulness in clinical settings is under evaluation [180]. It is suggested that this technique should be used only in dementia expert centers to confirm AD diagnosis in atypical cases or in the differential diagnosis between amyloid-associated dementia and non-amyloid pathology. Notably, this technique is not able to discriminate among stages of dementia progression and has the caveat in preclinical AD studies that 10–30% of cognitively normal individuals can have positive amyloid PET [176,181].

PET and Biomarkers of Synaptic Damage or Loss

The existence of a strong association of synapses and AD pathophysiology is well known. Thus, biomarkers of synapse damage or loss as proxies for synaptic and cognitive function in AD have been studied. Very recently, new PET ligands ([11C]UCB-J, [11C]UCB-A and [18F]UCB-H) have been developed labelling the synaptic vesicle glycoprotein 2A (SV2A) and allowing synapses to be visualized in the living brain [182–184]. In particular, the use in PET scanning of [11C]UCB-J demonstrated a reduction of approximately 40% of SV2A signal in the hippocampus of AD cases in respect to cognitively healthy aged cases [185]. Although very few studies have been reported using this approach to measure synapse loss longitudinally in AD, it appears that a direct measure of synapse density is a very promising biomarker to be used probably in combination with one of the previously reported biomarkers in CSF or MRI or FDG PET imaging biomarkers.

PET and Tau Biomarkers

The presence of NFTs, p-tau protein aggregates, in AD suggests the use of tau imaging as a surrogate marker to predict cognitive decline or disease progression in AD.

Since 2002, selective tau PET tracers have been developed. The first tracers were quinolone and benzimidazole derivatives with affinity to the PHF among tau aggregates. This target was optimal for the use of the ligand because it co-occurs with Aβ. This created an additional challenge as the ligand could bind both PHF and Aβ. However, a 25-fold selectivity for PHF over Aβ was achieved [186].

Among the first generation tau selective PET tracers, the two mostly studied were [18F]AV1451 and T-807. Beside the advantage of being able to replicate features of the Braak histopathological changes, these tracers had two important limitations: they had increased striatal retention and they had off-target binding to monoamine oxidase A/B [187]. Then, they were quickly substituted by a second generation of new targets without off-target binding such as [18F]PI-2620 and [18F]MK-6240 [188].

Through the use of these new ligands, a specific tau PET signature was reported in AD patients starting from the transentorhinal/entorhinal cortex to the hippocampus and then extending to the rest of the temporal lobe and neocortical regions [189]. This stepwise pattern typical in AD is not commonly detected in cognitively normal individuals and could be then be used in staging the disease.

In more recent years, additional different tau selective PET tracers have been engineered to be used for human studies: [18F]THK523, [18F]THK5117, [18F]THK5105, [18F]THK5351, [18F]AV1451(T807) and [11C]PBB3 [190]. Despite the reported results being promising up to now, the characteristics of tau intracellular aggregates, which are subjected to conformational changes and coexist with a high concentration of amyloid deposition, complicate the use of tau imaging for clinical use. For future use of this approach in the clinical setting, tau PET tracers must be developed with a higher selectivity and affinity to tau compared to Aβ and they must be able to cross the blood-brain barrier without being metabolized [186].

Novel PET Approaches: Inflammatory and Epigenetic Biomarkers

New imaging approaches to study AD are being developed. One of them is TSPO-PET, a PET technique performed using a specific ligand for the translocator protein TSPO (previously named peripheral benzodiazepine receptor), a mitochondrial membrane protein involved in several functions relevant to neurodegeneration and upregulated in neuroinflammation. Increased TSPO expression was detected in AD animal models, and it was demonstrated to correctly overlap with brain pathological areas as well as with regions of increased immunohistochemical staining of TSPO [191]. Moreover, the study of TSPO PET signals in preclinical trials of novel therapeutic AD models suggested that it could be used as a biomarker to monitor treatment progress in clinical trials [192]. However, new TSPO PET radioligands are needed due to the fact that those currently available lack high affinity to a prevalent polymorphic form of TSPO (A147T) [193].

Another novel PET approach to AD is imaging epigenetics ([11C]Martinostat PET). Epigenetics consists of several newly detected mechanisms involved in regulating gene expression. The two main epigenetic mechanisms, both modifying chromatin structure, are DNA methylation and histone acetylation. The latter, and in particular the involved histone deacetylase (HDACs), is the mechanism targeted by [11C]Martinostat PET. In fact, the Martinostat radiotracer was designed with high specific binding of a subset of class I HDAC enzymes (isoforms 1, 2 and 3) and thus is able to image HDAC density with favorable kinetics and high affinity [194]. The rationale for the use of this PET approach for studying AD resides in the observation that epigenetics has a role in the dynamic process of learning and memory and that it is altered by AD pathology. However, there is not a clear causal-consequence correlation between epigenetics and AD owing to the observation that some epigenetic changes arise after AD onset [195,196] while other changes arise before the disease presentation [197]. HDAC expression was reported to be lowest in the amygdala and hippocampus among gray matter regions in healthy adults [198]. A new epigenetic PET tracer has been recently developed (the fluorinated variant of [18F]MGS3) showing specific binding, comparable brain uptake and regional distribution to [11C]Martinostat, but due to technical limits in the radiosynthesis process no validation has been done [199]. Epigenetic imaging is still a new technique and will be further improved, but it might become useful in analyzing gene regulatory processes underlying AD and might be considered in the future as a biomarker for AD.

3.2.3. fMRI and Metabolic Markers

Functional MRI (fMRI) consists of the acquisition of brain images during a specific brain activity and in a basal state, and it measures the oxygen concentration of certain specific brain areas corresponding to particular stimuli or cognitive tasks [200,201]. fMRI studies showed a decrease in coordinated activity of AD patients in the hippocampus and inferior parietal lobes and cingulate cortex compared with healthy controls [202,203]. fMRI can be performed in resting state to study synaptic integrity and circuit connectivity or be performed in task-activated state to study reduced inhibition and hippocampal hyperactivity.

In preclinical AD, resting state fMRI (rsfMRI) has been linked to metabolic changes and precedes neurodegeneration (reviewed by [204]). Most rsfMRI analyses focused on the default mode network (DMN) that involves the medial prefrontal cortex, posterior cingulate cortex, precuneus, anterior cingulate cortex, parietal cortex, and the medial temporal lobe, including the hippocampus. These studies demonstrated widespread changes in DMN connectivity in MCI and AD [205–208]. Other studies pointed also to a disruption in the connectivity within the MTL (between the entorhinal cortex and the dentate and CA3 regions of the hippocampus) [209] and in other networks [210]. Functional connectivity is then considered an early marker of synaptic pathology due to isolation of the hippocampus from its cortical input.

Task-activated fMRI was also used in several studies and resulted in a number of observations: mild AD cases have a decrease in hippocampal activity similar to more impaired MCI patients [211]; the extent of hippocampal hyperactivation at baseline predicted cognitive decline as measured by the CDR-SB scores over four years after scanning [212]; the hyperactivation is specific to the DG/CA3 subregions of the hippocampus [213].

3.2.4. Arterial Spin Labelling ASL and Metabolic Markers

ASL perfusion MRI is a non-invasive technique for measuring cerebral blood flow (CBF) using magnetically labelled arterial blood water protons as endogenous tracers [214]. This technique has been studied to evaluate the possibility of using it instead of FDG PET in AD diagnosis, ASL being cheaper, with no radiation risk and more widely available. These studies revealed a good concordance of ASL and FDG PET results with comparable diagnostic accuracy [215]. Given the possibility of adding ASL to routine structural MRI protocols without additional cost, it appears a promising method for classifying the degree of neurodegeneration in individuals with prodromal AD. However, some limitations that are still present (e.g., the sensitivity to blood velocity and arterial transit time, the interferences occurring in the presence of steno-occlusive disease or other cerebrovascular pathology and the sensibility to head motion) are still reducing its clinical use [216].

4. Ocular Biomarkers of AD

Very recently, several groups started to consider the ocular examination as new non-invasive tool for AD screening and diagnosis. The rationale for the focus on eyes is that they have an origin that is similar to the brain's and that neurons in the cerebral cortex are similar to those in the retina [217–219]. Changes in the neural and non-neural ocular tissues have been reported in AD [220]. These include accumulation of Aβ and degeneration of retinal axonal and neural tissue.

In the cornea of AD patients, several abnormalities have been reported including corneal lattice dystrophies, degeneration of the corneal sub-basal nerve plexus, corneal nerve fiber pathology and decreased corneal sensation [221–223]. The detection in the aqueous humor (AH) of Aβ in patients with pseudoexfoliation syndrome and glaucoma suggested a link between these conditions and AD and the usefulness of searching for APP and Aβ in the AH as a biomarker of the disease [224]. Studies are needed to check whether AD has a direct or indirect neurotoxicity on the corneal nerve plexus. A correlation between AD and changes in the crystalline lens was also reported in a number of studies, making it a promising biomarker tissue for disease diagnosis. In particular, patients show a progressive loss in the transparency and cataractous changes of the otherwise transparent crystalline human lens [102].

Changes have also been reported in several aspects of the retina. In particular, retinal vessels of AD patients show reduced caliber of retinal veins, reduced arteriolar and venular fractal dimensions, smaller, more tortuous and sparse retinal vessels and reduced flow of blood [225,226]. Moreover, a thinning of the retinal nerve fiber layer and the degeneration of the retinal ganglion cell were also reported in AD patients [227,228]. An optical retinal imaging platform has also been developed to detect Aβ plaques in retina in AD [229].

The use of ocular biomarkers will be very helpful in the future in diagnosing AD at an early stage or to follow-up the patient during treatment with therapeutic agents. However, their implementation to the population for screening purpose remains a challenge.

5. Conclusions

Despite the high diagnostic accuracy of the established CSF biomarkers for AD, Aβ42, t-tau and p-tau, several other candidates in alternative non-invasive biological fluids have been recently investigated for their potential clinical use in support of early AD diagnosis and prognosis (Table 6). Advancements in ultrasensitive assays enabled the precise measurements of analytes and decreased the intra- and inter-laboratory variation, helping to identify novel candidate biomarkers as well as to support harmonization efforts for the core biomarkers for AD. An increasing number of molecules have been identified so far and the highest potential was reached for NfL in both CSF and blood, lactoferrin in saliva and $A\beta_{42}$ and p-tau in plasma. Nevertheless, there are several issues related to most of the presented potential AD fluidic biomarkers: they often come from a single study or there is significant inconsistency in the results from different studies. Additionally, pre-analytical sample processing should be standardized, analytical methods validated and the impact of other factors, such as age, presence of comorbidities and disease diagnosis/stage, must be thoroughly evaluated. Advanced imaging techniques partially overcome these limitations, allowing the identification of AD-related structural and functional biomarkers (Table 7). They provide easily interpretable data for determining AD stage, giving a very high accuracy for disease diagnosis. The useful in clinical settings of several imaging approaches and novel biomarkers is still under evaluation. In fact, the use of each technique shows both limits and advantages and additional studies are required to define which one could be the best. Moreover, brain imaging is expensive, time-consuming and the equipment is not always available, thus limiting access to most of population. In this context, the recent developed biosensors are emerging as promising alternatives for rapid, low-cost and simple diagnosis for AD, even in the early stages. With high sensitivity and selectivity, they represent excellent analytical tools which have applications in detecting AD biomarkers, being applicable to easily sampled biological fluids, including blood, urine and tears.

Table 6. Main advantages and disadvantages of biological fluid biomarkers for Alzheimer's disease.

Fluid	Advantages	Disadvantages
CSF	Reliability	Invasiveness
blood	accessibility, reproducibility	need for validation, low protein concentration, difficulty of detection due to the presence of proteases or protein carriers
saliva	accessibility, cost-effectiveness	requirement of normalization, influence by circadian rhythm, flow rate and time of sample collection
urine	ease of normalization on creatinine	need of validation in larger and longitudinal studies
tears	information on systemic pathophysiological processes	need for validation on large scale populations

It is now commonly ascertained that the best way to adequately make a diagnosis of AD and the staging of the disease progress is to combine two or more of the above-reported biomarkers, particularly mixing together fluidic molecular analysis and imaging studies. The choice of which specific biomarkers and techniques to combine depends on the case under evaluation. In particular, different combinations will be properly used for studying AD patients at different stages of the disease or in monitoring the course of AD during drug treatment and clinical trials or finally in differential diagnosis. This is due to the fact that each biomarker and technique has different efficacy in the diagnosis, prognosis and staging of the disease. However, great attention still needs to be paid to several aspects of study design, sample collection, sample measurement and data analysis, and international collaboration to standardize assays and study protocols, as well as to recruit sufficiently large cohorts, will facilitate future biomarker discovery and development.

Table 7. Summary of characteristics of main "dry" biomarkers for Alzheimer's disease.

Imaging Biomarker	Type of Information	Early Changes Predictive for AD	Limits
MRI Imaging	N		need patients' collaboration time-consuming analyses
Morphometric technique Hippocampal segmentation DTI		most studied, good predictive value promising tool, more studies are needed	
Functional technique fMRI ALS		promising tool, more studies are needed more studies are needed	
Nuclear Medicine Imaging			
Perfusion SPECT PET Imaging FDG PET	N		invasiveness, ionizing radiation invasiveness, ionizing radiation expansive,
Amyloid PET Tau PET	N P P		invasiveness, ionizing radiation expansive low specificity small studies

D, Aβ dysfunction; N, neurodegeneration; P, pathognomonic; AD, pathological deposition of amyloid or fibrillary tau.

Studies should still be addressed towards the identification of a non-invasive biomarker for predicting AD before the onset of symptoms. Despite extensive research worldwide, no diagnostic method is currently available for pre-clinical AD and the existing treatments are only symptomatic.

Author Contributions: C.V. and R.C. carried out the literature review and conceptualized and wrote the manuscript. M.L. and E.S. edited the manuscript. All authors have read and agreed to the published version of the manuscript.

Funding: This work was funded by the Italian Ministry for University and Scientific Research (MOLIM OncoBrain LAB project to M.L.).

Acknowledgments: All the figures were created with BioRender.com.

Conflicts of Interest: The authors declare no conflicts of interest.

References

1. Crous-Bou, M.; Minguillón, C.; Gramunt, N.; Molinuevo, J.L. Alzheimer's disease prevention: From risk factors to early intervention. *Alzheimers Res. Ther.* **2017**, *9*, 71. [CrossRef]
2. Anand, R.; Gill, K.D.; Mahdi, A.A. Therapeutics of Alzheimer's disease: Past, present and future. *Neuropharmacology* **2014**, *76 Pt A*, 27–50. [CrossRef]
3. Huang, Y.; Mucke, L. Alzheimer mechanisms and therapeutic strategies. *Cell* **2012**, *148*, 1204–1222. [CrossRef] [PubMed]
4. Karran, E.; Mercken, M.; De Strooper, B. The amyloid cascade hypothesis for Alzheimer's disease: An appraisal for the development of therapeutics. *Nat. Rev. Drug Discov.* **2011**, *10*, 698–712. [CrossRef] [PubMed]
5. Reddy, P.H. A Critical Assessment of Research on Neurotransmitters in Alzheimer's Disease. *J. Alzheimers Dis.* **2017**, *57*, 969–974. [CrossRef] [PubMed]
6. Calsolaro, V.; Edison, P. Neuroinflammation in Alzheimer's disease: Current evidence and future directions. *Alzheimers Dement.* **2016**, *12*, 719–732. [CrossRef] [PubMed]
7. Liu, C.C.; Liu, C.C.; Kanekiyo, T.; Xu, H.; Bu, G. Apolipoprotein E and Alzheimer disease: Risk, mechanisms and therapy. *Nat. Rev. Neurol.* **2013**, *9*, 106–118. [CrossRef]

8. D'Argenio, V.; Sartanaro, D. New insights into the molecular bases of familial Alzheimer's disease. *J. Pers. Med.* **2020**, *10*, 26. [CrossRef]
9. Bachurin, S.O.; Bovina, E.V.; Ustyugov, A.A. Drugs in Clinical Trials for Alzheimer's Disease: The Major Trends. *Med. Res. Rev.* **2017**, *37*, 1186–1225. [CrossRef]
10. Mattsson, N.; Carrillo, M.C.; Dean, R.A.; Devous, M.D., Sr.; Nikolcheva, T.; Pesini, P.; Salter, H.; Potter, W.Z.; Sperling, R.S.; Bateman, R.J.; et al. Revolutionizing Alzheimer's disease and clinical trials through biomarkers. *Alzheimers Dement (Amst.)* **2015**, *1*, 412–419. [CrossRef]
11. Hampel, H.; O'Bryant, S.E.; Durrleman, S.; Younesi, E.; Rojkova, K.; Escott-Price, V.; Corvol, J.C.; Broich, K.; Dubois, B.; Lista, S.; et al. A Precision Medicine Initiative for Alzheimer's disease: The road ahead to biomarker-guided integrative disease modeling. *Climacteric* **2017**, *20*, 107–118. [CrossRef] [PubMed]
12. Jack, C.R., Jr.; Bennett, D.A.; Blennow, K.; Carrillo, M.C.; Dunn, B.; Haeberlein, S.B.; Holtzman, D.M.; Jagust, W.; Jessen, F.; Karlawish, J.; et al. NIA-AA Research Framework: Toward a biological definition of Alzheimer's disease. *Alzheimers Dement.* **2018**, *14*, 535–562. [CrossRef] [PubMed]
13. Jack, C.R., Jr.; Bennett, D.A.; Blennow, K.; Carrillo, M.C.; Feldman, H.H.; Frisoni, G.B.; Hampel, H.; Jagust, W.J.; Johnson, K.A.; Knopman, D.S.; et al. A/T/N: An unbiased descriptive classification scheme for Alzheimer disease biomarkers. *Neurology* **2016**, *87*, 539–547. [CrossRef] [PubMed]
14. Forgrave, L.M.; Ma, M.; Best, J.R.; DeMarco, M.L. The diagnostic performance of neurofilament light chain in CSF and blood for Alzheimer's disease, frontotemporal dementia, and amyotrophic lateral sclerosis: A systematic review and meta-analysis. *Alzheimers Dement. (Amst.)* **2019**, *11*, 730–743. [CrossRef]
15. Osborn, K.E.; Khan, O.A.; Kresge, H.A.; Bown, C.W.; Liu, D.; Moore, E.E.; Gifford, K.A.; Acosta, L.M.Y.; Bell, S.P.; Hohman, T.J.; et al. Cerebrospinal fluid and plasma neurofilament light relate to abnormal cognition. *Alzheimers Dement. (Amst.)* **2019**, *11*, 700–709. [CrossRef]
16. Palmqvist, S.; Janelidze, S.; Stomrud, E.; Zetterberg, H.; Karl, J.; Zink, K.; Bittner, T.; Mattsson, N.; Eichenlaub, U.; Blennow, K.; et al. Performance of Fully Automated Plasma Assays as Screening Tests for Alzheimer Disease-Related β-Amyloid Status. *JAMA Neurol.* **2019**, *76*, 1060–1069. [CrossRef]
17. Marizzoni, M.; Ferrari, C.; Babiloni, C.; Albani, D.; Barkhof, F.; Cavaliere, L.; Didic, M.; Forloni, G.; Fusco, F.; Galluzzi, S.; et al. CSF cutoffs for MCI due to AD depend on APOEε4 carrier status. *Neurobiol. Aging* **2020**, *89*, 55–62. [CrossRef]
18. Dhiman, K.; Gupta, V.B.; Villemagne, V.L.; Eratne, D.; Graham, P.L.; Fowler, C.; Bourgeat, P.; Li, Q.X.; Collins, S.; Bush, A.I.; et al. Cerebrospinal fluid neurofilament light concentration predicts brain atrophy and cognition in Alzheimer's disease. *Alzheimers Dement. (Amst.)* **2020**, *12*, e12005. [CrossRef]
19. Schmidt, F.M.; Mergl, R.; Stach, B.; Jahn, I.; Gertz, H.J.; Schönknecht, P. Elevated levels of cerebrospinal fluid neuron-specific enolase (NSE) in Alzheimer's disease. *Neurosci. Lett.* **2014**, *570*, 81–85. [CrossRef]
20. Llorens, F.; Schmitz, M.; Knipper, T.; Schmidt, C.; Lange, P.; Fischer, A.; Hermann, P.; Zerr, I. Cerebrospinal Fluid Biomarkers of Alzheimer's Disease Show Different but Partially Overlapping Profile Compared to Vascular Dementia. *Front. Aging Neurosci.* **2017**, *9*, 289. [CrossRef]
21. Lista, S.; Hampel, H. Synaptic degeneration and neurogranin in the pathophysiology of Alzheimer's disease. *Expert Rev. Neurother.* **2017**, *17*, 47–57. [CrossRef] [PubMed]
22. Kvartsberg, H.; Duits, F.H.; Ingelsson, M.; Andreasen, N.; Öhrfelt, A.; Andersson, K.; Brinkmalm, G.; Lannfelt, L.; Minthon, L.; Hansson, O.; et al. Cerebrospinal fluid levels of the synaptic protein neurogranin correlates with cognitive decline in prodromal Alzheimer's disease. *Alzheimers Dement.* **2015**, *11*, 1180–1190. [CrossRef]
23. Blennow, K.; Zetterberg, H. The Past and the Future of Alzheimer's Disease Fluid Biomarkers. *J. Alzheimers Dis.* **2018**, *62*, 1125–1140. [CrossRef]
24. Kvartsberg, H.; Lashley, T.; Murray, C.E.; Brinkmalm, G.; Cullen, N.C.; Höglund, K.; Zetterberg, H.; Blennow, K.; Portelius, E. The intact postsynaptic protein neurogranin is reduced in brain tissue from patients with familial and sporadic Alzheimer's disease. *Acta Neuropathol.* **2019**, *137*, 89–102. [CrossRef] [PubMed]
25. Willemse, E.A.J.; De Vos, A.; Herries, E.M.; Andreasson, U.; Engelborghs, S.; Van der Flier, W.M.; Scheltens, P.; Crimmins, D.; Ladenson, J.H.; Vanmechelen, E.; et al. Neurogranin as Cerebrospinal Fluid Biomarker for Alzheimer Disease: An Assay Comparison Study. *Clin. Chem.* **2018**, *64*, 927–937. [CrossRef] [PubMed]

26. Portelius, E.; Zetterberg, H.; Skillbäck, T.; Törnqvist, U.; Andreasson, U.; Trojanowski, J.Q.; Weiner, M.W.; Shaw, L.M.; Mattsson, N.; Blennow, K. Alzheimer's Disease Neuroimaging Initiative. Cerebrospinal fluid neurogranin: Relation to cognition and neurodegeneration in Alzheimer's disease. *Brain* **2015**, *138 Pt 11*, 3373–3385. [CrossRef]
27. Wellington, H.; Paterson, R.W.; Portelius, E.; Törnqvist, U.; Magdalinou, N.; Fox, N.C.; Blennow, K.; Schott, J.M.; Zetterberg, H. Increased CSF neurogranin concentration is specific to Alzheimer disease. *Neurology* **2016**, *86*, 829–835. [CrossRef]
28. Brinkmalm, A.; Brinkmalm, G.; Honer, W.G.; Frölich, L.; Hausner, L.; Minthon, L.; Hansson, O.; Wallin, A.; Zetterberg, H.; Blennow, K.; et al. SNAP-25 is a promising novel cerebrospinal fluid biomarker for synapse degeneration in Alzheimer's disease. *Mol. Neurodegener.* **2014**, *9*, 53. [CrossRef]
29. Davidsson, P.; Blennow, K. Neurochemical dissection of synaptic pathology in Alzheimer's disease. *Int. Psychogeriatr.* **1998**, *10*, 11–23. [CrossRef]
30. Clarke, M.T.M.; Brinkmalm, A.; Foiani, M.S.; Woollacott, I.O.C.; Heller, C.; Heslegrave, A.; Keshavan, A.; Fox, N.C.; Schott, J.M.; Warren, J.D.; et al. CSF synaptic protein concentrations are raised in those with atypical Alzheimer's disease but not frontotemporal dementia. *Alzheimers Res. Ther.* **2019**, *11*, 105. [CrossRef]
31. Öhrfelt, A.; Brinkmalm, A.; Dumurgier, J.; Brinkmalm, G.; Hansson, O.; Zetterberg, H.; Bouaziz-Amar, E.; Hugon, J.; Paquet, C.; Blennow, K. The pre-synaptic vesicle protein synaptotagmin is a novel biomarker for Alzheimer's disease. *Alzheimers Res. Ther.* **2016**, *8*, 41. [CrossRef] [PubMed]
32. Atagi, Y.; Liu, C.C.; Painter, M.M.; Chen, X.F.; Verbeeck, C.; Zheng, H.; Li, X.; Rademakers, R.; Kang, S.S.; Xu, H.; et al. Apolipoprotein E Is a Ligand for Triggering Receptor Expressed on Myeloid Cells 2 (TREM2). *J. Biol. Chem.* **2015**, *290*, 26043–26050. [CrossRef]
33. Parhizkar, S.; Arzberger, T.; Brendel, M.; Kleinberger, G.; Deussing, M.; Focke, C.; Nuscher, B.; Xiong, M.; Ghasemigharagoz, A.; Katzmarski, N.; et al. Loss of TREM2 function increases amyloid seeding but reduces plaque-associated *ApoE*. *Nat. Neurosci.* **2019**, *22*, 191–204. [CrossRef] [PubMed]
34. Heslegrave, A.; Heywood, W.; Paterson, R.; Magdalinou, N.; Svensson, J.; Johansson, P.; Öhrfelt, A.; Blennow, K.; Hardy, J.; Schott, J.; et al. Increased cerebrospinal fluid soluble TREM2 concentration in Alzheimer's disease. *Mol. Neurodegener.* **2016**, *11*, 3. [CrossRef] [PubMed]
35. Suárez-Calvet, M.; Araque Caballero, M.Á.; Kleinberger, G.; Bateman, R.J.; Fagan, A.M.; Morris, J.C.; Levin, J.; Danek, A.; Ewers, M.; Haass, C.; et al. Early changes in CSF sTREM2 in dominantly inherited Alzheimer's disease occur after amyloid deposition and neuronal injury. *EMBO Mol. Med.* **2016**, *8*, 466–476. [CrossRef]
36. Piccio, L.; Deming, Y.; Del-Águila, J.L.; Ghezzi, L.; Holtzman, D.M.; Fagan, A.M.; Fenoglio, C.; Galimberti, D.; Borroni, B.; Cruchaga, C. Cerebrospinal fluid soluble TREM2 is higher in Alzheimer disease and associated with mutation status. *Acta Neuropathol.* **2016**, *131*, 925–933. [CrossRef] [PubMed]
37. Liu, D.; Cao, B.; Zhao, Y.; Huang, H.; McIntyre, R.S.; Rosenblat, J.D.; Zhou, H. Soluble TREM2 changes during the clinical course of Alzheimer's disease: A meta-analysis. *Neurosci. Lett.* **2018**, *686*, 10–16. [CrossRef] [PubMed]
38. Suárez-Calvet, M.; Morenas-Rodríguez, E.; Kleinberger, G.; Schlepckow, K.; Araque Caballero, M.Á.; Franzmeier, N.; Capell, A.; Fellerer, K.; Nuscher, B.; Erden, E.; et al. Early increase of CSF sTREM2 in Alzheimer's disease is associated with tau related-neurodegeneration but not with amyloid-β pathology. *Mol. Neurodegener.* **2019**, *14*, 1. [CrossRef]
39. Piccio, L.; Buonsanti, C.; Cella, M.; Tassi, I.; Schmidt, R.E.; Fenoglio, C.; Rinker, J., II; Naismith, R.T.; Panina-Bordignon, P.; Passini, N.; et al. Identification of soluble TREM-2 in the cerebrospinal fluid and its association with multiple sclerosis and CNS inflammation. *Brain* **2008**, *131 Pt 11*, 3081–3091. [CrossRef]
40. Henjum, K.; Quist-Paulsen, E.; Zetterberg, H.; Blennow, K.; Nilsson, L.N.G.; Watne, L.O. CSF sTREM2 in delirium-relation to Alzheimer's disease CSF biomarkers Aβ42, t-tau and p-tau. *J. Neuroinflamm.* **2018**, *15*, 304. [CrossRef]
41. Ishiki, A.; Kamada, M.; Kawamura, Y.; Terao, C.; Shimoda, F.; Tomita, N.; Arai, H.; Furukawa, K. Glial fibrillar acidic protein in the cerebrospinal fluid of Alzheimer's disease, dementia with Lewy bodies, and frontotemporal lobar degeneration. *J. Neurochem.* **2016**, *136*, 258–261. [CrossRef] [PubMed]
42. Craig-Schapiro, R.; Perrin, R.J.; Roe, C.M.; Xiong, C.; Carter, D.; Cairns, N.J.; Mintun, M.A.; Peskind, E.R.; Li, G.; Galasko, D.R.; et al. YKL-40: A novel prognostic fluid biomarker for preclinical Alzheimer's disease. *Biol. Psychiatry* **2010**, *68*, 903–912. [CrossRef] [PubMed]

43. Olsson, B.; Lautner, R.; Andreasson, U.; Öhrfelt, A.; Portelius, E.; Bjerke, M.; Hölttä, M.; Rosén, C.; Olsson, C.; Strobel, G.; et al. CSF and blood biomarkers for the diagnosis of Alzheimer's disease: A systematic review and meta-analysis. *Lancet Neurol.* **2016**, *15*, 673–684. [CrossRef]
44. Lleó, A.; Alcolea, D.; Martínez-Lage, P.; Scheltens, P.; Parnetti, L.; Poirier, J.; Simonsen, A.H.; Verbeek, M.M.; Rosa-Neto, P.; Slot, R.E.R.; et al. Longitudinal cerebrospinal fluid biomarker trajectories along the Alzheimer's disease continuum in the BIOMARKAPD study. *Alzheimers Dement.* **2019**, *15*, 742–753. [CrossRef]
45. Kester, M.I.; Teunissen, C.E.; Sutphen, C.; Herries, E.M.; Ladenson, J.H.; Xiong, C.; Scheltens, P.; Van der Flier, W.M.; Morris, J.C.; Holtzman, D.M.; et al. Cerebrospinal fluid VILIP-1 and YKL-40, candidate biomarkers to diagnose, predict and monitor Alzheimer's disease in a memory clinic cohort. *Alzheimers Res. Ther.* **2015**, *7*, 59. [CrossRef]
46. Wennström, M.; Surova, Y.; Hall, S.; Nilsson, C.; Minthon, L.; Hansson, O.; Nielsen, H.M. The Inflammatory Marker YKL-40 Is Elevated in Cerebrospinal Fluid from Patients with Alzheimer's but Not Parkinson's Disease or Dementia with Lewy Bodies. *PLoS ONE* **2015**, *10*, e0135458. [CrossRef]
47. Zetterberg, H.; Andreasson, U.; Hansson, O.; Wu, G.; Sankaranarayanan, S.; Andersson, M.E.; Buchhave, P.; Londos, E.; Umek, R.M.; Minthon, L.; et al. Elevated cerebrospinal fluid BACE1 activity in incipient Alzheimer disease. *Arch. Neurol.* **2008**, *65*, 1102–1107. [CrossRef]
48. Mulder, S.D.; Van der Flier, W.M.; Verheijen, J.H.; Mulder, C.; Scheltens, P.; Blankenstein, M.A.; Hack, C.E.; Veerhuis, R. BACE1 activity in cerebrospinal fluid and its relation to markers of AD pathology. *J. Alzheimers Dis.* **2010**, *20*, 253–260. [CrossRef]
49. Zhong, Z.; Ewers, M.; Teipel, S.; Bürger, K.; Wallin, A.; Blennow, K.; He, P.; McAllister, C.; Hampel, H.; Shen, Y. Levels of beta-secretase (BACE1) in cerebrospinal fluid as a predictor of risk in mild cognitive impairment. *Arch. Gen. Psychiatry* **2007**, *64*, 718–726. [CrossRef]
50. Perneczky, R.; Alexopoulos, P.; Alzheimer's Disease euroimaging Initiative. Cerebrospinal fluid BACE1 activity and markers of amyloid precursor protein metabolism and axonal degeneration in Alzheimer's disease. *Alzheimers Dement.* **2014**, *10* (Suppl. 4), S425–S429. [CrossRef]
51. Savage, M.J.; Holder, D.J.; Wu, G.; Kaplow, J.; Siuciak, J.A.; Potter, W.Z.; Foundation for National Institutes of Health (FNIH) Biomarkers Consortium CSF Proteomics Project Team for Alzheimer's Disease Neuroimaging Initiative. Soluble BACE-1 Activity and sAβPPβ Concentrations in Alzheimer's Disease and Age-Matched Healthy Control Cerebrospinal Fluid from the Alzheimer's Disease Neuroimaging Initiative-1 Baseline Cohort. *J. Alzheimers Dis.* **2015**, *46*, 431–440. [CrossRef]
52. Molinuevo, J.L.; Ayton, S.; Batrla, R.; Bednar, M.M.; Bittner, T.; Cummings, J.; Fagan, A.M.; Hampel, H.; Mielke, M.M.; Mikulskis, A.; et al. Current state of Alzheimer's fluid biomarkers. *Acta Neuropathol.* **2018**, *136*, 821–853. [CrossRef]
53. Blennow, K.; Zetterberg, H. Biomarkers for Alzheimer's Disease: Current Status and Prospects for the Future. *J. Intern. Med.* **2018**, *284*, 643–663. [CrossRef] [PubMed]
54. Blennow, K. A Review of Fluid Biomarkers for Alzheimer's Disease: Moving from CSF to Blood. *Neurol. Ther.* **2017**, *6* (Suppl. 1), 15–24. [CrossRef] [PubMed]
55. Guzman-Martinez, L.; Maccioni, R.B.; Farías, G.A.; Fuentes, P.; Navarrete, L.P. Biomarkers for Alzheimer's Disease. *Curr. Alzheimer Res.* **2019**, *16*, 518–528. [CrossRef] [PubMed]
56. Henriksen, K.; O'Bryant, S.E.; Hampel, H.; Trojanowski, J.Q.; Montine, T.J.; Jeromin, A.; Blennow, K.; Lönneborg, A.; Wyss-Coray, T.; Soares, H. The future of blood-based biomarkers for Alzheimer's disease. *Alzheimers Dement.* **2014**, *10*, 115–131. [CrossRef] [PubMed]
57. Zetterberg, H.; Burnham, S.C. Blood-based molecular biomarkers for Alzheimer's disease. *Mol. Brain* **2019**, *12*, 26. [CrossRef] [PubMed]
58. Jia, L.; Qiu, Q.; Zhang, H.; Chu, L.; Du, Y.; Zhang, J.; Zhou, C.; Liang, F.; Shi, S.; Wang, S.; et al. Concordance between the assessment of Aβ42, T-tau, and P-T181-tau in peripheral blood neuronal-derived exosomes and cerebrospinal fluid. *Alzheimers Dement.* **2019**, *15*, 1071–1080. [CrossRef] [PubMed]
59. Toledo, J.B.; Vanderstichele, H.; Figurski, M.; Aisen, P.S.; Petersen, R.C.; Weiner, M.W.; Jack, C.R., Jr.; Jagust, W.; Decarli, C.; Toga, A.W.; et al. Factors affecting Aβ plasma levels and their utility as biomarkers in ADNI. *Acta Neuropathol.* **2011**, *122*, 401–413. [CrossRef]
60. Rembach, A.; Faux, N.G.; Watt, A.D.; Pertile, K.K.; Rumble, R.L.; Trounson, B.O.; Fowler, C.J.; Roberts, B.R.; Perez, K.A.; Li, Q.X.; et al. Changes in plasma amyloid beta in a longitudinal study of aging and Alzheimer's disease. *Alzheimers Dement.* **2014**, *10*, 53–61. [CrossRef]

61. Hanon, O.; Vidal, J.S.; Lehmann, S.; Bombois, S.; Allinquant, B.; Tréluyer, J.M.; Gelé, P.; Delmaire, C.; Blanc, F.; Mangin, J.F.; et al. Plasma amyloid levels within the Alzheimer's process and correlations with central biomarkers. *Alzheimers Dement.* **2018**, *14*, 858–868. [CrossRef] [PubMed]
62. Janelidze, S.; Stomrud, E.; Palmqvist, S.; Zetterberg, H.; Van Westen, D.; Jeromin, A.; Song, L.; Hanlon, D.; Tan Hehir, C.A.; Baker, D.; et al. Plasma β-Amyloid in Alzheimer's disease and vascular disease. *Sci. Rep.* **2016**, *6*, 26801. [CrossRef] [PubMed]
63. Verberk, I.M.W.; Slot, R.E.; Verfaillie, S.C.J.; Heijst, H.; Prins, N.D.; Van Berckel, B.N.M.; Scheltens, P.; Teunissen, C.E.; Van der Flier, W.M. Plasma Amyloid as Prescreener for the Earliest Alzheimer Pathological Changes. *Ann. Neurol.* **2018**, *84*, 648–658. [CrossRef] [PubMed]
64. Ovod, V.; Ramsey, K.N.; Mawuenyega, K.G.; Bollinger, J.G.; Hicks, T.; Schneider, T.; Sullivan, M.; Paumier, K.; Holtzman, D.M.; Morris, J.C.; et al. Amyloid β concentrations and stable isotope labeling kinetics of human plasma specific to central nervous system amyloidosis. *Alzheimers Dement.* **2017**, *13*, 841–849. [CrossRef]
65. Nakamura, A.; Kaneko, N.; Villemagne, V.L.; Kato, T.; Doecke, J.; Doré, V.; Fowler, C.; Li, Q.X.; Martins, R.; Rowe, C.; et al. High performance plasma amyloid-β biomarkers for Alzheimer's disease. *Nature* **2018**, *554*, 249–254. [CrossRef]
66. Li, Q.X.; Berndt, M.C.; Bush, A.I.; Rumble, B.; Mackenzie, I.; Friedhuber, A.; Beyreuther, K.; Masters, C.L. Membrane-associated forms of the beta A4 amyloid protein precursor of Alzheimer's disease in human platelet and brain: Surface expression on the activated human platelet. *Blood* **1994**, *84*, 133–142. [CrossRef]
67. Fossati, S.; Ramos Cejudo, J.; Debure, L.; Pirraglia, E.; Sone, J.Y.; Li, Y.; Chen, J.; Butler, T.; Zetterberg, H.; Blennow, K.; et al. Plasma tau complements CSF tau and P-tau in the diagnosis of Alzheimer's disease. *Alzheimers Dement. (Amst.)* **2019**, *11*, 483–492. [CrossRef]
68. Mattsson, N.; Zetterberg, H.; Janelidze, S.; Insel, P.S.; Andreasson, U.; Stomrud, E.; Palmqvist, S.; Baker, D.; Tan Hehir, C.A.; Jeromin, A.; et al. Plasma tau in Alzheimer disease. *Neurology* **2016**, *87*, 1827–1835. [CrossRef]
69. Pase, M.P.; Beiser, A.S.; Himali, J.J.; Satizabal, C.L.; Aparicio, H.J.; DeCarli, C.; Chêne, G.; Dufouil, C.; Seshadri, S. Assessment of Plasma Total Tau Level as a Predictive Biomarker for Dementia and Related Endophenotypes. *JAMA Neurol.* **2019**, *76*, 598–606. [CrossRef]
70. Mielke, M.M.; Hagen, C.E.; Wennberg, A.M.V.; Airey, D.C.; Savica, R.; Knopman, D.S.; Machulda, M.M.; Roberts, R.O.; Jack, C.R., Jr.; Petersen, R.C.; et al. Association of Plasma Total Tau Level with Cognitive Decline and Risk of Mild Cognitive Impairment or Dementia in the Mayo Clinic Study on Aging. *JAMA Neurol.* **2017**, *74*, 1073–1080. [CrossRef]
71. Tatebe, H.; Kasai, T.; Ohmichi, T.; Kishi, Y.; Kakeya, T.; Waragai, M.; Kondo, M.; Allsop, D.; Tokuda, T. Quantification of plasma phosphorylated tau to use as a biomarker for brain Alzheimer pathology: Pilot case-control studies including patients with Alzheimer's disease and down syndrome. *Mol. Neurodegener.* **2017**, *12*, 63. [CrossRef] [PubMed]
72. Mattsson, N.; Andreasson, U.; Zetterberg, H.; Blennow, K.; Alzheimer's Disease Neuroimaging Initiative. Association of Plasma Neurofilament Light With Neurodegeneration in Patients with Alzheimer Disease. *JAMA Neurol.* **2017**, *74*, 557–566. [CrossRef]
73. Gisslén, M.; Price, R.W.; Andreasson, U.; Norgren, N.; Nilsson, S.; Hagberg, L.; Fuchs, D.; Spudich, S.; Blennow, K.; Zetterberg, H. Plasma Concentration of the Neurofilament Light Protein (NFL) is a Biomarker of CNS Injury in HIV Infection: A Cross-Sectional Study. *EBioMedicine* **2015**, *3*, 135–140. [CrossRef] [PubMed]
74. Rojas, J.C.; Bang, J.; Lobach, I.V.; Tsai, R.M.; Rabinovici, G.D.; Miller, B.L.; Boxer, A.L.; AL-108-231 Investigators. CSF neurofilament light chain and phosphorylated tau 181 predict disease progression in PSP. *Neurology* **2018**, *90*, e273–e281. [CrossRef]
75. Preische, O.; Schultz, S.A.; Apel, A.; Kuhle, J.; Kaeser, S.A.; Barro, C.; Gräber, S.; Kuder-Buletta, E.; LaFougere, C.; Laske, C.; et al. Serum neurofilament dynamics predicts neurodegeneration and clinical progression in presymptomatic Alzheimer's disease. *Nat. Med.* **2019**, *25*, 277–283. [CrossRef] [PubMed]
76. Mattsson, N.; Cullen, N.C.; Andreasson, U.; Zetterberg, H.; Blennow, K. Association between Longitudinal Plasma Neurofilament Light and Neurodegeneration in Patients with Alzheimer Disease. *JAMA Neurol.* **2019**, *76*, 791–799. [CrossRef]
77. Sánchez-Valle, R.; Heslegrave, A.; Foiani, M.S.; Bosch, B.; Antonell, A.; Balasa, M.; Lladó, A.; Zetterberg, H.; Fox, N.C. Serum neurofilament light levels correlate with severity measures and neurodegeneration markers in autosomal dominant Alzheimer's disease. *Alzheimers Res. Ther.* **2018**, *10*, 113. [CrossRef]

78. Weston, P.S.J.; Poole, T.; Ryan, N.S.; Nair, A.; Liang, Y.; Macpherson, K.; Druyeh, R.; Malone, I.B.; Ahsan, R.L.; Pemberton, H.; et al. Serum neurofilament light in familial Alzheimer disease: A marker of early neurodegeneration. *Neurology* **2017**, *89*, 2167–2175. [CrossRef]
79. Abdullah, M.; Kimura, N.; Akatsu, H.; Hashizume, Y.; Ferdous, T.; Tachita, T.; Iida, S.; Zou, K.; Matsubara, E.; Michikawa, M. Flotillin is a Novel Diagnostic Blood Marker of Alzheimer's Disease. *J. Alzheimers Dis.* **2019**, *72*, 1165–1176. [CrossRef]
80. Angelopoulou, E.; Paudel, Y.N.; Shaikh, M.F.; Piperi, C. Flotillin: A Promising Biomarker for Alzheimer's Disease. *J. Pers. Med.* **2020**, *10*, 20. [CrossRef]
81. Serpente, M.; Fenoglio, C.; Villa, C.; Cortini, F.; Cantoni, C.; Ridolfi, E.; Clerici, F.; Marcone, A.; Benussi, L.; Ghidoni, R.; et al. Role of OLR1 and its regulating hsa-miR369-3p in Alzheimer's disease: Genetics and expression analysis. *J. Alzheimers Dis.* **2011**, *26*, 787–793. [CrossRef] [PubMed]
82. Galimberti, D.; Villa, C.; Fenoglio, C.; Serpente, M.; Ghezzi, L.; Cioffi, S.M.; Arighi, A.; Fumagalli, G.; Scarpini, E. Circulating miRNAs as potential biomarkers in Alzheimer's disease. *J. Alzheimers Dis.* **2014**, *42*, 1261–1267. [CrossRef] [PubMed]
83. Cortini, F.; Roma, F.; Villa, C. Emerging roles of long non-coding RNAs in the pathogenesis of Alzheimer's disease. *Ageing Res. Rev.* **2019**, *50*, 19–26. [CrossRef] [PubMed]
84. Leidinger, P.; Backes, C.; Deutscher, S.; Schmitt, K.; Mueller, S.C.; Frese, K.; Haas, J.; Ruprecht, K.; Paul, F.; Stähler, C.; et al. A blood based 12-miRNA signature of Alzheimer disease patients. *Genome Biol.* **2013**, *14*, R78. [CrossRef]
85. Lugli, G.; Cohen, A.M.; Bennett, D.A.; Shah, R.C.; Fields, C.J.; Hernandez, A.G.; Smalheiser, N.R. Plasma Exosomal miRNAs in Persons with and without Alzheimer Disease: Altered Expression and Prospects for Biomarkers. *PLoS ONE* **2015**, *10*, e0139233. [CrossRef]
86. Müller, M.; Jäkel, L.; Bruinsma, I.B.; Claassen, J.A.; Kuiperij, H.B.; Verbeek, M.M. MicroRNA-29a Is a Candidate Biomarker for Alzheimer's Disease in Cell-Free Cerebrospinal Fluid. *Mol. Neurobiol.* **2016**, *53*, 2894–2899. [CrossRef]
87. Nagaraj, S.; Zoltowska, K.M.; Laskowska-Kaszub, K.; Wojda, U. microRNA diagnostic panel for Alzheimer's disease and epigenetic trade-off between neurodegeneration and cancer. *Ageing Res. Rev.* **2019**, *49*, 125–143. [CrossRef]
88. Swarbrick, S.; Wragg, N.; Ghosh, S.; Stolzing, A. Systematic Review of miRNA as Biomarkers in Alzheimer's Disease. *Mol. Neurobiol.* **2019**, *56*, 6156–6167. [CrossRef]
89. Gupta, P.; Bhattacharjee, S.; Sharma, A.R.; Sharma, G.; Lee, S.S.; Chakraborty, C. miRNAs in Alzheimer Disease—A Therapeutic Perspective. *Curr. Alzheimer Res.* **2017**, *14*, 1198–1206. [CrossRef]
90. Ashton, N.J.; Schöll, M.; Heurling, K.; Gkanatsiou, E.; Portelius, E.; Höglund, K.; Brinkmalm, G.; Hye, A.; Blennow, K.; Zetterberg, H. Update on biomarkers for amyloid pathology in Alzheimer's disease. *Biomark. Med.* **2018**, *12*, 799–812. [CrossRef]
91. Spielmann, N.; Wong, D.T. Saliva: Diagnostics and therapeutic perspectives. *Oral Dis.* **2011**, *17*, 345–354. [CrossRef] [PubMed]
92. Ashton, N.J.; Ide, M.; Zetterberg, H.; Blennow, K. Salivary Biomarkers for Alzheimer's Disease and Related Disorders. *Neurol. Ther.* **2019**, *8* (Suppl. 2), 83–94. [CrossRef] [PubMed]
93. Shi, M.; Sui, Y.T.; Peskind, E.R.; Li, G.; Hwang, H.; Devic, I.; Ginghina, C.; Edgar, J.S.; Pan, C.; Goodlett, D.R.; et al. Salivary tau species are potential biomarkers of Alzheimer's disease. *J. Alzheimers Dis.* **2011**, *27*, 299–305. [CrossRef] [PubMed]
94. Ashton, N.J.; Ide, M.; Schöll, M.; Blennow, K.; Lovestone, S.; Hye, A.; Zetterberg, H. No association of salivary total tau concentration with Alzheimer's disease. *Neurobiol. Aging* **2018**, *70*, 125–127. [CrossRef]
95. Carro, E.; Bartolomé, F.; Bermejo-Pareja, F.; Villarejo-Galende, A.; Molina, J.A.; Ortiz, P.; Calero, M.; Rabano, A.; Cantero, J.L.; Orive, G. Early diagnosis of mild cognitive impairment and Alzheimer's disease based on salivary lactoferrin. *Alzheimers Dement. (Amst.)* **2017**, *8*, 131–138. [CrossRef]
96. François, M.; Bull, C.F.; Fenech, M.F.; Leifert, W.R. Current State of Saliva Biomarkers for Aging and Alzheimer's Disease. *Curr. Alzheimer Res.* **2019**, *16*, 56–66. [CrossRef]
97. Zengi, O.; Karakas, A.; Ergun, U.; Senes, M.; Inan, L.; Yucel, D. Urinary 8-hydroxy-2′-deoxyguanosine level and plasma paraoxonase 1 activity with Alzheimer's disease. *Clin. Chem. Lab. Med.* **2011**, *50*, 529–534. [CrossRef]

98. Praticò, D.; Clark, C.M.; Lee, V.M.; Trojanowski, J.Q.; Rokach, J.; FitzGerald, G.A. Increased 8,12-iso-iPF2alpha-VI in Alzheimer's disease: Correlation of a noninvasive index of lipid peroxidation with disease severity. *Ann. Neurol.* **2000**, *48*, 809–812. [CrossRef]
99. Zhang, J.; Zhang, C.H.; Li, R.J.; Li, X.L.; Chen, Y.D.; Gao, H.Q.; Shi, S.L. Accuracy of urinary AD7c-NTP for diagnosing Alzheimer's disease: A systematic review and meta-analysis. *J. Alzheimers Dis.* **2014**, *40*, 153–159. [CrossRef]
100. Yao, F.; Hong, X.; Li, S.; Zhang, Y.; Zhao, Q.; Du, W.; Wang, Y.; Ni, J. Urine-Based Biomarkers for Alzheimer's Disease Identified Through Coupling Computational and Experimental Methods. *J. Alzheimers Dis.* **2018**, *65*, 421–431. [CrossRef] [PubMed]
101. An, H.J.; Ninonuevo, M.; Aguilan, J.; Liu, H.; Lebrilla, C.B.; Alvarenga, L.S.; Mannis, M.J. Glycomics analyses of tear fluid for the diagnostic detection of ocular rosacea. *J. Proteome Res.* **2005**, *4*, 1981–1987. [CrossRef] [PubMed]
102. Zhou, L.; Zhao, S.Z.; Koh, S.K.; Chen, L.; Vaz, C.; Tanavde, V.; Li, X.R.; Beuerman, R.W. In-depth analysis of the human tear proteome. *J. Proteomics* **2012**, *75*, 3877–3885. [CrossRef]
103. Goldstein, L.E.; Muffat, J.A.; Cherny, R.A.; Moir, R.D.; Ericsson, M.H.; Huang, X.; Mavros, C.; Coccia, J.A.; Faget, K.Y.; Fitch, K.A.; et al. Cytosolic beta-amyloid deposition and supranuclear cataracts in lenses from people with Alzheimer's disease. *Lancet* **2003**, *361*, 1258–1265. [CrossRef]
104. Kalló, G.; Emri, M.; Varga, Z.; Ujhelyi, B.; Tőzsér, J.; Csutak, A.; Csősz, É. Changes in the Chemical Barrier Composition of Tears in Alzheimer's Disease Reveal Potential Tear Diagnostic Biomarkers. *PLoS ONE* **2016**, *11*, e0158000. [CrossRef] [PubMed]
105. Westman, E.; Simmons, A.; Zhang, Y.; Muehlboeck, J.S.; Tunnard, C.; Liu, Y.; Collins, L.; Evans, A.; Mecocci, P.; Vellas, B.; et al. Multivariate analysis of MRI data for Alzheimer's disease, mild cognitive impairment and healthy controls. *Neuroimage* **2011**, *54*, 1178–1187. [CrossRef]
106. Babiloni, C.; Del Percio, C.; Lizio, R.; Noce, G.; Lopez, S.; Soricelli, A.; Ferri, R.; Nobili, F.; Arnaldi, D.; Famà, F.; et al. Abnormalities of resting-state functional cortical connectivity in patients with dementia due to Alzheimer's and Lewy body diseases: An EEG study. *Neurobiol. Aging* **2018**, *65*, 18–40. [CrossRef] [PubMed]
107. Morrison, C.; Rabipour, S.; Taler, V.; Sheppard, C.; Knoefel, F. Visual Event-Related Potentials in Mild Cognitive Impairment and Alzheimer's Disease: A Literature Review. *Curr. Alzheimer Res.* **2019**, *16*, 67–89. [CrossRef] [PubMed]
108. Cantone, M.; Di Pino, G.; Capone, F.; Piombo, M.; Chiarello, D.; Cheeran, B.; Pennisi, G.; Di Lazzaro, V. The contribution of transcranial magnetic stimulation in the diagnosis and in the management of dementia. *Clin. Neurophysiol.* **2014**, *125*, 1509–1532. [CrossRef] [PubMed]
109. Davis, P.C.; Gearing, M.; Gray, L.; Mirra, S.S.; Morris, J.C.; Edland, S.D.; Lin, T.; Heyman, A. The CERAD experience, Part VIII: Neuroimaging-neuropathology correlates of temporal lobe changes in Alzheimer's disease. *Neurology* **1995**, *45*, 178–179. [CrossRef]
110. Harper, L.; Barkhof, F.; Scheltens, P.; Schott, J.M.; Fox, N.C. An algorithmic approach to structural imaging in dementia. *J. Neurol. Neurosurg. Psychiatry* **2014**, *85*, 692–698. [CrossRef]
111. Dickerson, B.C.; Bakkour, A.; Salat, D.H.; Feczko, E.; Pacheco, J.; Greve, D.N.; Grodstein, F.; Wright, C.I.; Blacker, D.; Rosas, H.D.; et al. The cortical signature of Alzheimer's disease: Regionally specific cortical thinning relates to symptom severity in very mild to mild AD dementia and is detectable in asymptomatic amyloid positive individuals. *Cereb. Cortex* **2009**, *19*, 497–510. [CrossRef] [PubMed]
112. Frisoni, G.B.; Prestia, A.; Rasser, P.E.; Bonetti, M.; Thompson, P.M. In vivo mapping of incremental cortical atrophy from incipient to overt Alzheimer's disease. *J. Neurol.* **2009**, *256*, 916–924. [CrossRef] [PubMed]
113. Perrin, R.J.; Fagan, A.M.; Holtzman, D.M. Multimodal techniques for diagnosis and prognosis of Alzheimer's disease. *Nature* **2009**, *461*, 916–922. [CrossRef] [PubMed]
114. Wolz, R.; Julkunen, V.; Koikkalainen, J.; Niskanen, E.; Zhang, D.P.; Rueckert, D.; Soininen, H.; Lötjönen, J.; Alzheimer's Disease Neuroimaging Initiative. Multi-method analysis of MRI images in early diagnostics of Alzheimer's disease. *PLoS ONE* **2011**, *6*, e25446. [CrossRef]
115. Whitwell, J.L.; Shiung, M.M.; Przybelski, S.A.; Weigand, S.D.; Knopman, D.S.; Boeve, B.F.; Petersen, R.C.; Jack, C.R. MRI patterns of atrophy associated with progression to AD in amnestic mild cognitive impairment. *Neurology* **2008**, *70*, 512–520. [CrossRef]

116. Frisoni, G.B.; Jack, C.R., Jr.; Bocchetta, M.; Bauer, C.; Frederiksen, K.S.; Liu, Y.; Preboske, G.; Swihart, T.; Blair, M.; Cavedo, E.; et al. The EADC-ADNI harmonized protocol for manual hippocampal segmentation on magnetic resonance: Evidence of validity. *Alzheimers Dement.* **2015**, *11*, 111–125. [CrossRef]
117. Henneman, W.J.; Sluimer, J.D.; Barnes, J.; Van der Flier, W.M.; Sluimer, I.C.; Fox, N.C.; Scheltens, P.; Vrenken, H.; Barkhof, F. Hippocampal atrophy rates in Alzheimer disease: Added value over whole brain volume measures. *Neurology* **2009**, *72*, 999–1007. [CrossRef]
118. Risacher, S.L.; Shen, L.; West, J.D.; Kim, S.; McDonald, B.C.; Beckett, L.A.; Harvey, D.J.; Jack, C.R., Jr.; Weiner, M.W.; Saykin, A.J.; et al. Longitudinal MRI atrophy biomarkers: Relationship to conversion in the ADNI cohort. *Neurobiol. Aging* **2010**, *31*, 1401–1418. [CrossRef]
119. Rosen, A.C.; Prull, M.W.; Gabrieli, J.D.E.; Stoub, T.; O'Hara, R.; Friedman, L.; Yesavage, J.A.; deToledo-Morrell, L. Differential associations between entorhinal and hippocampal volumes and memory performance in older adults. *Behav. Neurosci.* **2003**, *117*, 1150–1160. [CrossRef]
120. Rodrigue, K.M.; Raz, N. Shrinkage of the entorhinal cortex over five years predicts memory performance in healthy adults. *J. Neurosci. Off. J. Soc. Neurosci.* **2004**, *24*, 956–963. [CrossRef]
121. Wei, H.; Kong, M.; Zhang, C.; Guan, L.; Ba, M.; Alzheimer's Disease Neuroimaging Initiative. The structural MRI markers and cognitive decline in prodromal Alzheimer's disease: A 2-year longitudinal study. *Quant. Imaging Med. Surg.* **2018**, *8*, 1004–1019. [CrossRef] [PubMed]
122. Risacher, S.L.; Anderson, W.H.; Charil, A.; Castelluccio, P.F.; Shcherbinin, S.; Saykin, A.J.; Schwarz, A.J.; Alzheimer's Disease Neuroimaging Initiative. Alzheimer disease brain atrophy subtypes are associated with cognition and rate of decline. *Neurology* **2017**, *89*, 2176–2186. [CrossRef] [PubMed]
123. Mishra, V.; Sreenivasan, K.; Banks, S.J.; Zhuang, X.; Yang, Z.; Cordes, D.; Bernick, C. Investigating structural and perfusion deficits due to repeated head trauma in active professional fighters. *NeuroImage Clin.* **2018**, *17*, 616–627. [CrossRef] [PubMed]
124. Okonkwo, O.C.; Xu, G.; Dowling, N.M.; Bendlin, B.B.; LaRue, A.; Hermann, B.P.; Koscik, R.; Jonaitis, E.; Rowley, H.A.; Carlsson, C.M.; et al. Family history of Alzheimer disease predicts hippocampal atrophy in healthy middle-aged adults. *Neurology* **2012**, *78*, e1769–e1776. [CrossRef]
125. O'Dwyer, L.; Lamberton, F.; Matura, S.; Tanner, C.; Scheibe, M.; Miller, J.; Rujescu, D.; Prvulovic, D.; Hampel, H. Reduced hippocampal volume in healthy young ApoE4 carriers: An MRI study. *PLoS ONE* **2012**, *7*, e48895. [CrossRef]
126. Weiner, M.W.; Veitch, D.P.; Aisen, P.S.; Beckett, L.A.; Cairns, N.J.; Cedarbaum, J.; Green, R.C.; Harvey, D.; Jack, C.R.; Jagust, W.; et al. 2014 Update of the Alzheimer's Disease Neuroimaging Initiative: A review of papers published since its inception. *Alzheimers Dement.* **2015**, *11*, e1–e120. [CrossRef] [PubMed]
127. Mueller, S.G.; Yushkevich, P.A.; Das, S.; Wang, L.; Van Leemput, K.; Iglesias, J.E.; Alpert, K.; Mezher, A.; Ng, P.; Paz, K.; et al. Systematic comparison of different techniques to measure hippocampal subfield volumes in ADNI2. *Neuroimage Clin.* **2018**, *17*, 1006–1018. [CrossRef]
128. Mechelli, A.; Price, C.J.; Friston, K.J.; Ashburner, J. Voxel-based morphometry of the human brain: Methods and applications. *Curr. Med. Imaging Rev.* **2005**, *1*, 1–9. [CrossRef]
129. Zandifar, A.; Fonov, V.S.; Pruessner, J.C.; Collins, D.L.; Alzheimer's Disease Neuroimaging Initiative. The EADC-ADNI harmonized protocol for hippocampal segmentation: A validation study. *Neuroimage* **2018**, *181*, 142–148. [CrossRef]
130. Holland, D.; McEvoy, L.K.; Dale, A.M. Unbiased comparison of sample size estimates from longitudinal structural measures in ADNI. *Hum. Brain Mapp.* **2012**, *33*, 2586–2602. [CrossRef]
131. Frenzel, S.; Wittfeld, K.; Habes, M.; Klinger-König, J.; Bülow, R.; Völzke, H.; Grabe, H.J. A Biomarker for Alzheimer's Disease Based on Patterns of Regional Brain Atrophy. *Front. Psychiatry* **2020**, *10*, 953. [CrossRef] [PubMed]
132. Wolk, D.A.; Das, S.R.; Mueller, S.G.; Weiner, M.W.; Yushkevich, P.A. Medial temporal lobe subregional morphometry using high resolution MRI in Alzheimer's disease. *Neurobiol. Aging* **2017**, *49*, 204–213. [CrossRef] [PubMed]
133. Taylor, W.D.; Hsu, E.; Krishnan, K.R.R.; MacFall, J.R. Diffusion tensor imaging: Background, potential, and utility in psychiatric research. *Biol. Psychiatry* **2004**, *55*, 201–207. [CrossRef] [PubMed]
134. Sajjadi, S.A.; Acosta-Cabronero, J.; Patterson, K.; Diaz-deGrenu, L.Z.; Williams, G.B.; Nestor, P.J. Diffusion tensor magnetic resonance imaging for single subject diagnosis in neurodegenerative diseases. *Brain* **2013**, *136 Pt 7*, 2253–2261. [CrossRef]

135. Agosta, F.; Galantucci, S.; Filippi, M. Advanced magnetic resonance imaging of neurodegenerative diseases. *Neurol. Sci.* **2017**, *38*, 41–51. [CrossRef]
136. Clerx, L.; Visser, P.J.; Verhey, F.; Aalten, P. New MRI markers for Alzheimer's disease: A meta-analysis of diffusion tensor imaging and a comparison with medial temporal lobe measurements. *J. Alzheimers Dis.* **2012**, *29*, 405429. [CrossRef]
137. Jones, D.K.; Leemans, A. Diffusion tensor imaging. *Methods Mol. Biol.* **2011**, *711*, 127–144. [CrossRef]
138. Lenglet, C.; Campbell, J.S.; Descoteaux, M.; Haro, G.; Savadjiev, P.; Wassermann, D.; Anwander, A.; Deriche, R.; Pike, G.B.; Sapiro, G.; et al. Mathematical Methods for Diffusion MRI Processing. *NeuroImage* **2009**, *45*, S111. [CrossRef]
139. Gao, F.; Barker, P.B. Various MRS application tools for Alzheimer disease and mild cognitive impairment. *AJNR Am. J. Neuroradiol.* **2014**, *35*, S4–S11. [CrossRef]
140. Shonk, T.K.; Moats, R.A.; Gifford, P.; Michaelis, T.; Mandigo, J.C.; Izumi, J.; Ross, B.D. Probable Alzheimer disease: Diagnosis with proton MR spectroscopy. *Radiology* **1995**, *195*, 65–72. [CrossRef]
141. Klunk, W.E.; Panchalingam, K.; Moossy, J.; McClure, R.J.; Pettegrew, J.W. N-acetyl-L-aspartate and other amino acid metabolites in Alzheimer's disease brain: A preliminary proton nuclear magnetic resonance study. *Neurology* **1992**, *42*, 1578–1585. [CrossRef] [PubMed]
142. Kantarci, K.; Jack, C.R., Jr.; Xu, Y.C.; Campeau, N.G.; O'Brien, P.C.; Smith, G.E.; Ivnik, R.J.; Boeve, B.F.; Kokmen, E.; Tangalos, E.G.; et al. Regional metabolic patterns in mild cognitive impairment and Alzheimer's disease: A 1H MRS study. *Neurology* **2000**, *55*, 210–217. [CrossRef] [PubMed]
143. Kantarci, K.; Reynolds, G.; Petersen, R.C.; Boeve, B.F.; Knopman, D.S.; Edland, S.D.; Smith, G.E.; Ivnik, R.J.; Tangalos, E.G.; Jack, C.R., Jr. Proton MR spectroscopy in mild cognitive impairment and Alzheimer disease: Comparison of 1.5 and 3 T. *AJNR Am. J. Neuroradiol.* **2003**, *24*, 843–849.
144. Schuff, N.; Capizzano, A.A.; Du, A.T.; Amend, D.L.; O'Neill, J.; Norman, D.; Kramer, J.; Jagust, W.; Miller, B.; Wolkowitz, O.M.; et al. Selective reduction of N-acetylaspartate in medial temporal and parietal lobes in AD. *Neurology* **2002**, *58*, 928–935. [CrossRef] [PubMed]
145. Schuff, N.; Amend, D.; Ezekiel, F.; Steinman, S.K.; Tanabe, J.; Norman, D.; Jagust, W.; Kramer, J.H.; Mastrianni, J.A.; Fein, G.; et al. Changes of hippocampal N-acetyl aspartate and volume in Alzheimer's disease. A proton MR spectroscopic imaging and MRI study. *Neurology* **1997**, *49*, 1513–1521. [CrossRef]
146. Meyerhoff, D.J.; MacKay, S.; Constans, J.M.; Norman, D.; Van Dyke, C.; Fein, G.; Weiner, M.W. Axonal injury and membrane alterations in Alzheimer's disease suggested by in vivo proton magnetic resonance spectroscopic imaging. *Ann. Neurol.* **1994**, *36*, 40–47. [CrossRef]
147. Moats, R.A.; Ernst, T.; Shonk, T.K.; Ross, B.D. Abnormal cerebral metabolite concentrations in patients with probable Alzheimer disease. *Magn. Reson. Med.* **1994**, *32*, 110–115. [CrossRef]
148. Pfefferbaum, A.; Adalsteinsson, E.; Spielman, D.; Sullivan, E.V.; Lim, K.O. In vivo spectroscopic quantification of the N-acetyl moiety, creatine, and choline from large volumes of brain gray and white matter: Effects of normal aging. *Magn. Reson. Med.* **1994**, *41*, 276–284. [CrossRef]
149. Rose, S.E.; De Zubicaray, G.I.; Wang, D.; Galloway, G.J.; Chalk, J.B.; Eagle, S.C.; Semple, J.; Doddrell, D.M. A 1H MRS study of probable Alzheimer's disease and normal aging: Implications for longitudinal monitoring of dementia progression. *Magn. Reson. Imaging* **1999**, *17*, 291–299. [CrossRef]
150. Krishnan, K.R.; Charles, H.C.; Doraiswamy, P.M.; Mintzer, J.; Weisler, R.; Yu, X.; Perdomo, C.; Ieni, J.R.; Rogers, S. Randomized, placebo-controlled trial of the effects of donepezil on neuronal markers and hippocampal volumes in Alzheimer's disease. *Am. J. Psychiatry* **2003**, *160*, 2003–2011. [CrossRef]
151. Kantarci, K.; Petersen, R.C.; Boeve, B.F.; Knopman, D.S.; Tang-Wai, D.F.; O'Brien, P.C.; Weigand, S.D.; Edland, S.D.; Smith, G.E.; Ivnik, R.J.; et al. 1H MR spectroscopy in common dementias. *Neurology* **2004**, *63*, 1393–1398. [CrossRef] [PubMed]
152. Antuono, P.G.; Jones, J.L.; Wang, Y.; Li, S.J. Decreased glutamate + glutamine in Alzheimer's disease detected in vivo with (1)H-MRS at 0.5 T. *Neurology* **2001**, *56*, 737–742. [CrossRef] [PubMed]
153. Stuhler, E.; Platsch, G.; Weih, M.; Kornhuber, J.; Kuwert, T.; Merhof, D. Multiple discriminant analysis of SPECT data for alzheimer's disease, frontotemporal dementia and asymptomatic controls. In Proceedings of the IEEE Nuclear Science Symposium and Medical Imaging Conference (NSS/MIC), Valencia, Spain, 23–29 October 2011; IEEE: Anaheim, CA, USA, 2011; Volume 3, pp. 4398–43401. [CrossRef]

154. O'Brien, J.T.; Firbank, M.J.; Davison, C.; Barnett, N.; Bamford, C.; Donaldson, C.; Olsen, K.; Herholz, K.; Williams, D.; Lloyd, J. 18F-FDG PET and perfusion SPECT in the diagnosis of Alzheimer and Lewy body dementias. *J. Nucl. Med.* **2014**, *55*, 1959–1965. [CrossRef] [PubMed]
155. Yeo, J.M.; Lim, X.X.; Khan, Z.; Pal, S. Systematic review of the diagnostic utility of SPECT imaging in dementia. *Eur. Arch. Psychiatry Clin. Neurosci.* **2013**, *263*, 539–552. [CrossRef] [PubMed]
156. Dougall, N.J.; Bruggink, S.; Ebmeier, K. Systematic review of the diagnostic accuracy of 99mTc-HMPAO-SPECT in dementia. *Am. J. Geriatr. Psychiatry* **2004**, *12*, 554–570. [CrossRef]
157. Phelps, M.E. PET: The merging of biology and imaging into molecular imaging. *J. Nucl. Med.* **2000**, *41*, 661–681.
158. García-Ribas, G.; Arbizu, J.; Carrió, I.; Garrastachu, P.; Martinez-Lage, P. Biomarcadores por tomografia por emisión de positrons (PET): Imagen de la patología de Alzheimer y la neurodegeneración al servicio del diagnóstico clínico. *Neurología* **2017**, *32*, 275–277. [CrossRef]
159. De Leon, M.J.; Convit, A.; Wolf, O.T.; Tarshish, C.Y.; DeSanti, S.; Rusinek, H.; Tsui, W.; Kandil, E.; Scherer, A.J.; Roche, A.; et al. Prediction of cognitive decline in normal elderly subjects with 2-[(18)F]fluoro-2-deoxy-D-glucose/positronemission tomography (FDG/PET). *Proc. Natl. Acad. Sci. USA* **2001**, *98*, 10966–10971. [CrossRef]
160. Mosconi, L. Brain glucose metabolism in the early and specific diagnosis of Alzheimer's disease. FDG-PET studies in MCI and AD. *Eur. J. Nucl. Med. Mol. Imaging* **2005**, *32*, 486–510. [CrossRef]
161. Walhovd, K.B.; Fjell, A.M.; Brewer, J.; McEvoy, L.K.; Fennema-Notestine, C.; Hagler, D.J., Jr.; Jennings, R.G.; Karow, D.; Dale, A.M.; Alzheimer's Disease Neuroimaging Initiative. Combining MR imaging, positron-emission tomography, and CSF biomarkers in the diagnosis and prognosis of Alzheimer disease. *Am. J. Neuroradiol.* **2010**, *31*, 347–354. [CrossRef]
162. Ou, Y.N.; Xu, W.; Li, J.Q.; Guo, Y.; Cui, M.; Chen, K.L.; Huang, Y.Y.; Dong, Q.; Tan, L.; Yu, J.T.; et al. FDG-PET as an independent biomarker for Alzheimer's biological diagnosis: A longitudinal study. *Alzheimers Res. Ther.* **2019**, *11*, 57. [CrossRef] [PubMed]
163. Engler, H.; Forsberg, A.; Almkvist, O.; Blomquist, G.; Larsson, E.; Savitcheva, I.; Wall, A.; Ringheim, A.; Langstrom, B.; Nordberg, A. Two-year follow-up of amyloid deposition in patients with Alzheimer's disease. *Brain* **2006**, *129*, 2856–2866. [CrossRef] [PubMed]
164. Mielke, R.; Herholz, K.; Grond, M.; Kessler, J.; Heiss, W.D. Clinical deterioration in probable Alzheimer's disease correlates with progressive metabolic impairment of association areas. *Dementia* **1994**, *5*, 36–41. [CrossRef] [PubMed]
165. Sanchez-Catasus, C.A.; Stormezand, G.N.; Van Laar, P.J.; De Deyn, P.P.; Sanchez, M.A.; Dierckx, R.A. FDG-PET for prediction of AD dementia in mild cognitive impairment. A review of the state of the art with particular emphasis on the comparison with other neuroimaging modalities (MRI and Perfusion SPECT). *Curr. Alzheimer Res.* **2017**, *14*, 127–142. [CrossRef]
166. Altomare, D.; Ferrari, C.; Caroli, A.; Galluzzi, S.; Prestia, A.; Van der Flier, W.M.; Ossenkoppele, R.; Van Berckel, B.; Barkhof, F.; Teunissen, C.E.; et al. Prognostic value of Alzheimer's biomarkers in mild cognitive impairment: The effect of age at onset. *J. Neurol.* **2019**, *266*, 2535–2545. [CrossRef]
167. Bloudek, L.M.; Spackman, D.E.; Blankenburg, M.; Sullivan, S.D. Review and metaanalysis of biomarkers and diagnostic imaging in Alzheimer's disease. *J. Alzheimers Dis.* **2011**, *26*, 627–645. [CrossRef]
168. Herholz, K. The role of PET quantification in neurological imaging: FDG and amyloid imaging in dementia. *Clin. Transl. Imaging* **2014**, *2*, 321–330. [CrossRef]
169. Nordberg, A. Amyloid imaging in early detection of Alzheimer's disease. *Neurodegener. Dis.* **2010**, *7*, 136–138. [CrossRef]
170. Huijbers, W.; Mormino, E.C.; Wigman, S.E.; Ward, A.M.; Vannini, P.; McLaren, D.G.; Becker, J.A.; Schultz, A.P.; Hedden, T.; Johnson, K.A.; et al. Amyloid Deposition Is Linked to Aberrant Entorhinal Activity among Cognitively Normal Older Adults. *J. Neurosci.* **2014**, *34*, 5200–5210. [CrossRef]
171. Villeneuve, S.; Reed, B.R.; Wirth, M.; Haase, C.M.; Madison, C.M.; Ayakta, N.; Mack, W.; Mungas, D.; Chui, H.C.; DeCarli, C.; et al. Cortical thickness mediates the effect of β-amyloid on episodic memory. *Neurology* **2014**, *82*, 761–767. [CrossRef]
172. Leal, S.L.; Lockhart, S.N.; Maass, A.; Bell, R.K.; Jagust, W.J. Subthreshold Amyloid Predicts Tau Deposition in Aging. *J. Neurosci. Off. J. Soc. Neurosci.* **2018**, *38*, 4482–4489. [CrossRef] [PubMed]

173. Klunk, W.E.; Engler, H.; Nordberg, A.; Wang, Y.; Blomqvist, G.; Holt, D.P.; Bergstrom, M.; Savitcheva, I.; Huang, G.F.; Estrada, S.; et al. Imaging brain amyloid in Alzheimer's disease with Pittsburgh Compound-B. *Ann. Neurol.* **2004**, *55*, 306–319. [CrossRef] [PubMed]
174. Klunk, W.E.; Wang, Y.; Huang, G.F.; Debnath, M.L.; Holt, D.P.; Shao, L.; Hamilton, R.L.; Ikonomovic, M.D.; DeKosky, S.T.; Mathis, C.A. The binding of 2-(4′methylaminophenyl)benzothiazole to postmortem brain homogenates is dominated by the amyloid component. *J. Neurosci.* **2003**, *23*, 2086–2092. [CrossRef] [PubMed]
175. Price, J.C.; Klunk, W.E.; Lopresti, B.J.; Lu, X.; Hoge, J.A.; Ziolko, S.K.; Holt, D.P.; Meltzer, C.C.; De Kosky, S.T.; Mathis, C.A. Kinetic modeling of amyloid binding in humans using PET imaging and Pittsburgh Compound-B. *J. Cereb. Blood Flow Metab.* **2005**, *25*, 1528–1547. [CrossRef]
176. Mintun, M.A.; Larossa, G.N.; Sheline, Y.I.; Dence, C.S.; Lee, S.Y.; Mach, R.H.; Klunk, W.E.; Mathis, C.A.; DeKosky, S.T.; Morris, J.C. [11C]PIB in a nondemented population: Potential antecedent marker of Alzheimer disease. *Neurology* **2006**, *67*, 446–452. [CrossRef]
177. Lockhart, A.; Lamb, J.R.; Osredkar, T.; Sue, L.I.; Joyce, J.N.; Ye, L.; Libri, V.; Leppert, D.; Beach, T.G. PIB is a non-specific imaging marker of amyloid-beta (Aβ) peptiderelated cerebral amyloidosis. *Brain* **2007**, *130*, 2607–2615. [CrossRef]
178. Selkoe, D.J. Soluble oligomers of the amyloid β-protein impair synaptic plasticity and behavior. *Behav. Brain Res.* **2008**, *192*, 106–113. [CrossRef]
179. O'Brien, J.T.; Herholz, K. Amyloid imaging for dementia in clinical practice. *BMC Med.* **2015**, *13*, 163. [CrossRef]
180. Boccardi, M.; Altomare, D.; Ferrari, C.; Festari, C.; Guerra, U.P.; Paghera, B.; Pizzocaro, C.; Lussignoli, G.; Geroldi, C.; Zanetti, O.; et al. Assessment of the Incremental Diagnostic Value of Florbetapir F 18 Imaging in Patients with Cognitive Impairment: The Incremental Diagnostic Value of Amyloid PET With [18F]-Florbetapir (INDIA-FBP) Study. *JAMA Neurol.* **2016**, *73*, 1417–1424. [CrossRef]
181. Forsberg, A.; Engler, H.; Almkvist, O.; Blomquist, G.; Hagman, G.; Wall, A.; Ringheim, A.; Langstrom, B.; Nordberg, A. PET imaging of amyloid deposition in patients with mild cognitive impairment. *Neurobiol. Aging* **2008**, *29*, 1456–1465. [CrossRef]
182. Finnema, S.J.; Nabulsi, N.B.; Eid, T.; Detyniecki, K.; Lin, S.; Chen, M.; Dhaher, R.; Matuskey, D.; Baum, E.; Holden, D.; et al. Imaging synaptic density in the living human brain. *Sci. Transl. Med.* **2016**, *8*, 348ra96. [CrossRef] [PubMed]
183. Constantinescu, C.C.; Tresse, C.; Zheng, M.; Gouasmat, A.; Carroll, V.M.; Mistico, L.; Alagille, D.; Sandiego, C.M.; Papin, C.; Marek, K.; et al. Development and in vivo preclinical imaging of fluorine-18-labeled synaptic vesicle protein 2A (SV2A) PET tracers. *Mol. Imaging Biol.* **2019**, *21*, 509–518. [CrossRef] [PubMed]
184. Li, S.; Cai, Z.; Wu, X.; Holden, D.; Pracitto, R.; Kapinos, M.; Gao, H.; Labaree, D.; Nabulsi, N.; Carson, R.E.; et al. Synthesis and in vivo evaluation of a novel PET radiotracer for imaging of synaptic vesicle glycoprotein 2A (SV2A) in nonhuman primates. *ACS Chem. Neurosci.* **2019**, *10*, 1544–1554. [CrossRef] [PubMed]
185. Chen, M.K.; Mecca, A.P.; Naganawa, M.; Finnema, S.J.; Toyonaga, T.; Lin, S.; Najafzadeh, S.; Ropchan, J.; Lu, Y.; McDonald, J.W.; et al. Assessing synaptic density in Alzheimer disease with synaptic vesicle glycoprotein 2A positron emission tomographic imaging. *JAMA Neurol.* **2018**, *8042*, 1–10. [CrossRef] [PubMed]
186. Villemagne, V.L.; Fodero-Tavoletti, M.T.; Masters, C.L.; Rowe, C.C. Tau imaging: Early progress and future directions. *Lancet Neurol.* **2015**, *14*, 114–124. [CrossRef]
187. Schwarz, A.J.; Yu, P.; Miller, B.B.; Shcherbinin, S.; Dickson, J.; Navitsky, M.; Joshi, A.D.; Devous, M.D., Sr.; Mintun, M.S. Regional profiles of the candidate tau PET ligand 18F-AV-1451 recapitulate key features of Braak histopathological stages. *Brain* **2016**, *139*, 1539–1550. [CrossRef]
188. Wilson, H.; Pagano, G.; Politis, M. Dementia spectrum disorders: Lessons learnt from decades with PET research. *J. Neural. Transm. Vienna* **2019**, *126*, 233–251. [CrossRef]
189. Vemuri, P.; Lowe, V.J.; Knopman, D.S.; Senjem, M.L.; Kemp, B.J.; Schwarz, C.G.; Przybelski, S.A.; Machulda, M.M.; Petersen, R.C.; Jack, C.R., Jr. Tau-PET uptake: Regional variation in average SUVR and impact of amyloid deposition. *Alzheimers Dement. (Amst.)* **2016**, *6*, 21–30. [CrossRef]
190. Okamura, N.; Harada, R.; Furukawa, K.; Furumoto, S.; Tago, T.; Yanai, K.; Arai, H.; Kudo, Y. Advances in the development of tau PET radiotracers and their clinical applications. *Ageing Res. Rev.* **2016**, *30*, 107–113. [CrossRef]

191. Mirzaei, N.; Tang, S.P.; Ashworth, S.; Coello, C.; Plisson, C.; Passchier, J.; Selvaraj, V.; Tyacke, R.J.; Nutt, D.J.; Sastre, M. In vivo imaging of microglial activation by positron emission tomography with [11C]PBR28 in the 5XFAD model of Alzheimer's disease. *Glia* **2016**, *64*, 993–1006. [CrossRef]
192. James, M.L.; Belichenko, N.P.; Shuhendler, A.J.; Hoehne, A.; Andrews, L.E.; Condon, C.; Nguyen, T.V.; Reiser, V.; Jones, P.; Trigg, W.; et al. [18F]GE-180 PET Detects Reduced Microglia Activation After LM11A-31 Therapy in a Mouse Model of Alzheimer's Disease. *Theranostics* **2017**, *7*, 1422–1436. [CrossRef] [PubMed]
193. Owen, D.R.; Yeo, A.J.; Gunn, R.N.; Song, K.; Wadsworth, G.; Lewis, A.; Rhodes, C.; Pulford, D.J.; Bennacef, I.; Parker, C.A.; et al. An 18-kDa translocator protein (TSPO) polymorphism explains differences in binding affinity of the PET radioligand PBR28. *J. Cereb. Blood Flow Metab.* **2012**, *32*, 1–5. [CrossRef] [PubMed]
194. Wey, H.Y.; Wang, C.; Schroeder, F.A.; Logan, J.; Price, J.C.; Hooker, J.M. Kinetic Analysis and Quantification of [11C] Martinostat for in Vivo HDAC Imaging of the Brain. *ACS Chem. Neurosci.* **2015**, *6*, 708–715. [CrossRef] [PubMed]
195. Mastroeni, D.; Grover, A.; Delvaux, E.; Whiteside, C.; Coleman, P.D.; Rogers, J. Epigenetic mechanisms in Alzheimer's disease. *Neurobiol. Aging* **2011**, *32*, 1161–1180. [CrossRef]
196. Cencioni, C.; Spallotta, F.; Martelli, F.; Valente, S.; Mai, A.; Zeiher, A.M.; Gaetano, C. Oxidative stress and epigenetic regulation in ageing and age-related diseases. *Int. J. Mol. Sci.* **2013**, *14*, 17643–17663. [CrossRef]
197. Bihaqi, S.W.; Schumacher, A.; Maloney, B.; Lahiri, D.K.; Zawia, N.H. Do epigenetic pathways initiate late onset Alzheimer disease (LOAD): Towards a new paradigm. *Curr. Alzheimer Res.* **2012**, *9*, 574–588. [CrossRef]
198. Wey, H.Y.; Gilbert, T.M.; Zürcher, N.R.; She, A.; Bhanot, A.; Taillon, B.D.; Schroeder, F.A.; Wang, C.; Haggarty, S.J.; Hooker, J.M. Insights into neuroepigenetics through human histone deacetylase PET imaging. *Sci. Transl. Med.* **2016**, *8*, 351ra106. [CrossRef]
199. Strebl, M.G.; Wang, C.; Schroeder, F.A.; Placzek, M.S.; Wey, H.Y.; Van de Bittner, G.C.; Neelamegam, R.; Hooker, J.M. Development of a Fluorinated Class-I HDAC Radiotracer Reveals Key Chemical Determinants of Brain Penetrance. *ACS Chem. Neurosci.* **2016**, *7*, 528–533. [CrossRef]
200. Wagner, A.D. Early detection of Alzheimer's disease: An fMRI marker for people at risk? *Nat. Neurosci.* **2000**, *3*, 973–974. [CrossRef]
201. Tripoliti, E.E.; Fotiadis, D.I.; Argyropoulou, M.; Manis, G. A six stage approach for the diagnosis of the Alzheimer's disease based on fMRI data. *J. Biomed. Inform.* **2010**, *43*, 307–320. [CrossRef]
202. Wang, K.; Liang, M.; Wang, L.; Tian, L.; Zhang, X.; Li, K.; Jiang, T. Altered functional connectivity in early Alzheimer's disease: A resting-state fMRI study. *Hum. Brain Mapp.* **2007**, *28*, 967–978. [CrossRef] [PubMed]
203. Tripoliti, E.E.; Fotiadis, D.I.; Argyropoulou, M.A. A supervised method to assist the diagnosis and classification of the status of Alzheimer's disease using data from an fMRI experiment. In Proceedings of the 2008 30th Annual International Conference of the IEEE Engineering in Medicine and Biology Society, Vancouver, BC, Canada, 20–25 August 2008; Volume 2008, pp. 4419–4422. [CrossRef] [PubMed]
204. Sheline, Y.I.; Raichle, M.E. Resting state functional connectivity in preclinical Alzheimer's disease. *Biol. Psychiatry* **2013**, *74*, 340–347. [CrossRef]
205. Supekar, K.; Menon, V.; Rubin, D.; Musen, M.; Greicius, M.D. Network Analysis of Intrinsic Functional Brain Connectivity in Alzheimer's Disease. *PLoS Comput. Biol.* **2008**, *4*, e1000100. [CrossRef] [PubMed]
206. Sanz-Arigita, E.J.; Schoonheim, M.M.; Damoiseaux, J.S.; Rombouts, S.A.; Maris, E.; Barkhof, F.; Scheltens, P.; Stam, C.J. Loss of "small-world" networks in Alzheimer's disease: Graph analysis of FMRI resting-state functional connectivity. *PLoS ONE* **2010**, *5*, e13788. [CrossRef] [PubMed]
207. Zhang, Z.; Lu, G.; Zhong, Y.; Tan, Q.; Liao, W.; Wang, Z.; Wang, Z.; Li, K.; Chen, H.; Liu, Y. Altered spontaneous neuronal activity of the default-mode network in mesial temporal lobe epilepsy. *Brain Res.* **2010**, *1323*, 152–160. [CrossRef] [PubMed]
208. Petrella, J.R.; Sheldon, F.C.; Prince, S.E.; Calhoun, V.D.; Doraiswamy, P.M. Default mode network connectivity in stable vs progressive mild cognitive impairment. *Neurology* **2011**, *76*, 511–517. [CrossRef]
209. Fredericks, C.A.; Sturm, V.E.; Brown, J.A.; Hua, A.Y.; Bilgel, M.; Wong, D.F.; Resnick, S.M.; Seeley, W.W. Early affective changes and increased connectivity in preclinical Alzheimer's disease. *Alzheimers Dement. (Amst.)* **2018**, *10*, 471–479. [CrossRef]
210. Dickerson, B.C.; Salat, D.H.; Greve, D.N.; Chua, E.F.; Rand-Giovannetti, E.; Rentz, D.M.; Bertram, L.; Mullin, K.; Tanzi, R.E.; Blacker, D.; et al. Increased hippocampal activation in mild cognitive impairment compared to normal aging and AD. *Neurology* **2005**, *65*, 404–411. [CrossRef]

211. Sperling, R. Functional MRI Studies of Associative Encoding in Normal Aging, Mild Cognitive Impairment, and Alzheimer's Disease. *Ann. N. Y. Acad. Sci.* **2007**, *1097*, 146–155. [CrossRef]
212. Miller, S.L.; Fenstermacher, E.; Bates, J.; Blacker, D.; Sperling, R.A.; Dickerson, B.C. Hippocampal activation in adults with mild cognitive impairment predicts subsequent cognitive decline. *J. Neurol. Neurosurg. Psychiatry* **2008**, *79*, 630–635. [CrossRef] [PubMed]
213. Yassa, M.A.; Stark, S.M.; Bakker, A.; Albert, M.S.; Gallagher, M.; Stark, C.E.L. Highresolution structural and functional MRI of hippocampal CA3 and dentate gyrus in patients with amnestic Mild Cognitive Impairment. *NeuroImage* **2010**, *51*, 1242–1252. [CrossRef] [PubMed]
214. Deibler, A.R.; Pollock, J.M.; Kraft, R.A.; Tan, H.; Burdette, J.H.; Maldjian, J.A. Arterial spin-labeling in routine clinical practice, part 1: Technique and artifacts. *AJNR Am. J. Neuroradiol.* **2008**, *29*, 1228–1234. [CrossRef] [PubMed]
215. Dolui, S.; Li, Z.; Nasrallah, I.M.; Detre, J.A.; Wolk, D.A. Arterial spin labeling versus 18F-FDG-PET to identify mild cognitive impairment. *Neuroimage Clin.* **2020**, *25*, 102146. [CrossRef]
216. Zhang, J. How far is arterial spin labeling MRI from a clinical reality? Insights from arterial spin labeling comparative studies in Alzheimer's disease and other neurological disorders. *J. Magn. Reson. Imaging* **2016**, *43*, 10201045. [CrossRef]
217. Ganmor, E.; Segev, R.; Schneidman, E. The architecture of functional interaction networks in the retina. *J. Neurosci.* **2011**, *31*, 3044–3054. [CrossRef]
218. Masland, R.H. The neuronal organization of the retina. *Neuron* **2012**, *76*, 266–280. [CrossRef]
219. London, A.; Benhar, I.; Schwartz, M. The retina as a window to the brain-from eye research to CNS disorders. *Nat. Rev. Neurol.* **2013**, *9*, 44–53. [CrossRef]
220. Van Wijngaarden, P.; Hadoux, X.; Alwan, M.; Keel, S.; Dirani, M. Emerging ocular biomarkers of Alzheimer disease. *Clin. Exp. Ophthalmol.* **2017**, *45*, 54–61. [CrossRef]
221. Stix, B.; Leber, M.; Bingemer, P.; Gross, C.; Rüschoff, J.; Fändrich, M.; Schorderet, D.F.; Vorwerk, C.K.; Zacharias, M.; Roessner, A.; et al. Hereditary lattice corneal dystrophy is associated with corneal amyloid deposits enclosing C-terminal fragments of keratoepithelin. *Invest. Ophthalmol. Vis. Sci.* **2005**, *46*, 1133–1139. [CrossRef]
222. Prakasam, A.; Muthuswamy, A.; Ablonczy, Z.; Greig, N.H.; Fauq, A.; Rao, K.J.; Pappolla, M.A.; Sambamurti, K. Differential accumulation of secreted AbetaPP metabolites in ocular fluids. *J. Alzheimers Dis.* **2010**, *20*, 1243–1253. [CrossRef] [PubMed]
223. Örnek, N.; Dağ, E.; Örnek, K. Corneal sensitivity and tear function in neurodegenerative diseases. *Curr. Eye Res.* **2015**, *40*, 423–428. [CrossRef] [PubMed]
224. Mancino, R.; Martucci, A.; Cesareo, M.; Giannini, C.; Corasaniti, M.T.; Bagetta, G.; Nucci, C. Glaucoma and Alzheimer Disease: One Age-Related Neurodegenerative Disease of the Brain. *Curr. Neuropharmacol.* **2018**, *16*, 971–977. [CrossRef] [PubMed]
225. Frost, S.; Kanagasingam, Y.; Sohrabi, H.; Vignarajan, J.; Bourgeat, P.; Salvado, O.; Villemagne, V.; Rowe, C.C.; Macaulay, S.L.; Szoeke, C.; et al. Retinal vascular biomarkers for early detection and monitoring of Alzheimer's disease. *Transl. Psychiatry* **2013**, *3*, e233. [CrossRef]
226. Feke, G.T.; Hyman, B.T.; Stern, R.A.; Pasquale, L.R. Retinal blood flow in mild cognitive impairment and Alzheimer's disease. *Alzheimers Dement. (Amst.)* **2015**, *1*, 144–151. [CrossRef]
227. Parisi, V.; Restuccia, R.; Fattapposta, F.; Mina, C.; Bucci, M.G.; Pierelli, F. Morphological and functional retinal impairment in Alzheimer's disease patients. *Clin. Neurophysiol.* **2001**, *112*, 1860–1867. [CrossRef]
228. Kesler, A.; Vakhapova, V.; Korczyn, A.D.; Naftaliev, E.; Neudorfer, M. Retinal thickness in patients with mild cognitive impairment and Alzheimer's disease. *Clin. Neurol. Neurosurg.* **2011**, *113*, 523–526. [CrossRef]
229. Koronyo, Y.; Biggs, D.; Barron, E.; Boyer, D.S.; Pearlman, J.A.; Au, W.J.; Kile, S.J.; Blanco, A.; Fuchs, D.T.; Ashfaq, A.; et al. Retinal amyloid pathology and proof-of-concept imaging trial in Alzheimer's disease. *JCI Insight* **2017**, *2*, 93621. [CrossRef]

Commentary

Biomarkers for Alzheimer's Disease (AD) and the Application of Precision Medicine

Walter J. Lukiw [1,2,3,4,*], Andrea Vergallo [5], Simone Lista [5,6,7], Harald Hampel [5] and Yuhai Zhao [1,2]

1 LSU Neuroscience Center, Louisiana State University Health Science Center, New Orleans, LA 70112, USA; yzhao4@lsuhsc.edu
2 Department of Cell Biology and Anatomy, LSU-HSC, New Orleans, LA 70112, USA
3 Department of Ophthalmology, LSU Neuroscience Center, LSU-HSC, New Orleans, LA 70112, USA
4 Department Neurology, LSU Neuroscience Center, LSU-HSC, New Orleans, LA 70112, USA
5 Sorbonne University, GRC no 21, Alzheimer Precision Medicine (APM), AP-HP, Pitié-Salpêtrière hospital, F-75013 Paris, France; a_vergallo@yahoo.com (A.V.); slista@libero.it (S.L.); harald.hampel@med.uni-muenchen.de (H.H.)
6 Brain & Spine Institute (ICM), INSERM U 1127, CNRS UMR 7225, Boulevard de l'Hôpital, F-75013 Paris, France
7 Department of Neurology, Institute of Memory and Alzheimer's Disease (IM2A), Pitié-Salpêtrière Hospital, AP-HP, F-75013 Paris, France
* Correspondence: wlukiw@lsuhsc.edu

Received: 18 August 2020; Accepted: 15 September 2020; Published: 21 September 2020

Abstract: An accurate diagnosis of Alzheimer's disease (AD) currently stands as one of the most difficult and challenging in all of clinical neurology. AD is typically diagnosed using an integrated knowledge and assessment of multiple biomarkers and interrelated factors. These include the patient's age, gender and lifestyle, medical and genetic history (both clinical- and family-derived), cognitive, physical, behavioral and geriatric assessment, laboratory examination of multiple AD patient biofluids, especially within the systemic circulation (blood serum) and cerebrospinal fluid (CSF), multiple neuroimaging-modalities of the brain's limbic system and/or retina, followed up in many cases by post-mortem neuropathological examination to finally corroborate the diagnosis. More often than not, prospective AD cases are accompanied by other progressive, age-related dementing neuropathologies including, predominantly, a neurovascular and/or cardiovascular component, multiple-infarct dementia (MID), frontotemporal dementia (FTD) and/or strokes or 'mini-strokes' often integrated with other age-related neurological and non-neurological disorders including cardiovascular disease and cancer. Especially over the last 40 years, enormous research efforts have been undertaken to discover, characterize, and quantify more effectual and reliable biological markers for AD, especially during the pre-clinical or prodromal stages of AD so that pre-emptive therapeutic treatment strategies may be initiated. While a wealth of genetic, neurobiological, neurochemical, neuropathological, neuroimaging and other diagnostic information obtainable for a single AD patient can be immense: (**i**) it is currently challenging to integrate and formulate a definitive diagnosis for AD from this multifaceted and multidimensional information; and (**ii**) these data are unfortunately not directly comparable with the etiopathological patterns of other AD patients even when carefully matched for age, gender, familial genetics, and drug history. Four decades of AD research have repeatedly indicated that diagnostic profiles for AD are reflective of an extremely heterogeneous neurological disorder. This commentary will illuminate the heterogeneity of biomarkers for AD, comment on emerging investigative approaches and discuss why ***'precision medicine'*** is emerging as our best paradigm yet for the most accurate and definitive prediction, diagnosis, and prognosis of this insidious and lethal brain disorder.

Keywords: Alzheimer's disease (AD); biomarkers; diagnostics; messenger RNA; microRNA; neuroimaging; neurotropic microbes; precision medicine; prognostics

1. Overview

Senile dementia is the progressive, age-related loss of memory and cognition sufficiently severe to irreversibly affect social, behavioral, perceptual, occupational, and functional capabilities. Recent statistics indicate that globally, about ~50 million people live with dementia, now costing an estimated one trillion dollars in annual healthcare. By 2050, the number of people with dementia is projected to increase to ~130 million. In the United States, Alzheimer disease (AD), the leading cause of senile dementia in the elderly, currently affects about ~6 million people age 65 and older; by 2050, the number of people aged 65 and older with AD will grow to a projected ~15 million if no medical breakthroughs occur to prevent, slow, or cure this incapacitating disorder of the human mind [1–3]. One hundred and fourteen years since its original description, despite immense research efforts and clinical trials employing multiple strategic therapeutic approaches, there is currently no adequate treatment or cure for this widely expanding socioeconomic and healthcare concern [2–8].

The discovery, characterization, and quantification of biomarkers as measurable substances or cognitive disruptions in the ***'prospective AD patient'***, whose presence are indicative of disease, are urgently needed so that: (**i**) AD may be more accurately diagnosed, especially at an earlier ***'prodromal'*** stage and with the goal of preventive and or targeted therapeutic strategies that may be implemented at the earliest signs of AD onset; and (**ii**) more effective and reliable integration of multi-modal biomarkers for AD that can streamline, support, and strengthen the diagnostic and therapeutic decision-making. Remarkably, peer-reviewed publications on biomarkers for AD have yielded almost ~53,000 original research reports and reviews since 1983 crossing the words ***'Alzheimer's disease'*** and ***'biomarkers'***; [2,3,8–10]. These include observations on the classical and established AD biomarkers [11–14], including altered genetics (incorporating genome-wide association studies or GWAS), genetic mutations and gene modifications (including methylation and post-transcriptional modifications), end-stage neurotoxic and pathogenic metabolic products that accumulate in AD brains, such as multiple forms of tau aggregates and amyloid-beta (Aβ) aggregation species and plaques. AD biomarkers also include protein, lipid, proteolipid, inflammatory cytokine, chemokine, carbohydrate, microRNA (miRNA), and messenger RNA (mRNA) abundance, speciation, and complexity, as well as an evolving assortment of neuro-radiological and neuroimaging technologies (Table 1). AD biomarkers are certainly useful in the detection of dementing illness during its progressive course, and their appearance and magnitude correlate with cognitive loss in a dynamic way, allowing clinicians to monitor responses to therapeutic intervention across a background of aging of the AD patient.

Table 1. Multiple interrelated factors contribute to AD. The considerable heterogeneity of Alzheimer's disease (AD) appears to be mediated in part by a highly interconnected network of biological factors, and each of these can be used as diagnostic biomarkers which appear to each have a variable potential to contribute to AD-type change. There is abundant evidence that all 23 of these biomarkers and/or factors (listed alphabetically) contribute to AD initiation, onset, or propagation, and there may be other important biological factors and other elements that may contribute to this complex neurological disease that we have not yet recognized or even considered. Data derived from each of these multiple biomarkers and factors combined is amenable to systems and network analysis, information integration, and the application of ***precision medicine*** that should ultimately yield a more accurate diagnosis of AD at any stage of the disease (see text for references; specific references to the biomarkers listed in Table 1 can be found in [15–29]. Abbreviations: BACE = β-amyloid cleavage enzyme; CRP = C-reactive protein (a blood-serum-based inflammatory biomarker); lncRNA = long non-coding RNA; PSI, PS2 = presenilin 1, presenilin 2; rRNA = ribosomal RNA; sncRNA = sall non-coding RNA.

age and age-related effects;
amyloid (Aβ40 and Aβ42 peptides);
compartmentalization of biomarkers [brain tissue, extracellular fluid (ECF), CSF, blood serum, urine];
cytokine storm (cytokines and chemokines);
environment and environmental effects;
epigenetics (methylation, mRNA and miRNA editing);
exosomes and extracellular micro-vesicles (EXs and EMVs);
gender and gender-related effects;
genetics (mutations in BACE, PS1, PS2, etc.,);
gastrointestinal (GI) tract microbiome;
innate immunity;
Neuro-inflammatory markers (CRP);
inter-current illness (cardiovascular disease);
lifestyle (diet, smoking);
messenger RNA (mRNA);
microbial contribution (viral, bacterial, fungal, other);
microbiome (oral, other);
microRNA (miRNA);
miRNA-mRNA linking patterns;
misdiagnosis;
oral microbiome and dental hygiene;
other RNA (sncRNA, lncRNA);
overlapping neurological disorders: [Downs syndrome (Trisomy 21), frontotemporal dementia (FTD), multi-infarct dementia (MID), neuro-vascular disease, prion disease (PrD), etc.,].

As improvements in AD diagnostics are based on advances in both AD biomarker acquisition and the technologies used to gain these data, below we briefly discuss some of the most recent advances contributing in a major way to the more accurate and comprehensive accrual of important AD biomarker data.

2. Novel, Emerging, and Advanced Diagnostic Biomarkers for AD

2.1. Analysis of Exosomes (EXs), Extracellular Microvesicles (EMVs), and Their Molecular Cargos

Currently, the complex molecular cargos of exosomes and extracellular microvesicles (EXs and EMVs) have emerged as one of the most representative, significant, dependable, and trustworthy of all AD biomarkers. Typically, EX and EMV cargos consist of various mixtures of protein, lipid, proteolipid, cytokine, chemokine, carbohydrate, microRNA (miRNA) and messenger RNA (mRNA), and other constituents including end-stage neurotoxic and pathogenic metabolic products. These in part, consist of tau proteins, amyloid beta (Aβ) peptides, alpha-synuclein, TAR DNA-binding protein 43 (TDP-43) and others. EXs and EMVs (**i**) have been analyzed in the cerebrospinal fluid (CSF), blood serum, and post-mortem tissues of AD patients; (**ii**) are derived from their cells of origin and typically contain hundreds of different signaling molecules, many of which are potentially pathogenic and may be involved in the horizontal spread of neurological disease from one brain region to another; (**iii**) may represent a defined class of plasma membrane-derived nanovesicles released by all cell lineages of the human central nervous system (CNS); and (**iv**) as potential biomarkers, may contribute to an additional element of certainty into the diagnostic assessment of AD [20,21,26,30–32].

2.2. The Evaluation of Neurotropic Microbes in AD as Potential Diagnostic Biomarkers

There is a wealth of data indicating that neurotropic microbes including both DNA and RNA viruses (such as *Herpes simplex 1* or *SARS-CoV-2*) or bacterial Phyla such as *Proteobacteria*, *Verrucomicrobia*, *Fusobacteria*, *Cyanobacteria*, *Actinobacteria*, and *Spirochetes* or microbe-derived viral, fungal, or prokaryotic cellular components or microbial neurotoxins have high affinity for neural cells of the human brain and CNS [24,33–41]. Multiple independent laboratories continue to report the detection of viral, bacterial, fungal, protozoal, or other microbially-derived nucleic acid sequences or neurotoxins, such as highly inflammatory bacterial amyloid peptides, lipopolysaccharide (LPS), and many microbe-derived endotoxins within AD affected brain tissues [24,29,34,36,39,40]. Microbial biomarkers and systems biology approaches to understand host–microbiome interactions have been suggested by multiple AD researchers that both: (**i**) predict the risk of developing AD well before the onset of cognitive decline; and (**ii**) stimulate and/or accelerate the development of classical AD neuropathology [24,34,39,41–44].

Whether these viral, bacterial, or other microbial DNA- or RNA-based nucleic acids or associated lipoproteins, liposaccharides, peptidoglycans, bacterial-derived amyloids, and/or neurotoxins originate from the gastrointestinal (GI) tract microbiome, a potential brain microbiome, or some dormant pathological microbiome is currently not well understood [24,35,36,40,43,45]. Since 1978, at least ~4400 peer-reviewed research articles provide convincing evidence that multiple species of microbes, including viruses, bacteria (especially Gram-negative bacteria), and other microorganisms or their secreted components are strongly associated with the onset and/or the development of AD-type change [24,29,33,34,41–43,46]. If microbial presence in the brain is involved in the early initiation or propagation of AD, as currently suspected, then specialized RNA-sequencing applications or nucleic acid-containing gene chips, electrochemical biosensors, or panels of microbial-derived 16S ribosomal RNA (rRNA) interrogated with nucleic acid probes derived from AD biofluids might be useful as novel AD biomarkers in the detection of microbial patterns of expression from human brain tissues at any stage or degree of AD neuropathology.

2.3. Linking microRNA-messenger RNA (miRNA-mRNA) Signaling Patterns in AD

DNA microfluidic array technologies, quantitative reverse transcription PCR (RT-qPCR), RNA sequencing, LED-Northern, Western immunoassay, and electrochemical biosensors, integrated by advanced bioinformatics tools have uncovered families of up-regulated human brain enriched microRNAs (miRNAs) and their down-regulated messenger RNA (mRNA) targets. These have been found in short post-mortem interval (PMI) sporadic AD brain, in transgenic animal models of AD (TgAD), in brain biopsies, and in biofluids from AD patients. Genome-wide association studies (GWAS), epigenetic evaluations, such as miRNA-mRNA linkage or association mapping for AD-relevant neurological pathways, should provide useful diagnostic approaches since it has recently become apparent that miRNA-mediated mRNA-targeted regulatory mechanisms involve a large number of pathogenic and highly integrated gene expression pathways in the CNS [25,32,45–48]. To cite one very recent example, the human-brain-resident, nuclear factor kappa B p50/p65 (NF-κB p50/p65)-regulated microRNA-146a (miRNA-146a) is an inducible, 22-nucleotide, single-stranded non-coding RNA (sncRNA) easily detected in CNS neurons and immunological cell types. An inducible miRNA-146a: (**i**) is significantly up-regulated in AD brain tissues, CSF, and blood serum [25,48]; (**ii**) is an important epigenetic modulator of inflammatory signaling and the innate-immune response in several neurological disorders; and (**iii**) is essential in the down-regulation of the innate-immune regulator complement factor H (CFH; [10,20,23,25,40,46,49]).

LPS- and NF-kB (p50/p65)-inducible microRNAs, such as miRNA-146a and miRNA-125b, appear to contribute to neuropathological, neuro-inflammatory, and altered neuro-immunological aspects of both AD and prion disease (PrD; [25,32,40,46,48–50]). Interestingly, NF-κB-sensitive up-regulated miRNAs and their down-regulated mRNA targets appear to constitute an integrated NF-κB-miRNA-mRNA signaling network implicated in multiple AD pathophysiological processes [10,40,45,48,50,51]. Hence, potential signaling pathways to the acquisition of the AD phenotype appear to occur in part via an integrated and highly complex system of multiple miRNA-mRNA interactions that define many key pathogenic and pro-inflammatory gene expression pathways. Genetic and epigenetic signaling via miRNA-mRNA networks in the brain may be one of the most useful as potential biomarkers for early AD detection as they can detect subtle failure in multiple AD-relevant brain signaling systems and metabolic pathways [10,25,45,49,51].

2.4. Recent Advances in Neuro-Radiological and Neuroimaging Technologies

A number of neuro-radiological and neuroimaging technologies are currently being used to view physical atrophy and structural change in specific anatomical regions of the human brain, such as the hippocampus, neocortex (gray matter), white matter, ventricles, and other brain regions for the purpose of acquiring real time data for the diagnosis of AD [19,28,52–61]. These neuroimaging technologies and structural and functional imaging techniques include computerized tomography (CT; including dual-energy CT), positron emission tomography (PET), scintigraphic neuroimaging (PET-SN), diffuse optical imaging (DOT), structural and functional magnetic resonance imaging (MRI), including ultra-high field MRI (UHF-MRI), magnetoencephalography (MEG), single-photon emission computed tomography (SPECT), cranial ultrasound, and functional ultrasound imaging (fUS) in the search for anatomically-based diagnostic biomarkers for AD with high accuracy and sensitivity [19,28,57–62]. Neuroimaging techniques, hardware and software design, and imaging resolution are being constantly improved, updated, and refined [28,60–62]. For example, with high signal-to-noise (S/N) ratios, improved contrast and unparalleled spatial resolution, ultra-high field MRI of ≥7 tesla (T) has been highly successful in imaging the neuroanatomy of highly focused brain regions targeted by AD pathophysiology while providing additional information on morphological, quantitative, and subtle metabolic changes associated with early AD-type pathological alterations [57,61]. In vivo biomarkers for AD performed by recently implemented scintigraphic neuroimaging and employing amyloid binding PET agents along with non-scintigraphic biomarkers from blood (plasma/serum) and CSF have provided unique and novel opportunities to investigate the pathogenesis, prodromal changes, and time course of the disease in living individuals across the AD continuum [19,28,61–63]. Imaging technologies have indicated that AD changes in brain tissues begin as much as ~25 years prior to the onset of clinical symptomatology [28,61,63,64]. The opportunities afforded by *in vivo* biomarkers of AD, whether by blood (plasma/serum) or CSF examination or imaging technologies, are beginning to transform the strategic design of AD therapeutic trials by shifting the focus to the preclinical stages of the disease and massively integrating both molecular-based and neuroimaging data [60,61,63–67].

3. AD Biomarkers and Post-Mortem Neuropathological Examination of the AD Brain

Classically, the diagnosis of AD was a clinic-pathological one and there was a considerable error rate in the clinical diagnosis, especially early in the course of the disease. The differential diagnosis for AD by exclusion was confounded by a great many clinically overlapping neurological disorders including, mainly, MID, FTD, prion disease, tumors, subdural hematomas, neurovascular disruption and disease, and others [4–6]. Early neurophysiological diagnostic observations of AD included a diffusely slow electroencephalogram (EEG) and reduced cerebral blood flow [4,5,7]. Early PET studies demonstrated that oxygen extraction in the AD brain was relatively normal, thus tentatively excluding ischemia as a potential pathogenic factor [4,5,64]. Morphological pathological changes including the appearance of amyloid-enriched senile plaques (SP) and neurofibrillary tangles (NFT), widely distributed in neocortex but excluded from the basal ganglia, thalamus, and substantia nigra, and a severe loss of large neocortical neurons, were 'classical' diagnostic characteristics of the AD patient [4–8].

Usually at the family or care-giver's request, post-mortem neuropathological examinations of the deceased AD patient's brain were routinely performed by qualified AD-trained specialists and neuropathologists and brain tissues were often subsequently provided to AD researchers for further molecular-genetic and biomarker research including the examination for the presence of AD-relevant brain lesions. Light microscopy, NFT and SP amyloid dyes (such as Congo red, Thioflavin S, Thioflavin T and methylene yellow), and antibody-based staining (such as 6E10 and 10G4), the evaluation, density, composition, and quantitation of NFT and SP amyloid, and the examination of blood (plasma/serum) or CSF amyloid were additional indicators of immune- or inflammatory-neuropathology in the individual AD patient, which often contributed to the confirmation of AD in the ***'prospective AD patient'***. Currently, in many medical schools in the US and Canada, the post-mortem examination of the AD brain still remains the classical exercise to certify and verify the existence of AD. It is generally appreciated that the application of ***'precision medicine'***, involving massively integrated data sets of multi-faceted AD biomarkers, data-driven analytical methodologies, and the application of systems theory, will challenge and may eventually supersede the need for the classical post-mortem neuropathological examination of the brain in order to verify and confirm the diagnosis of AD [10–16,22,59,64,65,68].

3.1. Challenges in the Validation of AD Biomarkers

There are inherent problems in current approaches to AD biomarker research: (**i**) as most early reports emphasized just one or at most a few AD molecular-genetic biomarkers, biophysical, and neuroimaging modalities without consideration of the other hundreds or thousands of proteins, peptides, carbohydrates, lipids, or DNA and multiple species of RNA that have been previously implicated as being 'probable' diagnostic markers for AD; (**ii**) very often the nature of the acquisition of AD biomarkers represented a '***snapshot in time***' of one specific portion of the AD continuum that does not take into consideration the time course of changes observed in AD and/or the progressive nature of the disorder; (**iii**) because easily accessible and non-invasive AD biomarkers are often limited in their diagnostic applicability because of their overlap with other neurological diseases related to AD, such as Down's syndrome, Parkinson's and Lewy body disease, prion disease, FTD, hippocampal sclerosis, and MID and stroke; and (**iv**) no single newly generated ***de novo*** biomarker has ever been found to be associated with AD; that is, fluctuations in the abundance of pre-existing AD biomarkers reflect significant and absolute differences in the quantity, speciation, and stoichiometric relationships of AD-related biomolecules, including indicators of pro-inflammatory and immune system dysfunction. Put another way, no 'specific' AD biomarker 'suddenly appears' at the earliest onset, or propagation, or throughout any time-point during the course of AD, or at any stage of cognitive failure for that matter. Rather, it is usually a quantifiable up-regulation or down-regulation of an already existing biomarker in a specific anatomical region or biofluid compartment that has been the most consistently observed in the progressively degenerating brain. To cite a recent example, over 50 susceptibility genes and gene loci have been associated with late-onset AD and multiple models have been proposed [27], however, these associations are relatively rare and non-penetrant, occur in a few but not all AD cases, adding to the complexity and heterogeneity in the diagnosis of AD [15,56,58,63,64,69,70]. To further confound the establishment of definitive AD biomarkers, AD is commonly associated with more than one single neuropathology, in the case of AD usually with cerebrovascular and/or neurovascular involvement, and every one of these ancillary neurological disorders can carry their own set of complex and often overlapping disease biomarkers [10,63,64,69,70].

Especially over the last 10 years, the progressive and steadily increasing accumulation of molecular, genetic, epigenetic, neuroimaging, clinical, and geriatric data acquired from multiple AD cohorts has significantly increased our appreciation and understanding of the intrinsic variability and heterogeneity of AD biomarkers associated with the continuum of AD and other forms of progressive age-related neurodegenerative disease. The generation of massive datasets integrating multiple genetic-, epigenetic-, molecular-, and neuroimaging-derived biomarkers is enabling the application of clustering techniques and the identification and stratification of AD subtypes that may further categorize the multiple aspects of AD heterogeneity [10–18,67–70]. These approaches hold great potential: **(i)** for improving both the diagnosis and prognosis of AD; **(ii)** for projecting the clinical and neurological evolution of AD for planning suitable directions in therapeutic mediation; **(iii)** in providing multiple opportunities for the more directed analysis of AD heterogeneity in a data driven manner; **(iv)** in providing strategic guidelines for more decisive therapeutic intervention and the more efficacious clinical management of AD; and **(v)** for advancing *'**precision medicine**'* not only for the individual AD patient, but also for other cases of inflammatory neurodegeneration and neurological disease.

3.2. Using Precision Medicine in the Diagnosis of AD

Multiple analytical molecular-genetic approaches, advances in geriatric psychiatry and clinical evaluation, advancements in neuro-radiological labelling techniques and neuroimaging technologies, integrated diagnostic and predictive strategies and methodological improvements, discoveries of the comprehensive pathophysiological profiles of complex multi-factorial neurodegenerative diseases: **(i)** are presently well within the capabilities and scope of contemporary clinical, medical, and diagnostic neurology; **(ii)** are currently yielding increasingly large volumes of biomarker data for both individual AD patients, large populations of AD cases and age- and gender-matched controls; and **(iii)** are providing a data-driven basis for the paradigm shift of using the *'**precision medicine**'* approach in AD prevention, diagnostics, prognostics, and therapeutics [10,13,14,16,22,27,64]. Less common clinical presentations of AD are becoming increasingly recognized, adding to the increasing volume of AD biomarker data [10,14,17,19,64]. Since one of the pillars of *'**precision medicine**'* is supported by biomarker-derived medical data, further improvements in the acquisition, integration, interpretation, and bioinformatics aspects of clinical data and the coordination and analysis of clinical, laboratory, molecular-genetic, and neuroimaging data, geriatric and psychological information and related healthcare resources should obtain significantly increased accuracy in the diagnostic synopsis for the *"**prospective AD patient**"*. The significant heterogeneity of the AD condition: **(i)** will certainly benefit from an equally wide variety of AD biomarker-derived *'**precision medicine**'*-oriented treatment approaches and/or data-driven pharmacological strategies; and **(ii)** whose biomarker-based therapeutic design will greatly improve the current situation regarding the healthcare, more effective and successful treatment, and the development of disease-modifying drugs for AD patients at any stage of the disease [10,22,65,68,71].

4. Summary

The ongoing search for valid biomarkers for AD is being carried out globally in at least a dozen major geriatric, bioinformatic, neurobiological, neuro-genetic and neurological bioscience arenas: **(i)** those involving the age, gender, and geriatrics of the *'**prospectiveAD patient**'*; **(ii)** in the genetics and epigenetics of the AD patient including messenger RNA (mRNA) and microRNA (miRNA) signaling patterns, complexity and genomic methylation research; **(iii)** in multiple biofluids from AD patients including the blood (plasma/serum) of the systemic circulation, the glymphatic system, the cerebrospinal fluid (CSF) and/or urine; **(iv)**; through the detailed analysis of molecular cargos from both biofluids and tissue-compartmentalized exosomes and extracellular microvesicles (EXs and EMVs); **(v)** throughout the peripheral nervous system (PNS; typically using skin biopsies); **(vi)** via clinically-based geriatric, psychiatric, and neurological assessment and testing; **(vii)** via advances in neuro-radiological labeling techniques and neuroimaging technologies including CAT, PET, PET-SN,

MRI, fMRI; UHF-MRI, DOT, MEG, SPECT, cranial ultrasound, functional ultrasound (fUS) imaging, and immunohistochemistry involving confocal laser scanning microscopy and other advanced microscopic and neuroimaging techniques; (**viii**) from the quantitation and characterization of the load of microbial and microbial-derived components in the AD-affected brain; (**ix**) via the identification, quantitation, and characterization of AD-specific lesions including amyloid peptide-enriched SPs and NFTs; (**x**) after post-mortem examination and biopsies of AD cases, again matched up against those same biomarkers in age- and gender-matched neurologically normal controls to corroborate the prospective diagnosis of AD; (**xi**) via the comprehensive analysis of the potential contribution of overlapping progressive, age-related neurological disorders to AD-type change; and lastly (**xii**), through the assessment of the socioeconomic, environmental, and lifestyle factors of the ***'prospectiveAD patient'*** (Table 1). The recent application of highly integrated data sets of these multiple AD biomarkers and analytical techniques on large cohorts of AD patients and involving systems-biology and ***'precision medicine'*** may well serve to unravel many of the more intricate aspects of AD heterogeneity and expand and build on current therapeutic strategies to more effectively address both the diagnosis and clinical management of this devastating neurological disease.

Author Contributions: Conceptualization, W.J.L. and H.H.; writing original draft preparation, W.J.L. and Y.Z.; Validation/formal analysis/writing review and editing/visualization/supervision/project administration/funding acquisitions W.J.L., H.H., A.V., S.L., Y.Z. All authors have read and agreed to the published version of the manuscript.

Funding: This research was funded through an unrestricted grant to the LSU Eye Center from Research to Prevent Blindness (RPB); the Louisiana Biotechnology Research Network (LBRN) and NIH grants NEI EY006311, NIA AG18031 and NIA AG038834 (W.J.L.).

Acknowledgments: A.V. is an employee of Eisai Inc. This work has been performed during his previous position at Sorbonne University, Paris, France. H.H. is an employee of Eisai Inc. This work has been performed during his previous position at Sorbonne University, Paris, France. At Sorbonne University he was supported by the AXA Research Fund, the "*Fondation partenariale Sorbonne Université*" and the "*Fondation pour la Recherche sur Alzheimer*", Paris, France. This work was presented in part at the Society for Neuroscience (SFN) Annual Meeting 19–23 October 2019 Chicago IL, USA. Sincere thanks are extended to L. Carver, E. Head, W. Poon, H. LeBlanc, F. Culicchia, C. Eicken and C. Hebel for short post-mortem interval (PMI) human brain and/or retinal tissues or extracts, miRNA array work and initial data interpretation, and to D Guillot and AI Pogue for expert technical assistance. Thanks are also extended to the many neuropathologists, physicians and researchers of Canada and the US who have provided high quality, short post-mortem interval (PMI), human CNS, retinal tissues or extracted total brain and retinal RNA and biofluids (ECF, CSF and blood serum sample) for scientific studies.

Conflicts of Interest: The authors declare no conflict of interest.

Appendix A

H.H. is an employee of Eisai Inc. and serves as Senior Associate Editor for the Journal Alzheimer's & Dementia and does not receive any fees or honoraria since May 2019; before May 2019 he had received lecture fees from Servier, Biogen and Roche, research grants from Pfizer, Avid, and MSD Avenir (paid to the institution), travel funding from Functional Neuromodulation, Axovant, Eli Lilly and company, Takeda and Zinfandel, GE-Healthcare and Oryzon Genomics, consultancy fees from Qynapse, Jung Diagnostics, Cytox Ltd., Axovant, Anavex, Takeda and Zinfandel, GE Healthcare and Oryzon Genomics, and Functional Neuromodulation, and participated in scientific advisory boards of Functional Neuromodulation, Axovant, Eisai, Eli Lilly and company, Cytox Ltd., GE Healthcare, Takeda and Zinfandel, Oryzon Genomics and Roche Diagnostics. H.H. is co-inventor in the following patents as a scientific expert and has received no royalties: In Vitro Multiparameter Determination Method for The Diagnosis and Early Diagnosis of Neurodegenerative Disorders Patent Number: 8916388; In Vitro Procedure for Diagnosis and Early Diagnosis of Neurodegenerative Diseases Patent Number: 8298784; Neurodegenerative Markers for Psychiatric Conditions Publication Number: 20120196300; In Vitro Multiparameter Determination Method for The Diagnosis and Early Diagnosis of Neurodegenerative Disorders Publication Number: 20100062463; In Vitro Method for The Diagnosis and Early Diagnosis of Neurodegenerative Disorders Publication Number: 20100035286; In Vitro Procedure for Diagnosis and Early Diagnosis of Neurodegenerative Diseases Publication Number:

20090263822; In Vitro Method for The Diagnosis of Neurodegenerative Diseases Patent Number: 7547553; CSF Diagnostic in Vitro Method for Diagnosis of Dementias and Neuroinflammatory Diseases Publication Number: 20080206797; In Vitro Method for The Diagnosis of Neurodegenerative Diseases Publication Number: 20080199966; Neurodegenerative Markers for Psychiatric Conditions Publication Number: 20080131921; A.V. is an employee of Eisai Inc. He does not receive any fees or honoraria since November 2019. Before November 2019 he had received lecture honoraria from Roche, MagQu LLC, and Servier. S.L. has received lecture honoraria from Roche and Servier. W.J.L. is the Bollinger Professor of Alzheimer's disease (AD) at the LSU School of Medicine and Health Sciences Center. He serves on 36 journal editorial boards and study sections and is senior academic editor for the journals *Frontiers in Genetics, Neurochemical Research, Folia Neuropathologica* and *PLoS One.* Research on miRNA and mRNA in the Lukiw laboratory involving biomarkers in AD and in other forms of neurological or retinal disease, prion disease, amyloidogenesis, synaptogenesis, neuro-inflammation and environmental neurotoxicology was supported through an unrestricted grant to the LSU Eye Center from Research to Prevent Blindness (RPB); the Louisiana Biotechnology Research Network (LBRN) and NIH grants NEI EY006311, NIA AG18031 and NIA AG038834 (W.J.L.).

References

1. Available online: https://www.alz.org/alzheimers-dementia/facts (accessed on 17 September 2020).
2. Arvanitakis, Z.; Shah, R.C.; Bennett, D.A. Diagnosis and management of dementia: Review. *JAMA* **2019**, *322*, 1589–1599. [CrossRef]
3. Patnode, C.D.; Perdue, L.A.; Rossom, R.C.; Rushkin, M.C.; Redmond, N.; Thomas, R.G.; Lin, J.S. *Screening for Cognitive Impairment in Older Adults: An Evidence Update for the U.S. Preventive Services Task Force*; Rockville, M.D., Ed.; Agency for Healthcare Research and Quality (US): Rockville, MD, USA, 2020.
4. Terry, R.D.; Katzman, R. Senile dementia of the Alzheimer type. *Ann. Neurol.* **1983**, *14*, 497–506. [CrossRef]
5. Thienhaus, O.J.; Hartford, J.T.; Skelly, M.F.; Bosmann, H.B. Biologic markers in Alzheimer's disease. *J. Am. Geriatr. Soc.* **1985**, *33*, 715–726. [CrossRef]
6. Alzheimer, A.; Stelzmann, R.A.; Schnitzlein, H.N.; Murtagh, F.R. An English translation of Alzheimer's 1907 paper, "Uber eine eigenartige Erkankung der Hirnrinde". *Clin. Anat.* **1995**, *8*, 429–431. [CrossRef] [PubMed]
7. Lukiw, W.J. 100 years of Alzheimer's disease research: Are we any closer to a cure? *Aging Health* **2007**, *3*. [CrossRef]
8. Available online: https://www.alz.org/alzheimers-dementia (accessed on 17 September 2020).
9. Available online: www.ncbi.nlm.nih.gov/pmc (accessed on 17 September 2020).
10. Zhao, Y.; Jaber, V.; Alexandrov, P.N.; Vergallo, A.; Lista, S.; Hampel, H.; Lukiw, W.J. microRNA-Based Biomarkers and Alzheimer's disease (AD) Frontiers in Neuroscience—Neurodegeneration—Special Research Topic 'Deciphering the Biomarkers of Alzheimer's disease'. **2020**, in press.
11. Blennow, K.; Hampel, H.; Weiner, M.; Zetterberg, H. Cerebrospinal fluid and plasma biomarkers in Alzheimer disease. *Nat. Rev. Neurol.* **2010**, *6*, 131–144. [CrossRef] [PubMed]
12. Hampel, H.; Frank, R.; Broich, K.; Teipel, S.J.; Katz, R.G.; Hardy, J.; Herholz, K.; Bokde, A.L.; Jessen, F.; Hoessler, Y.C.; et al. Biomarkers for Alzheimer's disease: Academic, industry and regulatory perspectives. *Nat. Rev. Drug Discov.* **2010**, *9*, 560–574. [CrossRef] [PubMed]
13. Hampel, H.; O'Bryant, S.E.; Molinuevo, J.L.; Zetterberg, H.; Masters, C.L.; Lista, S.; Kiddle, S.J.; Batrla, R.; Blennow, K. Blood-based biomarkers for Alzheimer disease: Mapping the road to the clinic. *Nat. Rev. Neurol.* **2018**, *14*, 639–652. [CrossRef] [PubMed]
14. Hampel, H.; Vergallo, A.; Afshar, M.; Akman-Anderson, L.; Arenas, J.; Benda, N.; Batrla, R.; Broich, K.; Caraci, F.; Claudio Cuello, A.; et al. Blood-based systems biology biomarkers for next-generation clinical trials in Alzheimer's disease. *Dialogues Clin Neurosci.* **2019**, *21*, 177–191. [CrossRef]
15. Kim, D.H.; Yeo, S.H.; Park, J.M.; Choi, J.Y.; Lee, T.H.; Park, S.Y.; Ock, M.S.; Eo, J.; Kim, H.S.; Cha, H.J. Genetic markers for diagnosis and pathogenesis of Alzheimer's disease. *Gene* **2014**, *545*, 185–193. [CrossRef] [PubMed]

16. Lista, S.; Khachaturian, Z.S.; Rujescu, D.; Garaci, F.; Dubois, B.; Hampel, H. Application of systems theory in longitudinal studies on the origin and progression of Alzheimer's disease. *Methods Mol. Biol.* **2016**, *1303*, 49–67. [CrossRef] [PubMed]
17. DeTure, M.A.; Dickson, D.W. The neuropathological diagnosis of Alzheimer's disease. *Mol. Neurodegener.* **2019**, *14*, 32. [CrossRef] [PubMed]
18. Toschi, N.; Lista, S.; Baldacci, F.; Cavedo, E.; Zetterberg, H.; Blennow, K.; Kilimann, I.; Teipel, S.J.; dos Santos, A.M.; Epelbaum, S.; et al. INSIGHT-preAD study group; Alzheimer Precision Medicine Initiative (APMI); Biomarker-guided clustering of Alzheimer's disease clinical syndromes. *Neurobiol. Aging* **2019**, *83*, 42–53. [CrossRef] [PubMed]
19. Veitch, D.P.; Weiner, M.W.; Aisen, P.S.; Beckett, L.A.; Cairns, N.J.; Green, R.C.; Harvey, D.; Jack Jr, C.R.; Jagust, W.; Morris, J.C.; et al. Understanding disease progression and improving Alzheimer's disease clinical trials: Recent highlights from the Alzheimer's Disease Neuroimaging Initiative. *Alzheimers Dement.* **2019**, *15*, 106–152. [CrossRef]
20. Aharon, A.; Spector, P.; Ahmad, R.S.; Horrany, N.; Sabbach, A.; Brenner, B.; Aharon-Peretz, J. Extracellular vesicles of Alzheimer's disease patients as a biomarker for disease progression. *Mol. Neurobiol.* **2020**. [CrossRef]
21. Ellegaard Nielsen, J.; Sofie Pedersen, K.; Vestergård, K.; Georgiana Maltesen, R.; Christiansen, G.; Lundbye-Christensen, S.; Moos, T.; Risom Kristensen, S.; Pedersen, S. Novel blood-derived extracellular vesicle-based biomarkers in Alzheimer's disease identified by proximity extension assay. *Biomedicines* **2020**, *8*, 199. [CrossRef]
22. Hampel, H.; Vergallo, A.; Caraci, F.; Cuello, A.C.; Lemercier, P.; Vellas, B.; Giudici, K.V.; Baldacci, F.; Hänisch, B.; Haberkamp, M.; et al. Alzheimer precision medicine initiative (APMI). Future avenues for Alzheimer's disease detection and therapy: Liquid biopsy, intracellular signaling modulation, systems pharmacology drug discovery. *Neuropharmacology* **2020**, 108081. [CrossRef]
23. Lu, G.; Liu, W.; Huang, X.; Zhao, Y. Complement factor H levels are decreased and correlated with serum C-reactive protein in late-onset Alzheimer's disease. *Arq. Neuropsiquiatr.* **2020**, *78*, 76–80. [CrossRef]
24. Lukiw, W.J. Gastrointestinal (GI) tract microbiome-derived neurotoxins-potent neuro-inflammatory signals from the GI tract via the systemic circulation into the brain. *Front. Cell Infect. Microbiol.* **2020**, *10*, 22. [CrossRef]
25. Lukiw, W.J. microRNA-146a Signaling in Alzheimer's disease (AD) and prion disease (PrD). *Front. Neurol.* **2020**, *11*, 462. [CrossRef] [PubMed]
26. Lukiw, W.J.; Pogue, A.I. Vesicular transport of encapsulated microRNA between glial and neuronal cells. *Int. J. Mol. Sci.* **2020**, *21*, 5078. [CrossRef] [PubMed]
27. Sims, R.; Hill, M.; Williams, J. The multiplex model of the genetics of Alzheimer's disease. *Nat. Neurosci.* **2020**, *23*, 311–322. [CrossRef] [PubMed]
28. Vernooij, M.W.; van Buchem, M.A. Neuroimaging in Dementia. In *Diseases of the Brain, Head and Neck, Spine 2020–2023: Diagnostic Imaging*; Hodler, J., Kubik-Huch, R.A., von Schulthess, G.K., Eds.; Springer: Cham, Switzerland, 2020; pp. 131–142.
29. Zhao, Y.; Lukiw, W.J. Bacteroidetes neurotoxins and inflammatory neurodegeneration. *Mol. Neurobiol.* **2018**, *55*, 9100–9107. [CrossRef] [PubMed]
30. Ruan, Z.; Ikezu, T. Tau Secretion. *Adv. Exp. Med. Biol.* **2019**, *1184*, 123–134. [CrossRef]
31. Sheng, L.; Stewart, T.; Yang, D.; Thorland, E.; Soltys, D.; Aro, P.; Khrisat, T.; Xie, Z.; Li, N.; Liu, Z.; et al. Erythrocytic α-synuclein contained in microvesicles regulates astrocytic glutamate homeostasis: A new perspective on Parkinson's disease pathogenesis. *Acta Neuropathol. Commun.* **2020**, *8*, 102. [CrossRef]
32. Zhao, Y.; Jaber, V.; Lukiw, W.J. Over-expressed pathogenic miRNAs in Alzheimer's disease (AD) and prion disease (PrD) drive deficits in TREM2-mediated Aβ42 peptide clearance. *Front. Aging Neurosci.* **2016**, *6*, 140. [CrossRef]
33. Bhattacharjee, S.; Lukiw, W.J. Alzheimer's disease and the microbiome. *Front. Cell Neurosci.* **2013**, *7*, 153. [CrossRef]
34. Emery, D.C.; Shoemark, D.K.; Batstone, T.E.; Waterfall, C.M.; Coghill, J.A.; Cerajewska, T.L.; Davies, M.; West, N.X.; Allen, S.J. 16S rRNA next generation sequencing analysis shows bacteria in Alzheimer's post-mortem brain. *Front. Aging Neurosci.* **2017**, *9*, 195. [CrossRef]

35. Pisa, D.; Alonso, R.; Fernández-Fernández, A.M.; Rábano, A.; Carrasco, L. Polymicrobial infections in brain tissue from Alzheimer's disease patients. *Sci. Rep.* **2017**, *7*, 5559. [CrossRef]
36. Zhan, X.; Stamova, B.; Sharp, F.R. Lipopolysaccharide associates with amyloid plaques, neurons and oligodendrocytes in Alzheimer's disease brain: A Review. *Front. Aging Neurosci.* **2018**, *10*, 42. [CrossRef] [PubMed]
37. Ceppa, F.A.; Izzo, L.; Sardelli, L.; Raimondi, I.; Tunesi, M.; Albani, D.; Giordano, C. Human gut-microbiota interaction in neurodegenerative disorders and current engineered tools for its modeling. *Front. Cell Infect. Microbiol.* **2020**, *10*, 297. [CrossRef] [PubMed]
38. Lukiw, W.J.; Pogue, A.I.; Hill, J.M. SARS-CoV-2 infectivity and neurological targets in the brain. *Cell. Mol. Neurobiol.* **2020**. [CrossRef] [PubMed]
39. Itzhaki, R.F. Corroboration of a major role for herpes simplex virus Type 1 in Alzheimer's disease. *Front. Aging Neurosci.* **2018**, *10*, 324. [CrossRef] [PubMed]
40. Alexandrov, P.; Zhao, Y.; Li, W.; Lukiw, W.J. Lipopolysaccharide-stimulated, NF-kB-, miRNA-146a- and miRNA-155-mediated molecular-genetic communication between the human gastrointestinal tract microbiome and the brain. *Folia Neuropathol.* **2019**, *57*, 211–219. [CrossRef] [PubMed]
41. Johnson, R.T.; ter Meulen, V. Slow infections of the nervous system. *Adv. Intern. Med.* **1978**, *23*, 353–383.
42. Hill, J.M.; Bhattacharjee, S.; Pogue, A.I.; Lukiw, W.J. The gastrointestinal tract microbiome and potential link to Alzheimer's disease. *Front. Neurol.* **2014**, *5*, 43. [CrossRef]
43. Naughton, S.X.; Raval, U.; Pasinetti, G.M. The viral hypothesis in Alzheimer's disease: Novel insights and pathogen-based biomarkers. *J. Pers. Med.* **2020**, *10*, 74. [CrossRef]
44. Rosario, D.; Boren, J.; Uhlen, M.; Proctor, G.; Aarsland, D.; Mardinoglu, A.; Shoaie, S. Systems biology approaches to understand the host-microbiome interactions in neurodegenerative diseases. *Front. Neurosci.* **2020**, *14*, 716. [CrossRef]
45. Jaber, V.; Zhao, Y.; Lukiw, W.J. Alterations in micro RNA-messenger RNA (miRNA-mRNA) coupled signaling networks in sporadic Alzheimer's disease (AD) hippocampal CA1. *J. Alzheimers Dis. Parkinsonism* **2017**, *7*, 312. [CrossRef]
46. Mai, H.; Fan, W.; Wang, Y.; Cai, Y.; Li, X.; Chen, F.; Chen, X.; Yang, J.; Tang, P.; Chen, H.; et al. Intranasal Administration of miR-146a Agomir Rescued the Pathological Process and Cognitive Impairment in an AD Mouse Model. *Mol. Ther. Nucleic Acids* **2019**, *18*, 681–695. [CrossRef] [PubMed]
47. Mujica, M.L.; Gallay, P.A.; Perrachione, F.; Montemerlo, A.E.; Tamborelli, L.A.; Vaschetti, V.; Reartes, D.; Bollo, S.; Rodríguez, M.C.; Dalmasso, P.D.; et al. New trends in the development of electrochemical biosensors for the quantification of microRNAs. *J. Pharm. Biomed. Anal.* **2020**, *189*, 113478. [CrossRef] [PubMed]
48. Alexandrov, P.N.; Dua, P.; Hill, J.M.; Bhattacharjee, S.; Zhao, Y.; Lukiw, W.J. microRNA (miRNA) speciation in Alzheimer's disease (AD) cerebrospinal fluid (CSF) and extracellular fluid (ECF). *Int. J. Biochem. Mol. Biol.* **2012**, *3*, 365–373. [PubMed]
49. Slota, J.A.; Booth, S.A. MicroRNAs in neuroinflammation: Implications in disease pathogenesis, biomarker discovery and therapeutic applications. *Noncoding RNA* **2019**, *5*, 35. [CrossRef]
50. Zhao, Y.; Bhattacharjee, S.; Jones, B.M.; Hill, J.; Dua, P.; Lukiw, W.J. Regulation of neurotropic signaling by the inducible, NF-kB-sensitive miRNA-125b in Alzheimer's disease (AD) and in primary human neuronal-glial (HNG) cells. *Mol. Neurobiol.* **2014**, *50*, 97–106. [CrossRef]
51. Jaber, V.R.; Zhao, Y.; Sharfman, N.M.; Li, W.; Lukiw, W.J. Addressing Alzheimer's Disease (AD) Neuropathology Using Anti-microRNA (AM) Strategies. *Mol. Neurobiol.* **2019**, *56*, 8101–8108. [CrossRef]
52. Kaipainen, A.; Jääskeläinen, O.; Liu, Y.; Haapalinna, F.; Nykänen, N.; Vanninen, R.; Koivisto, A.M.; Julkunen, V.; Remes, A.M.; Herukka, S.K. Cerebrospinal Fluid and MRI Biomarkers in Neurodegenerative Diseases: A Retrospective Memory Clinic-Based Study. *J. Alzheimers Dis.* **2020**, *75*, 751–765. [CrossRef]
53. Kamagata, K.; Andica, C.; Hatano, T.; Ogawa, T.; Takeshige-Amano, H.; Ogaki, K.; Akashi, T.; Hagiwara, A.; Fujita, S.; Aoki, S.; et al. Advanced diffusion magnetic resonance imaging in patients with Alzheimer's and Parkinson's diseases. *Neural Regen. Res.* **2020**, *15*, 1590–1600. [CrossRef]
54. Lo Buono, V.; Palmeri, R.; Corallo, F.; Allone, C.; Pria, D.; Bramanti, P.; Marino, S. Diffusion tensor imaging of white matter degeneration in early stage of Alzheimer's disease: A review. *Int. J. Neurosci.* **2020**, *130*, 243–250. [CrossRef]

55. Lombardi, G.; Crescioli, G.; Cavedo, E.; Lucenteforte, E.; Casazza, G.; Bellatorre, A.; Lista, C.; Costantino, G.; Frisoni, G.; Virgili, G.; et al. Structural magnetic resonance imaging for the early diagnosis of dementia due to Alzheimer's disease in people with mild cognitive impairment. *Cochrane Database Syst. Rev.* **2020**, *3*. [CrossRef]
56. Blennow, K.; Zetterberg, H. Biomarkers for Alzheimer's disease: Current status and prospects for the future. *J. Intern. Med.* **2018**, *284*, 643–663. [CrossRef]
57. Donatelli, G.; Ceravolo, R.; Frosini, D.; Tosetti, M.; Bonuccelli, U.; Cosottini, M. Present and future of ultra-high field MRI in neurodegenerative disorders. *Curr. Neurol. Neurosci. Rep.* **2018**, *18*, 31. [CrossRef] [PubMed]
58. Sherva, R.; Tripodis, Y.; Bennett, D.A.; Chibnik, L.B.; Crane, P.K.; de Jager, P.L.; Farrer, L.A.; Saykin, A.J.; Shulman, J.M.; Naj, A.; et al. Genome-wide association study of the rate of cognitive decline in Alzheimer's disease. *Alzheimers Dement.* **2014**, *10*, 45–52. [CrossRef] [PubMed]
59. Dong, A.; Toledo, J.B.; Honnorat, N.; Doshi, J.; Varol, E.; Sotiras, A.; Wolk, D.; Trojanowski, J.Q.; Davatzikos, C. Alzheimer's Disease Neuroimaging Initiative. Heterogeneity of neuroanatomical patterns in prodromal Alzheimer's disease: Links to cognition, progression and biomarkers. *Brain* **2017**, *140*, 735–747. [CrossRef] [PubMed]
60. Gibney, B.; Redmond, C.E.; Byrne, D.; Mathur, S.; Murray, N. A review of the applications of dual-energy CT in acute neuroimaging. *Can. Assoc. Radiol. J.* **2020**, *71*, 253–265. [CrossRef] [PubMed]
61. Koychev, I.; Hofer, M.; Friedman, N.C. Correlation of Alzheimer's disease neuropathologic staging with amyloid and tau scintigraphic imaging biomarkers. *J. Nucl. Med.* **2020**. [CrossRef] [PubMed]
62. Tetreault, A.M.; Phan, T.; Orlando, D.; Lyu, I.; Kang, H.; Landman, B.; Darby, R. Network localization of clinical, cognitive, and neuropsychiatric symptoms in Alzheimer's disease. *Brain* **2020**, *143*, 1249–1260. [CrossRef]
63. Emrani, S.; Lamar, M.; Price, C.C.; Wasserman, V.; Matusz, E.; Au, R.; Swenson, R.; Nagele, R.; Heilman, K.M.; Libon, D.J. Alzheimer's/vascular spectrum dementia: Classification in addition to diagnosis. *J. Alzheimers Dis.* **2020**, *73*, 63–71. [CrossRef] [PubMed]
64. Habes, M.; Grothe, M.J.; Tunc, B.; McMillan, C.; Wolk, D.A.; Davatzikos, C. Disentangling heterogeneity in Alzheimer's disease and related dementias using data-driven methods. *Biol. Psychiatry* **2020**, *88*, 70–72. [CrossRef]
65. Hampel, H.; Goetzl, E.J.; Kapogiannis, D.; Lista, S.; Vergallo, A. Biomarker-Drug and Liquid Biopsy Co-development for Disease Staging and Targeted Therapy: Cornerstones for Alzheimer's Precision Medicine and Pharmacology. *Front. Pharmacol.* **2019**, *10*, 310. [CrossRef]
66. Lewczuk, P.; Łukaszewicz-Zając, M.; Mroczko, P.; Kornhuber, J. Clinical significance of fluid biomarkers in Alzheimer's disease. *Pharmacol. Rep.* **2020**. [CrossRef] [PubMed]
67. Pulliam, L.; Sun, B.; Mustapic, M.; Chawla, S.; Kapogiannis, D. Plasma neuronal exosomes serve as biomarkers of cognitive impairment in HIV infection and Alzheimer's disease. *J. Neurovirol.* **2019**, *25*, 702–709. [CrossRef] [PubMed]
68. Hampel, H.; Vergallo, A.; Perry, G.; Lista, S.; Alzheimer Precision Medicine Initiative (APMI). The Alzheimer precision medicine initiative. *J. Alzheimers Dis.* **2019**, *68*, 1–24. [CrossRef] [PubMed]
69. Guerreiro, R.J.; Gustafson, D.R.; Hardy, J. The genetic architecture of Alzheimer's disease: Beyond APP, PSENs and APOE. *Neurobiol. Aging.* **2012**, *33*, 437–456. [CrossRef]
70. Cole, M.A.; Seabrook, G.R. On the horizon-the value and promise of the global pipeline of Alzheimer's disease therapeutics. *Alzheimers Dement. (N. Y.)* **2020**, *6*, e12009. [CrossRef]
71. Jellinger, K.A. Neuropathological assessment of the Alzheimer spectrum. *J. Neural. Transm.* **2020**, *127*, 1229–1256. [CrossRef]

Journal of *Personalized Medicine*

Review

New Insights into the Molecular Bases of Familial Alzheimer's Disease

Valeria D'Argenio [1,2,*] and Daniela Sarnataro [3,*]

1 CEINGE-Biotecnologie Avanzate scarl, via G. Salvatore 486, 80145 Naples, Italy
2 Department of Human Sciences and Quality of Life Promotion, San Raffaele Open University, via di val Cannuta 247, 00166 Rome, Italy
3 Department of Molecular Medicine and Medical Biotechnology, Federico II University, via S. Pansini 5, 80131 Naples, Italy
* Correspondence: dargenio@ceinge.unina.it (V.D.); daniela.sarnataro@unina.it (D.S.); Tel.: +39-081-3737909 (V.D.); +39-081-7464575 (D.S.)

Received: 25 March 2020; Accepted: 17 April 2020; Published: 19 April 2020

Abstract: Like several neurodegenerative disorders, such as Prion and Parkinson diseases, Alzheimer's disease (AD) is characterized by spreading mechanism of aggregated proteins in the brain in a typical "prion-like" manner. Recent genetic studies have identified in four genes associated with inherited AD (amyloid precursor protein-*APP*, Presenilin-1, Presenilin-2 and Apolipoprotein E), rare mutations which cause dysregulation of APP processing and alterations of folding of the derived amyloid beta peptide (Aβ). Accumulation and aggregation of Aβ in the brain can trigger a series of intracellular events, including hyperphosphorylation of tau protein, leading to the pathological features of AD. However, mutations in these four genes account for a small of the total genetic risk for familial AD (FAD). Genome-wide association studies have recently led to the identification of additional AD candidate genes. Here, we review an update of well-established, highly penetrant FAD-causing genes with correlation to the protein misfolding pathway, and novel emerging candidate FAD genes, as well as inherited risk factors. Knowledge of these genes and of their correlated biochemical cascade will provide several potential targets for treatment of AD and aging-related disorders.

Keywords: Alzheimer's disease; *APP* mutations; *APOE* alleles; *PSEN1*; *PSEN2*; germline mutations; late onset AD; early onset AD; familial AD; genetics of AD

---

1. Introduction

Alzheimer's disease (AD) is responsible for about 60–70% of cases of dementia, equivalent to an estimated population of 40–50 million persons worldwide. This more than doubled from 1990 to 2016 [1]. Typically, AD is featured by a progressive neurodegeneration that lead to a gradual loss of memory and alterations affecting other cognitive functions, such as spatial cognition, reasoning, word-finding, judgment and problem solving [2]. Aging is the most important AD risk factor and, even if it may occur at any age, more often AD appears in older individuals (later than 65 years of age) [3,4]. However, early onset AD cases, characterized by the occurrence of clinical signs between 30 and 65 years old, have been also reported [3,4]. Even if atypical phenotypes have been described, clinical and pathological features seem to be the same between early and late onset AD, so that it may be difficult to distinguish these 2 groups [4]. Interestingly, while late onset AD is usually sporadic and doesn't show any segregation within the families, it has been highlighted that early onset AD is featured by a high recurrence rate within the affected families, thus suggesting the presence of inherited forms of AD [3].

In particular, it has been estimated that 15–25% of total AD accounts for familial Alzheimer's disease (FAD) (Figure 1) [5].

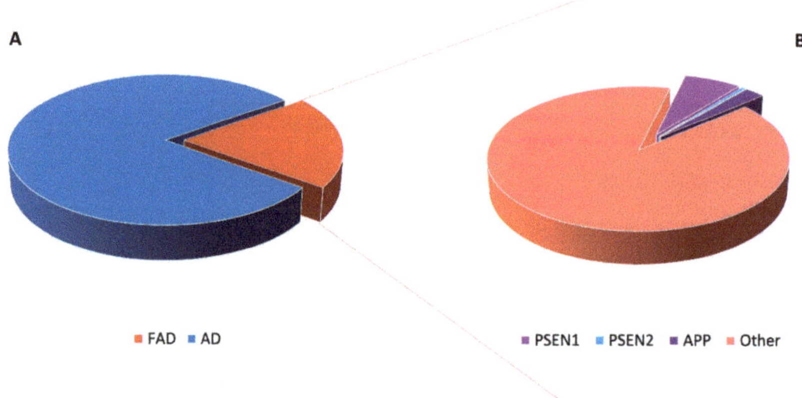

Figure 1. Prevalence and genetic causes of familial Alzheimer's disease (FAD). FAD represents only a small fraction (about 20%) of all Alzheimer's disease (AD) cases (**A**). In addition, within FAD, mutations the amyloid precursor protein (*APP*), presenilin 1 (*PSEN1*) or presenilin 2 (*PSEN2*) genes account for a small proportion of cases, underlying that the molecular bases of the larger fraction still remain to be unveiled (**B**).

In this category, it is included the autosomal dominant Alzheimer's disease (ADAD), associated to the presence of known causative gene mutations, mainly in the amyloid precursor protein (*APP*), presenilin 1 (*PSEN1*) or presenilin 2 (*PSEN2*) genes, that have been described as highly penetrant, disease-causing FAD genes [3–5]. However, mutations in these genes are able to explain just a small percentage of all FAD cases, suggesting the existence of other, inherited, disease-predisposing genes [4]. Since FAD typically shows a sequential, clinical progression from pre-dementia to dementia stage, thus it is crucial to recognize and diagnose it before symptoms onset in order to begin treatments as soon as possible [6,7]. As a consequence, a better understanding of FAD molecular bases, i.e., the identification of the causative genes, may ameliorate the management and the clinical outcome of these patients and of their families.

Recent advances in genomics, i.e., the availability of highly performing next generation sequencing (NGS)-based methods to accurately analyze single genes [8,9] or a subset of genes of interest [10–12], up to the whole exome [13,14] or the whole genome [13,14], has prompted the study of the molecular bases of human diseases and is a promising tool to discover novel FAD-related genes [4].

Here, we will review the current knowledge regarding the genetic etiology of FAD. In particular, we will focus first on well-established, highly penetrant, FAD-causing genes. In this context, the relationship between mutations affecting AD-related genes and proteins' trafficking, folding and aggregation properties will be highlighted, with special attention to APP. Next, novel, emerging and candidate FAD genes, as well as inherited risk factors will be also discussed, suggesting that enlarged genetic testing may be useful in FAD families in order to improve the identification and management of the at-risk subjects.

2. Methods

Indexed articles in English were searched in PubMed using the following keywords: "Alzheimer's disease molecular bases", "Alzheimer's disease mutations", "Alzheimer's disease germline mutations", "Alzheimer's disease genes", "inherited Alzheimer's disease", "familial Alzheimer's disease",

"*APP* mutations", "*PSEN1* mutations", "*PSEN2* mutations" and "novel Alzheimer's disease genes". In the attempt to focus on the most recent and updated papers on these topics, we fixed the time within 2010 and 2020 as temporal window; however, a manual search for oldest references mentioned in the found articles was also carried out. Papers in the search results reporting somatic mutations or describing genetic risk factors related to sporadic, late onset AD were not included since are out of the topic of the present review (Figure 2).

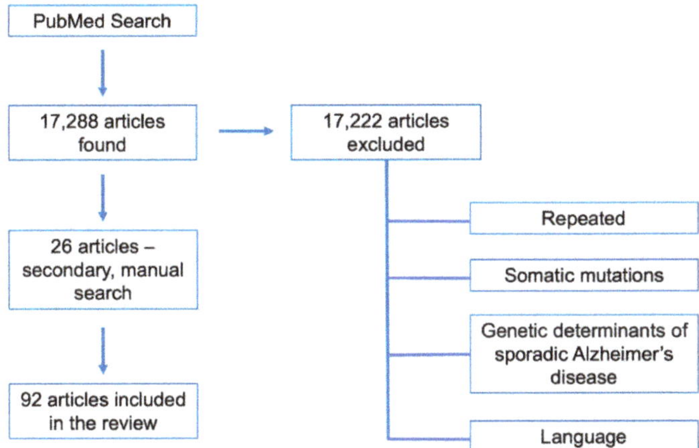

Figure 2. Flow-diagram summarizing the steps for the selection of the articles reviewed herein.

3. Highly Penetrant Familial Alzheimer Disease-Causing Genes

Genetic analysis of large FAD families allowed the discovery of the three well established and high-penetrant genes related to this disease, namely, *APP*, *PSEN1* and *PSEN2* (Figure 3).

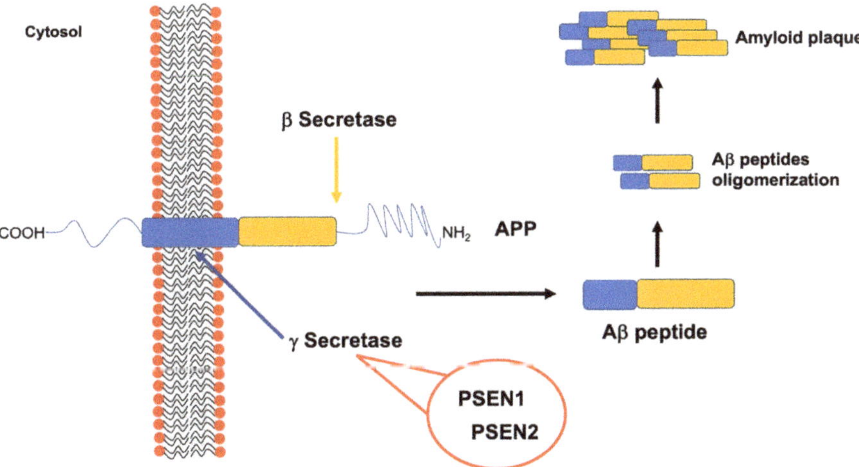

Figure 3. Amyloid precursor protein (APP) structure and amyloid beta (Aβ) peptide production. The cleavage of APP by specific secretases is required to produce the Aβ peptide. One of these secretases, namely the γ secretase, is a multimeric complex involving also presenilin 1 (PSEN1) and presenilin 2 (PSEN2). Mutations affecting *APP*, as well as *PSEN1* or *PSEN2* genes, have been associated to familial Alzheimer's disease (FAD) because of their ability to increase Aβ peptides production and, consequently, their aggregation up to amyloid plaques formation.

The amyloid precursor protein gene (*APP*, OMIM #104760, chromosome 21q21.3) encodes for an integral type 1 membrane glycoprotein that is almost ubiquitously expressed. APP is sequentially processed for proteolytic cleavage to produce amyloid β (Aβ), by β- and γ-secretases (the latter composed by PSEN1, PSEN2, Nicastrin and Aph-1). Because APP processing by γ-secretase is not restricted to a single site, it gives rise to different Aβ species, $A\beta_{42}$ being more prone to aggregate [15].

Recently it has become evident that the AD-typical Aβ assemblies are able to adopt alternative conformations and become self-propagating, like prions [16–19].

To date, 32 pathogenic mutations in *APP* were reported within or flanking the Aβ sequence (https://www.alzforum.org/mutations/app), being located mostly near the β- and γ-secretase sites [20]. As a consequence, *APP* mutations seem to be able to increase Aβ aggregation rate, thus causing FAD (Figure 3) [21].

The p.K670N and p.M671L Swedish AD-related pathogenic mutations are localized near to the γ-secretase site and lead to increased absolute levels of $A\beta_{42}$ (without changing $A\beta_{42}$ to $A\beta_{40}$ ratio) [22]. Interestingly, Del Prete et al., found that in a Swedish cell culture model of AD, APP and its catabolites are present in mitochondrial-ER associated membranes (MAMs) and β- and γ-secretases harbor APP processing activities in MAMs [23]. This finding is extremely interesting, considering the fact that the localization of APP and its enzymes plays a critical role in the Aβ generation and its signaling inside the cell [24]. The p.T714I, p.V715M, p.V715A, p.I716V, p.V717I and p.V717L *APP* mutations are all located in proximity of the γ-secretase site, affect the cleavage and, contrary to the double Swedish mutations, cause an increase of the $A\beta_{42}$ to $A\beta_{40}$ ratio and are able to influence the stability of APP C-terminal fragments [20,25]. Very recently, the p.I716T *APP* mutation [25], as well as the p.L723P featured by the local unfolding of the C-terminal turn [26], was described to affect the efficiency of the γ-secretase ε-cleavage and to induce a major $A\beta_{42}$ to $A\beta_{40}$ ratio. This effect is due to the additional H-bond between the T716 side chain and the transmembrane backbone, which can affect the cleavage domain dynamic [25].

Mutations within the Aβ sequence have been predicted not only to affect APP processing, thus regulating the amount of total Aβ production, but were also thought to affect the aggregation properties of the resulting Aβ peptide [25]. Indeed, the *APP* p.A692G, p.E693Q and p.D694N mutations, all located inside the Aβ sequence, have been described to have an increased aggregative ability and neurotoxicity respect to the wild type Aβ [21].

Although a common effect of these mutations is the dysregulation of the production of different Aβ forms of APP, a recent paper form Lumsden et al. proposed also a dysregulation of iron homeostasis as a common effect of mutations related to early onset AD [27].

Table 1 summarizes *APP* mutations, affecting protein stability, folding, processing, from the N- to the C-terminal.

Despite the above, it has been reported that *APP* dominant mutations account for about 16% of all ADAD [3]. Some years ago, 2 *APP* mutations, namely p.A673V and p.E693del, have been found to be able to cause FAD only in the homozygous status supporting the existence also of a recessive pattern of inheritance [28,29]. More recently, Conidi et al., described an Italian family carrying the *APP* p.A713T in homozygous status; surprisingly the clinical phenotype was not more severe respect to the heterozygous carriers [30]. This finding not only highlighted that the homozygosity for *APP* dominant mutations is not lethal, but also suggested that other, independently inherited genetic factors, may exert a protective effect and modify the clinical presentation of the disease.

Finally, in addition to single nucleotide variants and small insertion/deletions, dominantly inherited duplications of the *APP* locus have been also described and associated to AD [31–35]. In particular, small chromosomal duplications with different genomic coordinates, but all including the *APP* locus, have been described in some French FAD families [31,32]. Next, *APP* duplications were reported also in Finnish and Dutch FAD cases [33–35]. The clinical consequences of these duplications are not yet clearly defined and also their frequency as FAD cause is variable in the different studies [31–35]. Totally, 25 duplications have been identified so far and, respect to missense mutations, these duplications seem to have a reduced penetrance and a variable age of onset [36].

Table 1. Amyloid precursor protein (*APP*) pathogenic mutations and their effects at protein level.

Mutation and Protein Region	Protein Stability/Folding/Processing	Reference
N-terminal		
p.K670N/p.M671L	Leads to increased absolute levels of $A\beta_{42}$, doesn't alter $A\beta_{42}/A\beta_{40}$ ratio Present in MAMs	Kumar-Singh_2009 [22] Del Prete_2017 [23]
Amyloid-beta domain		
p.A692G		Murakami_2002 [21]
p.E693Q		
p.E693K	Potent aggregative	
p.E693G		
p.D694N		
p.A713T p.T714I	Increase $A\beta_{42}/A\beta_{40}$ ratio, affect stability of *APP* CTFs	Kumar-Singh_2009 [22]
Transmembrane/C-terminal		
p.V715M		
p.I716V	Increase $A\beta_{42}/A\beta_{40}$ ratio, affect stability of *APP* CTFs	De Jonghe_2001 [20]
p.V717L		
Pp.I716T	Increases $A\beta_{42}/A\beta_{40}$ ratio reducing the efficiency of the γ-secretase ε-cleavage	Götz_2019 [25]
p.L723P	Causes unfolding of C-terminal turn of *APP* TM domain helix	Bocharov_2019 [26]

Interestingly, Jonsson et al., identified a rare *APP* variant in the Icelandic population showing a protective effect [37]. Finally, a number of variants of unknown clinical significance have been also detected and further functional tests are required to establish their pathogenicity.

The presenilin 1 (*PSEN1*, OMIM #104311, chromosome 14q24.3) encodes for a protein that is a subunit of γ-secretase, i.e., one of the 2 enzymes responsible for APP proteolytic cleavage. As a consequence, mutations in *PSEN1*, impairing the activity of γ-secretase complex, may lead to the production of more aggregation-prone forms of the Aβ peptide, that is a typical hallmark of AD, thus inducing the disease development (Figure 3) [3,4].

PSEN1 is the most common gene related to FAD. To date, 221 *PSEN1* pathogenic mutations have been described, accounting for up to 70% of ADAD cases (http://www.alzforum.org/mutations). These mutations can be both single nucleotide variants and small insertions/deletions; in addition, a deletion able to cause *PSEN1* exon 9 skipping, has been also described [38]. Typically, FAD onset in *PSEN1* mutations carriers ranges from 30 to 50 years, the mutations showing an autosomal dominant inheritance and almost always a complete penetrance [39]. Interestingly, *de novo* mutations, featured by a very early age of onset (before 30 years) have been also described [40–42].

As in the case of *APP*, Kosik et al., described a Colombian family carrying the *PSEN1* mutation p.E280A in homozygous status; also, in this case, the severity of the disease was not influenced by the homozygosity of the mutation [43].

The presenilin 2 (*PSEN2*, OMIM #600759, chromosome 1q31-q42) gene has a genetic structure very similar to *PSEN1*, sharing a sequence homology of 67% [3]. It encodes for another component of the γ-secretase complex; thus, its impairment, due to pathogenic mutations, is able to increase the $A\beta_{42}/A\beta_{40}$ ratio too (Figure 3).

To date, 19 different *PSEN2* pathogenic mutations have been reported (http://www.alzforum.org/mutations). As for *PSEN1*, these pathogenic mutations are scattered along the entire gene sequence with a higher frequency in the transmembrane domains [44,45]. The ADAD age of onset ranges from 40 to 70 years in *PSEN2* mutations carriers, the penetrance of these mutations being still controversial to assess, due to the few numbers of families reported to date and showing also a so wide age-range of onset [39].

It has been estimated that totally, *APP*, *PSEN1* and *PSEN2* mutations account only for about 5–10% of all FAD (Figure 1) [46,47]. Within the FAD, considering only the ADAD forms the contribution of these genes is highly heterogeneous based on the population studied, 23% up to 88% of patients remaining without a genetic diagnosis [47,48]. Since the clinical features of FAD can be variable, the diagnosis is often difficult and delayed, underlying the importance of identifying other molecular alterations responsible for the currently unexplained FAD cases.

4. Apolipoprotein E ε4 Risk Allele and Familial Alzheimer Disease

The apolipoprotein E gene (*APOE*, OMIM #107741, chromosome 19q13.2) encodes a glycoprotein involved in the mobilization of peripheral cholesterol, also during neuronal growth and regeneration [49,50]. Three APOE isoforms are known, namely ApoE2, ApoE3 and ApoE4, differing at level of 2 aminoacidic residues (the 112 and 158) and coded by 3 alleles, i.e., ε2, ε3 and ε4, whose frequency varies among different populations [3].

To date, an association between the ε4 allele and the sporadic, late-onset AD has been reported [3,4]. In particular, it has been assessed an increased risk up to 3-fold in the heterozygous carriers and up to 15-fold in the homozygous [51]. Interestingly, the ε2 allele has been reported as a protective factor, reducing the AD risk and also positively impacting longevity [52]. These different features have been related to ta different binding affinity of the encoded proteins for the Aβ peptide; in particular, ApoE4 shows the highest affinity leading to the creation of monofibrils that are able to produce dense precipitates [3]. However, it is important to underline that the ε4 allele is not a cause but an AD risk factor; thus, other genetic or environmental factors are required for disease development. A working model, attempting to explain the relationship between ApoE and Alzheimer's disease, has been proposed (Figure 4) [53].

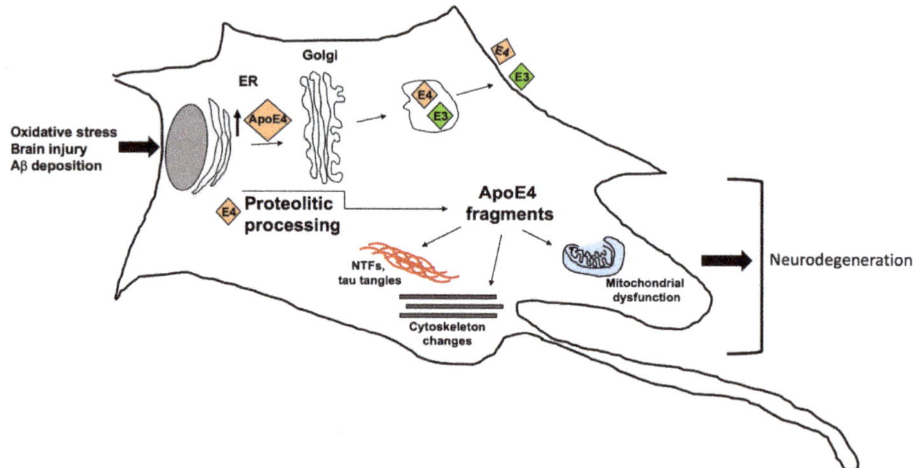

Figure 4. Working model for Apolipoprotein E4 (ApoE4) contribution to Alzheimer's disease (AD) development. Different stimuli can induce ApoE4 and ApoE3 overexpression, with E4 contribution to neurodegeneration in AD. ApoE4, undergoing to proteolytic cleavage, can generate different fragments which can contribute to mitochondrial dysfunction, cytoskeletal disorganization, neurofibrillary tangles and, consequently to neurodegeneration.

In particular, ApoE4, unlike ApoE3, contributes to AD by interacting with different factors through various pathways. In response to oxidative stress, aging, brain damage or Aβ deposition, neurons synthesize increasing amount of ApoE, that in turn undergoes proteolytic processing generating fragments which cause mitochondrial dysfunction, cytoskeletal changes, NFT (neurofibrillary tangles) formation, leading to neurodegeneration.

Interestingly, it has been shown that the *APOE* ε4 allele is able to increase also the risk for early onset AD in presence of familiarity for the disease [54]. In particular, in the ε4 homozygous carriers the risk was independent from other genetic factors, while in the heterozygous no, suggesting that it may act as disease phenotype-modifier in presence of other genetic mutations [4]. However, Genin et al., reported for *APOE* ε4 allele an AD risk comparable to that of other genetic factors [55]. Some studies, evaluating the effects of the *APOE* genotype on AD clinical features in families carrying pathogenic mutations in *APP*, *PSEN1* or *PSEN2* genes, showed that the ε4 allele is associated to an earlier age of onset in the mutations' carriers, while the carriers of the ε2 allele had a later onset [56–58]. Nevertheless, it is important to underline that the significance of *APOE* testing in clinical practice is still under debate and it has been recently reviewed to not significantly impact diagnostic and prognostic evaluations [59]. Indeed, being a risk factor, the *APOE* ε4 allele is common in the general population, i.e., also in healthy individual without a positive family history of AD. Further large longitudinal studies are required to assess the contribution of *APOE* to AD risk and its possible use in clinical routine settings.

5. Novel, Emerging and Candidate Genes Associated to Familial Alzheimer Disease

Mutations in *APP*, *PSEN1* and *PSEN2* genes, as well as the *APOE* ε4 risk allele, explain only a small percentage of all FAD cases, suggesting that other genes may play a role. In the last years, NGS based studies, through the analysis of large pedigrees, are allowing the detection of novel genes potentially related to FAD.

Guerreiro et al., analyzing a Turkish FAD family, identified a pathogenic mutation in the *NOTCH3* gene [60]. Interestingly, the same mutation was previously associated with a dementia disorder similar to AD and the proband belongs to a consanguineous family with a complex history of neurological disorders [60]. *NOTCH3* (OMIM# 600276, chromosome 19p13.12) encodes a transmembrane receptor involved in cellular signaling and embryonic development. More than 130 mutations have been reported in this gene and related to the rare syndrome cerebral arteriopathy autosomal dominant with subcortical infarcts and leukoencephalopathy; a role also in FAD has been recently proposed [61].

The finding of a shared gene between degenerative and vascular dementias suggests the presence of a similar neurovascular unit dysfunction. Accordingly, a consensus paper by Bordet et al., based on the observation that most of patients currently seem to be affected by mixed forms, proposed that also therapeutic strategies should be common [62]. Indeed, therapeutic approaches should be oriented towards an integrated strategy, including antioxidants, anti-inflammatory, modulation of proteins aggregation and neuronal plasticity. Since in older patients, vascular cognitive impairment (VCI) leading to vascular dementia is often mixed to AD and VCI is rarely "pure", a disease modifying strategy seems to be justified [62]. It is noteworthy that a mitochondrial dysfunction may play a role in this context as an additional cause of cognitive impairment, either of vascular, degenerative or both natures [63]. In particular, a reduction of respiratory chain complex I activity, related to mitochondrial dysfunction, has been reported in a group of patients with vascular dementia and several mitochondrial mechanisms have been invoked in Aβ-related cerebrovascular degeneration [63].

Pottier et al., carried out the WES of 29 probands from FAD families resulted negative for mutations in the 3 main FAD genes and found 7 mutations in the *SORL1* gene [64]. In particular, one of these mutations, i.e., the p.G511R, has been shown to be able to reduce the ability of the protein to bind the Aβ peptide, thus inducing its accumulation [65]. The sortilin-related receptor (*SORL1*, OMIM# 602005, chromosome 11q24.1) encodes for a mosaic protein that is the receptor of neuronal ApoE. Accordingly, *SORL-1* mutations have been described in 2 families with early onset AD. In particular, the *SORL-1* variants where shown to be able to weaken the interaction with APP, interfering with APP trafficking and altering the Aβ levels [66]. Other studies have confirmed the role of *SORL1* mutations in FAD and also in late onset AD [67,68]. Li et al., have recently reported the case of a patient with early onset AD and cognitive impairment, carrying a heterozygous mutation in the *SORL1* gene [69]. Taken together, these items of evidence suggest that *SORL1* mutations may be involved in FAD, its contribution being probably underestimated, and that this gene should be tested in the affected families in addition to *APP*, *PSEN1* and *PSEN2* genes. Indeed, the analysis of *SORL1* in larger cohorts of patients may allow to better clarify its contribution to FAD.

Interestingly, genome-wide association studies identified about 30 additional risk factors/alleles for late onset AD [70,71]. Among these, variants affecting *CLU* (*APOJ*) or *CR1* (complement component 3b/4b receptor 1), being involved in the clearance of Aβ have been associated to AD [72] and heterozygous missense mutations in *TREM2* (triggering receptor myeloid 2 cells) have been described to increase by 3-fold the risk of AD [73]. It has been proposed that these genes may be responsible also for FAD cases.

In particular, an increased frequency of *CLU* gene rare coding mutations has been highlighted in AD patients, predominantly affecting the β chain of the protein [74]. Clusterin (*CLU*, OMIM# 185430, chromosome 8p21.1) encodes a protein involved in synapsis turnover. Most of the CLU variants described so far are able to promote CLU degradation, thus reducing its activity [75].

Two NGS-based studies identified a rare variant (p.Arg47His) in *TREM2* gene [73,76] *TREM2* (OMIM# 605086, chromosome 6p21.1) encodes a type I transmembrane protein belonging to the immunoglobulin receptor superfamily and involved in immune responses activation. Interestingly, TREM2 has been found to be able to bind ApoE, thus increasing the phagocytosis of ApoE-bound apoptotic neurons [77]. Some *TREM2* variants may increase AD risk by reducing the affinity for ApoE, and thus decreasing Aβ peptide clearance. Additionally, it has been showed that *TREM2* mutations in its extracellular domain impair protein maturation and its phagocytic activity [78,79]. The *TREM2* p.Arg47His variant has been reported in different population as associated to AD, including some cases of FAD [73,76,80–82]. The role of other *TREM2* variants is still poorly understood.

Three independent studies identified loss-of-function mutations in the *ABCA7* gene in AD patients [83–85]. The ATP-binding cassette, subfamily A, member 7 (*ABCA7*, OMIM# 605414, chromosome 19p13.3) encodes a transporter protein able to move lipids across the membranes. It has been reported that the inhibition of ABCA7 expression is able to increase β secretase cleavage of APP, thus increasing the production of Aβ peptide [86]. In particular, Cuyvers et al., identified an *ABCA7* frameshift mutation as a founder mutation in a Belgian population, since it was detected in several FAD families showing a dominant pattern of inheritance [83].

Vardarajan et al., by sequencing 76 AD-related loci, identified a rare missense mutation in the *EPHA1* gene (p.P460L) segregating within a large Caribbean FAD family [85]. The Ephrin receptor (*EPHA1*, OMIM # 179610, chromosome 7q34-q35) encodes a tyrosine kinase receptor implicated in neuronal development.

The main features of novel FAD candidate genes are summarized in Table 2.

Table 2. Novel candidate genes and inherited risk factors associated to familial Alzheimer's disease (FAD).

Gene Name (Acronym)	Proposed Function in FAD	References
Apolipoprotein E (*APOE*)	Contribution of ApoE4 to mitochondrial dysfunction, cytoskeletal disorganization and neurofibrillary tangles	Huang_2006 [53]
Neurogenic Locus Notch Homolog Protein 3 (*NOTCH3*)	Cellular signaling impairment	Patel_2019 [61]
Sortilin-related receptor (*SORL1*)	Interference with APP trafficking and alteration of the Aβ levels	Cuccaro_2016 [66]
Complement component 3b/4b receptor 1 (*CR1*)	Aβ peptide clearance reduction	Shen 2016 [72]
Clusterin (*CLU*)	Synapsis turnover reduction	Bettens_2015 [75]
Triggering Receptor Expressed on Myeloid Cells 2 (*TREM2*)	Aβ peptide clearance reduction	Bailey_2015 [77] Lue_2015 [78] Kleinberger_2014 [79]
The ATP-binding cassette, subfamily a, member 7 (*ABCA7*)	Aβ peptide production increase	Satoh_2015 [86]
Ephrin receptor (*EPHA1*)	Alteration of neuronal development	Vardarajan_2016 [85]

It is noticeable that the same genes identified as risk factors for sporadic and late onset AD, harbor also rare variants segregating with FAD. Even if data in this field are still inconclusive since they are often based on isolated findings, however they suggest that some familial cases may be due the combination of rare variants and other risk factors. Recent NGS-based screening including FAD cases, identified established risk alleles with moderate penetrance and one or more variants of uncertain significance, thus suggesting the hypothesis, in presence of no mutations in the 3 main FAD genes, of a polygenic inheritance [87,88]. The use of NGS-based methods for the analysis of large genomic regions in several patients simultaneously may provide further insights to improve the diagnosis of disorders featured by high genetic and phenotypic variability, such as AD. However, the interpretation of NGS data, due to the large number of variants of unknown significance, is often challenging and inconclusive in clinical settings [14,89].

6. Conclusions

AD incidence is showing an increasing trend worldwide, so its early and accurate diagnosis has become mandatory. Indeed, the earlier the diagnosis is made, the sooner the treatments may begin, the latter being an important prognostic factor to ameliorate patients' clinical outcome. It is becoming evident that within the "AD" definition are included different entities, whose correct identification may be important to drive treatments choices. While most of AD cases are sporadic, featured by late onset and by the presence of polygenic risk factors, a small percentage of AD is familial, often featured by early age of onset and related to the presence of rare, pathogenic mutations segregating within the affected families. To date, 3 main genes (*APP*, *PSEN1* and *PSEN2*) have been related to autosomal dominant FAD, accounting just for a small percentage of cases. Novel candidate genes are being identified. Larger studies on large group of patients are required to better address their contribution to the disease and discover other potential candidates. This will allow a better prognostic classification of the patients and a better management of the probands and of their families, by the identification of the -at risk individuals. These genetic data, together with novel clues coming from other "omics" [90–92], will allow also the development of novel and even more personalized therapies for AD and FAD treatment.

Author Contributions: V.D. and D.S.: conceptualization; writing—original draft preparation; writing—review and editing. All authors have read and agreed to the published version of the manuscript.

Funding: This research received no external funding.

Conflicts of Interest: The authors declare no conflict of interest.

References

1. Collaborators GBDD. Global, regional, and national burden of Alzheimer's disease and other dementias, 1990–2016: A systematic analysis for the Global Burden of Disease Study 2016. *Lancet Neurol.* **2019**, *181*, 88–106.
2. McKhann, G.M.; Knopman, D.S.; Chertkow, H.; Hyman, B.T.; Jack, C.R., Jr.; Kawas, C.H.; Klunk, W.E.; Koroshetz, W.J.; Manly, J.J.; Mayeux, R.; et al. The diagnosis of dementia due to Alzheimer's disease: Recommendations from the National Institute on Aging-Alzheimer's Association workgroups on diagnostic guidelines for Alzheimer's disease. *Alzheimers Dement.* **2011**, *7*, 263–269. [CrossRef]
3. Guerreiro, R.J.; Gustafson, D.R.; Hardy, J. The genetic architecture of Alzheimer's disease: Beyond APP, PSENs and APOE. *Neurobiol. Aging* **2012**, *33*, 437–456. [CrossRef]
4. Cacace, R.; Sleegers, K.; Van Broeckhoven, C. Molecular genetics of early-onset Alzheimer's disease revisited. *Alzheimers Dement.* **2016**, *12*, 733–748. [CrossRef]
5. Goldman, J.S.; Hahn, S.E.; Catania, J.W.; LaRusse-Eckert, S.; Butson, M.B.; Rumbaugh, M.; Strecker, M.N.; Roberts, J.S.; Burke, W.; Mayeux, R.; et al. American College of Medical Genetics and the National Society of Genetic Counselors. Genetic counseling and testing for Alzheimer disease: Joint practice guidelines of the American College of Medical Genetics and the National Society of Genetic Counselors. *Genet. Med.* **2011**, *136*, 597–605. [CrossRef]
6. Reiman, E.M.; Langbaum, J.B.; Tariot, P.N.; Lopera, F.; Bateman, R.J.; Morris, J.C.; Sperling, R.A.; Aisen, P.S.; Roses, A.D.; Welsh-Bohmer, K.A.; et al. CAP-advancing the evaluation of preclinical Alzheimer disease treatments. *Nat. Rev. Neurol.* **2016**, *121*, 56–61. [CrossRef]

7. Tariot, P.N.; Lopera, F.; Langbaum, J.B.; Thomas, R.G.; Hendrix, S.; Schneider, L.S.; Rios-Romenets, S.; Giraldo, M.; Acosta, N.; Tobon, C.; et al. The Alzheimer's Prevention Initiative Autosomal-Dominant Alzheimer's Disease Trial: A study of crenezumab versus placebo in preclinical PSEN1 E280A mutation carriers to evaluate efficacy and safety in the treatment of autosomal-dominant Alzheimer's disease, including a placebo-treated noncarrier cohort. *Alzheimers Dement. (N. Y.)* **2018**, *4*, 150–160.
8. Bergougnoux, A.; D'Argenio, V.; Sollfrank, S.; Verneau, F.; Telese, A.; Postiglione, I.; Lackner, K.J.; Claustres, M.; Castaldo, G.; Rossman, H.; et al. Multicenter validation study for the certification of a CFTR gene scanning method using next generation sequencing technology. *Clin. Chem. Lab. Med.* **2018**, *56*, 1046–1053. [CrossRef]
9. D'Argenio, V.; Esposito, M.V.; Telese, A.; Precone, V.; Starnone, F.; Nunziato, M.; Cantiello, P.; Iorio, M.; Evangelista, E.; D'Aiuto, M.; et al. The molecular analysis of BRCA1 and BRCA2: Next-generation sequencing supersedes conventional approaches. *Clin. Chim. Acta* **2015**, *446*, 221–225. [CrossRef]
10. D'Argenio, V.; Frisso, G.; Precone, V.; Boccia, A.; Fienga, A.; Pacileo, G.; Limongelli, G.; Paolella, G.; Calabrò, R.; Salvatore, F. DNA sequence capture and next-generation sequencing for the molecular diagnosis of genetic cardiomyopathies. *J. Mol. Diagn.* **2014**, *16*, 32–44. [CrossRef]
11. Cariati, F.; D'Argenio, V.; Tomaiuolo, R. The evolving role of genetic tests in reproductive medicine. *J. Transl. Med.* **2019**, *17*, 267. [CrossRef]
12. Nunziato, M.; Esposito, M.V.; Starnone, F.; Diroma, M.A.; Calabrese, A.; Del Monaco, V.; Buono, P.; Frasci, G.; Botti, G.; D'Aiuto, M.; et al. A multi-gene panel beyond BRCA1/BRCA2 to identify new breast cancer-predisposing mutations by a picodroplet PCR followed by a next-generation sequencing strategy: A pilot study. *Anal. Chim. Acta* **2019**, *1046*, 154–162. [CrossRef]
13. Precone, V.; Del Monaco, V.; Esposito, M.V.; De Palma, F.D.; Ruocco, A.; Salvatore, F.; D'Argenio, V. Cracking the code of human diseases using next-generation sequencing: Applications, challenges, and perspectives. *Biomed. Res. Int.* **2015**, 161648. [CrossRef]
14. D'Argenio, V. The high-throughput analyses era: are we ready for the data struggle? *High. Throughput* **2018**, *7*, 8. [CrossRef]
15. Belinova, I.; Karran, E.; De Strooper, B. The toxic Ab oligomer and Alzheimer's disease: An emperor in need of clothes. *Nat. Neurosci.* **2012**, *15*, 349–357.
16. Goedert, M.; Clavaguera, F.; Tolnay, M. The propagation of prion-like protein inclusions in neurodegenerative diseases. *Trends Neurosci.* **2010**, *33*, 317–325. [CrossRef]
17. Jucker, M.; Walker, L.C. Self-propagation of pathogenic protein aggregates in neurodegenerative diseases. *Nature* **2013**, *501*, 45–51. [CrossRef]
18. Sarnataro, D.; Pepe, A.; Zurzolo, C. Cell biology of prion protein. *Prog. Mol. Biol. Transl. Sci.* **2017**, *150*, 57–82.
19. Sarnataro, D. Attempt to untangle the prion-like misfolding mechanism for neurodegenerative diseases. *Int. J. Mol. Sci.* **2018**, *19*, 3081. [CrossRef]
20. De Jonghe, C.; Esselens, C.; Kumar-Singh, S.; Craessaerts, K.; Serneela, S.; Checler, F.; Annaert, W.; Van Broeckhoven, C.; De Strooper, B. Pathogenic APP mutations near the γ-secretase cleavage site differentially affect Aβ secretion and APP C-terminal fragment stability. *Hum. Mol. Gen.* **2001**, *16*, 1665–1671. [CrossRef]
21. Murakami, K.; Irie, K.; Morimoto, A.; Ohigashi, H.; Shindo, M.; Nagao, M.; Shimizu, T.; Shirasawa, T. Synthesis, aggregation, neurotoxicity, and secondary structure of various Aβ1-42 mutants of familial Alzheimer's disease at positions 21–23. *Biochem. Biophys. Res. Commun.* **2002**, *294*, 5–10. [CrossRef]
22. Kumar-Singh, S. Hereditary and sporadic forms of Aβ-cerebrovascular amyloidosis and relevant transgenic mouse models. *Int. J. Mol. Sci.* **2009**, *10*, 1872–1895. [CrossRef] [PubMed]
23. Del Prete, D.; Suski, J.M.; Oulès, B.; Debayle, D.; Gay, A.S.; Lacas-Gervais, S.; Bussiere, R.; Bauer, C.; Pinton, P.; Paterlini-Bréchot, P.; et al. Localization and processing of the Amyloid-β protein precursor in mitochondria-associated membranes. *J. Alzheimers Dis.* **2017**, *55*, 1549–1570. [CrossRef] [PubMed]
24. Bhattacharya, A.; Limone, A.; Napolitano, F.; Cerchia, C.; Parisi, S.; Minopoli, G.; Montuori, N.; Lavecchia, A.; Sarnataro, D. APP maturation and intracellular localization are controlled by a specific inhibitor of 37/67 kDa laminin-1 receptor in neuronal cells. *Int. J. Mol. Sci.* **2020**, *21*, 1738. [CrossRef] [PubMed]
25. Götz, A.; Hogel, P.; Silber, M.; Chaitoglou, I.; Luy, B.; Muhle-Goll, C.; Scharnagl, C.; Langosch, D. Increased H-bond stability relates to altered e-cleavage efficiency and Ab levels in the I45T familial Alzheimer's disease mutant of APP. *Sci. Rep.* **2019**, *9*, 5321. [CrossRef]

26. Bocharov, E.V.; Nadezhdin, K.D.; Urban, A.S.; Volynsky, P.E.; Pavlov, K.V., Efremov, R.G.; Arseniev, A.S.; Bocharova, O.V. Familial L723P mutation can shift the distribution between the alternative APP transmembrane domain cleavage cascades by local unfolding of the E-cleavage site suggesting a straightforward mechanism of Alzheimer's disease pathogenesis. *ACS Chem. Biol.* **2019**, *4*, 1573–1582. [CrossRef]
27. Lumsden, A.L.; Rogers, J.T.; Majd, S.; Newman, M.; Sutherland, G.T.; Verdile, G.; Lardelli, M. Dysregulation of neuronal iron homeostasis as an alternative unifying effect of mutations causing familial Alzheimer's disease. *Front. Neurosci.* **2018**, *12*, 533. [CrossRef]
28. Di Fede, G.; Catania, M.; Morbin, M.; Rossi, G.; Suardi, S.; Mazzoleni, G.; Merlin, M.; Giovagnoli, A.R.; Prioni, S.; Erbetta, A.; et al. A recessive mutation in the APP gene with dominant-negative effect on amyloidogenesis. *Science* **2009**, *323*, 1473–1477. [CrossRef]
29. Tomiyama, T.; Nagata, T.; Shimada, H.; Teraoka, R.; Fukushima, A.; Kanemitsu, H.; Takuma, H.; Kuwano, R.; Imagawa, M.; Ataka, S.; et al. A new amyloid beta variant favoring oligomerization in Alzheimer'stype dementia. *Ann. Neurol.* **2008**, *63*, 377–387. [CrossRef]
30. Conidi, M.E.; Bernardi, L.; Puccio, G.; Smirne, N.; Muraca, M.G.; Curcio, S.A.; Colao, R.; Piscopo, P.; Gallo, M.; Anfossi, M.; et al. Homozygous carriers of APP A713T mutation in an autosomal dominant Alzheimer disease family. *Neurology* **2015**, *84*, 2266–2273. [CrossRef]
31. Cabrejo, L.; Guyant-Marechal, L.; Laquerriere, A.; Vercelletto, M.; De la Fourniere, F.; Thomas-Anterion, C.; Verny, C.; Letournel, F.; Pasquier, F.; Vital, A.; et al. Phenotype associated with APP duplication in five families. *Brain* **2006**, *129*, 2966–2976. [CrossRef] [PubMed]
32. Rovelet-Lecrux, A.; Hannequin, D.; Raux, G.; Le Meur, N.; Laquerriere, A.; Vital, A.; Dumanchin, C.; Feuillette, S.; Brice, A.; Vercelletto, M.; et al. APP locus duplication causes autosomal dominant early-onset Alzheimer disease with cerebral amyloid angiopathy. *Nat. Genet.* **2006**, *38*, 24–26. [CrossRef] [PubMed]
33. Remes, A.M.; Finnila, S.; Mononen, H.; Tuominen, H.; Takalo, R.; Herva, R.; Majamaa, K. Hereditary dementia with intracerebral hemorrhages and cerebral amyloid angiopathy. *Neurology* **2004**, *63*, 234–240. [CrossRef] [PubMed]
34. Rovelet-Lecrux, A.; Frebourg, T.; Tuominen, H.; Majamaa, K.; Campion, D.; Remes, A.M. APP locus duplication in a Finnish family with dementia and intracerebral haemorrhage. *J. Neurol. Neurosurg. Psychiatry* **2007**, *78*, 1158–1159. [CrossRef] [PubMed]
35. Sleegers, K.; Brouwers, N.; Gijselinck, I.; Theuns, J.; Goossens, D.; Wauters, J.; Del-Favero, J.; Cruts, M.; van Duijn, C.M.; Van Broeckhoven, C. APP duplication is sufficient to cause early onset Alzheimer's dementia with cerebral amyloid angiopathy. *Brain* **2006**, *129*, 2977–2983. [CrossRef] [PubMed]
36. Hooli, B.V.; Mohapatra, G.; Mattheisen, M.; Parrado, A.R.; Roehr, J.T.; Shen, Y.; Gusella, J.F.; Moir, R.; Saunders, A.J.; Lange, C.; et al. Role of common and rare APP DNA sequence variants in Alzheimer disease. *Neurology* **2012**, *78*, 1250–1257. [CrossRef]
37. Jonsson, T.; Atwal, J.K.; Steinberg, S.; Snaedal, J.; Jonsson, P.V.; Bjornsson, S.; Stefansson, H.; Sulem, P.; Gudbjartsson, D.; Maloney, J.; et al. A mutation in APP protects against Alzheimer's disease and age-related cognitive decline. *Nature* **2012**, *488*, 96–99. [CrossRef]
38. Cruts, M.; Theuns, J.; Van Broeckhoven, C. Locus-specific mutation databases for neurodegenerative brain diseases. *Hum. Mutat.* **2012**, *33*, 1340–1344. [CrossRef]
39. Pilotto, A.; Padovani, A.; Borroni, B. Clinical, biological, and imaging features of monogenic Alzheimer's Disease. *Biomed. Res. Int.* **2013**, 689591. [CrossRef]
40. Portet, F.; Dauvilliers, Y.; Campion, D.; Raux, G.; Hauw, J.J.; Lyon-Caen, O.; Camu, W.; Touchon, J. Very early onset AD with a de novo mutation in the presenilin 1 gene (Met 233 Leu). *Neurology* **2003**, *61*, 1136–1137. [CrossRef]
41. Golan, M.P.; Styczynska, M.; Jozwiak, K.; Walecki, J.; Maruszak, A.; Pniewski, J.; Lugiewicz, R.; Filipek, S.; Zekanowski, C.; Barcikowska, M. Early-onset Alzheimer's disease with a de novo mutation in the presenilin 1 gene. *Exp. Neurol.* **2007**, *208*, 264–268. [CrossRef] [PubMed]
42. Lanoiselèe, H.M.; Nicolas, G.; Wallon, D.; Rovelet-Lecrux, A.; Lacour, M.; Rousseau, S.; Richard, A.C.; Pasquier, F.; Rollin-Sillaire, A.; Martinaud, O.; et al. Collaborators of the CNR-MAJ project. APP, PSEN1, and PSEN2 mutations in early-onset Alzheimer disease: A genetic screening study of familial and sporadic cases. *PLoS Med.* **2017**, *14*, e1002270. [CrossRef] [PubMed]

43. Kosik, K.S.; Munoz, C.; Lopez, L.; Arcila, M.L.; Garcia, G.; Madrigal, L.; Moreno, S.; Ríos Romenets, S.; Lopez, H.; Gutierrez, M.; et al. Homozygosity of the autosomal dominant Alzheimer disease presenilin 1 E280A mutation. *Neurology* **2015**, *84*, 206–208. [CrossRef]
44. Hardy, J.; Crook, R. Presenilin mutations line up along transmembrane alpha-helices. *Neurosci. Lett.* **2001**, *306*, 203–205. [CrossRef]
45. Guerreiro, R.J.; Beck, J.; Gibbs, J.R.; Santana, I.; Rossor, M.N.; Schott, J.M.; Nalls, M.A.; Ribeiro, H.; Santiago, B.; Fox, N.C.; et al. Genetic Variability in CLU and Its Association with Alzheimer's Disease. *PLoS ONE* **2010**, *5*, e9510. [CrossRef]
46. Brouwers, N.; Sleegers, K.; Van Broeckhoven, C. Molecular genetics of Alzheimer's disease: An update. *Ann. Med.* **2008**, *40*, 562–583. [CrossRef]
47. Wingo, T.S.; Lah, J.J.; Levey, A.I.; Cutler, D.J. Autosomal recessive causes likely in early-onset Alzheimer disease. *Arch. Neurol.* **2012**, *69*, 59–64. [CrossRef]
48. Jarmolowicz, A.I.; Chen, H.Y.; Panegyres, P.K. The Patterns of Inheritance in Early-Onset Dementia: Alzheimer's Disease and Frontotemporal Dementia. *Am. J. Alzheimers Dis. Other Dement.* **2014**, *30*, 299–306. [CrossRef]
49. Mahley, R.W. Apolipoprotein E: Cholesterol transport protein with expanding role in cell biology. *Science* **1988**, *240*, 622–630. [CrossRef]
50. Mahley, R.W.; Rall, S.C., Jr. Apolipoprotein E: Far more than a lipid transport protein. *Annu Rev. Genomics Hum. Genet.* **2000**, *1*, 507–537. [CrossRef]
51. Ashford, J.W. APOE genotype effects on Alzheimer's disease onset and epidemiology. *J. Mol. Neurosci.* **2004**, *23*, 157–165. [CrossRef]
52. Corder, E.H.; Saunders, A.M.; Risch, N.J.; Strittmatter, W.J.; Schmechel, D.E.; Gaskell, P.C., Jr.; Rimmler, J.B.; Locke, P.A.; Conneally, P.M.; Schmader, K.E. Protective effect of apolipoprotein E type 2 allele for late onset Alzheimer disease. *Nat. Genet.* **1994**, *7*, 180–184. [CrossRef]
53. Huang, Y. Apolipoprotein E and Alzheimer disease. *Neurology* **2006**, *66*, S79–S85. [CrossRef] [PubMed]
54. Van Duijn, C.M.; de Knijff, P.; Cruts, M.; Wehnert, A.; Havekes, L.M.; Hofman, A.; Van Broeckhoven, C. Apolipoprotein E4 allele in a population-based study of early-onset Alzheimer's disease. *Nat. Genet.* **1994**, *7*, 74–78. [CrossRef] [PubMed]
55. Genin, E.; Hannequin, D.; Wallon, D.; Sleegers, K.; Hiltunen, M.; Combarros, O.; Bullido, M.J.; Engelborghs, S.; De Deyn, P.; Berr, C.; et al. APOE and Alzheimer disease: A major gene with semi-dominant inheritance. *Mol. Psychiatry* **2011**, *16*, 903–907. [CrossRef] [PubMed]
56. Sorbi, S.; Nacmias, B.; Forleo, P.; Piacentini, S.; Latorraca, S.; Amaducci, L. Epistatic effect of APP717 mutation and apolipoprotein E genotype in familial Alzheimer's disease. *Ann. Neurol.* **1995**, *38*, 124–127. [CrossRef] [PubMed]
57. Pastor, P.; Roe, C.M.; Villegas, A.; Bedoya, G.; Chakraverty, S.; Garcia, G.; Bedoya, G.; Chakraverty, S.; García, G.; Tirado, V.; et al. Apolipoprotein Eepsilon4 modifies Alzheimer's disease onset in an E280A PS1 kindred. *Ann. Neurol.* **2003**, *54*, 163–169. [CrossRef]
58. Wijsman, E.M.; Daw, E.W.; Yu, X.; Steinbart, E.J.; Nochlin, D.; Bird, T.D.; Schellenberg, G.D. APOE and other loci affect age-at-onset in Alzheimer's disease families with PS2 mutation. *Am. J. Med. Genet. Part B Neuropsychiatr. Genet.* **2005**, *132*, 14–20. [CrossRef]
59. Petersen, R.C. Mild Cognitive Impairment. *Continuum (Minneap Minn)* **2016**, *22*, 404–418. [CrossRef]
60. Guerreiro, R.J.; Lohmann, E.; Kinsella, E.; Bras, J.M.; Luu, N.; Gurunlian, N.; Dursun, B.; Bilgic, B.; Santana, I.; Hanagasi, H.; et al. Exome sequencing reveals an unexpected genetic cause of disease: NOTCH3 mutation in a Turkish family with Alzheimer's disease. *Neurobiol. Aging* **2012**, *33*, 1008–1023. [CrossRef]
61. Patel, D.; Mez, J.; Vardarajan, B.N.; Staley, L.; Chung, J.; Zhang, X.; Farrell, J.J.; Rynkiewicz, M.J.; Cannon-Albright, L.A.; Teerlink, C.C.; et al. Alzheimer's Disease Sequencing Project. Association of Rare Coding Mutations with Alzheimer Disease and Other Dementias Among Adults of European Ancestry. *JAMA Netw. Open* **2019**, *2*, 191350. [CrossRef] [PubMed]
62. Bordet, R.; Ihl, R.; Korczyn, A.D.; Lanza, G.; Jansa, J.; Hoerr, R.; Guekht, A. Towards the concept of disease-modifier in post-stroke or vascular cognitive impairment: A consensus report. *BMC Med.* **2017**, *15*, 107. [CrossRef] [PubMed]
63. Lanza, G.; Cantone, M.; Musso, S.; Borgione, E.; Scuderi, C.; Ferri, R. Early-onset subcortical ischemic vascular dementia in an adult with mtDNA mutation 3316G>A. *J. Neurol.* **2018**, *265*, 968–969. [CrossRef] [PubMed]

64. Pottier, C.; Hannequin, D.; Coutant, S.; Rovelet-Lecrux, A.; Wallon, D.; Rousseau, S.; Legallic, S.; Paquet, C.; Bombois, S.; Pariente, J.; et al. PHRC GMAJ Collaborators. High frequency of potentially pathogenic SORL1 mutations in autosomal dominant early-onset Alzheimer disease. *Mol. Psychiatry* **2012**, *17*, 875–879. [CrossRef]
65. Caglayan, S.; Takagi-Niidome, S.; Liao, F.; Carlo, A.S.; Schmidt, V.; Burgert, T.; Kitago, Y.; Füchtbauer, E.M.; Füchtbauer, A.; Holtzman, D.M.; et al. Lysosomal sorting of amyloid-beta by the SORLA receptor is impaired by a familial Alzheimer's disease mutation. *Sci. Transl. Med.* **2014**, *6*, 223. [CrossRef]
66. Cuccaro, M.L.; Camey, R.M.; Zhang, Y.; Bohm, C.; Kunkle, B.W.; Vardarajan, B.N.; Whitehead, P.L.; Cukier, H.N.; Mayeux, R.; George-Hyslop, P.; et al. SORL1 mutations in early- and late-onset Alzheimer disease. *Neurol. Genet.* **2016**, *2*, e116.
67. Nicolas, G.; Charbonnier, C.; Wallon, D.; Quenez, O.; Bellenguez, C.; Grenier-Boley, B.; Rousseau, S.; Richard, A.C.; Rovelet-Lecrux, A.; Le Guennec, K.; et al. CNR-MAJ collaborators. SORL1 rare variants: A major risk factor for familial early-onset Alzheimer's disease. *Mol. Psychiatry* **2016**, *21*, 831–836. [CrossRef]
68. Vardarajan, B.N.; Zhang, Y.; Lee, J.H.; Cheng, R.; Bohm, C.; Ghani, M.; Cheng, R.; Bohm, C.; Ghani, M.; Reitz, C.; et al. Coding mutations in SORL1 and Alzheimer disease. *Ann. Neurol.* **2015**, *77*, 215–227. [CrossRef]
69. Li, X.; Xiong, Z.; Liu, Y.; Yuan, Y.; Deng, J.; Xiang, W.; Li, Z. Case report of first—Episode psychotic symptoms in a patient with early-onset Alzheimer's disease. *BMC Psychiatry* **2020**, *20*, 128. [CrossRef]
70. Cuyvers, E.; Sleegers, K. Genetic variations underlying Alzheimer's disease: Evidence from genome-wide association studies and beyond. *Lancet Neurol.* **2016**, *15*, 857–868. [CrossRef]
71. Shen, L.; Jia, J. An overview of Genome-Wide Association Studies in Alzheimer's Disease. *Neurosci. Bull.* **2016**, *32*, 183–190. [CrossRef] [PubMed]
72. Lambert, J.C.; Heath, S.; Even, G.; Campion, D.; Sleegers, K.; Hiltunen, M.; Combarros, O.; Zelenika, D.; Bullido, M.J.; Tavernier, B.; et al. Genome-wide association study identifies variants at CLU and CR1 associated with Alzheimer's disease. *Nat. Gen.* **2009**, *41*, 1094–1099. [CrossRef] [PubMed]
73. Jonsson, T.; Stefansson, H.; Steinberg, S.; Jonsdottir, I.; Jonsson, P.V.; Snaedal, J.; Bjornsson, S.; Huttenlocher, J.; Levey, A.I.; Lah, J.J.; et al. Variant of TREM2 associated with the risk of Alzheimer's disease. *N. Engl. J. Med.* **2013**, *368*, 107–116. [CrossRef] [PubMed]
74. Bettens, K.; Brouwers, N.; Engelborghs, S.; Lambert, J.C.; Rogaeva, E.; Vandenberghe, R.; Le Bastard, N.; Pasquier, F.; Vermeulen, S.; Van Dongen, J.; et al. Both common variations and rare non-synonymous substitutions and small insertion/deletions in CLU are associated with increased Alzheimer risk. *Mol. Neurodegener.* **2012**, *7*, 3. [CrossRef] [PubMed]
75. Bettens, K.; Vermeulen, S.; Van Cauwenberghe, C.; Heeman, B.; Asselbergh, B.; Robberecht, C.; Engelborghs, S.; Vandenbulcke, M.; Vandenberghe, R.; De Deyn, P.P.; et al. Reduced secreted clusterin as a mechanism for Alzheimer-associated CLU mutations. *Mol. Neurodegener.* **2015**, *10*, 30. [CrossRef]
76. Guerreiro, R.; Wojtas, A.; Bras, J.; Carrasquillo, M.; Rogaeva, E.; Majounie, E.; Cruchaga, C.; Sassi, C.; Kauwe, J.S.; Younkin, S.; et al. Alzheimer Genetic Analysis Group. TREM2 variants in Alzheimer's disease. *Engl. J. Med.* **2013**, *368*, 117–127. [CrossRef]
77. Bailey, C.C.; DeVaux, L.B.; Farzan, M. The triggering receptor expressed on myeloid cells 2 binds apolipoprotein E. *J. Biol. Chem.* **2015**, *290*, 26033–26042. [CrossRef]
78. Lue, L.F.; Schmitz, C.; Walker, D.G. What happens to microglial TREM2 in Alzheimer's disease: Immunoregulatory turned into immunopathogenic? *Neuroscience* **2015**, *302*, 138–150. [CrossRef] [PubMed]
79. Kleinberger, G.; Yamanishi, Y.; Suarez-Calvet, M.; Czirr, E.; Lohmann, E.; Cuyvers, E.; Struyfs, H.; Pettkus, N.; Wenninger Weinzierl, A.; Mazaheri, F.; et al. TREM2 mutations implicated in neurodegeneration impair cell surface transport and phagocytosis. *Sci. Transl. Med.* **2014**, *6*, 243ra86. [CrossRef] [PubMed]
80. Cuyvers, E.; Bettens, K.; Philtjens, S.; Van Langenhove, T.; Gijselinck, I.; van der Zee, J.; Engelborghs, S.; Vandenbulcke, M.; Van Dongen, J.; Geerts, N.; et al. BELNEU consortium. Investigating the role of rare heterozygous TREM2 variants in Alzheimer's disease and frontotemporal dementia. *Neurobiol. Aging* **2014**, *35*, 726–729. [CrossRef] [PubMed]
81. Pottier, C.; Wallon, D.; Rousseau, S.; Rovelet-Lecrux, A.; Richard, A.C.; Rollin-Sillaire, A.; Frebourg, T.; Campion, D.; Hannequin, D. TREM2 R47H variant as a risk factor for early-onset Alzheimer's disease. *J. Alzheimers Dis.* **2013**, *35*, 45–49. [CrossRef] [PubMed]

82. Cruchaga, C.; Karch, C.M.; Jin, S.C.; Benitez, B.A.; Cai, Y.; Guerreiro, R.; Harari, O.; Norton, J.; Budde, J.; Bertelsen, S.; et al. Rare coding variants in the phospholipase D3 gene confer risk for Alzheimer's disease. *Nature* **2014**, *505*, 550–554. [CrossRef] [PubMed]
83. Cuyvers, E.; De Roeck, A.; Van den Bossche, T.; Van Cauwenberghe, C.; Bettens, K.; Vermeulen, S.; Mattheijssens, M.; Peeters, K.; Engelborghs, S.; Vandenbulcke, M.; et al. Mutations in 74 ABCA7 in a Belgian cohort of Alzheimer's disease patients: A targeted resequencing study. *Lancet Neurol.* **2015**, *14*, 814–822. [CrossRef]
84. Steinberg, S.; Stefansson, H.; Jonsson, T.; Johannsdottir, H.; Ingason, A.; Helgason, H.; Sulem, P.; Magnusson, O.T.; Gudjonsson, S.A.; Unnsteinsdottir, U.; et al. Loss-of-function variants in ABCA7 confer risk of Alzheimer's disease. *Nat. Genet.* **2015**, *47*, 445–447. [CrossRef]
85. Vardarajan, B.N.; Ghani, M.; Kahn, A.; Sheikh, S.; Sato, C.; Barral, S.; Lee, J.H.; Cheng, R.; Reitz, C.; Lantigua, R.; et al. Rare coding mutations identified by sequencing of Alzheimer disease genome-wide 76 association studies loci. *Ann. Neurol.* **2015**, *78*, 487–498. [CrossRef] [PubMed]
86. Satoh, K.; Abe-Dohmae, S.; Yokoyama, S.; St George-Hyslop, P.; Fraser, P.E. ATP-binding cassette transporter A7 (ABCA7) loss of function alters Alzheimer amyloid processing. *J. Biol. Chem.* **2015**, *290*, 24152–24165. [CrossRef]
87. Bonvicini, C.; Scassellati, C.; Benussi, L.; Di Maria, E.; Maj, C.; Ciani, M.; Fostinelli, S.; Mega, A.; Bocchetta, M.; Lanzi, G.; et al. Next Generation Sequencing Analysis in Early Onset Dementia Patients. *J. Alzheimers Dis.* **2019**, *67*, 243–256. [CrossRef]
88. Cochran, J.N.; McKinley, E.C.; Cochran, M.; Amaral, M.D.; Moyers, B.A.; Lasseigne, B.N.; Gray, D.E.; Lawlor, J.M.J.; Prokop, J.W.; Geier, E.G.; et al. Genome sequencing for early-onset or atypical dementia: High diagnostic yield and frequent observation of multiple contributory alleles. *Cold Spring Harb. Mol. Case Stud.* **2019**, *5*, a003491. [CrossRef] [PubMed]
89. Cali, F.; Cantone, M.; Cosentino, F., II; Lanza, G.; Ruggeri, G.; Chiavetta, V.; Salluzzo, R.; Ragalmuto, A.; Vinci, M.; Ferri, R. Interpreting Genetic Variants: Hints from a Family Cluster of Parkinson's Disease. *J. Parkinsons Dis.* **2019**, *9*, 203–206. [CrossRef] [PubMed]
90. Qazi, T.J.; Quan, Z.; Mir, A.; Qing, H. Epigenetics in Alzheimer's Disease: Perspective of DNA Methylation. *Mol. Neurobiol.* **2018**, *55*, 1026–1044. [CrossRef]
91. D'Argenio, V.; Sarnataro, D. Microbiome influence in the pathogenesis of prion and Alzheimer's Diseases. *Int. J. Mol. Sci.* **2019**, *20*, 4704. [CrossRef] [PubMed]
92. Lee, S.Y.; Song, M.Y.; Kim, D.; Park, C.; Park, D.K.; Kim, D.G.; Yoo, J.S.; Kim, Y.H. A Proteotranscriptomic-based computational drug-repositioning method for Alzheimer's Disease. *Front. Pharmacol.* **2020**, *10*, 1653. [CrossRef] [PubMed]

Article

Influence of *ApoE* Genotype and *Clock* T3111C Interaction with Cardiovascular Risk Factors on the Progression to Alzheimer's Disease in Subjective Cognitive Decline and Mild Cognitive Impairment Patients

Valentina Bessi [1,*], Juri Balestrini [1], Silvia Bagnoli [1], Salvatore Mazzeo [1], Giulia Giacomucci [1], Sonia Padiglioni [1], Irene Piaceri [1], Marco Carraro [1], Camilla Ferrari [1], Laura Bracco [1], Sandro Sorbi [1,2] and Benedetta Nacmias [1]

1 Department of Neuroscience, Psychology, Drug Research and Child Health-University of Florence–Viale Pieraccini 6, 50139 Florence, Italy; juri.balestrini@gmail.com (J.B.); silvia.bagnoli@unifi.it (S.B.); salvatore.mazzeo@unifi.it (S.M.); giuliagiacomucci.md@gmail.com (G.G.); sonia_padiglioni@libero.it (S.P.); irene.piaceri@gmail.com (I.P.); marco.carraro@unifi.it (M.C.); camilla.ferrari@unifi.it (C.F.); bracco.neuro@gmail.com (L.B.); sandro.sorbi@unifi.it (S.S.); benedetta.nacmias@unifi.it (B.N.)

2 IRCCS Fondazione Don Carlo Gnocchi, via di Scandicci 269, 50143 Florence, Italy

* Correspondence: valentina.bessi@unifi.it; Tel.: +39-05-7948660; Fax: +39-05-7947484

Received: 11 May 2020; Accepted: 27 May 2020; Published: 29 May 2020

Abstract: Background: Some genes could interact with cardiovascular risk factors in the development of Alzheimer's disease. We aimed to evaluate the interaction between *ApoE* ε4 status, *Clock* T3111C and *Per2* C111G polymorphisms with cardiovascular profile in Subjective Cognitive Decline (SCD) and Mild Cognitive Impairment (MCI). Methods: We included 68 patients who underwent clinical evaluation; neuropsychological assessment; *ApoE*, *Clock* and *Per2* genotyping at baseline; and neuropsychological follow-up every 12–24 months for a mean of 13 years. We considered subjects who developed AD and non-converters. Results: *Clock* T3111C was detected in 47% of cases, *Per2* C111G in 19% of cases. *ApoE* ε4 carriers presented higher risk of heart disease; *Clock* C-carriers were more frequently smokers than non C-carriers. During the follow-up, 17 patients progressed to AD. Age at baseline, *ApoE* ε 4 and dyslipidemia increased the risk of conversion to AD. *ApoE* ε4 carriers with history of dyslipidemia showed higher risk to convert to AD compared to *ApoE* ε4− groups and *ApoE* ε4+ without dyslipidemia patients. *Clock* C-carriers with history of blood hypertension had a higher risk of conversion to AD. Conclusions: *ApoE* and *Clock* T3111C seem to interact with cardiovascular risk factors in SCD and MCI patients influencing the progression to AD.

Keywords: Alzheimer's disease; subjective cognitive decline; mild cognitive impairment; clock genes; *Clock*; *ApoE*; cardiovascular risk factors

1. Introduction

Alzheimer's disease (AD) is characterized by a slow but progressive trend, with a presymptomatic phase that can last from years to decades [1]. Subjective Cognitive Decline (SCD) is defined as a self-experienced persistent decline in cognitive capacity in comparison with the subject's previously status, during which the subject has normal performance on standardized cognitive tests [2]. Mild cognitive impairment (MCI) concerns an objective cognitive impairment with minimal impact on instrumental activity of daily living [3], and it is considered an intermediate phase between normal cognition and dementia. MCI is associated with an increased risk of positive AD biomarkers and with an annual conversion rate of 5%–17% to AD [4,5]. For SCD, the annual conversion rate (ACR) to MCI is 3.6%–6.6%, while it is 1.5%–2.3% to dementia [4,6].

A growing amount of evidence has underlined the importance of cardiovascular health on the risk of developing AD [7–9]. It has been reported that approximately one-third of AD cases worldwide may be attributable to cardiovascular risk factors, including hypertension, obesity, diabetes, smoking, and physical inactivity [10].

In addition, it is well known that genetic aspects play a central role in development of AD [11]. Apolipoprotein E e4 carrier status (*ApoE* e4) is a well-defined genetic risk factor, and recently, current research has been focused on clock genes. *Clock* (*Circadian Locomotor Output Cycle Kaput*, chromosome 4q12) and *Per2* (*Period2*, chromosome 2q37.3) are part of the transcriptional-translational feedback loops regulating circadian rhythm [12]. Several polymorphisms of these genes have been recently studied to elucidate their role in sleep-wake cycle alterations, aging, psychiatric disturbance and neurodegeneration [13,14]. Moreover, *Clock* and *Per2* polymorphisms have been associated with overweight and glucose and lipid metabolism impairments [15,16]. Some studies have investigated the possible role of the *Clock* T3111C polymorphism on the quality of aging in very elderly patients [17], while other studies focused on influence of *Per2* C111G polymorphism on lipid metabolism in adults with metabolic syndrome [16]. On the basis of the above-mentioned initial findings about of the influence of these polymorphisms on cardiovascular profile, the aim of the present study was to define the interaction between *Clock* T3111C and *Per2* C111G, *ApoE* ε4, and cognitive function, in relation to cardiovascular risk factors, in SCD and MCI patients and in the progression to AD.

2. Materials and Methods

2.1. Participants and Clinical Assessment

As part of a longitudinal, clinical-neuropsychological-genetic survey on SCD and MCI, we included 74 consecutive spontaneous patients who self-referred to the Centre for Alzheimer's Disease and Adult Cognitive Disorders of Careggi Hospital in Florence between April 1996 and May 2014. All participants underwent a comprehensive family and clinical history, general and neurological examination, extensive neuropsychological investigation, estimation of premorbid intelligence, as well as assessment of depression. A positive family history was defined as one or more first-degree relatives with documented cognitive decline. Sleep quality was assessed according to anamnestic data: we considered as "poor sleepers" patients who had difficulties in falling asleep or woke up early or have frequent sleep interruptions; patients that did not report sleep disturbances were classified as "good sleepers". For this study, inclusion criteria were: (1) complaining of cognitive decline with a duration of ≥6 months; (2) normal functioning on the Activities of Daily Living and the Instrumental Activities of Daily Living scales; (3) unsatisfied criteria for dementia at baseline [18]; (4) attainment of the clinical endpoint, i.e., conversion to AD according to the NIA-AA [18] criteria during follow up, regardless of follow-up duration; (5) a follow-up time of more than 2 years from the baseline visit for those patients who did not develop AD. Exclusion criteria were: (1) history of head injury, current neurological and/or systemic disease, symptoms of psychosis, major depression, alcoholism or other substance abuse; (2) the complete data loss of patients' follow-up; (3) progression to dementia other than AD.

From the initial sample, we excluded six patients: two patients had a follow-up shorter than 2 years; two diagnosed with psychiatric disturbance, and one with Fronto-Temporal Dementia, according to Neary criteria [19]; one patient received a diagnosis of Vascular Dementia [20]. Therefore, in the end 68 patients were included.

We divided this sample into two groups: 41 patients classified as SCD, according to the terminology proposed by the Subjective Cognitive Decline Initiative (SCD-I) Working Group [2] (i.e., presence of a self-experienced persistent decline in cognitive capacities with normal performance on standardized cognitive tests); 27 patients classified as MCI according to (NIA-AA) criteria for the diagnosis of MCI [3] (i.e., evidence of lower performance in one or more cognitive domains with preserved independence of function in daily life).

All patients underwent clinical and neuropsychological follow-up every 12 or 24 months. All of them were genotyped for ApoE (Apolipoprotein E), Clock and Per2.

On the basis of progression to AD during the follow-up, patients were classified respectively into converters and non-converters. All subjects gave their informed consent for inclusion before they participated in the study. The study was conducted in accordance with the Declaration of Helsinki, and the protocol was approved by the Ethics Committee (DSM study).

2.2. Neuropsychological Assessment

All patients were evaluated by means of an extensive neuropsychological battery [21]. The battery consisted of global measurements [Mini-Mental State Examination (MMSE)], tasks exploring verbal and spatial short- term memory (Digit Span; Corsi Tapping Test) and verbal long-term memory [Five Words and Paired Words Acquisition (FWA, PWA); recall after 10min (FWR10, PWR10); recall after 24-h (FWR24, PWR24); Babcock Short Story Immediate and Delayed Recall (BS, BSR)]; language (Token Test; Category Fluency Task); and visuo-motor functions (Copying Drawings) [21]. Visuospatial abilities were also evaluated by Rey–Osterrieth Complex Figure copy, and visuospatial long-term memory was assessed by means of recall of Rey–Osterrieth Complex Figure test [22]; attention/executive function was explored by means of Dual Task [23], Phonemic Fluency Test [24], and Trail Making Test [25]. Everyday memory was assessed by means of Rivermead Behavioral Memory Test (RBMT) [26]. All raw test scores were adjusted for age, education and gender according to the correction factor reported in validation studies for the Italian population [21–26]. In order to estimate pre-morbid intelligence, all patients were given the TIB ("Test di Intelligenza Breve") [27], an Italian version of the National Adult Reading Test [28]. The presence and severity of depressive symptoms was evaluated by means of the 22-item Hamilton Depression Rating Scale (HRSD) [29].

2.3. Apolipoprotein E ε4, Clock T3111C and Per2 C111G Genotyping

A standard automated method (QIAcube, QIAGEN) was used to isolate DNA from peripheral blood samples. ApoE genotypes were investigated by high resolution melting analysis (HRMA). Two sets of PCR primers were designed to amplify the regions encompassing rs7412 [NC_000019.9:g.45412079C>T] and rs429358 (NC_000019.9:g.45411941T>C). The samples with known ApoE genotypes, which had been validated by DNA sequencing, were used as standard references. The ApoE genotype was coded as ApoE ε4− (no ApoE ε4 alleles) and ApoE ε4+ (presence of one or two ApoE ε4 alleles).

The analyses of Clock and Per2 were performed using HRMA in order to detect the 3111T/C Clock polymorphism using primers as reported [30] and the Per2 C111G polymorphism with the following primers Forward 5′-ACAGAAAGAGTCAAATGGGTGC-3′, Reverse 5′-TGTCCACATCTTCCTGCAGT-3′ with Annealing temperature 60 °C.

2.4. Statistical Analysis

Patient groups were characterized using means and standard deviations (SD). We tested for the normality distribution of the data using the Kolmogorov–Smirnov test. Depending on the distribution of our data, we used t-test or non-parametric Mann–Whitney U Tests for between-groups' comparisons. We used chi-square test to compare categorical data. We analyzed survival curves using the Kaplan–Meier estimator. Finally, we used logistic regression to analyze the role of some cardiovascular risk factors in worsening cognition. All statistical analyses were performed with SPSS software v.25 (SPSS Inc., Chicago, IL, USA). The significance level was set at $p < 0.05$.

3. Results

3.1. Participants and Clinical Assessment

In the whole cohort, 32 of 68 patients (47%) were *Clock* C carriers (29 TC, 3 CC), while 13 of 68 (19%) were *Per2* G carriers (13 CG, 0 GG); 7 of 68 (10%) carried both *Clock* C and *Per2* G alleles. The genotypic distributions of the *Clock* and *Per2* genes in this sample were in Hardy–Weinberg equilibrium (*Clock* T3111C $\chi^2 = 0.91$, $p > 0.05$; *Per2* C111G $\chi^2 = 0.77$, $p > 0.05$). The prevalence of *Clock* T3111C and *Per2* C111G polymorphisms did not significantly differ between SCD and MCI (*Clock* T3111C: 51.2% in SCD and 40.7% in MCI; *Per2* C111G: 19.5% in SCD and 18.5% in MCI); moreover, there were not any differences in the prevalence of both *Clock* C and *Per2* G carriers in SCD and MCI groups. (Table 1).

Table 1. Comparison between prevalence of *Clock* and *Per2* polymorphisms in SCD and MCI individuals.

Features	SCD	MCI	*p*
Per2 G carriers–units (%)	8 (19.5%)	5 (18.5%)	0.919
Clock C carriers-units (%)	21 (51.2%)	11 (40.7%)	0.397
Per2 G carriers and *Clock* C carriers-units (%)	4 (9.75%)	3 (8.10%)	0.843

p indicates level of significance for comparison between groups (statistical significance at $p < 0.05$, in bold characters).

There were no differences between *Clock* C carriers and non C carriers with regards to age at onset of symptoms, age at baseline visit, disease duration, sex, family history of AD, years of education, TIB, MMSE, and *ApoE* ε4 allele status. With respect to CV risk factors, there was a higher proportion of smokers in *Clock* C carriers than in non C carriers (19.40% vs. 2.80%, $\chi^2 = 4{,}892$, $p = 0.027$) (Table 2).

Comparing *Per2* G carriers and non G carriers, there were no differences in age at onset of symptoms, age at baseline visit, disease duration, sex, family, MMSE, and *ApoE* ε4 allele status. *Per2* G carriers had lower premorbid intelligence score on TIB (104.29 ± 10.74 vs. 109.98 ± 8.06, $p = 0.049$), less years of education (7 ± 3.05 vs 10.73 ± 4.51, $p = 0.007$), and lower frequency of family history of AD (15.38% vs. 60%, $\chi^2 = 8.37$, $p = 0.004$) (Table 2). There were no differences in CV risk factors proportion between G and non G carriers.

We did not find any differences in sleep quality between *Clock* C carriers and *Clock* non C carriers ($\chi^2 = 0.136$, $p = 0.454$), neither between *Per2* G carriers and *Per2* non G carriers ($\chi^2 = 0.879$, $p = 0.273$) (Table 2).

ApoE ε4+ patients have a higher proportion of history of heart disease than *ApoE* ε4− (8.70% vs. 0.00%, $\chi^2 = 3.944$, $p = 0.047$). We did not find any statistically significant difference between *ApoE* ε4+ and ApoE ε4− as far age at onset of symptoms, age at baseline evaluation, disease duration (time from onset of symptoms and baseline evaluation), follow-up time, familiarity, sex, education, and MMSE (Table 2).

Table 2. Demographic data according to *Per2*, *Clock*, *ApoE4*.

Features	*Per2*			*Clock*			*ApoE*		
	non G (n = 55)	G (n = 13)	*p*	non C (n = 36)	C (n = 32)	*p*	ε4− (n = 45)	ε4+ (n = 23)	*p*
SCD (% within SCD)	80.5%	19.5%	-	48.8%	51.2%	-	68.3%	31.7%	-
MCI (% within MCI)	81.5%	18.5%	-	59.3%	40.7%	-	63%	37%	-
Conversion to AD (SCD) (% in consideration of carrier status)	0.0%	25.0%	-	5.0%	4.8%	-	3.6%	7.7%	-
Conversion to AD (MCI) (% in consideration of carrier status)	59.1%	40.0%	-	62.5%	45.5%	-	35.3%	90%	-
Age at baseline (±SD) in years	64.03 ± 9.12	65.22 ± 7.26	0.739	64.29 ± 9.33	63.69 ± 8.60	0.777	63.38 ± 9.08	65.38 ± 8.75	0.354
Age at onset (±SD) in years	59.69 ± 10.02	62.38 ± 7.96	0.562	60.56 ± 10.87	59.25 ± 8.64	0.694	59.38 ± 10.12	61.83 ± 9.17	0.223
Follow-up time (±SD) in years	10.87 ± 4.23	12.60 ± 4.68	0.234	11.34 ± 4.31	10.90 ± 4.45	0.815	13.00 ± 4.63	10.54 ± 4.71	0.065
Disease duration (±SD) in years	4.36 ± 3.63	2.83 ± 2.33	0.090	3.73 ± 3.65	4.44 ± 3.18	0.195	4.00 ± 4.85	3.55 ± 2.00	0.800
Sex (females, males)	37. 17	10. 3	0.552	26. 10	22. 10	0.754	33, 12	15, 8	0.487
Family history of AD (%)	61.11%	15.38%	**0.003 ***	41.66%	62.5%	0.086	43.18%	65.22%	0.105
Education in years (±SD)	10.72 ± 4.55	7 ± 3.05	**0.007 ***	9.86 ± 4.80	10.19 ± 4.21	0.737	9.73 ± 4.51	10.57 ± 4.54	0.455
MMSE (±SD)	28.46 ± 1.66	28.31 ± 1.49	0.583	28.09 ± 1.78	28.81 ± 1.33	0.064	28.34 ± 1.78	28.595 ± 1.26	0.839
***ApoE* ε4 (%)**	35.18%	23.07%	0.404	30.55%	37.5%	0.546	-	-	-
Diabetes (%) *	5.60%	7.70%	0.770	8.30%	3.10%	0.362	2.20%	13.00%	0.730
Hypertension (%) *	22.20%	23.10%	0.947	25.00%	18.8%	0.535	22.20%	21.70%	0.964
Dyslipidemia (%) *	30.80%%	23.10%	0.585	30.60%	26.70%	0.728	34.10%	18.20%	0.178
Heart Disease (%) *	3.80%	0.00%	0.477	2.90%	3.10%	0.949	0.00%	8.70%	**0.047**
Smoking (%) *	11.30%	7.70%	0.703	2.80%	19.40%	**0.027**	6.80%	17.40%	0.179

Values quoted in the table are mean (±SD) or (%) or units. *p* indicates level of significance for comparison between groups (statistical significance at $p < 0.05$, in bold characters). * every risk factor was evaluated at the baseline for statistical purpose.

3.2. Description of Sample at Follow-Up

During the follow-up, 17 patients (25%, 2 SCD and 15 MCI) converted to AD (converters) while 51 patients did not progress to AD (non-converters). Mean conversion time was 4.73 ± 3.91 years (range: 1.41–14.01, IQR = 3.71 years). Mean follow-up time of non-converters was 13.03 ± 4.48 years (range: 4.06–23.74, IQR = 6.73 years). There were no differences between converters and non-converters with respect to disease duration, sex, family history of AD, years of education, and MMSE at baseline. Converters had higher age at the onset of symptoms (68.20 ± 7.49 vs. 61.25 ± 6.21, $p = 0.016$), age at baseline (71.60 ± 6.19 vs. 63.42 ± 6.96, $p = 0.003$) and greater proportion of ApoE ε4 (56.25% vs. 25.49%, $\chi^2 = 5.22$, $p = 0.022$). (Table 3).

Table 3. Comparison of demographic and clinical data between converters and non-converters.

Features	Converters (n = 17)	Non Converters (n = 51)	*p*
Prevalence *Per2* G (%)	23.5%	17.6%	0.593
Prevalence *Clock* C (%)	35.3%	51.0%	0.262
Age at baseline (±SD) in years	70.7 (± 6.3)	61.8 (±8.6)	**<0.01**
Age at onset * (±SD) in years	67.4 (± 7.5)	57.9 (±9.4)	**<0.01**
Follow-up time (±SD) in years	9.6 (± 4.7)	13.03 (±4.5)	**<0.01**
Disease duration (±SD) in years	3.3 (± 3.6)	4.0 (±4.3)	0.602
Sex (females, males)	12,5	36,15	1.000
Family history of AD (%)	52.9%	51.0%	0.889
Education in years (±SD)	8.4 (±3.9)	10.6 (±4.6)	0.059
MMSE (±SD)	28.1 (±1.1)	28.5 (±1.8)	0.399
***ApoE* ε4 (%)**	58.8%	25.5%	**0.012**
Hypertension (%)	29.4%	19.6%	0.399
Diabetes (%)	11.8%	3.9%	0.234
Dyslipidemia (%)	47.1%	22.4%	0.053
Heart disease (%)	11.8%	0.0%	**0.014**
Smoking (current) (%)	0.0%	14.0%	0.103
Chronic kidney disease (%)	0.0%	0.0%	-

Values quoted in the table are mean (± SD) or (%). *p* indicates level of significance for comparison between groups (statistical significance at $p < 0.05$, in bold characters). * onset of memory problems (not overt dementia).

There were no significant differences between converters and non-converters in the prevalence of Clock T3111C (35.3% vs. 51.0%, $\chi^2 = 1.26$, $p = 0.262$) and Per2 C111G (23.5% vs. 17.6%, $\chi^2 = 0.285$ $p = 0.593$) polymorphism (Table 3).

In order to ascertain the effects of cardiovascular risk factors on the conversion to AD, we performed a proportional hazards regression analysis considering conversion time as time and "conversion to AD" as dependent variable. We considered as covariates age at onset, age at baseline, ApoE, Clock and Per2 genotype, hypertension, diabetes, dyslipidemia, heart disease, and smoking habit. Dyslipidemia ($p = 0.041$, HR = 3.08, 95% I.C. = 1.05: 9.09), age at baseline ($p = 0.001$, HR = 1.16, 95% I.C. = 1.07:1.27) and ApoE ε4 ($p = 0.001$, HR = 6.21, 95% I.C. = 2.04:18.9) were statistically significantly associated with an increased likelihood of conversion to AD (Table 4).

Table 4. Proportional hazards regression analysis.

	B	*p*	HR	95% C.I.	
				Lower	Upper
Whole sample					
ApoE e4	1.826	0.001	6.212	2.045	18.872
Age at baseline	0.151	0.001	1.163	1.067	1.269
Dyslipidemia	0.126	0.041	3.083	1.045	9.099
***Clock* C carriers**					
Hypertension	3.265	0.025	26.18	1.510	454.039
***Clock* non C carriers**					
Age at baseline	0.204	0.002	1.23	1.078	1.394
ApoE	2.111	0.009	8.25	1.712	39.791

Regression Coefficients (B), *p*-value (*p*), Hazard Ratio (OR) and 95% Confidence Intervals (95% CI) for covariates included in the proportional hazards regression model are reported (significant differences at $p < 0.05$).

3.3. Relationship between ApoE and Dyslipidemia

In order to explore the relationship between dyslipidemia and ApoE genotype, we divided the sample according to ApoE ε4 status (ε4+ and ε4−). In the ε4+ sample, a Kaplan–Meier survival analysis showed significant difference in survival distributions between patients with history of dyslipidemia and patients without history of dyslipidemia ($\chi^2 = 4.42$, $p = 0.036$), as 100% of dyslipidemic patients and 29.4% of non-dyslipidemic patients converted to AD (Figure 1a). When we performed the same analysis on the ε4− sample, we found no statistically significant effect of dyslipidemia on rate of progression to AD ($\chi^2 = 1.92$, $p = 0.166$) (Figure 1b).

Finally, we ranked the whole sample according to history of dyslipidemia and ApoE genotype (non-dyslipidemic/ε4−, n = 29; non-dyslipidemic/ε4+, n = 17; dyslipidemic/ε4−, n = 15; dyslipidemic/ε4, n = 4) and Kaplan–Meier survival analysis was conducted to compare the proportions of conversions in the four different groups. Patients in dyslipidemic/ε4+ group had a higher rate of conversion to AD compared to non-dyslipidemic/ ε4− ($\chi^2 = 25.47$, $p < 0.001$), non-dyslipidemic/ε4+ ($\chi^2 = 4.42$, $p = 0.036$) and dyslipidemic/ε4− ($\chi^2 = 7.64$, $p = 0.006$). Proportion of conversion in non-dyslipidemic/ε4+ was higher as compared to non-dyslipidemic/ε4− ($\chi^2 = 3.73$, $p = 0.05$). There was no significant difference between non-dyslipidemic/ε4−, dyslipidemic/ε4− and dyslipidemic/ε4− (Figure 2).

3.4. Relationship between Clock and Per2 and Risk Factors on Progression to AD

In order to explore the relationship between CV risk factors and Clock polymorphism, we divided the sample according to Clock genotype status (C carriers and non C carriers). For each sample, we performed a proportional hazards regression analysis considering conversion time as time; "conversion to AD" as dependent variable; and age at onset, age at baseline, ApoE genotype, hypertension, diabetes, dyslipidemia, heart disease, and smoking habit as covariates. In the C carriers sample, hypertension at baseline ($p = 0.025$, HR = 3.625) was statistically significantly associated with an increased likelihood of conversion to AD (Table 4). In the Clock non C carriers, none of the CV factors were included in the model as only age at baseline was statistically significantly associated with high risk of progression to AD.

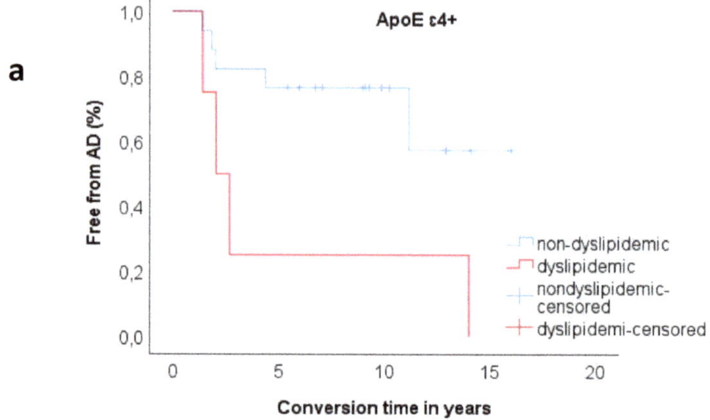

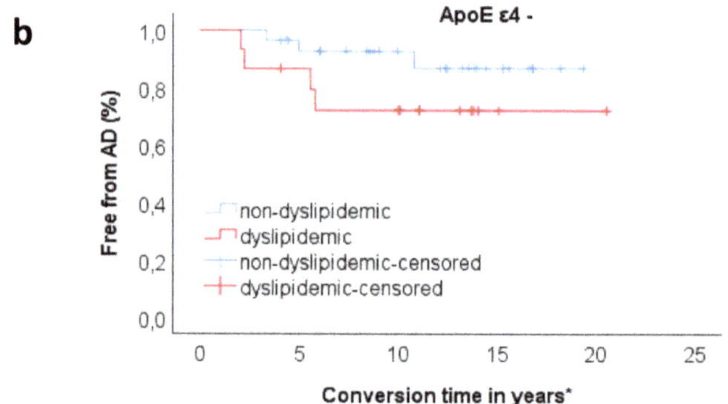

Figure 1. (**a**) Kaplan–Meier survival analysis for comparisons of proportion of progression to AD between dyslipidemic (n = 4) and non-dyslipidemic (n = 17) patients in *ApoE ε4+* carrier group. Proportion of progression was higher in dyslipidemic group (100.00%) compared to non-dyslipidemic (29.40%). The pairwise log rank comparisons showed significant difference in survival distributions for the dyslipidemic vs. non-dyslipidemic ($\chi^2 = 4.42$, $p = 0.036$). (**b**) Kaplan–Meier survival analysis for comparisons of proportion of progression to AD between dyslipidemic (n = 15) and non-dyslipidemic (n = 29) patients in *ApoE ε4−* carrier group. The pairwise log rank comparisons showed no significant difference in survival distributions for the dyslipidemic vs. non-dyslipidemic ($\chi^2 = 1.92$, $p = 0.166$). * For censored cases (non-converters), conversion time indicates follow-up time.

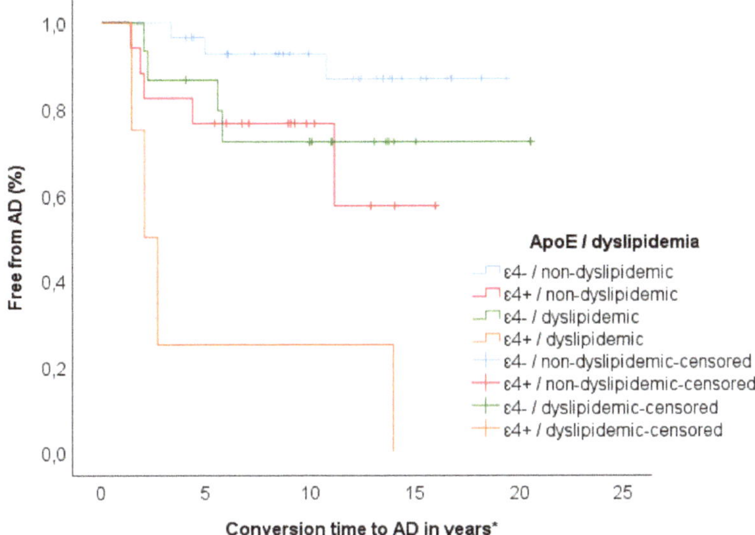

Figure 2. Kaplan–Meier survival analysis for comparisons of patients ranked according to history of dyslipidemia and ApoE genotype (non-dylipidemic/ε4−, n = 29; dyslipidemic/ε4−, n = 17; non-dylipidemic/high/ε4+, n = 15; dylipidemic/ε4+, n = 4). Proportion of progression was higher in dylipidemic/ε4+ (100.00%) compared to non-dylipidemic/ε4− (10.30%, χ^2 = 25.47, p < 0.001), non-dylipidemic/ε4+ (29.40%, χ^2 = 4.42, p = 0.036), and dylipidemic/ε4− (26.7%, χ^2 = 7.64, p = 0.006). Proportion of progression in non-dylipidemic/ε4+ group (26.7%) was higher than non-dylipidemic/ε4− (10.3%, χ^2 = 3.73, p = 0.05). * For censored cases (non converters) conversion time indicates follow-up time.

A Kaplan–Meier survival analysis showed significant difference in survival distributions between patients with history of hypertension and patients without history of hypertension (χ^2 = 4.42, p = 0.036) only in the Clock C carriers sample (Figure 3a). In this group, 50% of hypertensive patients and 11.5% of non-hypertensive patients converted to AD. When we performed the same analysis on the Clock non C carriers sample, we found no statistically significant effect of hypertension on rate of progression to AD; (Figure 3b).

We performed the proportional hazard regression analysis ranking patients according to Per2 polymorphism, including as covariates age at onset, age at baseline, ApoE genotype, hypertension, diabetes, dyslipidemia, heart disease, and smoking. In the G carriers sample, none of the covariates showed a statistically significant effect on risk of conversion to AD. In the Per2 non G carriers, none of the CV factors were included in the model as only age at baseline was statistically significantly associated with high risk of progression to AD (p = 0.010, HR = 1.185).

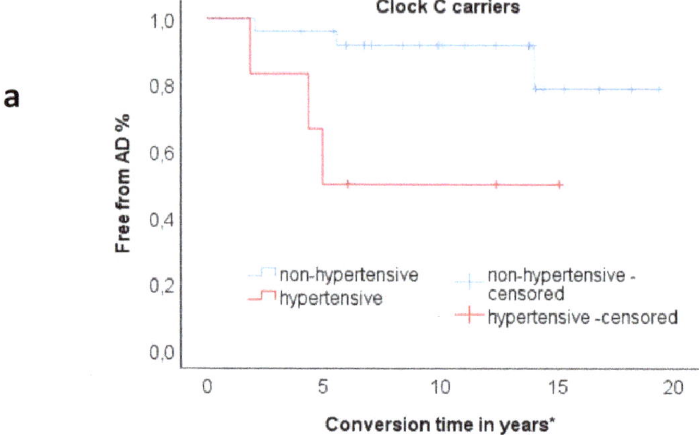

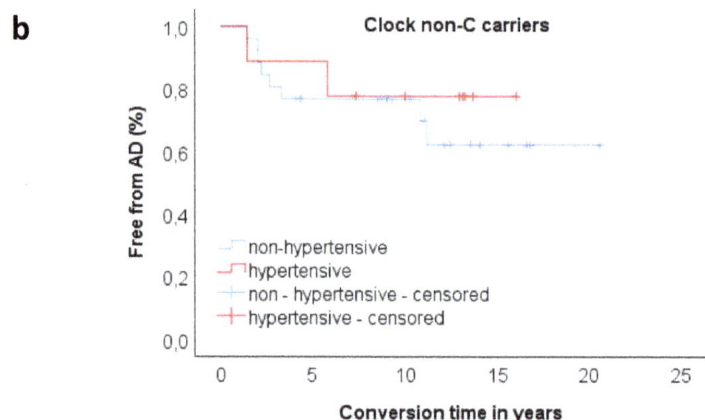

Figure 3. (**a**) Kaplan–Meier survival analysis for comparisons of proportion of progression to AD between hypertensive (n = 6) and non-hypertensive (n = 26) patients in *Clock* C carriers group. Proportion of progression was higher in hypertensive group (50.00%) compared to non-hypertensive (11.50%). The pairwise log rank comparisons showed significant difference in survival distributions for the hypertensive vs. non-hypertensive ($\chi^2 = 5.77$, $p = 0.017$). (**b**) Kaplan–Meier survival analysis for comparisons of proportion of progression to AD between hypertensive (n = 9) and non-hypertensive (n = 26) patients in *Clock* non C carriers group. The pairwise log rank comparisons showed no significant difference in survival distributions for the hypertensive vs. non-hypertensive ($\chi^2 = 0.323$, $p = 0.570$). * For censored cases (non converters) conversion time indicates follow-up time.

4. Discussion

This study investigated the interaction between genetic features (*Clock* T3111C, *Per2* C111G polymorphisms, and *ApoE* genotype) and cardiovascular risk factors in a sample of SCD and MCI patients. We are not aware of any previous studies exploring this topic on these groups of patients.

The frequency of these polymorphisms in our cohort was similar to prevalence data reported in a previous study on healthy Italian population [31].

We found that *Clock* T3111C carriers were more frequent smokers compared to non-carriers of the polymorphism. Our result may suggest an implication also of Clock T3111C on nicotine dependence. Other researches have shown that clock genes are associated with substance abuse, including alcohol, cocaine and cannabis [32]. In fact, circadian genes have a direct role in in the regulation of dopaminergic transmission, especially in reward circuitry. This evidence could represent the biological substrate for the role of Clock genes in the development of addicted behaviours [33,34].

We found a correlation between *Per2* C111G polymorphism with years of education and family history of AD. This result confirms previous evidence by our group [35], but it is difficult to interpret this finding due to the absence of previous works on this issue.

For future researches on wider sample, we will aim to clarify our current findings. With regard to cardiovascular risk factors, we did not find any association with *Per2* C111G.

We found that *ApoE* ε4 carriers had more frequent history of heart disease than ε4 non carriers. Our result is supported by previous studies. In particular, two different meta-analysis [36,37] showed a different distribution of coronary disease risk according to *ApoE* genotype. Furthermore, this association could be independent from other cardiovascular risk factors, as we did not find any correlation with diabetes, dyslipidemia, smoking, or hypertension.

A multivariate analysis showed that age at baseline, *ApoE* e4 and dyslipidemia increase the risk of progression to AD.

With regard to age, this is not surprising as it is recognized to be the major risk factor for AD [38,39].

The effect of *ApoE* ε4 allele on risk of progression to AD has been widely demonstrated by previous studies [40–43].

Last, lipid disorders have been said to have a role in cognitive impairment and their treatment has been studied as a prevention tool, but evidence about this topic is not yet conclusive [44].

The interaction between *ApoE* ε4 and dyslipidemia on cognition is not yet completely understood [36]. In order to explore this point, we ranked the patients according to *ApoE* genotype. The effect of dyslipidemia on progression to AD was confirmed only in *ApoE* ε4 carriers. Furthermore, we showed that patients who were *ApoE* ε4 carriers and had history of dyslipidemia showed higher risk to convert to AD both compared to *ApoE*ε4− groups and *ApoE* ε4+ without dyslipidemia patients. According to this analysis, dyslipidemia could be synergistic with E4 carrier status in contributing to AD pathogenesis as reported by other authors [44].

Previous studies suggested an association between *Clock* gene polymorphisms with different cardiovascular risk factors, as well as with the cognitive state [15–17]. In order to investigate a possible interaction between *Clock* T3111C and cardiovascular risk factors, we ranked patients according to *Clock* genotype. We found that *Clock* C carriers with history of blood hypertension had a higher risk of conversion to AD than *Clock* C carriers without hypertension. This difference was not detected in the *Clock* non C carrier group. No other work, to the best of our knowledge, has previously studied the possible influence of *Clock* gene polymorphisms on the effect of hypertension in conversion to AD in this particular group of patients.

A limitation of this study is the small size of our cohort. In future we aim to expand our sample, also including a healthy control group, to support our present results.

Secondly, the lack of quantitative data about dyslipidemia and hypertension did not allow us to understand if our results might be different according to the level of blood lipids and blood pressure. Another limitation of this study is that AD diagnosis was not supported by AD biomarkers. Future researches including cerebrospinal fluid amyloid beta, tau and p-tau levels or neuroimaging data, as amyloid PET, could provide interesting and additional information. Finally, as it is a single-center study, there may be biases with regard to assessment and diagnosis procedures. On the other hand, this study has some remarkable strengths. First of all, to the best of our knowledge, this is the first prospective study that assessed the interaction of these clock genes polymorphisms with cardiovascular factors on the risk or progression to AD in cohort of well-defined SCD and MCI patients. The second strength is the very long mean follow-up time of 13 years. Moreover, non-converter patients had a longer follow-up time than patients who converted to AD. This is crucial information as follow-up time could influence rate of conversion to AD. The long follow-up time in our sample allows us to minimize the risk of classifying as stable subjects carrying an Alzheimer pathology who will convert later. Indeed, the present study suggests an association between *ApoE* genotype and *Clock* T3111C with different cardiovascular risk factors in a cohort of SCD and MCI patients. This interaction could influence the progression to AD in this group of patients. Understanding the mechanisms by which genetic and cardiovascular risk factors contribute to AD could inspire the development of new personalized therapeutic approaches for this disease.

Author Contributions: Conceptualization, V.B. and B.N.; methodology, V.B., S.M., B.N.; software, S.M.; validation, V.B., J.B., S.M., G.G., S.B., S.P., I.P., M.C., L.B., S.S., B.N.; formal analysis, V.B., S.M., J.B., S.B., I.P.; investigation, V.B., J.B., S.M., G.G., S.B., S.P., I.P., M.C., L.B., S.S., B.N.; resources, V.B., J.B., S.M., G.G., S.B., S.P., I.P., M.C., L.B., S.S., B.N.; data curation, V.B., J.B., S.M., G.G., S.B., S.P., I.P., M.C., C.F., L.B., S.S., B.N.; writing—original draft preparation, V.B., J.B., S.M., G.G.; writing—review and editing, V.B., S.M., G.G., S.S., B.N.; visualization, V.B., J.B., S.M., G.G., S.B., S.P., I.P., M.C., C.F., L.B., S.S., B.N.; supervision, V.B., S.S., B.N.; project administration, V.B., S.S., B.N.; funding acquisition, V.B., S.S. All authors have read and agreed to the published version of the manuscript.

Funding: This research was funded by Ministero della Salute and Regione Toscana (grants n° GR-2010-2316359-Longitudinal clinical-neuropsychological study of subjective memory complaints) and by Fondazione Cassa di Risparmio di Firenze (ECRF16-CONTRIBUTO 2015.0713, SORBI).

Conflicts of Interest: No authors report any conflicts of interest for this study.

References

1. Sperling, R.A.; Aisen, P.S.; Beckett, L.A.; Bennett, D.A.; Craft, S.; Fagan, A.M.; Iwatsubo, T.; Jack, C.R.; Kaye, J.; Montine, T.J.; et al. Toward defining the preclinical stages of Alzheimer's disease: Recommendations from the National Institute on Aging-Alzheimer's Association workgroups on diagnostic guidelines for Alzheimer's disease. *Alzheimer's Dement.* **2011**, *7*, 280–292.
2. Jessen, F.; Amariglio, R.E.; van Boxtel, M.; Breteler, M.; Ceccaldi, M.; Chételat, G.; Dubois, B.; Dufouil, C.; Ellis, K.A.; van der Flier, W.M.; et al. A conceptual framework for research on subjective cognitive decline in preclinical Alzheimer's disease. *Alzheimer's Dement.* **2014**, *10*, 844–852.
3. Albert, M.S.; DeKosky, S.T.; Dickson, D.; Dubois, B.; Feldman, H.H.; Fox, N.C.; Gamst, A.; Holtzman, D.M.; Jagust, W.J.; Petersen, R.C.; et al. The diagnosis of mild cognitive impairment due to Alzheimer's disease: Recommendations from the National Institute on Aging-Alzheimer's Association workgroups on diagnostic guidelines for Alzheimer's disease. *Alzheimer's Dement.* **2011**, *7*, 270–279. [CrossRef]
4. Bessi, V.; Mazzeo, S.; Padiglioni, S.; Piccini, C.; Nacmias, B.; Sorbi, S.; Bracco, L. From Subjective Cognitive Decline to Alzheimer's Disease: The Predictive Role of Neuropsychological Assessment, Personality Traits, and Cognitive Reserve. A 7-Year Follow-Up Study. *J. Alzheimer's Dis.* **2018**, *63*, 1523–1535.
5. Brodaty, H.; Heffernan, M.; Kochan, N.A.; Draper, B.; Trollor, J.N.; Reppermund, S.; Slavin, M.J.; Sachdev, P.S. Mild cognitive impairment in a community sample: The Sydney Memory and Ageing Study. *Alzheimer's Dement.* **2013**, *9*, 310–317.e1. [CrossRef]
6. Mitchell, A.J.; Beaumont, H.; Ferguson, D.; Yadegarfar, M.; Stubbs, B. Risk of dementia and mild cognitive impairment in older people with subjective memory complaints: Meta-analysis. *Acta Psychiatr. Scand.* **2014**, *130*, 439–451. [CrossRef]
7. Raichlen, D.A.; Bharadwaj, P.K.; Nguyen, L.A.; Franchetti, M.K.; Zigman, E.K.; Solorio, A.R.; Alexander, G.E. Effects of simultaneous cognitive and aerobic exercise training on dual-task walking performance in healthy older adults: Results from a pilot randomized controlled trial. *BMC Geriatr.* **2020**, *20*, 83. [CrossRef]
8. Reiter, K.; Nielson, K.A.; Smith, T.J.; Weiss, L.R.; Alfini, A.J.; Smith, J.C. Improved Cardiorespiratory Fitness Is Associated with Increased Cortical Thickness in Mild Cognitive Impairment. *J. Int. Neuropsychol. Soc. JINS* **2015**, *21*, 757–767.
9. World Health Organization. *Risk Reduction of Cognitive Decline and Dementia: WHO Guidelines*; WHO: Geneva, Switzerland, 2019; p. 401.
10. Norton, S.; Matthews, F.E.; Barnes, D.E.; Yaffe, K.; Brayne, C. Potential for primary prevention of Alzheimer's disease: An analysis of population-based data. *Lancet Neurol.* **2014**, *13*, 788–794. [CrossRef]

11. Cazaly, E.; Charlesworth, J.; Dickinson, J.L.; Holloway, A.F. Genetic Determinants of Epigenetic Patterns: Providing Insight into Disease. *Mol. Med.* **2015**, *21*, 400–409. [CrossRef]
12. Ko, C.H.; Takahashi, J.S. Molecular components of the mammalian circadian clock. *Hum. Mol. Genet.* **2006**, *15*, R271–R277.
13. Hood, S.; Amir, S. Neurodegeneration and the Circadian Clock. *Front. Aging Neurosci.* **2017**, *9*, 170. [CrossRef]
14. Leng, Y.; Musiek, E.S.; Hu, K.; Cappuccio, F.P.; Yaffe, K. Association between circadian rhythms and neurodegenerative diseases. *Lancet Neurol.* **2019**, *18*, 307–318. [CrossRef]
15. Bonney, S.; Kominsky, D.; Brodsky, K.; Eltzschig, H.; Walker, L.; Eckle, T. Cardiac Per2 functions as novel link between fatty acid metabolism and myocardial inflammation during ischemia and reperfusion injury of the heart. *PLoS ONE* **2013**, *8*, e71493. [CrossRef]
16. Gómez-Abellán, P.; Hernández-Morante, J.J.; Luján, J.A.; Madrid, J.A.; Garaulet, M. Clock genes are implicated in the human metabolic syndrome. *Int. J. Obes.* **2008**, *32*, 121–128. [CrossRef]
17. Pagliai, G.; Sofi, F.; Dinu, M.; Sticchi, E.; Vannetti, F.; Molino Lova, R.; Ordovàs, J.M.; Gori, A.M.; Marcucci, R.; Giusti, B.; et al. CLOCK gene polymorphisms and quality of aging in a cohort of nonagenarians—The MUGELLO Study. *Sci. Rep.* **2019**, *9*, 1–7. [CrossRef]
18. McKhann, G.M.; Knopman, D.S.; Chertkow, H.; Hyman, B.T.; Jack, C.R.; Kawas, C.H.; Klunk, W.E.; Koroshetz, W.J.; Manly, J.J.; Mayeux, R.; et al. The diagnosis of dementia due to Alzheimer's disease: Recommendations from the National Institute on Aging-Alzheimer's Association workgroups on diagnostic guidelines for Alzheimer's disease. *Alzheimer's Dement.* **2011**, *7*, 263–269. [CrossRef]
19. Neary, D.; Snowden, J.S.; Gustafson, L.; Passant, U.; Stuss, D.; Black, S.; Freedman, M.; Kertesz, A.; Robert, P.H.; Albert, M.; et al. Frontotemporal lobar degeneration: A consensus on clinical diagnostic criteria. *Neurology* **1998**, *51*, 1546–1554.
20. Gorelick, P.B.; Scuteri, A.; Black, S.E.; Decarli, C.; Greenberg, S.M.; Iadecola, C.; Launer, L.J.; Laurent, S.; Lopez, O.L.; Nyenhuis, D.; et al. Vascular contributions to cognitive impairment and dementia: A statement for healthcare professionals from the american heart association/american stroke association. *Stroke* **2011**, *42*, 2672–2713. [CrossRef]
21. Bracco, L.; Amaducci, L.; Pedone, D.; Bino, G.; Lazzaro, M.P.; Carella, F.; D'Antona, R.; Gallato, R.; Denes, G. Italian Multicentre Study on Dementia (SMID): A neuropsychological test battery for assessing Alzheimer's disease. *J. Psychiatr. Res.* **1990**, *24*, 213–226. [CrossRef]
22. Caffarra, P.; Vezzadini, G.; Dieci, F.; Zonato, F.; Venneri, A. Rey-Osterrieth complex figure: Normative values in an Italian population sample. *Neurol. Sci.* **2002**, *22*, 443–447. [CrossRef]
23. Baddeley, A.; Della Sala, S.; Papagno, C.; Spinnler, H. Dual-task performance in dysexecutive and nondysexecutive patients with a frontal lesion. *Neuropsychology* **1997**, *11*, 187–194.
24. Spinnler, H.; Tognoni, G. *Standardizzazione e Taratura Italian di Test Neuropsicologici: Gruppo Italiano per lo Studio Neuropsicologico Dell'invecchiamento*; Masson Italia Periodici: Milano, Italy, 1987.
25. Giovagnoli, A.R.; Del Pesce, M.; Mascheroni, S.; Simoncelli, M.; Laiacona, M.; Capitani, E. Trail making test: Normative values from 287 normal adult controls. *Ital. J. Neurol. Sci.* **1996**, *17*, 305–309. [CrossRef]
26. Brazzelli, M.; Della Sala, S.; Laiacona, M. Taratura della versione italiana del Rivermead Behavioural Memory Test: Un test di valutazione ecologica della memoria. *Giunti Organ. Spec.* **1993**, *206*, 33–42.
27. Colombo, L.; Sartori, G.; Brivio, C. Stima del quoziente intellettivo tramite l'applicazione del TIB (Test Breve di Intelligenza). *Giornale Italiano di Psicologia* **2002**, *29*, 613–638.
28. Nelson, H. *National Adult Reading Test (NART): For the Assessment of Premorbid Intelligence in Patients with Dementia: Test Manual*; Windsor: Windsor, UK, 1982.
29. Hamilton, M. A rating scale for depression. *J. Neurol. Neurosurg. Psychiatry* **1960**, *23*, 56–62. [CrossRef]
30. Mishima, K.; Tozawa, T.; Satoh, K.; Saitoh, H.; Mishima, Y. The 3111T/C polymorphism of hClock is associated with evening preference and delayed sleep timing in a Japanese population sample. *Am. J. Med Genet. Part B Neuropsychiatr. Genet.* **2005**, *133*, 101–104.
31. Choub, A.; Mancuso, M.; Coppedè, F.; LoGerfo, A.; Orsucci, D.; Petrozzi, L.; DiCoscio, E.; Maestri, M.; Rocchi, A.; Bonanni, E.; et al. Clock T3111C and Per2 C111G SNPs do not influence circadian rhythmicity in healthy Italian population. *Neurol. Sci.* **2011**, *32*, 89–93. [CrossRef]
32. Saffroy, R.; Lafaye, G.; Desterke, C.; Ortiz-Tudela, E.; Amirouche, A.; Innominato, P.; Pham, P.; Benyamina, A.; Lemoine, A. Several clock genes polymorphisms are meaningful risk factors in the development and severity of cannabis addiction. *Chronobiol. Int.* **2019**, *36*, 122–134. [CrossRef]

33. Becker-Krail, D.; McClung, C. Implications of circadian rhythm and stress in addiction vulnerability. *F1000Research* **2016**, *5*, 59.
34. Perreau-Lenz, S.; Spanagel, R. Clock genes × stress × reward interactions in alcohol and substance use disorders. *Alcohol* **2015**, *49*, 351–357. [CrossRef] [PubMed]
35. Bessi, V.; Giacomucci, G.; Mazzeo, S.; Bagnoli, S.; Padiglioni, S.; Balestrini, J.; Tomaiuolo, G.; Piaceri, I.; Carraro, M.; Bracco, L.; et al. Per2 C111G polymorphism influences cognition in Subjective Cognitive Decline and Mild Cognitive Impairment. A 10-year follow-up study. *Eur. J. Neurol.*. under review.
36. Bennet, A.M.; Di Angelantonio, E.; Ye, Z.; Wensley, F.; Dahlin, A.; Ahlbom, A.; Keavney, B.; Collins, R.; Wiman, B.; de Faire, U.; et al. Association of apolipoprotein E genotypes with lipid levels and coronary risk. *JAMA* **2007**, *298*, 1300–1311. [CrossRef] [PubMed]
37. Tang, G.; Wang, D.; Long, J.; Yang, F.; Si, L. Meta-analysis of the association between whole grain intake and coronary heart disease risk. *Am. J. Cardiol.* **2015**, *115*, 625–629. [CrossRef]
38. Gorelick, P.B. Risk factors for vascular dementia and Alzheimer disease. *Stroke* **2004**, *35*, 2620–2622. [CrossRef]
39. van der Flier, W.M.; Scheltens, P. Epidemiology and risk factors of dementia. *J. Neurol. Neurosurg. Psychiatry* **2005**, *76*, v2–v7. [CrossRef]
40. Rawle, M.J.; Davis, D.; Bendayan, R.; Wong, A.; Kuh, D.; Richards, M. Apolipoprotein-E (Apoe) ε4 and cognitive decline over the adult life course. *Transl. Psychiatry* **2018**, *8*, 1–8. [CrossRef]
41. Xu, W.-L.; Caracciolo, B.; Wang, H.-X.; Santoni, G.; Winblad, B.; Fratiglioni, L. Accelerated progression from mild cognitive impairment to dementia among APOE ε4ε4 carriers. *J. Alzheimers Dis. JAD* **2013**, *33*, 507–515. [CrossRef]
42. Hong, Y.J.; Yoon, B.; Shim, Y.S.; Kim, S.-O.; Kim, H.J.; Choi, S.H.; Jeong, J.H.; Yoon, S.J.; Yang, D.W.; Lee, J.-H. Predictors of Clinical Progression of Subjective Memory Impairment in Elderly Subjects: Data from the Clinical Research Centers for Dementia of South Korea (CREDOS). *Dement. Geriatr. Cogn. Disord.* **2015**, *40*, 158–165. [CrossRef]
43. Mazzeo, S.; Padiglioni, S.; Bagnoli, S.; Bracco, L.; Nacmias, B.; Sorbi, S.; Bessi, V. The dual role of cognitive reserve in subjective cognitive decline and mild cognitive impairment: A 7-year follow-up study. *J. Neurol.* **2019**, *266*, 487–497. [CrossRef]
44. Salameh, T.S.; Rhea, E.M.; Banks, W.A.; Hanson, A.J. Insulin resistance, dyslipidemia, and apolipoprotein E interactions as mechanisms in cognitive impairment and Alzheimer's disease. *Exp. Biol. Med.* **2016**, *241*, 1676–1683.

Journal of *Personalized Medicine*

Article

Behavioral and Psychological Symptoms of Dementia (BPSD): Clinical Characterization and Genetic Correlates in an Italian Alzheimer's Disease Cohort

Catia Scassellati [1,2,†], Miriam Ciani [3,†], Carlo Maj [1,4], Cristina Geroldi [5], Orazio Zanetti [5], Massimo Gennarelli [1,6] and Cristian Bonvicini [3,*]

1 Genetics Unit, IRCCS Istituto Centro San Giovanni di Dio Fatebenefratelli, 25123 Brescia, Italy; c.scassellati@fatebenefratelli.eu (C.S.); cmaj@uni-bonn.de (C.M.); gennarelli@fatebenefratelli.eu (M.G.)
2 Biological Psychiatry Unit, IRCCS Istituto Centro San Giovanni di Dio Fatebenefratelli, 25123 Brescia, Italy
3 Molecular Markers Laboratory, IRCCS Istituto Centro San Giovanni di Dio Fatebenefratelli, 25123 Brescia, Italy; mciani@fatebenefratelli.eu
4 Institute of Genomic Statistics and Bioinformatics, University of Bonn, 53127 Bonn, Germany
5 Alzheimer's Research Unit-Memory Clinic, IRCCS Istituto Centro San Giovanni di Dio Fatebenefratelli, 25123 Brescia, Italy; cgeroldi@fatebenefratelli.eu (C.G.); ozanetti@fatebenefratelli.eu (O.Z.)
6 Section of Biology and Genetic, Department of Molecular and Translational Medicine, University of Brescia, 25123 Brescia, Italy
* Correspondence: cbonvicini@fatebenefratelli.eu
† These authors contributed equally to this work.

Received: 25 June 2020; Accepted: 12 August 2020; Published: 14 August 2020

Abstract: Background: The occurrence of Behavioral and Psychological Symptoms of Dementia (BPSD) in Alzheimer's Disease (AD) patients hampers the clinical management and exacerbates the burden for caregivers. The definition of the clinical distribution of BPSD symptoms, and the extent to which symptoms are genetically determined, are still open to debate. Moreover, genetic factors that underline BPSD symptoms still need to be identified. Purpose. To characterize our Italian AD cohort according to specific BPSD symptoms as well as to endophenotypes. To evaluate the associations between the considered BPSD traits and *COMT*, *MTHFR*, and *APOE* genetic variants. Methods. AD patients (n = 362) underwent neuropsychological examination and genotyping. BPSD were assessed with the Neuropsychiatric Inventory scale. Results. *APOE* and *MTHFR* variants were significantly associated with specific single BPSD symptoms. Furthermore, "Psychosis" and "Hyperactivity" resulted in the most severe endophenotypes, with *APOE* and *MTHFR* implicated as both single risk factors and "gene*x*gene" interactions. Conclusions. We strongly suggest the combined use of both BPSD single symptoms/endophenotypes and the "gene*x*gene" interactions as valid strategies for expanding the knowledge about the BPSD aetiopathogenetic mechanisms.

Keywords: behavioral and psychological symptoms of dementia (BPSD); Alzheimer's disease (AD); neuropsychiatry inventory scale (NPI); endophenotypes; CART analysis; *MTHFR*; *APOE*; *COMT*; genetic variants

1. Introduction

The core clinical criteria for Alzheimer's Disease (AD), the most common neurodegenerative dementing illness, focus on the presence of memory disturbance or other cognitive symptoms that interfere with the ability to function at work or in usual daily activities [1].

Although cognitive symptoms are the actual hallmark of disease, patients often show also a broad range of "non-cognitive" disturbances, more commonly known with the term "Behavioral and Psychological Symptoms of Dementia (BPSD)". Highlighting the importance of these symptoms is pivotal because BPSD represents the leading reason for loss of independence and institutionalization for AD *patients* [2–4]. About up to 90% of people with AD can present neuropsychiatric disturbances including agitation, aggression, irritation, disinhibition, anxiety, depression, apathy, delusions, and hallucinations at onset or later in the course of the disease [5,6].

Furthermore, BPSD can be recognized individually, but more often they occur in association. Indeed, 50% of subjects with AD show at least four neuropsychiatric symptoms simultaneously [7]. Their clusterization into distinct domains based on their frequency of co-occurrence allowed the definition of distinct behavioral endophenotypes [8]. In systematic reviews of studies that applied unbiased approaches to cluster BPSD, the following domains were identified: Affective (anxiety and depression), Disinhibition/Hyperactivity (aggression, impulsivity, and motor hyperactivity), Apathy and Psychosis (hallucinations, delusions, and paranoia) [3,8,9]. However, the debate about the definition of an appropriate clusterization in dementia is still ongoing [3]. More importantly, the studies available to date are characterized by a high heterogeneity, also because a certain number of symptoms (i.e., apathy, sleep disorders, eating disturbances) are not adequately grouped [3].

The susceptibility to BPSD is somewhat unclear as well as the molecular link between BPSD and AD. Within this complex of aetiological interplay, genetic background has been considered one of the key players involved in predisposing patients to specific behavioral and psychological manifestations in AD [10]. Multiple genes, prevalently involved in the processes of neurotransmission/neurodevelopment, have been assessed for their putative influence on BPSD risk, whose findings have been often inconsistent (for review [10]). For instance, the Catechol-O-Methyltransferase (*COMT*) gene was investigated in five published papers, four of which were conducted by the same authors [11–14], and the association found with "frontal"/cognitive and "psychosis" endophenotypes was weak. COMT protein is one of the major enzymes involved in the synaptic dopamine catabolism and, thus, has a crucial role in the prefrontal cortex. The well-known functional single nucleotide polymorphism (SNP) is characterized by a G to A transition at codon 108/158 (soluble/membrane-bound COMT) resulting in a valine-to-methionine substitution, giving rise to a significant, three-to-four-fold reduction in its enzymatic activity. The presence of valine (H allele = high activity) in the coding sequence corresponds dose-dependently with reduced prefrontal dopamine levels, leading subsequently to the upregulation of striatal dopamine activity [15].

On the other hand, the Apolipoprotein E (*APOE*) gene, the main recognized genetic risk factor for late-onset AD (LOAD) (allele ε4) [16], seems to give more consistent results, with the well-known alleles ε4 and ε2 associated with specific BPSD symptoms (for review [10]).

Thus, further new biological and molecular mechanisms could be involved in BPSD genetics architecture. For instance, *COMT* also regulates the folate pathway [17]. Folate is a cofactor in one-carbon metabolism, where it promotes the remethylation of homocysteine (Hcy). Numerous studies associate the folate deficiency and the resultant increase of Hcy levels with AD [18]. Interestingly, circulating Hcy levels in AD patients with and without BPSD were higher compared to the control subjects, and the plasma Hcy concentration in AD patients with BPSD was the highest among the three considered groups [19]. Hyperhomocysteinemia has been also correlated with psychosis and depression [20–22]. The Methylene tetrahydrofolate reductase (*MTHFR*) gene, whose genetic defects lead to hyperhomocysteinemia, could be a good and new candidate in the aetiopathogenetic mechanisms of BPSD. Within this gene, the two most commonly studied SNPs are the C677T (rs1801133) and the A1298C (rs1801131). The T allele of the C677T polymorphism provokes the synthesis of an athermolabile variant of the enzyme leading to a reduced enzymatic activity, which in turn produce an increase in the blood Hcy levels [23,24]. Recent findings supported that this allele was associated with an increased risk of LOAD [25,26], and correlated with a significant increase of Hcy levels [27,28]. Concerning the A1298C, the C allele results in a reduced enzymatic activity, but does not influence its thermolability [29,30].

Based on this rationale, in this study we analyzed an Italian cohort of AD patients in order to: 1. define the clinical distribution of BPSD symptoms through the characterization of single BPSD symptoms as well as of endophenotype clusters; 2. confirm and clarify the aetiopathogenetic effect of *APOE* and *COMT* variants in relation to both single symptoms and behavioral endophenotypes, selecting functional polymorphisms such as rs429358 (Cys130Arg) and rs7412 (Arg176Cys) and Val108/158Met, respectively [15,16]; 3. evaluate for the first time the putative involvement of *MTHFR* in single BPSD symptoms and likewise behavioral endophenotypes, selecting functional polymorphisms such as C677T (rs1801133) and the A1298C (rs1801131) [23,24,29,30]; 4. highlight for individual symptoms and endophenotypes a "single gene" involvement and/or a "gene×gene" interaction of *APOE*, *COMT*, and *MTHFR*.

2. Materials and Methods

2.1. Sample and Clinical Evaluation

The study was approved by the Local Ethics Committee (IRCCS Istituto Centro San Giovanni di Dio Fatebenefratelli, Brescia, Italy, n. 6/2006) and conducted in accordance with local clinical research regulations. Written informed consent was obtained from all participants.

AD subjects (n = 362) were recruited from the Alzheimer Unit of the IRCCS Istituto Centro San Giovanni di Dio Fatebenefratelli, Brescia, Italy. All patients were unrelated Caucasian subjects residing in Northern Italy with an Italian origin descent for at least two generations. All of them were assessed at their admittance with a complete sociodemographic and clinical data collection (cognitive, behavioral, neurological, functional, and physical abilities). A multidisciplinary clinical examination was performed and the diagnosis of probable AD was established based on criteria of the "National Institute of Neurological and Communicative Disorders" and the "Stroke-Alzheimer's disease and Related Disorders Association" [31]. To assess cognitive decline, the Mini Mental State Examination (MMSE) test was used [32], in addition to the Cumulative Illness Rating Scale for Geriatrics (CIRS-G) [33]; function abilities were evaluated with the Basic Activity Daily Living (BADL), and the Instrumental Activity Daily Living (IADL) scales [34,35].

BPSD were assessed with the Neuropsychiatric Inventory (NPI) scale [36], a fully structured interview exploring 12 behavioral and neuropsychiatric domains (i.e., delusions, hallucinations, agitation/aggression, disphoria/depression, anxiety, apathy, irritability, euphoria, disinhibition, aberrant motor behavior, sleep behavior disturbances, besides appetite and eating abnormalities). It specifically provides an individual score for each explored cognitive domain specifically obtained by multiplying the severity of symptoms (1 = mild; 2 = moderate; 3 = severe) with their frequencies (4-point scale from 1 = occasionally to 4 = very frequently, more than once a day). The NPI then yields a domain rating of frequency times severity (range = 0–12). In agreement with a group of expert clinicians, each of the 12 behavioral/neuropsychiatric symptoms was classified in three groups, according to the "severity*frequency" score: 1—symptom-free (NPI = 0); 2—with low "severity*frequencies" score (NPI from 1 to 4); 3—with high "severity*frequencies" score (NPI from 6 to 12). Moreover, NPI provides a total score that globally evaluates the 12 domains for a complete clinical picture given by the sum of the individual scores (defining a range from 0 to 144).

The adopted exclusionary criteria for subjects were: (a) A history of schizophrenia, schizoaffective disorder, delusional disorder, mood disorder with psychotic features, major depressive disorder, substance use disorder, or mental retardation according to The Diagnostic and Statistical Manual of Mental Disorders (DSM-IV) criteria; (b) severe cerebrovascular disorders, hydrocephalus and intra-cranial mass, documented by *Computed Tomography* or *Magnetic Resonance Imaging* within the past 12 months; (c) abnormalities in serum folate and Vitamin B12, syphilis serology or altered thyroid hormone levels; (d) a history of traumatic brain injury or other neurological diseases (e.g., Parkinson's Disease, Huntington Disease, seizure disorders); (e) current acute or poorly controlled medical problems (e.g., poorly controlled diabetes or hypertension; cancer within the past five years; clinically significant hepatic, renal, cardiac or pulmonary disorders); (f) absence of knowledgeable informant who could properly report information regarding the patient's behavior.

2.2. Genetic Analysis

Genomic DNA (gDNA) was extracted from peripheral blood with the PureGene genomic DNA isolation kit (Gentra Systems, Minneapolis, MN, USA). The obtained gDNA was quantified by spectrophotometric quantification using the NanoDrop microvolume sample retention system (Thermo Fisher Scientific, Waltham, MA, USA) [37]. In order to verify the possible degradation, all samples were analyzed on a 0.8% agarose gel electrophoresis and a long *Polymerase Chain Reaction* (PCR) protocol was developed for exploring genetic variability in *MTHFR*, *COMT*, and *APOE*.

In all samples, the *MTHFR* [rs1801133 (C677T) and rs1801131 (A1298C)], *COMT* [rs4680 (Val158Met)], and *APOE* [rs429358 (Cys130Arg) and rs7412 (Arg176Cys)] polymorphisms were genotyped by using the SNaPshot assay [38]. Briefly, 100 ng of gDNA of each sample were amplified in the GeneAmp PCR System 9700 (Applied Biosystems, Foster City, CA, USA), and the PCR-amplification products were used as a template in a SNaPshot Multiplex assay performed according to the manufacturer's instructions. Finally, samples were analyzed, and the allele peak determination was performed on an ABI 3130xl Genetic Analyzer. Electrophoresis results were analyzed using the GeneMapper ID software v4.0 (Applied Biosystems, Foster City, CA, USA).

2.3. Statistical Analyses

All statistical analyses were conducted by using the SPSS version 23.0 (SPSS Inc., Chicago, IL, USA). The Hardy-Weinberg equilibrium (HWE) was calculated using an online calculator (http://www.husdyr.kvl.dk/htm/kc/popgen/genetik/applets/kitest.htm) for the presence of multiallelic genotypes.

2.3.1. Principal Component Analysis (PCA)

In the first exploratory phase, the set of symptoms from the NPI scale was factorially analyzed to identify possible underlying latent variables. The Principal Component Analysis (PCA) method and Varimax rotation were performed. The number of factors was determined on the basis of eigenvalues greater than one of the Pearson's correlation matrix, and by sharp breaks in the size of the eigenvalues using a scree plot [39]. Symptom-factor correlations, (i.e., factor loading) being greater than 0.40 in absolute value were chosen to identify a simple factor structure (i.e., factors with non-overlapping clusters of symptoms).

2.3.2. Classification and Regression Tree (CART) Analysis for Single BPSD Symptoms and Endophenotypes

The Classification and Regression Tree (CART) analysis using the exhaustive Chi-squared Automatic Interaction Detector (CHAID) algorithm was performed to explore the interaction between single BPSD symptoms as well as endophenotypes and all polymorphisms in *APOE*, *MTHFR*, and *COMT* genes in the BPSD cohort [40,41]. Specifically, we performed the analyses considering ε4 carriers (ε2ε4 + ε3ε4 + ε4ε4 genotypes) versus the others indicated as *APOE* ε4 non-carriers (ε2ε3 + ε3ε3 genotypes). *APOE* allele ε4 carriers, the carriers of both alleles in 677C/T *MTHFR*, in 1298A/C *MTHFR*, and in Val108/158Met *COMT* genes polymorphisms were used as a dominant model of inheritance. Moreover, the comparisons were performed incorporating "Free group" (NPI 0) along with "Low group" (NPI 1–4) versus "High group" (NPI 6–12). This is due to the evidence that "Low group" is present in our population with a frequency < 20%. As this study does not have a longitudinal design, we assumed causality in our CART model [https://www.epa.gov/caddis-vol4/caddis-volume-4-data-analysis-classification-and-regression-tree-cart-analysis].

The exhaustive CHAID data mining algorithm is a nonparametric procedure that makes no assumptions of the underlying data. In the CHAID analysis, nominal, ordinal, and continuous data can be used, where continuous predictors are split into categories with approximately an equal number of observations. CHAID creates all possible cross tabulations for each categorical predictor until the best outcome is achieved and no further splitting can be performed [42]. In the CHAID technique, we can visually see the relationships between the split variables and the associated related factor within the tree. The development of the decision, or classification tree, starts with identifying the target variable or dependent variable, which would be considered the root. CHAID analysis splits the target into two or more categories that are called the initial, or parent nodes, and then the nodes are split using statistical algorithms into child nodes.

The exhaustive CHAID data mining algorithm automatically pruning insignificant nodes in a decision tree was constructed through the IBM SPSS 23 statistical package program. It works on the basis of F test if a continuous dependent variable is used as in our study (endophenotypes), whereas, if the predictor variable has only two categories (single items), the χ^2-test for independence is performed for each pair of categories of the predictor variable in relation to the binary target variable. For non-significant outcomes, those paired categories are merged.

Minimum subject numbers for the parent and child nodes were fixed at 20 and 10 for constructing an optimal decision tree structure and improving the predictive performance of the algorithms. A ten-fold cross validation was activated in the study.

Since the analyses were conducted by using decision tree models, it was not needed to have transformed data.

SPSS automatically made a Bonferroni adjustment to calculate the adjusted p-values for the merged categories to control for the Type I error rate.

To exclude the influence of age and gender, we performed the univariate analyses (ANOVA) related to BPSD endophenotypes, including the variables age or gender. Selecting ENDOPHENOTYPES ("Psychosis", "Hyperactivity", "Mood", "Frontal") as a dependent variable, AGE as covariate, SEX as an independent variable, we did not find any significant association in separate analyses (Table S1—gender and Table S2—age, Supplementary Materials). The only significant association was observed in the "Mood" endophenotype (Table S1, F = 4.7; p = 0.032); data not confirmed by the exhaustive CHAID data mining algorithm. Thus, given the non-significant analyses, age and sex were not included in the calculations. Medications as well as comorbidity were not included in our analyses.

The Hosmer-Lemeshow goodness of fit test was used to confirm the suitability of the trees. The interactions were given by calculating the sum of the scores relative to "severity*frequencies" of the single BSPD symptoms components as a mean risk ± SD.

The stepwise multiple logistic regression was used to confirm the classification trees.

3. Results

Socio-demographic and clinical features of our Italian AD cohort, as well as the alleles, genotypes, and carriers frequencies of the all polymorphisms in *APOE*, *MTHFR*, and *COMT* genes are shown in Tables 1 and 2. Table 1 describes also the clinical features according to stratification for gender; whereas in Table 2 we reported the HWE results where no HWE deviation was observed.

3.1. Clinical Characterization according to Single NPI Scores

Three hundred and eight patients were characterized by the NPI scale. Considering the NPI scores and the relative proposed ranges (Table 3), our population was mainly characterized by agitation, irritability, night-time behavior disturbances, and aberrant motor behavior. Within each group, we found that: 52% of AD patients showed higher severity in agitation symptomatology, 48% in irritability, 42% in night-time behavior disturbances and, finally, 39% showed higher severity in aberrant motor behavior.

Slightly less representative in our sample, there were: Apathy (28% higher severity), delusions (29% higher severity), anxiety (25% higher severity), depression (22% higher severity), and hallucinations (21% higher severity).

On the contrary, appetite and eating disturbances were referred in only a minority of patients, while no one presented disinhibition or euphoria according to their caregivers' reports.

Table 1. Clinical features of our Alzheimer's Disease Italian cohort, stratified according to gender.

Gender (F/M)	241/121	Female	Male
Age (years)	80.5 ± 7.0	80.9 ± 7.1	79.7 ± 6.7
Schooling (years)	6.0 ± 3.6	5.4 ± 2.9	7.4 ± 4.4
MMSE, total score	11.7 ± 7.3	12.0 ± 6.9	10.9 ± 8.1
Onset (years)	76.1 ± 7.9	76.5 ± 8.1	75.4 ± 7.6
Illness Duration (years)	4.5 ± 2.9	4.4 ± 2.6	4.5 ± 3.5
BMI	22.3 ± 4.1	22.2 ± 4.2	22.5 ± 3.8
IADL	7.2 ± 1.5	7.4 ± 1.2	6.8 ± 1.9
BADL	3.7 ± 1.9	3.6 ± 1.9	3.8 ± 1.8
Tinetti	19.8 ± 5.4	19.8 ± 5.2	19.6 ± 5.9
CIRS	12.7 ± 9.7	12.2 ± 9.5	13.8 ± 10.0
Hypertension	27.9%	29.5%	24.8%
Cardiopathy	28.7%	26.1%	33.9%
Hypercholesterolemia	4.7%	5.8%	2.5%
Diabetes	6.6%	5.4%	9.1%
Psychiatric diseases	26.0%	27.4%	23.1%
AChEIs	20.2%	21.6%	17.4%
Neuroleptics	45.5%	41.1%	51.2%
Antidepressants	45.6%	46.9%	43.0%
Benzodiazepines/hypnotics	41.4%	46.1%	32.2%

Note: The measurement data are expressed as mean ± standard deviation (mean ± SD) or in percentual (%). MMSE: Mini-Mental State Examination; BMI: Body max index; IADL: Instrumental activities daily living; BADL: Basic activities of daily living; Tinetti; CIRS: Cumulative illness rating scales; AChEIs: Acetylcholinesterase inhibitors.

Table 2. Genetic distribution of our Alzheimer's Disease Italian cohort for *APOE*, *MTHFR*, and *COMT* genes polymorphisms.

***APOE* Haplotypes (rs429358 and rs7412)**			***MTHFR* (rs1801133: C677T)**			***MTHFR* (rs1801131: A1298C)**			***COMT* (rs4680: V158M)**		
HWE	χ^2 = 0.91	p = 0.92	HWE	χ^2 = 1.70	p = 0.19	HWE	χ^2 = 0.53	p = 0.47	HWE	χ^2 = 1.22	p = 0.27
Alleles	N	freq	*Alleles*	N	freq	*Alleles*	N	freq	*Alleles*	N	freq
ε2	20	0.03	T	314	0.43	A	499	0.69	A	330	0.46
ε3	510	0.70	C	410	0.57	C	225	0.31	G	394	0.54
ε4	194	0.27									
total	724	1.00	total	724	1.00	total	724	1.00	total	724	1.00
Genotypes	N	freq	*Genotypes*	N	freq	*Genotypes*	N	freq	*Genotypes*	N	freq
ε2 ε3	16	0.04	TT	62	0.17	AA	169	0.47	AA	70	0.19
ε2 ε4	4	0.01	TC	190	0.52	AC	161	0.44	AG	190	0.52
ε3 ε3	178	0.49	CC	110	0.30	CC	32	0.09	GG	102	0.28
ε3 ε4	138	0.38									
ε4 ε4	26	0.07									
total	362	1.00	total	362	1.00	total	362	1.00	total	362	1.00
ε2 carriers	20	0.06	T carriers	252	0.70	A carriers	330	0.91	A carriers	260	0.72
ε3 carriers	332	0.92	C carriers	300	0.83	C carriers	193	0.53	G carriers	292	0.81
ε4 carriers	168	0.46									

Note: *APOE*: Apolipoprotein E; *MTHFR*: Methylenetetrahydrofolate Reductase; *COMT*: Catechol-O-Methyltransferase; HWE: Hardy-Weinberg equilibrium.

Table 3. Clinical distributions of single Behavioral and Psychological Symptoms of Dementia (BPSD) in our Italian Alzheimer's Disease cohort.

		"Severity*Frequencies" Groups N Subjects (%) &		
Symptoms ("Severity*Frequencies" = 0–12)	**Mean ± SD §**	**Free (NPI 0)**	**Low (NPI 1–4)**	**High (NPI 6–12)**
Agitation	4.96 ± 4.32	102 (33.0)	47 (15.2)	160 (51.8)
Irritability	4.57 ± 4.25	109 (35.4)	52 (16.9)	147 (47.7)
Night-time behavior disturbances	4.22 ± 4.55	136 (44.0)	44 (14.2)	129 (41.7)
Aberrant motor behavior	4.12 ± 4.72	148 (48.2)	39 (12.7)	120 (39.1)
Apathy	3.03 ± 4.19	178 (58.0)	42 (13.7)	87 (28.3)
Delusions	3.00 ± 4.00	166 (54.1)	52 (16.9)	89 (29.0)
Anxiety	2.53 ± 4.10	200 (65.4)	30 (9.8)	76 (24.8)
Depression	2.50 ± 3.90	185 (60.3)	53 (17.3)	69 (22.5)
Hallucination	2.30 ± 3.55	187 (60.9)	55 (17.9)	65 (21.2)
Appetite and eating disturbances	1.94 ± 3.59	223 (72.2)	28 (9.1)	58 (18.8)
Disinhibition	0.93 ± 2.44	253 (82.4)	28 (9.1)	26 (8.5)
Euphoria	0.12 ± 0.86	299 (97.4)	6 (2.0)	2 (0.7)
NPI, total score (0–144)	34.1 ± 22.6			
NPI, n symptoms	4.9 ± 2.4			

Note: § Mean ± standard deviation (SD) of the scores relative to "severity*frequencies" of the single items. & Number (N) and percentage (%) of subjects present in each group (Free, Low, and High) compared to individual symptoms. Free: "severity*frequencies" score (NPI 0) for all the individual symptoms; Low: "severity*frequencies" score (NPI 1–4) for all the individual symptoms; High: "severity*frequencies" score (NPI 6–12) for all the individual symptoms.

3.2. Clinical Characterization according to Behavioral Endophenotypes

The results of the exploratory factor analysis on BPSD in AD, conducted by PCA and Varimax rotation, are shown in Table 4. Using the criterion of eigenvalues greater than 1 and the scree plot, PCA allowed grouping the 12 explored BPSD symptoms in four endophenotypes. These factors explained the 56% of the total variance of data. Specifically, thanks to this approach, the following clinical categories were established in our cohort: (a) "Psychosis" characterized by delusions, hallucinations, agitation, aberrant motor behavior, and night-time behavioral symptoms; (b) "Hyperactivity" with agitation, irritability, appetite and eating disturbances; (c) "Mood" associated with high loadings on anxiety, depression, and apathy; (d) "Frontal" including disinhibition and euphoria.

The endophenotypes more representative in our population were "Psychosis" (37% higher severity), and "Hyperactivity" (39% higher severity). These were followed by the "Mood" and "Frontal" endophenotypes (Table 5).

3.3. *APOE*, *MTHFR*, and *COMT* Genetic Correlates in Endophenotype Clusterization and in Single BPSD Symptoms

All polymorphisms investigated in this study within the *APOE*, *MTHFR*, and *COMT* genes showed a prevalence comparable to that reported in https://www.alzforum.org/, considering Caucasian Populations (Table 2). Globally, the prevalence of ε2, ε3, and ε4 alleles for the *APOE* gene was estimated to be 8%, 78%, and 14%, respectively; for rs1801133 *MTHFR* gene to be 39% (allele T) and 61% (allele C); for rs1801131 *MTHFR* gene to be 34% (allele C) and 66% (allele A); for Val108/158Met *COMT* gene to be 49% (allele A) and 51% (allele G).

In order to explore the genetic correlates in both approaches, we exploited the potential of genetic analysis by using the exhaustive CHAID data mining algorithm to reveal "gene*x*gene" interaction as aetiopathogenetic mechanisms in BPSD. The Hosmer-Lemeshow goodness of fit test confirmed the suitability of the trees (Table S3 Supplementary Materials). Moreover, the stepwise multiple logistic regression validated the classification trees (Tables S4 and S5 Supplementary Materials).

Table 4. Principal component analysis (PCA) of the Behavioral and Psychological Symptoms of Dementia (BPSD) in our Italian Alzheimer's Disease cohort.

Symptoms	Endophenotype Components			
	Factor 1	Factor 2	Factor 3	Factor 4
	Mood	Hyperactivity	Psychosis	Frontal
Delusions	0.276	0.372	*0.463*	0.144
Hallucination	0.022	0.011	*0.690*	0.248
Agitation	0.058	*0.654*	*0.468*	0.068
Depression	*0.855*	0.058	0.166	0.089
Anxiety	*0.820*	0.062	0.156	0.075
Euphoria	−0.040	−0.109	0.186	*0.766*
Apathy	*0.579*	0.292	−0.212	−0.184
Disinhibition	0.131	0.376	−0.166	*0.687*
Irritability	0.028	*0.645*	0.367	0.119
Aberrant motor behavior	0.104	0.373	*0.402*	−0.157
Night-time behavior disturbances	0.029	0.056	*0.621*	−0.087
Appetite and eating disturbances	0.191	*0.671*	−0.157	0.014

Note: Extraction method: Principal component analysis. Rotation Method: Varimax with Kaiser normalization. Coefficients with values over 0.4 are italicized and in bold.

Table 5. Clinical distributions of behavioral endophenotypes in our Italian Alzheimer's Disease cohort.

		Mean of N Subjects ± SD £ (% Mean of N Subjects ± SD) $		
Endophenotypes (Components; "Severity*Frequencies" Range)	**Mean ± SD §**	**Free**	**Low**	**High**
Psychosis (Delusions, Hallucination, Agitation, Aberrant motor behavior, Night-time behavior disturbances; 0–60)	18.5 ± 13.3	148 ± 32 (48.0 ± 10.5)	47 ± 6 (15.4 ± 2.1)	113 ± 37 (36.6 ± 11.8)
Hyperactivity (Agitation, Irritability, Appetite and eating disturbances; 0–36)	11.4 ± 9.2	145 ± 68 (46.9 ± 22.0)	42 ± 13 (13.7 ± 4.1)	122 ± 56 (39.4 ± 18.0)
Mood (Depression, Anxiety, Apathy; 0–36)	8.1 ± 9.4	188 ± 11 (61.2 ± 3.8)	42 ± 12 (13.6 ± 3.7)	77 ± 9 (25.0 ± 3.0)
Frontal (Euphoria, Disinhibition; 0–24)	1.0 ± 2.7	276 ± 33 (89.9 ± 10.6)	17 ± 16 (5.5 ± 5.1)	14 ± 17 (4.6 ± 5.5)

Note: § Mean ± standard deviation (SD) of the scores relative to "severity*frequencies" of the endophenotype components. Free: NPI = 0 for all the individual components involved in the endophenotype; Low: Sum of the "severity*frequencies" score (NPI 1–4) for all the individual components involved in the endophenotype; High: Sum of the "severity*frequencies" score (NPI 6–12) for all the individual components involved in the endophenotype. £ Mean number of subjects N ± SD present in each group (Free, Low, and High) compared to individual endophenotypes. $ Percentage (%) of the mean of the number of subjects (N ± SD) present in each group (Free, Low, and High) compared to individual endophenotypes.

3.3.1. "Genexgene" Interactions in "Psychosis" Endophenotype and the Relative Single BPSD Symptoms

Table 6 reports the main significant results obtained in the genetic analyses performed on single BPSD symptoms. Starting by the agitation, we found that the *APOE* ε4 allele carriers showed a high risk to develop more severe symptoms (odd ratio (OR) = 1.87, 95% CI: 1.19–2.95). Concerning the aberrant motor behavior, we evidenced similar results with *APOE* ε4 allele carriers showing higher risk to develop high severity in this symptomatology (OR = 1.91, 95% CI: 1.20–3.04). In relation to *MTHFR*, we found that the homozygotes CC (C677T) showed a high risk to develop more severe delusions symptoms (OR = 1.75, 95% CI: 1.04–2.94).

Table 6. Genetic correlates related to single BPSD symptoms in our Italian Alzheimer's Disease cohort according to different values of "severity*frequencies" (exhaustive CHAID data mining algorithm).

	Gene	Free + Low [&]	High [&]	
		N (Freq.)	N (Freq.)	
Agitation	*APOE*			
	ε4-carriers	58 (0.39)	86 (0.54)	χ^2 = 7.435, $p_{adjusted}$ = 0.006 OR = 1.87; 95% CI: 1.19–2.95
	ε4-non-carriers	91 (0.61)	72(0.46)	
	total	149 (1.00)	158 (1.00)	
Aberrant motor behavior	*APOE*			
	4-carriers	76 (0.41)	68 (0.57)	χ^2 = 7.553, $p_{adjusted}$ = 0.006 OR = 1.91; 95% CI: 1.20–3.04
	ε4-non-carriers	111 (0.59)	52 (0.43)	
	total	187 (1.00)	120 (1.00)	
Delusions	*MTHFR*			
	CC	59 (0.27)	35 (0.39)	χ^2 = 4.363, $p_{adjusted}$ = 0.037 OR = 1.75; 95% CI: 1.04–2.94
	Genotypes_C677T	159 (0.73)	54 (0.61)	
	T-carriers_C677T	218 (1.00)	89 (1.00)	
Appetite/eating abnormalities	*APOE*			
	ε4-carriers	109 (0.44)	35 (0.60)	χ^2 = 5.922, $p_{adjusted}$ = 0.015 OR = 2.06; 95% CI: 1.14–3.71
	ε4-non-carriers	141 (0.56)	22 (0.40)	
	total	250 (1.00)	57 (1.00)	

Note: [&] Free + Low: "severity*frequencies" scores (NPI 0–4) for all individual symptoms; High: "severity*frequencies" scores (NPI 6–12) for all the individual symptoms. OR: Odd ratio.

On the contrary, no genetic correlates were observed for night-time behavior disturbances. These results were further represented as decision model trees in Figure S1 (Supplementary Materials).

When we considered the "Psychosis" endophenotype as whole, we confirmed the effect of *APOE* ε4 allele alone (risk mean = 20.7, 95% CI: 18.5–22.9), but also a "gene*x*gene" interaction with *MTHFR* for both polymorphisms. In particular, *APOE* ε4 non-carriers along with the homozygotes CC (C677T) showed a high risk to develop psychosis (risk mean = 20.2, 95% CI: 18.9–29.3), but also with the homozygotes CC (C677T) and with the 1298A allele carriers (risk mean = 23.1, 95% CI: 18.7–27.6) (Figure 1A).

3.3.2. "Genexgene" Interactions in "Hyperactivity" Endophenotype and the Relative Single BPSD Symptoms

For irritability symptoms, no genetic correlates were observed, whereas for agitation, the results are reported above. Although in an unrepresentative sample, a risk role of *APOE* ε4 allele was found with higher severity in appetite/eating abnormalities (OR = 2.06, 95% CI: 1.14–3.71) (Table 6). These results were further represented in Figure S1 (Supplementary Materials) as decision model trees.

When we consider the "Hyperactivity" endophenotype as a whole, the most significant result is represented by the risk role of *APOE* ε4 allele (risk mean = 13.2, 95% CI: 11.8–14.7). Moreover, an interaction with the *MTHFR* C677T CC genotype and *APOE* ε4 non-carriers was observed (risk mean = 12.1, 95% CI: 9.7–14.4) (Figure 1B).

3.3.3. "Genexgene" Interactions in "Mood"/"Frontal" Endophenotypes and the Relative Single BPSD Symptoms

No significant results were found for these endophenotype as well as for the relative single BPSD symptoms.

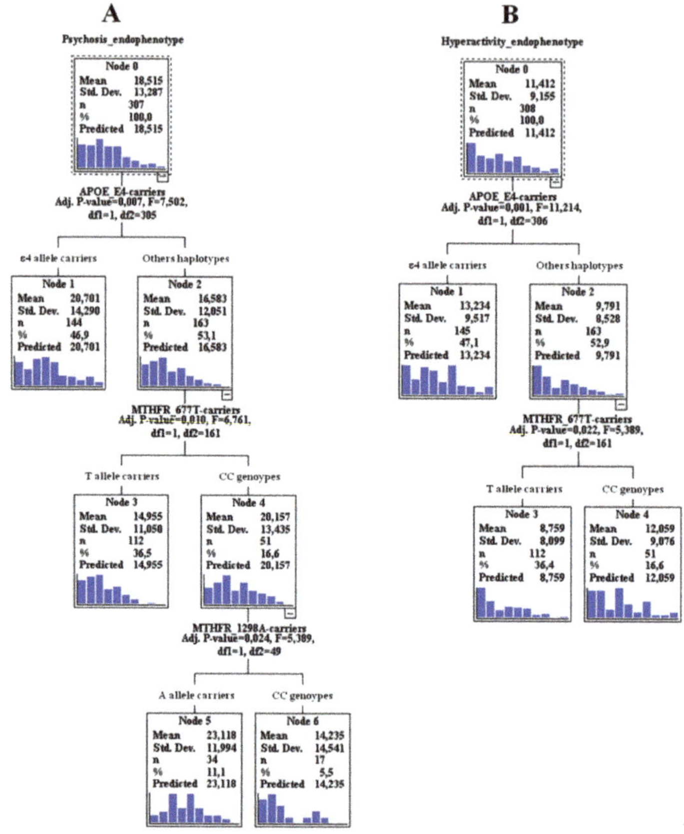

Figure 1. Exhaustive Chi-squared Automatic Interaction Detector (CHAID) data mining algorithm analysis results ("genexgene" interaction) for "Psychosis" (**A**) and "Hyperactivity" (**B**) endophenotypes. Note: Mean ± SD is given by the sum of the scores relative to "severity*frequencies" of the endophenotype components. For other haplotypes, we meant "*APOE* ε4 non-carriers".

4. Discussion

This study wants to deeply define the clinical distribution of BPSD symptoms in an Italian cohort of AD patients by the NPI scale. To the best of our knowledge, this is the largest BPSD cohort investigated until now. In this population, "Psychosis" and "Hyperactivity" endophenotypes as well as "agitation" as single symptoms were revealed to be associated with a high severity in symptomatology. In addition, "Mood" was associated to lower mean values, and, for last there was the "Frontal" endophenotype. As major clinical information, we suggest two combinatorial approaches to characterize a BPSD cohort: through 12 individual symptoms as well as four specific behavioral and neuropsychiatric endophenotypes. This choice allows us to overcome the limitations linked to a single methodological approach, bringing to light the interconnected aspects of the complex pathogenic mechanisms of BPSD. Moreover, we strongly propose analyzing the genetic correlates also in relation to single BPSD symptoms as well as to BPSD endophenotypes. This represents a valid strategy for expanding the knowledge about the aetiopathogenetic mechanisms of BPSD. There is more. We strongly suggest the use of both strategies also in relation to the genetic approach "single gene" involvement/"gene*x*gene" interaction, because this contributes to clarifying deeply the complex genetic architecture underlying BPSD. The main results obtained in this work are represented by the involvement of *APOE* as a "single gene" in the modulation of the severity in agitation, also reflected in "Psychosis" as well as in "Hyperactivity" endophenotypes, and in aberrant motor behavior. Moreover, we found the involvement of *APOE* also in the "gene×gene" interaction with *MTHFR* (both polymorphisms) as risk factors for "Psychosis". *MTHFR* as a "single gene" was associated to higher severity of delusions. This implicates that specific functional variants in *APOE* and *MTHFR* genes can impact on the relative proteins' translation and thus on their activity [16,23,24,29,30], to influence specific BPSD traits. No associations or minor relevant results include the "Frontal" endophenotype (euphoria and disinhibition) and the relation between *APOE* and appetite/eating abnormalities.

Neuropsychiatric symptoms are frequent in dementia and contribute significantly for burden caregiver and illness costs. Correct identification and evaluation of these symptoms is a crucial part of the clinical approach to dementia. Despite the added value and tentative efforts to group different symptoms into clusters (to facilitate clinical/diagnostic/treatments investigations), there is not yet an established model. Cerejeira et al., 2012 [3] reviewed different studies showing the heterogeneity in the clusterization of behavioral endophenotypes, even though a certain concordance can be found. Delusions and hallucinations have been consistently grouped in "Psychosis". A distinct "Mood" or "Affective" cluster (depression and anxiety) has been suggested, in accordance with our study. On the contrary, some studies support that apathy and depression are distinct phenomena and belong to different neuropsychiatric syndromes, whereas others group both symptoms in the same factor [3], as in our case study. The frequencies of BPSD symptoms in the AD population can change [43]. As reported in Deardorff and Grossberg, 2019 [43], apathy is the most common BPSD symptom in AD (49%), followed by depression (42%), agitation and aggression (40%), psychosis (delusions, 31% and hallucinations, 16%), and sleep disturbances (39%). Such a distribution is different as compared to our BPSD/AD cohort.

These discrepancies could be related to different factors. First of all, the distribution of BPSD symptoms is strongly linked to differences in patient populations, their ethnicity, and composition (for instance whether the patients are affected by AD or by other neurodegenerative disorders). A small sample size is an important limitation. This study represents the larger cohort available to date, that permits performing a better stratification according to BPSD symptoms, where the range of symptoms has been precisely defined, besides being crucial to detect significant genetic associations. Indeed, genome-wide association studies and studies examining copy number variations performed in AD and psychosis have detected only suggestive underpowered associations in intronic SNPs (review in [44,45]). Moreover, individual symptoms evolve differently over the course of dementia. Notwithstanding the general impression that the overall level of psychopathology increases with dementia severity, they have a tendency to wax and wane, their severity fluctuating over time. For this reason, longitudinal studies are required to achieve significant insights into the evolution of BPSD during the course of disease [3]. Finally, the choice of setting for patient recruitment is a further important selection bias: Nursing home dwellers, inpatients, or patients treated in an ambulatory setting. The most inclusive population-based studies recruit "real-life" patients, making the results much more practical, although at a price of numerous medical, environmental, and drug-related confounders [10]. In our case, we have recruited patients from the department of our hospital, who in the majority of cases, enter for a control of BPSD. Thus, for definition, they are affected by disorders with a more prevalent "agitated" symptomatology rather than "apathy".

If age is the greatest risk factor for AD, the ε4 allele of the *APOE* is the greatest genetic risk factor for LOAD [16]. ε4 carriers showed higher severity in agitation symptoms, grouping in "Hyperactivity"/"Psychosis" endophenotypes. The single association with agitation is confirmed in different studies, including recent reviews [10,46,47]. Based on neuroimaging studies, agitation is associated with volume loss in the frontal cortex, circulate cortex, insula, amygdala, and hippocampus [48]. Neurochemically, agitation is associated with decreased cholinergic activity in the frontal and temporal cortex. Interestingly, AD patients carrying the *APOE* ε4 allele show a more profound loss in cholinergic activity in the hippocampus and the cortex, and in neuroimaging studies the presence of the *APOE* ε4 allele was associated with a greater rate of hippocampal, cortical, and whole-brain atrophy [10]. On the other hand, there are also negative studies, reporting no associations of this gene with agitation [49]. These findings, thus, further emphasize the importance of sub-grouping BPSD in distinct neuropsychiatric syndromes. In fact, we found an association of *APOE* also within "Hyperactivity"/"Psychosis" endophenotypes. We therefore support the evidence on the role played by the allele ε4 as a risk factor in these endophenotypes, along with the importance of the BPSD clusterization.

The ε4 carriers showed also higher severity in aberrant motor behavior, the finding was confirmed also in other different studies [50–52], where they demonstrated that the *APOE* ε4 status increased the tendency toward this symptom.

Another more significant result of this work regarding the "gene*x*gene" interactions between *APOE* and *MTHFR* (both polymorphisms) as risk factors for higher severity in the "Psychosis" endophenotype, underlies the important role played by *MTHFR* as a single gene in delusions (genotype CC of C677T). This interaction was observed also in "Hyperactivity". What is highlighted from the CART analysis, is that also the other *APOE* ε4 non carriers can play a role, and this is performed through the interactions with the CC genotype (C677T) and/or 1298A allele carriers. Different studies supported the epistatic effect between *MTHFR* and *APOE* on cognitive performance in the elderly but also in AD [53–55]. Moreover, plasma homocysteine concentrations are associated with increased amyloid-β (Aβ) deposition in the brain [28]. Although this effect was supported, no studies are available to confirm this interaction also in the psychosis/hyperactivity dimension. The most probable hypotheses behind this result could be that *MTHFR* contributes to the regulation of relin [56], whose expression is decreased in the hippocampus of subjects suffering from schizophrenia and major depressive disorder (MDD) [57]. Relin binds to the receptor for APOE, and might thus similarly modulate glutamatergic neurotransmission, altered in psychosis. Moreover, a significant positive correlation of serum homocysteine levels was found with delusions [58].

In relation to *COMT* as a "single gene", we did not confirm the results reported by Borroni et al., 2006 [12,13]. Further studies are needed to clarify the role of this gene, that seems to be less relevant as compared to the findings found for *APOE* and *MTHFR* genes, also in their interaction.

If we select in PubMed the following keywords "BPSD-gene-Alzheimer", we found 32 results where the studies have been actively published until 2013, from 2013 to 2020 there are only seven works performed on the genetics of BPSD. To date, the classic limitations linked to genetic studies remain and there is still the necessity of: larger cohorts with appropriate assessment tools and statistical methods with correction for multiple testing. There are limitations regarding inconsistent data in genome-wide association studies with prevalently intronic SNPs detection and presence of confounders influencing the associations. In this study, appropriate statistical tools such as the exhaustive CHAID data mining algorithm were used allowing to obtain the adjusted *p*-values for multiple testing, and where age and sex did not influence the calculations. The only variable not included in our analyses was "Medications". Medications as well as comorbidity are clinical information extremely heterogeneous in a BPSD population, including our cohort (Table 1). To include "Medications" or "Comorbidity" as variables in the statistical analyses would result in being overly complex, especially in a sample of sample sizes such as ours. Thus, we suggest that future studies should consider these variables, but starting from homogeneous groups for medications and for comorbidity, on which then to investigate the genetic correlates.

Although aware that further studies are still needed, we want to ignite the enthusiasm and encourage more research to deepen the etiopathogenetic mechanisms of BPSD in AD, starting by a well-defined BPSD population, appropriate statistical methods with multiple corrections, larger sample, and "gene*x*gene" and possibly "gene*x*environment" interactions. This permits not only to facilitate a clinical/diagnostic assessment, but also and mainly to define appropriate treatments chosen that, to date, represent the major concern for the clinicals.

5. Conclusions

We strongly suggest the combined use of both BPSD single symptoms/endophenotypes and the "gene$x$gene" interactions as valid strategies for expanding the knowledge about the BPSD aetiopathogenetic mechanisms.

Supplementary Materials: The following are available online at http://www.mdpi.com/2075-4426/10/3/90/s1, Table S1: Univariate Analysis of Variance in our BPSD cohort: Tests of Between-Subjects Effects for Gender variable and BPSD endophenotypes, Table S2: Univariate Analysis of Variance in our BPSD cohort: Tests of Between-Subjects Effects for Age variable and BPSD endophenotypes, Table S3: Hosmer-Lemeshow goodness of fit test in single BSPD symptoms. Table S4: Stepwise multiple logistic regression for "Psychosis" endophenotype, Table S5: Stepwise multiple logistic regression for "Hyperactivity" endophenotype, Figure S1: Exaustive CHAID data mining algorithm analysis results ("gene × gene" interaction) for single BPSD symptoms according to different values of "severity*frequencies".

Author Contributions: Conceptualization, C.S., M.C., and C.B.; data curation, M.C. and C.B.; formal analysis, M.C., C.M., and C.B.; investigation, C.B.; project administration, C.B.; validation, C.S. and C.B.; writing—review and editing, C.S., M.C., C.M., C.G., O.Z., M.G., and C.B. All authors have read and agreed to the published version of the manuscript.

Funding: This research was supported by grants from the Italian Ministry of Health [Ricerca Corrente].

Conflicts of Interest: The authors declare no conflict of interest.

References

1. McKhann, G.M.; Knopman, D.S.; Chertkow, H.; Hyman, B.T.; Jack, C.R., Jr.; Kawas, C.H.; Klunk, W.E.; Koroshetz, W.J.; Manly, J.J.; Mayeux, R.; et al. The diagnosis of dementia due to Alzheimer's disease: Recommendations from the National Institute on Aging-Alzheimer's Association workgroups on diagnostic guidelines for Alzheimer's disease. *Alzheimers Dement.* **2011**, *7*, 263–269. [CrossRef]
2. Haupt, M.; Kurz, A.; Janner, M. A 2-year follow-up of behavioural and psychological symptoms in Alzheimer's disease. *Dement. Geriatr. Cogn. Disord.* **2000**, *11*, 147–152. [CrossRef] [PubMed]
3. Cerejeira, J.; Lagarto, L.; Mukaetova-Ladinska, E.B. Behavioral and psychological symptoms of dementia. *Front. Neurol.* **2012**, *3*, 73. [CrossRef] [PubMed]
4. Cloak, N.; Al Khalili, Y. *Behavioral and Psychological Symptoms in Dementia (BPSD)*; StatPearls Publishing LLC: Treasure Island, FL, USA, 2020.
5. Angelucci, F.; Spalletta, G.; di Iulio, F.; Ciaramella, A.; Salani, F.; Colantoni, L.; Varsi, A.E.; Gianni, W.; Sancesario, G.; Caltagirone, C.; et al. Alzheimer's disease (AD) and Mild Cognitive Impairment (MCI) patients are characterized by increased BDNF serum levels. *Curr. Alzheimer Res.* **2010**, *7*, 15–20. [CrossRef] [PubMed]
6. Finkel, S.I.; Costa e Silva, J.; Cohen, G.; Miller, S.; Sartorius, N. Behavioral and psychological signs and symptoms of dementia: A consensus statement on current knowledge and implications for research and treatment. *Int. Psychogeriatr.* **1996**, *8*, 497–500. [CrossRef] [PubMed]
7. Frisoni, G.B.; Rozzini, L.; Gozzetti, A.; Binetti, G.; Zanetti, O.; Bianchetti, A.; Trabucchi, M.; Cummings, J.L. Behavioral syndromes in Alzheimer's disease: Description and correlates. *Dement. Geriatr. Cogn. Disord.* **1999**, *10*, 130–138. [CrossRef]
8. van der Linde, R.M.; Dening, T.; Matthews, F.E.; Brayne, C. Grouping of behavioural and psychological symptoms of dementia. *Int. J. Geriatr. Psychiatry* **2014**, *29*, 562–568. [CrossRef]
9. Mushtaq, R.; Pinto, C.; Tarfarosh, S.F.; Hussain, A.; Shoib, S.; Shah, T.; Shah, S.; Manzoor, M.; Bhat, M.; Arif, T. A Comparison of the Behavioral and Psychological Symptoms of Dementia (BPSD) in Early-Onset and Late-Onset Alzheimer's Disease-A Study from South East Asia (Kashmir, India). *Cureus* **2016**, *8*, e625. [CrossRef]
10. Flirski, M.; Sobow, T.; Kloszewska, I. Behavioural genetics of Alzheimer's disease: A comprehensive review. *Arch. Med. Sci.* **2011**, *7*, 195–210. [CrossRef]
11. Borroni, B.; Agosti, C.; Archetti, S.; Costanzi, C.; Bonomi, S.; Ghianda, D.; Lenzi, G.L.; Caimi, L.; Di Luca, M.; Padovani, A. Catechol-O-methyltransferase gene polymorphism is associated with risk of psychosis in Alzheimer Disease. *Neurosci. Lett.* **2004**, *370*, 127–129. [CrossRef]

12. Borroni, B.; Grassi, M.; Agosti, C.; Archetti, S.; Costanzi, C.; Cornali, C.; Caltagirone, C.; Caimi, L.; Di Luca, M.; Padovani, A. Cumulative effect of COMT and 5-HTTLPR polymorphisms and their interaction with disease severity and comorbidities on the risk of psychosis in Alzheimer disease. *Am. J. Geriatr. Psychiatry* **2006**, *14*, 343–351. [CrossRef] [PubMed]
13. Borroni, B.; Grassi, M.; Agosti, C.; Costanzi, C.; Archetti, S.; Franzoni, S.; Caltagirone, C.; Di Luca, M.; Caimi, L.; Padovani, A. Genetic correlates of behavioral endophenotypes in Alzheimer disease: Role of COMT, 5-HTTLPR and APOE polymorphisms. *Neurobiol. Aging* **2006**, *27*, 1595–1603. [CrossRef] [PubMed]
14. Borroni, B.; Grassi, M.; Costanzi, C.; Zanetti, M.; Archetti, S.; Franzoni, S.; Caimi, L.; Padovani, A. Haplotypes in cathechol-O-methyltransferase gene confer increased risk for psychosis in Alzheimer disease. *Neurobiol. Aging* **2007**, *28*, 1231–1238. [CrossRef] [PubMed]
15. Akil, M.; Kolachana, B.S.; Rothmond, D.A.; Hyde, T.M.; Weinberger, D.R.; Kleinman, J.E. Catechol-O-methyltransferase genotype and dopamine regulation in the human brain. *J. Neurosci.* **2003**, *23*, 2008–2013. [CrossRef]
16. Lewandowski, C.T.; Maldonado Weng, J.; LaDu, M.J. Alzheimer's disease pathology in APOE transgenic mouse models: The Who, What, When, Where, Why, and How. *Neurobiol. Dis.* **2020**, *139*, 104811. [CrossRef]
17. Muller, T. Catechol-O-methyltransferase enzyme: Cofactor S-adenosyl-L-methionine and related mechanisms. *Int. Rev. Neurobiol.* **2010**, *95*, 49–71.
18. Seshadri, S.; Beiser, A.; Selhub, J.; Jacques, P.F.; Rosenberg, I.H.; D'Agostino, R.B.; Wilson, P.W.; Wolf, P.A. Plasma homocysteine as a risk factor for dementia and Alzheimer's disease. *N. Engl. J. Med.* **2002**, *346*, 476–483. [CrossRef]
19. Zheng, Z.; Wang, J.; Yi, L.; Yu, H.; Kong, L.; Cui, W.; Chen, H.; Wang, C. Correlation between behavioural and psychological symptoms of Alzheimer type dementia and plasma homocysteine concentration. *BioMed Res. Int.* **2014**, *2014*, 383494. [CrossRef]
20. Reif, A.; Pfuhlmann, B.; Lesch, K.P. Homocysteinemia as well as methylenetetrahydrofolate reductase polymorphism are associated with affective psychoses. *Prog. Neuro-psychopharmacol. Biol. Psychiatry* **2005**, *29*, 1162–1168. [CrossRef]
21. Monji, A.; Yanagimoto, K.; Maekawa, T.; Sumida, Y.; Yamazaki, K.; Kojima, K. Plasma folate and homocysteine levels may be related to interictal "schizophrenia-like" psychosis in patients with epilepsy. *J. Clin. Psychopharmacol.* **2005**, *25*, 3–5. [CrossRef]
22. Vogel, T.; Dali-Youcef, N.; Kaltenbach, G.; Andres, E. Homocysteine, vitamin B12, folate and cognitive functions: A systematic and critical review of the literature. *Int. J. Clin. Pract.* **2009**, *63*, 1061–1067. [CrossRef] [PubMed]
23. Frosst, P.; Blom, H.J.; Milos, R.; Goyette, P.; Sheppard, C.A.; Matthews, R.G.; Boers, G.J.; den Heijer, M.; Kluijtmans, L.A.; van den Heuvel, L.P. A candidate genetic risk factor for vascular disease: A common mutation in methylenetetrahydrofolate reductase. *Nat. Genet.* **1995**, *10*, 111–113. [CrossRef] [PubMed]
24. Rozen, R. Genetic predisposition to hyperhomocysteinemia: Deficiency of methylenetetrahydrofolate reductase (MTHFR). *Thromb. Haemost.* **1997**, *78*, 523–526. [CrossRef] [PubMed]
25. Rai, V. Methylenetetrahydrofolate Reductase (MTHFR) C677T Polymorphism and Alzheimer Disease Risk: A Meta-Analysis. *Mol. Neurobiol.* **2017**, *54*, 1173–1186. [CrossRef]
26. Stoccoro, A.; Tannorella, P.; Salluzzo, M.G.; Ferri, R.; Romano, C.; Nacmias, B.; Siciliano, G.; Migliore, L.; Coppede, F. The Methylenetetrahydrofolate Reductase C677T Polymorphism and Risk for Late-Onset Alzheimer's disease: Further Evidence in an Italian Multicenter Study. *J. Alzheimers Dis.* **2017**, *56*, 1451–1457. [CrossRef]
27. Hu, Q.; Teng, W.; Li, J.; Hao, F.; Wang, N. Homocysteine and Alzheimer's Disease: Evidence for a Causal Link from Mendelian Randomization. *J. Alzheimers Dis.* **2016**, *52*, 747–756. [CrossRef]
28. Roussotte, F.F.; Hua, X.; Narr, K.L.; Small, G.W.; Thompson, P.M. Alzheimer's Disease Neuroimaging Initiative The C677T variant in MTHFR modulates associations between brain integrity, mood, and cognitive functioning in old age. *Biol. Psychiatry Cogn. Neurosci. Neuroimaging* **2017**, *2*, 280–288.
29. van der Put, N.M.; Gabreels, F.; Stevens, E.M.; Smeitink, J.A.; Trijbels, F.J.; Eskes, T.K.; van den Heuvel, L.P.; Blom, H.J. A second common mutation in the methylenetetrahydrofolate reductase gene: An additional risk factor for neural-tube defects? *Am. J. Hum. Genet.* **1998**, *62*, 1044–1051. [CrossRef]

30. Weisberg, I.; Tran, P.; Christensen, B.; Sibani, S.; Rozen, R. A second genetic polymorphism in methylenetetrahydrofolate reductase (MTHFR) associated with decreased enzyme activity. *Mol. Genet. Metab.* **1998**, *64*, 169–172. [CrossRef]
31. McKhann, G.; Drachman, D.; Folstein, M.; Katzman, R.; Price, D.; Stadlan, E.M. Clinical diagnosis of Alzheimer's disease: Report of the NINCDS-ADRDA Work Group under the auspices of Department of Health and Human Services Task Force on Alzheimer's Disease. *Neurology* **1984**, *34*, 939–944. [CrossRef]
32. Folstein, M.F.; Folstein, S.E.; McHugh, P.R. "Mini-mental state". A practical method for grading the cognitive state of patients for the clinician. *J. Psychiatr. Res.* **1975**, *12*, 189–198. [CrossRef]
33. Miller, M.D.; Paradis, C.F.; Houck, P.R.; Mazumdar, S.; Stack, J.A.; Rifai, A.H.; Mulsant, B.; Reynolds, C.F., 3rd. Rating chronic medical illness burden in geropsychiatric practice and research: Application of the Cumulative Illness Rating Scale. *Psychiatry Res.* **1992**, *41*, 237–248. [CrossRef]
34. Katz, S.; Downs, T.D.; Cash, H.R.; Grotz, R.C. Progress in development of the index of ADL. *Gerontologist* **1970**, *10*, 20–30. [CrossRef] [PubMed]
35. Lawton, M.P.; Brody, E.M. Assessment of older people: Self-maintaining and instrumental activities of daily living. *Gerontologist* **1969**, *9*, 179–186. [CrossRef] [PubMed]
36. Cummings, J.L.; Mega, M.; Gray, K.; Rosenberg-Thompson, S.; Carusi, D.A.; Gornbein, J. The Neuropsychiatric Inventory: Comprehensive assessment of psychopathology in dementia. *Neurology* **1994**, *44*, 2308–2314. [CrossRef] [PubMed]
37. Desjardins, P.; Conklin, D. NanoDrop microvolume quantitation of nucleic acids. *J. Vis. Exp.* **2010**, *45*, 2565. [CrossRef]
38. Ingelsson, M.; Shin, Y.; Irizarry, M.C.; Hyman, B.T.; Lilius, L.; Forsell, C.; Graff, C. Genotyping of apolipoprotein E: Comparative evaluation of different protocols. *Curr. Protoc. Hum. Genet.* **2003**, *38*, 9–14. [CrossRef]
39. Gorsuch, R. *Factor Analysis*; Erlbaum: Hillsdale, NJ, USA, 1983.
40. Breiman, L.; Friedman, J.; Stone, C.J.; Olshen, R.A. *Classification and Regression Trees*; CRC Press: Boca Raton, FL, USA, 1984.
41. Zhang, H.; Bonney, G. Use of classification trees for association studies. *Genet. Epidemiol.* **2000**, *19*, 323–332. [CrossRef]
42. Biggs, D.; De Ville, B.; Suen, E. A method of choosing multiway partitions for classification and decision trees. *J. Appl. Stat.* **1991**, *18*, 49–62. [CrossRef]
43. Deardorff, W.J.; Grossberg, G.T. Behavioral and psychological symptoms in Alzheimer's dementia and vascular dementia. *Handb. Clin. Neurol.* **2019**, *165*, 5–32.
44. Shah, C.; DeMichele-Sweet, M.A.; Sweet, R.A. Genetics of psychosis of Alzheimer disease. *Am. J. Med. Genet. B Neuropsychiatr. Genet.* **2017**, *174*, 27–35. [CrossRef] [PubMed]
45. DeMichele-Sweet, M.A.A.; Weamer, E.A.; Klei, L.; Vrana, D.T.; Hollingshead, D.J.; Seltman, H.J.; Sims, R.; Foroud, T.; Hernandez, I.; Moreno-Grau, S.; et al. Genetic risk for schizophrenia and psychosis in Alzheimer disease. *Mol. Psychiatry* **2018**, *23*, 963–972. [CrossRef] [PubMed]
46. Panza, F.; Frisardi, V.; Seripa, D.; D'Onofrio, G.; Santamato, A.; Masullo, C.; Logroscino, G.; Solfrizzi, V.; Pilotto, A. Apolipoprotein E genotypes and neuropsychiatric symptoms and syndromes in late-onset Alzheimer's disease. *Ageing Res. Rev.* **2012**, *11*, 87–103. [CrossRef] [PubMed]
47. Ruthirakuhan, M.; Lanctot, K.L.; Di Scipio, M.; Ahmed, M.; Herrmann, N. Biomarkers of agitation and aggression in Alzheimer's disease: A systematic review. *Alzheimers Dement.* **2018**, *14*, 1344–1376. [CrossRef]
48. Rosenberg, P.B.; Nowrangi, M.A.; Lyketsos, C.G. Neuropsychiatric symptoms in Alzheimer's disease: What might be associated brain circuits? *Mol. Asp. Med.* **2015**, *43–44*, 25–37. [CrossRef]
49. Banning, L.C.P.; Janssen, E.P.C.J.; Hamel, R.E.G.; de Vugt, M.; Kohler, S.; Wolfs, C.A.G.; Oosterveld, S.M.; Melis, R.J.F.; Olde Rikkert, M.G.M.; Kessels, R.P.C.; et al. Determinants of Cross-Sectional and Longitudinal Health-Related Quality of Life in Memory Clinic Patients Without Dementia. *J. Geriatr. Psychiatry Neurol.* **2019**, *33*, 256–264. [CrossRef]
50. Garcia-Segura, M.E.; Fischer, C.E.; Schweizer, T.A.; Munoz, D.G. APOE varepsilon4/varepsilon4 Is Associated with Aberrant Motor Behavior Through Both Lewy Body and Cerebral Amyloid Angiopathy Pathology in High Alzheimer's Disease Pathological Load. *J. Alzheimers Dis.* **2019**, *72*, 1077–1087. [CrossRef]
51. Oliveira, F.F.; Chen, E.S.; Smith, M.C.; Bertolucci, P.H. Associations of cerebrovascular metabolism genotypes with neuropsychiatric symptoms and age at onset of Alzheimer's disease dementia. *Braz. J. Psychiatry.* **2017**, *39*, 95–103. [CrossRef]

52. Steinberg, M.; Corcoran, C.; Tschanz, J.T.; Huber, C.; Welsh-Bohmer, K.; Norton, M.C.; Zandi, P.; Breitner, J.C.; Steffens, D.C.; Lyketsos, C.G. Risk factors for neuropsychiatric symptoms in dementia: The Cache County Study. *Int. J. Geriatr. Psychiatry* **2006**, *21*, 824–830. [CrossRef]
53. Polito, L.; Poloni, T.E.; Vaccaro, R.; Abbondanza, S.; Mangieri, M.; Davin, A.; Villani, S.; Guaita, A. High homocysteine and epistasis between MTHFR and APOE: Association with cognitive performance in the elderly. *Exp. Gerontol.* **2016**, *76*, 9–16. [CrossRef]
54. Durmaz, A.; Kumral, E.; Durmaz, B.; Onay, H.; Aslan, G.I.; Ozkinay, F.; Pehlivan, S.; Orman, M.; Cogulu, O. Genetic factors associated with the predisposition to late onset Alzheimer's disease. *Gene* **2019**, *707*, 212–215. [CrossRef] [PubMed]
55. Yi, J.; Xiao, L.; Zhou, S.Q.; Zhang, W.J.; Liu, B.Y. The C677T Polymorphism of the Methylenetetrahydrofolate Reductase Gene and Susceptibility to Late-onset Alzheimer's Disease. *Open Med.* **2019**, *14*, 32–40. [CrossRef] [PubMed]
56. Chen, Z.; Schwahn, B.C.; Wu, Q.; He, X.; Rozen, R. Postnatal cerebellar defects in mice deficient in methylenetetrahydrofolate reductase. *Int. J. Dev. Neurosci.* **2005**, *23*, 465–474. [CrossRef] [PubMed]
57. Fatemi, S.H.; Earle, J.A.; McMenomy, T. Reduction in Reelin immunoreactivity in hippocampus of subjects with schizophrenia, bipolar disorder and major depression. *Mol. Psychiatry* **2000**, *5*, 654–663. [CrossRef] [PubMed]
58. Soni, R.M.; Tiwari, S.C.; Mahdi, A.A.; Kohli, N. Serum Homocysteine and Behavioral and Psychological Symptoms of Dementia: Is There Any Correlation in Alzheimer's Disease? *Ann. Neurosci.* **2019**, *25*, 152–159. [CrossRef]

Review

Obesity and Diabetes Mediated Chronic Inflammation: A Potential Biomarker in Alzheimer's Disease

Md Shahjalal Hossain Khan and Vijay Hegde *

Obesity and Metabolic Health Laboratory, Department of Nutritional Sciences, Texas Tech University, Lubbock, TX 79409, USA; Md-Shahjalal.Khan@ttu.edu

* Correspondence: vijay.hegde@ttu.edu; Tel.: +1-(806)-834-6695

Received: 17 April 2020; Accepted: 19 May 2020; Published: 22 May 2020

Abstract: Alzheimer's disease (AD) is the sixth leading cause of death and is correlated with obesity, which is the second leading cause of preventable diseases in the United States. Obesity, diabetes, and AD share several common features, and inflammation emerges as the central link. High-calorie intake, elevated free fatty acids, and impaired endocrine function leads to insulin resistance and systemic inflammation. Systemic inflammation triggers neuro-inflammation, which eventually hinders the metabolic and regulatory function of the brain mitochondria leading to neuronal damage and subsequent AD-related cognitive decline. As an early event in the pathogenesis of AD, chronic inflammation could be considered as a potential biomarker in the treatment strategies for AD.

Keywords: obesity; diabetes; inflammation; Alzheimer's disease; Amyloid Beta; Tau; biomarker; mitochondrial dysfunction

---

1. Introduction

Alzheimer's disease (AD) is a progressive and irreversible brain disorder that begins well before the clinical symptoms appear. The actual symptoms may appear only after several years of changes in the brain due to damage or destruction of brain cells (neurons) in the area involved in cognitive function such as memory, thinking, and learning [1]. AD slowly abolishes brain function and hinders thinking ability. AD is recognized as a common cause of an estimated 60–80% of cases of dementia [2]. The early clinical symptoms of AD have been described as difficulty in conversations and depression and is followed by disorientation, confusion, impaired communication, behavioral changes, and, eventually, trouble in speaking and walking. At the critical stage, consequences are fatal and the patients are bed-bound and need special attention such as round-the-clock care [3].

AD prevalence is escalating rapidly worldwide especially among the population aged 65 years and older. In the United States, it has been reported that, in 2019, an estimated 5.8 million Americans were living with AD-related dementia and, among them, 81% were aged 75 or older and 200,000 individuals were under 65 years of age [3]. The numbers have been projected to grow from the current 55 million to 88 million [3], approximately doubling by 2050 [4]. This alone depicts the magnitude of the burden of AD on health care and the overall society in the future. Considering disability-adjusted life years (DALYs) as a primary measure of disease burden, in the United States, AD has risen from the 12th most burdensome disease in 1990 to the 6th most burdensome disease in 2016 [5]. It has been estimated that, in 2019, the healthcare burden will be around $290 million [3]. Collectively, the above statistics indicate that AD does not only affect morbidity or mortality, but AD is also affecting the socio-economic and healthcare burden in the USA. There have been numerous studies elucidating the pathogenesis, molecular and clinical mechanisms, the association of diabetes and metabolic syndromes, and consequences in a broader range. Yet no single or combined treatment has shown to have satisfactory levels of efficiency to delay or prevent AD pathogenesis.

In this review, we will focus on a much narrower context of AD pathogenesis considering inflammation as the central mechanistic link among obesity, diabetes, and AD-related dementia. In obesity, the elevated circulating free fatty acids (FFA) attributes to inflammation, initiated by Toll-like receptors (TLR-4) and release of pro-inflammatory cytokines, which is also associated with insulin resistance and diabetes [6,7] via nuclear factor-kappa B (NF-kB) mediated inflammation that might mediate the intracellular signaling impairments [8]. In the brain, pro-inflammatory cytokines activated by microglia can induce oxidative stress and compromised antioxidant defense [9]. Collectively, they contribute to impaired insulin signaling, synapses loss, reduced mitochondrial axonal transport [10], mitochondrial fragmentation, dynamics, and eventual dysfunction [11]. Mitochondrial dysfunction has been considered as an early event for AD pathogenesis and is also associated with a metabolic syndrome including obesity, diabetes, insulin resistance, and cardiovascular diseases [8]. Thus, it is possible that mitochondrial dysfunction has an evident role in the initiation and development of the metabolic disorder, and it is worthy to investigate to what extent inflammation and mitochondrial health affect the progression of AD.

2. Pathogenesis of AD

The characteristic features of AD involve two major fundamental processes: extracellular beta amyloid (Aβ) deposition and intracellular tau protein hyper accumulation. Aβ is insoluble and is a major component for senile plaques. Insoluble tau is the major component of neurofibrillary tangles (NFT) [12]. Aβ is a 36 to 43 amino acid peptide, which is a part of the large transmembrane protein, Amyloid Precursor Protein (APP), and derived from cleavage of APP by β- and γ–secretase enzymes. Defective clearance of Aβ during the cleaving process of APP results in the accumulation of insoluble Aβ [13]. Initially, the Aβ monomer polymerizes into soluble oligomers and then into larger insoluble fragments like Aβ42 that can precipitate as amyloid fibrils [14]. On the other hand, tau is a protein associated with the microtubule and helps in modulating the axonal microtubule stability [15]. In an AD patient's brain, the tau protein gets hyper-phosphorylated and causes the protein to lose microtubule-binding ability and dissociate from the microtubules, which can progressively disrupt the transport structure and result in starvation of neurons and, ultimately, neuronal cell death [16]. Deposition of neuronal bodies and processing of insoluble phosphorylated tau protein into paired helical filaments can cause neurofibrillary degeneration. These deposits interfere with the spacing between microtubules and hinder the axonal terminals and dendrite nutrient transport [17].

3. Obesity, Diabetes, and AD

Obesity is a chronic and multifactorial disease characterized by excessive body fat accumulation. It is considered a risk factor for many diseases and disorders including hypertension, type 2 diabetes mellitus, coronary heart disease, and AD [18]. Adipose tissue possesses endocrine function secreting adipokines, inflammatory cytokines, and other bioactive mediators that influence energy homeostasis in metabolically active organs like adipose tissue, liver, pancreas, and even brain [19]. During a positive energy balance, adipocyte size and proliferation increases result in the expansion of adipose tissue to accommodate the excess energy, which leads to an increase in adipose tissue mass. This is followed by adipose tissue dysfunction promoting chronic low-grade inflammation [20], elevated oxidative stress [21,22], and altered mitochondrial dysfunction [23–25]. Although how obesity contributes to the mitochondrial dysfunction is still not clear, it is postulated that the induced inflammation and metabolic alteration implicated by impaired insulin function could be a possible factor [26,27].

Altered glucose homeostasis due to hyperinsulinemia and insulin resistance in diabetes is also one of the most prominent features of obesity. Obesity and a high fat diet (HFD) can induce insulin resistance, which, subsequently, impairs insulin signaling in the periphery as well as in the brain [28]. It is well-established that insulin has a significant role in the central nervous system (CNS) [29–33].

In the brain, insulin along with insulin growth factors (IGF) modulate neuronal growth, differentiation, survival, migration, metabolism, protein synthesis, gene expression, synapse formation, and synaptic plasticity [34]. Moreover, insulin regulates myelin production and oligodendrocyte maintenance [34]. Defective insulin signaling is also associated with impaired cognition and AD-related dementia [35]. Severely decreased phosphorylation of insulin receptors has been found in patient brains from both AD and diabetes [36]. The disturbance in insulin signaling could contribute to making the CNS environment more vulnerable to metabolic stress and, therefore, accelerate the neuronal dysfunction [37]. Diabetes is associated with islet amyloid polypeptide (IAPP) accumulation in the pancreatic islets. The IAPP is concurrently secreted with insulin from the pancreas. In diabetic and AD patients, the IAPP is found to be misfolded and elevated [38], accompanied by elevated Aβ accumulation, hyper-phosphorylated tau, and impaired fasting glucose as comorbidities [39].

Substantial evidence indicates that patients with obesity and diabetes are more susceptible to develop AD-related cognitive degeneration. Several longitudinal [40,41] and epidemiological studies [42–45] have established a significant association between midlife large waist-hip ratios to AD-related dementia and decreased hippocampal volume in later life. Moreover, the common pathways of neurodegeneration in the AD brain resemble similar pathologies observed in the brains of individuals with diabetes. Obesity, AD, and diabetes concomitantly share common features like brain atrophy, reduced cerebral glucose, and CNS insulin resistance [46].

4. Obesity, Diabetes, and Subsequent Systemic and Neuro-Inflammation

Obesity, diabetes, and AD attributes to a common shared chronic inflammatory process. Both epidemiological and observational studies suggested that neuro-inflammation is an early-stage marker of AD pathogenesis [47]. A higher plasma and CNS levels of inflammatory markers known as Interleukin-6 (IL-6), Interleukin 1β (IL-1β), transforming growth factor-β (TGF-β), and tumor necrosis factor-α (TNF-α) has been reported in AD patients, which indicates the potential role of inflammatory markers in AD pathogenesis [48]. Insulin resistance and inflammation have a bidirectional relation with AD. In the chronic peripheral inflammatory process in obesity and diabetes, the production of inflammatory cytokines can lead to serine phosphorylation of insulin receptor substrate-1 (IRS-1), which can inhibit the downstream signaling pathways like kappa B kinase (IKK), c-Jun N-terminal kinase (JNK), and extracellular signal regulated kinase 2 (ERK2). These can block the intracellular insulin signaling by downregulating the insulin receptor-mediated signaling [49]. Additionally, the systemic inflammation can damage and cross the blood-brain barrier (BBB) and enter into the brain, which might trigger the brain-specific inflammatory response [50]. These circulating cytokines can increase the apoptosis (cell death) rate, reduce synaptic function, and inhibit the neurogenesis and, thus, causing neuronal death [51]. Moreover, systemic inflammatory processes, e.g., IL-6, TNF-α, and C-reactive proteins (CRP), inhibit the transfer of Aβ from the CNS to the periphery [52]. Thus, Aβ oligomers accumulation can activate brain microglia that secrete the pro-inflammatory cytokines (IL-1β, IL-6, TNF-α), which can phosphorylate insulin receptor substrate (IRS) in multiple sites by activating the IRS-1 serine kinase upon binding with the respective receptors and alter brain insulin signaling [53]. Although the precise mechanisms are yet to be understood, the putative mechanisms for systemic and neuro-inflammation caused by obesity and diabetes are as follows.

4.1. Systemic Inflammation and AD

The circulating pro-inflammatory molecules either increase the permeability of the BBB or gain access via the area that lacks effective BBB to initiate neuro-inflammation in the hypothalamus [54,55]. The mechanism of BBB disruption is multifaceted and includes changes in the structural components like pericyte dysfunction, tight junction, and elevated endothelial oxidative stress [56]. HFD can elevate the expression and activation of pro-inflammatory cytokines and corresponding transcription factors, such as NF-*k*B, in the hypothalamus [57]. This pro-inflammatory action (a) will increase the microglial (brain's residing macrophage) infiltration and activation in the hypothalamus resulting in the local inflammatory mediators such as cytokines [57,58]. In addition, the elevated free fatty acids (FFA) enters the arcuate nucleus and increases TLR-4, a molecular pattern recognition receptor, on microglia and astrocytes and initiates the inflammatory response centrally [59]. (b) The circulating cytokines have limited spread and activate the hypothalamic cytokine receptors. This action can augment the brain inflammation [28]. (c) The direct entry of cytokines, chemokines, and FFA in systemic circulation can also propagate neuro-inflammation by initiating pro-inflammatory cytokines and prostaglandin production [60]. Collectively, this pro-inflammatory milieu can disrupt the hypothalamic function by inducing synaptic remodeling, neuronal cell apoptosis, and disturbed neurogenesis [61]. The hypothalamus has a potential role in cognitive function related to feeding, metabolism, stress regulation, and cardiovascular function along with cognition, attention, and memory function [62]. The remodeling of the hypothalamic circuit leads to dysregulation in the hypothalamic-pituitary-adrenal (HPA) axis resulting in increased production of glucocorticoids, which are related to impaired cognition and memory including depression, Cushing's syndrome, and AD [63]. Moreover, the chronic HPA axis activation and elevated glucocorticoid level is also associated with dendritic atrophy, hippocampal volume reduction, and reduced synaptic plasticity [64]. One of the characteristic features of AD is the hippocampus and cerebral cortex atrophy [65]. All these possible events collectively can lead to neurodegeneration and eventually AD.

4.2. Neuro-Inflammation in AD

The brain was thought to be unaffected by systemic inflammation and, thus, regarded as an "immune-privileged" organ not susceptible to inflammation for many years. However, this notion changed after extensive neuro-immune research that explored the central nervous system interaction with the peripheral system through hormonal and paracrine action [66]. In the case of the AD brain, Aβ deposition triggered by systemic inflammation initiates a series of immune-responses intended to reduce the Aβ plaques and the aggregates by activating the innate immunity to elicit the inflammatory response from microglia and astrocytes [67,68]. These pathological adaptations incite the release of pro-inflammatory cytokines including IL-6, IL-1β, or TNF-α along with other pro-inflammatory molecules including macrophage inflammatory proteins, monocyte chemoattractant proteins, coagulating factors, reactive oxygen species (ROS), nitric oxide, proteases, protease inhibitors, prostaglandins, thromboxanes, leukotrienes, and CRP from glial cells [52,69–71]. The aggravated environment induces additional phosphorylation of tau, accumulation of Aβ, and pro-inflammatory molecules [52], which, consequently, release reactive substances like nitric oxide, proteolytic enzymes, excitatory amino acids, and complementary factors, and damage the adjacent healthy neurons [72]. Therefore, the well-intended initial response to assist Aβ clearance, in turn, also secretes mediators that cause damage and leads to neurodegeneration [73]. An elevated level of IL-1β was found in serum, cerebrospinal fluids, and in the brain region [74,75], particularly astrocytes of the cortex and hippocampus [76] of patients with AD. The IL-1β secreted from astrocytes increases the neurotoxic Aβ and APP production [76,77]. The IL-1β also activates microglia to induce pro-inflammatory cytokine release that can lead to neurotoxicity [78]. Additionally, oxidative damage and alteration in tau phosphorylation in neurons, along with chronic low-grade inflammatory states, collectively lead to compromised brain integrity [79].

4.3. RAGE-Mediated Inflammation in AD

Dyslipidemia and chronic hyperglycemia caused by disturbance of insulin signaling in obesity can lead to glucolipotoxicity, which can play a pivotal role in AD. Hyperglycemia also generates advanced glycation end products (AGE), which is a senescent macro-protein derivative that can be identified in both senile plaques and NFTs as a possible link between AD and diabetes. AGEs are the derivatives of proteins, lipids, or nucleic acids modified by non-enzymatic glycosylation and tend to increase the production and accumulation during aging, diabetes, or obesity [80,81]. Advanced glycation end products (AGE) and its pattern recognition receptors (RAGE) induce a chronic, persistent inflammatory response that propagates vascular injuries [81–83]. RAGE is also a putative receptor for Aβ [84] and is found to be expressed in neurons, microglia, and astrocytes [85,86]. Aβ is a ligand for RAGE that interacts with the N-terminal domain. Neurons and microglia of the hippocampus and inferior frontal cortex area of the AD brain have reported expressing an elevated level of RAGE [85,87].

The Aβ-RAGE interaction in the neuronal cells leads to an ROS mediated cellular response and activation of NF-kB that induce elevated inflammatory milieu [87]. RAGE dependent Aβ-induced migration has been shown in both in-vitro Aβ plaque models and postmortem human microglia cell culture experiments [85]. It was reported that Aβ-RAGE dependent microglial activation elevates the RAGE and microglial colony-stimulating factors (M-CSF), while inhibition of RAGE using anti-RAGE F(ab')$_2$ inhibits the microglial chemotactic response against Aβ_{42} [85]. Moreover, increased production of IL-1β or TNF-α, enhanced microglial infiltration in Aβ plaques, reduced acetylcholine esterase activity, and impaired spatial and learning memory observed in double transgenic mouse models when compared to APP or RAGE mouse models [88]. Elevated levels of AGE in the brain and plasma are associated with the cognitive dysfunction in AD patients [89,90]. AGE, along with increasing Aβ cytotoxicity, supports fibrillary tangles and Aβ plaque formation, which contributes to the pathogenesis of AD [91].

5. Inflammation Mediated Impaired Mitochondrial Health in AD

Mitochondria, which is the powerhouse of cells that provides energy by utilizing ATP, is also responsible for cellular processes including energy metabolism, ROS generation, calcium ion (Ca^{2+}) homeostasis, cell survival, and apoptosis [92]. Mitochondria maintain the normal functioning through regulated quality control even in the presence of persistent insults. However, any failure of quality control results in mechanical defects and damages that cause mitochondrial dysfunction [93]. It can lead to metabolic disorders. The major metabolic organs like liver, muscle, and adipose tissue are the active contributors in this process. During obesity and diabetes, excess lipid accumulation and hyperglycemia induced insulin resistance can lead to abnormal mitochondrial function via altering the ATP synthesis activity of mitochondria, beta-oxidation, and elevated oxidative stress [94]. Adipocyte differentiation drastically increases the activity and biogenesis of mitochondria, pointing towards a link between mitochondria and obesity. Mitochondrial dysfunction is correlated with a reduced size and number of mitochondria [95], oxidative capacity [96], reduced oxygen consumption, and oxidative phosphorylation related gene expression [97], which have been observed in individuals with obesity [90]. In mature adipocytes, mitochondrial dysfunction has been linked to fatty acid oxidation [98], adipokine secretion [99], and impaired glucose homeostasis [100]. Additionally, in obese mice, the expression of mitochondrial DNA (mtDNA), mtDNA transcription factor A (Tfam), and respiratory proteins are also significantly reduced [92]. HFD increases the production of ROS and oxidative stress in mouse adipocytes [101]. HFD consumption or excess calorie intake results in an excess mitochondrial load that hinders the effective dissipation of the proton gradient, which increases ROS generation, mtDNA mutation, and apoptosis [102].

There are several ways obesity and diabetes-induced inflammation affects mitochondrial functioning.

5.1. *Inflammation and Energy Metabolism*

Neuronal glucose metabolism consists of mechanisms that regulate uptake of glucose, insulin, and insulin signaling pathways, glucose transporters (GLUTs), and entry of glycolytic end-products into mitochondria that eventually metabolize and generate ATP via oxidative phosphorylation [103]. Mitochondria participates and modulates several metabolic signaling pathways including cytosolic signaling, redox-sensitive signaling, JNK, and 5' AMP-activated protein kinase (AMPK) signaling [104]. These signaling pathways along with the associated metabolites, transporters, receptors, and enzymes ensure proper neuronal energy metabolism. Mitochondria-centered altered glucose metabolism manifested by impaired insulin signaling, altered receptors activity, and reduced glucose uptake is one of the key features of AD [104]. Mitochondria regulate the tricarboxylic acid (TCA) cycle, which is the principal regulator of cellular respiration. The TCA cycle generates adenosine triphosphate (ATP) via a series of enzyme-catalyzed chemical reactions and any malfunction in the enzymes involved in this cycle leads to impaired cellular respiration and ATP production [104].

Inflammatory processes mediated by infiltrated immune cells from the periphery and activated microglia initiate an intracellular signaling cascade that modifies the mitochondrial energy metabolism [105]. The activated microglia and astrocytes release pro-inflammatory cytokines, particularly TNF-α, and induce oxidative phosphorylation impairment, ATP production, and ROS production [105]. In one study, loss of mitochondrial transmembrane potential and reduction in intracellular ATP upon treatment with IL-1β in the retinal neuronal cells has also been reported [106]. Moreover, IL-1β and TNF-α are reported in decreasing the pyruvate dehydrogenase (PDH) enzyme activity in vivo [107], particularly in cardiomyocytes [108]. Reduced glucose oxidation in cultured human dermal fibroblast [109] reduced skeletal muscle [110,111], hepatic PDH activity [107], and complex I and II activity reported in the presence of inflammatory cytokines [105]. Impaired PDH activity has been found in the postmortem AD brain along with increased IL-6, IL-1β, or TNF-α, which suggests a compromised TCA cycle activity in the inflamed CNS of mild cognitive impairment and AD patients [105]. A significant reduction of neuronal complex I and III along with several other nuclear-encoded subunits of the Electron Transport Chain (ETC) has also been reported [112]. A reduction of ATP production and basal respiration upon TNF-α exposure has been shown in a dose-dependent manner in the mouse hippocampal cell line and primary neuronal cell culture [113]. TNF-α administration also reduces the peroxisome proliferator-activated receptor (PPAR)-γ co-activator 1α (PGC-1α) in myoblasts [114] and human cardiomyocytes [115] even though the effect of TNF-α on neuronal cells of various neurodegenerative diseases has not yet been reported. PGC-1α plays an important role in many cellular and metabolic processes including energy metabolism, cardiovascular diseases, and neurodegenerative diseases [116].

It has been reported that, in the AD brain, the levels of mitochondrial enzymes including PDH, cytochrome oxidase (COX), and α-ketoglutarate dehydrogenase complex are all decreased [117]. Proteomics studies on AD brains reveal that enzymes involved in metabolic pathways of the Kreb's cycle and glycolysis including malate dehydrogenase, glyceraldehyde 3-phosphate dehydrogenase (GAPDH), fructose-bis-phosphate-enolase, alpha-enolase (ENO1), and ATP synthase are oxidized [46]. The oxidation of these enzymes is related to dysfunctional cerebral glucose metabolism and reduction in ATP synthesis, which leads to loss of synaptic function [118]. These oxidative changes can lead to inflammation and eventually compromise metabolic function. The exact mechanism of how neuro-inflammation affects the metabolic function of mitochondria is yet to be understood. However, cellular and biochemical studies revealed that APP, Aβ, tau, and presenilin are associated with impaired mitochondrial energy metabolism [119,120]. It has been proposed that Aβ interacts with mitochondrial proteins disrupting ETC, increases ROS production, and free superoxide radicals that preclude the cellular ATP generation [121]. It has also been proposed that the hyper-phosphorylated tau and Aβ aggregates can block mitochondria and other cell organelles from nerve terminals, synapsis, and other brain regions with high ATP demands [122], which might lead to starvation of dendritic spines and synapsis due to severe mitochondrial ATP depletion [123].

5.2. *Inflammation and Altered Mitochondrial Dynamics*

Mitochondria are organelles that exist in dynamic networks, migrate throughout the cell, continuously fuse, divide, and undergo regulated turnover as the metabolic environment demands [124]. Mitochondrial dynamics can change the size and shape and adapt to the challenges [125]. Mitochondrial fission and fusion are two distinct processes that mediate the mitochondrial dynamic morphology and integrity [126]. In healthy neurons, mitochondria need to maintain the fusion and fission mechanism in a balanced way for the normal functioning of the synapses [125]. The enzymes involved in the fusion process are mitofusion 1 (Mfn1), mitofusion 2 (Mfn2), and optic atrophy protein 1 (OPA1). The proteins involved in the fission process are dynamin-related protein-1 (Drp1) and fission 1 (Fis1).

Mitochondrial biogenesis is the cellular process of producing new mitochondria and increasing the mitochondrial mass [127]. The regulation of biogenesis is modulated by nuclear factors including a set of nuclear transcription factors 1 and 2 (NRF 1 and NRF2), which control the cytochrome c and COX gene expression [128]. Any dysregulation in the mitochondrial dynamics and biogenesis might lead to excessive mitochondrial fragmentation and altered mitochondrial function resulting in mitochondrial dysfunction and eventually contribute to AD progression [129,130].

In astrocytes, impaired respiration rate and increased mitochondrial fragmentation (Drp1) have been reported upon IL-1β exposure [131]. In 3T3L1 adipocytes, TNF-α treatment alters mitochondrial morphology [132], along with smaller condensed mitochondria attributed by an increased level of Fis1 and decreased level of mitochondrial OPA1 [133,134]. In another in-vivo study, IL-6 downregulated Mfn1/Mfn2 and Fis1, which suggests neurotoxic effects of IL-6 [135]. In HFD mice, a decreased level of mitochondrial fusion genes (OPA1, Mfn1) and elevated expression of fission genes (Fis1, Drp1) along with increased pro-inflammatory cytokines has been reported [136]. Perturbation in mitochondrial function and release of mitochondrial contents in extracellular milieu activates the innate immune system, which can exacerbate the inflammatory response and alter the mitochondrial function. The production of ROS can provoke inflammation, and further mitochondrial dysfunction can lead to the production of pro-inflammatory IL-1β [137]. These studies suggested that inflammation and mitochondrial dysfunction operate in a synergic autotoxic feedback loop.

5.3. Inflammation and Mitochondrial Oxidative Stress

As a major source of ROS, mitochondria regulate the oxidative stress process [138]. Oxidative stress results from the imbalance between ROS production and detoxification in biological systems. ROS production is an important physiological by-product of the ETC. In the respiratory chain, while transferring the electron to molecular oxygen, about 0.4% to 5% of electrons lose their way and generate superoxide radicals ($O_2\bullet^-$), which, in turn, activates the mitochondrial permeability transition pore and results in apoptosis [139]. The brain is a highly susceptible organ to oxidative stress when compared to other organs due to the high demand for energy and oxygen, high levels of peroxidizable polyunsaturated fatty acids, the relative paucity of antioxidants, and anti-oxidant defense mechanisms and relative abundance of potent ROS catalyst iron [140]. During obesity and HFD consumption, adipose tissue propagates the inflammation and secretes pro-inflammatory cytokines including TNF-α, IL-6, and IL-1β, which also induce ROS production [141]. The activation of these cytokines promotes nitric oxide and ROS generation by macrophages and monocytes. HFD consumption propagates the lipid peroxidation in the brain by elevated levels of ROS via a similar mechanism that has been found in the non-neuronal tissue [142]. In mice, upon HFD consumption, elevated expression of ERK, and inducible nitric oxide synthase (i-NOS) are associated with oxidative stress [143,144]. The occurrence of oxidative stress has been shown as one of the early events in AD development and plays a major role in the pathogenesis of AD [118,145,146], whereas oxidative damage has been estimated to occur before other prognoses including the onset of Aβ aggregates, tau pathologies, or inflammation in the AD brain [147,148]. The activation of these cytokines promotes nitrogen and ROS generation by macrophages and monocytes. It has also been shown that oxidative stress is related to the hippocampal dysfunction in obese mice [146,149]. In the AD brain, excessive levels of ROS production indicated by the presence of an increased level of oxidative stress markers including oxidized lipid, protein, and DNA has been noticed [118,146].

Aβ generates hydrogen peroxide (H_2O_2) by reducing metal ions and increasing the free radical production by zinc, iron, and copper, which is highly concentrated in the core and periphery of Aβ deposits [150]. Aβ plaques can lead to a cytosolic calcium ions overload by depleting Ca^{2+} storage in the endoplasmic reticulum (ER). Increased Ca^{2+} leads to decreased endogenous glutathione (GSH) levels and increased accumulation of ROS inside the cell [151]. ROS potentially can mediate the JNK/stress-activated protein kinase pathways that can be associated with tau hyper-phosphorylation [152]. Aβ can also initiate free radical formation upon Nicotinamide adenine dinucleotide phosphate hydrogen (NADPH) oxidase activation, which leads to over-accumulation of ROS. The Aβ-induced ROS can initiate tau hyper-phosphorylation and modify cellular signaling via p38 mitogen-activated protein kinase (MAPK) activation [153]. Aβ can generate free radicals when interacting with metal ions. Cu^{2+}/Zn^{2+}-bound Aβ has been found to possess a similar structure to the antioxidant superoxide dismutase (SOD) [154].

Figure 1 provides an overview of how obesity and diabetes associated with elevated free fatty acids and impaired endocrine function leads to insulin resistance and systemic inflammation. This systemic inflammation triggers neuro-inflammation, which eventually hinders the metabolic and regulatory function of the brain mitochondria and leads to neuronal damage.

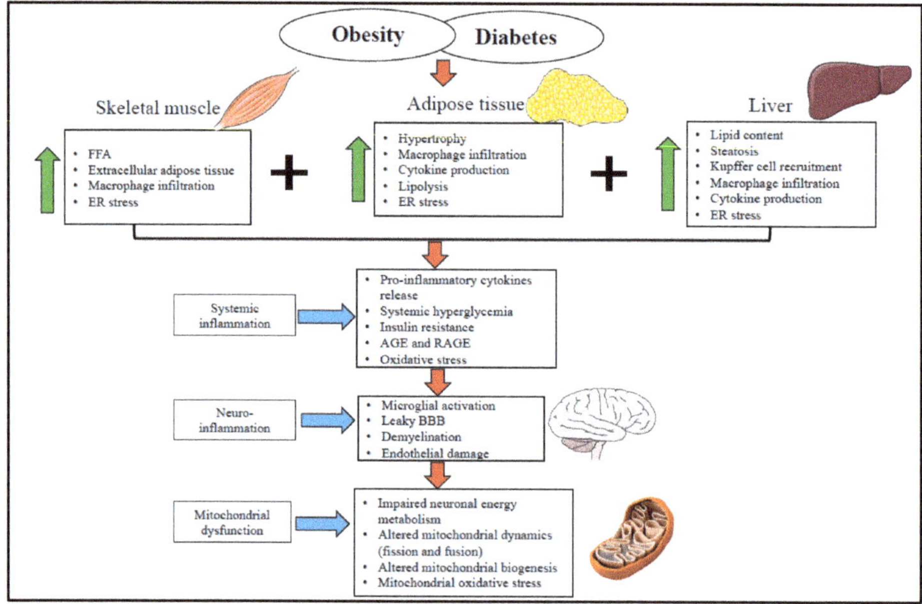

Figure 1. Overview of the diverse mechanism involved in obesity and diabetes-induced systemic and neuro-inflammation in mitochondrial and brain health. Skeletal muscle, adipose tissue, and liver are involved in glucose and lipid metabolism and play a vital role in developing adipose tissue dysfunction and insulin resistance during the obese condition. In obesity and diabetes, elevated free fatty acid, increased secretion of pro-inflammatory cytokines and macrophage infiltration, AGE and RAGE generation, and ER stress induce systemic inflammation. Systemic inflammation progressively initiates microglial activation, endothelial damage, and BBB disruption, which eventually leads to neuro-inflammation. These unfavorable inflammatory events alter the mitochondrial energy metabolism, mitochondrial dynamics, and biogenesis to escalate the mitochondrial oxidative stress collectively and trigger the mitochondrial assault. This gradually results in synapsis loss and neuronal death. FFA—Free Fatty Acid. ER—Endoplasmic Reticulum. AGE—Advanced Glycation End Products. RAGE—pattern recognition receptors of AGE. BBB—Blood-Brain Barrier.

Collectively, it has been proposed that the development of AD in the brain is the consequence of increased oxidative stress derived from several mechanisms: (a) elevated lipid peroxidation, (b) increased protein and DNA oxidation, (c) reduced energy metabolism and activity of COX, (d) ability of Aβ to generate free radicals in the form of increased accumulation of some metal ions like mercury, iron, and aluminum, which can stimulate free radical formation by the Fenton and Haber-Weiss pathway, and increased advanced glycation end products, SOD-1, and malonaldehyde in NFT and senile plaques [155].

5.4. Inflammation and Cognitive Impairment

The association between inflammation and cognitive decline has been reported in several cross-sectional studies [156–158]. However, the data were limited by the number of subjects [156–158], lack of adequate follow-up duration [156], or enough inflammatory markers [156] or subjects who were more than 80 years old. Elevated IL-6 in midlife is associated with cognitive decline whereas increased CRP levels did not predict the concurrent cognitive decline [159]. One neuroimaging study found that IL-6 is consistently associated with cognitive decline. In a follow-up study, increased circulating IL-6 level was found to be associated with accelerated cognitive decline after a 10-year follow up [160]. In another study, IL-6 was found strongly associated with the higher volume of white matter hyper-intensities, decreased gray matter, and reduced hippocampal volume whereas, in the case of CRP, the association was weaker than IL-6. This suggests that an inflammatory process could be associated with alteration of the brain [161]. Animal studies indicate declined cognition associated with elevated inflammation and Aβ deposition [162]. However, some studies found a minor association between cognition and inflammation [156–158,163]. In the Rotterdam Study, elevated levels of pro-inflammatory cytokines including IL-6 and CRP were found to be associated with cognitive decline and executive function. In the Leiden 85-plus study, a systematic level of IL-6 was found to be associated with declined cognition and memory function after adjustment for Apolipoprotein E $\varepsilon 4$ (APOE $\varepsilon 4$) carriers [164]. On the other hand, the Amsterdam Longitudinal Aging Study did not find any association between inflammation and cognitive decline after adjusting for APOE $\varepsilon 4$ [157]. Furthermore, cerebral small vessel disease (SVD) refers to intracranial vascular disease associated with clinical manifestation and neuroimaging features caused by any changes in the morphology of cerebral vessels, which are crucial for adequate cerebral blood flow and brain parenchyma [165]. SVD causes decreased cerebral blood flow, increased BBB permeability, impaired cerebral autoregulation, brain functionality loss, and cognitive decline in the elderly [165]. SVD has often detected the parenchymal alteration based on four different features in magnetic resonance imaging (MRI), which include white matter hyperintensities, lacunes, cerebral microbleeds, and enlarged perivascular spaces [166–168]. A detailed systematic review suggested a robust association of inflammation with SVD represented by the presence of increased vascular inflammatory markers, especially among patients with stroke, which indicates an alteration in the endothelium and BBB [169]. Mounting evidence also indicates the potential role of inflammation and cerebral SVD in neurodegeneration and related disorders in recent years [168,170].

6. Conclusions and Future Direction

In this review, we discussed how obesity and diabetes can lead to systemic and neuro-inflammation, and whether inflammation affects the overall mitochondrial functioning, dynamics, oxidative stress, and cognitive decline in patients with AD. After cautiously considering the facts, we can suggest (but not conclude) that obesity or diabetes-induced inflammation is associated with impaired mitochondrial health. However, the question remains regarding to what extent the systemic or neuro-inflammation directly affect the brain mitochondrial health and subsequent AD pathogenesis. The association between inflammation and cognitive decline could be the consequence of underlying disease conditions triggered by diabetes, obesity, or cardiovascular disease and their consequent comorbidities. If the answer to this question can be addressed, it would be easier to decide whether targeting anti-inflammatory agents alone will be an effective and valid approach to combat AD. It seems even an anti-inflammatory approach is not the central focus for fighting AD but will support the mainstream treatment options including inhibiting Aβ production and tau hyper-phosphorylation. If the anti-inflammatory approach effectively improves the systemic neuro-inflammation, which is one of the events in AD that happens before memory loss, it is possible to improve the progressive Aβ deposition and tau hyper-phosphorylation. However, it is impossible to differentiate the effects of inflammation from other pathologies that occur concomitantly. One of the known modulators for the immune system is diet [171]. Diet high in sugar and saturated fat (SFA) is associated with cognitive impairment especially in learning and memory function both in humans and animal models, independent of obesity or associated metabolic changes [172–175]. High consumption of sugar and SFA results in substantial memory deficits in animals and humans, however, the impact of relative consumption remains unclear due to variability between human consumption and the controlled diet feeding in animal studies [176–178]. The ketogenic diet could be one way to counteract the inflammation and improve mitochondrial biogenesis [179]. The ketogenic diet was tested on subjects with a mild cognitive impairment syndrome and found to be associated with reduced inflammation, enhanced energy metabolism, and potentially improved neurocognitive function [180]. Dietary administration of a medium-chain triglyceride supplement was found to improve cognitive performance [181,182]. The ketogenic diet has been reported to be helpful in reducing the frequency of epilepsy in children [183]. Similar findings have been reported in animal studies where mice fed a ketogenic diet improved neuro-inflammation represented by reduced brain mRNA levels of TNFα, IL-6, and IL-1β and increased the PGC1β mRNA levels indicating ketone bodies related changes [184,185]. Moreover, the Mediterranean diet (MD) has been used as a potential dietary intervention to combat AD-related dementia due to its anti-inflammatory potential [186,187]. Numerous neuroimaging studies reported that the MD has protective effects on the neuronal structure and AD-related morphological changes [188–192].

Notably, several epidemiological studies effectively demonstrate the protective effects of NSAIDs in combating AD via its anti-inflammatory actions along with Aβ lowering properties whereas clinical studies did not support such an improvement [193–197]. More research is needed to conclude if obesity and diabetes associated with chronic inflammation and induced mitochondrial dysfunction can be identified as a biomarker that affects AD pathogenesis and whether the anti-inflammatory approach is a suitable therapeutic option to combat AD.

Author Contributions: Conceptualization, M.S.H.K. and V.H. Writing—original draft preparation, M.S.H.K. Writing—review and editing, V.H. Supervision, V.H. All authors have read and agreed to the published version of the manuscript.

Funding: This research received no external funding.

Conflicts of Interest: The authors declare no conflict of interest.

References

1. Schapira, A.H. Mitochondria in the etiology of Parkinson's disease. *Handb. Clin. Neurol.* **2007**, *83*, 479–491.
2. Alzheimer's Association. 2018 Alzheimer's disease facts and figures. *Alzheimers Dement.* **2018**, *14*, 367–429.
3. Gaugler, J.; James, B.; Johnson, T.; Marin, A.; Weuve, J. 2019 Alzheimer's disease facts and figures. *Alzheimers Dement.* **2019**, *15*, 321–387.
4. Hebert, L.E.; Beckett, L.A.; Scherr, P.A.; Evans, D.A. Annual incidence of Alzheimer disease in the United States projected to the years 2000 through 2050. *Alzheimer Dis. Assoc. Disord.* **2001**, *15*, 169–173. [CrossRef]
5. Mokdad, A.H.; Ballestros, K.; Echko, M.; Glenn, S.; Olsen, H.E.; Mullany, E.; Lee, A.; Khan, A.R.; Ahmadi, A.; Ferrari, A.J. The state of US health, 1990-2016: Burden of diseases, injuries, and risk factors among US states. *JAMA* **2018**, *319*, 1444–1472.
6. Lei, X.; Seldin, M.M.; Little, H.C.; Choy, N.; Klonisch, T.; Wong, G.W. C1q/TNF-related protein 6 (CTRP6) links obesity to adipose tissue inflammation and insulin resistance. *J. Biol. Chem.* **2017**, *292*, 14836–14850. [CrossRef]
7. Naznin, F.; Toshinai, K.; Waise, T.Z.; NamKoong, C.; Moin, A.S.M.; Sakoda, H.; Nakazato, M. Diet-induced obesity causes peripheral and central ghrelin resistance by promoting inflammation. *J. Endocrinol.* **2015**, *226*, 81. [CrossRef]
8. Rigotto, G.; Basso, E. Mitochondrial Dysfunctions: A Thread Sewing Together Alzheimer's Disease, Diabetes, and Obesity. *Oxidative Med. Cell. Longev.* **2019**, *2019*. [CrossRef]
9. Yoo, M.-H.; Gu, X.; Xu, X.-M.; Kim, J.-Y.; Carlson, B.A.; Patterson, A.D.; Cai, H.; Gladyshev, V.N.; Hatfield, D.L. Delineating the role of glutathione peroxidase 4 in protecting cells against lipid hydroperoxide damage and in Alzheimer's disease. *Antioxid. Redox Signal.* **2010**, *12*, 819–827. [CrossRef]

10. Riemer, J.; Kins, S. Axonal transport and mitochondrial dysfunction in Alzheimer's disease. *Neurodegener. Dis.* **2013**, *12*, 111–124. [CrossRef]
11. Reddy, P.H.; Yin, X.; Manczak, M.; Kumar, S.; Pradeepkiran, J.A.; Vijayan, M.; Reddy, A.P. Mutant APP and amyloid beta-induced defective autophagy, mitophagy, mitochondrial structural and functional changes and synaptic damage in hippocampal neurons from Alzheimer's disease. *Hum. Mol. Genet.* **2018**, *27*, 2502–2516. [CrossRef]
12. Jiang, Y.; Mullaney, K.A.; Peterhoff, C.M.; Che, S.; Schmidt, S.D.; Boyer-Boiteau, A.; Ginsberg, S.D.; Cataldo, A.M.; Mathews, P.M.; Nixon, R.A. Alzheimer's-related endosome dysfunction in Down syndrome is Aβ-independent but requires APP and is reversed by BACE-1 inhibition. *Proc. Natl. Acad. Sci. USA* **2010**, *107*, 1630–1635. [CrossRef]
13. Zetterberg, H.; Blennow, K.; Hanse, E. Amyloid β and APP as biomarkers for Alzheimer's disease. *Exp. Gerontol.* **2010**, *45*, 23–29. [CrossRef]
14. Younkin, S.G. The role of Aβ42 in Alzheimer's disease. *J. Physiol.-Paris* **1998**, *92*, 289–292. [CrossRef]
15. Kneynsberg, A.; Combs, B.; Christensen, K.; Morfini, G.; Kanaan, N.M. Axonal degeneration in tauopathies: Disease relevance and underlying mechanisms. *Front. Neurosci.* **2017**, *11*, 572. [CrossRef]
16. Mandelkow, E.-M.; Biernat, J.; Drewes, G.; Gustke, N.; Trinczek, B.; Mandelkow, E. Tau domains, phosphorylation, and interactions with microtubules. *Neurobiol. Aging* **1995**, *16*, 355–362. [CrossRef]
17. Kocahan, S.; Doğan, Z. Mechanisms of Alzheimer's disease pathogenesis and prevention: The brain, neural pathology, N-methyl-D-aspartate receptors, tau protein and other risk factors. *Clin. Psychopharmacol. Neurosci.* **2017**, *15*, 1. [CrossRef]
18. Greenberg, A.S.; Obin, M.S. Obesity and the role of adipose tissue in inflammation and metabolism. *Am. J. Clin. Nutr.* **2006**, *83*, 461S–465S. [CrossRef]
19. D O'Brien, P.; Hinder, L.M.; Callaghan, B.C.; Feldman, E.L. Neurological consequences of obesity. *Lancet Neurol.* **2017**, *16*, 465–477. [CrossRef]
20. Mraz, M.; Haluzik, M. The role of adipose tissue immune cells in obesity and low-grade inflammation. *J. Endocrinol.* **2014**, *222*, R113–R127. [CrossRef]
21. Noeman, S.A.; Hamooda, H.E.; Baalash, A.A. Biochemical study of oxidative stress markers in the liver, kidney and heart of high fat diet induced obesity in rats. *Diabetol. Metab. Syndr.* **2011**, *3*, 17. [CrossRef]
22. Matsuda, M.; Shimomura, I. Increased oxidative stress in obesity: Implications for metabolic syndrome, diabetes, hypertension, dyslipidemia, atherosclerosis, and cancer. *Obes. Res. Clin. Pract.* **2013**, *7*, e330–e341. [CrossRef]
23. Heinonen, S.; Buzkova, J.; Muniandy, M.; Kaksonen, R.; Ollikainen, M.; Ismail, K.; Hakkarainen, A.; Lundbom, J.; Lundbom, N.; Vuolteenaho, K. Impaired mitochondrial biogenesis in adipose tissue in acquired obesity. *Diabetes* **2015**, *64*, 3135–3145. [CrossRef]
24. De Pauw, A.; Tejerina, S.; Raes, M.; Keijer, J.; Arnould, T. Mitochondrial (dys) function in adipocyte (de) differentiation and systemic metabolic alterations. *Am. J. Pathol.* **2009**, *175*, 927–939. [CrossRef]
25. Altshuler-Keylin, S.; Kajimura, S. Mitochondrial homeostasis in adipose tissue remodeling. *Sci. Signal.* **2017**, *10*, eaai9248. [CrossRef]
26. Montgomery, M.K.; Turner, N. Mitochondrial dysfunction and insulin resistance: An update. *Endocr. Connect.* **2015**, *4*, R1–R15. [CrossRef]
27. Vernochet, C.; Damilano, F.; Mourier, A.; Bezy, O.; Mori, M.A.; Smyth, G.; Rosenzweig, A.; Larsson, N.-G.; Kahn, C.R. Adipose tissue mitochondrial dysfunction triggers a lipodystrophic syndrome with insulin resistance, hepatosteatosis, and cardiovascular complications. *FASEB J.* **2014**, *28*, 4408–4419. [CrossRef]
28. Cai, D.; Liu, T. Inflammatory cause of metabolic syndrome via brain stress and NF-κB. *Aging* **2012**, *4*, 98. [CrossRef]
29. Ho, L.; Qin, W.; Pompl, P.N.; Xiang, Z.; Wang, J.; Zhao, Z.; Peng, Y.; Cambareri, G.; Rocher, A.; Mobbs, C.V. Diet-induced insulin resistance promotes amyloidosis in a transgenic mouse model of Alzheimer's disease. *FASEB J.* **2004**, *18*, 902–904. [CrossRef]
30. Schubert, M.; Gautam, D.; Surjo, D.; Ueki, K.; Baudler, S.; Schubert, D.; Kondo, T.; Alber, J.; Galldiks, N.; Küstermann, E. Role for neuronal insulin resistance in neurodegenerative diseases. *Proc. Natl. Acad. Sci. USA* **2004**, *101*, 3100–3105. [CrossRef]
31. Stanley, M.; Macauley, S.L.; Holtzman, D.M. Changes in insulin and insulin signaling in Alzheimer's disease: Cause or consequence? *J. Exp. Med.* **2016**, *213*, 1375–1385. [CrossRef] [PubMed]

32. Arnold, S.E.; Arvanitakis, Z.; Macauley-Rambach, S.L.; Koenig, A.M.; Wang, H.-Y.; Ahima, R.S.; Craft, S.; Gandy, S.; Buettner, C.; Stoeckel, L.E. Brain insulin resistance in type 2 diabetes and Alzheimer disease: Concepts and conundrums. *Nat. Rev. Neurol.* **2018**, *14*, 168. [CrossRef] [PubMed]
33. Ferreira, S.T.; Lourenco, M.V.; Oliveira, M.M.; De Felice, F.G. Soluble amyloid-b oligomers as synaptotoxins leading to cognitive impairment in Alzheimer's disease. *Front. Cell. Neurosci.* **2015**, *9*, 191. [CrossRef]
34. D'Ercole, A.J.; Ye, P.; Calikoglu, A.S.; Gutierrez-Ospina, G. The role of the insulin-like growth factors in the central nervous system. *Mol. Neurobiol.* **1996**, *13*, 227–255. [CrossRef]
35. Neumann, K.F.; Rojo, L.; Navarrete, L.P.; Farías, G.; Reyes, P.; Maccioni, R.B. Insulin resistance and Alzheimer's disease: Molecular links & clinical implications. *Curr. Alzheimer Res.* **2008**, *5*, 438–447.
36. Liu, Y.; Liu, F.; Grundke-Iqbal, I.; Iqbal, K.; Gong, C.X. Deficient brain insulin signalling pathway in Alzheimer's disease and diabetes. *J. Pathol.* **2011**, *225*, 54–62. [CrossRef]
37. Kim, B.; Feldman, E.L. Insulin resistance in the nervous system. *Trends Endocrinol. Metab.* **2012**, *23*, 133–141. [CrossRef]
38. Jackson, K.; Barisone, G.A.; Diaz, E.; Jin, L.w.; DeCarli, C.; Despa, F. Amylin deposition in the brain: A second amyloid in Alzheimer disease? *Ann. Neurol.* **2013**, *74*, 517–526. [CrossRef]
39. Miklossy, J.; Qing, H.; Radenovic, A.; Kis, A.; Vileno, B.; Làszló, F.; Miller, L.; Martins, R.N.; Waeber, G.; Mooser, V. Beta amyloid and hyperphosphorylated tau deposits in the pancreas in type 2 diabetes. *Neurobiol. Aging* **2010**, *31*, 1503–1515. [CrossRef]
40. Whitmer, R.; Gustafson, D.; Barrett-Connor, E.; Haan, M.; Gunderson, E.; Yaffe, K. Central obesity and increased risk of dementia more than three decades later. *Neurology* **2008**, *71*, 1057–1064. [CrossRef]
41. Jagust, W.; Harvey, D.; Mungas, D.; Haan, M. Central obesity and the aging brain. *Arch. Neurol.* **2005**, *62*, 1545–1548. [CrossRef] [PubMed]
42. Fitzpatrick, A.L.; Kuller, L.H.; Lopez, O.L.; Diehr, P.; O'Meara, E.S.; Longstreth, W.; Luchsinger, J.A. Midlife and late-life obesity and the risk of dementia: Cardiovascular health study. *Arch. Neurol.* **2009**, *66*, 336–342. [CrossRef] [PubMed]
43. Hassing, L.B.; Dahl, A.K.; Thorvaldsson, V.; Berg, S.; Gatz, M.; Pedersen, N.L.; Johansson, B. Overweight in midlife and risk of dementia: A 40-year follow-up study. *Int. J. Obes.* **2009**, *33*, 893–898. [CrossRef]
44. Whitmer, R.A.; Gunderson, E.P.; Barrett-Connor, E.; Quesenberry, C.P.; Yaffe, K. Obesity in middle age and future risk of dementia: A 27 year longitudinal population based study. *BMJ* **2005**, *330*, 1360. [CrossRef]
45. Qizilbash, N.; Gregson, J.; Johnson, M.E.; Pearce, N.; Douglas, I.; Wing, K.; Evans, S.J.; Pocock, S.J. BMI and risk of dementia in two million people over two decades: A retrospective cohort study. *Lancet Diabetes Endocrinol.* **2015**, *3*, 431–436. [CrossRef]
46. Verdile, G.; Fuller, S.J.; Martins, R.N. The role of type 2 diabetes in neurodegeneration. *Neurobiol. Dis.* **2015**, *84*, 22–38. [CrossRef]
47. Bhamra, M.S.; Ashton, N.J. Finding a pathological diagnosis for A lzheimer's disease: Are inflammatory molecules the answer? *Electrophoresis* **2012**, *33*, 3598–3607. [CrossRef]
48. Swardfager, W.; Lanctôt, K.; Rothenburg, L.; Wong, A.; Cappell, J.; Herrmann, N. A meta-analysis of cytokines in Alzheimer's disease. *Biol. Psychiatry* **2010**, *68*, 930–941. [CrossRef]
49. Nakamura, M.; Watanabe, N. Ubiquitin-like protein MNSFβ/endophilin II complex regulates Dectin-1-mediated phagocytosis and inflammatory responses in macrophages. *Biochem. Biophys. Res. Commun.* **2010**, *401*, 257–261. [CrossRef]
50. Banks, W.A.; Kastin, A.J.; Broadwell, R.D. Passage of cytokines across the blood-brain barrier. *Neuroimmunomodulation* **1995**, *2*, 241–248. [CrossRef]
51. Rosenberg, P.B. Clinical aspects of inflammation in Alzheimer's disease. *Int. Rev. Psychiatry* **2005**, *17*, 503–514. [CrossRef]
52. Chiroma, S.M.; Baharuldin, M.T.H.; Taib, C.; Amom, Z.; Jagadeesan, S.; Moklas, M. Inflammation in Alzheimer's disease: A friend or foe? *Biomed. Res. Ther.* **2018**, *5*, 2552–2564. [CrossRef]
53. Ma, Q.-L.; Yang, F.; Rosario, E.R.; Ubeda, O.J.; Beech, W.; Gant, D.J.; Chen, P.P.; Hudspeth, B.; Chen, C.; Zhao, Y. β-amyloid oligomers induce phosphorylation of tau and inactivation of insulin receptor substrate via c-Jun N-terminal kinase signaling: Suppression by omega-3 fatty acids and curcumin. *J. Neurosci.* **2009**, *29*, 9078–9089. [CrossRef]
54. Pickup, J.; Crook, M. Is type II diabetes mellitus a disease of the innate immune system? *Diabetologia* **1998**, *41*, 1241–1248. [CrossRef]

55. Weisberg, S.P.; McCann, D.; Desai, M.; Rosenbaum, M.; Leibel, R.L.; Ferrante, A.W. Obesity is associated with macrophage accumulation in adipose tissue. *J. Clin. Investig.* **2003**, *112*, 1796–1808. [CrossRef]
56. Tucsek, Z.; Toth, P.; Sosnowska, D.; Gautam, T.; Mitschelen, M.; Koller, A.; Szalai, G.; Sonntag, W.E.; Ungvari, Z.; Csiszar, A. Obesity in aging exacerbates blood–brain barrier disruption, neuroinflammation, and oxidative stress in the mouse hippocampus: Effects on expression of genes involved in beta-amyloid generation and Alzheimer's disease. *J. Gerontol. Ser. A Biomed. Sci. Med Sci.* **2014**, *69*, 1212–1226. [CrossRef]
57. De Souza, C.T.; Araujo, E.P.; Bordin, S.; Ashimine, R.; Zollner, R.L.; Boschero, A.C.; Saad, M.J.; Velloso, L.C.A. Consumption of a fat-rich diet activates a proinflammatory response and induces insulin resistance in the hypothalamus. *Endocrinology* **2005**, *146*, 4192–4199. [CrossRef]
58. Thaler, J.P.; Yi, C.-X.; Schur, E.A.; Guyenet, S.J.; Hwang, B.H.; Dietrich, M.O.; Zhao, X.; Sarruf, D.A.; Izgur, V.; Maravilla, K.R. Obesity is associated with hypothalamic injury in rodents and humans. *J. Clin. Investig.* **2012**, *122*, 153–162. [CrossRef]
59. Milanski, M.; Degasperi, G.; Coope, A.; Morari, J.; Denis, R.; Cintra, D.E.; Tsukumo, D.M.; Anhe, G.; Amaral, M.E.; Takahashi, H.K. Saturated fatty acids produce an inflammatory response predominantly through the activation of TLR4 signaling in hypothalamus: Implications for the pathogenesis of obesity. *J. Neurosci.* **2009**, *29*, 359–370. [CrossRef]
60. Blatteis, C.M. The onset of fever: New insights into its mechanism. *Prog. Brain Res.* **2007**, *162*, 3–14.
61. Miller, A.A.; Spencer, S.J. Obesity and neuroinflammation: A pathway to cognitive impairment. *Brain Behav. Immun.* **2014**, *42*, 10–21. [CrossRef]
62. Koessler, S.; Engler, H.; Riether, C.; Kissler, J. No retrieval-induced forgetting under stress. *Psychol. Sci.* **2009**, *20*, 1356–1363. [CrossRef]
63. Raber, J. Detrimental effects of chronic hypothalamic—Pituitary—Adrenal axis activation. *Mol. Neurobiol.* **1998**, *18*, 1–22. [CrossRef]
64. MacQueen, G.; Frodl, T. The hippocampus in major depression: Evidence for the convergence of the bench and bedside in psychiatric research? *Mol. Psychiatry* **2011**, *16*, 252–264. [CrossRef]
65. Alves, L.; Correia, A.S.A.; Miguel, R.; Alegria, P.; Bugalho, P. Alzheimer's disease: A clinical practice-oriented review. *Front. Neurol.* **2012**, *3*, 63. [CrossRef]
66. Wärnberg, J.; Gomez-Martinez, S.; Romeo, J.; Díaz, L.E.; Marcos, A. Nutrition, inflammation, and cognitive function. *Ann. N. Y. Acad. Sci.* **2009**, *1153*, 164–175.
67. Eikelenboom, P.; Veerhuis, R.; Van Exel, E.; Hoozemans, J.J.M.; Rozemuller, A.J.M.; van Gool, W.A. The early involvement of the innate immunity in the pathogenesis of lateonset Alzheimer's disease: Neuropathological, epidemiological and genetic evidence. *Curr. Alzheimer Res.* **2011**, *8*, 142–150. [CrossRef]
68. Morales, I.; Guzmán-Martínez, L.; Cerda-Troncoso, C.; Farías, G.A.; Maccioni, R.B. Neuroinflammation in the pathogenesis of Alzheimer's disease. A rational framework for the search of novel therapeutic approaches. *Front. Cell. Neurosci.* **2014**, *8*, 112. [CrossRef]
69. Tuppo, E.E.; Arias, H.R. The role of inflammation in Alzheimer's disease. *Int. J. Biochem. Cell Biol.* **2005**, *37*, 289–305. [CrossRef]
70. Akiyama, H.; Barger, S.; Barnum, S.; Bradt, B.; Bauer, J.; Cole, G.M.; Cooper, N.R.; Eikelenboom, P.; Emmerling, M.; Fiebich, B.L. Inflammation and Alzheimer's disease. *Neurobiol. Aging* **2000**, *21*, 383–421. [CrossRef]
71. Mrak, R.E.; Sheng, J.G.; Griffin, W.S.T. Glial cytokines in Alzheimer's disease: Review and pathogenic implications. *Hum. Pathol.* **1995**, *26*, 816–823. [CrossRef]
72. Halliday, G.; Robinson, S.R.; Shepherd, C.; Kril, J. Alzheimer's disease and inflammation: A review of cellular and therapeutic mechanisms. *Clin. Exp. Pharmacol. Physiol.* **2000**, *27*, 1–8. [CrossRef] [PubMed]
73. Minter, M.R.; Taylor, J.M.; Crack, P.J. The contribution of neuroinflammation to amyloid toxicity in Alzheimer's disease. *J. Neurochem.* **2016**, *136*, 457–474. [CrossRef] [PubMed]
74. Déniz-Naranjo, M.; Muñoz-Fernandez, C.; Alemany-Rodríguez, M.; Pérez-Vieitez, M.; Aladro-Benito, Y.; Irurita-Latasa, J.; Sánchez-García, F. Cytokine IL-1 beta but not IL-1 alpha promoter polymorphism is associated with Alzheimer disease in a population from the Canary Islands, Spain. *Eur. J. Neurol.* **2008**, *15*, 1080–1084. [CrossRef]
75. Li, C.; Zhao, R.; Gao, K.; Wei, Z.; Yaoyao Yin, M.; Ting Lau, L.; Chui, D.; Cheung Hoi Yu, A. Astrocytes: Implications for neuroinflammatory pathogenesis of Alzheimer's disease. *Curr. Alzheimer Res.* **2011**, *8*, 67–80. [CrossRef]

76. Wyss-Coray, T. Inflammation in Alzheimer disease: Driving force, bystander or beneficial response? *Nat. Med.* **2006**, *12*, 1005–1015.
77. Griffin, W.S.T.; Liu, L.; Li, Y.; Mrak, R.E.; Barger, S.W. Interleukin-1 mediates Alzheimer and Lewy body pathologies. *J. Neuroinflammation* **2006**, *3*, 5. [CrossRef]
78. Liu, L.; Chan, C. The role of inflammasome in Alzheimer's disease. *Ageing Res. Rev.* **2014**, *15*, 6–15. [CrossRef]
79. Milionis, H.J.; Florentin, M.; Giannopoulos, S. Metabolic syndrome and Alzheimer's disease: A link to a vascular hypothesis? *CNS Spectr.* **2008**, *13*, 606–613. [CrossRef]
80. Chellan, P.; Nagaraj, R.H. Protein crosslinking by the Maillard reaction: Dicarbonyl-derived imidazolium crosslinks in aging and diabetes. *Arch. Biochem. Biophys.* **1999**, *368*, 98–104. [CrossRef]
81. Thorpe, S.R.; Baynes, J.W. Role of the Maillard reaction in diabetes mellitus and diseases of aging. *Drugs Aging* **1996**, *9*, 69–77. [CrossRef]
82. Hori, O.; Yan, S.D.; Ogawa, S.; Kuwabara, K.; Matsumoto, M.; Stern, D.; Schmidt, A.M. The receptor for advanced glycation end-products has a central role in mediating the effects of advanced glycation end-products on the development of vascular disease in diabetes mellitus. *Nephrol. Dial. Transplant.* **1996**, *11*, 13–16. [CrossRef]
83. Ulrich, P.; Cerami, A. Protein glycation, diabetes, and aging. *Recent Prog. Horm. Res.* **2001**, *56*, 1–22. [CrossRef]
84. Srikanth, V.; Maczurek, A.; Phan, T.; Steele, M.; Westcott, B.; Juskiw, D.; Münch, G. Advanced glycation endproducts and their receptor RAGE in Alzheimer's disease. *Neurobiol. Aging* **2011**, *32*, 763–777. [CrossRef]
85. Lue, L.-F.; Walker, D.G.; Brachova, L.; Beach, T.G.; Rogers, J.; Schmidt, A.M.; Stern, D.M.; Du Yan, S. Involvement of microglial receptor for advanced glycation endproducts (RAGE) in Alzheimer's disease: Identification of a cellular activation mechanism. *Exp. Neurol.* **2001**, *171*, 29–45. [CrossRef]
86. Sasaki, N.; Toki, S.; Chowei, H.; Saito, T.; Nakano, N.; Hayashi, Y.; Takeuchi, M.; Makita, Z. Immunohistochemical distribution of the receptor for advanced glycation end products in neurons and astrocytes in Alzheimer's disease. *Brain Res.* **2001**, *888*, 256–262. [CrossRef]
87. Lue, L.-F.; Andrade, C.; Sabbagh, M.; Walker, D. Is there inflammatory synergy in type II diabetes mellitus and Alzheimer's disease? *Int. J. Alzheimer's Dis.* **2012**, *2012*. [CrossRef]
88. Fang, F.; Lue, L.-F.; Yan, S.; Xu, H.; Luddy, J.S.; Chen, D.; Walker, D.G.; Stern, D.M.; Yan, S.; Schmidt, A.M. RAGE-dependent signaling in microglia contributes to neuroinflammation, Aβ accumulation, and impaired learning/memory in a mouse model of Alzheimer's disease. *FASEB J.* **2010**, *24*, 1043–1055. [CrossRef]
89. Beeri, M.S.; Moshier, E.; Schmeidler, J.; Godbold, J.; Uribarri, J.; Reddy, S.; Sano, M.; Grossman, H.T.; Cai, W.; Vlassara, H. Serum concentration of an inflammatory glycotoxin, methylglyoxal, is associated with increased cognitive decline in elderly individuals. *Mech. Ageing Dev.* **2011**, *132*, 583–587. [CrossRef]
90. Pugazhenthi, S.; Qin, L.; Reddy, P.H. Common neurodegenerative pathways in obesity, diabetes, and Alzheimer's disease. *Biochim. Et Biophys. Acta BBA-Mol. Basis Dis.* **2017**, *1863*, 1037–1045. [CrossRef]
91. Kasten, T.; Zhang, L.; Goate, A.; Morris, J.C.; Holtzman, D.; Bateman, R.J. Age and Amyloid Effects on Human CNS Amyloid-Beta Kinetics. *Ann. Neurol.* **2015**, *78*, 439–453. [CrossRef]
92. Bhatti, J.; Kumar, S.; Vijayan, M.; Bhatti, G.; Reddy, P.H. Therapeutic strategies for mitochondrial dysfunction and oxidative stress in age-related metabolic disorders. In *Progress in Molecular Biology and Translational Science*; Elsevier: Amsterdam, The Netherlands, 2017; Volume 146, pp. 13–46.
93. Yoo, S.-M.; Park, J.; Kim, S.-H.; Jung, Y.-K. Emerging perspectives on mitochondrial dysfunction and inflammation in Alzheimer's disease. *BMB Rep.* **2020**, *53*, 35. [CrossRef]
94. Bournat, J.C.; Brown, C.W. Mitochondrial dysfunction in obesity. *Curr. Opin. Endocrinol. Diabetes Obes.* **2010**, *17*, 446. [CrossRef]
95. Ritov, V.B.; Menshikova, E.V.; He, J.; Ferrell, R.E.; Goodpaster, B.H.; Kelley, D.E. Deficiency of subsarcolemmal mitochondria in obesity and type 2 diabetes. *Diabetes* **2005**, *54*, 8–14. [CrossRef]
96. Mogensen, M.; Sahlin, K.; Fernström, M.; Glintborg, D.; Vind, B.F.; Beck-Nielsen, H.; Højlund, K. Mitochondrial respiration is decreased in skeletal muscle of patients with type 2 diabetes. *Diabetes* **2007**, *56*, 1592–1599. [CrossRef]
97. Richardson, D.K.; Kashyap, S.; Bajaj, M.; Cusi, K.; Mandarino, S.J.; Finlayson, J.; DeFronzo, R.A.; Jenkinson, C.P.; Mandarino, L.J. Lipid infusion decreases the expression of nuclear encoded mitochondrial genes and increases the expression of extracellular matrix genes in human skeletal muscle. *J. Biol. Chem.* **2005**, *280*, 10290–10297. [CrossRef]

98. Bogacka, I.; Xie, H.; Bray, G.A.; Smith, S.R. Pioglitazone induces mitochondrial biogenesis in human subcutaneous adipose tissue in vivo. *Diabetes* **2005**, *54*, 1392–1399. [CrossRef]
99. Gao, C.-L.; Zhu, C.; Zhao, Y.-P.; Chen, X.-H.; Ji, C.-B.; Zhang, C.-M.; Zhu, J.-G.; Xia, Z.-K.; Tong, M.-L.; Guo, X.-R. Mitochondrial dysfunction is induced by high levels of glucose and free fatty acids in 3T3-L1 adipocytes. *Mol. Cell. Endocrinol.* **2010**, *320*, 25–33. [CrossRef]
100. Koh, E.H.; Park, J.-Y.; Park, H.-S.; Jeon, M.J.; Ryu, J.W.; Kim, M.; Kim, S.Y.; Kim, M.-S.; Kim, S.-W.; Park, I.S. Essential role of mitochondrial function in adiponectin synthesis in adipocytes. *Diabetes* **2007**, *56*, 2973–2981. [CrossRef]
101. Sparks, L.M.; Xie, H.; Koza, R.A.; Mynatt, R.; Hulver, M.W.; Bray, G.A.; Smith, S.R. A high-fat diet coordinately downregulates genes required for mitochondrial oxidative phosphorylation in skeletal muscle. *Diabetes* **2005**, *54*, 1926–1933. [CrossRef]
102. Lin, Y.; Berg, A.H.; Iyengar, P.; Lam, T.K.; Giacca, A.; Combs, T.P.; Rajala, M.W.; Du, X.; Rollman, B.; Li, W. The hyperglycemia-induced inflammatory response in adipocytes the role of reactive oxygen species. *J. Biol. Chem.* **2005**, *280*, 4617–4626. [CrossRef]
103. Yin, F.; Sancheti, H.; Patil, I.; Cadenas, E. Energy metabolism and inflammation in brain aging and Alzheimer's disease. *Free Radic. Biol. Med.* **2016**, *100*, 108–122. [CrossRef]
104. Yin, F.; Boveris, A.; Cadenas, E. Mitochondrial energy metabolism and redox signaling in brain aging and neurodegeneration. *Antioxid. Redox Signal.* **2014**, *20*, 353–371. [CrossRef]
105. van Horssen, J.; van Schaik, P.; Witte, M. Inflammation and mitochondrial dysfunction: A vicious circle in neurodegenerative disorders? *Neurosci. Lett.* **2019**, *710*, 132931. [CrossRef]
106. Abcouwer, S.F.; Shanmugam, S.; Gomez, P.F.; Shushanov, S.; Barber, A.J.; Lanoue, K.F.; Quinn, P.G.; Kester, M.; Gardner, T.W. Effect of IL-1β on survival and energy metabolism of R28 and RGC-5 retinal neurons. *Investig. Ophthalmol. Vis. Sci.* **2008**, *49*, 5581–5592. [CrossRef]
107. Tredget, E.E.; Yu, Y.M.; Zhong, S.; Burini, R.; Okusawa, S.; Gelfand, J.A.; Dinarello, C.A.; Young, V.; Burke, J. Role of interleukin 1 and tumor necrosis factor on energy metabolism in rabbits. *Am. J. Physiol.-Endocrinol. Metab.* **1988**, *255*, E760–E768. [CrossRef]
108. Zell, R.; Geck, P.; Werdan, K.; Boekstegers, P. TNF-α and IL-1α inhibit both pyruvate dehydrogenase activity and mitochondrial function in cardiomyocytes: Evidence for primary impairment of mitochondrial function. *Mol. Cell. Biochem.* **1997**, *177*, 61–67. [CrossRef]
109. Taylor, D.; Faragher, E.; Evanson, J. Inflammatory cytokines stimulate glucose uptake and glycolysis but reduce glucose oxidation in human dermal fibroblasts in vitro. *Circ. Shock* **1992**, *37*, 105–110.
110. Vary, T.C.; Siegel, J.H.; Nakatani, T.; Sato, T.; Aoyama, H. Effect of sepsis on activity of pyruvate dehydrogenase complex in skeletal muscle and liver. *Am. J. Physiol.-Endocrinol. Metab.* **1986**, *250*, E634–E640. [CrossRef]
111. Vary, T.; Martin, L. Potentiation of decreased pyruvate dehydrogenase activity by inflammatory stimuli in sepsis. *Circ. Shock* **1993**, *39*, 299–305.
112. Dutta, R.; McDonough, J.; Yin, X.; Peterson, J.; Chang, A.; Torres, T.; Gudz, T.; Macklin, W.B.; Lewis, D.A.; Fox, R.J. Mitochondrial dysfunction as a cause of axonal degeneration in multiple sclerosis patients. *Ann. Neurol.* **2006**, *59*, 478–489. [CrossRef]
113. Doll, D.N.; Rellick, S.L.; Barr, T.L.; Ren, X.; Simpkins, J.W. Rapid mitochondrial dysfunction mediates TNF-alpha-induced neurotoxicity. *J. Neurochem.* **2015**, *132*, 443–451. [CrossRef]
114. Tang, K.; Wagner, P.D.; Breen, E.C. TNF-α-mediated reduction in PGC-1α may impair skeletal muscle function after cigarette smoke exposure. *J. Cell. Physiol.* **2010**, *222*, 320–327. [CrossRef]
115. Palomer, X.; Álvarez-Guardia, D.; Rodríguez-Calvo, R.; Coll, T.; Laguna, J.C.; Davidson, M.M.; Chan, T.O.; Feldman, A.M.; Vázquez-Carrera, M. TNF-α reduces PGC-1 α expression through NF-κB and p38 MAPK leading to increased glucose oxidation in a human cardiac cell model. *Cardiovasc. Res.* **2009**, *81*, 703–712. [CrossRef]
116. Zhang, Y.; Chen, C.; Jiang, Y.; Wang, S.; Wu, X.; Wang, K. PPARγ coactivator-1α (PGC-1α) protects neuroblastoma cells against amyloid-beta (Aβ) induced cell death and neuroinflammation via NF-κB pathway. *BMC Neurosci.* **2017**, *18*, 1–8. [CrossRef]
117. Gibson, G.; Sheu, K.-F.; Blass, J. Abnormalities of mitochondrial enzymes in Alzheimer disease. *J. Neural Transm.* **1998**, *105*, 855–870. [CrossRef]

118. Butterfield, D.A.; Poon, H.F.; Clair, D.S.; Keller, J.N.; Pierce, W.M.; Klein, J.B.; Markesbery, W.R. Redox proteomics identification of oxidatively modified hippocampal proteins in mild cognitive impairment: Insights into the development of Alzheimer's disease. *Neurobiol. Dis.* **2006**, *22*, 223–232. [CrossRef]
119. Caspersen, C.; Wang, N.; Yao, J.; Sosunov, A.; Chen, X.; Lustbader, J.W.; Xu, H.W.; Stern, D.; McKhann, G.; Yan, S.D. Mitochondrial Aβ: A potential focal point for neuronal metabolic dysfunction in Alzheimer's disease. *FASEB J.* **2005**, *19*, 2040–2041. [CrossRef]
120. Lustbader, J.W.; Cirilli, M.; Lin, C.; Xu, H.W.; Takuma, K.; Wang, N.; Caspersen, C.; Chen, X.; Pollak, S.; Chaney, M. ABAD directly links Aß to mitochondrial toxicity in Alzheimer's disease. *Science* **2004**, *304*, 448–452. [CrossRef]
121. Reddy, P.H.; Beal, M.F. Amyloid beta, mitochondrial dysfunction and synaptic damage: Implications for cognitive decline in aging and Alzheimer's disease. *Trends Mol. Med.* **2008**, *14*, 45–53. [CrossRef]
122. Reddy, P.H.; Beal, M.F. Are mitochondria critical in the pathogenesis of Alzheimer's disease? *Brain Res. Rev.* **2005**, *49*, 618–632. [CrossRef] [PubMed]
123. Thies, E.; Mandelkow, E.-M. Missorting of tau in neurons causes degeneration of synapses that can be rescued by the kinase MARK2/Par-1. *J. Neurosci.* **2007**, *27*, 2896–2907. [CrossRef] [PubMed]
124. Chan, D.C. Mitochondria: Dynamic organelles in disease, aging, and development. *Cell* **2006**, *125*, 1241–1252. [CrossRef] [PubMed]
125. Suen, D.-F.; Norris, K.L.; Youle, R.J. Mitochondrial dynamics and apoptosis. *Genes Dev.* **2008**, *22*, 1577–1590. [CrossRef] [PubMed]
126. J Bonda, D.; A Smith, M.; Perry, G.; Lee, H.-G.; Wang, X.; Zhu, X. The mitochondrial dynamics of Alzheimer's disease and Parkinson's disease offer important opportunities for therapeutic intervention. *Curr. Pharm. Des.* **2011**, *17*, 3374–3380. [CrossRef]
127. Valero, T. Editorial (thematic issue: Mitochondrial biogenesis: Pharmacological approaches). *Curr. Pharm. Des.* **2014**, *20*, 5507–5509. [CrossRef]
128. Wenz, T. Regulation of mitochondrial biogenesis and PGC-1α under cellular stress. *Mitochondrion* **2013**, *13*, 134–142. [CrossRef]
129. Wang, X.; Su, B.; Siedlak, S.L.; Moreira, P.I.; Fujioka, H.; Wang, Y.; Casadesus, G.; Zhu, X. Amyloid-β overproduction causes abnormal mitochondrial dynamics via differential modulation of mitochondrial fission/fusion proteins. *Proc. Natl. Acad. Sci. USA* **2008**, *105*, 19318–19323. [CrossRef]
130. Wang, X.; Su, B.; Lee, H.-G.; Li, X.; Perry, G.; Smith, M.A.; Zhu, X. Impaired balance of mitochondrial fission and fusion in Alzheimer's disease. *J. Neurosci.* **2009**, *29*, 9090–9103. [CrossRef]
131. Motori, E.; Puyal, J.; Toni, N.; Ghanem, A.; Angeloni, C.; Malaguti, M.; Cantelli-Forti, G.; Berninger, B.; Conzelmann, K.-K.; Götz, M. Inflammation-induced alteration of astrocyte mitochondrial dynamics requires autophagy for mitochondrial network maintenance. *Cell Metab.* **2013**, *18*, 844–859. [CrossRef]
132. Chen, X.-H.; Zhao, Y.-P.; Xue, M.; Ji, C.-B.; Gao, C.-L.; Zhu, J.-G.; Qin, D.-N.; Kou, C.-Z.; Qin, X.-H.; Tong, M.-L. TNF-α induces mitochondrial dysfunction in 3T3-L1 adipocytes. *Mol. Cell. Endocrinol.* **2010**, *328*, 63–69. [CrossRef]
133. Anusree, S.; Nisha, V.; Priyanka, A.; Raghu, K. Insulin resistance by TNF-α is associated with mitochondrial dysfunction in 3T3-L1 adipocytes and is ameliorated by punicic acid, a PPARγ agonist. *Mol. Cell. Endocrinol.* **2015**, *413*, 120–128. [CrossRef] [PubMed]
134. Hahn, W.S.; Kuzmicic, J.; Burrill, J.S.; Donoghue, M.A.; Foncea, R.; Jensen, M.D.; Lavandero, S.; Arriaga, E.A.; Bernlohr, D.A. Proinflammatory cytokines differentially regulate adipocyte mitochondrial metabolism, oxidative stress, and dynamics. *Am. J. Physiol.-Endocrinol. Metab.* **2014**, *306*, E1033–E1045. [CrossRef] [PubMed]
135. White, J.P.; Puppa, M.J.; Sato, S.; Gao, S.; Price, R.L.; Baynes, J.W.; Kostek, M.C.; Matesic, L.E.; Carson, J.A. IL-6 regulation on skeletal muscle mitochondrial remodeling during cancer cachexia in the Apc Min/+ mouse. *Skelet. Muscle* **2012**, *2*, 14. [CrossRef] [PubMed]
136. Miotto, P.M.; LeBlanc, P.J.; Holloway, G.P. High-fat diet causes mitochondrial dysfunction as a result of impaired ADP sensitivity. *Diabetes* **2018**, *67*, 2199–2205. [CrossRef] [PubMed]
137. Zhou, R.; Yazdi, A.S.; Menu, P.; Tschopp, J. A role for mitochondria in NLRP3 inflammasome activation. *Nature* **2011**, *469*, 221–225. [CrossRef]
138. Brookes, P.S.; Yoon, Y.; Robotham, J.L.; Anders, M.; Sheu, S.-S. Calcium, ATP, and ROS: A mitochondrial love-hate triangle. *Am. J. Physiol.-Cell Physiol.* **2004**, *287*, C817–C833. [CrossRef]

139. Reddy, P.H. Amyloid precursor protein-mediated free radicals and oxidative damage: Implications for the development and progression of Alzheimer's disease. *J. Neurochem.* **2006**, *96*, 1–13. [CrossRef]
140. Wang, X.; Wang, W.; Li, L.; Perry, G.; Lee, H.-G.; Zhu, X. Oxidative stress and mitochondrial dysfunction in Alzheimer's disease. *Biochim. Et Biophys. Acta BBA-Mol. Basis Dis.* **2014**, *1842*, 1240–1247. [CrossRef]
141. Fernández-Sánchez, A.; Madrigal-Santillán, E.; Bautista, M.; Esquivel-Soto, J.; Morales-González, Á.; Esquivel-Chirino, C.; Durante-Montiel, I.; Sánchez-Rivera, G.; Valadez-Vega, C.; Morales-González, J.A. Inflammation, oxidative stress, and obesity. *Int. J. Mol. Sci.* **2011**, *12*, 3117–3132. [CrossRef]
142. Amin, K.A.; Nagy, M.A. Effect of Carnitine and herbal mixture extract on obesity induced by high fat diet in rats. *Diabetol. Metab. Syndr.* **2009**, *1*, 17. [CrossRef]
143. Brown, G.C. Nitric oxide and neuronal death. *Nitric Oxide* **2010**, *23*, 153–165. [CrossRef] [PubMed]
144. Zhang, L.; Bruce-Keller, A.J.; Dasuri, K.; Nguyen, A.; Liu, Y.; Keller, J.N. Diet-induced metabolic disturbances as modulators of brain homeostasis. *Biochim. Et Biophys. Acta (BBA)-Mol. Basis Dis.* **2009**, *1792*, 417–422. [CrossRef] [PubMed]
145. Nunomura, A.; Perry, G.; Aliev, G.; Hirai, K.; Takeda, A.; Balraj, E.K.; Jones, P.K.; Ghanbari, H.; Wataya, T.; Shimohama, S. Oxidative damage is the earliest event in Alzheimer disease. *J. Neuropathol. Exp. Neurol.* **2001**, *60*, 759–767. [CrossRef] [PubMed]
146. Morrison, C.D.; Pistell, P.J.; Ingram, D.K.; Johnson, W.D.; Liu, Y.; Fernandez-Kim, S.O.; White, C.L.; Purpera, M.N.; Uranga, R.M.; Bruce-Keller, A.J. High fat diet increases hippocampal oxidative stress and cognitive impairment in aged mice: Implications for decreased Nrf2 signaling. *J. Neurochem.* **2010**, *114*, 1581–1589. [CrossRef]
147. Reddy, P.H.; McWeeney, S.; Park, B.S.; Manczak, M.; Gutala, R.V.; Partovi, D.; Jung, Y.; Yau, V.; Searles, R.; Mori, M. Gene expression profiles of transcripts in amyloid precursor protein transgenic mice: Up-regulation of mitochondrial metabolism and apoptotic genes is an early cellular change in Alzheimer's disease. *Hum. Mol. Genet.* **2004**, *13*, 1225–1240. [CrossRef]
148. Manczak, M.; Park, B.S.; Jung, Y.; Reddy, P.H. Differential expression of oxidative phosphorylation genes in patients with Alzheimer's disease. *Neuromolecular Med.* **2004**, *5*, 147–162. [CrossRef]
149. Pistell, P.J.; Morrison, C.D.; Gupta, S.; Knight, A.G.; Keller, J.N.; Ingram, D.K.; Bruce-Keller, A.J. Cognitive impairment following high fat diet consumption is associated with brain inflammation. *J. Neuroimmunol.* **2010**, *219*, 25–32. [CrossRef]
150. Butterfield, D.A.; Drake, J.; Pocernich, C.; Castegna, A. Evidence of oxidative damage in Alzheimer's disease brain: Central role for amyloid β-peptide. *Trends Mol. Med.* **2001**, *7*, 548–554. [CrossRef]
151. Ferreiro, E.; Oliveira, C.R.; Pereira, C.M. The release of calcium from the endoplasmic reticulum induced by amyloid-beta and prion peptides activates the mitochondrial apoptotic pathway. *Neurobiol. Dis.* **2008**, *30*, 331–342. [CrossRef]
152. Shelat, P.B.; Chalimoniuk, M.; Wang, J.H.; Strosznajder, J.B.; Lee, J.C.; Sun, A.Y.; Simonyi, A.; Sun, G.Y. Amyloid beta peptide and NMDA induce ROS from NADPH oxidase and AA release from cytosolic phospholipase A2 in cortical neurons. *J. Neurochem.* **2008**, *106*, 45–55. [CrossRef] [PubMed]
153. Liu, Z.; Zhou, T.; Ziegler, A.C.; Dimitrion, P.; Zuo, L. Oxidative stress in neurodegenerative diseases: From molecular mechanisms to clinical applications. *Oxidative Med. Cell. Longev.* **2017**, *2017*. [CrossRef] [PubMed]
154. Curtain, C.C.; Ali, F.; Volitakis, I.; Cherny, R.A.; Norton, R.S.; Beyreuther, K.; Barrow, C.J.; Masters, C.L.; Bush, A.I.; Barnham, K.J. Alzheimer's disease amyloid-β binds copper and zinc to generate an allosterically ordered membrane-penetrating structure containing superoxide dismutase-like subunits. *J. Biol. Chem.* **2001**, *276*, 20466–20473. [CrossRef] [PubMed]
155. Rojas-Gutierrez, E.; Muñoz-Arenas, G.; Treviño, S.; Espinosa, B.; Chavez, R.; Rojas, K.; Flores, G.; Díaz, A.; Guevara, J. Alzheimer's disease and metabolic syndrome: A link from oxidative stress and inflammation to neurodegeneration. *Synapse* **2017**, *71*, e21990. [CrossRef] [PubMed]
156. Weaver, J.; Huang, M.-H.; Albert, M.; Harris, T.; Rowe, J.; Seeman, T.E. Interleukin-6 and risk of cognitive decline: MacArthur studies of successful aging. *Neurology* **2002**, *59*, 371–378. [CrossRef] [PubMed]
157. Dik, M.; Jonker, C.; Hack, C.; Smit, J.; Comijs, H.; Eikelenboom, P. Serum inflammatory proteins and cognitive decline in older persons. *Neurology* **2005**, *64*, 1371–1377. [CrossRef]
158. Teunissen, C.; Van Boxtel, M.; Bosma, H.; Bosmans, E.; Delanghe, J.; De Bruijn, C.; Wauters, A.; Maes, M.; Jolles, J.; Steinbusch, H. Inflammation markers in relation to cognition in a healthy aging population. *J. Neuroimmunol.* **2003**, *134*, 142–150. [CrossRef]

159. Singh-Manoux, A.; Dugravot, A.; Brunner, E.; Kumari, M.; Shipley, M.; Elbaz, A.; Kivimaki, M. Interleukin-6 and C-reactive protein as predictors of cognitive decline in late midlife. *Neurology* **2014**, *83*, 486–493. [CrossRef]
160. Ozawa, M.; Shipley, M.; Kivimaki, M.; Singh-Manoux, A.; Brunner, E.J. Dietary pattern, inflammation and cognitive decline: The Whitehall II prospective cohort study. *Clin. Nutr.* **2017**, *36*, 506–512. [CrossRef]
161. Satizabal, C.; Zhu, Y.; Mazoyer, B.; Dufouil, C.; Tzourio, C. Circulating IL-6 and CRP are associated with MRI findings in the elderly: The 3C-Dijon Study. *Neurology* **2012**, *78*, 720–727. [CrossRef]
162. Kyrkanides, S.; Tallents, R.H.; Jen-nie, H.M.; Olschowka, M.E.; Johnson, R.; Yang, M.; Olschowka, J.A.; Brouxhon, S.M.; O'Banion, M.K. Osteoarthritis accelerates and exacerbates Alzheimer's disease pathology in mice. *J. Neuroinflammation* **2011**, *8*, 112. [CrossRef] [PubMed]
163. Ravaglia, G.; Forti, P.; Maioli, F.; Brunetti, N.; Martelli, M.; Servadei, L.; Bastagli, L.; Bianchin, M.; Mariani, E. Serum C-reactive protein and cognitive function in healthy elderly Italian community dwellers. *J. Gerontol. Ser. A Biol. Sci. Med Sci.* **2005**, *60*, 1017–1021. [CrossRef] [PubMed]
164. Schram, M.T.; Euser, S.M.; De Craen, A.J.; Witteman, J.C.; Frölich, M.; Hofman, A.; Jolles, J.; Breteler, M.M.; Westendorp, R.G. Systemic markers of inflammation and cognitive decline in old age. *J. Am. Geriatr. Soc.* **2007**, *55*, 708–716. [CrossRef] [PubMed]
165. Li, Q.; Yang, Y.; Reis, C.; Tao, T.; Li, W.; Li, X.; Zhang, J.H. Cerebral small vessel disease. *Cell Transplant.* **2018**, *27*, 1711–1722. [CrossRef] [PubMed]
166. Staals, J.; Booth, T.; Morris, Z.; Bastin, M.E.; Gow, A.J.; Corley, J.; Redmond, P.; Starr, J.M.; Deary, I.J.; Wardlaw, J.M. Total MRI load of cerebral small vessel disease and cognitive ability in older people. *Neurobiol. Aging* **2015**, *36*, 2806–2811. [CrossRef] [PubMed]
167. Staals, J.; Makin, S.D.; Doubal, F.N.; Dennis, M.S.; Wardlaw, J.M. Stroke subtype, vascular risk factors, and total MRI brain small-vessel disease burden. *Neurology* **2014**, *83*, 1228–1234. [CrossRef]
168. Wardlaw, J.M.; Smith, C.; Dichgans, M. Mechanisms of sporadic cerebral small vessel disease: Insights from neuroimaging. *Lancet Neurol.* **2013**, *12*, 483–497. [CrossRef]
169. Low, A.; Mak, E.; Rowe, J.B.; Markus, H.S.; O'Brien, J.T. Inflammation and cerebral small vessel disease: A systematic review. *Ageing Res. Rev.* **2019**, 100916. [CrossRef]
170. Vemuri, P.; Lesnick, T.G.; Przybelski, S.A.; Knopman, D.S.; Preboske, G.M.; Kantarci, K.; Raman, M.R.; Machulda, M.M.; Mielke, M.M.; Lowe, V.J. Vascular and amyloid pathologies are independent predictors of cognitive decline in normal elderly. *Brain* **2015**, *138*, 761–771. [CrossRef]
171. Calder, P.C.; Bosco, N.; Bourdet-Sicard, R.; Capuron, L.; Delzenne, N.; Doré, J.; Franceschi, C.; Lehtinen, M.J.; Recker, T.; Salvioli, S. Health relevance of the modification of low grade inflammation in ageing (inflammageing) and the role of nutrition. *Ageing Res. Rev.* **2017**, *40*, 95–119. [CrossRef]
172. Benito-León, J.; Mitchell, A.; Hernández-Gallego, J.; Bermejo-Pareja, F. Obesity and impaired cognitive functioning in the elderly: A population-based cross-sectional study (NEDICES). *Eur. J. Neurol.* **2013**, *20*, 899-e877. [CrossRef] [PubMed]
173. Darling, J.; Ross, A.; Bartness, T.; Parent, M. Predicting the effects of a high-energy diet on fatty liver and hippocampal-dependent memory in male rats. *Obesity* **2013**, *21*, 910–917. [CrossRef]
174. Kanoski, S.E.; Zhang, Y.; Zheng, W.; Davidson, T.L. The effects of a high-energy diet on hippocampal function and blood-brain barrier integrity in the rat. *J. Alzheimer's Dis.* **2010**, *21*, 207–219. [CrossRef] [PubMed]
175. Kanoski, S.E.; Davidson, T.L. Different patterns of memory impairments accompany short-and longer-term maintenance on a high-energy diet. *J. Exp. Psychol. Anim. Behav. Process.* **2010**, *36*, 313. [CrossRef] [PubMed]
176. Francis, H.; Stevenson, R. The longer-term impacts of Western diet on human cognition and the brain. *Appetite* **2013**, *63*, 119–128. [CrossRef]
177. Freeman, L.R.; Haley-Zitlin, V.; Rosenberger, D.S.; Granholm, A.-C. Damaging effects of a high-fat diet to the brain and cognition: A review of proposed mechanisms. *Nutr. Neurosci.* **2014**, *17*, 241–251. [CrossRef]
178. Martin, A.A.; Davidson, T.L. Human cognitive function and the obesogenic environment. *Physiol. Behav.* **2014**, *136*, 185–193. [CrossRef]
179. M Wilkins, H.; H Swerdlow, R. Relationships between mitochondria and neuroinflammation: Implications for Alzheimer's disease. *Curr. Top. Med. Chem.* **2016**, *16*, 849–857. [CrossRef]

180. Krikorian, R.; Shidler, M.D.; Dangelo, K.; Couch, S.C.; Benoit, S.C.; Clegg, D.J. Dietary ketosis enhances memory in mild cognitive impairment. *Neurobiol. Aging* **2012**, *33*, 425.e19–425.e27. [CrossRef]
181. Reger, M.A.; Henderson, S.T.; Hale, C.; Cholerton, B.; Baker, L.D.; Watson, G.; Hyde, K.; Chapman, D.; Craft, S. Effects of β-hydroxybutyrate on cognition in memory-impaired adults. *Neurobiol. Aging* **2004**, *25*, 311–314. [CrossRef]
182. Henderson, S.T.; Vogel, J.L.; Barr, L.J.; Garvin, F.; Jones, J.J.; Costantini, L.C. Study of the ketogenic agent AC-1202 in mild to moderate Alzheimer's disease: A randomized, double-blind, placebo-controlled, multicenter trial. *Nutr. Metab.* **2009**, *6*, 31. [CrossRef]
183. Neal, E.G.; Chaffe, H.; Schwartz, R.H.; Lawson, M.S.; Edwards, N.; Fitzsimmons, G.; Whitney, A.; Cross, J.H. The ketogenic diet for the treatment of childhood epilepsy: A randomised controlled trial. *Lancet Neurol.* **2008**, *7*, 500–506. [CrossRef]
184. Selfridge, J.E.; Wilkins, H.M.; Lezi, E.; Carl, S.M.; Koppel, S.; Funk, E.; Fields, T.; Lu, J.; Tang, E.P.; Slawson, C. Effect of one month duration ketogenic and non-ketogenic high fat diets on mouse brain bioenergetic infrastructure. *J. Bioenerg. Biomembr.* **2015**, *47*, 1–11. [CrossRef]
185. Rutledge, A.C.; Adeli, K. Fructose and the metabolic syndrome: Pathophysiology and molecular mechanisms. *Nutr. Rev.* **2007**, *65*, S13–S23. [CrossRef]
186. Casas, R.; Sacanella, E.; Urpí-Sardà, M.; Corella, D.; Castaner, O.; Lamuela-Raventos, R.-M.; Salas-Salvadó, J.; Martínez-González, M.-A.; Ros, E.; Estruch, R. Long-term immunomodulatory effects of a mediterranean diet in adults at high risk of cardiovascular disease in the PREvención con DIeta MEDiterránea (PREDIMED) randomized controlled trial. *J. Nutr.* **2016**, *146*, 1684–1693. [CrossRef]
187. Casas, R.; Urpi-Sardà, M.; Sacanella, E.; Arranz, S.; Corella, D.; Castañer, O.; Lamuela-Raventós, R.-M.; Salas-Salvadó, J.; Lapetra, J.; Portillo, M.P. Anti-inflammatory effects of the Mediterranean diet in the early and late stages of atheroma plaque development. *Mediat. Inflamm.* **2017**, *2017*. [CrossRef]
188. Berti, V.; Walters, M.; Sterling, J.; Quinn, C.G.; Logue, M.; Andrews, R.; Matthews, D.C.; Osorio, R.S.; Pupi, A.; Vallabhajosula, S. Mediterranean diet and 3-year Alzheimer brain biomarker changes in middle-aged adults. *Neurology* **2018**, *90*, e1789–e1798. [CrossRef]
189. Luciano, M.; Corley, J.; Cox, S.R.; Hernández, M.C.V.; Craig, L.C.; Dickie, D.A.; Karama, S.; McNeill, G.M.; Bastin, M.E.; Wardlaw, J.M. Mediterranean-type diet and brain structural change from 73 to 76 years in a Scottish cohort. *Neurology* **2017**, *88*, 449–455. [CrossRef]
190. Staubo, S.C.; Aakre, J.A.; Vemuri, P.; Syrjanen, J.A.; Mielke, M.M.; Geda, Y.E.; Kremers, W.K.; Machulda, M.M.; Knopman, D.S.; Petersen, R.C. Mediterranean diet, micronutrients and macronutrients, and MRI measures of cortical thickness. *Alzheimer's Dement.* **2017**, *13*, 168–177. [CrossRef]
191. Rainey-Smith, S.R.; Gu, Y.; Gardener, S.L.; Doecke, J.D.; Villemagne, V.L.; Brown, B.M.; Taddei, K.; Laws, S.M.; Sohrabi, H.R.; Weinborn, M. Mediterranean diet adherence and rate of cerebral Aβ-amyloid accumulation: Data from the Australian Imaging, Biomarkers and Lifestyle Study of Ageing. *Transl. Psychiatry* **2018**, *8*, 1–7. [CrossRef]
192. Gu, Y.; Scarmeas, N.; Stern, Y.; Manly, J.J.; Schupf, N.; Mayeux, R.; Brickman, A.M. Mediterranean diet is associated with slower rate of hippocampal atrophy: A longitudinal study in cognitively normal older adults. *Alzheimer's Dement. J. Alzheimer's Assoc.* **2016**, *12*, P193–P194. [CrossRef]
193. Beher, D.; Clarke, E.E.; Wrigley, J.D.; Martin, A.C.; Nadin, A.; Churcher, I.; Shearman, M.S. Selected Non-steroidal Anti-inflammatory Drugs and Their Derivatives Target γ-Secretase at a Novel Site EVIDENCE FOR AN ALLOSTERIC MECHANISM. *J. Biol. Chem.* **2004**, *279*, 43419–43426. [CrossRef]
194. Eriksen, J.L.; Sagi, S.A.; Smith, T.E.; Weggen, S.; Das, P.; McLendon, D.; Ozols, V.V.; Jessing, K.W.; Zavitz, K.H.; Koo, E.H. NSAIDs and enantiomers of flurbiprofen target γ-secretase and lower Aβ42 in vivo. *J. Clin. Investig.* **2003**, *112*, 440–449. [CrossRef]
195. Jantzen, P.T.; Connor, K.E.; DiCarlo, G.; Wenk, G.L.; Wallace, J.L.; Rojiani, A.M.; Coppola, D.; Morgan, D.; Gordon, M.N. Microglial activation and β-amyloid deposit reduction caused by a nitric oxide-releasing nonsteroidal anti-inflammatory drug in amyloid precursor protein plus presenilin-1 transgenic mice. *J. Neurosci.* **2002**, *22*, 2246–2254. [CrossRef]

196. Weggen, S.; Eriksen, J.L.; Sagi, S.A.; Pietrzik, C.U.; Ozols, V.; Fauq, A.; Golde, T.E.; Koo, E.H. Evidence that nonsteroidal anti-inflammatory drugs decrease amyloid β42 production by direct modulation of γ-secretase activity. *J. Biol. Chem.* **2003**, *278*, 31831–31837. [CrossRef]
197. Weggen, S.; Eriksen, J.L.; Sagi, S.A.; Pietrzik, C.U.; Golde, T.E.; Koo, E.H. Aβ42-lowering nonsteroidal anti-inflammatory drugs preserve intramembrane cleavage of the amyloid precursor protein (APP) and ErbB-4 receptor and signaling through the APP intracellular domain. *J. Biol. Chem.* **2003**, *278*, 30748–30754. [CrossRef]

Review

The Other Side of Alzheimer's Disease: Influence of Metabolic Disorder Features for Novel Diagnostic Biomarkers

Chiara Argentati, Ilaria Tortorella, Martina Bazzucchi, Carla Emiliani, Francesco Morena and Sabata Martino *

Department of Chemistry, Biology and Biotechnologies, University of Perugia, 06123 Perugia, Italy; chiara.argentati89@gmail.com (C.A.); tortorella.i@hotmail.it (I.T.); marti89.b@libero.it (M.B.); carla.emiliani@unipg.it (C.E.); francesco.morena@unipg.it (F.M.)

* Correspondence: sabata.martino@unipg.it; Tel.: +39-075-585-7442

Received: 3 August 2020; Accepted: 4 September 2020; Published: 6 September 2020

Abstract: Nowadays, the amyloid cascade hypothesis is the dominant model to explain Alzheimer's disease (AD) pathogenesis. By this hypothesis, the inherited genetic form of AD is discriminated from the sporadic form of AD (SAD) that accounts for 85–90% of total patients. The cause of SAD is still unclear, but several studies have shed light on the involvement of environmental factors and multiple susceptibility genes, such as Apolipoprotein E and other genetic risk factors, which are key mediators in different metabolic pathways (e.g., glucose metabolism, lipid metabolism, energetic metabolism, and inflammation). Furthermore, growing clinical evidence in AD patients highlighted the presence of affected systemic organs and blood similarly to the brain. Collectively, these findings revise the canonical understating of AD pathogenesis and suggest that AD has metabolic disorder features. This review will focus on AD as a metabolic disorder and highlight the contribution of this novel understanding on the identification of new biomarkers for improving an early AD diagnosis.

Keywords: amyloid cascade hypothesis; glucose metabolism; adipose tissue dysfunction; energetic metabolism; lysosomes dysfunction; Type-3-Diabetes; neuroinflammation; neurodegeneration

1. Alzheimer's Disease: State of the Art

Alzheimer's disease (AD) is the most widespread neurodegenerative disease and the commonest cause of dementia. Data from The World Health Organization indicate that AD is the fifth cause of death worldwide and predict a dramatic increase in AD incidence in the next years, reaching 152 million people affected by 2050 [1]. This scenario is worsened by the absence of an effective therapy as well as by the absence of an early diagnosis, as it is usually made after symptoms manifestations, when neural impairment and brain injury damages are already severe [2–5]. One of the major obstacles for the early diagnosis is related to the complexity of AD etiology and pathogenesis [2–6]. It is acknowledged that the amyloid cascade hypothesis, the best model explaining AD development and progression, is not sufficiently clearly described [2–6]. It is also accepted that pathological signs similar to those found in the brain are present in the systemic organs and blood of AD patients [2–5,7].

In this review, we discuss the novel knowledge evidencing AD as a metabolic disorder. In particular, we will describe the contribution to AD pathophysiology of dysfunction in (i) glucose metabolism, (ii) adipose tissue, (iii) mitochondria, (iv) lysosomal compartment functionality, (v) metabolic syndrome, and their correlation with neuroinflammation and neurodegeneration. The influence of these discoveries in the identification of new biomarkers for AD diagnosis is also highlighted.

To be comprehensive, this work starts with the description of the amyloid cascade hypothesis correlating consolidated and new findings.

2. The Amyloid Cascade Hypothesis

The amyloid cascade hypothesis is based on the observation that at least 10–15% of AD patients have an inherited genetic background, while most of the cases (85–90%) are sporadic [4,8].

2.1. Familial Form of Alzheimer's Disease

The hereditary form of AD, also known as FAD (familial form of Alzheimer's disease) usually has an early-onset AD (EOAD), around 30 years of age. FAD is very rare and is caused by dominant inherited mutations in three genes, Amyloid Precursor Protein (APP), Presenilin 1 (PSEN1), and Presenilin 2 (PSEN2), which are implicated in the same metabolic pathway [9,10] (Table 1). APP missense mutations account for 10–15% of EOAD, while more than 272 different missense mutations were found for PSEN1 gene that represent 18% to 50% of the autosomal dominant cases of familial EOAD. PSEN2 gene mutations are the rarest cause of familial EOAD [10–12]. The APP gene encodes for the β-amyloid precursor protein, while PSEN1 and PSEN2 encode respectively for presenilin 1 and presenilin 2 subunits of the γ-secretase complex, which are involved in the cleavage of APP to generate the β-amyloid peptide. The alteration of one of these three proteins may steer the physiological non-amyloidogenic pathway to the pathogenic amyloidogenic pathway with consequent accumulation of the amyloid peptide and formation of amyloid plaques, the main hallmark of AD [10–12].

Table 1. The dominant inherited Alzheimer's disease (AD) genes Amyloid Precursor Protein (APP), Presenilin 1 (PSEN1), and Presenilin 2 (PSEN2).

Gene	Chr.	Protein Length	Protein Domains	N° of Mutations			Reference
				Pathogenic	Non Pathogenic	Protective	
APP	21q21.3	770 aa	Extracellular	10	15	1	Alzaforum Database [13]
			Transmembrane	16	-	-	
			Intracellular	1	-	-	
PSEN1	14q24.2	467 aa	Extracellular	16	1	1	Alzaforum Database [13]
			9 Transmembrane	226	-	-	
			Intracellular	30	-	-	
PSEN2	1q42.13	448 aa	Extracellular	4	-	-	Alzaforum Database [13]
			9 Transmembrane	7	5	-	
			Intracellular	3	5	-	

2.2. Sporadic Form of Alzheimer's Disease

The sporadic form of Alzheimer's disease (SAD) has prevalently a late-onset AD (LOAD) on average over 65 years of age [12,14]. SAD represents the majority of the cases of AD and has a more complex pathogenesis than FAD as it may have different potential causes not yet fully understood. During the past decade, clinical and experimental studies have identified many genetic and non-genetic risk factors for SAD [14].

Among the genetic risk factors, the presence of mutations on the Apolipoprotein E (APOE) gene correlates with a high probability of developing SAD [14,15]. The APOE gene has three variants, namely ε2, ε3, and ε4 that have different roles in AD pathogenesis. While ε2 and ε3 variants are not involved in AD onset, the ε4 variant is considered the single biggest risk factor for SAD [16]. It has been proposed that APOE-ε4 (APOE4) mediates β-Amyloid (Aβ) aggregation and Tau hyperphosphorylation [3,17]. Moreover, the ApoE protein has immunomodulatory functions [18] by its binding with the microglia Triggering Receptor Expressed on Myeloid Cells 2 (TREM2) receptor [19].

Genetic alteration causing Down Syndrome (DS) is another well-established risk factor for AD. As the APP gene is located on chromosome 21, patients with DS that carry an extra copy of this chromosome have a high probability, nearly 90% after 60 years of age, of developing Aβ deposit and AD-related pathological changes [20].

Additionally, more than 20 loci associated with SAD have been identified by Genome-Wide Association Studies (GWAS) [21,22]. These genes are involved in different pathways such as glucose metabolism, lipid metabolism, inflammatory pathways, endosomal vesicle recycling, regulation of autophagy, phagocytosis, cholesterol efflux, axon guidance, and cytokine-mediated signaling pathway [2,8,21,23–25] (Table 2).

Table 2. Genetic variant, max magnitude, chromosome position, and clinical characteristics of definite AD cases carrying risk variants of principal sporadic form of Alzheimer's disease (SAD)-associated genes. ApoE: Apolipoprotein E.

Gene	Variant	Max Magnitude	Chr. Position	Clinical Features
TOMM40	rs10524523	3	44,899,792	Higher risk for late-onset Alzheimer's disease
	rs157582	2.1	44,892,962	Weaker memory performance
	rs2075650	2	44,892,362	Possibly 2–4× higher Alzheimer's risk
APOE	rs199768005	2.1	44,909,057	Marked reduced risk of Alzheimer's disease
	rs429358	3	44,908,684	>3× increased risk for Alzheimer's
	rs449647	2	44,905,307	Lower levels of ApoE
TREM2	rs104894002	6	41,161,557	Alzheimer's, late-onset, possible/predicted
	rs143332484	2	41,161,469	Moderate increase (1.7×) in risk for Alzheimer's disease
	rs75932628	3.5	41,161,514	Risk of Alzheimer's disease
ABCA7	rs113809142	3	1,056,245	≈2× higher risk for Alzheimer's disease
	rs115550680	2.5	1,050,421	Increased risk (≈2.2×) of Alzheimer's, observed for African-Americans
	rs200538373	3	1,061,893	≈3× higher risk for Alzheimer's disease
	rs72973581	2.5	1,043,104	Slightly lower risk (0.57×) for Alzheimer's, according to one study
	rs78117248	2	1,052,854	Risk factor for Alzheimer disease (odds ratio ≈2×)
CLU	rs11136000	1.5	27,607,002	0.84× decreased risk for Alzheimer's disease
CR1	rs6656401	1.5	207,518,704	1.18× increased risk for late-onset Alzheimer's
	rs3818361	1.2	207,611,623	1.2× increased risk for late-onset Alzheimer's
CD33	rs3865444	1.6	51,224,706	Slight reduction in risk for Alzheimer's disease
MS4A6A	rs610932	1.5	60,171,834	An allele associated with reduced risk of Alzheimer's in East Asian populations
BIN1	rs6733839	NA	127,135,234	This SNP has a population attributable fraction for AD of 8.1 which is second only to APOE4's of 27.3
PICALM	rs3851179	1.5	86,157,598	0.85× decreased risk for Alzheimer's disease
SORL1	rs10892759	1.01	121,593,379	Reduced risk for Alzheimer's
	rs1784931	1.01	121,612,229	Reduced risk for Alzheimer's
PLD3	rs145999145	2	40,371,688	2× higher risk for Alzheimer's disease
CTNNA3	rs2306402	2.1	67,175,727	1.2× increased risk for late-onset Alzheimer's disease
DNMBP	rs3740057	NA	99,898,828	Increased risk for late-onset Alzheimer's disease in both Japanese and Belgian populations
	rs10883421	NA	99,912,584	Increased risk for late-onset Alzheimer's disease in both Japanese and Belgian populations
BACE1	rs638405	2	117,293,108	2× increased Alzheimer's risk in ApoE4 carriers
	rs4938369	NA	117,317,404	1.6× increased risk for Alzheimer's
GAB2	rs7101429	2	78,281,921	0.70× reduced risk for Alzheimer's risk
ADAM10	rs145518263	4	58,665,141	Rare mutation increasing risk for late-onset Alzheimer's disease
	rs61751103	4	58,665,172	Rare mutation increasing risk for late-onset Alzheimer's disease
ATP8B4	rs10519262	NA	50,140,297	1.9× risk for AD
ABCA2	rs908832	NA	137,018,032	3.8× increased risk for early-onset Alzheimer's
OLR1	rs1050283	NA	10,159,690	Increased risk for Alzheimer's
A2M	rs669	NA	9,079,672	3.8× or higher increased risk for Alzheimer's
OTC	rs5963409	NA	38,351,716	1.19× increased risk for Alzheimer's disease

Table 2 summarizes the main risk genetic variants correlated with AD. Only the entry with reported clinical futures are selected from the SNPedia database [26]. The full gene name is reported in the Abbreviation list. SNP: Single Nucleotide Polymorphism; NA: Not Assigned.

Among the non-genetic risk factors for AD, age, gender, comorbidities, and lifestyle are strictly monitored [27–29]. Gender seems to deeply affect the incidence, clinical manifestation, development, and prognosis of the disease, resulting in a higher AD risk for females [30–32]. This observation is also confirmed by epidemiologic studies showing that clinical symptoms and neurodegeneration occur more rapidly in women after diagnosis, therefore suggesting that the faster disease progression could be due to neurobiological vulnerability in postmenopausal females [31].

A critical risk is also represented by AD comorbidities such as cerebrovascular diseases (e.g., subcortical leukoencephalopathy, ischemic and hemorrhagic strokes), cardiovascular risk factors (e.g., hypertension, insulin resistance, hyperlipidemia, chronic inflammatory diseases, vitamin D deficiency, alcohol consumption, and smoking) [4] and metabolic alterations (see Section 3). Hypertension may also increase the risk of AD through the alteration of vascular integrity of the Blood–Brain Barrier (BBB), resulting in protein extravasation into brain tissue causing cell damage, reduction in neuronal or synaptic function, apoptosis, increase of Aβ accumulation, and cognitive impairment [33].

2.3. Molecular Mechanism of β-Amyloid Cleavage

As above mentioned, AD pathogenesis is mostly explained by the mechanisms causing the accumulation of oligomeric β-amyloid-peptide-42 (Aβ42) or related peptides in the brain and their consequent toxic effects on neurons and glia, culminating with programmed neuronal death [34,35].

This pathway, called 'Amyloid Cascade Hypothesis' is schematized in Figure 1 and is described below.

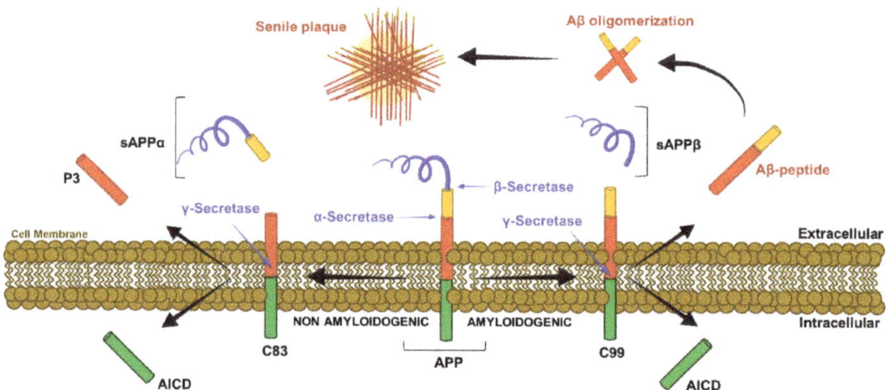

Figure 1. Molecular mechanism of APP cleavage. Cartoon shows the mechanistic events leading the non-amyloidogenic and amyloidogenic process. Details of proteolytic steps are in the text. sAPPα, Soluble Amyloid Precursor Protein α; P3, residues from 17 to 40/42 of the APP; AICD, APP Intracellular Domain; C83, αAPP COOH-terminal fragment 83; APP, Amyloid Precursor Protein; C99, βAPP COOH-terminal fragment 99.

The Aβ peptides are the result of incorrect cleavage of the integral membrane protein Type I APP. In the physiological non-amyloidogenic pathway, Type I APP is first cleaved at the *n*-terminal domain by the α-secretase, which triggers the release of Soluble Amyloid Precursor Protein (sAPP) α, which is a large soluble fragment of ectodomain, while the C-terminal domain (the αAPP COOH-terminal fragment (αAPP-CTF), C83) remains in the plasma membrane. Afterwards, the C83 fragment is cleaved by the γ-secretase, generating the soluble non-aggregating and non-neurotoxic peptide P3 (residues from 17 to 40/42 of the APP) that is released in the extracellular space and the APP Intracellular Domain (AICD) fragment, which is released in the intracellular space (Figure 1).

The toxic amyloidogenic pathway occurs when the β-secretase Beta-site of beta-Amyloid precursor protein Cleaving Enzyme (BACE) cleaves APP at the *n*-terminal domain, releasing the sAPPβ ectodomain and the C-terminal fragments C99 and C100 residues called βAPP COOH-terminal fragments (βAPP-CTFs). After the cleavage by γ-secretase, the Aβ-peptide is released. The γ-secretase may act with two different cuts: the first, at the Valine40, generates the Aβ40 (the most abundant

Aβ-peptide isoform), while the second generates the Aβ42 isoform, which accounts for 10% of the total Aβ peptides [36–38] (Figure 1).

Aβ is thought to damage neurons directly by increasing the toxicity and promoting hypoglycemia and oxidative stress [5]. Aβ can also act indirectly on neuronal loss by the activation of microglia, leading to the release of toxic and inflammatory mediators such as nitric oxide and cytokines including interleukin (IL)-1β, tumor necrosis factor α (TNFα), and interferon-γ [5,39]. It has also been shown that Aβ oligomers can also induce Tau hyperphosphorylation and, in cultured neurons, neuritic dystrophy [5,39].

The presence and the accumulation of Aβ peptides is necessary for the diagnosis of AD, but, as a wide proportion of AD patients die without significant Aβ deposition, it is suggested that this event is not sufficient to completely explain AD pathophysiology [38,40].

The Role of Tau Protein

A corollary of the amyloid cascade hypothesis is the accumulation of Tau Neurofibrillary Tangles (NFTs) [41]. NFTs are intracellular clumps that are mainly composed of paired hyperphosphorylated Tau protein leading to the generation of helical filaments commonly found in neurodegenerative disorders known as "tauopathies" [41–43].

The Tau protein is encoded by the Microtubule-Associated Protein Tau gene located on chromosome 17q21. In normal conditions, it binds microtubules to promote their assembly and stability through a mechanism of phosphorylation and dephosphorylation [44,45] (Figure 2). In AD, Tau hyperphosphorylation causes the loss of its microtubule binding capacity; it also induces the formation of intracellular NFTs and the consequent microtubule depolymerization and defective function [45] (Figure 2).

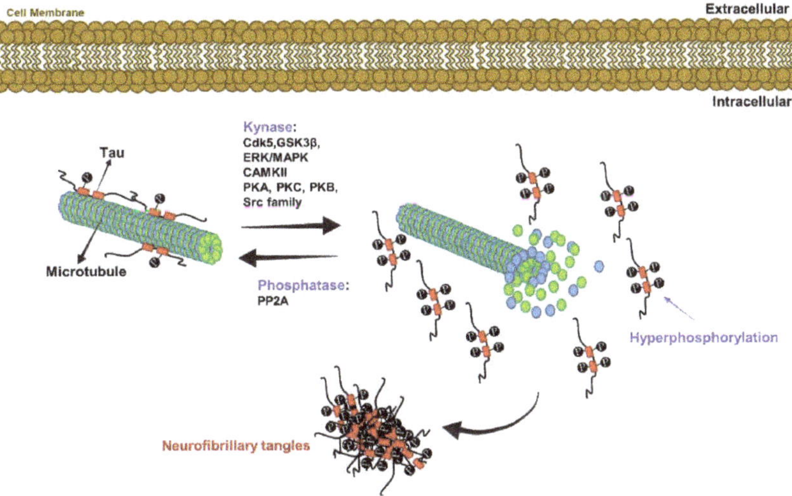

Figure 2. Molecular mechanism of Neurofibrillary Tangles formation. Cartoon shows the mechanistic event leading to altered Tau hyperphosphorylation and the consequent Neurofibrillary Tangles formation as described in the text. CAMKII, Calcium/Calmodulin-dependent Protein Kinase Type II; Cdk5, Cyclin-dependent kinase 5; ERK, Extracellular signal-regulated kinase; GSK3β, Glycogen Synthase Kinase 3 β; MAPK, Mitogen-Activated Protein Kinase; PKA, Protein Kinase A; PKB, Protein Kinase B; PKC, Protein Kinase C; PP2A, Protein Phosphatase 2A; SRC, Proto-oncogene tyrosine-protein kinase Src.

After the first report describing and identifying Tau as a phosphoprotein (p-Tau), at least 85 known sites of phosphorylation (mostly serine and threonine, but also tyrosine) are now classified [46,47].

Tau kinases and phosphatase reciprocally balance their activity (Figure 2) [46,47]. For example, Glycogen Synthase Kinase 3 β (GSK3β) activation represses Protein Phosphatase 2A (PP2A), while the repression of PP2A leads to GSK3β activation [48]. Impairment of the Protein kinase B (Akt)/Mammalian Target of Rapamycin (mTOR) pathway may alter the physiological phosphorylation balance between GSK3β and PP2A, as Akt inhibits GSK3β, which in turn inhibits PP2A [45]. Interestingly, it was proposed that the Aβ peptide promotes GSK3β activation, resulting in Tau phosphorylation [49]. Finally, other studies have shown that the acetylation of certain Tau residues (for example, Lys280) can promote Tau autophosphorylation, exacerbate the aggregate formation, and ultimately lead to tauopathy [50].

2.4. Biomarkers from Amyloid Cascade Hypothesis for AD Diagnosis

The Food and Drug Administration defines an ideal biomarker as specific, sensitive, predictive, robust, simple, accurate, inexpensive, and measurable in peripheral, easily accessible districts [51,52]. In accordance with the amyloid cascade hypothesis, the National Institute on Aging and Alzheimer's Association has recognized three categories of biomarkers on the basis of the pathological mechanism in which are involved: Aβ deposits, hyperphosphorylated Tau aggregates, and neurodegeneration [6].

Currently, among the various imaging methods (structural Magnetic Resonance Imaging (MRI), functional MRI, amyloid-Positron Emission Tomography (PET) and 18F-fluorodeoxyglucose (FDG)-PET), amyloid-PET is the most reliable diagnostic tool for AD diagnosis because of its ability to highlight aggregated Aβ peptides within the brain by using amyloid tracers [53]. Similarly, recent new ligands for Tau have been developed such as [18F] PI-2620 and [18F] MK-6240 [54–57]. Other innovative imaging methods are being developed, including the detection of neuroinflammation process or microglia activation through Translocator Protein-PET [58] and epigenetic modifications and synaptic density/loss [59].

Among fluid biomarkers for AD, the gold standard is the Cerebrospinal Fluid (CSF), because it directly interacts with the brain interstitial fluid and it could reflect pathological changes in AD [60,61]. The typical AD CSF composition is characterized by halved Aβ42-peptide concentration (due to its accumulation in the brain) and an increase of p-Tau and total Tau [60–63]. In particular, total plasma Aβ42/40-peptide levels were shown to be correlated with amyloid and Tau deposits on a PET scan [64,65].

3. Insight into Alzheimer's Disease as a Metabolic Disorder

As reported above, the current understanding documents the widespread dysfunctions of peripheral organs and blood in AD patients and the contribution of some metabolic alterations to the disease.

In the following paragraphs, we will discuss in detail the contribution of metabolic alterations associated with AD pathogenesis and the consequent emerged new biomarkers suitable for AD early diagnosis. The section includes (i) aberrant glucose metabolism, (ii) adipose tissue dysfunction, (iii) the alteration of mitochondria, (iv) dysfunction of lysosomal compartment, and (v) metabolic syndrome, and it describes how each of them may improve the identification of new markers for AD.

3.1. Glucose Metabolism and AD

Dysregulation of the glucose metabolism is a prominent feature in AD patients' brains [66,67]. Glucose is an essential energy substrate for the maintenance of neuronal activity and is uptaken by glucose transporters expressed in astrocytes, in neurons, and in the cerebral endothelium. A reduction in glucose transporters in both neurons and endothelial cells of the BBB has been documented during AD progression [66] (Figure 3). This includes the glucose transporter 1 (GLUT-1) in glia and endothelial cells and glucose transporter 3 (GLUT-3) in neurons [68]. Experimental evidence showed that GLUT-1 deficiency in endothelial cells in mouse models accelerates AD progression, promoting neurodegeneration, Aβ deposition, and cognitive decline [69].

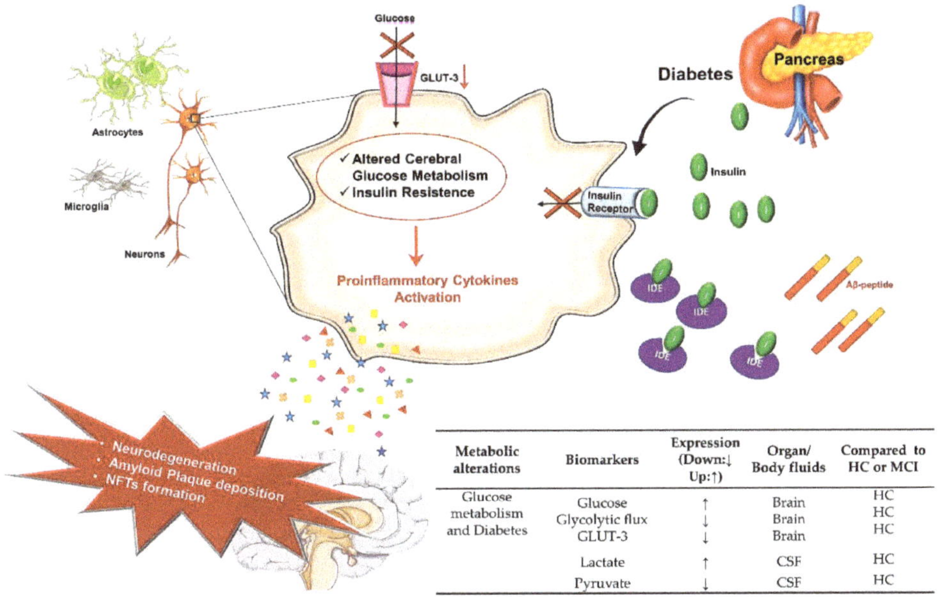

Metabolic alterations	Biomarkers	Expression (Down:↓ Up:↑)	Organ/ Body fluids	Compared to HC or MCI
Glucose metabolism and Diabetes	Glucose	↑	Brain	HC
	Glycolytic flux	↓	Brain	HC
	GLUT-3	↓	Brain	HC
	Lactate	↑	CSF	HC
	Pyruvate	↓	CSF	HC

Figure 3. Alteration of glucose metabolism and AD. Cartoon schematizes the main altered events leading to the alteration of glucose homeostasis and reports the table with potentials AD biomarkers. Details are in the text. The arrow direction (up ↑, down ↓) indicates the higher and lower expression levels of the related biomarker in AD with respect to healthy controls (HC). Aβ, β-Amyloid; GLUT-3, Glucose Transporter 3; IDE, Insulin Degrading Enzyme; MCI, Mild Cognitive Impairment; NFTs, Tau Neurofibrillary Tangles.

In a recent study, Yan An and co-authors measured the levels of glucose, GLUT-3, and GLUT-1 in the autopsy brain of AD patients samples that were part of the Baltimore Longitudinal Study of Aging [67,70]. They found a significant higher glucose concentration in the brain of AD patients that correlated with the severity of the disease and symptoms' onset. Interestingly, they also measured a reduced cerebral glycolytic flux and lower levels of GLUT-3 in the same AD patients [67].

The correlation of altered glucose metabolism in the brain and AD is mostly confirmed by clinical and experimental studies establishing that diabetes, due to insulin resistance, is one of the main risk factors for the development of AD. It is estimated that 65% of diabetic patients have an increased risk of developing AD [71–74].

The role of insulin in brain homeostasis has been investigated in physiology [75,76] and pathology, including AD [77] and diabetes [78]. Insulin is a peptide hormone secreted by pancreatic beta cells that has well-characterized functions in glucose/lipid metabolism and cell growth [79]. It can cross the BBB from the periphery to the Central Nervous System (CNS) competing with the Aβ peptide for the Insulin Degrading Enzyme (IDE) in the human brain. In the hippocampus, it appears to be involved in the regulation of GSK3β signaling and in the maintenance of neuroplasticity, neurotrophic, and neuroendocrine functions [80]. Altogether, insulin and Insulin-like Growth Factor 1 (IGF1) resistance in AD results in a reduced catabolism of cerebral glucose and promotes oxidative stress, mitochondrial dysfunction, pro-inflammatory cytokines activation, and impaired energy metabolism [81–83] (Figure 3). Furthermore, cardiovascular disorders, oxidative stress, inflammation, high level of cholesterol, and Aβ deposition are also common risk factors for AD and diabetes [73].

Collectively, these findings have driven the scientific community to define as 'type 3 diabetes' or "brain diabetes" the common molecular and cellular features of type 1 diabetes, type 2 diabetes, and insulin resistance associated with the development of the neurodegeneration [84–87]. In fact, compared to type 2 and type 1 diabetes, together with insulin resistance or deficiency (canonical hallmarks of diabetes), type 3 diabetes is also characterized by other relevant symptoms such as cognitive decline, impairments in visuospatial function and psychomotor speed flexibility, and loss of attention and memory. Of course, the amyloid aggregation and deposition are also present (see for review [88]).

Potential Glucose Metabolism Biomarkers for AD Diagnosis

Even though today, the role of altered glucose metabolism in AD is well established, the use of related biomarkers is still very moderate because it is difficult to restrict them to AD pathogenesis. Nevertheless, reduced FDG-PET brain metabolism was recognized as a biomarker of neurodegeneration. In fact, FDG-PET detects functions of glucose metabolism, recognizing areas of reduced brain activity and neuronal injury [89,90]. The technique has excellent sensitivity and specificity in discriminating AD from healthy controls [91].

In a recent meta-analysis of CSF marker, data highlighted an increased anaerobic glycolysis in AD patients. In particular, it was observed a relevant increase in lactate and a decrease in pyruvate, whereas the levels of glucose and glutamate in the CSF of AD patients were comparable to control subjects [92] (Figure 3). Conversely, as mentioned above, the glucose level increased in the brain of AD patients, while the glycolytic flux and GLUT-3 levels decreased.

3.2. *Adipose Tissue Dysfunction and AD*

Several studies have associated the excess of body weight with an increased risk of AD [82,93–95]. Therapies for reinstating metabolic homeostasis could improve cognitive functions in AD patients [96–98], while a high-fat diet might exacerbate cognitive function in animal models of AD [99–101].

There is a number of potential mechanisms linking high adiposity to AD (Figure 4). For example, the increase of free fatty acids, which is common in obesity, contributes to the onset of insulin resistance [102]. In addition, adiposity is a risk factor for diabetes, hypertension, and cardiovascular changes, all conditions contributing themselves to a significantly increased risk of AD [93,103].

The adipose tissue plays a central role in regulating the body energy and the homeostasis of glucose, both at organs and systemic levels [104–108]. In particular, adipose white tissue stores energy in the form of lipids and controls the mobilization and distribution of lipids in the body, while adipose brown tissue maintains body temperature and acts as an endocrine organ, producing numerous bioactive factors such as adipokines (e.g., Leptin, Adiponectin, Apelin, Resistin, Monocyte Chemoattractant Protein-1 (MCP-1), IL-1β, IL-6, IL-10, TNFα, Transforming Growth Factor β (TGF-β)) and lipokines (such as Lysophosphatidic Acid, LPA), which control many metabolic pathways including in the brain [104–106,109,110]. In particular, Leptin has pro-inflammatory activities, maintains neurogenesis and synaptogenesis, and is involved both in neuroprotection and neuroinflammation. Adiponectin controls the proliferation of hippocampal neural stem cells as well as the release of somatotropins and gonadotropins, and it partakes in the neurodegeneration process. Resistin is necessary for cognitive function performance and hypothalamic insulin resistance. LPA is critical for brain development, neurogenesis, the proliferation and differentiation of neural progenitor cells, and synaptic transmission. TNFα controls neurogenesis, neuroprotection, and the survival of neural progenitor cells [111]. The role of adipokines/lipokines able to cross the BBB and enter in the CNS is further confirmed by the fact that their misexpression might alter or disrupt directly/indirectly brain's homeostasis and functions [105,106,109]. As a matter of fact, alterations of the BBB structure such as in inflammatory conditions cause an increased permeability to adipokines and LPA into the CNS and an increase of oxidative stress and neurodegeneration [111]. Interestingly, adipokines can also activate endothelial cell receptors, resulting in the modulation of tight junctions expression and BBB permeability [111]. Finally, adipokine dysregulation and oxidative stress were also involved in the remodeling of blood vessels and arterial stiffness in high-fat diet mice [112].

Altogether, these findings suggest that in pathological conditions, adipokines promote Reactive Oxygen Species (ROS) overproduction and inflammatory processes, which are involved in BBB disruption and could potentially act on different brain regions such as the hippocampus (Figure 4). This could explain why metabolic dysfunctions are associated with hippocampus atrophy and with an increased risk of developing dementia and AD [113]. In this regard, systemic alterations have been correlated to chronic inflammation caused by adiposity [114–116].

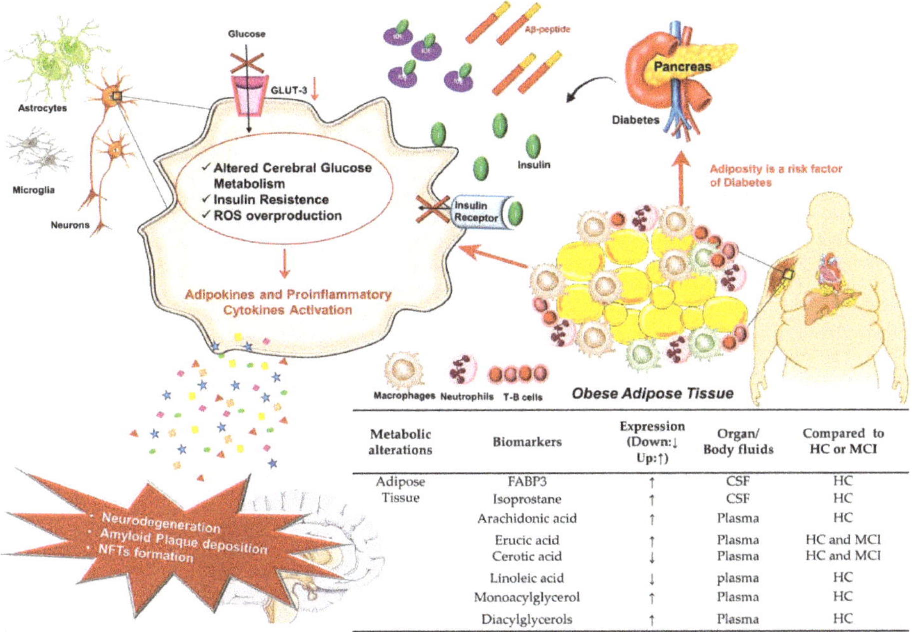

Metabolic alterations	Biomarkers	Expression (Down:↓ Up:↑)	Organ/ Body fluids	Compared to HC or MCI
Adipose Tissue	FABP3	↑	CSF	HC
	Isoprostane	↑	CSF	HC
	Arachidonic acid	↑	Plasma	HC
	Erucic acid	↑	Plasma	HC and MCI
	Cerotic acid	↓	Plasma	HC and MCI
	Linoleic acid	↓	plasma	HC
	Monoacylglycerol	↑	Plasma	HC
	Diacylglycerols	↑	Plasma	HC

Figure 4. Adipose tissue dysfunction and AD. Cartoon schematizes the main altered events leading to the alteration of adipose tissue homeostasis and reports the table with potentials AD biomarkers. Details are in the text. The arrow direction (up ↑, down ↓) indicates the higher and lower expression levels of the related biomarker in AD with respect to healthy controls (HC) or MCI (Mild Cognitive Impairment). Aβ, β-Amyloid; FABP3, Fatty Acid-Binding Protein; GLUT-3, Glucose Transporter 3; IDE, Insulin Degrading Enzyme; NFTs, Tau Neurofibrillary Tangles.

Potential Adipose Tissue Biomarkers for AD Diagnosis

Considering the important role that adipose tissue plays in AD pathogenesis, several studies were carried out to monitor lipids and lipid metabolism-related molecules in the peripheral districts of AD and Mild Cognitive Impairment (MCI) patients (Figure 4). For example, it has been observed a higher CSF level of Fatty Acid-Binding Protein 3 (FABP3) compared to healthy subjects [117,118] (Figure 4). Moreover, higher CSF levels of FABP3 and isoprostane were observed in MCI patients that evolved toward AD [117–120]. Different types of lipids such as arachidonic acid, erucic acid, monoacylglycerol, and diacylglycerol were shown to be higher in AD compared to MCI and healthy subjects, whereas other lipids were found to be lower, including cerotic acid and linoleic acid [119–125] (Figure 4).

In the last years, several studies aimed to understand if alterations in adipokines expression could be used as a biomarker for AD progression. Current data are contradictory, perhaps due to the intricate interplay in which these biomolecules are involved [126–136]. Thus, the measurement of adiponectin levels in CSF or plasma led to divergent results: reduced levels of adiponectin have been observed in the serum [137] and CSF [138] of AD patients, while in other studies, plasma and CSF adiponectin were significantly higher in MCI and AD compared to controls [139]. Wennberg et al. found that serum adiponectin is positively correlated to amyloid levels but inversely correlated to the hippocampal volume in women with MCI [140].

3.3. *Energetic Metabolism, Mitochondria Dysfunction, and AD*

Over the years, age-related alterations of mitochondria functioning have been observed in different neurodegenerative diseases such as AD [141–143]. An impairment of cellular bioenergetics has been observed in AD patients [144,145], and mitochondrial dysfunction appears to be an early event in AD pathogenesis [146,147] that is deeply correlated to the initiation of neuroinflammation [148]. Different observations have been made in an attempt to explain the mechanisms behind mitochondrial dysfunctions observed in AD (Figure 5).

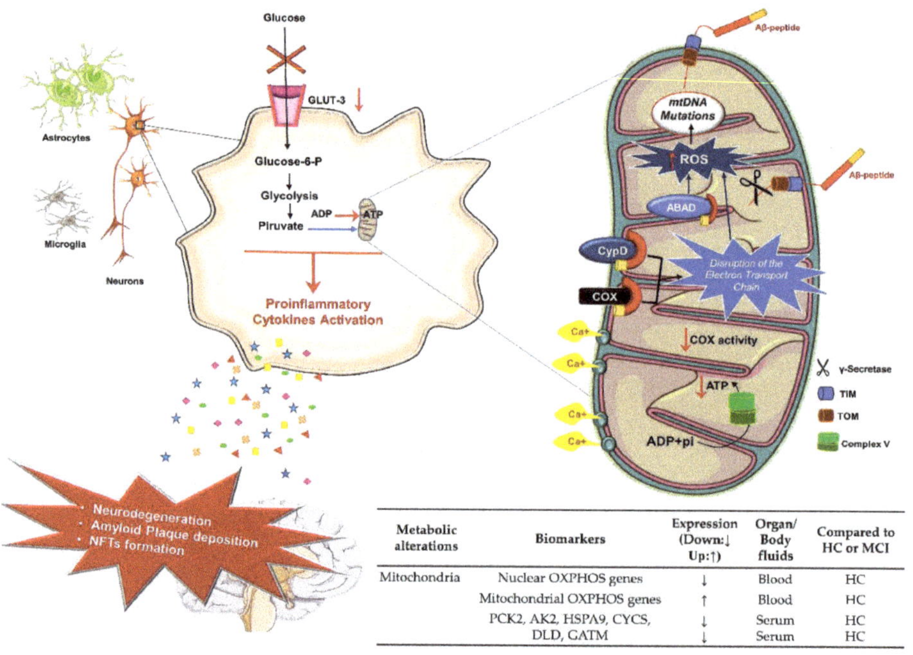

Metabolic alterations	Biomarkers	Expression (Down:↓ Up:↑)	Organ/ Body fluids	Compared to HC or MCI
Mitochondria	Nuclear OXPHOS genes	↓	Blood	HC
	Mitochondrial OXPHOS genes	↑	Blood	HC
	PCK2, AK2, HSPA9, CYCS,	↓	Serum	HC
	DLD, GATM	↓	Serum	HC

Figure 5. Energetic metabolism, mitochondria dysfunction, and AD. Cartoon schematizes the main altered events leading to the alteration of energetic metabolism and mitochondria dysfunction in AD described in the text and the table reports the potential AD biomarkers. The arrow direction (up ↑, down ↓) indicates the higher and lower expression levels of the related biomarker in AD with respect to healthy controls (HC). ABAD, Aβ Binding Alcohol Dehydrogenase; ADP, Adenosine Diphosphate; AK2, Adenylate Kinase 2; ATP, Adenosine Triphosphate; Aβ, β-Amyloid; COX, Cytochrome C Oxidase; CypD, Cyclophilin D; CYCS, Cytochrome C; DLD, Dihydrolipoyl dehydrogenase; GATM, Glycine Amidinotransferase; Glucose-6-P, Glucose 6-Phosphate; GLUT-3, Glucose Transporter 3; HSPA9, Stress-70 protein; MCI, Mild Cognitive Impairment; mtDNA, mitochondrial DNA; NFTs, Tau Neurofibrillary Tangles; OXPHOS: Oxidative Phosphorylation; PCK2, Phosphoenolpyruvate Carboxykinase [GTP]; Pi, Inorganic Phosphate; ROS, Reactive Oxygen Species; TIM, Translocase of the Inner Membrane; TOM, Translocase of the Outer Membrane.

First, neuronal mitochondria in AD are different in number and shape. The mitochondria number per neuron is reduced as the physiological balance between fission and fusion is altered, leading to a decreased biogenesis [149], and they appear to be swollen with misshapen cristae [150]. It has been observed in the AD brain of transgenic mice that there is a physical association of Aβ peptide and mitochondria [151] occurring through the involvement of the translocase of the outer membrane machinery [152]. It has also been shown that the Aβ peptide interacts directly with intracellular proteins such as the mitochondrial enzyme Aβ Binding Alcohol Dehydrogenase (ABAD), thus mediating mitochondrial enzymes insufficiency, oxidative stress, and cell death [153,154] (Figure 5).

Even though the exact mechanism of how Aβ peptide damages neuronal cells is still unknown, several studies elucidated that the interaction of Aβ peptide with mitochondrial proteins, such as Cyclophilin D (CypD) [155] and Cytochrome C Oxidase (COX) [156], leads to a disruption of the Electron Transport Chain with a consequent increase in ROS production and loss of Adenosine Triphosphate (ATP) generation [157]. Interestingly, COX activity was reported to be reduced also in platelet mitochondria derived from AD patients and in animal models of AD [158,159] (Figure 5). Moreover, the Aβ peptide appears to interfere with antioxidant enzymes (i.e., superoxide dismutase [Cu-Zn], catalase, and glutathione) and Kreb's cycle enzymes (pyruvate dehydrogenase, malate dehydrogenase, and aconitase) [160]. On the other hand, other reports suggested that the mitochondrial dysfunction is independent of Aβ-peptide deposition and hypothesized the 'Mitochondrial cascade hypothesis' for AD pathogenesis, by which the impairment of mitochondrial activity (mostly due to the respiratory chain and bioenergetic mechanisms failures) is upstream the Aβ cascade formation [161,162].

Mitochondria normal homeostasis appears to be compromised in the AD brain at different levels, including Ca^{2+} homeostasis, which regulates various neuronal functions such as impulse transmission, synaptic plasticity, and neuronal death. It has been observed that Aβ/Tau mitochondrial binding impairs also the mitochondrial Ca^{2+} handling capacity, resulting in oxidative stress, decreased mitochondrial membrane potential, decreased mitochondrial permeability transition pores formation, and deficient ATP synthesis [163]. On the one hand, impaired intraneuronal Ca^{2+} signaling is known to promote the abnormal aggregation of Aβ peptide, while on the other hand, Aβ can lead to a cytosolic Ca^{2+} overload by reducing its storage in the endoplasmic reticulum [144].

As a result of its proximity to the principal intracellular source of ROS and the lack of protective histone and limited repair mechanisms, mitochondrial DNA (mtDNA) is particularly susceptible to DNA damage caused by oxidative stress. As a matter of fact, several studies observed that AD-mtDNA differs from age-matched control subjects. Brain mtDNA nucleotide oxidative damage, as well as the frequency of point mutations, appear to be higher in AD subjects [164,165]. Moreover, association studies highlighted that particular mtDNA haplogroups and specific mtDNA SNPs are statistically over or underrepresented in AD patients [166–169].

Interestingly, recent studies have demonstrated a correlation between APOE4 expression and the mitochondrial function [170,171]. In particular, Schmukler and co-author have demonstrated that mitochondria have reduced fission and mitophagy in APOE4-carrier astrocytes compared to APOE3-carrier astrocytes [170]. These results were also confirmed in the hippocampus of ApoE4 mice [171].

Potential Mitochondria Biomarkers for AD Diagnosis

Considering the central role of mitochondria in AD pathophysiology, a major issue for AD early diagnosis is the identification of new candidate biomarkers related to mitochondrial enzymes and metabolism in peripheral districts such as CSF and blood. Even though the establishment of reliable and consistent mitochondrial biomarkers for AD is still not achieved, studies have been carried out to shed light on the possible use of mitochondrial markers for this purpose [172]. For example, a recent study has demonstrated the possibility of detecting alterations in the expression of nuclear and mitochondrial encoded Oxidative Phosphorylation (OXPHOS) genes in blood samples from AD and MCI patients [173] (Figure 5). Interestingly, a recent serum proteome profiling study identified deregulated proteins in AD samples, in particular, 12 proteins associated with mitochondrial function including PCK2, AK2, HSPA9, CYCS, DLD, and GATM (see Abbreviation List) [174] (Figure 5).

In another study aiming at identifying differentially expressed genes in blood samples of AD patients, it has been suggested that mitochondrial dysfunction, nuclear factor-κB (NF-κB) signaling and inducible Nitric Oxide Synthase (iNOS) signaling pathways are all dysregulated in AD pathophysiology [175]. However, these results still need to be confirmed. Finally, giving the importance of mitochondria in AD pathogenesis, Ridge et al. recently created a new extended dataset of mitochondrial genomes to investigate the impact of mitochondrial genetic variation on the risk for AD, with the aim of helping further research on mitochondrial peripheral biomarkers [176].

3.4. Lysosomes Dysfunction and AD

In post-mitotic neurons, endocytosis and macroautophagy processes are particularly important for maintaining the correct homeostasis and transducing signals through axons and synapses; thus, the correct functioning of the lysosomal system is crucial for the nervous system [177–179].

Many reports have associated autophagic pathway dysfunction to neurodegeneration development [180]. Alterations of this pathway have been identified at different levels and resulted in defective autophagic proteolysis of the macromolecules/molecular complexes within the lysosomal compartment [180–182]. The correlation of abnormal autophagic activity in AD has been described by Nixon and co-authors [180], who emphasized the impairment of the autophagolysosomes in human AD fibroblasts and in several models of the disease [183,184]. For instance, they have demonstrated the role of presenilin-1 for targeting v-ATPase to lysosomes as well as for their acidification and autophagic activity. These functions were absent in cells from psen1-null mice [184]. The abnormal autophagosomes accumulation was also recently described in the distal region of axons in neurons of AD patients and animal models, and it was correlated with the impairment of retrograde transport along the axons and with changes in the autophagic clearance of the Aβ peptide [179]. This work is an example of many others showing the autophagy deficit in the "neuronal housekeeper" function by which lysosomes lose the capability to degrade aberrant proteins, thereby participating in the accumulation Aβ peptide and Tau [185–193].

Interestingly, the enlargement of early endosomes/lysosomes compartments in neurons of AD brains as well as the extracellular presence of lysosomal hydrolases and their co-localization with amyloid plaques have been also observed before amyloid deposition [194–197]. During AD progression and in primary tauopathies, lysosomes become compromised, as there is an accumulation of endo/autolysosomal structures and intermediates in dystrophic neurons around amyloid plaques [197–200]. Many other laboratories have reported marked lysosomal system alterations in mouse models of AD (both amyloidosis and tauopathy models [201–203]). which had been noted also in brain regions affected only at very late stages of the disease [204,205], thereby suggesting that lysosomal alterations could precede neurodegeneration and encouraging further studies focused on the initial stages of AD [206]. Moreover, lysosomal impairment seems to be accompanied by the activation of particular secretory routes to remove misfolded protein, including Tau and Aβ peptide [207–210], as observed in Neuroaminidase 1 (Neu1) null mutant mice [211]. This could diminish the intracellular proteins accumulation but facilitate altered cell-to-cell transmission of the pathology.

Overall, the connection between AD and lysosomal dysfunction appears to be a vicious cycle in which lysosomal impairment contributes to Aβ peptide and Tau accumulation which, in turn, contributes to improper functioning of the lysosomal pathways [212].

In particular, Aβ peptide aggregates were found in endosomes, autophagic vesicles, multivesicular bodies, and lysosome [196,213–215], where the presence of APP in the outer lysosomal membrane and γ-secretase activity in lysosomal membranes was shown [216–218]. In addition to being a potential site of Aβ-peptide production, lysosomes are responsible for the complete hydrolysis of APP, APP-CTFs, and Aβ [219,220]. A recent study in APP/*psen1* mice highlighted the importance of lysosomal degradation of APP in preventing its availability to the canonical amyloidogenic pathway [220]. Moreover, over-activation of the lysosomal system is emphasized in EOAD caused by mutations of *psen1* and in transgenic mice overexpressing the PS L146M mutation [221]. Lysosomal system activation progressively worsens as neurons become metabolically compromised. Loss of function gene mutations of Cathepsin D (CatD), a ubiquitous lysosomal protease [222], have been shown to cause progressive neurodegeneration [206,223–226] and, additionally, CatD was found to be decreased in AD patients' fibroblasts [227] and mouse models [228]. Cathepsin B (CatB) was also found to be decreased in a mouse model [228], and the evidence of its important role in reducing Aβ-peptide levels reinforced the interest in cathepsins role in AD [229]. Further evidence of lysosomal dysfunction in AD came from alterations in lysosomal enzymes expression and activity. In particular, activities of the lysosomal glycohydrolases β-Galactosidase (Gal), β-Hexosaminidase (Hex), and α-Mannosidase were found to be increased in fibroblasts from AD patients and presymptomatic individuals with FAD [230] and in the cortex of mouse model of AD, where there was an increase in Gal and Hex activity [231]. Studies conducted in our laboratory led to the observation that alterations in lysosomal glycohydrolases (Gal, Hex, β-Galactosylcerebrosidase, CatB and Cathepsin S (CatS)) were detectable also in peripheral districts such as blood plasma and the Peripheral Blood Mononuclear Cells (PBMCs) of AD patients, and that some of these alterations could discriminate AD from MCI [232,233]. In addition, gangliosides, which are substrates for some lysosomal enzymes [234,235], are altered in AD: monosialo-gangliosides are reported to increase in AD, while more complex gangliosides tend to decrease [202]. It was proposed that APP pathological processing could contribute to gangliosides alteration as AICD appears to down-regulate ganglioside GD3-synthase [236]. Moreover, GM1 and GM3 gangliosides are both capable of binding Aβ-peptide, and these interactions appear to occur early in disease progression, as they have been detected in brains that showed only the earliest signs of AD [202,236,237]. Other evidence of the involvement of gangliosides in AD pathophysiology came from the observation that the inhibition of glycosylceramide synthase, which catalyzes the first step in glycosphingolipid biosynthesis, was correlated with a reduction of amyloidogenic processing of APP [238].

Of note, AD patients showed different exosome profiles compared to MCI patients [239] and differences in synaptic proteins and autolysosomal markers (including CatD and Heat Shock Protein 70) have been found in AD patients' exosome before symptoms onset [240,241], suggesting the potential use of exosomes as diagnostic biomarkers [212].

Potential Lysosomal Biomarkers for AD Diagnosis

The increasing evidence demonstrating the involvement of lysosomal system alterations in AD pathogenesis suggest that monitoring levels of the autophagy proteins as well as lysosomal enzymes and their products may be used as a biomarker for early diagnostic purposes (Figure 6).

For instance, the correlation of AD and Lysosomal Storage Disorders (LSDs) (a group of inherited diseases caused by mutations in lysosomal enzymes [242,243]) was confirmed by the detection of altered sphingolipid metabolism-related molecules in the CSF and serum of AD patients, thereby highlighting their biological relevance as possible biomarkers of the lysosomal proteins [244].

High baseline plasma levels of Ceramide (Cer) 16:0 and Cer 24:0 correlated with an increased risk of AD in older women, and increased Cer 22:0 and Cer 24:0 levels suggested hippocampal volume loss and cognitive decline [125]. Moreover, although a major decrease of CSF sulfatide was observed in the early stages of AD, very little change in its concentration was observed at the advanced stages [125,245,246] (Figure 6).

The levels of lysosomal enzyme Hex in plasma, Gal, and CatB in PBMCs from AD patients allowed discriminating AD versus healthy subjects and AD versus MCI [232,233].

Increased levels of lysosomal proteins (i.e., CatD and Lysosomal-Associated Membrane Protein 1, LAMP1) and decreased levels of synaptic proteins (synaptophysin, synaptopodin, synaptotagmin-2, and neurogranin) were also observed in the neural-derived plasma exosomes of AD patients [247] (Figure 6). Finally, neural-derived exosomes have been proposed as peripheral AD biomarkers. These exosomes, isolated from AD patients, showed significantly higher levels of Aβ1-42-peptide,

Tau, p-Thr181 Tau, and p-Ser396 Tau (compared to controls), which provided a high predictability of disease development in the preclinical stage [247].

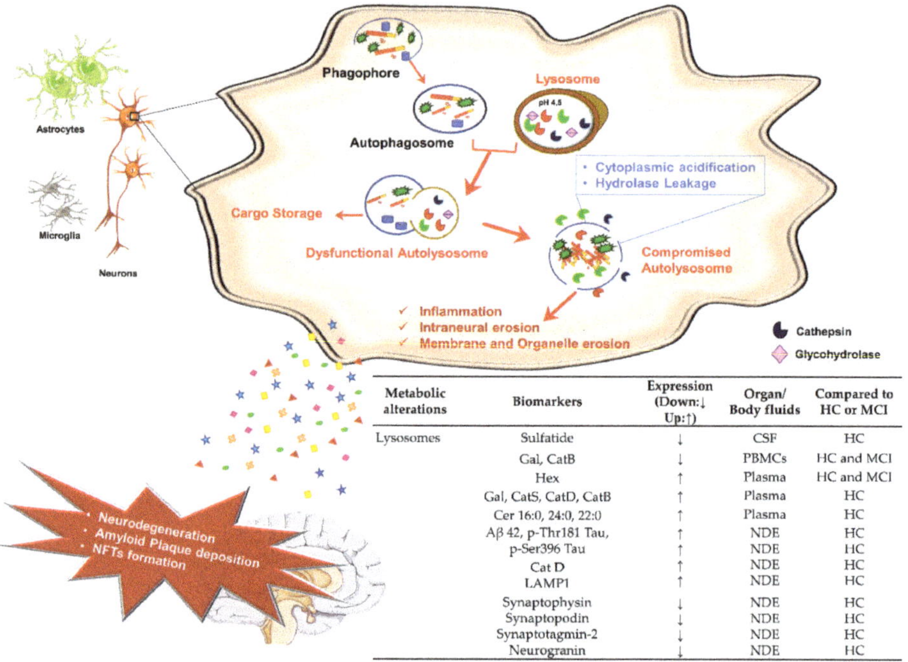

Metabolic alterations	Biomarkers	Expression (Down:↓ Up:↑)	Organ/ Body fluids	Compared to HC or MCI
Lysosomes	Sulfatide	↓	CSF	HC
	Gal, CatB	↓	PBMCs	HC and MCI
	Hex	↑	Plasma	HC and MCI
	Gal, CatS, CatD, CatB	↑	Plasma	HC
	Cer 16:0, 24:0, 22:0	↑	Plasma	HC
	Aβ 42, p-Thr181 Tau,	↑	NDE	HC
	p-Ser396 Tau	↑	NDE	HC
	Cat D	↑	NDE	HC
	LAMP1	↑	NDE	HC
	Synaptophysin	↓	NDE	HC
	Synaptopodin	↓	NDE	HC
	Synaptotagmin-2	↓	NDE	HC
	Neurogranin	↓	NDE	HC

Figure 6. Lysosomes dysfunction and AD. Cartoon schematizes the main altered event leading to the alteration of lysosomal compartment and the table reports potential useful AD biomarkers as described in the text. The arrow direction (up ↑, down ↓) indicates the higher and lower expression levels of the related biomarker in AD with respect to healthy controls (HC) or MCI (Mild Cognitive Impairment). Aβ, β-Amyloid; CatB, Cathepsin B; CatD, Cathepsin D; CatS, Cathepsin S; Cer, Ceramide; Gal, β-Galactosidase; Hex, β-Hexosaminidase; LAMP1, Lysosomal Associated Membrane Protein 1; NDE, Neural-derived exosomes; NFTs, Tau Neurofibrillary Tangles.

3.5. Metabolic Syndrome

In addition to the above-described metabolic alterations, metabolic syndrome also includes systemic organ dysfunction, hyperglycemia, insulin resistance, hypertriglyceridemia, low High-Density Lipoprotein (HDL) cholesterol, obesity, and hypertension [248–252]. As there is a strict correlation with systemic dysfunction, the identification of specific biomolecules as AD biomarkers is still under evaluation. However, FDG-PET imaging has evidenced that a gradual decrease in the cerebral glucose metabolic rate could discriminate patients with MCI that will develop AD from those who will not, suggesting that metabolic dysfunction may have a key role in the early mechanisms of AD [253].

A lipidomic study in the post-mortem human brain highlighted 34 metabolites that might differentiate AD patients from healthy controls [254]. These metabolites belong to pathways of some amino acids (alanine, aspartate, glutamate; arginine, proline, cysteine, methionine, glycine, serine, and threonine), purine metabolism, pantothenate, and coenzyme A biosynthesis [254], which were all altered in AD. This was also confirmed by metabolomic analysis of human plasma indicating the alterations of polyamine and arginine metabolism in MCI patients who afterwards developed AD [255].

4. Cross-Talk between Metabolic Dysfunctions, Neuroinflammation, and Neurodegeneration in AD

The altered metabolic pathways contribute to the chronic status of neuroinflammation and neurodegeneration present in AD patients. In the next paragraphs, we highlight this cross-talk (Figure 7).

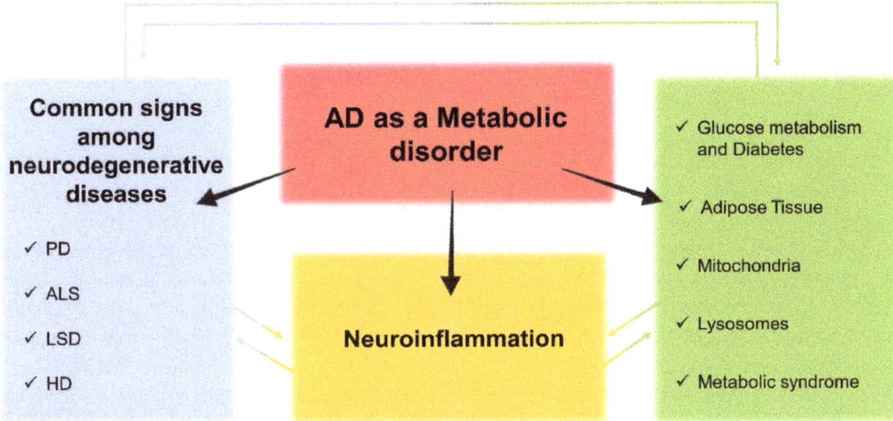

Figure 7. Cross-talk between metabolic dysfunctions, neuroinflammation, and neurodegeneration in AD. Schematic representation. AD, Alzheimer's Disease; ALS, Amyotrophic Lateral Sclerosis; HD, Huntington's disease; LSD, Lysosomal Storage Disorder; PD, Parkinson's Disease.

4.1. *Neuroinflammation, Metabolic Alteration, and AD*

The metabolic alterations occurring in AD described in the previous paragraphs all contribute to the worsening of the neuroinflammation state that characterizes this disease [256,257]. In AD, impaired glucose metabolism and insulin resistance are deeply correlated to the chronic inflammation state [79] as well as to adipose tissue that releases numerous pro-inflammatory adipokines [104–106,110]. Moreover, mitochondria dysfunctions promote oxidative stress and energy metabolism impairment [144,146], while lysosomal alterations lead to an impairment of the autophagic pathway and of its general degradation activity [197,200,212]. All the above events partake in the progress of AD neuroinflammation state [197,200,212].

Indeed, the CNS microglia and astrocytes have a central role in the neuroinflammation process occurring in AD [258–261]. In particular, the microglia has a phagocytic role that helps clear damaged neurons and remove pathogens other than facilitating tissue repair [262], while astrocytes are involved in many functions such as neurotransmitter uptake and recycling, the modulation of synaptic activity, and maintenance of the correct permeability of the BBB [260,261,263]. The integrity of the BBB in AD is compromised, and it is likely that both Tau and Aβ peptide may be involved in the loss of BBB integrity, exacerbating the neurodegenerative process and associated inflammatory responses [264]. In addition, further loosening of the tight junctions may be caused by the excessive release by the microglia of pro-inflammatory cytokines observed in AD, such as TNFα, IL-1β, and IL-17A [265,266]. Microglia appears to directly interact with soluble Aβ oligomers and Aβ fibrils via specific receptors including class A scavenger receptor A1, Cluster of differentiation (CD) 36, CD14, α6β1 integrin, CD47, and Toll-like receptors (TLR2, TLR4, TLR6, and TLR9) [267,268]. This binding leads to microglia activation that results in pro-inflammatory cytokines and chemokines production [267,268]. Moreover, astrocytes are highly activated in AD in response to excess Aβ peptide, which also leads to the upregulation of pro-inflammatory factors [269,270]. Among the different transcription factors, NF-κB is considered a primary regulator of inflammatory responses because its activation enhances

BACE1 expression, thus stimulating the cleavage of APP and Aβ-peptide production [270]. In turn, BACE1 promotes the activity of secretases that enhance the production of Aβ peptide from APP. In detail, the reactive astrocytes express BACE and PSEN1 genes [270].

Finally, in a recent work, Ising et al. [271] have demonstrated that the inhibition of the NLRP3 (see the Abbreviation List) inflammasome activity in mouse models significantly reduced Tau phosphorylation in different brain regions and prevented cognitive decline, thus indicating that the NLRP3 inflammasome may have an important role in AD pathophysiology [271].

Marker of Neuroinflammation for AD Diagnosis

The molecules that partake in AD-related neuroinflammation processes are extensively studied as potential markers that could be monitored for early AD diagnosis. For example, a recent meta-analysis reported high CSF concentration of soluble TREM2, MCP-1, Chitinase 3-like 1 (YKL-40), and TGF-β in AD patients compared to healthy controls and increased tumor necrosis factor receptor 1 and 2 in peripheral blood of AD patients but not MCI patients [272].

TREM2 is a membrane receptor that plays a key role in mediating the phagocytic clearance during apoptosis, including the microglial phagocytosis of apoptotic neurons, and in modulating the inflammatory response caused by damaged myelin and amyloid plaques [273–275]. The tailored proteolytic cleavage of TREM2 at the H157–S158 peptide bond by A Disintegrin and metalloproteinase domain-containing protein (ADAM) 10 and ADAM17 proteases produces the soluble TREM2 [276,277], which generally is highly released in the CSF, where its levels are considered positive markers for neuronal injury [278,279]. For instance, the levels of soluble TREM2 increased in patients with autosomal dominant AD, and this increase was measured in the CSF of 218 subjects, 127 mutation carriers (MCs) and 91 noncarriers (NCs). In particular, the CSF levels of soluble TREM2 increased in MCs with respect to NCs 5 years before the appearance of symptoms, and these levels remained significantly higher until 5 years after symptoms onset [280].

In addition, peripheral cytokines levels such as IL-1β, IL-2, IL-6, IL-12, IL-18, IL-23, and TGF-β were seen to be altered in AD patients and could be used as candidate biomarkers [281–284].

However, when considering using peripheral inflammatory biomarkers for detecting asymptomatic phases of AD, it has to be taken into account that the alteration of these molecules is common to diverse neurodegenerative disease; thus, it is necessary to correlate their levels with other markers [282,285,286]. Nevertheless, these markers together with other characteristics of AD could serve as a peripheral panel that is useful to distinguish early AD from other neurodegenerative diseases.

4.2. *Neurodegeneration, Metabolic Alteration, and AD*

The numerous altered biological systems described so far in AD highlight the concept that the characterization of this neurodegenerative disease requires a deep understanding of many diverse pathological events. In this regard, it is crucial to underline that AD shares different pathological aspects with other neurodegenerative diseases, which could make early diagnosis difficult [287]. As a matter of fact, AD shares several features with Parkinson's Disease (PD), Huntington's disease (HD), and Amyotrophic Lateral Sclerosis (ALS) [286–292]. These common general pathways involve protein misfolding and aggregation, mitochondrial dysfunctions, oxidative stress and ROS production, neuroinflammation and phosphorylation impairment, and microRNA-altered expression [290,293,294], which appear to be changed concurrently [287,295–299].

The toxicity of misfolded and aggregated proteins is well-established, as it was observed in different neurodegenerative diseases (i.e., Aβ peptide in AD, α-Synuclein in PD, and Huntingtin in HD), and even though it occurs in different brain regions for each disease, confirms a crucial role of the toxicity of macromolecules accumulation in these disorders [300]. In this context, LSDs, due to their improper accumulation of disease-specific metabolites, share neural cell death, neurodegeneration, and other common interplay with the above-mentioned neurodegenerative diseases [287,301,302]. For example, cholesterol accumulates in late endosomes and lysosomes in the juvenile form of Niemann

Pick type C disease characterized by progressive neurodegeneration similar to AD, including NFTs formation and increased APP amyloidogenic processing [206,303,304]. An extensive study reported APP-CTFs, Aβ peptide, and α-Synuclein accumulation in the Sandhoff mouse model [305–307] and ganglioside-bound Aβ-peptide in post-mortem human brains of patients with GM1 gangliosidosis and GM2 gangliosidosis [308]. Other evidence came from the presence of soluble Aβ(1-40)-peptide in post-mortem brains of patients with Mucopolysaccharidosis type 1 [309], in the mouse models of Mucopolysaccharidosis type IIIB [310], and the neuropathic form of Gaucher disease [311]. The accumulation of APP and APP-CTFs in the hippocampus of Neu1-deficient mice, a model of the LSD sialidosis, has also been demonstrated [202,211]. Common signs among AD, PD, HD, and ALS came also from the pathways regulating the clearance of misfolded and abnormal proteins that are associated with neurodegenerative diseases [312–316].

The toxic accumulation of proteins and metabolites is also partially responsible for the neuroinflammation process that is common to almost all neurodegenerative diseases. Today, innate immunity, with the particular relevance of glial cells, is considered to have a central role in brain homeostasis neurodegenerative events [317]. Indeed, sequencing and GWAS studies contributed to highlight the role of immunity and microglia as important contributors to the pathological events occurring in neurodegenerative diseases. This strengthened the idea of reversing degenerative events by reducing inflammation, even though it is still not clear at which time point this approach could be determinant [318].

Another common feature in neurodegenerative diseases is the oxidative stress caused by ROS and Reactive Nitrogen Species (RNS) brain accumulation [319–322]. The oxidative damage of different molecules within neurons (such as lipids, DNA, and proteins) was observed in AD, PD, HD, and ALS. In particular, ROS derived from Nicotinamide Adenine Dinucleotide Phosphate Oxidase 2 [320], which were detected in endothelial cells, platelets, neurons, astrocytes, and microglia, were suggested to be critical in mediating inflammation and apoptotic pathways in the CNS [261,321,323].

Neurodegeneration and Biomarkers for AD

The common pathways involved in neurodegenerative diseases, such as protein aggregation, neuroinflammation, and dysfunction in the metabolic and autophagic system, make the diagnosis of each specific disease more difficult. Currently, it is possible to clearly discriminate one neurodegenerative disease among others by combining several common and genetic-specific markers, whether possible. These include AD rare mutations in APP, PSEN1, and PSEN2 genes [324]; PD-specific genetic markers (e.g., in SNCA, LRRK2, PRKN, UCHL1, PINK1, DJ-1, NR4A, see Abbreviation List) [324]; ALS genetic mutations in more than a dozen genes that have been found to cause familial ALS (ALS2, NEFH, FUS, TARDBP, C9orf72, and SOD1 [325], see Abbreviation List); and the HD characteristic mutation in the Huntingtin gene [324]. Finally, the LSDs can be identified by disease-specific genetic mutations that lead to enzyme deficiencies within the lysosomes, resulting in an accumulation of undegraded substrate [326].

5. Concluding Remarks

In this review, we have documented the involvement of metabolic alterations in AD pathogenesis and highlighted the contribution of these findings to improving a suitable biomarkers' panel for the early diagnosis of AD. Indeed, the field is under constant update due to the metanalysis investigations that integrate old and new findings from proteomic, metabolomic, and transcriptomic studies comparing AD versus healthy subjects or AD versus MCI or other neurodegenerative disorders [327–329]. Advancing is also from an available innovative in vitro cell model of AD or other neurogenerative diseases [330–335].

The overall findings provide a portfolio of biomolecules with potential therapeutic AD activity. This may help overcoming the absence of an effective therapeutic strategy for AD, as the available therapeutic agents are up to now just symptomatic treatments focused on improving cognitive symptoms: 3 acetylocholinesterase inhibitors (donepezil, galantamine, and rivastigmine) that help maintain high acetylcholine levels, and the noncompetitive *n*-Methyl-D-Aspartate (NMDA) receptor antagonist memantine used for counteracting glutamate excitatory neurotoxicity [336–338].

However, due to the discovery of the new AD biomarkers, there are currently 136 active trials involving 121 therapeutic agents at different stages. The disease-modifying drugs that have been studied in the last years that eventually reached phase 3 are mainly focused on counteracting (i) the deposition of extracellular amyloid β plaques mainly consisting of immunotherapy approaches and (ii) the synaptic plasticity or neuroprotection small molecules [339].

Small molecules in phase 3 are targeted to (i) Synaptic plasticity/Neuroprotection: AGB101, ANAVEX2-73, BHV4157, Icosapent Ethyl; (ii) Inflammation/Infection/Immunity: ALZT-OP1a, Azeliragon, COR388, Masitinib; (iii) Metabolism and bioenergetics: Metformin, Tricaprilin; (iv) Vasculature: Losartan+Amlodipine+Atorvastan and (v) Tau: TRx0237 [339]. In particular, TRx0237 (LMTX), which is a Tau aggregation inhibitor that decreases the level of aggregated Tau proteins [340], is a promising agent, and the ongoing phase 3 trial should be completed in 2022 [341]. Of note, 18F-FDG-PET is used as an outcome measure to monitor the efficacy of the treatment [341,342].

Other than small molecules, important immunotherapy strategies have been developed and represent another encouraging approach that may lead to the finding of some effective treatment for AD. Monoclonal antibodies have been developed to target Aβ and Tau oligomers, and the immunotherapy agents already in phase 3 are all targeted to amyloid: Aducanumab (Monoclonal antibody directed at plaques and oligomers), BAN2401 (Monoclonal antibody directed at protofibrils), Gantenerumab (Monoclonal antibody directed at plaques and oligomers), and Solanezumab (Monoclonal antibody directed at monomers) [339].

The active vaccination agent CAD106 (Amyloid Vaccine consisting of multiple copies of the Aβ1–6 peptide coupled to a carrier containing bacteriophage Qβ coat proteins) was a promising therapy that entered phase 2/3 in 2015 [343] and continued phase 3 in 2019, but in September 2019, Novartis noted that the CAD106 project had been retired [344,345].

Finally, the recent findings on the regulatory role of microRNAs in AD pathophysiology encouraged the possibility of modulating mRNA transcription using Antisense Oligonucleotides (ASOs). For example, ASOs targeting GSK-3β in Senescence-Accelerated Prone mice P8 mice improved memory and learning and decreased oxidative stress [346], while ASOs targeting human Tau expressed in one AD mouse model were able to ameliorate pathological Tau deposition [347].

In conclusion, the overall findings here reviewed envisage the great molecular complexity of AD pathogenesis, validate the difficulty to establish an early decision diagnosis, and also indicate a new roadmap that collects canonical AD hallmarks and new biomolecules coming from altered metabolic pathways.

Author Contributions: Conceptualization, C.A., I.T., M.B., F.M., and S.M.; Writing—review and editing, C.A., I.T., M.B., C.E., F.M., and S.M.; Supervision, S.M. All authors have read and agreed to the published version of the manuscript.

Funding: This research received no external funding.

Conflicts of Interest: The authors declare no conflict of interest.

Abbreviations

A2M	Alpha-2-Macroglobulin
ABCA2	Adenosine Triphosphate Binding Cassette Subfamily A Member 2
ABCA7	Adenosine Triphosphate Binding Cassette Subfamily A Member 7
ADAM10	A Disintegrin And Metalloprotease Domain 10
AK2	Adenylate Kinase 2
ALS2	Alsin Rho Guanine Nucleotide Exchange Factor Amyotrophic Lateral Sclerosis 2 APOE
ATP8B4	Adenosine Triphosphatase Phospholipid Transporting 8B4
BACE1	Beta-Secretase 1

BIN1	Bridging Integrator-1
C9orf72	C9orf72-SMCR8 complex subunit
CD33	Sialic Acid-Binding Ig-Like Lectin 3
CLU	Clusterin
CR1	Complement C3b/C4b receptor 1 (Knops blood group)
CTNNA3	Catenin Alpha 3
CYCS	Cytochrome C
DJ-1	Parkinsonism-associated deglycase
DLD	Dihydrolipoyl dehydrogenase
DNMBP	Dynamin Binding Protein
FUS	Fusion RNA binding protein
GAB2	Growth Factor Receptor Bound Protein 2 Associated Binding Protein 2
GATM	Glycine Amidinotransferase
HSPA9	Stress-70 protein
LRRK2	Leucine-Rich Repeat Kinase2
MS4A6A	Membrane Spanning 4-Domains A6A
NEFH	Neurofilament Heavy
NLRP3	Nucleotide-Binding Oligomerization Domain, Leucine-Rich Repeat and Pyrin Domain Containing 3
NR4A2	Nuclear Receptor Subfamily 4 Group A Member 2
OLR1	Oxidized Low Density Lipoprotein Receptor 1
OTC	Ornithine Carbamoyltransferase
PCK2	Phosphoenolpyruvate Carboxykinase 2
PICALM	Phosphatidylinositol Binding Clathrin Assembly Protein
PINK1	Phosphatase and Tensin Homolog-induced Kinase 1
PLD3	Phospholipase D family member 3
PRKN	Parkin RBR E3 ubiquitin protein ligase
SNCA	Synuclein alpha
SOD1	Superoxide Dismutase 1
SORL1	Sortilin Related Receptor 1
TARDBP	TAR DNA Binding Protein
TOMM40	Translocase of Outer Mitochondrial Membrane 40
TREM2	Triggering Receptor Expressed on Myeloid Cells 2
UCHL1	Ubiquitin C-Terminal Hydrolase L1

References

1. Dementia. Available online: https://www.who.int/news-room/fact-sheets/detail/dementia (accessed on 28 July 2020).
2. Jansen, I.E.; Savage, J.E.; Watanabe, K.; Bryois, J.; Williams, D.M.; Steinberg, S.; Sealock, J.; Karlsson, I.K.; Hägg, S.; Athanasiu, L.; et al. Genome-wide meta-analysis identifies new loci and functional pathways influencing Alzheimer's disease risk. *Nat. Genet.* **2019**, *51*, 404–413. [CrossRef] [PubMed]
3. Sala Frigerio, C.; De Strooper, B. Alzheimer's disease mechanisms and emerging roads to novel therapeutics. *Annu. Rev. Neurosci.* **2016**, *39*, 57–79. [CrossRef] [PubMed]
4. Reitz, C.; Brayne, C.; Mayeux, R. Epidemiology of Alzheimer disease. *Nat. Rev. Neurol.* **2011**, *7*, 137–152. [CrossRef]
5. Selkoe, D.J.; Hardy, J. The amyloid hypothesis of Alzheimer's disease at 25 years. *EMBO Mol. Med.* **2016**, *8*, 595–608. [CrossRef] [PubMed]
6. Jack, C.R.; Bennett, D.A.; Blennow, K.; Carrillo, M.C.; Dunn, B.; Haeberlein, S.B.; Holtzman, D.M.; Jagust, W.; Jessen, F.; Karlawish, J.; et al. NIA-AA Research framework: Toward a biological definition of Alzheimer's disease. *Alzheimer's Dement.* **2018**, *14*, 535–562. [CrossRef]
7. Maarouf, C.L.; Walker, J.E.; Sue, L.I.; Dugger, B.N.; Beach, T.G.; Serrano, G.E. Impaired hepatic amyloid-beta degradation in Alzheimer's disease. *PLoS ONE* **2018**, *13*, e0203659. [CrossRef]
8. Bertram, L.; Tanzi, R.E. Genome-wide association studies in Alzheimer's disease. *Hum. Mol. Genet.* **2009**, *18*, R137–R145. [CrossRef]

9. Tcw, J.; Goate, A.M. Genetics of β-amyloid precursor protein in Alzheimer's disease. *Cold Spring Harb. Perspect. Med.* **2017**, *7*, a024539. [CrossRef]
10. Lanoiselée, H.-M.M.; Nicolas, G.; Wallon, D.; Rovelet-Lecrux, A.; Lacour, M.; Rousseau, S.; Richard, A.-C.C.; Pasquier, F.; Rollin-Sillaire, A.; Martinaud, O.; et al. APP, PSEN1, and PSEN2 mutations in early-onset Alzheimer disease: A genetic screening study of familial and sporadic cases. *PLoS Med.* **2017**, *14*, e1002270. [CrossRef]
11. Batelli, S.; Albani, D.; Prato, F.; Polito, L.; Franceschi, M.; Gavazzi, A.; Forloni, G. Early-onset Alzheimer disease in an Italian family with presenilin-1 double mutation E318G and G394V. *Alzheimer Dis. Assoc. Disord.* **2008**, *22*, 184–187. [CrossRef]
12. Chávez-Gutiérrez, L.; Szaruga, M. Mechanisms of neurodegeneration—Insights from familial Alzheimer's disease. In *Seminars in Cell & Developmental Biology*; Academic Press: Cambridge, MA, USA, 2020.
13. Mutations | ALZFORUM. Available online: https://www.alzforum.org/mutations (accessed on 30 July 2020).
14. Dorszewska, J.; Prendecki, M.; Oczkowska, A.; Dezor, M.; Kozubski, W. Molecular basis of familial and sporadic Alzheimer's disease. *Curr. Alzheimer Res.* **2016**, *13*, 952–963. [CrossRef] [PubMed]
15. Armstrong, R. Risk factors for Alzheimer's disease. *Folia Neuropathol.* **2019**, *57*, 87–105. [CrossRef] [PubMed]
16. Lane, C.A.; Hardy, J.; Schott, J.M. Alzheimer's disease. *Eur. J. Neurol.* **2018**, *25*, 59–70. [CrossRef] [PubMed]
17. Van der Lee, S.J.; Wolters, F.J.; Ikram, M.K.; Hofman, A.; Ikram, M.A.; Amin, N.; van Duijn, C.M. The effect of APOE and other common genetic variants on the onset of Alzheimer's disease and dementia: A community-based cohort study. *Lancet Neurol.* **2018**, *17*, 434–444. [CrossRef]
18. Shi, Y.; Yamada, K.; Liddelow, S.A.; Smith, S.T.; Zhao, L.; Luo, W.; Tsai, R.M.; Spina, S.; Grinberg, L.T.; Rojas, J.C.; et al. ApoE4 markedly exacerbates tau-mediated neurodegeneration in a mouse model of tauopathy. *Nature* **2017**, *549*, 523–527. [CrossRef] [PubMed]
19. Shi, Y.; Holtzman, D.M. Interplay between innate immunity and Alzheimer disease: APOE and TREM2 in the spotlight. *Nat. Rev. Immunol.* **2018**, *18*, 759–772. [CrossRef]
20. Hartley, S.L.; Handen, B.L.; Devenny, D.A.; Hardison, R.; Mihaila, I.; Price, J.C.; Cohen, A.D.; Klunk, W.E.; Mailick, M.R.; Johnson, S.C.; et al. Cognitive functioning in relation to brain amyloid-β in healthy adults with Down syndrome. *Brain* **2014**, *137*, 2556–2563. [CrossRef]
21. Bertram, L.; Tanzi, R.E. Alzheimer disease risk genes: 29 and counting. *Nat. Rev. Neurol.* **2019**, *15*, 191–192. [CrossRef]
22. Misra, A.; Chakrabarti, S.S.; Gambhir, I.S. New genetic players in late-onset Alzheimer's disease: Findings of genome-wide association studies. *Indian J. Med. Res.* **2018**, *148*, 135–144.
23. Giri, M.; Zhang, M.; Lü, Y. Genes associated with Alzheimer's disease: An overview and current status. *Clin. Interv. Aging* **2016**, *11*, 665–681. [CrossRef]
24. Lambert, J.C.; Ibrahim-Verbaas, C.A.; Harold, D.; Naj, A.C.; Sims, R.; Bellenguez, C.; Jun, G.; DeStefano, A.L.; Bis, J.C.; Beecham, G.W.; et al. Meta-analysis of 74,046 individuals identifies 11 new susceptibility loci for Alzheimer's disease. *Nat. Genet.* **2013**, *45*, 1452–1458. [CrossRef] [PubMed]
25. Kunkle, B.W.; Grenier-Boley, B.; Sims, R.; Bis, J.C.; Damotte, V.; Naj, A.C.; Boland, A.; Vronskaya, M.; van der Lee, S.J.; Amlie-Wolf, A.; et al. Genetic meta-analysis of diagnosed Alzheimer's disease identifies new risk loci and implicates Aβ, tau, immunity and lipid processing. *Nat. Genet.* **2019**, *51*, 414–430. [CrossRef] [PubMed]
26. SNPedia. Available online: https://www.snpedia.com/index.php/SNPedia (accessed on 30 July 2020).
27. Sala Frigerio, C.; Wolfs, L.; Fattorelli, N.; Thrupp, N.; Voytyuk, I.; Schmidt, I.; Mancuso, R.; Chen, W.T.; Woodbury, M.E.; Srivastava, G.; et al. The major risk factors for Alzheimer's disease: Age, sex, and genes modulate the microglia response to Aβ plaques. *Cell Rep.* **2019**, *27*, 1293–1306. [CrossRef] [PubMed]
28. Guerreiro, R.; Bras, J. The age factor in Alzheimer's disease. *Genome Med.* **2015**, *7*, 1–3. [CrossRef]
29. Verhaar, B.J.H.; Leeuw, F.A.; Doorduijn, A.S.; Fieldhouse, J.L.P.; Rest, O.; Teunissen, C.E.; Berckel, B.N.M.; Barkhof, F.; Visser, M.; Schueren, M.A.E.; et al. Nutritional status and structural brain changes in Alzheimer's disease: The NUDAD project. *Alzheimer's Dement. Diagn. Assess. Dis. Monit.* **2020**, *12*, e12063.
30. Kim, S.; Kim, M.J.; Kim, S.; Kang, H.S.; Lim, S.W.; Myung, W.; Lee, Y.; Hong, C.H.; Choi, S.H.; Na, D.L.; et al. Gender differences in risk factors for transition from mild cognitive impairment to Alzheimer's disease: A CREDOS study. *Compr. Psychiatry* **2015**, *62*, 114–122. [CrossRef]
31. Podcasy, J.; Epperson, C. Considering sex and gender in Alzheimer disease and other dementias. *Dialogues Clin. Neurosci.* **2016**, *18*, 437–446.

32. Vasan, R.S.; Beiser, A.; Seshadri, S.; Larson, M.G.; Kannel, W.B.; D'Agostino, R.B.; Levy, D. Residual lifetime risk for developing hypertension in middle-aged women and men: The Framingham heart study. *J. Am. Med. Assoc.* **2002**, *287*, 1003–1010. [CrossRef]
33. Kalaria, R.N. Vascular basis for brain degeneration: Faltering controls and risk factors for dementia. *Nutr. Rev.* **2010**, *68*, S74–S87. [CrossRef]
34. Swerdlow, R.H. Pathogenesis of Alzheimer's disease. *Clin. Interv. Aging* **2007**, *2*, 347–359.
35. Suzhen, D.; Yale, D.; Feng, G.; Yinghe, H.; Zheng, Z. Advances in the pathogenesis of Alzheimer's disease: A re-evaluation of amyloid cascade hypothesis. *Transl. Neurodegener.* **2012**, *1*, 18. [CrossRef] [PubMed]
36. De Strooper, B.; Vassar, R.; Golde, T. The secretases: Enzymes with therapeutic potential in Alzheimer disease. *Nat. Rev. Neurol.* **2010**, *6*, 99–107. [CrossRef] [PubMed]
37. Lee, S.J.C.; Nam, E.; Lee, H.J.; Savelieff, M.G.; Lim, M.H. Towards an understanding of amyloid-β oligomers: Characterization, toxicity mechanisms, and inhibitors. *Chem. Soc. Rev.* **2017**, *46*, 310–323. [CrossRef] [PubMed]
38. Siddiqi, M.K.; Malik, S.; Majid, N.; Alam, P.; Khan, R.H. Cytotoxic species in amyloid-associated diseases: Oligomers or mature fibrils. *Adv. Protein Chem. Struct. Biol.* **2019**, *118*, 333–369. [PubMed]
39. Jin, M.; Shepardson, N.; Yang, T.; Chen, G.; Walsh, D.; Selkoe, D.J. Soluble amyloid β-protein dimers isolated from Alzheimer cortex directly induce Tau hyperphosphorylation and neuritic degeneration. *Proc. Natl. Acad. Sci. USA* **2011**, *108*, 5819–5824. [CrossRef] [PubMed]
40. Musiek, E.S.; Holtzman, D.M. Three dimensions of the amyloid hypothesis: Time, space and "wingmen". *Nat. Neurosci.* **2015**, *18*, 800–806. [CrossRef]
41. Fitzpatrick, A.W.P.; Falcon, B.; He, S.; Murzin, A.G.; Murshudov, G.; Garringer, H.J.; Crowther, R.A.; Ghetti, B.; Goedert, M.; Scheres, S.H.W. Cryo-EM structures of tau filaments from Alzheimer's disease. *Nature* **2017**, *547*, 185–190. [CrossRef]
42. Falcon, B.; Zhang, W.; Schweighauser, M.; Murzin, A.G.; Vidal, R.; Garringer, H.J.; Ghetti, B.; Scheres, S.H.W.; Goedert, M. Tau filaments from multiple cases of sporadic and inherited Alzheimer's disease adopt a common fold. *Acta Neuropathol.* **2018**, *136*, 699–708. [CrossRef]
43. Cicognola, C.; Brinkmalm, G.; Wahlgren, J.; Portelius, E.; Gobom, J.; Cullen, N.C.; Hansson, O.; Parnetti, L.; Constantinescu, R.; Wildsmith, K.; et al. Novel tau fragments in cerebrospinal fluid: Relation to tangle pathology and cognitive decline in Alzheimer's disease. *Acta Neuropathol.* **2019**, *137*, 279–296. [CrossRef]
44. Badiola, N.; Suarez-Calvet, M.; Lleo, A. Tau phosphorylation and aggregation as a therapeutic target in tauopathies. *CNS Neurol. Disord. Drug Targets* **2012**, *9*, 727–740. [CrossRef]
45. Guo, T.; Noble, W.; Hanger, D.P. Roles of tau protein in health and disease. *Acta Neuropathol.* **2017**, *133*, 665–704. [CrossRef] [PubMed]
46. Wang, Y.; Mandelkow, E. Tau in physiology and pathology. *Nat. Rev. Neurosci.* **2016**, *17*, 5–21. [CrossRef] [PubMed]
47. Hanger, D.P.; Anderton, B.H.; Noble, W. Tau phosphorylation: The therapeutic challenge for neurodegenerative disease. *Trends Mol. Med.* **2009**, *15*, 112–119. [CrossRef] [PubMed]
48. Wang, J.Z.; Xia, Y.Y.; Grundke-Iqbal, I.; Iqbal, K. Abnormal hyperphosphorylation of tau: Sites, regulation, and molecular mechanism of neurofibrillary degeneration. *J. Alzheimer's Dis.* **2013**, *33*, S123–S139. [CrossRef] [PubMed]
49. Hernández, F.; Gómez de Barreda, E.; Fuster-Matanzo, A.; Lucas, J.J.; Avila, J. GSK3: A possible link between beta amyloid peptide and tau protein. *Exp. Neurol.* **2010**, *223*, 322–325. [CrossRef]
50. Min, S.W.; Cho, S.H.; Zhou, Y.; Schroeder, S.; Haroutunian, V.; Seeley, W.W.; Huang, E.J.; Shen, Y.; Masliah, E.; Mukherjee, C.; et al. Acetylation of tau inhibits its degradation and contributes to tauopathy. *Neuron* **2010**, *67*, 953–966. [CrossRef]
51. FDA-NIH Biomarker Working Group. *BEST (Biomarkers, EndpointS, and other Tools) Resource*; FDA-NIH Biomarker Working Group: Silver Spring, MD, USA, 2017.
52. Karley, D.; Gupta, D.; Tiwari, A. Biomarker for Cancer: A great Promise for Future. *World J. Oncol.* **2011**, *2*, 151–157.
53. Johnson, K.A.; Fox, N.C.; Sperling, R.A.; Klunk, W.E. Brain imaging in Alzheimer disease. *Cold Spring Harb. Perspect. Med.* **2012**, *2*, a006213. [CrossRef]
54. Wilson, H.; Pagano, G.; Politis, M. Dementia spectrum disorders: Lessons learnt from decades with PET research. *J. Neural Transm.* **2019**, *126*, 233–251. [CrossRef]

55. Hanseeuw, B.J.; Betensky, R.A.; Jacobs, H.I.L.; Schultz, A.P.; Sepulcre, J.; Becker, J.A.; Cosio, D.M.O.; Farrell, M.; Quiroz, Y.T.; Mormino, E.C.; et al. Association of Amyloid and Tau with cognition in preclinical Alzheimer disease: A longitudinal study. *JAMA Neurol.* **2019**, *76*, 915–924. [CrossRef]
56. Villemagne, V.L.; Fodero-Tavoletti, M.T.; Masters, C.L.; Rowe, C.C. Tau imaging: Early progress and future directions. *Lancet Neurol.* **2015**, *14*, 114–124. [CrossRef]
57. Leuzy, A.; Chiotis, K.; Lemoine, L.; Gillberg, P.G.; Almkvist, O.; Rodriguez-Vieitez, E.; Nordberg, A. Tau PET imaging in neurodegenerative tauopathies—Still a challenge. *Mol. Psychiatry* **2019**, *24*, 1112–1134. [CrossRef] [PubMed]
58. Werry, E.L.; Bright, F.M.; Piguet, O.; Ittner, L.M.; Halliday, G.M.; Hodges, J.R.; Kiernan, M.C.; Loy, C.T.; Kril, J.J.; Kassiou, M. Recent developments in TSPO PET imaging as a biomarker of neuroinflammation in neurodegenerative disorders. *Int. J. Mol. Sci.* **2019**, *20*, 3161. [CrossRef] [PubMed]
59. Márquez, F.; Yassa, M.A. Neuroimaging biomarkers for Alzheimer's disease. *Mol. Neurodegener.* **2019**, *14*, 1–14. [CrossRef]
60. Blennow, K.; Dubois, B.; Fagan, A.M.; Lewczuk, P.; De Leon, M.J.; Hampel, H. Clinical utility of cerebrospinal fluid biomarkers in the diagnosis of early Alzheimer's disease. *Alzheimer's Dement.* **2015**, *11*, 58–69. [CrossRef]
61. Biscetti, L.; Salvadori, N.; Farotti, L.; Cataldi, S.; Eusebi, P.; Paciotti, S.; Parnetti, L. The added value of Aβ42/Aβ40 in the CSF signature for routine diagnostics of Alzheimer's disease. *Clin. Chim. Acta* **2019**, *494*, 71–73. [CrossRef]
62. Forlenza, O.V.; Radanovic, M.; Talib, L.L.; Aprahamian, I.; Diniz, B.S.; Zetterberg, H.; Gattaz, W.F. Cerebrospinal fluid biomarkers in Alzheimer's disease: Diagnostic accuracy and prediction of dementia. *Alzheimer's Dement. Diagn. Assess. Dis. Monit.* **2015**, *1*, 455–463. [CrossRef]
63. De Souza, L.C.; Chupin, M.; Lamari, F.; Jardel, C.; Leclercq, D.; Colliot, O.; Lehéricy, S.; Dubois, B.; Sarazin, M. CSF tau markers are correlated with hippocampal volume in Alzheimer's disease. *Neurobiol. Aging* **2012**, *33*, 1253–1257. [CrossRef]
64. Risacher, S.L.; Fandos, N.; Romero, J.; Sherriff, I.; Pesini, P.; Saykin, A.J.; Apostolova, L.G. Plasma amyloid beta levels are associated with cerebral amyloid and tau deposition. *Alzheimer's Dement. Diagn. Assess. Dis. Monit.* **2019**, *11*, 510–519. [CrossRef]
65. Ovod, V.; Ramsey, K.N.; Mawuenyega, K.G.; Bollinger, J.G.; Hicks, T.; Schneider, T.; Sullivan, M.; Paumier, K.; Holtzman, D.M.; Morris, J.C.; et al. Amyloid β concentrations and stable isotope labeling kinetics of human plasma specific to central nervous system amyloidosis. *Alzheimer's Dement.* **2017**, *13*, 841–849. [CrossRef]
66. Duran-Aniotz, C.; Hetz, C. Glucose metabolism: A sweet relief of Alzheimer's disease. *Curr. Biol.* **2016**, *26*, R806–R809. [CrossRef]
67. An, Y.; Varma, V.R.; Varma, S.; Casanova, R.; Dammer, E.; Pletnikova, O.; Chia, C.W.; Egan, J.M.; Ferrucci, L.; Troncoso, J.; et al. Evidence for brain glucose dysregulation in Alzheimer's disease. *Alzheimer's Dement.* **2018**, *14*, 318–329. [CrossRef] [PubMed]
68. Chen, Z.; Zhong, C. Decoding Alzheimer's disease from perturbed cerebral glucose metabolism: Implications for diagnostic and therapeutic strategies. *Prog. Neurobiol.* **2013**, *108*, 21–43. [CrossRef] [PubMed]
69. Winkler, E.A.; Nishida, Y.; Sagare, A.P.; Rege, S.V.; Bell, R.D.; Perlmutter, D.; Sengillo, J.D.; Hillman, S.; Kong, P.; Nelson, A.R.; et al. GLUT1 reductions exacerbate Alzheimer's disease vasculo-neuronal dysfunction and degeneration. *Nat. Neurosci.* **2015**, *18*, 521–530. [CrossRef] [PubMed]
70. Ferrucci, L. The Baltimore longitudinal study of aging (BLSA): A 50-year-long journey and plans for the future. *J. Gerontol. Ser. A* **2008**, *63*, 1416–1419. [CrossRef] [PubMed]
71. Willette, A.A.; Bendlin, B.B.; Starks, E.J.; Birdsill, A.C.; Johnson, S.C.; Christian, B.T.; Okonkwo, O.C.; La Rue, A.; Hermann, B.P.; Koscik, R.L.; et al. Association of insulin resistance with cerebral glucose uptake in late middle-aged adults at risk for Alzheimer disease. *JAMA Neurol.* **2015**, *72*, 1013–1020. [CrossRef]
72. Alexander, C.M.; Landsman, P.B.; Teutsch, S.M.; Haffner, S.M. NCEP-defined metabolic syndrome, diabetes, and prevalence of coronary heart disease among NHANES III participants age 50 years and older. *Diabetes* **2003**, *52*, 1210–1214. [CrossRef]
73. van Duinkerken, E.; Ryan, C.M. Diabetes mellitus in the young and the old: Effects on cognitive functioning across the life span. *Neurobiol. Dis.* **2020**, *134*, 104608. [CrossRef]
74. Zhu, Y.; Ding, X.; She, Z.; Bai, X.; Nie, Z.; Wang, F.; Wang, F.; Geng, X. Exploring shared pathogenesis of Alzheimer's disease and type 2 diabetes mellitus via co-expression networks analysis. *Curr. Alzheimer Res.* **2020**, *17*, 566–575. [CrossRef]

75. Dodd, G.T.; Tiganis, T. Insulin action in the brain: Roles in energy and glucose homeostasis. *J. Neuroendocrinol* **2017**, *29*, e12513. [CrossRef]
76. Nakabeppu, Y. Origins of brain insulin and its function. In *Advances in Experimental Medicine and Biology*; Springer: New York, NY, USA, 2019; Volume 1128, pp. 1–11.
77. Frazier, H.N.; Ghoweri, A.O.; Anderson, K.L.; Lin, R.L.; Porter, N.M.; Thibault, O. Broadening the definition of brain insulin resistance in aging and Alzheimer's disease. *Exp. Neurol.* **2019**, *313*, 79–87. [CrossRef]
78. Al Haj Ahmad, R.M.; Al-Domi, H.A. Thinking about brain insulin resistance. *Diabetes Metab. Syndr. Clin. Res. Rev.* **2018**, *12*, 1091–1094. [CrossRef] [PubMed]
79. Neth, B.J.; Craft, S. Insulin resistance and Alzheimer's disease: Bioenergetic linkages. *Front. Aging Neurosci.* **2017**, *9*, 345. [CrossRef] [PubMed]
80. Ferreira, L.S.S.; Fernandes, C.S.; Vieira, M.N.N.; De Felice, F.G. Insulin resistance in Alzheimer's disease. *Front. Neurosci.* **2018**, *12*, 830. [CrossRef] [PubMed]
81. De La Monte, S.M.; Re, E.; Longato, L.; Tong, M. Dysfunctional pro-ceramide, ER stress, and insulin/IGF signaling networks with progression of Alzheimer's disease. *J. Alzheimer's Dis.* **2012**, *30*, S217–S229. [CrossRef] [PubMed]
82. De La Monte, S.M.; Tong, M. Brain metabolic dysfunction at the core of Alzheimer's disease. *Biochem. Pharmacol.* **2014**, *88*, 548–559. [CrossRef]
83. Rivera, E.J.; Goldin, A.; Fulmer, N.; Tavares, R.; Wands, J.R.; De La Monte, S.M. Insulin and insulin-like growth factor expression and function deteriorate with progression of Alzheimer's disease: Link to brain reductions in acetylcholine. *J. Alzheimer's Dis.* **2005**, *8*, 247–268. [CrossRef] [PubMed]
84. Kandimalla, R.; Thirumala, V.; Reddy, P.H. Is Alzheimer's disease a type 3 diabetes? A critical appraisal. *Biochim. Biophys. Acta Mol. Basis Dis.* **2017**, *1863*, 1078–1089. [CrossRef] [PubMed]
85. Rojas-Gutierrez, E.; Muñoz-Arenas, G.; Treviño, S.; Espinosa, B.; Chavez, R.; Rojas, K.; Flores, G.; Díaz, A.; Guevara, J. Alzheimer's disease and metabolic syndrome: A link from oxidative stress and inflammation to neurodegeneration. *Synapse* **2017**, *71*, e21990. [CrossRef]
86. De la Monte, S.M. The full spectrum of Alzheimer's disease is rooted in metabolic derangements that drive type 3 diabetes. In *Advances in Experimental Medicine and Biology*; Springer: New York, NY, USA, 2019; Volume 1128, pp. 45–83.
87. Arnold, S.E.; Arvanitakis, Z.; Macauley-Rambach, S.L.; Koenig, A.M.; Wang, H.Y.; Ahima, R.S.; Craft, S.; Gandy, S.; Buettner, C.; Stoeckel, L.E.; et al. Brain insulin resistance in type 2 diabetes and Alzheimer disease: Concepts and conundrums. *Nat. Rev. Neurol.* **2018**, *14*, 168–181. [CrossRef]
88. Stanciu, G.D.; Bild, V.; Ababei, D.C.; Rusu, R.N.; Cobzaru, A.; Paduraru, L.; Bulea, D. Link between diabetes and Alzheimer's disease due to the shared amyloid aggregation and deposition involving both neurodegenerative changes and neurovascular damages. *J. Clin. Med.* **2020**, *9*, 1713. [CrossRef] [PubMed]
89. Ou, Y.N.; Xu, W.; Li, J.Q.; Guo, Y.; Cui, M.; Chen, K.L.; Huang, Y.Y.; Dong, Q.; Tan, L.; Yu, J.T. FDG-PET as an independent biomarker for Alzheimer's biological diagnosis: A longitudinal study. *Alzheimer's Res. Ther.* **2019**, *11*, 57. [CrossRef] [PubMed]
90. Anchisi, D.; Borroni, B.; Franceschi, M.; Kerrouche, N.; Kalbe, E.; Beuthien-Beumann, B.; Cappa, S.; Lenz, O.; Ludecke, S.; Marcone, A.; et al. Heterogeneity of brain glucose metabolism in mild cognitive impairment and clinical progression to Alzheimer disease. *Arch. Neurol.* **2005**, *62*, 1728–1733. [CrossRef] [PubMed]
91. Rice, L.; Bisdas, S. The diagnostic value of FDG and amyloid PET in Alzheimer's disease—A systematic review. *Eur. J. Radiol.* **2017**, *94*, 16–24. [CrossRef] [PubMed]
92. Manyevitch, R.; Protas, M.; Scarpiello, S.; Deliso, M.; Bass, B.; Nanajian, A.; Chang, M.; Thompson, S.M.; Khoury, N.; Gonnella, R.; et al. Evaluation of metabolic and synaptic dysfunction hypotheses of Alzheimer's disease (AD): A meta-analysis of CSF markers. *Curr. Alzheimer Res.* **2018**, *15*, 164–181. [CrossRef] [PubMed]
93. Luchsinger, J.A.; Gustafson, D.R. Adiposity and Alzheimer's disease. *Curr. Opin. Clin. Nutr. Metab. Care* **2009**, *12*, 15–21. [CrossRef]
94. Tharp, W.G.; Gupta, D.; Smith, J.; Jones, K.P.; Jones, A.M.; Pratley, R.E. Effects of glucose and insulin on secretion of amyloid-β by human adipose tissue cells. *Obesity* **2016**, *24*, 1471–1479. [CrossRef]
95. Khan, M.S.H.; Hegde, V. Obesity and diabetes mediated chronic inflammation: A potential biomarker in Alzheimer's disease. *J. Pers. Med.* **2020**, *10*, 42. [CrossRef]
96. He, D.; Xi, B.; Xue, J.; Huai, P.; Zhang, M.; Li, J. Association between leisure time physical activity and metabolic syndrome: A meta-analysis of prospective cohort studies. *Endocrine* **2014**, *46*, 231–240. [CrossRef]

97. Misiak, B.; Leszek, J.; Kiejna, A. Metabolic syndrome, mild cognitive impairment and Alzheimer's disease—The emerging role of systemic low-grade inflammation and adiposity. *Brain Res. Bull.* **2012**, *89*, 144–149. [CrossRef]
98. Horie, N.C.; Serrao, V.T.; Simon, S.S.; Gascon, M.R.P.; Dos Santos, A.X.; Zambone, M.A.; Del Bigio De Freitas, M.M.; Cunha-Neto, E.; Marques, E.L.; Halpern, A.; et al. Cognitive effects of intentional weight loss in elderly obese individuals with mild cognitive impairment. *J. Clin. Endocrinol. Metab.* **2016**, *101*, 1104–1112. [CrossRef] [PubMed]
99. Ho, L.; Qin, W.; Pompl, P.N.; Xiang, Z.; Wang, J.; Zhao, Z.; Peng, Y.; Cambareri, G.; Rocher, A.; Mobbs, C.V.; et al. Diet-induced insulin resistance promotes amyloidosis in a transgenic mouse model of Alzheimer's disease. *FASEB J.* **2004**, *18*, 902–904. [CrossRef]
100. Julien, C.; Tremblay, C.; Phivilay, A.; Berthiaume, L.; Émond, V.; Julien, P.; Calon, F. High-fat diet aggravates amyloid-beta and tau pathologies in the 3xTg-AD mouse model. *Neurobiol. Aging* **2010**, *31*, 1516–1531. [CrossRef] [PubMed]
101. Sah, S.K.; Lee, C.; Jang, J.H.; Park, G.H. Effect of high-fat diet on cognitive impairment in triple-transgenic mice model of Alzheimer's disease. *Biochem. Biophys. Res. Commun.* **2017**, *493*, 731–736. [CrossRef]
102. Mazon, J.N.; de Mello, A.H.; Ferreira, G.K.; Rezin, G.T. The impact of obesity on neurodegenerative diseases. *Life Sci.* **2017**, *182*, 22–28. [CrossRef]
103. Pegueroles, J.; Pané, A.; Vilaplana, E.; Montal, V.; Bejanin, A.; Videla, L.; Carmona-Iragui, M.; Barroeta, I.; Ibarzabal, A.; Casajoana, A.; et al. Obesity impacts brain metabolism and structure independently of amyloid and tau pathology in healthy elderly. *Alzheimer's Dement. Diagn. Assess. Dis. Monit.* **2020**, *12*, e12052.
104. Letra, L.; Santana, I. The influence of adipose tissue on brain development, cognition, and risk of neurodegenerative disorders. In *Advances in Neurobiology*; Springer: New York, NY, USA, 2017; Volume 19, pp. 151–161.
105. Ouchi, N.; Parker, J.L.; Lugus, J.J.; Walsh, K. Adipokines in inflammation and metabolic disease. *Nat. Rev. Immunol.* **2011**, *11*, 85–97. [CrossRef] [PubMed]
106. Pichiah, P.B.T.; Sankarganesh, D.; Arunachalam, S.; Achiraman, S. Adipose-derived molecules—Untouched horizons in Alzheimer's disease biology. *Front. Aging Neurosci.* **2020**, *12*, 17. [CrossRef]
107. Argentati, C.; Morena, F.; Bazzucchi, M.; Armentano, I.; Emiliani, C.; Martino, S. Adipose stem cell translational applications: From bench-to-bedside. *Int. J. Mol. Sci.* **2018**, *19*, 3475. [CrossRef]
108. Farruggia, M.C.; Small, D.M. Effects of adiposity and metabolic dysfunction on cognition: A review. *Physiol. Behav.* **2019**, *208*, 112578. [CrossRef]
109. Beall, C.; Hanna, L.; Ellacott, K.L.J. CNS targets of adipokines. *Compr. Physiol.* **2017**, *7*, 1359–1406. [PubMed]
110. Forny-Germano, L.; De Felice, F.G.; Do Nascimento Vieira, M.N. The role of leptin and adiponectin in obesity-associated cognitive decline and Alzheimer's disease. *Front. Neurosci.* **2019**, *13*, 1027. [CrossRef]
111. Parimisetty, A.; Dorsemans, A.C.; Awada, R.; Ravanan, P.; Diotel, N.; Lefebvre d'Hellencourt, C. Secret talk between adipose tissue and central nervous system via secreted factors—An emerging frontier in the neurodegenerative research. *J. Neuroinflamm.* **2016**, *13*, 1–13. [CrossRef] [PubMed]
112. Gil-Ortega, M.; Martín-Ramos, M.; Arribas, S.M.; González, M.C.; Aránguez, I.; Ruiz-Gayo, M.; Somoza, B.; Fernández-Alfonso, M.S. Arterial stiffness is associated with adipokine dysregulation in non-hypertensive obese mice. *Vascul. Pharmacol.* **2016**, *77*, 38–47. [CrossRef] [PubMed]
113. Hiscox, L.V.; Johnson, C.L.; McGarry, M.D.J.; Marshall, H.; Ritchie, C.W.; van Beek, E.J.R.; Roberts, N.; Starr, J.M. Mechanical property alterations across the cerebral cortex due to Alzheimer's disease. *Brain Commun.* **2020**, *2*, fcz049. [CrossRef] [PubMed]
114. Weisberg, S.P.; McCann, D.; Desai, M.; Rosenbaum, M.; Leibel, R.L.; Ferrante, A.W. Obesity is associated with macrophage accumulation in adipose tissue. *J. Clin. Investig.* **2003**, *112*, 1796–1808. [CrossRef] [PubMed]
115. Zeyda, M.; Stulnig, T.M. Obesity, inflammation, and insulin resistance—A mini-review. *Gerontology* **2009**, *55*, 379–386. [CrossRef]
116. Thaler, J.P.; Schwartz, M.W. Minireview: Inflammation and obesity pathogenesis: The hypothalamus heats up. *Endocrinology* **2010**, *151*, 4109–4115. [CrossRef] [PubMed]
117. Park, S.A.; Han, S.M.; Kim, C.E. New fluid biomarkers tracking non-amyloid-β and non-tau pathology in Alzheimer's disease. *Exp. Mol. Med.* **2020**, *52*, 556–568. [CrossRef]
118. Sepe, F.N.; Chiasserini, D.; Parnetti, L. Role of FABP3 as biomarker in Alzheimer's disease and synucleinopathies. *Future Neurol.* **2018**, *13*, 199–207. [CrossRef]

119. De Leon, M.J.; Mosconi, L.; Li, J.; De Santi, S.; Yao, Y.; Tsui, W.H.; Pirraglia, E.; Rich, K.; Javier, E.; Brys, M.; et al. Longitudinal CSF isoprostane and MRI atrophy in the progression to AD. *J. Neurol.* **2007**, *254*, 1666–1675. [CrossRef] [PubMed]
120. Praticò, D. The neurobiology of isoprostanes and Alzheimer's disease. *Biochim. Biophys. Acta Mol. Cell Biol. Lipids* **2010**, *1801*, 930–933. [CrossRef] [PubMed]
121. Goozee, K.; Chatterjee, P.; James, I.; Shen, K.; Sohrabi, H.R.; Asih, P.R.; Dave, P.; Ball, B.; Manyan, C.; Taddei, K.; et al. Alterations in erythrocyte fatty acid composition in preclinical Alzheimer's disease. *Sci. Rep.* **2017**, *7*, 1–9. [CrossRef] [PubMed]
122. Iuliano, L.; Pacelli, A.; Ciacciarelli, M.; Zerbinati, C.; Fagioli, S.; Piras, F.; Orfei, M.D.; Bossù, P.; Pazzelli, F.; Serviddio, G.; et al. Plasma fatty acid lipidomics in amnestic mild cognitive impairment and Alzheimer's disease. *J. Alzheimer's Dis.* **2013**, *36*, 545–553. [CrossRef] [PubMed]
123. Wood, P.L.; Medicherla, S.; Sheikh, N.; Terry, B.; Phillipps, A.; Kaye, J.A.; Quinn, J.F.; Woltjer, R.L. Targeted lipidomics of fontal cortex and plasma diacylglycerols (DAG) in mild cognitive impairment and Alzheimer's disease: Validation of DAG accumulation early in the pathophysiology of Alzheimer's disease. *J. Alzheimer's Dis.* **2015**, *48*, 537–546. [CrossRef]
124. Proitsi, P.; Kim, M.; Whiley, L.; Simmons, A.; Sattlecker, M.; Velayudhan, L.; Lupton, M.K.; Soininen, H.; Kloszewska, I.; Mecocci, P.; et al. Association of blood lipids with Alzheimer's disease: A comprehensive lipidomics analysis. *Alzheimer's Dement.* **2017**, *13*, 140–151. [CrossRef]
125. Kao, Y.C.; Ho, P.C.; Tu, Y.K.; Jou, I.M.; Tsai, K.J. Lipids and Alzheimer's disease. *Int. J. Mol. Sci.* **2020**, *21*, 1505. [CrossRef]
126. Bonda, D.J.; Stone, J.G.; Torres, S.L.; Siedlak, S.L.; Perry, G.; Kryscio, R.; Jicha, G.; Casadesus, G.; Smith, M.A.; Zhu, X.; et al. Dysregulation of leptin signaling in Alzheimer disease: Evidence for neuronal leptin resistance. *J. Neurochem.* **2014**, *128*, 162–172. [CrossRef]
127. Maioli, S.; Lodeiro, M.; Merino-Serrais, P.; Falahati, F.; Khan, W.; Puerta, E.; Codita, A.; Rimondini, R.; Ramirez, M.J.; Simmons, A.; et al. Alterations in brain leptin signalling in spite of unchanged CSF leptin levels in Alzheimer's disease. *Aging Cell* **2015**, *14*, 122–129. [CrossRef]
128. Oania, R.; McEvoy, L.K. Plasma leptin levels are not predictive of dementia in patients with mild cognitive impairment. *Age Ageing* **2014**, *44*, 53–58. [CrossRef]
129. Teunissen, C.E.; Van Der Flier, W.M.; Scheltens, P.; Duits, A.; Wijnstok, N.; Nijpels, G.; Dekker, J.M.; Blankenstein, R.M.A.; Heijboer, A.C. Serum leptin is not altered nor related to cognitive decline in Alzheimer's disease. *J. Alzheimer's Dis.* **2015**, *44*, 809–813. [CrossRef] [PubMed]
130. Gustafson, D.R.; Bäckman, K.; Lissner, L.; Carlsson, L.; Waern, M.; Östling, S.; Guo, X.; Bengtsson, C.; Skoog, I. Leptin and dementia over 32 years—The prospective population study of women. *Alzheimer's Dement.* **2012**, *8*, 272–277. [CrossRef] [PubMed]
131. Littlejohns, T.J.; Kos, K.; Henley, W.E.; Cherubini, A.; Ferrucci, L.; Lang, I.A.; Langa, K.M.; Melzer, D.; Llewellyn, D.J. Serum leptin and risk of cognitive decline in elderly Italians. *J. Alzheimer's Dis.* **2015**, *44*, 1231–1239. [CrossRef] [PubMed]
132. Johnston, J.; Hu, W.; Fardo, D.; Greco, S.; Perry, G.; Montine, T.; Trojanowski, J.; Shaw, L.; Ashford, J.; Tezapsidis, N. Low plasma leptin in cognitively impaired ADNI subjects: Gender differences and diagnostic and therapeutic potential. *Curr. Alzheimer Res.* **2014**, *11*, 165–174. [CrossRef] [PubMed]
133. Horvath, A.; Salman, Z.; Quinlan, P.; Wallin, A.; Svensson, J. Patients with Alzheimer's disease have increased levels of insulin-like growth factor-I in serum but not in cerebrospinal fluid. *J. Alzheimer's Dis.* **2020**, *75*, 289–298. [CrossRef]
134. Suzuki, K.; Suzuki, S.; Ishii, Y.; Fujita, H.; Matsubara, T.; Okamura, M.; Sakuramoto, H.; Hirata, K. Serum insulin-like growth factor-1 levels in neurodegenerative diseases. *Acta Neurol. Scand.* **2019**, *139*, 563–567. [CrossRef]
135. Galle, S.A.; van der Spek, A.; Drent, M.L.; Brugts, M.P.; Scherder, E.J.A.; Janssen, J.A.M.J.L.; Ikram, M.A.; van Duijn, C.M. Revisiting the role of insulin-like growth factor-I receptor stimulating activity and the apolipoprotein E in Alzheimer's disease. *Front. Aging Neurosci.* **2019**, *11*, 20. [CrossRef]
136. Ostrowski, P.P.; Barszczyk, A.; Forstenpointner, J.; Zheng, W.; Feng, Z.-P. Meta-analysis of serum insulin-like growth factor 1 in Alzheimer's disease. *PLoS ONE* **2016**, *11*, e0155733. [CrossRef]

137. Teixeira, A.L.; Diniz, B.S.; Campos, A.C.; Miranda, A.S.; Rocha, N.P.; Talib, L.L.; Gattaz, W.F.; Forlenza, O.V. Decreased levels of circulating adiponectin in mild cognitive impairment and Alzheimer's disease. *Neuromol. Med.* **2013**, *15*, 115–121. [CrossRef]
138. Waragai, M.; Adame, A.; Trinh, I.; Sekiyama, K.; Takamatsu, Y.; Une, K.; Masliah, E.; Hashimoto, M. Possible involvement of adiponectin, the anti-diabetes molecule, in the pathogenesis of Alzheimer's disease. *J. Alzheimer's Dis.* **2016**, *52*, 1453–1459. [CrossRef]
139. Une, K.; Takei, Y.A.; Tomita, N.; Asamura, T.; Ohrui, T.; Furukawa, K.; Arai, H. Adiponectin in plasma and cerebrospinal fluid in MCI and Alzheimer's disease. *Eur. J. Neurol.* **2011**, *18*, 1006–1009. [CrossRef] [PubMed]
140. Wennberg, A.M.V.; Gustafson, D.; Hagen, C.E.; Roberts, R.O.; Knopman, D.; Jack, C.; Petersen, R.C.; Mielke, M.M. Serum adiponectin levels, neuroimaging, and cognition in the mayo clinic study of aging. *J. Alzheimer's Dis.* **2016**, *53*, 573–581. [CrossRef]
141. Golpich, M.; Amini, E.; Mohamed, Z.; Azman Ali, R.; Mohamed Ibrahim, N.; Ahmadiani, A. Mitochondrial dysfunction and biogenesis in neurodegenerative diseases: Pathogenesis and treatment. *CNS Neurosci. Ther.* **2017**, *23*, 5–22. [CrossRef] [PubMed]
142. Swerdlow, R.H.; Koppel, S.; Weidling, I.; Hayley, C.; Ji, Y.; Wilkins, H.M. Mitochondria, cybrids, aging, and Alzheimer's disease. In *Progress in Molecular Biology and Translational Science;* Elsevier B.V.: Amsterdam, The Netherlands, 2017; Volume 146, pp. 259–302.
143. Rossi, A.; Rigotto, G.; Valente, G.; Giorgio, V.; Basso, E.; Filadi, R.; Pizzo, P. Defective mitochondrial pyruvate flux affects cell bioenergetics in Alzheimer's disease-related models. *Cell Rep.* **2020**, *30*, 2332–2348. [CrossRef] [PubMed]
144. Adiele, R.C.; Adiele, C.A. Mitochondrial regulatory pathways in the pathogenesis of Alzheimer's disease. *J. Alzheimer's Dis.* **2016**, *53*, 1257–1270. [CrossRef] [PubMed]
145. Swerdlow, R.H.; Burns, J.M.; Khan, S.M. The Alzheimer's disease mitochondrial cascade hypothesis. *J. Alzheimer's Dis.* **2010**, *20*, S265–S279. [CrossRef] [PubMed]
146. Swerdlow, R.H.; Burns, J.M.; Khan, S.M. The Alzheimer's disease mitochondrial cascade hypothesis: Progress and perspectives. *Biochim. Biophys. Acta Mol. Basis Dis.* **2014**, *1842*, 1219–1231. [CrossRef]
147. Moreira, P.I.; Carvalho, C.; Zhu, X.; Smith, M.A.; Perry, G. Mitochondrial dysfunction is a trigger of Alzheimer's disease pathophysiology. *Biochim. Biophys. Acta Mol. Basis Dis.* **2010**, *1802*, 2–10. [CrossRef]
148. Caldwell, C.C.; Yao, J.; Brinton, R.D. Targeting the prodromal stage of Alzheimer's disease: Bioenergetic and mitochondrial opportunities. *Neurotherapeutics* **2015**, *12*, 66–80. [CrossRef]
149. Sheng, B.; Wang, X.; Su, B.; Lee, H.G.; Casadesus, G.; Perry, G.; Zhu, X. Impaired mitochondrial biogenesis contributes to mitochondrial dysfunction in Alzheimer's disease. *J. Neurochem.* **2012**, *120*, 419–429. [CrossRef]
150. Baloyannis, S.J. Mitochondrial alterations in Alzheimer's disease. *J. Alzheimer's Dis.* **2006**, *9*, 119–126. [CrossRef] [PubMed]
151. Yao, J.; Du, H.; Yan, S.; Fang, F.; Wang, C.; Lue, L.F.; Guo, L.; Chen, D.; Stern, D.M.; Gunn Moore, F.J.; et al. Inhibition of amyloid-β(Aβ) peptide-binding alcohol dehydrogenase-Aβ interaction reduces Aβ accumulation and improves mitochondrial function in a mouse model of Alzheimer's disease. *J. Neurosci.* **2011**, *31*, 2313–2320. [CrossRef] [PubMed]
152. Hansson Petersen, C.A.; Alikhani, N.; Behbahani, H.; Wiehager, B.; Pavlov, P.F.; Alafuzoff, I.; Leinonen, V.; Ito, A.; Winblad, B.; Glaser, E.; et al. The amyloid β-peptide is imported into mitochondria via the TOM import machinery and localized to mitochondrial cristae. *Proc. Natl. Acad. Sci. USA* **2008**, *105*, 13145–13150. [CrossRef] [PubMed]
153. Lustbader, J.W.; Cirilli, M.; Lin, C.; Xu, H.W.; Takuma, K.; Wang, N.; Caspersen, C.; Chen, X.; Pollak, S.; Chaney, M.; et al. ABAD directly links Aβ to mitochondrial toxicity in Alzheimer's disease. *Science* **2004**, *304*, 448–452. [CrossRef]
154. Lim, Y.A.; Grimm, A.; Giese, M.; Mensah-Nyagan, A.G.; Villafranca, J.E.; Ittner, L.M.; Eckert, A.; Götz, J. Inhibition of the mitochondrial enzyme ABAD restores the amyloid-β-mediated deregulation of estradiol. *PLoS ONE* **2011**, *6*, e28887. [CrossRef]
155. Du, H.; Guo, L.; Fang, F.; Chen, D.A.; Sosunov, A.; M. McKhann, G.; Yan, Y.; Wang, C.; Zhang, H.; Molkentin, J.D.; et al. Cyclophilin D deficiency attenuates mitochondrial and neuronal perturbation and ameliorates learning and memory in Alzheimer's disease. *Nat. Med.* **2008**, *14*, 1097–1105. [CrossRef]

156. Crouch, P.J.; Blake, R.; Duce, J.A.; Ciccotosto, G.D.; Li, Q.X.; Barnham, K.J.; Curtain, C.C.; Cherny, R.A.; Cappai, R.; Dyrks, T.; et al. Copper-dependent inhibition of human cytochrome c oxidase by a dimeric conformer of amyloid-β1-42. *J. Neurosci.* **2005**, *25*, 672–679. [CrossRef]
157. Reddy, P.H.; Beal, M.F. Amyloid beta, mitochondrial dysfunction and synaptic damage: Implications for cognitive decline in aging and Alzheimer's disease. *Trends Mol. Med.* **2008**, *14*, 45–53. [CrossRef]
158. Cardoso, S.M.; Proença, M.T.; Santos, S.; Santana, I.; Oliveira, C.R. Cytochrome c oxidase is decreased in Alzheimer's disease platelets. *Neurobiol. Aging* **2004**, *25*, 105–110. [CrossRef]
159. Mancuso, M.; Filosto, M.; Bosetti, F.; Ceravolo, R.; Rocchi, A.; Tognoni, G.; Manca, M.L.; Solaini, G.; Siciliano, G.; Murri, L. Decreased platelet cytochrome c oxidase activity is accompanied by increased blood lactate concentration during exercise in patients with Alzheimer disease. *Exp. Neurol.* **2003**, *182*, 421–426. [CrossRef]
160. Shaerzadeh, F.; Motamedi, F.; Minai-Tehrani, D.; Khodagholi, F. Monitoring of neuronal loss in the hippocampus of Aβ-injected rat: Autophagy, mitophagy, and mitochondrial biogenesis stand against apoptosis. *Neuromol. Med.* **2014**, *16*, 175–190. [CrossRef]
161. Swerdlow, R.H. Mitochondria and mitochondrial cascades in Alzheimer's disease. *J. Alzheimer's Dis.* **2018**, *62*, 1403–1416. [CrossRef]
162. Cunnane, S.C.; Trushina, E.; Morland, C.; Prigione, A.; Casadesus, G.; Andrews, Z.B.; Beal, M.F.; Bergersen, L.H.; Brinton, R.D.; de la Monte, S.; et al. Brain energy rescue: An emerging therapeutic concept for neurodegenerative disorders of ageing. *Nat. Rev. Drug Discov.* **2020**, *19*, 609–633. [CrossRef]
163. Panel, M.; Ghaleh, B.; Morin, D. Mitochondria and aging: A role for the mitochondrial transition pore? *Aging Cell* **2018**, *17*, e12793. [CrossRef]
164. Santos, R.X.; Correia, S.C.; Zhu, X.; Smith, M.A.; Moreira, P.I.; Castellani, R.J.; Nunomura, A.; Perry, G. Mitochondrial DNA oxidative damage and repair in aging and Alzheimer's disease. *Antioxid. Redox Signal.* **2013**, *18*, 2444–2457. [CrossRef]
165. Casoli, T.; Spazzafumo, L.; Di Stefano, G.; Conti, F. Role of diffuse low-level heteroplasmy of mitochondrial DNA in Alzheimer's disease neurodegeneration. *Front. Aging Neurosci.* **2015**, *7*, 142. [CrossRef]
166. Wang, Y.; Brinton, R.D. Triad of risk for late onset Alzheimer's: Mitochondrial haplotype, apoe genotype and chromosomal sex. *Front. Aging Neurosci.* **2016**, *8*, 232. [CrossRef]
167. Ridge, P.G.; Koop, A.; Maxwell, T.J.; Bailey, M.H.; Swerdlow, R.H.; Kauwe, J.S.K.; Honea, R.A. Mitochondrial haplotypes associated with biomarkers for Alzheimer's disease. *PLoS ONE* **2013**, *8*, e74158. [CrossRef]
168. Maruszak, A.; Safranow, K.; Branicki, W.; Gawda-Walerych, K.; Poṡpiech, E.; Gabryelewicz, T.; Canter, J.A.; Barcikowska, M.; Ekanowski, C. The impact of mitochondrial and nuclear DNA variants on late-onset Alzheimer's disease risk. *J. Alzheimer's Dis.* **2011**, *27*, 197–210. [CrossRef]
169. Ridge, P.G.; Kauwe, J.S.K. Mitochondria and Alzheimer's disease: The role of mitochondrial genetic variation. *Curr. Genet. Med. Rep.* **2018**, *6*, 1–10. [CrossRef]
170. Schmukler, E.; Solomon, S.; Simonovitch, S.; Goldshmit, Y.; Wolfson, E.; Michaelson, D.M.; Pinkas-Kramarski, R. Altered mitochondrial dynamics and function in APOE4-expressing astrocytes. *Cell Death Dis.* **2020**, *11*, 1–13. [CrossRef]
171. Simonovitch, S.; Schmukler, E.; Masliah, E.; Pinkas-Kramarski, R.; Michaelson, D.M. The effects of APOE4 on mitochondrial dynamics and proteins in vivo. *J. Alzheimer's Dis.* **2019**, *70*, 861–875. [CrossRef]
172. Peng, Y.; Gao, P.; Shi, L.; Chen, L.; Liu, J.; Long, J. Central and peripheral metabolic defects contribute to the pathogenesis of Alzheimer's disease: Targeting mitochondria for diagnosis and prevention. *Antioxid. Redox Signal.* **2020**, *32*, 1188–1236. [CrossRef]
173. Lunnon, K.; Keohane, A.; Pidsley, R.; Newhouse, S.; Riddoch-Contreras, J.; Thubron, E.B.; Devall, M.; Soininen, H.; Kłoszewska, I.; Mecocci, P.; et al. Mitochondrial genes are altered in blood early in Alzheimer's disease. *Neurobiol. Aging* **2017**, *53*, 36–47. [CrossRef]
174. Dey, K.K.; Wang, H.; Niu, M.; Bai, B.; Wang, X.; Li, Y.; Cho, J.H.; Tan, H.; Mishra, A.; High, A.A.; et al. Deep undepleted human serum proteome profiling toward biomarker discovery for Alzheimer's disease. *Clin. Proteom.* **2019**, *16*, 16. [CrossRef]
175. Li, X.; Wang, H.; Long, J.; Pan, G.; He, T.; Anichtchik, O.; Belshaw, R.; Albani, D.; Edison, P.; Green, E.K.; et al. Systematic analysis and biomarker study for Alzheimer's disease. *Sci. Rep.* **2018**, *8*, 17394. [CrossRef]

176. Ridge, P.G.; Wadsworth, M.E.; Miller, J.B.; Saykin, A.J.; Green, R.C.; Kauwe, J.S.K. Assembly of 809 whole mitochondrial genomes with clinical, imaging, and fluid biomarker phenotyping. *Alzheimer's Dement.* **2018**, *14*, 514–519. [CrossRef]
177. Liang, Y. Emerging concepts and functions of autophagy as a regulator of synaptic components and plasticity. *Cells* **2019**, *8*, 34. [CrossRef]
178. Nixon, R.A. The aging lysosome: An essential catalyst for late-onset neurodegenerative diseases. *Biochim. Biophys. Acta Proteins Proteom.* **2020**, *1868*, 140443. [CrossRef]
179. Zhou, F.; Xiong, X.; Li, S.; Liang, J.; Zhang, X.; Tian, M.; Li, X.; Gao, M.; Tang, L.; Li, Y. Enhanced autophagic retrograde axonal transport by dynein intermediate chain upregulation improves Aβ clearance and cognitive function in APP/PS1 double transgenic mice. *Aging* **2020**, *12*, 12142–12159. [CrossRef]
180. Nixon, R.A. The role of autophagy in neurodegenerative disease. *Nat. Med.* **2013**, *19*, 983–997. [CrossRef]
181. Wong, Y.C.; Holzbaur, E.L.F. Autophagosome dynamics in neurodegeneration at a glance. *J. Cell Sci.* **2015**, *128*, 1259–1267. [CrossRef]
182. Yan, X.; Wang, B.; Hu, Y.; Wang, S.; Zhang, X. Abnormal mitochondrial quality control in neurodegenerative diseases. *Front. Cell. Neurosci.* **2020**, *14*, 138. [CrossRef]
183. Rubinsztein, D.C.; DiFiglia, M.; Heintz, N.; Nixon, R.A.; Qin, Z.H.; Ravikumar, B.; Stefanis, L.; Tolkovsky, A. Autophagy and its possible roles in nervous system diseases, damage and repair. *Autophagy* **2005**, *1*, 11–22. [CrossRef]
184. Lee, J.H.; Yu, W.H.; Kumar, A.; Lee, S.; Mohan, P.S.; Peterhoff, C.M.; Wolfe, D.M.; Martinez-Vicente, M.; Massey, A.C.; Sovak, G.; et al. Lysosomal proteolysis and autophagy require presenilin 1 and are disrupted by Alzheimer-related PS1 mutations. *Cell* **2010**, *141*, 1146–1158. [CrossRef]
185. Yuan, Z.; Yidan, Z.; Jian, Z.; Xiangjian, Z.; Guofeng, Y. Molecular mechanism of autophagy: Its role in the therapy of Alzheimer's disease. *Curr. Neuropharmacol.* **2020**, *18*, 720–739. [CrossRef]
186. Malampati, S.; Song, J.-X.; Chun-Kit Tong, B.; Nalluri, A.; Yang, C.-B.; Wang, Z.; Gopalkrishnashetty Sreenivasmurthy, S.; Zhu, Z.; Liu, J.; Su, C.; et al. Targeting aggrephagy for the treatment of Alzheimer's disease. *Cells* **2020**, *9*, 311. [CrossRef]
187. Norambuena, A.; Wallrabe, H.; Cao, R.; Wang, D.B.; Silva, A.; Svindrych, Z.; Periasamy, A.; Hu, S.; Tanzi, R.E.; Kim, D.Y.; et al. A novel lysosome-to-mitochondria signaling pathway disrupted by amyloid-β oligomers. *EMBO J.* **2018**, *37*, e100241. [CrossRef]
188. Nilsson, P.; Sekiguchi, M.; Akagi, T.; Izumi, S.; Komori, T.; Hui, K.; Sörgjerd, K.; Tanaka, M.; Saito, T.; Iwata, N.; et al. Autophagy-related protein 7 deficiency in amyloid β (Aβ) precursor protein transgenic mice decreases Aβ in the multivesicular bodies and induces Aβ accumulation in the golgi. *Am. J. Pathol.* **2015**, *185*, 305–313. [CrossRef]
189. Reddy, K.; Cusack, C.L.; Nnah, I.C.; Khayati, K.; Saqcena, C.; Huynh, T.B.; Noggle, S.A.; Ballabio, A.; Dobrowolski, R. Dysregulation of nutrient sensing and CLEARance in presenilin deficiency. *Cell Rep.* **2016**, *14*, 2166–2179. [CrossRef]
190. De Kimpe, L.; van Haastert, E.S.; Kaminari, A.; Zwart, R.; Rutjes, H.; Hoozemans, J.J.M.; Scheper, W. Intracellular accumulation of aggregated pyroglutamate amyloid beta: Convergence of aging and Aβ pathology at the lysosome. *Age* **2013**, *35*, 673–687. [CrossRef]
191. Yang, D.-S.; Stavrides, P.; Mohan, P.S.; Kaushik, S.; Kumar, A.; Ohno, M.; Schmidt, S.D.; Wesson, D.; Bandyopadhyay, U.; Jiang, Y.; et al. Reversal of autophagy dysfunction in the TgCRND8 mouse model of Alzheimer's disease ameliorates amyloid pathologies and memory deficits. *Brain* **2010**, *134*, 258–277. [CrossRef]
192. Lafay-Chebassier, C.; Paccalin, M.; Page, G.; Barc-Pain, S.; Perault-Pochat, M.C.; Gil, R.; Pradier, L.; Hugon, J. mTOR/p70S6k signalling alteration by Aβ exposure as well as in APP-PS1 transgenic models and in patients with Alzheimer's disease. *J. Neurochem.* **2005**, *94*, 215–225. [CrossRef]
193. Wang, Y.; Martinez-Vicente, M.; Krüger, U.; Kaushik, S.; Wong, E.; Mandelkow, E.-M.; Cuervo, A.M.; Mandelkow, E. Tau fragmentation, aggregation and clearance: The dual role of lysosomal processing. *Hum. Mol. Genet.* **2009**, *18*, 4153–4170. [CrossRef]
194. Cataldo, A.M.; Barnett, J.L.; Pieroni, C.; Nixon, R.A. Increased neuronal endocytosis and protease delivery to early endosomes in sporadic Alzheimer's disease: Neuropathologic evidence for a mechanism of increased β-amyloidogenesis. *J. Neurosci.* **1997**, *17*, 6142–6151. [CrossRef]

195. Cataldo, A.M.; Peterhoff, C.M.; Troncoso, J.C.; Gomez-Isla, T.; Hyman, B.T.; Nixon, R.A. Endocytic pathway abnormalities precede amyloid β deposition in sporadic Alzheimer's disease and down syndrome: Differential effects of APOE genotype and presenilin mutations. *Am. J. Pathol.* **2000**, *157*, 277–286. [CrossRef]
196. Cataldo, A.M.; Petanceska, S.; Terio, N.B.; Peterhoff, C.M.; Durham, R.; Mercken, M.; Mehta, P.D.; Buxbaum, J.; Haroutunian, V.; Nixon, R.A. Aβ localization in abnormal endosomes: Association with earliest Aβ elevations in AD and Down syndrome. *Neurobiol. Aging* **2004**, *25*, 1263–1272. [CrossRef]
197. Nixon, R.A.; Wegiel, J.; Kumar, A.; Yu, W.H.; Peterhoff, C.; Cataldo, A.; Cuervo, A.M. Extensive involvement of autophagy in Alzheimer disease: An immuno-electron microscopy study. *J. Neuropathol. Exp. Neurol.* **2005**, *64*, 113–122. [CrossRef]
198. Gowrishankar, S.; Yuan, P.; Wu, Y.; Schrag, M.; Paradise, S.; Grutzendler, J.; De Camilli, P.; Ferguson, S.M. Massive accumulation of luminal protease-deficient axonal lysosomes at Alzheimer's disease amyloid plaques. *Proc. Natl. Acad. Sci. USA* **2015**, *112*, E3699–E3708. [CrossRef]
199. Köhler, C. Granulovacuolar degeneration: A neurodegenerative change that accompanies tau pathology. *Acta Neuropathol.* **2016**, *132*, 339–359. [CrossRef]
200. Piras, A.; Collin, L.; Grüninger, F.; Graff, C.; Rönnbäck, A. Autophagic and lysosomal defects in human tauopathies: Analysis of post-mortem brain from patients with familial Alzheimer disease, corticobasal degeneration and progressive supranuclear palsy. *Acta Neuropathol. Commun.* **2016**, *4*, 22. [CrossRef] [PubMed]
201. Avrahami, L.; Farfara, D.; Shaham-Kol, M.; Vassar, R.; Frenkel, D.; Eldar-Finkelman, H. Inhibition of glycogen synthase kinase-3 ameliorates β-amyloid pathology and restores lysosomal acidification and mammalian target of rapamycin activity in the Alzheimer disease mouse model: In vivo and in vitro studies. *J. Biol. Chem.* **2013**, *288*, 1295–1306. [CrossRef] [PubMed]
202. Whyte, L.S.; Lau, A.A.; Hemsley, K.M.; Hopwood, J.J.; Sargeant, T.J. *Endo-Lysosomal and Autophagic Dysfunction: A Driving Factor in Alzheimer's Disease?* Blackwell Publishing Ltd.: Oxford, UK, 2017; Volume 140, pp. 703–717.
203. Hung, C.O.Y.; Livesey, F.J. Altered γ-secretase processing of APP disrupts lysosome and autophagosome function in monogenic Alzheimer's disease. *Cell Rep.* **2018**, *25*, 3647–3660. [CrossRef] [PubMed]
204. Cataldo, A.M.; Barnett, J.L.; Mann, D.M.A.; Nixon, R.A. Colocalization of lysosomal hydrolase and β-amyloid in diffuse plaques of the cerebellum and striatum in Alzheimer's disease and Down's syndrome. *J. Neuropathol. Exp. Neurol.* **1996**, *55*, 704–715. [CrossRef]
205. Cataldo, A.M.; Hamilton, D.J.; Nixon, R.A. Lysosomal abnormalities in degenerating neurons link neuronal compromise to senile plaque development in Alzheimer disease. *Brain Res.* **1994**, *640*, 68–80. [CrossRef]
206. 2Nixon, R.A.; Cataldo, A.M. Lysosomal system pathways: Genes to neurodegeneration in Alzheimer's disease. *J. Alzheimer's Dis.* **2006**, *9*, 277–289.
207. Mohamed, N.V.; Plouffe, V.; Rémillard-Labrosse, G.; Planel, E.; Leclerc, N. Starvation and inhibition of lysosomal function increased tau secretion by primary cortical neurons. *Sci. Rep.* **2014**, *4*, 1–11. [CrossRef]
208. Lee, J.G.; Takahama, S.; Zhang, G.; Tomarev, S.I.; Ye, Y. Unconventional secretion of misfolded proteins promotes adaptation to proteasome dysfunction in mammalian cells. *Nat. Cell Biol.* **2016**, *18*, 765–776. [CrossRef]
209. Fontaine, S.N.; Zheng, D.; Sabbagh, J.J.; Martin, M.D.; Chaput, D.; Darling, A.; Trotter, J.H.; Stothert, A.R.; Nordhues, B.A.; Lussier, A.; et al. DnaJ/Hsc70 chaperone complexes control the extracellular release of neurodegenerative-associated proteins. *EMBO J.* **2016**, *35*, 1537–1549. [CrossRef]
210. Nilsson, P.; Loganathan, K.; Sekiguchi, M.; Matsuba, Y.; Hui, K.; Tsubuki, S.; Tanaka, M.; Iwata, N.; Saito, T.; Saido, T.C. Aβ secretion and plaque formation depend on autophagy. *Cell Rep.* **2013**, *5*, 61–69. [CrossRef]
211. Annunziata, I.; Patterson, A.; Helton, D.; Hu, H.; Moshiach, S.; Gomero, E.; Nixon, R.; D'Azzo, A. Lysosomal NEU1 deficiency affects amyloid precursor protein levels and amyloid-β secretion via deregulated lysosomal exocytosis. *Nat. Commun.* **2013**, *4*, 1–12. [CrossRef] [PubMed]
212. Van Weering, J.R.T.; Scheper, W. Endolysosome and autolysosome dysfunction in Alzheimer's disease: Where intracellular and extracellular meet. *CNS Drugs* **2019**, *33*, 639–648. [CrossRef] [PubMed]
213. Yu, W.H.; Kumar, A.; Peterhoff, C.; Shapiro Kulnane, L.; Uchiyama, Y.; Lamb, B.T.; Cuervo, A.M.; Nixon, R.A. Autophagic vacuoles are enriched in amyloid precursor protein-secretase activities: Implications for β-amyloid peptide over-production and localization in Alzheimer's disease. *Int. J. Biochem. Cell Biol.* **2004**, *36*, 2531–2540. [CrossRef]

214. Morel, E.; Chamoun, Z.; Lasiecka, Z.M.; Chan, R.B.; Williamson, R.L.; Vetanovetz, C.; Dall'Armi, C.; Simoes, S.; Point Du Jour, K.S.; McCabe, B.D.; et al. Phosphatidylinositol-3-phosphate regulates sorting and processing of amyloid precursor protein through the endosomal system. *Nat. Commun.* **2013**, *4*, 2250. [CrossRef]
215. Liu, R.Q.; Zhou, Q.H.; Ji, S.R.; Zhou, Q.; Feng, D.; Wu, Y.; Sui, S.F. Membrane localization of β-amyloid 1-42 in lysosomes: A possible mechanism for lysosome labilization. *J. Biol. Chem.* **2010**, *285*, 19986–19996. [CrossRef]
216. Pasternak, S.H.; Bagshaw, R.D.; Guiral, M.; Zhang, S.; Ackerleyll, C.A.; Pak, B.J.; Callahan, J.W.; Mahuran, D.J. Presenilin-1, nicastrin, amyloid precursor protein, and γ-secretase activity are co-localized in the lysosomal membrane. *J. Biol. Chem.* **2003**, *278*, 26687–26694. [CrossRef] [PubMed]
217. Oikawa, N.; Walter, J. Presenilins and γ-secretase in membrane proteostasis. *Cells* **2019**, *8*, 209. [CrossRef]
218. Tam, J.H.K.; Seah, C.; Pasternak, S.H. The amyloid precursor protein is rapidly transported from the golgi apparatus to the lysosome and where it is processed into beta-amyloid. *Mol. Brain* **2014**, *7*, 54. [CrossRef]
219. Saido, T.; Leissring, M.A. Proteolytic degradation of amyloid β-protein. *Cold Spring Harb. Perspect. Med.* **2012**, *2*, a006379. [CrossRef]
220. Xiao, Q.; Yan, P.; Ma, X.; Liu, H.; Perez, R.; Zhu, A.; Gonzales, E.; Tripoli, D.L.; Czerniewski, L.; Ballabio, A.; et al. Neuronal-targeted TFEB accelerates lysosomal degradation of app, reducing Aβ generation and amyloid plaque pathogenesis. *J. Neurosci.* **2015**, *35*, 12137–12151. [CrossRef]
221. Nixon, R.A.; Cataldo, A.M.; Mathews, P.M. The Endosomal-Lysosomal System of Neurons in Alzheimer's Disease Pathogenesis: A Review. *Neurochem. Res.* **2000**, *25*, 1161–1172. [CrossRef] [PubMed]
222. Martino, S.; Tiribuzi, R.; Ciraci, E.; Makrypidi, G.; D'Angelo, F.; Di Girolamo, I.; Gritti, A.; De Angelis, G.M.C.; Papaccio, G.; Sampaolesi, M.; et al. Coordinated involvement of cathepsins S, D and cystatin C in the commitment of hematopoietic stem cells to dendritic cells. *Int. J. Biochem. Cell Biol.* **2011**, *43*, 775–783. [CrossRef] [PubMed]
223. Koike, M.; Nakanishi, H.; Saftig, P.; Ezaki, J.; Isahara, K.; Ohsawa, Y.; Schulz-Schaeffer, W.; Watanabe, T.; Waguri, S.; Kametaka, S.; et al. Cathepsin D deficiency induces lysosomal storage with ceroid lipofuscin in mouse CNS neurons. *J. Neurosci.* **2000**, *20*, 6898–6906. [CrossRef] [PubMed]
224. Myllykangas, L.; Tyynelä, J.; Page-McCaw, A.; Rubin, G.M.; Haltia, M.J.; Feany, M.B. Cathepsin D-deficient Drosophila recapitulate the key features of neuronal ceroid lipofuscinoses. *Neurobiol. Dis.* **2005**, *19*, 194–199. [CrossRef] [PubMed]
225. Wang, Y.; Wu, Q.; Anand, B.G.; Karthivashan, G.; Phukan, G.; Yang, J.; Thinakaran, G.; Westaway, D.; Kar, S. Significance of cytosolic cathepsin D in Alzheimer's disease pathology: Protective cellular effects of PLGA nanoparticles against β-amyloid-toxicity. *Neuropathol. Appl. Neurobiol.* **2020**. [CrossRef] [PubMed]
226. Suire, C.N.; Abdul-Hay, S.O.; Sahara, T.; Kang, D.; Brizuela, M.K.; Saftig, P.; Dickson, D.W.; Rosenberry, T.L.; Leissring, M.A. Cathepsin D regulates cerebral Aβ42/40 ratios via differential degradation of Aβ42 and Aβ40. *Alzheimer's Res. Ther.* **2020**, *12*, 80. [CrossRef] [PubMed]
227. Urbanelli, L.; Emiliani, C.; Massini, C.; Persichetti, E.; Orlacchio, A.A.; Pelicci, G.; Sorbi, S.; Hasilik, A.; Bernardi, G.; Orlacchio, A.A. Cathepsin D expression is decreased in Alzheimer's disease fibroblasts. *Neurobiol. Aging* **2008**, *29*, 12–22. [CrossRef]
228. Torres, M.; Jimenez, S.; Sanchez-Varo, R.; Navarro, V.; Trujillo-Estrada, L.; Sanchez-Mejias, E.; Carmona, I.; Davila, J.C.; Vizuete, M.; Gutierrez, A.; et al. Defective lysosomal proteolysis and axonal transport are early pathogenic events that worsen with age leading to increased APP metabolism and synaptic Abeta in transgenic APP/PS1 hippocampus. *Mol. Neurodegener.* **2012**, *7*, 1–15. [CrossRef]
229. Wang, C.; Sun, B.; Zhou, Y.; Grubb, A.; Gan, L. Cathepsin B degrades amyloid-β in mice expressing wild-type human amyloid precursor protein. *J. Biol. Chem.* **2012**, *287*, 39834–39841. [CrossRef]
230. Emiliani, C.; Urbanelli, L.; Racanicchi, L.; Orlacchio, A.; Pelicci, G.; Sorbi, S.; Bernardi, G.; Orlacchio, A. Up-regulation of glycohydrolases in Alzheimer's disease fibroblasts correlates with Ras activation. *J. Biol. Chem.* **2003**, *278*, 38453–38460. [CrossRef]
231. Magini, A.; Polchi, A.; Tozzi, A.; Tancini, B.; Tantucci, M.; Urbanelli, L.; Borsello, T.; Calabresi, P.; Emiliani, C. Abnormal cortical lysosomal β-hexosaminidase and β-galactosidase activity at post-synaptic sites during Alzheimer's disease progression. *Int. J. Biochem. Cell Biol.* **2015**, *58*, 62–70. [CrossRef] [PubMed]

232. Tiribuzi, R.; Orlacchio, A.; Crispoltoni, L.; Maiotti, M.; Zampolini, M.; De Angelis, M.; Mecocci, P.; Cecchetti, R.; Bernardi, G.; Datti, A.; et al. Lysosomal β-galactosidase and β-hexosaminidase activities correlate with clinical stages of dementia associated with Alzheimer's disease and type 2 diabetes mellitus. *J. Alzheimer's Dis.* **2011**, *24*, 785–797. [CrossRef] [PubMed]
233. Morena, F.; Argentati, C.; Trotta, R.; Crispoltoni, L.; Stabile, A.; Pistilli, A.; di Baldassarre, A.; Calafiore, R.; Montanucci, P.; Basta, G.; et al. A comparison of lysosomal enzymes expression levels in peripheral blood of mild- and severe-Alzheimer's disease and MCI patients: Implications for regenerative medicine approaches. *Int. J. Mol. Sci.* **2017**, *18*, 1806. [CrossRef]
234. Martino, S.; Emiliani, C.; Tancini, B.; Severini, G.M.; Chigorno, V.; Bordignon, C.; Sonnino, S.; Orlacchio, A. Absence of metabolic cross-correction in Tay-Sachs cells: Implications for gene therapy. *J. Biol. Chem.* **2002**, *277*, 20177–20184. [CrossRef] [PubMed]
235. Martino, S.; Cavalieri, C.; Emiliani, C.; Dolcetta, D.; De Cusella Angelis, M.G.; Chigorno, V.; Severini, G.M.; Sandhoff, K.; Bordignon, C.; Sonnino, S.; et al. Restoration of the GM2 ganglioside metabolism in bone marrow-derived stromal cells from Tay-Sachs disease animal model. *Neurochem. Res.* **2002**, *27*, 793–800. [CrossRef] [PubMed]
236. Grimm, M.O.W.; Zinser, E.G.; Grösgen, S.; Hundsdörfer, B.; Rothhaar, T.L.; Burg, V.K.; Kaestner, L.; Bayer, T.A.; Lipp, P.; Müller, U.; et al. Amyloid precursor protein (APP) mediated regulation of ganglioside homeostasis linking Alzheimer's disease pathology with ganglioside metabolism. *PLoS ONE* **2012**, *7*, e34095. [CrossRef]
237. Calamai, M.; Pavone, F.S. Partitioning and confinement of GM1 ganglioside induced by amyloid aggregates. *FEBS Lett.* **2013**, *587*, 1385–1391. [CrossRef]
238. Tamboli, I.Y.; Prager, K.; Barth, E.; Heneka, M.; Sandhoff, K.; Walter, J. Inhibition of glycosphingolipid biosynthesis reduces secretion of the β-amyloid precursor protein and amyloid β-peptide. *J. Biol. Chem.* **2005**, *280*, 28110–28117. [CrossRef]
239. Winston, C.N.; Goetzl, E.J.; Akers, J.C.; Carter, B.S.; Rockenstein, E.M.; Galasko, D.; Masliah, E.; Rissman, R.A. Prediction of conversion from mild cognitive impairment to dementia with neuronally derived blood exosome protein profile. *Alzheimer's Dement. Diagn. Assess. Dis. Monit.* **2016**, *3*, 63–72. [CrossRef]
240. Goetzl, E.J.; Kapogiannis, D.; Schwartz, J.B.; Lobach, I.V.; Goetzl, L.; Abner, E.L.; Jicha, G.A.; Karydas, A.M.; Boxer, A.; Miller, B.L. Decreased synaptic proteins in neuronal exosomes of frontotemporal dementia and Alzheimer's disease. *FASEB J.* **2016**, *30*, 4141–4148. [CrossRef]
241. Goetzl, E.J.; Boxer, A.; Schwartz, J.B.; Abner, E.L.; Petersen, R.C.; Miller, B.L.; Kapogiannis, D. Altered lysosomal proteins in neural-derived plasma exosomes in preclinical Alzheimer disease. *Neurology* **2015**, *85*, 40–47. [CrossRef] [PubMed]
242. Platt, F.M.; d'Azzo, A.; Davidson, B.L.; Neufeld, E.F.; Tifft, C.J. Lysosomal storage diseases. *Nat. Rev. Dis. Prim.* **2018**, *4*, 27. [CrossRef] [PubMed]
243. Bonam, S.R.; Wang, F.; Muller, S. Lysosomes as a therapeutic target. *Nat. Rev. Drug Discov.* **2019**, *18*, 923–948. [CrossRef] [PubMed]
244. Varma, V.R.; Oommen, A.M.; Varma, S.; Casanova, R.; An, Y.; Andrews, R.M.; O'Brien, R.; Pletnikova, O.; Troncoso, J.C.; Toledo, J.; et al. Brain and blood metabolite signatures of pathology and progression in Alzheimer disease: A targeted metabolomics study. *PLoS Med.* **2018**, *15*, e1002482. [CrossRef] [PubMed]
245. Han, X.; Fagan, A.M.; Cheng, H.; Morris, J.C.; Xiong, C.; Holtzman, D.M. Cerebrospinal fluid sulfatide is decreased in subjects with incipient dementia. *Ann. Neurol.* **2003**, *54*, 115–119. [CrossRef]
246. Han, X. The pathogenic implication of abnormal interaction between apolipoprotein e isoforms, amyloid-beta peptides, and sulfatides in Alzheimer's disease. In *Proceedings of the Molecular Neurobiology*; Humana Press: Totowa, NJ, USA, 2010; Volume 41, pp. 97–106.
247. Lee, S.; Mankhong, S.; Kang, J.H. Extracellular vesicle as a source of Alzheimer's biomarkers: Opportunities and challenges. *Int. J. Mol. Sci.* **2019**, *20*, 1728. [CrossRef]
248. Tune, J.D.; Goodwill, A.G.; Sassoon, D.J.; Mather, K.J. Cardiovascular consequences of metabolic syndrome. *Transl. Res.* **2017**, *183*, 57–70. [CrossRef]
249. Atti, A.R.; Valente, S.; Iodice, A.; Caramella, I.; Ferrari, B.; Albert, U.; Mandelli, L.; De Ronchi, D. Metabolic syndrome, mild cognitive impairment, and dementia: A meta-analysis of longitudinal studies. *Am. J. Geriatr. Psychiatry* **2019**, *27*, 625–637. [CrossRef]
250. Eckel, R.H.; Grundy, S.M.; Zimmet, P.Z. The metabolic syndrome. In *Proceedings of the Lancet*; Elsevier Limited: London, UK, 2005; Volume 365, pp. 1415–1428.

251. Bangen, K.J.; Armstrong, N.M.; Au, R.; Gross, A.L. Metabolic syndrome and cognitive trajectories in the Framingham offspring study. *J. Alzheimer's Dis.* **2019**, *71*, 931–943. [CrossRef]
252. Campos-Peña, V.; Toral-Rios, D.; Becerril-Pérez, F.; Sánchez-Torres, C.; Delgado-Namorado, Y.; Torres-Ossorio, E.; Franco-Bocanegra, D.; Carvajal, K. Metabolic syndrome as a risk factor for Alzheimer's disease: Is Aβ a crucial factor in both pathologies? *Antioxid. Redox Signal.* **2017**, *26*, 542–560. [CrossRef]
253. Pagani, M.; Nobili, F.; Morbelli, S.; Arnaldi, D.; Giuliani, A.; Öberg, J.; Girtler, N.; Brugnolo, A.; Picco, A.; Bauckneht, M.; et al. Early identification of MCI converting to AD: A FDG PET study. *Eur. J. Nucl. Med. Mol. Imaging* **2017**, *44*, 2042–2052. [CrossRef]
254. Paglia, G.; Stocchero, M.; Cacciatore, S.; Lai, S.; Angel, P.; Alam, M.T.; Keller, M.; Ralser, M.; Astarita, G. Unbiased metabolomic investigation of Alzheimer's disease brain points to dysregulation of mitochondrial aspartate metabolism. *J. Proteome Res.* **2016**, *15*, 608–618. [CrossRef] [PubMed]
255. Graham, S.F.; Chevallier, O.P.; Elliott, C.T.; Hölscher, C.; Johnston, J.; McGuinness, B.; Kehoe, P.G.; Passmore, A.P.; Green, B.D. Untargeted metabolomic analysis of human plasma indicates differentially affected polyamine and L-arginine metabolism in mild cognitive impairment subjects converting to Alzheimer's disease. *PLoS ONE* **2015**, *10*, e0119452. [CrossRef]
256. Finneran, D.J.; Nash, K.R. Neuroinflammation and fractalkine signaling in Alzheimer's disease. *J. Neuroinflamm.* **2019**, *16*, 30. [CrossRef] [PubMed]
257. Chaney, A.; Williams, S.R.; Boutin, H. In vivo molecular imaging of neuroinflammation in Alzheimer's disease. *J. Neurochem.* **2019**, *149*, 438–451. [CrossRef] [PubMed]
258. Walters, A.; Phillips, E.; Zheng, R.; Biju, M.; Kuruvilla, T. Evidence for neuroinflammation in Alzheimer's disease. *Prog. Neurol. Psychiatry* **2016**, *20*, 25–31. [CrossRef]
259. Hansen, D.V.; Hanson, J.E.; Sheng, M. Microglia in Alzheimer's disease. *J. Cell Biol.* **2018**, *217*, 459–472. [CrossRef]
260. Fakhoury, M. Microglia and astrocytes in Alzheimer's disease: Implications for therapy. *Curr. Neuropharmacol.* **2017**, *15*, 508–518. [CrossRef]
261. Subhramanyam, C.S.; Wang, C.; Hu, Q.; Dheen, S.T. Microglia-mediated neuroinflammation in neurodegenerative diseases. *Semin. Cell Dev. Biol.* **2019**, *94*, 112–120. [CrossRef]
262. Hemonnot, A.L.; Hua, J.; Ulmann, L.; Hirbec, H. Microglia in Alzheimer disease: Well-known targets and new opportunities. *Front. Cell. Infect. Microbiol.* **2019**, *9*, 233. [CrossRef]
263. González-Reyes, R.E.; Nava-Mesa, M.O.; Vargas-Sánchez, K.; Ariza-Salamanca, D.; Mora-Muñoz, L. Involvement of astrocytes in Alzheimer's disease from a neuroinflammatory and oxidative stress perspective. *Front. Mol. Neurosci.* **2017**, *10*, 427. [CrossRef] [PubMed]
264. Zenaro, E.; Piacentino, G.; Constantin, G. The blood-brain barrier in Alzheimer's disease. *Neurobiol. Dis.* **2017**, *107*, 41–56. [CrossRef] [PubMed]
265. Zhang, F.; Jiang, L. Neuroinflammation in Alzheimer's disease. *Neuropsychiatr. Dis. Treat.* **2015**, *11*, 243–256. [CrossRef] [PubMed]
266. Pasqualetti, G.; Brooks, D.J.; Edison, P. The role of neuroinflammation in dementias. *Curr. Neurol. Neurosci. Rep.* **2015**, *15*, 17. [CrossRef]
267. Heneka, M.T.; Carson, M.J.; El Khoury, J.; Landreth, G.E.; Brosseron, F.; Feinstein, D.L.; Jacobs, A.H.; Wyss-Coray, T.; Vitorica, J.; Ransohoff, R.M.; et al. Neuroinflammation in Alzheimer's disease. *Lancet Neurol.* **2015**, *14*, 388–405. [CrossRef]
268. Lee, C.Y.D.; Landreth, G.E. The role of microglia in amyloid clearance from the AD brain. *J. Neural Transm.* **2010**, *117*, 949–960. [CrossRef]
269. Malik, M.; Parikh, I.; Vasquez, J.B.; Smith, C.; Tai, L.; Bu, G.; Ladu, M.J.; Fardo, D.W.; Rebeck, G.W.; Estus, S. Genetics ignite focus on microglial inflammation in Alzheimer's disease. *Mol. Neurodegener.* **2015**, *10*, 52. [CrossRef]
270. Frost, G.R.; Li, Y.M. The role of astrocytes in amyloid production and Alzheimer's disease. *Open Biol.* **2017**, *7*, 170228. [CrossRef]
271. Ising, C.; Venegas, C.; Zhang, S.; Scheiblich, H.; Schmidt, S.V.; Vieira-Saecker, A.; Schwartz, S.; Albasset, S.; McManus, R.M.; Tejera, D.; et al. NLRP3 inflammasome activation drives tau pathology. *Nature* **2019**, *575*, 669–673. [CrossRef]

272. Shen, X.N.; Niu, L.D.; Wang, Y.J.; Cao, X.P.; Liu, Q.; Tan, L.; Zhang, C.; Yu, J.T. Inflammatory markers in Alzheimer's disease and mild cognitive impairment: A meta-analysis and systematic review of 170 studies. *J. Neurol. Neurosurg. Psychiatry* **2019**, *90*, 590–598. [CrossRef]
273. Zhao, Y.; Wu, X.; Li, X.; Jiang, L.L.; Gui, X.; Liu, Y.; Sun, Y.; Zhu, B.; Piña-Crespo, J.C.; Zhang, M.; et al. TREM2 is a receptor for β-amyloid that mediates microglial function. *Neuron* **2018**, *97*, 1023–1031.e7. [CrossRef] [PubMed]
274. Carmona, S.; Zahs, K.; Wu, E.; Dakin, K.; Bras, J.; Guerreiro, R. The role of TREM2 in Alzheimer's disease and other neurodegenerative disorders. *Lancet Neurol.* **2018**, *17*, 721–730. [CrossRef]
275. Filipello, F.; Morini, R.; Corradini, I.; Zerbi, V.; Canzi, A.; Michalski, B.; Erreni, M.; Markicevic, M.; Starvaggi-Cucuzza, C.; Otero, K.; et al. The microglial innate immune receptor TREM2 is required for synapse elimination and normal brain connectivity. *Immunity* **2018**, *48*, 979–991. [CrossRef] [PubMed]
276. Schlepckow, K.; Kleinberger, G.; Fukumori, A.; Feederle, R.; Lichtenthaler, S.F.; Steiner, H.; Haass, C. An Alzheimer-associated TREM2 variant occurs at the ADAM cleavage site and affects shedding and phagocytic function. *EMBO Mol. Med.* **2017**, *9*, 1356–1365. [CrossRef]
277. Thornton, P.; Sevalle, J.; Deery, M.J.; Fraser, G.; Zhou, Y.; Ståhl, S.; Franssen, E.H.; Dodd, R.B.; Qamar, S.; Gomez Perez-Nievas, B.; et al. TREM 2 shedding by cleavage at the H157-S158 bond is accelerated for the Alzheimer's disease-associated H157Y variant. *EMBO Mol. Med.* **2017**, *9*, 1366–1378. [CrossRef]
278. Rauchmann, B.S.; Sadlon, A.; Perneczky, R. Soluble TREM2 and Inflammatory Proteins in Alzheimer's disease cerebrospinal fluid. *J. Alzheimer's Dis.* **2020**, *73*, 1615–1626. [CrossRef]
279. Zhong, L.; Xu, Y.; Zhuo, R.; Wang, T.; Wang, K.; Huang, R.; Wang, D.; Gao, Y.; Zhu, Y.; Sheng, X.; et al. Soluble TREM2 ameliorates pathological phenotypes by modulating microglial functions in an Alzheimer's disease model. *Nat. Commun.* **2019**, *10*, 1–16.
280. Suárez-Calvet, M.; Caballero, M.Á.A.; Kleinberger, G.; Bateman, R.J.; Fagan, A.M.; Morris, J.C.; Levin, J.; Danek, A.; Ewers, M.; Haass, C. Early changes in CSF sTREM2 in dominantly inherited Alzheimer's disease occur after amyloid deposition and neuronal injury. *Sci. Transl. Med.* **2016**, *8*, 369. [CrossRef]
281. Hesse, R.; Wahler, A.; Gummert, P.; Kirschmer, S.; Otto, M.; Tumani, H.; Lewerenz, J.; Schnack, C.; von Arnim, C.A.F. Decreased IL-8 levels in CSF and serum of AD patients and negative correlation of MMSE and IL-1β. *BMC Neurol.* **2016**, *16*, 185. [CrossRef]
282. Hampel, H.; Caraci, F.; Cuello, A.C.; Caruso, G.; Nisticò, R.; Corbo, M.; Baldacci, F.; Toschi, N.; Garaci, F.; Chiesa, P.A.; et al. A Path Toward precision medicine for neuroinflammatory mechanisms in Alzheimer's disease. *Front. Immunol.* **2020**, *11*, 456. [CrossRef]
283. Molinuevo, J.L.; Ayton, S.; Batrla, R.; Bednar, M.M.; Bittner, T.; Cummings, J.; Fagan, A.M.; Hampel, H.; Mielke, M.M.; Mikulskis, A.; et al. Current state of Alzheimer's fluid biomarkers. *Acta Neuropathol.* **2018**, *136*, 821–853. [CrossRef] [PubMed]
284. Hampel, H.; Vergallo, A.; Aguilar, L.F.; Benda, N.; Broich, K.; Cuello, A.C.; Cummings, J.; Dubois, B.; Federoff, H.J.; Fiandaca, M.; et al. Precision pharmacology for Alzheimer's disease. *Pharmacol. Res.* **2018**, *130*, 331–365. [CrossRef] [PubMed]
285. Bekris, L.M.; Khrestian, M.; Dyne, E.; Shao, Y.; Pillai, J.; Rao, S.; Bemiller, S.M.; Lamb, B.; Fernandez, H.H.; Leverenz, J.B. Soluble TREM2 and biomarkers of central and peripheral inflammation in neurodegenerative disease. *J. Neuroimmunol.* **2018**, *319*, 19–27. [CrossRef] [PubMed]
286. Chen, X.; Hu, Y.; Cao, Z.; Liu, Q.; Cheng, Y. Cerebrospinal fluid inflammatory cytokine aberrations in Alzheimer's disease, Parkinson's disease and amyotrophic lateral sclerosis: A systematic review and meta-analysis. *Front. Immunol.* **2018**, *9*, 2122. [CrossRef] [PubMed]
287. Bicchi, I.; Emiliani, C.; Vescovi, A.; Martino, S. The big bluff of amyotrophic lateral sclerosis diagnosis: The role of neurodegenerative disease mimics. *Neurodegener. Dis.* **2015**, *15*, 313–321. [CrossRef]
288. Cummings, J. Disease modification and neuroprotection in neurodegenerative disorders. *Transl. Neurodegener.* **2017**, *6*, 25. [CrossRef]
289. Pihlstrøm, L.; Wiethoff, S.; Houlden, H. Genetics of neurodegenerative diseases: An overview. In *Handbook of Clinical Neurology*; Elsevier B.V.: Amsterdam, The Netherlands, 2018; Volume 145, pp. 309–323.
290. Morena, F.; Argentati, C.; Bazzucchi, M.; Emiliani, C.; Martino, S. Above the epitranscriptome: RNA modifications and stem cell identity. *Genes* **2018**, *9*, 329. [CrossRef]
291. Bartel, D.P. MicroRNAs: Genomics, biogenesis, mechanism, and function. *Cell* **2004**, *116*, 281–297. [CrossRef]

292. Martino, S.; Montesano, S.; Di Girolamo, I.; Tiribuzi, R.; Di Gregorio, M.; Orlacchio, A.; Datti, A.; Calabresi, P.; Sarchielli, P.; Orlacchio, A. Expression of cathepsins S and D signals a distinctive biochemical trait in CD34+ hematopoietic stem cells of relapsing-remitting multiple sclerosis patients. *Mult. Scler. J.* **2013**, *19*, 1443–1453. [CrossRef]
293. Bicchi, I.; Morena, F.; Montesano, S.; Polidoro, M.; Martino, S. MicroRNAs and molecular mechanisms of neurodegeneration. *Genes* **2013**, *4*, 244–263. [CrossRef]
294. Orlacchio, A.; Bernardi, G.; Orlacchio, A.; Martino, S. RNA interference as a tool for Alzheimers disease therapy. *Mini Rev. Med. Chem.* **2007**, *7*, 1166–1176. [CrossRef] [PubMed]
295. Sheikh, S.; Safia; Haque, E.; Mir, S.S. Neurodegenerative diseases: Multifactorial conformational diseases and their therapeutic interventions. *J. Neurodegener. Dis.* **2013**, *2013*, 563481. [CrossRef] [PubMed]
296. Orlacchio, A.; Bernardi, G.; Martino, S. Stem cells: An overview of the current status of therapies for central and peripheral nervous system diseases. *Curr. Med. Chem.* **2010**, *17*, 595–608. [CrossRef] [PubMed]
297. Orlacchio, A.; Bernardi, G.; Orlacchio, A.; Martino, S. Stem cells and neurological diseases. *Discov. Med.* **2010**, *9*, 546–553. [PubMed]
298. Nuzziello, N.; Liguori, M. The MicroRNA centrism in the orchestration of neuroinflammation in neurodegenerative diseases. *Cells* **2019**, *8*, 1193. [CrossRef]
299. Gaudet, A.D.; Fonken, L.K.; Watkins, L.R.; Nelson, R.J.; Popovich, P.G. MicroRNAs: Roles in regulating neuroinflammation. *Neuroscientist* **2018**, *24*, 221–245. [CrossRef] [PubMed]
300. Thibaudeau, T.A.; Anderson, R.T.; Smith, D.M. A common mechanism of proteasome impairment by neurodegenerative disease-associated oligomers. *Nat. Commun.* **2018**, *9*, 1–14. [CrossRef]
301. Poswar, F.; Vairo, F.; Burin, M.; Michelin-Tirelli, K.; Brusius-Facchin, A.; Kubaski, F.; Desouza, C.; Baldo, G.; Giugliani, R. Lysosomal diseases: Overview on current diagnosis and treatment. *Genet. Mol. Biol.* **2019**, *42*, 165–177. [CrossRef]
302. Elmonem, M.A.; Abdelazim, A.M. Novel biomarkers for lysosomal storage disorders: Metabolomic and proteomic approaches. *Clin. Chim. Acta* **2020**, *509*, 195–209. [CrossRef]
303. Jin, L.W.; Maezawa, I.; Vincent, I.; Bird, T. Intracellular accumulation of amyloidogenic fragments of amyloid-β precursor protein in neurons with niemann-pick type C defects is associated with endosomal abnormalities. *Am. J. Pathol.* **2004**, *164*, 975–985. [CrossRef]
304. Nixon, R.A. Niemann-Pick Type C Disease and Alzheimer's disease: The APP-endosome connection fattens up. *Am. J. Pathol.* **2004**, *164*, 757–761. [CrossRef]
305. Ornaghi, F.; Sala, D.; Tedeschi, F.; Maffia, M.C.; Bazzucchi, M.; Morena, F.; Valsecchi, M.; Aureli, M.; Martino, S.; Gritti, A. Novel bicistronic lentiviral vectors correct β-Hexosaminidase deficiency in neural and hematopoietic stem cells and progeny: Implications for in vivo and ex vivo gene therapy of GM2 gangliosidosis. *Neurobiol. Dis.* **2020**, *134*, 104667. [CrossRef] [PubMed]
306. Lattanzi, A.; Neri, M.; Maderna, C.; di Girolamo, I.; Martino, S.; Orlacchio, A.; Amendola, M.; Naldini, L.; Gritti, A. Widespread enzymatic correction of CNS tissues by a single intracerebral injection of therapeutic lentiviral vector in leukodystrophy mouse models. *Hum. Mol. Genet.* **2010**, *19*, 2208–2227. [CrossRef] [PubMed]
307. Morena, F.; Oikonomou, V.; Argentati, C.; Bazzucchi, M.; Emiliani, C.; Gritti, A.; Martino, S. Integrated computational analysis highlights unique miRNA signatures in the subventricular zone and striatum of GM2 gangliosidosis animal models. *Int. J. Mol. Sci.* **2019**, *20*, 3179. [CrossRef]
308. Keilani, S.; Lun, Y.; Stevens, A.C.; Williams, H.N.; Sjoberg, E.R.; Khanna, R.; Valenzano, K.J.; Checler, F.; Buxbaum, J.D.; Yanagisawa, K.; et al. Lysosomal dysfunction in a mouse model of Sandhoff disease leads to accumulation of ganglioside-bound amyloid-β peptide. *J. Neurosci.* **2012**, *32*, 5223–5236. [CrossRef]
309. Ginsberg, S.D.; Galvin, J.E.; Lee, V.M.Y.; Rorke, L.B.; Dickson, D.W.; Wolfe, J.H.; Jones, M.Z.; Trojanowski, J.Q. Accumulation of intracellular amyloid-β peptide (Aβ 1-40) in mucopolysaccharidosis brains. *J. Neuropathol. Exp. Neurol.* **1999**, *58*, 815–824. [CrossRef]
310. Ohmi, K.; Zhao, H.-Z.; Neufeld, E.F. Defects in the medial entorhinal cortex and dentate gyrus in the mouse model of Sanfilippo syndrome type B. *PLoS ONE* **2011**, *6*, e27461. [CrossRef]
311. Xu, Y.; Xu, K.; Sun, Y.; Liou, B.; Quinn, B.; Li, R.; Xue, L.; Zhang, W.; Setchell, K.D.R.; Witte, D.; et al. Multiple pathogenic proteins implicated in neuronopathic Gaucher disease mice. *Hum. Mol. Genet.* **2014**, *23*, 3943–3957. [CrossRef]

312. Fujikake, N.; Shin, M.; Shimizu, S. Association between autophagy and neurodegenerative diseases. *Front. Neurosci.* **2018**, *12*, 255. [CrossRef]
313. Erie, C.; Sacino, M.; Houle, L.; Lu, M.L.; Wei, J. Altered lysosomal positioning affects lysosomal functions in a cellular model of Huntington's disease. *Eur. J. Neurosci.* **2015**, *42*, 1941–1951. [CrossRef]
314. Frakes, A.E.; Ferraiuolo, L.; Haidet-Phillips, A.M.; Schmelzer, L.; Braun, L.; Miranda, C.J.; Ladner, K.J.; Bevan, A.K.; Foust, K.D.; Godbout, J.P.; et al. Microglia induce motor neuron death via the classical NF-κB pathway in amyotrophic lateral sclerosis. *Neuron* **2014**, *81*, 1009–1023. [CrossRef] [PubMed]
315. Wolfe, D.M.; Lee, J.-H.; Kumar, A.; Lee, S.; Orenstein, S.J.; Nixon, R.A. Autophagy failure in Alzheimer's disease and the role of defective lysosomal acidification. *Eur. J. Neurosci.* **2013**, *37*, 1949–1961. [CrossRef] [PubMed]
316. Guo, F.; Liu, X.; Cai, H.; Le, W. Autophagy in neurodegenerative diseases: Pathogenesis and therapy. *Brain Pathol.* **2018**, *28*, 3–13. [CrossRef] [PubMed]
317. Novellino, F.; Saccà, V.; Donato, A.; Zaffino, P.; Spadea, M.F.; Vismara, M.; Arcidiacono, B.; Malara, N.; Presta, I.; Donato, G. Innate immunity: A common denominator between neurodegenerative and neuropsychiatric diseases. *Int. J. Mol. Sci.* **2020**, *21*, 1115. [CrossRef] [PubMed]
318. Scheiblich, H.; Trombly, M.; Ramirez, A.; Heneka, M.T. Neuroimmune connections in aging and neurodegenerative diseases. *Trends Immunol.* **2020**, *41*, 300–312. [CrossRef] [PubMed]
319. Niedzielska, E.; Smaga, I.; Gawlik, M.; Moniczewski, A.; Stankowicz, P.; Pera, J.; Filip, M. Oxidative stress in neurodegenerative diseases. *Mol. Neurobiol.* **2016**, *53*, 4094–4125. [CrossRef]
320. Sorce, S.; Krause, K.H. NOX enzymes in the central nervous system: From signaling to disease. *Antioxid. Redox Signal.* **2009**, *11*, 2481–2504. [CrossRef]
321. Cahill-Smith, S.; Li, J.M. Oxidative stress, redox signalling and endothelial dysfunction in ageing-related neurodegenerative diseases: A role of NADPH oxidase 2. *Br. J. Clin. Pharmacol.* **2014**, *78*, 441–453. [CrossRef]
322. Singh, A.; Kukreti, R.; Saso, L.; Kukreti, S. Oxidative stress: A key modulator in neurodegenerative diseases. *Molecules* **2019**, *24*, 1583. [CrossRef]
323. Tarafdar, A.; Pula, G. The role of NADPH oxidases and oxidative stress in neurodegenerative disorders. *Int. J. Mol. Sci.* **2018**, *19*, 3824. [CrossRef]
324. Agrawal, M.; Biswas, A. Molecular diagnostics of neurodegenerative disorders. *Front. Mol. Biosci.* **2015**, *2*, 54. [CrossRef] [PubMed]
325. Robelin, L.; Gonzalez De Aguilar, J.L. Blood biomarkers for amyotrophic lateral sclerosis: Myth or reality? *Biomed Res. Int.* **2014**, *2014*, 525097. [CrossRef] [PubMed]
326. Sun, A. Lysosomal storage disease overview. *Ann. Transl. Med.* **2018**, *6*, 476. [CrossRef] [PubMed]
327. Van Giau, V.; Bagyinszky, E.; Yang, Y.S.; Youn, Y.C.; An, S.S.A.; Kim, S.Y. Genetic analyses of early-onset Alzheimer's disease using next generation sequencing. *Sci. Rep.* **2019**, *9*, 1–10. [CrossRef] [PubMed]
328. Sancesario, G.M.; Bernardini, S. Alzheimer's disease in the omics era. *Clin. Biochem.* **2018**, *59*, 9–16. [CrossRef] [PubMed]
329. Johnson, E.C.B.; Dammer, E.B.; Duong, D.M.; Ping, L.; Zhou, M.; Yin, L.; Higginbotham, L.A.; Guajardo, A.; White, B.; Troncoso, J.C.; et al. Large-scale proteomic analysis of Alzheimer's disease brain and cerebrospinal fluid reveals early changes in energy metabolism associated with microglia and astrocyte activation. *Nat. Med.* **2020**, *26*, 769–780. [CrossRef]
330. Wu, Y.Y.; Chiu, F.L.; Yeh, C.S.; Kuo, H.C. Opportunities and challenges for the use of induced pluripotent stem cells in modelling neurodegenerative disease. *Open Biol.* **2019**, *9*, 180177. [CrossRef]
331. Israel, M.A.; Yuan, S.H.; Bardy, C.; Reyna, S.M.; Mu, Y.; Herrera, C.; Hefferan, M.P.; Van Gorp, S.; Nazor, K.L.; Boscolo, F.S.; et al. Probing sporadic and familial Alzheimer's disease using induced pluripotent stem cells. *Nature* **2012**, *482*, 216–220. [CrossRef]
332. Frati, G.; Luciani, M.; Meneghini, V.; De Cicco, S.; Ståhlman, M.; Blomqvist, M.; Grossi, S.; Filocamo, M.; Morena, F.; Menegon, A.; et al. Human iPSC-based models highlight defective glial and neuronal differentiation from neural progenitor cells in metachromatic leukodystrophy. *Cell Death Dis.* **2018**, *9*, 1–15. [CrossRef]
333. Argentati, C.; Tortorella, I.; Bazzucchi, M.; Morena, F.; Martino, S. Harnessing the potential of stem cells for disease modeling: Progress and promises. *J. Pers. Med.* **2020**, *10*, 8. [CrossRef]

334. Martino, S.; di Girolamo, I.; Cavazzin, C.; Tiribuzi, R.; Galli, R.; Rivaroli, A.; Valsecchi, M.; Sandhoff, K.; Sonnino, S.; Vescovi, A.; et al. Neural precursor cell cultures from GM2 gangliosidosis animal models recapitulate the biochemical and molecular hallmarks of the brain pathology. *J. Neurochem.* **2009**, *109*, 135–147. [CrossRef] [PubMed]
335. Meneghini, V.; Frati, G.; Sala, D.; De Cicco, S.; Luciani, M.; Cavazzin, C.; Paulis, M.; Mentzen, W.; Morena, F.; Giannelli, S.; et al. Generation of human induced pluripotent stem cell-derived bona fide neural stem cells for ex vivo gene therapy of metachromatic leukodystrophy. *Stem Cells Transl. Med.* **2017**, *6*, 352–368. [CrossRef] [PubMed]
336. Cummings, J.; Lee, G.; Ritter, A.; Sabbagh, M.; Zhong, K. Alzheimer's disease drug development pipeline: 2019. *Alzheimer's Dement. Transl. Res. Clin. Interv.* **2019**, *5*, 272–293. [CrossRef] [PubMed]
337. Atri, A. Current and future treatments in Alzheimer's disease. *Semin. Neurol.* **2019**, *39*, 227–240. [CrossRef] [PubMed]
338. Hampel, H.; Mesulam, M.-M.; Cuello, A.C.; Farlow, M.R.; Giacobini, E.; Grossberg, G.T.; Khachaturian, A.S.; Vergallo, A.; Cavedo, E.; Snyder, P.J.; et al. The cholinergic system in the pathophysiology and treatment of Alzheimer's disease. *Brain* **2018**, *141*, 1917–1933. [CrossRef] [PubMed]
339. Cummings, J.; Lee, G.; Ritter, A.; Sabbagh, M.; Zhong, K. Alzheimer's disease drug development pipeline: 2020. *Alzheimer's Dement. Transl. Res. Clin. Interv.* **2020**, *6*, e12050. [CrossRef]
340. LMTM | ALZFORUM. Available online: https://www.alzforum.org/therapeutics/lmtm (accessed on 26 July 2020).
341. Safety and Efficacy of TRx0237 in Subjects with Alzheimer's Disease Followed by Open-Label Treatment—Full Text View—ClinicalTrials.gov. Available online: https://clinicaltrials.gov/ct2/show/NCT03446001 (accessed on 26 July 2020).
342. Huang, L.K.; Chao, S.P.; Hu, C.J. Clinical trials of new drugs for Alzheimer disease. *J. Biomed. Sci.* **2020**, *27*, 18. [CrossRef]
343. Vandenberghe, R.; Riviere, M.E.; Caputo, A.; Sovago, J.; Maguire, R.P.; Farlow, M.; Marotta, G.; Sanchez-Valle, R.; Scheltens, P.; Ryan, J.M.; et al. Active Aβ immunotherapy CAD106 in Alzheimer's disease: A phase 2b study. *Alzheimer's Dement. Transl. Res. Clin. Interv.* **2017**, *3*, 10–22. [CrossRef]
344. Amilomotide | ALZFORUM. Available online: https://www.alzforum.org/therapeutics/amilomotide (accessed on 26 July 2020).
345. A Study of CAD106 and CNP520 Versus Placebo in Participants at Risk for the Onset of Clinical Symptoms of Alzheimer's Disease—Full Text View—ClinicalTrials.gov. Available online: https://clinicaltrials.gov/ct2/show/study/NCT02565511 (accessed on 26 July 2020).
346. Farr, S.A.; Ripley, J.L.; Sultana, R.; Zhang, Z.; Niehoff, M.L.; Platt, T.L.; Murphy, M.P.; Morley, J.E.; Kumar, V.; Butterfield, D.A. Antisense oligonucleotide against GSK-3β in brain of SAMP8 mice improves learning and memory and decreases oxidative stress: Involvement of transcription factor Nrf2 and implications for Alzheimer disease. *Free Radic. Biol. Med.* **2014**, *67*, 387–395. [CrossRef]
347. DeVos, S.L.; Miller, R.L.; Schoch, K.M.; Holmes, B.B.; Kebodeaux, C.S.; Wegener, A.J.; Chen, G.; Shen, T.; Tran, H.; Nichols, B.; et al. Tau reduction prevents neuronal loss and reverses pathological tau deposition and seeding in mice with tauopathy. *Sci. Transl. Med.* **2017**, *9*, eaag0481. [CrossRef]

Journal of
Personalized
Medicine

Article

Deficits in Mitochondrial Spare Respiratory Capacity Contribute to the Neuropsychological Changes of Alzheimer's Disease

Simon M. Bell, Matteo De Marco, Katy Barnes, Pamela J. Shaw, Laura Ferraiuolo, Daniel J. Blackburn, Heather Mortiboys *,† and Annalena Venneri *,†

Sheffield Institute for Translational Neuroscience (SITraN), University of Sheffield, 385a Glossop Road, Sheffield S10 2HQ, UK; s.m.bell@sheffield.ac.uk (S.M.B.); m.demarco@sheffield.ac.uk (M.D.M.); kabarnes1@sheffield.ac.uk (K.B.); pamela.shaw@sheffield.ac.uk (P.J.S.); l.ferraiuolo@shef.ac.uk (L.F.); d.blackburn@sheffield.ac.uk (D.J.B.)

* Correspondence: h.mortiboys@sheffield.ac.uk (H.M.); a.venneri@sheffield.ac.uk (A.V.); Tel.: +44-(0)114-222-2259 (H.M.); +44-(0)114-271-3430 (A.V.)

† These two authors are joint senior authors.

Received: 30 March 2020; Accepted: 24 April 2020; Published: 29 April 2020

Abstract: Alzheimer's disease (AD) is diagnosed using neuropsychological testing, supported by amyloid and tau biomarkers and neuroimaging abnormalities. The cause of neuropsychological changes is not clear since they do not correlate with biomarkers. This study investigated if changes in cellular metabolism in AD correlate with neuropsychological changes. Fibroblasts were taken from 10 AD patients and 10 controls. Metabolic assessment included measuring total cellular ATP, extracellular lactate, mitochondrial membrane potential (MMP), mitochondrial respiration and glycolytic function. All participants were assessed with neuropsychological testing and brain structural MRI. AD patients had significantly lower scores in delayed and immediate recall, semantic memory, phonemic fluency and Mini Mental State Examination (MMSE). AD patients also had significantly smaller left hippocampal, left parietal, right parietal and anterior medial prefrontal cortical grey matter volumes. Fibroblast MMP, mitochondrial spare respiratory capacity (MSRC), glycolytic reserve, and extracellular lactate were found to be lower in AD patients. MSRC/MMP correlated significantly with semantic memory, immediate and delayed episodic recall. Correlations between MSRC and delayed episodic recall remained significant after controlling for age, education and brain reserve. Grey matter volumes did not correlate with MRSC/MMP. AD fibroblast metabolic assessment may represent an emergent disease biomarker of AD.

Keywords: Alzheimer's disease; mitochondrial spare respiratory capacity; mitochondrial; membrane potential; glycolytic reserve; semantic memory; phonemic fluency; episodic memory; neuropsychology; neuroimaging

---

1. Background

Alzheimer's disease (AD) is the most common cause of dementia worldwide and in 2018 was estimated to cost the global economy 1 trillion US dollars [1]. The clinical symptoms of the disease are the progressive loss of different aspects of cognitive function until a patient becomes completely dependent on the care of family members and healthcare workers [2]. Median survival after diagnosis is 7 to 10 years for people in their 60s to 70s, and 3 years for people in their 90s [3].

The disease is characterized pathologically by the presence of extracellular amyloid plaques comprising mainly of the amyloid beta protein; and intracellular neurofibrillary tangles (NFT) made mainly of the cytoskeletal protein tau [4]. To date the cause of AD still remains poorly understood.

As the buildup of amyloid appears to be a key step in the development of both familial and sporadic forms of the disease, the amyloid cascade hypothesis has become the leading theory for the cause of the condition [5]. In brief, this hypothesis states that the key step in developing AD is the accumulation of amyloid beta through reduced breakdown and clearance, and/or increased production. Strategies aimed at reducing the amyloid load in the brain, however, have failed to control the disease [6] and have resulted in a large number of clinical trials that have failed to achieve primary outcome measures [7]. Even in pre-clinical carriers of dominantly inherited AD mutations, amyloid removal therapies have not slowed disease progression [8]. Furthermore, brain amyloid load does not correlate with clinical symptoms [9]. This has led researchers to investigate alternative pathophysiological mechanisms [10].

AD is clinically defined by distinctive changes in cognitive status identified by neuropsychological assessment. Brain imaging changes and amyloid and tau protein levels in the cerebrospinal fluid are used to confirm the diagnosis in vivo [11]. The changes seen in a patient's ability to perform a cognitive task are often difficult to explain from a cellular perspective. Tau deposition does explain elements of the observed neuropsychological abnormalities [12], but does not fully account for all cognitive changes observed in AD patients [13,14].

Cognitive processing, such as that required while performing memory tasks, puts an increased metabolic demand on the brain [15]. This is evidenced by neuroimaging studies of the brain which use tracers of metabolism such as 2-[18F]fluoro-2-Deoxy-D-glucose (FDG) that have shown poor glucose utilization in patients who perform poorly on memory tasks [16,17]. Positron-emission tomography (PET) imaging studies that use oxygen-15 labelled water also show reduced uptake when AD patients perform cognitive tasks [18]. These imaging studies suggest that any deficit in metabolic function, such as deficits in mitochondrial respiration or glycolysis, are likely to affect an individual's performance on cognitive tasks. It is therefore possible that mitochondrial respiration or glycolytic dysfunction might contribute to cognitive deficits in AD, making these cellular processes suitable pharmacological targets to improve the cognitive symptoms of AD.

Cellular metabolic changes within the brains of patients with AD and in peripheral cell populations are seen very early in the condition, and often precede the development of both amyloid plaques and NFT. Abnormalities have been shown in many metabolic pathways in AD [19]. Mounting evidence suggests that deficits in glycolysis and the function of mitochondria, specifically how they control oxidative phosphorylation, are likely to be key in the development and establishment of AD [20–22].

A mitochondrial cascade hypothesis has been suggested for the aetiology of AD, and states that people who inherit mitochondrial genes that predispose them to lower mitochondrial respiration rates may be more likely to develop the condition [23]. In animal models of AD, changes in mitochondrial function are seen prior to amyloid deposition [20,24], and cell models show changes in mitochondrial function and oxidative stress without the presence of amyloid [25], giving further evidence for the key role of mitochondrial dysfunction in AD.

Impairment of glycolysis is also seen early in patients with AD. A 2-[18F]fluoro-2-Deoxy-D-glucose positron-emission tomography (FDG-PET) imaging of the brain shows a reduction in glucose metabolism [26]. In particular, there is a reduction in aerobic glycolysis in brain areas susceptible to amyloid deposition [27], and in those regions where high levels of tau accumulation are seen [28].

We have previously shown that fibroblasts from sporadic AD (sAD) patients have multiple mitochondrial structural and functional abnormalities, and that these can be ameliorated by treatment with ursodeoxycholic acid (UDCA) [29]. Other studies have shown that glycyl-l-histidyl-l-lysine (GHK-Cu), by increasing gene expression, has an effect on improving mitochondrial activity and influencing cognitive decline [30]. In our previous study we showed sAD fibroblasts to have deficits in mitochondrial membrane potential (MMP) and mitochondrial spare respiratory capacity (MSRC). MSRC refers to the difference in oxidative phosphorylation rates between the basal level of mitochondrial respiration and the maximal level a cell can achieve [31]. In essence, MSRC measures the cellular reserve respiratory capacity. In animal models of AD it has been shown that deficits in MSRC cause cognitive deficits that treatment with the antioxidant pyrroloquinoline quinone can improve [32]. It has not been previously shown, however, whether changes in MSRC in human cell lines obtained from sAD patients correlate with their performance on neuropsychological tests.

As deficits in both glycolysis and mitochondrial function have been shown in AD patients and models, in this proof of concept study we explored metabolic function in fibroblasts from sporadic AD patients and its role in the cognitive decline experienced by these patients. To this end, we assessed mitochondrial functional changes in a larger cohort of patients compared to the findings described in our previous study [29]. We have then assessed additional metabolic parameters in the sAD fibroblasts including glycolytic function and cellular ATP levels, which have previously not been described. Finally, we investigated whether these metabolic abnormalities detected in sAD fibroblasts correlated with neuropsychological and neuroimaging features typical of the early stages of this disease.

2. Results

2.1. Patient Demographic Details

Skin biopsies were taken from ten sAD patients (mean age 61.3 years, 6 male) and ten controls (mean age 66.7 years, 5 male). Body mass index (BMI) did not differ significantly between the groups (sAD mean 27.8 kg/m^2 SD 5.37 kg/m^2 Controls mean 28.2 kg/m^2 SD 4.00 kg/m^2 t-test $p = 0.44$). Table 1 shows patient and control demographic data.

Table 1. Patient Demographic Information and contemporary treatment status.

Patient Number	Age (Years)	Sex	MMSE	Length of Education (Years)	AD Treatment (in Disease Cohort, at Time of Biopsy)
1	63	Male	20	16	None
2	57	Male	18	15	Donepezil
3	53	Male	14	11	Donepezil
4 *	60	Male	18	9	None
5 *	59	Female	23	11	Galantamine
6 *	63	Female	26	10	Donepezil
7 *	60	Male	18	11	Memantine
8	60	Male	18	11	None
9	79	Female	28	15	None
10	61	Female	25	11	Donepezil
Group Mean (Standard Dev)	61.33 (7.19)		20.8 (4.4)	12.0 (2.4)	
Controls					
1	66	Male	29	11	NA
2 *	54	Male	27	17	NA
3 *	53	Male	29	16	NA
4 *	56	Male	24	11	NA
5	61	Female	30	NA	NA
6	54	Female	29	17	NA
7 *	100	Female	24 #	14	NA
8	75	Female	28	18	NA
9	73	Female	26	12	NA
10	75	Male	27	10	NA
Group Mean (Standard Dev)	65.77 (14.7)		27.6 (1.8)	14 (3.1)	

For control 7 some items of the Mini Mental State Examination (MMSE) could not be tested due to sensory impairment * indicates fibroblast lines in which Mitochondrial Membrane Potential (MMP) and Mitochondrial Spare Respiratory Capacity (MRSC) values were published in our previous paper [29].

2.2. Neuropsychological Profiles

Neuropsychological profiling of controls and patients used in this study was performed as part of the VPH-DARE research project (http://www.vph-dare.eu/), (see Methods section for additional details). For the present study, a subgroup of neuropsychological tests was selected to be correlated with metabolic findings. The tests which can detect the earliest typical cognitive impairments in mild sAD were selected. These included assessment of semantic memory, immediate and delayed episodic recall. Phonemic fluency was also selected as dysfunction on this cognitive test is seen later in AD but not in its earlier stages [33]. All 10 controls and 10 patients with AD were assessed. The Mini-Mental

State Examination (MMSE) [34] was also used, as this test is useful in staging disease severity in sAD [34,35].

Patients with sAD performed at a lower level than controls on all tests (see Figure 1). The most significant differences were seen in the tests of semantic memory (mean score in controls 42.6 points, mean score in sAD 17.5 points, difference between means 25.1 points, SD 4.774 $p < 0.0001$); immediate (mean score in controls 15.8 points, mean score in sAD 6.4 points, difference between means 25.1 points, SD 1.468 $p < 0.0001$); and delayed recall of the Prose Memory test (mean score in controls 18.6 points, mean score in sAD 5.5 points, difference between means 13.4 points, SD 1.488 $p < 0.0001$). The phonemic fluency test performance was also significantly different between the 2 groups (mean score in controls 50.5 points, mean score in sAD 27.4 points, difference between means 23.1 points, SD 6.837 $p = 0.0033$), but to a lesser extent than the other neuropsychological tests. As expected, MMSE scores of sAD patients were significantly lower than those of controls (mean score in controls 27.3 points, mean score in sAD 21.3 points, mean difference 6.0 points, SD 1.54 $p = 0.0011$).

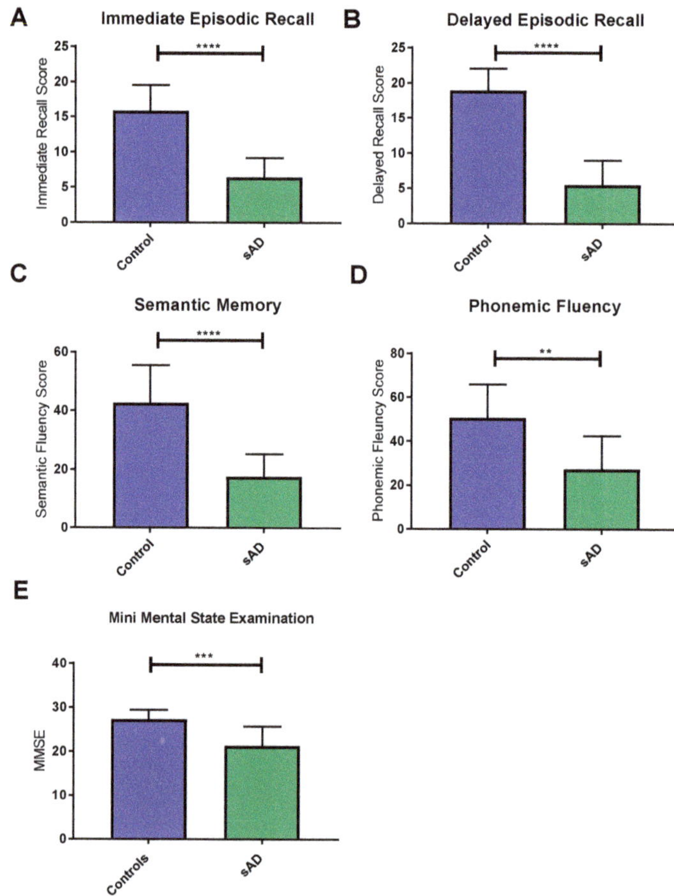

Figure 1. Mean scores of sporadic AD (sAD) patients and controls on cognitive tests included in the neuropsychological assessment. Graphs show mean with error bars indicating standard deviation. **** $p < 0.0001$; *** $p < 0.001$; ** $p < 0.01$. Controls are indicated by blue bars and sAD patients indicated by green bars. Significant reductions were seen in sAD immediate recall (**A**), delayed recall (**B**), semantic memory (**C**), phonemic fluency (**D**) and Mini Mental State Examination (**E**) when compared to controls.

2.3. Neuroimaging Profiles

Volumetric structural MRI scans were acquired on 9 controls and 10 patients with sAD. One control did not complete their MRI scan as an incidental finding (of no diagnostic significance for this study) on initial scans meant the participant could no longer take part in the VPH-DARE@IT original study. For the remaining participants, the left and right parietal lobes, anterior medial pre-frontal cortex and posterior cingulate gyrus, and left hippocampal grey matter volumes were extracted as these brain areas are affected early in AD [36].

Comparisons between the 2 groups showed no significant differences between the grey matter volumes of the preselected areas when performing a *t*-test. Brain grey matter volumes can be affected by age, education and brain reserve [37,38]. For these reasons a further analysis of the 2 groups was completed controlling for these covariates. A significant difference emerged in the volume of the left hippocampus, left parietal, right parietal and anterior medial prefrontal cortex in patients with sAD after controlling for confounding variables. Table 2 highlights the F-test and *p*-values for these differences. No significant difference was seen in the posterior cingulate cortex grey matter volume between the 2 groups.

Table 2. Differences in Control and sAD brain volumes. This table shows the significance of the differences in brain volumes between controls and patients with sAD when controlling for age, years of education and brain reserve.

Brain Volume	*F*-Test	*p*-Value
Left Hippocampal Volume	9.420	0.001
Left Parietal Volume	7.882	0.002
Right Parietal Volume	10.051	<0.0001
Anterior Medial Prefrontal Cortex	0.056	0.017
Posterior Cingulate Cortex	0.752	0.575

2.4. Fibroblast Metabolic Assessment

Mitochondrial function was investigated in both sAD patients and controls. We measured elements of both mitochondrial function and glycolysis. Five metabolic parameters were chosen that best describe different factors influencing the ability of the fibroblast to meet its energy demands. Mitochondrial parameters included: total cellular ATP, as this is a global marker of the energetic status of the cell; MMP and MSRC. MMP and MSRC were selected as these parameters are important in maintaining the rate of ATP production as cellular energy demand changes. These are also the two parameters which we have previously identified as being relevant to the mitochondrial phenotype in sAD fibroblasts [29]. Measures of glycolysis included glycolytic reserve and extracellular lactate levels. Glycolytic reserve, like MSRC and MMP for mitochondrial function, is a measure of cellular metabolic flexibility and deficits in this parameter are likely to contribute to an inability to maintain cellular energy production. Extracellular lactate is the final breakdown product of glycolysis and contributes information about how well the cell can utilize glucose as a metabolite.

First, we assessed MMP and MRSC as we have done in our previous work [29]. In this cohort, MRSC was significantly reduced (36% reduction $p < 0.0001$) in sAD when compared to controls (Figure 2A). MMP was also significantly reduced in the sAD fibroblast lines (14% reduction $p = 0.011$) (See Figure 2B). Figure 2C shows a representative oxygen consumption trace for the mitochondrial stress test experiment.

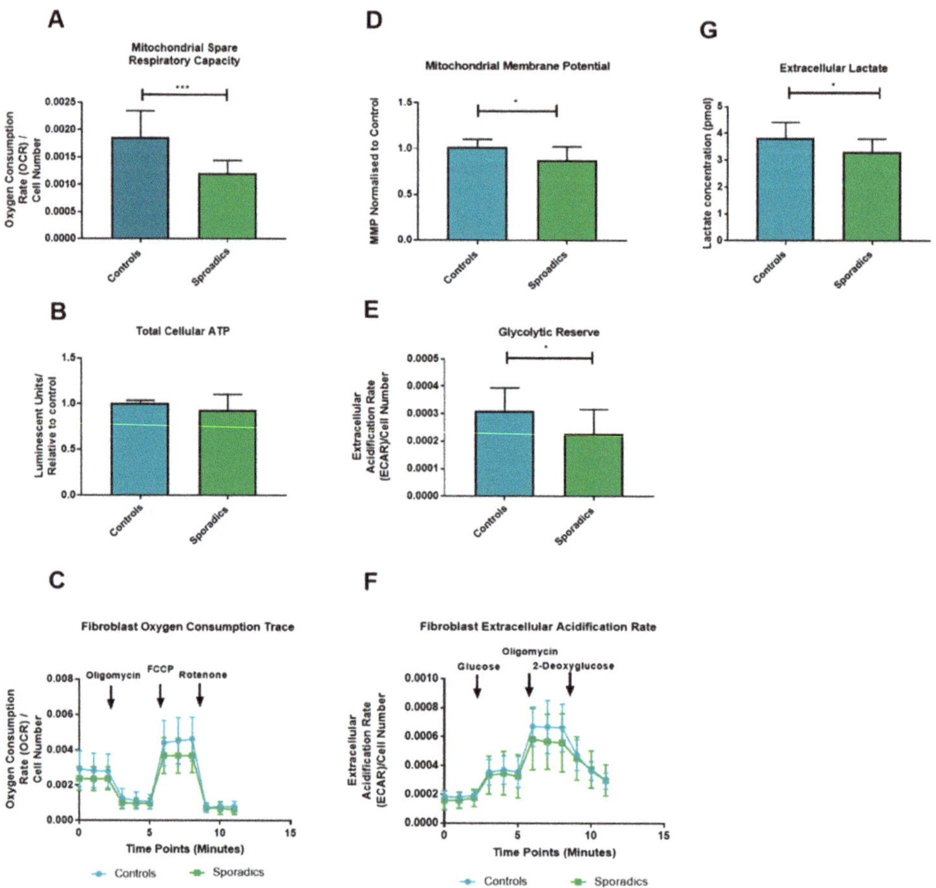

Figure 2. Oxidative phosphorylation and Glycolytic fibroblast Assessment: *** = $p < 0.001$; * = $p < 0.05$. Significant reductions in mitochondrial spare respiratory capacity (**A**), Mitochondrial Membrane potential (**D**), Glycolytic Reserve (**E**) and Extracellular lactate (**G**) was detected in sAD when compared to controls. No significant difference was seen in total cellular ATP (**B**). OCR and ECAR traces for mitochondrial stress test and glycolysis stress test are displayed in panels (**C**) and (**F**) respectively. For all panels controls are represented in blue and sAD are represented in green. Graphs represent mean with error bars indicating standard deviation.

Next, we assessed total cellular ATP, which showed no significant difference between sAD patient fibroblasts and controls when comparing mean values (p = 0.165, Figure 2D). The final metabolic assessment of the fibroblast lines involved assessment of glucose metabolism using the glycolysis stress test programme on the Seahorse analyzer and extracellular lactate measurement. A significantly lower glycolytic reserve was found in sAD fibroblasts when compared to controls (see Figure 1E, 25.8% reduction, p = 0.031). No significant differences were seen in maximum glycolytic rate (p = 0.792), or basal glycolytic rate (p = 0.381). Figure 1F shows the extracellular acidification rate (ECAR) trace for the glycolysis stress test, comparing the group of controls against the sAD fibroblast group. Measurement of extracellular lactate levels revealed significantly lower lactate levels released from sAD fibroblasts (14% reduction, p = 0.0227, Figure 2G).

2.5. Neuropsychological/Metabolic Correlations

Next, we sought to correlate the neuropsychological scores of the sAD patients and controls combined with their metabolic parameters measured in fibroblasts. As five neuropsychological measures and five fibroblast metabolic markers had been assessed to identify differences in the control and sAD fibroblast groups, we controlled for the effect of multiple comparisons by adjusting what we deemed to be a significant association to $p \leq 0.01$ (0.05/5). Using this new statistical threshold, only MSRC and MMP metabolic tests and immediate, delayed and semantic memory neuropsychological assessments were identified as significantly different between control and sAD groups. Correlations were, therefore, performed only for these data sets.

MSRC had a significant positive correlation with immediate episodic recall (r = 0.612, $p = 0.004$), delayed episodic recall ($r = 0.669$, $p = 0.001$) and semantic memory scores ($r = 0.614$, $p = 0.003$). MMP correlated only with semantic memory scores ($r = 0.565$, $p = 0.009$). Immediate episodic recall ($r = 0.552$, $p = 0.0134$) and delayed episodic recall correlated positively with MMP, but at a lower level of significance ($r = 0.540$, $p = 0.015$). Figure 3 displays these correlations graphically.

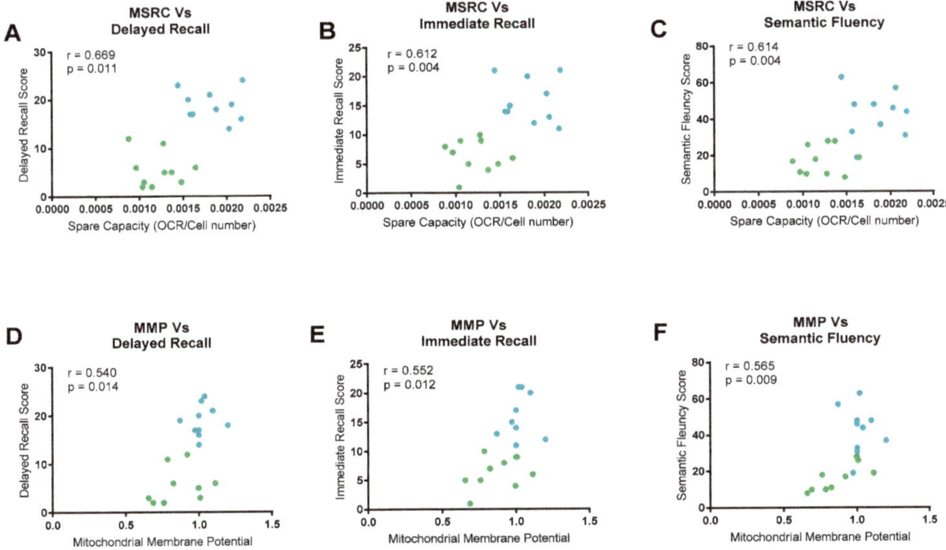

Figure 3. Neuropsychological Fibroblast Metabolism Correlations. Significant correlations were seen between mitochondrial spare respiratory capacity (MSRC) and delayed recall (**A**), Immediate recall (**B**) and semantic fluency (**C**). Mitochondrial membrane potential (MMP) correlated significantly with semantic fluency (**F**), and with delayed recall (**D**) and immediate recall (**E**) but to a lesser extent. For all panels controls are represented in blue and sAD are represented in green. Correlation coefficients and *p*-values for each correlation are displayed.

Age, years of education and brain reserve can potentially confound the correlations between neuropsychological measurements and the fibroblast metabolic markers, as all three factors have been shown to affect either performance on neuropsychological tests [39,40] or mitochondrial function [41]. To control for the effect of these three covariates, a further analysis was performed taking these parameters into account. This further analysis showed that the correlation between delayed episodic recall and MSRC was still significant at the more conservative significance threshold. Both immediate episodic recall ($p = 0.013$) and semantic memory ($p = 0.012$) correlations with MSRC were no longer significant once covariates were included in the analysis at the new more stringent *p*-value. All correlations seen between the neuropsychological tests and MMP were no longer significant after controlling for age, education and brain reserve. Table 3 displays the new correlation coefficients and p-values for the neuropsychological data.

Table 3. Neuropsychological fibroblast metabolism correlations corrected for age, years of education and grey matter reserve.

Psychological Test	Fibroblast Marker	R Value	p-Value
Immediate Episodic Recall			
	MSRC	0.605	0.013
	MMP	0.196	0.466
Delayed Episodic Recall			
	MSRC	0.695	0.003
	MMP	0.204	0.448
Semantic Memory			
	MSRC	0.610	0.012
	MMP	0.345	0.191

2.6. *Neuroimaging/Metabolic Correlations*

Volumetric imaging for areas of the brain known to be affected by AD were correlated with metabolic markers of the disease. Only the grey matter volumes that had shown significant differences between the sAD and control groups (with a p-value equal to or less than 0.01) were included. None of the grey matter volumes showed a significant correlation with changes in cellular metabolism. Table 4 shows the results of the correlations.

Table 4. Grey Matter Volume Fibroblast metabolic correlations: No significant correlations were seen between MMP nor MSRC and any grey matter volume when controlling for age, length of disease or brain reserve.

Grey Matter Volume	Fibroblast Marker	R Value	p-Value
Left Hippocampal Volume			
	MSRC	0.371	0.157
	MMP	0.041	0.881
Left Parietal Volume			
	MSRC	0.341	0.196
	MMP	−0.047	0.862
Right Parietal Volume			
	MSRC	0.418	0.107
	MMP	−0.043	0.875

3. Discussion

This proof of concept study shows that fibroblast MSRC correlates with established neuropsychological abnormalities that are affected early in AD. This is one of the first studies to show a functional biomarker of AD correlating with a marker of fibroblast mitochondrial function. MSRC is a measure of the ability of mitochondria within cells to increase the production of ATP in response to an increased energy demand. The correlation of MSRC with neuropsychological scores potentially helps to describe the cellular pathology underlying these neuropsychological changes seen in AD. This is the first study in humans to show that peripheral cell MSRC correlates with neuropsychological profiles. Correcting the abnormality in MSRC could be a future therapeutic approach in AD. The fact that this abnormality in mitochondrial function correlates with a clinical biomarker of AD also opens the avenue for monitoring drug response in future clinical trials.

MSRC has been shown to be abnormal in multiple diseases including acute myeloid leukaemia (AML) [42], Parkinson's disease [43,44] and motor neuron disease (MND) [45]. MSRC deficits may not be disease specific, but the combination of cellular metabolic changes seen here may represent a cellular metabolic profile specific to sAD. This metabolic phenotype could potentially represent a biomarker that can stratify and define a specific subset of sAD patients that might then respond to a personalised therapeutic approach. Previous research has shown that the metabolic profile identified in fibroblasts from sporadic Parkinson's disease patients differs from that identified here in fibroblasts from sAD [46,47].

Reductions in MSRC and MMP were seen in all sAD fibroblast lines when compared to controls at a group level. These data reinforce and extend the findings reported in our previous paper [29], and show that the finding of reduced MSRC and MMP is reproducible and more robust in sporadic AD fibroblasts when sample size is increased. Extracellular lactate levels and cellular glycolytic reserves, both markers of glycolytic function, were significantly reduced in all sAD fibroblasts. These measures, however, did not meet the statistical significance for correlation with neuropsychological markers of the disease. The deficits in markers of mitochondrial function and glycolysis in the fibroblasts reflect a lack of flexibility of metabolic pathways in sAD. This lack of flexibility is likely to be very important when performing cognitive tasks that depend on the coordination of multiple brain regions. Taken together, these results suggest that sAD fibroblasts may have limited ability to respond to increased cellular energy demand.

MMP did correlate with neuropsychological markers, but correlations did not survive the correction for confounding factors unlike MSRC, for which correlation with delayed episodic recall (a core feature in the clinical diagnostic criteria for sAD) remained significant. Semantic memory and immediate episodic memory did not significantly correlate with MSRC at the more conservative significance threshold after confounder consideration, but p-values for both of these correlations were less than 0.02. Potentially, these correlations would reach significance level with an increased sample size. The difference in MMP between control and sAD groups was much smaller than the difference seen in MSRC. MMP has a key role in maintaining many pathways in the mitochondria outside that of ATP generation, such as apoptosis fates [48]. It is likely that several mechanisms not affected by sAD in the fibroblast cell help to maintain MMP levels, which may explain the reduced variability seen in this parameter in sAD and the lack of a significant correlation with neuropsychological scores.

Baseline ATP levels were not significantly different between the two groups. These results suggest that, in a controlled environment when cells are not stressed, they can maintain ATP levels similar to that of controls. Interestingly though, the functional capacity of mitochondria is impaired, as described above, suggesting a potential inability to respond appropriately to increased ATP demand. The measurements of ATP we performed in this study only gives information about the total cellular ATP level. We did not investigate the rate of ATP breakdown nor how ATP levels change when the cell is under stressed conditions or the ratio of ATP/ADP in the cell. This may explain why no significant correlations were seen with ATP measurements, but were seen with the other markers of mitochondrial respiration.

It is interesting that the correlation between scores on phonemic fluency, a neuropsychological test not affected early in sAD, and cell metabolism markers did not reach significance threshold. It. The medial temporal lobe is important in semantic memory processing [49], an area of cognition that is impaired very early in sAD [50]. This is not the case for phonemic fluency which is supported by processes associated with frontal executive regions [51]. It could be that the correlations we see between neuropsychological tests and cellular metabolic function may reflect which areas of the brain are more susceptible to metabolic failure. It has been previously shown that areas of the brain that are affected early in sAD express lower levels of electron transport chain (ETC) genes when compared to controls [52], suggesting that this might be the case. Data from FDG-PET studies also support the concept of focal brain hypometabolism in sAD, with areas such as the medial temporal lobes, precuneus and lateral parietal lobes all preferentially affected [53].

We did not see a significant correlation between mitochondrial function or glycolysis and brain structural imaging markers of sAD. This may be explained by the fact that multiple factors can affect grey matter volume such as age, levels of education and brain reserve. It has been previously shown that smaller grey matter volumes in the frontal lobes associated with ageing are associated with worse performance on frontal lobe cognitive tests [54]. The neuropsychological assessment performed in this study mainly focused on temporal lobe cognitive functioning and not frontal lobe function. Potentially, the effect of brain ageing on grey matter volumes may mask any effect metabolic function may have.

The limitations of our study include a small sample size, and the fact that we have used fibroblasts to measure metabolic function. It could also be argued that a correlation between metabolic function of central nervous system cells and neuropsychological parameters would be more meaningful. This is not the first study, however, to show that the metabolism of peripheral cells outside the nervous system is affected in sAD. White blood cells [55], platelets [56] and fibroblasts have been shown to have metabolic abnormalities in multiple studies [46,47,57,58]. Identifying that fibroblast mitochondrial abnormalities correlate with neuropsychological markers of AD opens up avenues to use fibroblast metabolic parameters as biomarkers of sAD, and also as a high throughput drug screening model, as well as potential outcome measures of therapeutic efficacy. Replication of the findings of this study in larger cohorts and by other groups is needed, however, to consolidate the evidence that fibroblast metabolic function is a reliable biomarker of sAD.

Our patient cohort were selected at an early stage of sAD to try and reduce group variability. To understand the use of fibroblast metabolic abnormalities as a biomarker, further work investigating cohorts of patients with amnestic mild cognitive impairment (MCI), the prodromal stage of sAD, would be advantageous as well as investigating these parameters in more advanced sAD patients.

Future work could extend the findings of metabolic-neuropsychological correlations by creating neuronal lineage cells via cellular reprogramming methods such as that described by Takahashi et al. [59]. This type of model would allow for direct comparison between human nervous system cells and human nervous system function in a context where reprogrammed cells would maintain their ageing phenotype [60].

4. Methods

4.1. Patient Details

Skin biopsies and fibroblast metabolic assessments were performed as part of the MODEL-AD research study (Yorkshire and Humber Research and Ethics Committee number: 16/YH/0155). Due to the initial success of fibroblast metabolic assessment, the initial cohort of four controls and four patients (previously reported in [29]) with sAD was expanded to ten healthy controls (mean age: 65.77 years, SD 14.70 years) and 10 patients with sAD (mean age: 61.33 years, SD 7.19 years). All patients had been involved in the EU-funded Framework Programme 7 Virtual Physiological Human: Dementia Research Enabled by IT (VPH-DARE@IT) initiative (http://www.vph-dare.eu/). A diagnosis of sAD was made based on clinical criteria [11]. Ethical approval for the neuropsychological and neuroimaging measures collected was gained from Yorkshire and Humber Regional Ethics Committee, Reference number: 12/YH/0474. Informed written consent was obtained from each participant. Investigations were carried out following the rules of the Declaration of Helsinki of 1975 (https://www.wma.net/what-we-do/medical-ethics/declaration-of-helsinki/), revised in 2013.

4.2. Neuropsychological Testing

A battery of tests was devised to detect impaired performance in the cognitive domains most susceptible to sAD neurodegeneration. This included tests of short- and long-term memory, attention and executive functioning, language and semantics, and visuoconstructive skills. A detailed description of each task is provided elsewhere [61]. Most tests were exclusively used as part of the clinical assessment of patients and controls. Three tests were also included as part of the experimental protocol. Immediate and delayed recall on the Logical Memory test were used as measures of episodic memory, the cognitive domain centrally defined by diagnostic criteria for a diagnosis of sAD. The performance on the Category Fluency test was used as an index of semantic memory, the cognitive domain affected by the accumulation of neurofibrillary pathology in the transentorhinal cortex [50]. Finally, the Letter Fluency test (phonemic fluency) was used as a methodological control, since performance on this task is often within normal age limits in patients with mild sAD [33,62].

4.3. MRI Acquisition and Processing

An MRI protocol of anatomical scans was acquired on a Philips Ingenia 3 T scanner. Several acquisitions (including T1-weighted, T2-weighted, FLAIR and diffusion-weighted sequences) were used for diagnostic purposes. Three-dimensional T1-weighted images (voxel size: 0.94 mm × 0.94 mm × 1.00 mm; repetition time: 8.2 s; echo delay time: 3.8 s; field of view: 256 mm; matrix size: 256 × 256 × 170) were also used for the calculation of hippocampal volumes. These were processed with the Similarity and Truth Estimation for Propagated Segmentations routine [63]. This procedure, available at niftyweb.cs.ucl.ac.uk, allows automated segmentation of the left and right hippocampus in the brain's native space using multiple reference templates. Hippocampal volumes were then quantified using Matlab (version R2014a) and the "get totals" script (http://www0.cs.ucl.ac.uk/staff/G.Ridgway/vbm/get_totals.m). Fractional measures were also obtained dividing hippocampal volumes by the volume of the total intracranial space. The left hippocampal volume and left hippocampal ratio were chosen as structural markers of AD as evidence suggests the left hippocampus is affected early in the progression of the disease [64]. Native space maps created to extract hippocampal volume from MRI images are shown in Supplementary Figure S1. Additional regions of interest (listed in Table 2) were then defined as binary image masks. Segmented grey matter maps were registered to the Montreal Neurological Institute space, and mask-constrained volumes were extracted.

4.4. Tissue Culture

Fibroblasts were cultured as described previously [29]. In brief, all 10 control and all 10 sAD lines were cultured in EMEM-based media (Corning) incubated at 37 °C in a 5% carbon dioxide atmosphere. Sodium pyruvate (1%) (Sigma Aldrich), non-essential amino acids (1%) (Lonza), penicillin and streptomycin (1%) (Sigma Aldrich), multi-vitamins (1%) (Lonza), Fetal Bovine Serum (10%) (Biosera) and 50μg/mL uridine (Sigma Aldrich) were added to the media. All experiments described were performed on all 20 cell lines. Fibroblasts were plated at a density of 5000 cells per well in a white 96 well plate for ATP assays. For MMP assays fibroblasts were plated at a density of 2500 cells per well in a black 96 well plate. Each assay was performed on three separate passages of fibroblasts; cells between passages 5–10 were used. All experiments were performed on passage-matched cells.

4.5. Intracellular ATP levels

Cellular adenosine triphosphate (ATP) levels were measured using the ATPlite kit (Perkin Elmer) as described previously [65]. ATP levels were corrected for cell number by using CyQuant (ThermoFisher) measurements, as previously described [29]. These values were then normalised to control levels.

4.6. Mitochondrial Membrane Potential

MMP was measured using tetramethlyrhodamine (TMRM) staining of live fibroblasts, as previously described [29]. Cells were plated in a black 96 well plate, incubated at 37 °C for 48 h. TMRM dye was added one hour prior to imaging on an InCell Analyzer 2000 high-content imager (GE Healthcare).

4.7. Extracellular Lactate Measurement

Extracellular lactate was measured from confluent flasks of fibroblasts using an L-Lactate assay kit (Abcam ab65331). The assay was performed according to the manufacturer's instructions.

4.8. Metabolic Flux Assay

4.8.1. Mito Stress Test

Oxygen consumption rates (OCR) were measured using a 24-well Agilent Seahorse XF analyzer as described previously [29]. In brief, cells were plated at a density of 65,000 cells per well 48 h prior to measurement. Three measurements were taken at the basal point: after the addition of oligomycin (0.5 μM), after the addition of carbonyl cyanide-4-(trifluoromethoxy)phenylhydrazone (FCCP) (0.5 μM) and after the addition of rotenone (1 μM). OCR measurements were normalised to cell count as described in [29]. Measurements of basal mitochondrial respiration, maximal mitochondrial respiration and MSRC were also calculated.

4.8.2. Glycolysis Stress Test

Cells were plated in a Seahorse XF24 Cell Culture Microplate (Agilent) at a density of 65,000 cells per well 48 h prior to measurement. The glycolysis stress test standard protocol was used [66]. This is a completely separate assay to the mitochondrial stress test, and allows for the thorough interrogation of glycolysis via the addition of supplements and inhibitors of the glycolytic pathway. Three measurements were taken at the basal point: after the addition of glucose (10 μM), after the addition of oligomycin (1.0 μM) and after the addition of 2-deoxyglucose (50 μM). The glycolysis stress test can measure several aspects of cellular glycolysis. These include the basal glycolytic rate of the cell, the maximum level of glycolysis the cell can achieve and the glycolytic reserve which refers to the difference between maximum level of glycolysis and the basal level of glycolysis. The different aspects of glycolysis are calculated by measuring the extracellular acidification rate (ECAR). ECAR rates were normalised to cell count as described above.

4.9. Statistical Analysis

For comparing each neuropsychological, neuroimaging and metabolic functional markers, a Student's t-test was used, comparing the means of the control group for each parameter to the disease group. Statistics were calculated using GraphPad Prism Software (V7.02). Analyses of covariance were also used for group comparisons including confounding variables. For covariate analysis when comparing control and disease group grey matter volumes, or when correlating neuropsychological and neuroimaging sAD markers with fibroblast functional markers, the IBM SPSS Statistics suite (Version 26; https://www.ibm.com/uk-en/products/spss-statistics) was used as this function was not available in the GraphPad Prism Software. Significance levels were adjusted to account for multiple comparisons.

5. Conclusions

These data highlight how in-depth analysis of mitochondrial and glycolytic function in sAD fibroblasts identifies metabolic abnormalities that parallel changes seen in neuropsychological features distinctive of the early stages of sporadic Alzheimer's disease. This model system could be used to develop biomarkers useful in early detection as well as in the development of novel therapeutic approaches for sAD.

Supplementary Materials: The following are available online at http://www.mdpi.com/2075-4426/10/2/32/s1, Figure S1: Quantification of hippocampal volumes from T1-weighted MRI scans.

Author Contributions: Conceptualization, S.M.B., D.J.B., H.M., L.F., A.V., M.D.M. and P.J.S.; Methodology, S.M.B., M.D.M., K.B., H.M. and A.V.; Software, S.M.B., and M.D.M.; Validation, S.M.B., D.J.B., K.B., H.M., L.F., A.V., M.D.M. and P.J.S.; Formal Analysis, S.M.B., D.J.B., K.B., H.M., L.F., A.V., M.D.M. and P.J.S.; Investigation, S.M.B., M.D.M., and K.B.; Resources, S.M.B., D.J.B., K.B., H.M., L.F., A.V., M.D.M. and P.J.S.; Data Curation, S.M.B., D.J.B., K.B., H.M., L.F., A.V., M.D.M. and P.J.S.; Writing—Original Draft Preparation, S.M.B.; Writing—Review and Editing, S.M.B., D.J.B., K.B., H.M., L.F., A.V., M.D.M. and P.J.S.; Visualization, S.M.B., D.J.B., K.B., H.M., L.F., A.V., M.D.M. and P.J.S.; Supervision, H.M., P.J.S., A.V., L.F. and D.J.B.; Project Administration, S.M.B.; Funding Acquisition, S.M.B., D.J.B., K.B., H.M., L.F., A.V. and P.J.S. All authors have read and agreed to the published version of the manuscript.

Funding: This research was funded by Wellcome 4ward north (216340/Z/19/Z), ARUK Yorkshire Network Centre Small Grant Scheme (Ref: ARUK-PCRF2016A-1), Parkinson's UK (F1301), European Union Seventh Framework Programme (FP7/2007 – 2013) under grant agreement no. 601055, NIHR Sheffield Biomedical Research Centre award.

Acknowledgments: We would like to thank all research participants involved in this study. This research was supported by the Wellcome 4ward North Academy and ARUK Yorkshire Network Centre Small Grant Scheme (to SMB), Parkinson's UK (to HM, F1301) and the NIHR Sheffield Biomedical Research Centre. This is a summary of independent research carried out at the NIHR Sheffield Biomedical Research Centre (Translational Neuroscience). The views expressed are those of the author(s) and not necessarily those of the NHS, the NIHR or the Department of Health and Social Care (DHSC). Support of the European Union Seventh Framework Programme (FP7/2007–2013) under grant agreement no. 601055, VPH-DARE@IT to AV is acknowledged. PJS is supported by an NIHR Senior Investigator award. The support of the NIHR Clinical Research Facility—Sheffield Teaching Hospital is also acknowledged.

Conflicts of Interest: The authors have no financial or personal relationships with other people/organizations that could inappropriately influence (bias) their work.

References

1. Wimo, A.; Guerchet, M.; Ali, G.-C.; Wu, Y.-T.; Prina, A.M.; Winblad, B.; Jönsson, L.; Liu, Z.; Prince, M. The worldwide costs of dementia 2015 and comparisons with 2010. *Alzheimer's Dement.* **2017**, *13*, 1–7. [CrossRef]
2. Scheltens, P.; Blennow, K.; Breteler, M.M.B.; de Strooper, B.; Frisoni, G.B.; Salloway, S.; Van der Flier, W.M. Alzheimer's disease. *Lancet* **2016**, *388*, 505–517. [CrossRef]
3. Brookmeyer, R.; Corrada, M.M.; Curriero, F.C.; Kawas, C. Survival following a diagnosis of Alzheimer disease. *Arch. Neurol.* **2002**, *59*, 1764–1767. [CrossRef]
4. Perl, D.P. Neuropathology of Alzheimer's Disease. *Mt. Sinai J. Med.* **2010**, *77*, 32–42. [CrossRef] [PubMed]
5. Hardy, J.; Selkoe, D.J. The Amyloid Hypothesis of Alzheimer's Disease: Progress and Problems on the Road to Therapeutics. *Science* **2002**, *297*, 353. [CrossRef] [PubMed]
6. Honig, L.S.; Vellas, B.; Woodward, M.; Boada, M.; Bullock, R.; Borrie, M.; Hager, K.; Andreasen, N.; Scarpini, E.; Liu-Seifert, H.; et al. Trial of Solanezumab for Mild Dementia Due to Alzheimer's Disease. *N. Engl. J. Med.* **2018**, *378*, 321–330. [CrossRef] [PubMed]
7. Cummings, J.L.; Morstorf, T.; Zhong, K. Alzheimer's disease drug-development pipeline: Few candidates, frequent failures. *Alzheimers Res. Ther.* **2014**, *6*, 37. [CrossRef]
8. ALZFORUM. Topline Result for First DIAN-TU Clinical Trial: Negative on Primary. Available online: https://www.alzforum.org/news/research-news/topline-result-first-dian-tu-clinical-trial-negative-primary (accessed on 29 March 2020).
9. Savva, G.M.; Wharton, S.B.; Ince, P.G.; Forster, G.; Matthews, F.E.; Brayne, C. Age, neuropathology, and dementia. *N. Engl. J. Med.* **2009**, *360*, 2302–2309. [CrossRef]
10. Le Couteur, D.G.; Hunter, S.; Brayne, C. Solanezumab and the amyloid hypothesis for Alzheimer's disease. *BMJ* **2016**, *355*, i6771. [CrossRef]
11. McKhann, G.M.; Knopman, D.S.; Chertkow, H.; Hyman, B.T.; Jack, C.R., Jr.; Kawas, C.H.; Klunk, W.E.; Koroshetz, W.J.; Manly, J.J.; Mayeux, R.; et al. The diagnosis of dementia due to Alzheimer's disease: Recommendations from the National Institute on Aging-Alzheimer's Association workgroups on diagnostic guidelines for Alzheimer's disease. *Alzheimers Dement. J. Alzheimers Assoc.* **2011**, *7*, 263–269. [CrossRef]
12. Braak, H.; Braak, E. Neuropathological stageing of Alzheimer-related changes. *Acta Neuropathol.* **1991**, *82*, 239–259. [CrossRef] [PubMed]
13. Haroutunian, V.; Schnaider-Beeri, M.; Schmeidler, J.; Wysocki, M.; Purohit, D.P.; Perl, D.P.; Libow, L.S.; Lesser, G.T.; Maroukian, M.; Grossman, H.T. Role of the neuropathology of Alzheimer disease in dementia in the oldest-old. *Arch. Neurol.* **2008**, *65*, 1211–1217. [CrossRef] [PubMed]
14. Weintraub, S.; Wicklund, A.H.; Salmon, D.P. The neuropsychological profile of Alzheimer disease. *Cold Spring Harb. Perspect. Med.* **2012**, *2*, a006171. [CrossRef] [PubMed]
15. Magistretti, P.J.; Allaman, I. A Cellular Perspective on Brain Energy Metabolism and Functional Imaging. *Neuron* **2015**, *86*, 883–901. [CrossRef]

16. Ewers, M.; Brendel, M.; Rizk-Jackson, A.; Rominger, A.; Bartenstein, P.; Schuff, N.; Weiner, M.W. Alzheimer's Disease Neuroimaging I Reduced FDG-PET brain metabolism and executive function predict clinical progression in elderly healthy subjects. *NeuroImage. Clin.* **2013**, *4*, 45–52. [CrossRef]
17. Kalpouzos, G.; Eustache, F.; de la Sayette, V.; Viader, F.; Chételat, G.; Desgranges, B. Working memory and FDG–PET dissociate early and late onset Alzheimer disease patients. *J. Neurol.* **2005**, *252*, 548–558. [CrossRef]
18. Ishii, K.; Sasaki, M.; Yamaji, S.; Sakamoto, S.; Kitagaki, H.; Mori, E. Demonstration of decreased posterior cingulate perfusion in mild Alzheimer's disease by means of H215O positron emission tomography. *Eur. J. Nucl. Med.* **1997**, *24*, 670–673.
19. Morgen, K.; Frolich, L. The metabolism hypothesis of Alzheimer's disease: From the concept of central insulin resistance and associated consequences to insulin therapy. *J. Neural Transm.* **2015**, *122*, 499–504. [CrossRef]
20. Yao, J.; Irwin, R.W.; Zhao, L.; Nilsen, J.; Hamilton, R.T.; Brinton, R.D. Mitochondrial bioenergetic deficit precedes Alzheimer's pathology in female mouse model of Alzheimer's disease. *Proc. Natl. Acad. Sci. USA* **2009**, *106*, 14670. [CrossRef]
21. Blass, J.P.; Sheu, R.K.; Gibson, G.E. Inherent abnormalities in energy metabolism in Alzheimer disease. Interaction with cerebrovascular compromise. *Ann. N. Y. Acad. Sci.* **2000**, *903*, 204–221. [CrossRef]
22. Baik, S.H.; Kang, S.; Lee, W.; Choi, H.; Chung, S.; Kim, J.I.; Mook-Jung, I. A Breakdown in Metabolic Reprogramming Causes Microglia Dysfunction in Alzheimer's Disease. *Cell Metab.* **2019**, *30*, 493–507.e496. [CrossRef] [PubMed]
23. Swerdlow, R.H.; Khan, S.M. A "mitochondrial cascade hypothesis" for sporadic Alzheimer's disease. *Med. Hypotheses* **2004**, *63*, 8–20. [CrossRef]
24. Hartl, D.; Schuldt, V.; Forler, S.; Zabel, C.; Klose, J.; Rohe, M. Presymptomatic Alterations in Energy Metabolism and Oxidative Stress in the APP23 Mouse Model of Alzheimer Disease. *J. Proteome Res.* **2012**, *11*, 3295–3304. [CrossRef]
25. Gan, X.Q.; Huang, S.B.; Wu, L.; Wang, Y.F.; Hu, G.; Li, G.Y.; Zhang, H.J.; Yu, H.Y.; Swerdlow, R.H.; Chen, J.X.; et al. Inhibition of ERK-DLP1 signaling and mitochondrial division alleviates mitochondrial dysfunction in Alzheimer's disease cybrid cell. *Biochim. Biophys. Acta Mol. Basis Dis.* **2014**, *1842*, 220–231. [CrossRef] [PubMed]
26. Mosconi, L.; Tsui, W.H.; De Santi, S.; Li, J.; Rusinek, H.; Convit, A.; Li, Y.; Boppana, M.; de Leon, M.J. Reduced hippocampal metabolism in MCI and AD: Automated FDG-PET image analysis. *Neurology* **2005**, *64*, 1860–1867. [CrossRef] [PubMed]
27. Vlassenko, A.G.; Vaishnavi, S.N.; Couture, L.; Sacco, D.; Shannon, B.J.; Mach, R.H.; Morris, J.C.; Raichle, M.E.; Mintun, M.A. Spatial correlation between brain aerobic glycolysis and amyloid-beta (Abeta) deposition. *Proc. Natl. Acad. Sci. USA* **2010**, *107*, 17763–17767. [CrossRef]
28. Vlassenko, A.G.; Gordon, B.A.; Goyal, M.S.; Su, Y.; Blazey, T.M.; Durbin, T.J.; Couture, L.E.; Christensen, J.J.; Jafri, H.; Morris, J.C.; et al. Aerobic glycolysis and tau deposition in preclinical Alzheimer's disease. *Neurobiol. Aging* **2018**, *67*, 95–98. [CrossRef]
29. Bell, S.M.; Barnes, K.; Clemmens, H.; Al-Rafiah, A.R.; Al-Ofi, E.A.; Leech, V.; Bandmann, O.; Shaw, P.J.; Blackburn, D.J.; Ferraiuolo, L.; et al. Ursodeoxycholic Acid Improves Mitochondrial Function and Redistributes Drp1 in Fibroblasts from Patients with Either Sporadic or Familial Alzheimer's Disease. *J. Mol. Biol.* **2018**, *430*, 3942–3953. [CrossRef]
30. Pickart, L.; Vasquez-Soltero, J.M.; Margolina, A. The Effect of the Human Peptide GHK on Gene Expression Relevant to Nervous System Function and Cognitive Decline. *Brain Sci.* **2017**, *7*, 20. [CrossRef]
31. Desler, C.; Hansen, T.L.; Frederiksen, J.B.; Marcker, M.L.; Singh, K.K.; Juel Rasmussen, L. Is There a Link between Mitochondrial Reserve Respiratory Capacity and Aging? *J. Aging Res.* **2012**, *2012*, 192503. [CrossRef]
32. Martino Adami, P.V.; Quijano, C.; Magnani, N.; Galeano, P.; Evelson, P.; Cassina, A.; Do Carmo, S.; Leal, M.C.; Castaño, E.M.; Cuello, A.C.; et al. Synaptosomal bioenergetic defects are associated with cognitive impairment in a transgenic rat model of early Alzheimer's disease. *Off. J. Int. Soc. Cereb. Blood Flow Metab.* **2017**, *37*, 69–84. [CrossRef] [PubMed]
33. Capitani, E.; Rosci, C.; Saetti, M.C.; Laiacona, M. Mirror asymmetry of Category and Letter fluency in traumatic brain injury and Alzheimer's patients. *Neuropsychologia* **2009**, *47*, 423–429. [CrossRef] [PubMed]
34. Folstein, M.F.; Folstein, S.E.; McHugh, P.R. "Mini-mental state". A practical method for grading the cognitive state of patients for the clinician. *J. Psychiatr. Res.* **1975**, *12*, 189–198. [CrossRef]

35. Doody, R.S.; Massman, P.; Dunn, J.K. A method for estimating progression rates in Alzheimer disease. *Arch Neurol.* **2001**, *58*, 449–454. [CrossRef] [PubMed]
36. Scahill, R.I.; Schott, J.M.; Stevens, J.M.; Rossor, M.N.; Fox, N.C. Mapping the evolution of regional atrophy in Alzheimer's disease: Unbiased analysis of fluid-registered serial MRI. *Proc. Natl. Acad. Sci. USA* **2002**, *99*, 4703–4707. [CrossRef] [PubMed]
37. Boller, B.; Mellah, S.; Ducharme-Laliberte, G.; Belleville, S. Relationships between years of education, regional grey matter volumes, and working memory-related brain activity in healthy older adults. *Brain Imaging Behav.* **2017**, *11*, 304–317. [CrossRef]
38. Rzezak, P.; Squarzoni, P.; Duran, F.L.; de Toledo Ferraz Alves, T.; Tamashiro-Duran, J.; Bottino, C.M.; Ribeiz, S.; Lotufo, P.A.; Menezes, P.R.; Scazufca, M.; et al. Relationship between Brain Age-Related Reduction in Gray Matter and Educational Attainment. *PLoS ONE* **2015**, *10*, e0140945. [CrossRef]
39. Harada, C.N.; Natelson Love, M.C.; Triebel, K.L. Normal cognitive aging. *Clin. Geriatr. Med.* **2013**, *29*, 737–752. [CrossRef]
40. Stern, Y. Cognitive reserve: Implications for assessment and intervention. *Folia Phoniatr. Logop. Off. Organ. Int. Assoc. Logop. Phoniatr. IALP* **2013**, *65*, 49–54. [CrossRef]
41. Schniertshauer, D.; Gebhard, D.; Bergemann, J. Age-Dependent Loss of Mitochondrial Function in Epithelial Tissue Can Be Reversed by Coenzyme Q10. *J. Aging Res.* **2018**, *2018*, 6354680. [CrossRef] [PubMed]
42. Sriskanthadevan, S.; Jeyaraju, D.V.; Chung, T.E.; Prabha, S.; Xu, W.; Skrtic, M.; Jhas, B.; Hurren, R.; Gronda, M.; Wang, X.; et al. AML cells have low spare reserve capacity in their respiratory chain that renders them susceptible to oxidative metabolic stress. *Blood* **2015**, *125*, 2120–2130. [CrossRef] [PubMed]
43. Yadava, N.; Nicholls, D.G. Spare Respiratory Capacity Rather Than Oxidative Stress Regulates Glutamate Excitotoxicity after Partial Respiratory Inhibition of Mitochondrial Complex I with Rotenone. *J. Neurosci.* **2007**, *27*, 7310–7317. [CrossRef] [PubMed]
44. Mortiboys, H.; Johansen, K.K.; Aasly, J.O.; Bandmann, O. Mitochondrial impairment in patients with Parkinson disease with the G2019S mutation in LRRK2. *Neurology* **2010**, *75*, 2017–2020. [CrossRef] [PubMed]
45. Tan, W.; Pasinelli, P.; Trotti, D. Role of mitochondria in mutant SOD1 linked amyotrophic lateral sclerosis. *Biochim. Biophys. Acta* **2014**, *1842*, 1295–1301. [CrossRef]
46. Carling, P.J.; Mortiboys, H.; Green, C.; Mihaylov, S.; Sandor, C.; Schwartzentruber, A.; Taylor, R.; Wei, W.; Hastings, C.; Wong, S.; et al. Deep phenotyping of peripheral tissue facilitates mechanistic disease stratification in sporadic Parkinson's disease. *Prog. Neurobiol.* **2020**, *187*, 101772. [CrossRef]
47. Milanese, C.; Payán-Gómez, C.; Galvani, M.; Molano González, N.; Tresini, M.; Nait Abdellah, S.; van Roon-Mom, W.M.; Figini, S.; Marinus, J.; van Hilten, J.J.; et al. Peripheral mitochondrial function correlates with clinical severity in idiopathic Parkinson's disease. *Mov. Disord.* **2019**, *34*, 1192–1202. [CrossRef] [PubMed]
48. Gottlieb, E.; Armour, S.M.; Harris, M.H.; Thompson, C.B. Mitochondrial membrane potential regulates matrix configuration and cytochrome c release during apoptosis. *Cell Death Differ.* **2003**, *10*, 709–717. [CrossRef]
49. Grogan, A.; Green, D.W.; Ali, N.; Crinion, J.T.; Price, C.J. Structural correlates of semantic and phonemic fluency ability in first and second languages. *Cereb. Cortex* **2009**, *19*, 2690–2698. [CrossRef]
50. Venneri, A.; Jahn-Carta, C.; De Marco, M.; Quaranta, D.; Marra, C. Diagnostic and prognostic role of semantic processing in preclinical Alzheimer's disease. *Biomark. Med.* **2018**, *12*, 637–651. [CrossRef]
51. Baldo, J.V.; Schwartz, S.; Wilkins, D.; Dronkers, N.F. Role of frontal versus temporal cortex in verbal fluency as revealed by voxel-based lesion symptom mapping. *J. Int. Neuropsychol. Soc.* **2006**, *12*, 896–900. [CrossRef]
52. Liang, W.S.; Reiman, E.M.; Valla, J.; Dunckley, T.; Beach, T.G.; Grover, A.; Niedzielko, T.L.; Schneider, L.E.; Mastroeni, D.; Caselli, R.; et al. Alzheimer's disease is associated with reduced expression of energy metabolism genes in posterior cingulate neurons. *Proc. Natl. Acad. Sci. USA* **2008**, *105*, 4441–4446. [CrossRef] [PubMed]
53. Mosconi, L. Brain glucose metabolism in the early and specific diagnosis of Alzheimer's disease. FDG-PET studies in MCI and AD. *Eur. J. Nucl. Med. Mol. Imaging* **2005**, *32*, 486–510. [CrossRef] [PubMed]
54. Ramanoël, S.; Hoyau, E.; Kauffmann, L.; Renard, F.; Pichat, C.; Boudiaf, N.; Krainik, A.; Jaillard, A.; Baciu, M. Gray Matter Volume and Cognitive Performance During Normal Aging. A Voxel-Based Morphometry Study. *Front. Aging Neurosci.* **2018**, *10*, 235. [CrossRef] [PubMed]

55. Maynard, S.; Hejl, A.M.; Dinh, T.S.T.; Keijzers, G.; Hansen, Å.M.; Desler, C.; Moreno-Villanueva, M.; Bürkle, A.; Rasmussen, L.J.; Waldemar, G.; et al. Defective mitochondrial respiration, altered dNTP pools and reduced AP endonuclease 1 activity in peripheral blood mononuclear cells of Alzheimer's disease patients. *Aging* **2015**, *7*, 793. [CrossRef]
56. Fisar, Z.; Hroudová, J.; Hansíková, H.; Lelková, P.; Wenchich, L.; Jirák, R.; Zeman, J.; Martásek, P.; Raboch, J. Mitochondrial Respiration in the Platelets of Patients with Alzheimer's Disease. *Curr. Alzheimer Res.* **2016**, *13*, 930–941. [CrossRef]
57. Martin-Maestro, P.; Gargini, R.; Garcia, E.; Perry, G.; Avila, J. Slower Dynamics and Aged Mitochondria in Sporadic Alzheimer's Disease. *Oxid. Med. Cell. Longev.* **2017**, *2017*, 9302761. [CrossRef]
58. Sonntag, K.C.; Ryu, W.I.; Amirault, K.M.; Healy, R.A.; Siegel, A.J.; McPhie, D.L.; Forester, B.; Cohen, B.M. Late-onset Alzheimer's disease is associated with inherent changes in bioenergetics profiles. *Sci. Rep.* **2017**, *7*, 14038. [CrossRef]
59. Takahashi, K.; Yamanaka, S. Induction of pluripotent stem cells from mouse embryonic and adult fibroblast cultures by defined factors. *Cell* **2006**, *126*, 663–676. [CrossRef]
60. Mertens, J.; Paquola, A.C.; Ku, M.; Hatch, E.; Böhnke, L.; Ladjevardi, S.; McGrath, S.; Campbell, B.; Lee, H.; Herdy, J.R.; et al. Directly Reprogrammed Human Neurons Retain Aging-Associated Transcriptomic Signatures and Reveal Age-Related Nucleocytoplasmic Defects. *Cell Stem Cell* **2015**, *17*, 705–718. [CrossRef]
61. Wakefield, S.J.; McGeown, W.J.; Shanks, M.F.; Venneri, A. Differentiating normal from pathological brain ageing using standard neuropsychological tests. *Curr. Alzheimer Res.* **2014**, *11*, 765–772. [CrossRef]
62. De Marco, M.; Duzzi, D.; Meneghello, F.; Venneri, A. Cognitive Efficiency in Alzheimer's Disease is Associated with Increased Occipital Connectivity. *J. Alzheimers Dis.* **2017**, *57*, 541–556. [CrossRef] [PubMed]
63. Cardoso, M.J.; Leung, K.; Modat, M.; Keihaninejad, S.; Cash, D.; Barnes, J.; Fox, N.C.; Ourselin, S. STEPS: Similarity and Truth Estimation for Propagated Segmentations and its application to hippocampal segmentation and brain parcelation. *Med. Image Anal.* **2013**, *17*, 671–684. [CrossRef] [PubMed]
64. Vijayakumar, A.; Vijayakumar, A. Comparison of hippocampal volume in dementia subtypes. *ISRN Radiol.* **2013**, *2013*, 174524. [CrossRef] [PubMed]
65. Mortiboys, H.; Thomas, K.J.; Koopman, W.J.; Klaffke, S.; Abou-Sleiman, P.; Olpin, S.; Wood, N.W.; Willems, P.H.; Smeitink, J.A.; Cookson, M.R. Mitochondrial function and morphology are impaired in parkin-mutant fibroblasts. *Ann. Neurol.* **2008**, *64*, 555–565. [CrossRef] [PubMed]
66. Agilent Technologies, Agilent Seahorse XF Glycolysis Stress Test Kit. Available online: https://www.agilent.com/cs/library/usermanuals/public/XF_Glycolysis_Stress_Test_Kit_User_Guide.pdf (accessed on 29 March 2020).

Journal of *Personalized Medicine*

Article

The Brain Metabolic Correlates of the Main Indices of Neuropsychological Assessment in Alzheimer's Disease †

Agostino Chiaravalloti [1,2], Maria Ricci [1,*], Daniele Di Biagio [3], Luca Filippi [4], Alessandro Martorana [5,6] and Orazio Schillaci [1,2]

1 Department of Biomedicine and Prevention, University Tor Vergata, 00133 Rome, Italy; agostino.chiaravalloti@uniroma2.it (A.C.); orazio.schillaci@uniroma2.it (O.S.)
2 IRCCS Neuromed, 86077 Pozzilli, Italy
3 UOC Nuclear Medicine, Policlinico Tor Vergata, 00133 Rome, Italy; danieledibiagio@gmail.com
4 UOC Nuclear Medicine, Santa Maria Goretti Hospital, 04100 Latina, Italy; lucfil@hotmail.com
5 UOSD Centro Demenze Policlinico Tor Vergata, 00133 Rome, Italy; martorana@med.uniroma2.it
6 Department of System Medicine, University Tor Vergata, 00133 Rome, Italy
* Correspondence: maria.ricci28@gmail.com
† This paper is a revised version of our paper published as conference paper in the 29th European Association of Nuclear Medicine congress.

Received: 11 March 2020; Accepted: 15 April 2020; Published: 18 April 2020

Abstract: Background: The study aimed to investigate the relationships between F-18 fluorodeoxyglucose (18F)FDG uptake and neuropsychological assessment in Alzheimer's disease (AD). Methods: We evaluated 116 subjects with AD according to the NINCDS-ADRDA criteria. All the subjects underwent a brain PET/CT with (18F)FDG, cerebrospinal fluid (CSF) assay, mini-mental state examination (MMSE) and further neuropsychological tests: Rey auditory verbal learning test, immediate recall (RAVLT immediate); Rey auditory verbal learning test, delayed recall (RAVLT, delayed); Rey complex figure test, copy (RCFT, copy); Rey complex figure test, delayed recall (RCFT, delayed); Raven's colored progressive matrices (RCPM); phonological word fluency test (PWF) and Stroop test. We performed the statistical analysis by using statistical parametric mapping (SPM12; Wellcome Department of Cognitive Neurology, London, UK). Results: A significant relationship has been reported between (18F)FDG uptake and RAVLT immediate test in Brodmann area (BA)37 and BA22 and with RCFT, copy in BA40, and BA7. We did not find any significant relationships with other tests. Conclusion: In the AD population, brain (18F)FDG uptake is moderately related to the neuropsychological assessment, suggesting a limited impact on statistical data analysis of glucose brain metabolism.

Keywords: Alzheimer's disease; PET/CT; (18F)FDG; neuropsychological assessment

1. Introduction

Subjective memory deficit, together with evidence of memory decline and cognitive impairment during the past months or few years, are key features for Alzheimer's disease (AD) already at the prodromal (or mild cognitive impairment; MCI) stage [1]. Although cognitive tests are frequently used as outcome measures in clinical trials, there are a number of limitations associated with their use [2].

Neurodegenerative AD-related changes are known to accumulate progressively 15 to 20 years before dementia, and even years before the detection of clinical deficits [3]. Cognitive decline in AD, generally assessed by neuropsychological battery [4], is postulated as consequential from neurological pathology as portrayed by biomarkers [5], but the specificity of these associations is still a matter of debate [6]. Particularly, the correspondence between the neuropsychological assessment and standard neuroimaging biomarkers remains incomplete, and the relationship between neuropsychological performance and individual variation in clinical AD neuroimaging markers is likely to be both convergent and unique [5,7]. Although it is tempting to presume neurological substrates causing poor performance at a common neuropsychological assessment, the substrates underlying the cognitive process may be reorganized in the diseased brain of patients with MCI, possibly due to neurodegenerative changes or functional compensation. It is important to understand how these biomarkers, interact to influence cognitive change to isolate the combination of pathologies that contribute most to decline, in order to inform the clinical utility and validation of cognitive tests [5] and the neuroimaging correlates to test performance.

In particular, data concerning the relationship between glucose metabolism and common neuropsychological assessment are poor, even if the results of the neuropsychological tests may be considered as statistical parameters in research papers that include brain data analysis, by using computer-aided metrics such as statistical parametric mapping (SPM) [8,9].

The present study aimed to evaluate the brain metabolic correlates of the main indices of neuropsychological assessment by studying their relationships to cortical and subcortical F-18 fluorodeoxyglucose (18F)FDG uptake in a cohort of subjects with AD. Moreover, this paper aimed to evaluate the impact of neuropsychological assessment as parameters in SPM brain data analysis. For this purpose, all the subjects underwent a brain PET/CT scan using (18F)FDG, a complete neuropsychological assessment that included mini-mental state examination (MMSE). Moreover, the following indices of neuropsychological assessment were determined in the AD population: Rey auditory verbal learning test, immediate recall (RAVLT immediate); Rey auditory verbal learning test, delayed recall (RAVLT, delayed); Rey complex figure test, copy (RCFT, copy); Rey complex figure test, delayed recall (RCFT, delayed); Raven colored progressive matrices (RCPM); phonological word fluency test (PWF) and Stroop test. A cerebrospinal fluid (CSF) assay for amyloid, total tau, and phosphorylated tau was performed in all patients.

An initial version of this paper was presented as a conference paper at the 29th European Association of Nuclear Medicine Congress.

2. Materials and Methods

We evaluated 116 subjects with a new clinical diagnosis of AD (males = 66; females = 50) according to the NINCDS-ADRDA criteria [10]. The mean age was 71.4 ± 6 years old. A complete clinical investigation was performed in all patients, including medical history, mini-mental state examination (MMSE), a complete blood screening (including routine exams, thyroid hormones, and level of B12). Moreover, in all patients, a neurologist examination, neuropsychological examination, a complete neuropsychiatric evaluation, and neuroimaging consisting of magnetic resonance imaging (1.5 T MRI) was performed. Exclusion criteria were the following: isolated deficits and/or unmodified MMSE (<25/30) on revisit (6, 12, and 18 months follow-up), clinically manifest acute stroke in the last 6 months (Hachinsky scale >4, and radiological evidence of subcortical lesions), as described in previous papers [11–13]. None of the patients enrolled had pyramidal and/or extrapyramidal signs reported at the neurological examination. At the time of enrollment, in the 30 days before participating in this study, none of the patients had been treated with drugs that might have modulated cerebral cortex excitability, including antidepressants, neuroactive drugs (i.e., benzodiazepines, antiepileptic drugs or neuroleptics), or cholinesterase inhibitors. The study was performed according to the Declaration of Helsinki and approved by the local ethics committee of the Tor Vergata University in Rome. All participants or their legal guardians gave the written informed consent after receiving an extensive disclosure of the study. A cognitive profile consistent with mild dementia (according to neuropsychological evaluation, including the MMSE) has been described in all AD patients.

In order to improve the diagnostic accuracy of the AD patients, a lumbar puncture and CSF sampling was performed in all patients. The first 12 mL of CSF was collected in a polypropylene tube, then directly transported to the local laboratory for centrifugation at 2000× g at +4 °C for 10 min. The supernatant was pipetted off, gently stirred and mixed to avoid potential gradient effects, and aliquoted in 1 mL portions in polypropylene tubes that were stored at −80 °C pending biochemical analyses, without being thawed and re-frozen. Then, CSF total Tau (T-Tau) and phosphorylated Tau (Thr181, p-Tau) concentration was evaluated using a sandwich ELISA (Innotest hTAU-Ag, Innogenetics, Gent, Belgium). CSF Aβ1–42 levels were determined using a sandwich ELISA (Innotest® ß- amyloid (1–42), Innogenetics, Gent, Belgium), specifically elaborated to measure Aβ containing both the first and 42nd amino acid, as described in previous papers [11–15].

All the subjects underwent a brain PET/CT scan using (18F)FDG, mini-mental state examination (MMSE). Several indices of neuropsychological assessment were explored: Rey auditory verbal learning test, immediate recall (RAVLT immediate); Rey auditory verbal learning test, delayed recall (RAVLT, delayed); Rey complex figure test, copy (RCFT, copy); Rey complex figure test, delayed recall (RCFT, delayed); Raven's colored progressive matrices (RCPM); phonological word fluency Test (PWF) and Stroop test. The relationship between brain uptake of (18F)FDG and CSF biomarkers were analyzed using statistical parametric mapping (SPM12; Wellcome Department of Cognitive Neurology, London, UK) implemented in Matlab R2018a using sex, age, and CSF biomarkers as covariates.

2.1. PET/CT Scanning

The PET/CT system Discovery VCT (GE Medical Systems, Tennessee, TN, USA) has been used to assess (18F)FDG brain distribution in all patients using a 3D-mode standard technique in a 256 × 256 matrix. Reconstruction was performed using the 3-dimensional reconstruction method of ordered-subsets expectation maximization (OSEM) with 20 subsets and with 4 iterations. The system combines a high-speed ultra 16-detector-row (912 detectors per row) CT unit and a PET scanner with 10,080 bismuth germanate crystals in 24 rings (axial full width at half-maximum 1 cm radius, 5.2 mm in 3D mode, an axial field of view 157 mm). A low-amperage CT scan of the head for attenuation correction (40 mA; 120 Kv) was performed before PET image acquisition. All subjects fasted for at least 5 h before intravenous injection of (18F)FDG ; the serum glucose level was up to 95 mg/mL in all of them. All the subjects were injected intravenously with 185–210 MBq of (18F)FDG and hydrated with 500 mL of saline (0.9% sodium chloride). The scan started 30 min after the injection in all the subjects according to standard guidelines, as described in previous papers [13,16,17].

2.2. Statistical Analysis

Differences in brain (18F)FDG uptake were analyzed using statistical parametric mapping (SPM12, Wellcome Department of Cognitive Neurology, London, UK) implemented in Matlab R2018a (Mathworks, Natick, Massachusetts, USA). PET data were subjected to affine and nonlinear spatial normalization into the MNI space. Then, the spatially set of images was smoothed with an 8 mm isotropic Gaussian filter to blur individual variations in gyral anatomy and to increase the signal-to-noise ratio. Images were globally normalized using proportional scaling to remove confounding effects to global CBF changes, with a threshold masking of 0.8. The statistical parametric maps were transformed into a normal distribution unit. Correction of SPM coordinates to match the Talairach coordinates was achieved by the subroutine implemented by Matthew Brett (http://www.mrc-cbu.cam.ac.uk/Imaging). Brodmann areas (BA) were then identified at a range of 0 to 3 mm from the corrected Talairach coordinates of the SPM output isocentres, after importing them by Talairach client (http://www.talairach.org/index.html). A statistical height threshold was equal to or lower than $p < 0.001$ at both clusters, and voxel-level was accepted as significant. We considered significant a cluster extension of more than 125 (5 × 5 × 5 voxels, i.e., 11 × 11 × 11 mm) contiguous voxels, based on the calculation of the partial volume effect resulting from the spatial resolution of the PET camera (about the double of full width at half maximum), as described in previous papers of the same research group [13].The resulting SPM data was correlated to each index of neuropsychological assessment, in order to study their relationships to cortical and subcortical (18F)FDG uptake.

3. Results

The values of CSF amyloid, total tau, and phosphorylated tau were respectively 363.6 ± 162, 689 ± 338.1, and 92.4 ± 70.7 pg/mL. A general overview of the patient population is reported in Table 1. Most of the cases were sporadic, but in about 15% of enrolled patients, there was a family link.

Table 1. A general overview of age, mini-mental state examination (MMSE), and Cerebral spinal fluid (CSF) parameters in the whole population.

Whole Population (n = 116)	Mean ± SD (Range)
Age (years)	71.4 ± 6 (48–82)
MMSE	20.3 ± 4.9 (9–30)
Amyloid (pg/mL)	363.6 ± 162 (70–913)
T-Tau (pg/mL)	689 ± 338.1 (173–1749)
P-Tau (pg/mL)	92.4 ± 70.7 (34–661)

Neuropsychological assessment resulted in 22.6 ± 8.6 for RAVLT, immediate; 71.4 ± 5,9 for RAVLT, delayed; 18.2 ± 10.4 for RCFT, copy; 7.6 ± 6 for RCFT, delayed; 18.8 ± 9 (RCPM); 22.2 ± 10.1 for PWF and 44.6 ± 36.2 for Stroop test (Table 2).

Table 2. A general overview of Neuropsychological tests results in the whole population: Rey auditory verbal learning test, immediate recall (RAVLT immediate); Rey auditory verbal learning test, delayed recall (RAVLT, delayed); Rey complex figure test, copy (RCFT, copy); Rey complex figure test, delayed recall (RCFT, delayed); Raven's colored progressive matrices (RCPM); phonological word fluency test (PWF) and Stroop test.

Neuropsychological Test	Mean ± SD (Range)
RAVLT immediate	22.6 ± 8.6 (0–43)
RAVLT delayed	71.4 ± 5.9 (48–82)
RCFT copy	18.2 ± 10.4 (0–36)
RCFT delayed	7.6 ± 6 (0–26)
RCPM	18.8 ± 9 (0–36)
PWF	22.2 ± 10.1 (0–55)
Stroop Test	44.6 ± 36.2 (0–162)

A positive correlation was reported in the statistical analysis between the PET data and performance in RAVLT immediate. We found a significant relationship between (18F)FDG uptake and performance in RAVLT immediate in a large portion of the left temporal lobe, in a cluster of 1141 voxels that included left temporal middle and superior gyrus, as shown in Figure 1 and Table 3 (positive correlation in Brodmann area 37 and Brodmann area 22).

Furthermore, a significant correlation was described between the PET data and performance in RCFT, copy. We found a significant positive relationship between (18F)FDG uptake and performance in RCFT, copy in a cluster of 1177 voxels that included left parietal inferior lobule and precuneus, and in a cluster of 804 voxels that include right parietal superior and inferior lobules (left and right BA40 and left and right BA7), as reported in Figure 2 and Table 4. We did not find any significant relationships between (18F)FDG uptake and other tests performed.

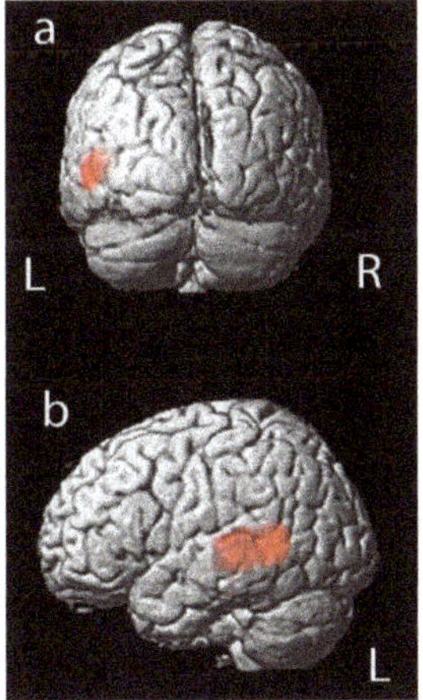

Figure 1. 3D rendering of the results of statistical parametric mapping analyses between (18F)FDG uptake and performance in RAVLT immediate that shows a positive correlation in a large portion of the left temporal lobe (Brodmann area 37 and Brodmann area BA22): (**a**) posterior view; (**b**) left lateral view.

Table 3. A general overview of statistical parametric mapping analyses that detected a positive correlation in Brodmann area (BA) 37 and Brodmann area 22 of FDG uptake and performance in RAVL immediate test. The Brodmann areas were then identified at a range of 0 to 3 mm from the corrected Talairach coordinates of the SPM output isocentres. A statistical height threshold was equal to or lower than $p < 0.001$ at both clusters, and voxel-level was accepted as significant. We considered as significant a cluster extension of more than 125 contiguous voxels.

Analysis		Cluster Level				Voxel Level	
Positive Correlation	p(FWE-corr)	q(FDRcorr)	Extent	Cortical Region	Z Score of Maximum	Talairach Coordinates	Cortical Region
	0.003	0.007	1141	Left temporal, middle temporal gyrus	4.55	−48, −52, 4	BA 37
				Left temporal, middle temporal gyrus	3.68	−54, −36, 8	BA 22
				Left temporal, superior temporal gyrus	3.60	−56, −28, 6	BA 22

Table 4. A general overview of statistical parametric mapping analyses that detected a positive correlation in left and right Brodmann area (BA) 7 and left and right Brodmann area 40 of FDG uptake and performance in RCFT, copy test. The Brodmann areas were then identified at a range of 0 to 3 mm from the corrected Talairach coordinates of the SPM output isocentres. A statistical height threshold was equal to or lower than $p < 0.001$ at both clusters, and voxel-level was accepted as significant. We considered as significant a cluster extension of more than 125 contiguous voxels.

Analysis		Cluster Level				Voxel Level	
Positive Correlation	p(FWE-corr)	q(FDRcorr)	Extent	Cortical Region	Z Score of Maximum	Talairach Coordinates	Cortical Region
	0.002	0.005	1177	Left parietal, inferior parietal lobule	4.84	−34, −50, 36	BA 40
				left parietal, precuneus	4.25	−24, −76, 42	BA 7
				left parietal, precuneus	4.08	−16, −74, 48	BA 7
	0.011	0.013	804	Right parietal, superior parietal lobule	4.52	34, −56, 48	BA 7
				right parietal, superior parietal lobule	3.90	28, −66, 48	BA 7
				right parietal, inferior parietal lobule	3.70	42, −40, 42	BA 40

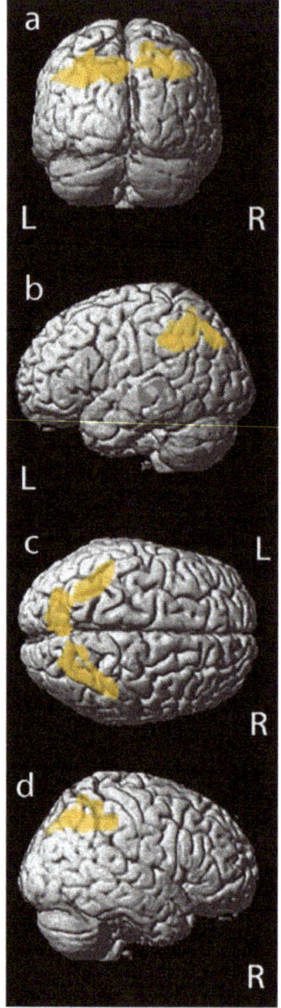

Figure 2. 3D rendering of the results of statistical parametric mapping analyses between (18F)FDG uptake and performance in RCFT copy that shows a positive correlation in left and right Brodmann area 40 and left and right Brodmann area 7: (**a**) posterior view; (**b**) left lateral view; (**c**) superior view; (**d**) right lateral view.

4. Discussion

Cognitive performance is an important outcome measure in AD diagnosis and progression. Nevertheless, the symptomatic significance of improvement or decline in clinical tests has not been well established, making it difficult to set a standard for what is meant by meaningful improvement [6]. Cognitive decline in AD is generally assessed by a neuropsychological battery [4]. An optimal outcome measure would reflect clinically-significant patient function, provide reliable measurements with minimal variability, and track a physiologically relevant disease process. However, the cognitive assessment also has several limitations, such as a large inter and intraindividual variability and floor and ceiling effects [2].

The neuropsychological test results may be considered as statistical parameters in research papers that include brain data analysis by using computer-aided metrics, such as statistical parametric mapping (SPM) [8]. For each cognitive measure, difference scores were entered into a regression analysis in SPM as the independent variable [9] or as a covariate.

The goal of the present study was to examine the brain metabolic correlates of the main indices of neuropsychological assessment tests in order to evaluate their impact on SPM brain data analysis.

We found a significant relationship between (18F)FDG uptake and performance in RAVLT immediate in a large portion of the left temporal lobe (positive correlation in BA37 and BA22) and with RCFT, copy (positive correlation in left and right BA40 and left and right BA7). We did not find any significant relationships with other tests.

A previous paper evaluated the brain metabolic correlates of the main indices of RAVLT, showing a significant correlation between the delayed recall score and metabolism in the posterior cingulate gyrus of both hemispheres and left precuneus, as well as between a score of long-term percent retention and metabolism in the left posterior cingulate gyrus, precuneus, and orbitofrontal areas. No correlation was found between immediate or total recall scores and glucose metabolism [1]. Nevertheless, the sample analyzed was a group of 54 elderly subjects with memory complaints. Therefore these differences, with our results, may be both due to the different sized sample and to the worse clinical conditions of our patients (elderly subjects with memory complaints vs. clinically diagnosed AD) with subsequent cortical hypometabolism according to the AD pattern. Neurological substrates underlying the tests widely used, such as delayed recall and executive function, were explored in MCI, whereas it was only partially explored in diagnosed AD.

A paper concerning MCI performances on delayed recall and executive function suggested that hypometabolism in the right medial temporal cortex, right prefrontal cortex, left superior parietal cortex, and bilateral posterior cingulate reflect impairments in delayed recall, while hypometabolism in the right prefrontal cortex mirrors deficits in executive function in MCI [18]. The differences with our results may be due to differences in both the methods and in the stage of disease of the examined population. A further paper evaluated the relationships of FDG-PET metabolism with cognition in MCI: the composite score predicted variation in cortical metabolism, and TMT B was significantly correlated with PET metabolism. The results indicated that RAVLT and TMT B are sensitive to variation in AD neuroimaging markers in MCI but, contrary to our paper focused on SPM brain analysis; the Statistical Package for the Social Sciences (SPSS version 22.0) was used [5]. Even if both statistical analysis and enrolled populations are different from our methods, these findings partially support our results concerning the correlation of RAVLT test performance with FDG uptake in AD.

Furthermore, a previous study compared glucose metabolism and clinical measurements, reporting that baseline and longitudinal FDG-ROI measures are sensitive to poor performance at neuropsychological assessment and validate the cognitive and functional relevance of longitudinal changes in (18F)FDG measurement. Nevertheless, the indices used in this paper were The Functional Activities Questionnaire (FAQ) and Alzheimer's Disease Assessment Scale—cognitive subscale (ADAS-cog) [6].

The patients and methods applied in the literature are as heterogeneous as the results obtained. According to our results concerning diagnosed AD, glucose metabolism of the left temporal lobe correlated to RAVLT, whereas glucose metabolism of the bilateral parietal lobe correlated to RCFT, copy. We did not find any significant relationships with other tests. Therefore, cortical and subcortical glucose consumption appear moderately related to the neuropsychological assessment. Even if the clinical usefulness of neuropsychological assessment is well established, our results suggest a limited impact on the data analysis of brain metabolism in patients with AD. Therefore, we suggest using these parameters in the statistical analysis of neuroimaging biomarkers by dedicated software, especially FDG uptake, just in the particular analysis concerning specific clinical aspects. Generally, considering the heterogeneous data about the influence of neuropsychological assessment on PET data, the use of neuropsychological indices as a covariate or independent variables in SPM analysis can be avoided in these patients.

Author Contributions: A.C., conceptualization, data curation, writing, supervision; M.R., data curation, formal analysis, writing; D.D.B., data curation, formal analysis; L.F., data curation, formal analysis; A.M., supervision, validation; O.S., supervision, validation, writing. All authors have read and agreed to the published version of the manuscript.

Funding: This research received no external funding.

Acknowledgments: The authors wish to thank Tiziana Martino (IRCCS Neuromed) for data collection. This study was supported by the PICASO project (grant no. 689209). An initial version of this paper has been presented as a conference paper at the 29th European Association of Nuclear Medicine congress https://link.springer.com/article/10.1007%2Fs00259-016-3484-4.

Conflicts of Interest: The authors declare no conflict of interest.

References

1. Brugnolo, A.; Morbelli, S.; Arnaldi, D.; De Carli, F.; Accardo, J.; Bossert, I.; Dessi, B.; Famà, F.; Ferrara, M.; Girtler, N.; et al. Metabolic Correlates of Rey Auditory Verbal Learning Test in Elderly Subjects with Memory Complaints. *J. Alzheimer's Dis.* **2014**, *39*, 103–113. [CrossRef] [PubMed]
2. Visser, P.J. Role of cognitive testing in disease modifying AD trials. *J. Nutr. Heal. Aging* **2006**, *10*, 132–133.
3. Tomadesso, C.; De La Sayette, V.; De Flores, R.; Bourgeat, P.; Villemagne, V.L.; Egret, S.; Eustache, F.; Chételat, G. Neuropsychology and neuroimaging profiles of amyloid-positive versus amyloid-negative amnestic mild cognitive impairment patients. *Alzheimer's Dementia Diagn. Assess. Dis. Monit.* **2018**, *10*, 269–277. [CrossRef] [PubMed]
4. Carlesimo, G.A.; Fadda, L.; Sabbadini, M.; Caltagirone, C. Recency effect in Alzheimer's disease: A reappraisal. *Q. J. Exp. Psychol. A* **1996**, *49*, 315–325. [CrossRef] [PubMed]
5. Eliassen, C.F.; Reinvang, I.; Selnes, P.; Fladby, T.; Hessen, E. Convergent Results from Neuropsychology and from Neuroimaging in Patients with Mild Cognitive Impairment. *Dement. Geriatr. Cogn. Disord.* **2017**, *43*, 144–154. [CrossRef] [PubMed]
6. Landau, S.M.; Harvey, D.; Madison, C.M.; Koeppe, R.A.; Reiman, E.M.; Foster, N.L.; Weiner, M.W.; Jagust, W.J.; Initiative, T.A.D.N. Associations between cognitive, functional, and FDG-PET measures of decline in AD and MCI. *Neurobiol. Aging* **2009**, *32*, 1207–1218. [CrossRef] [PubMed]
7. Hedden, T.; Schultz, A.; Rieckmann, A.; Mormino, E.C.; Johnson, K.A.; Sperling, R.A.; Buckner, R.L. Multiple Brain Markers are Linked to Age-Related Variation in Cognition. *Cereb. Cortex* **2014**, *26*, 1388–1400. [CrossRef] [PubMed]
8. Chiaravalloti, A.; Fiorentini, A.; Francesco, U.; Martorana, A.; Koch, G.; Belli, L.; Torniolo, S.; Di Pietro, B.; Motta, C.; Schillaci, O. Is cerebral glucose metabolism related to blood–brain barrier dysfunction and intrathecal IgG synthesis in Alzheimer disease? *Medicine* **2016**, *95*, e4206. [CrossRef] [PubMed]
9. Racine, A.; Adluru, N.; Alexander, A.L.; Christian, B.T.; Okonkwo, O.C.; Oh, J.; Cleary, C.A.; Birdsill, A.; Hillmer, A.T.; Murali, D.; et al. Associations between white matter microstructure and amyloid burden in preclinical Alzheimer's disease: A multimodal imaging investigation. *NeuroImage Clin.* **2014**, *4*, 604–614. [CrossRef] [PubMed]
10. Dubois, B.; Feldman, H.H.; Jacova, C.; DeKosky, S.T.; Barberger-Gateau, P.; Cummings, J.; Delacourte, A.; Galasko, D.; Gauthier, S.; Jicha, G.; et al. Research criteria for the diagnosis of Alzheimer's disease: Revising the NINCDS-ADRDA criteria. *Lancet Neurol.* **2007**, *6*, 734–746. [CrossRef]
11. Chiaravalloti, A.; Barbagallo, G.; Martorana, A.; Castellano, A.E.; Ursini, F.; Schillaci, O. Brain metabolic patterns in patients with suspected non-alzheimer's pathophysiology (SNAP) and alzheimer's disease (AD): Is [(18)F] FDG a specific biomarker in these patients? *Eur. J. Nucl. Med. Mol. Imaging* **2019**, *46*, 1796–1805. [CrossRef] [PubMed]
12. Chiaravalloti, A.; Barbagallo, G.; Ricci, M.; Martorana, A.; Ursini, F.; Sannino, P.; Karalis, G.; Schillaci, O. Brain metabolic correlates of CSF Tau protein in a large cohort of Alzheimer's disease patients: A CSF and FDG PET study. *Brain Res.* **2018**, *1678*, 116–122. [CrossRef] [PubMed]
13. Ricci, M.; Chiaravalloti, A.; Martorana, A.; Koch, G.; De Lucia, V.; Barbagallo, G.; Schillaci, O. The role of epsilon phenotype in brain glucose consumption in Alzheimer's disease. *Ann. Nucl. Med.* **2020**, *34*, 254–262. [CrossRef] [PubMed]

14. Liguori, C.; Chiaravalloti, A.; Sancesario, G.; Stefani, A.; Sancesario, G.; Mercuri, N.B.; Schillaci, O.; Pierantozzi, M. Cerebrospinal fluid lactate levels and brain [18F]FDG PET hypometabolism within the default mode network in Alzheimer's disease. *Eur. J. Nucl. Med. Mol. Imaging* **2016**, *43*, 2040–2049. [CrossRef] [PubMed]
15. Chiaravalloti, A.; Martorana, A.; Koch, G.; Toniolo, S.; Di Biagio, D.; Di Pietro, B.; Schillaci, O. Functional correlates of t-Tau, p-Tau and Aβ1–42 amyloid cerebrospinal fluid levels in Alzheimer's disease. *Nucl. Med. Commun.* **2015**, *36*, 1. [CrossRef] [PubMed]
16. Chiaravalloti, A.; Pagani, M.; Cantonetti, M.; Di Pietro, B.; Tavolozza, M.; Travascio, L.; Di Biagio, D.; Danieli, R.; Schillaci, O. Brain metabolic changes in Hodgkin disease patients following diagnosis and during the disease course: An 18F-FDG PET/CT study. *Oncol. Lett.* **2014**, *9*, 685–690. [CrossRef] [PubMed]
17. Chiaravalloti, A.; Koch, G.; Toniolo, S.; Belli, L.; Di Lorenzo, F.; Gaudenzi, S.; Schillaci, O.; Bozzali, M.; Sancesario, G.; Martorana, A. Comparison between Early-Onset and Late-Onset Alzheimer's Disease Patients with Amnestic Presentation: CSF and (18)F-FDG PET Study. *Dement. Geriatr. Cogn. Disord. Extra* **2016**, *6*, 108–119. [CrossRef] [PubMed]
18. Nishi, H.; Sawamoto, N.; Namiki, C.; Yoshida, H.; Thuy, D.H.D.; Ishizu, K.; Hashikawa, K.; Fukuyama, H. Correlation between Cognitive Deficits and Glucose Hypometabolism in Mild Cognitive Impairment. *J. Neuroimaging* **2010**, *20*, 29–36. [CrossRef] [PubMed]

Protocol

The "develOpment of metabolic and functional markers of Dementia IN Older people" (ODINO) Study: Rationale, Design and Methods

Anna Picca [1], Daniela Ronconi [2], Hélio J. Coelho-Junior [2], Riccardo Calvani [1], Federico Marini [3], Alessandra Biancolillo [4], Jacopo Gervasoni [1,2], Aniello Primiano [1,2], Cristina Pais [1], Eleonora Meloni [1], Domenico Fusco [1], Maria Rita Lo Monaco [1], Roberto Bernabei [1,2,*], Maria Camilla Cipriani [1], Emanuele Marzetti [1,2,*] and Rosa Liperoti [1,2]

1 Fondazione Policlinico Universitario "Agostino Gemelli" IRCCS, 00168 Rome, Italy; anna.picca@guest.policlinicogemelli.it (A.P.); riccardo.calvani@guest.policlinicogemelli.it (R.C.); jacopo.gervasoni@policlinicogemelli.it (J.G.); anielloprim@gmail.com (A.P.); cristina.pais@guest.policlinicogemelli.it (C.P.); meloni.eleonora@hotmail.it (E.M.); domenico.fusco@policlinicogemelli.it (D.F.); mariarita.lomonaco@policlinicogemelli.it (M.R.L.M.); mariacamilla.cipriani@policlinicogemelli.it (M.C.C.); rosa.liperoti@policlinicogemelli.it (R.L.)

2 Università Cattolica del Sacro Cuore, 00168 Rome, Italy; danielaronconi88@gmail.com (D.R.); coelhojunior@hotmail.com.br (H.J.C.-J.)

3 Department of Chemistry, Sapienza Università di Roma, 00185 Rome, Italy; federico.marini@uniroma1.it

4 Department of Physical and Chemical Sciences, Università degli Studi dell'Aquila, 67100 L'Aquila, Italy; alessandra.biancolillo@univaq.it

* Correspondence: roberto.bernabei@unicatt.it (R.B.); emanuele.marzetti@policlinicogemelli.it (E.M.); Tel.: +39-(06)-3015-5559 (R.B. & E.M.); Fax: +39-(06)-3051-911 (R.B. & E.M.)

Received: 11 March 2020; Accepted: 4 April 2020; Published: 9 April 2020

Abstract: Mild cognitive impairment (MCI), also termed mild neurocognitive disorder, includes a heterogeneous group of conditions characterized by declines in one or more cognitive domains greater than that expected during "normal" aging but not severe enough to impair functional abilities. MCI has been associated with an increased risk of developing dementia and even considered an early stage of it. Therefore, noninvasively accessible biomarkers of MCI are highly sought after for early identification of the condition. Systemic inflammation, metabolic perturbations, and declining physical performance have been described in people with MCI. However, whether biological and functional parameters differ across MCI neuropsychological subtypes is presently debated. Likewise, the predictive value of existing biomarkers toward MCI conversion into dementia is unclear. The "develOpment of metabolic and functional markers of Dementia IN Older people" (ODINO) study was conceived as a multi-dimensional investigation in which multi-marker discovery will be coupled with innovative statistical approaches to characterize patterns of systemic inflammation, metabolic perturbations, and physical performance in older adults with MCI. The ultimate aim of ODINO is to identify potential biomarkers specific for MCI subtypes and predictive of MCI conversion into Alzheimer's disease or other forms of dementia over a three-year follow-up. Here, we describe the rationale, design, and methods of ODINO.

Keywords: aging; biomarkers; cytokines; cognitive decline; Alzheimer's disease; metabolomics; neuroinflammation; multivariate analysis; physical performance; person-tailored

1. Introduction

Mild cognitive impairment (MCI), also termed mild neurocognitive disorder according to the Diagnostic and Statistical Manual of Mental Disorders, Fifth Edition (DSM-5) [1], is a condition characterized by a decline in cognitive function that is greater than that expected for the individual's age and education level but not severe enough to compromise engagement in daily activities [2]. From a neuropsychological perspective, MCI involves reduced cognitive abilities (non-amnestic subtype) in one (single domain) or more (multiple domains) domains or a reduced ability to recall stored information (amnestic subtype) [3]. Other cognitive domains including language, visuospatial function, complex attention, and executive functions may also be affected [3]. Noticeably, people diagnosed with MCI, especially those with the amnestic subtype (aMCI), have a 10-fold increased risk of progression toward Alzheimer's disease (AD) or other forms of dementia [4]. Therefore, establishing whether MCI is indeed a prodromal stage of dementia and understanding the mechanisms of its progression are necessary for the early take-in-charge of affected individuals and for the implementation of preventive interventions [5,6]. To this aim, biomarkers capable of identifying persons with MCI, especially those at higher risk of developing AD or other forms of dementia, are highly sought after.

Signs of neuroinflammation (e.g., detection of inflammatory cytokines in the proximity of ß-amyloid deposits and neurofibrillary tangles), whole-body metabolic perturbations, and declining physical performance (e.g., slow gait speed, impaired performance on dual-task tests) have been documented in people with MCI [7,8]. However, the existence of specific patterns of biological and functional markers across MCI subtypes and their predictive value for the conversion of MCI into dementia are debated.

Multi-marker approaches covering different physiological domains are increasingly implemented for the appraisal of complex and dynamic conditions [9–11]. These approaches are conceptualized within the notion of allostatic load, that is the exposure of cells and biological systems to recurrent or chronic stressors inflicting cumulative damage [12]. Within this paradigm, biomarkers represent endophenotypes of physiological dysregulation that may support diagnosis, tracking, clinical and therapeutic decision-making, and verification of the efficacy of an intervention before it is clinically detectable. In this scenario, the quantification of specific parameters coupled with ad hoc multivariate statistical approaches may allow identifying patterns of biomarkers that can be applied at the individual level as a measure of departure from his/her "normal operating conditions" [13].

The "develOpment of metabolic and functional markers of Dementia IN Older people" (ODINO) study has been conceived as an innovative multi-dimensional investigation in which clinical, neuropsychological, functional, and biological parameters will be analyzed through ad hoc statistical analyses to provide a comprehensive characterization of MCI subtypes. Biological and functional markers will be tested for their ability to predict MCI conversion into AD and other forms of dementia over a three-year follow-up.

2. Materials and Methods

2.1. Study Design and Population

The protocol of this observational study was approved by the Ethics Committee of the Fondazione Policlinico Universitario "Agostino Gemelli" IRCCS (Rome, Italy) (protocol #230/19). A convenience sample of 120 participants will be enrolled in ODINO. The study will be carried out through a two-step analytical process: (1) collection of clinical data, evaluation of cognitive and physical performance, and analysis of biological markers to evaluate patterns of physical performance, systemic inflammation, and metabolic perturbations in older adults with MCI, and (2) a longitudinal three-year follow-up to obtain indications of biomarkers associated with the conversion of MCI into dementia.

Participant recruitment will take place at the Fondazione Policlinico Universitario "Agostino Gemelli" IRCCS under the coordination of the outpatient clinic of the Department of Geriatrics. Participants will be recruited by convenience and asked about their willingness to participate in the study. Candidates aged ≥65 and ≤85 years with a diagnosis of MCI according to the criteria of the National Institute on Aging-Alzheimer's Association [14] will be considered eligible for enrolment. After obtaining written informed consent, participants will be stratified in MCI subtypes (i.e., amnestic single domain, non-amnestic single domain, amnestic multi-domains, and non-amnestic multi-domains).

Exclusion criteria will be: active treatment for cancer or cancer diagnosis (except for non-melanoma skin cancer), severe knee or hip osteoarthritis limiting mobility, inflammatory diseases (e.g., rheumatoid arthritis, vasculitis, autoimmune disorders, inflammatory bowel disease), stroke with upper and/or lower extremity involvement, Parkinson's disease or other neurological disorders likely to interfere with physical function, major psychiatric illnesses, sleep disorders, heart failure New York Heart Association (NYHA) class III–IV, respiratory insufficiency requiring supplemental oxygen, and use of long-acting benzodiazepines or antipsychotic drugs. Temporary exclusion criteria will be acute illnesses (e.g., infections, re-exacerbation of chronic obstructive pulmonary disease), major surgery, and traumata.

Participant assessment will be carried out at the geriatric outpatient clinic of the Fondazione Policlinico Universitario "Agostino Gemelli" IRCCS during four visits at baseline and every 12 months over a three-year follow-up. The activities expected at each visit will be completed over three days within one week (Table 1).

Table 1. Visit schedule and related activities.

Activity	Visit 1 (Baseline)	V2 (12 Months)	V3 (24 Months)	V4 (36 Months)
Informed Consent	x			
Sociodemographic Characteristics	x	x	x	x
Medical History	x	x	x	x
Medication Inventory	x	x	x	x
CDR Scale	x	x	x	x
MMSE	x	x	x	x
RAVLT	x	x	x	x
Digit Span	x	x	x	x
Corsi Span	x	x	x	x
Visuospatial Abilities and Praxis	x	x	x	x
Language	x	x	x	x
Attention and Executive Functions	x	x	x	x
GDS-15	x	x	x	x
Anthropometry	x	x	x	x
Body Composition	x	x	x	x
Muscle Strength	x	x	x	x
TUG	x	x	x	x
SPPB	x	x	x	x
6MWT	x	x	x	x
ADL	x	x	x	x
IADL	x	x	x	x
7-Day Activity and EE Monitor	x	x	x	x
Blood Draw	x	x	x	x

The activities of each visit will be completed over three days within one week. Abbreviations: CDR, clinical dementia rating; MMSE, mini-mental state examination; RAVLT, Rey auditory verbal learning test; GDS-15, geriatric depression scale 15 items; TUG, timed-up-and-go; SPPB, short physical performance battery; 6MWT, six-min walking test; ADL, activities of daily living; IADL, instrumental activities of daily living; EE, energy expenditure.

2.2. General Characteristics

Information on age, sex, smoking habit, alcohol consumption, comorbid conditions, and medications will be recorded by an attending physician through a structured interview and careful review of medical records. Vital signs will also be assessed. Standard blood biochemistry will be carried out by the centralized diagnostic laboratory of the Fondazione Policlinico Universitario "Agostino Gemelli" IRCCS.

2.3. Neuropsychological Assessment and Cognitive Evaluation

Participants will receive a comprehensive neuropsychological assessment and cognitive evaluation by means of the clinical dementia rating (CDR) scale [15] and a battery of neuropsychological tests exploring global cognitive function (mini-mental state examination, MMSE) [16], verbal learning and episodic memory (Rey auditory verbal learning test, RAVLT) [17], verbal short-term memory and verbal working memory (digit span) [18], visuospatial short-term memory and visuospatial working memory (Corsi span) [19,20], visuospatial abilities and praxis (copying drawings; copying drawings with landmarks; Rey-Osterrieth complex figure copy) [21], language (phonological verbal fluency task, semantic verbal fluency task, nouns naming test, and verbs naming test) [22], and attention and executive functions (Stroop color word interference test, multiple features target cancellation test, MFTC) [23,24]. Mood will be assessed through the 15-item geriatric depression scale (GDS-15) [25].

2.4. Anthropometry and Body Composition

Body mass and height will be measured by means of a medical graded weight scale with a stadiometer. Body mass index (BMI) will be calculated as the ratio between body mass (kg) and the square of height (m^2).

A flexible and inextensible anthropometric tape will be used to measure waist circumference (WC), hip circumference (HC), and mid-arm circumference (MAC). The waist-to-hip ratio will be calculated. For these measurements, participants will be requested to wear light clothes and to stay in standing position, head held erect, eyes forward, with arms relaxed at the sides of the body, and feet kept together. WC will be taken at the mid-point between the last floating rib and the highest point of the iliac crest. HC will be measured at the highest point of the buttocks. MAC will be taken at the mid-point between the elbow and the deltoid muscle [26].

Bone mineral density and appendicular lean mass will be measured by dual X-ray absorptiometry (DXA) on a Hologic® Discovery A (Hologic, Inc., Bedford, MA, USA), as previously described [27].

2.5. Assessment of Muscle Strength, Physical Function, and Disability Status

Upper and lower-limb muscle strength will be measured by isometric handgrip strength (IHG) and isokinetic analysis, respectively. IHG will be measured using a Jamar handheld hydraulic dynamometer (Patterson Medical Products, Inc., Cincinnati, OH, USA) [28]. For the test, participants will remain seated on a standard chair with shoulders abducted, and the elbow flexed at 90° and near to the trunk, and the wrist in a neutral position (thumb up). The contralateral arm will remain relaxed under the thigh [29]. IHG will be measured during four s. Participants will be familiarized and warmed up before performing three maximal efforts with a one-min rest period. Encouragement to perform the test as quickly and forcefully as possible will be provided during the entire experiment. The maximal concentric isokinetic strength of knee extensors of the dominant side will be measured on a REV9000 isokinetic dynamometer (Technogym, Gambettola, Italy), as previously described [30]. Briefly, participants will be asked to produce their maximum force while extending the knee from 90° to 0° of flexion at 60°/s with a hip angle of 90°–100°. Two practice repetitions will be completed prior to three test repetitions. The maximal peak torque achieved will be used for the analysis.

The timed-up-and-go (TUG) test involves getting up from a chair, walking three meters around a marker placed on the floor, coming back to the same position, and sitting back on the chair. Participants will begin the test while wearing their regular footwear, with their back against the chair, arms resting on the chair's arms, and with the feet in contact with the ground. A researcher will instruct them to, on the word "go", get up and walk at a normal pace through the demarcation of three meters on the ground, turn, return to the chair, and sit down again. Timing will start when participants get up from the chair and will stop when their back touches the backrest of the chair again [31]. After completing the standard TUG test, participants will be asked to perform TUG combined with a verbal fluency task (i.e., naming as many animals as they can remember), with a motor task (i.e., carrying a full cup of water), and with both cognitive and motor tasks (i.e., performing the verbal fluency test while carrying a full cup of water) [32].

The short physical performance battery (SPPB) is composed of three subtasks: standing balance, usual gait speed, and the five-repetition chair-stand test [33]. For the standing balance test, participants will be asked to stand in three progressively more difficult positions for 10 s each: a side-by-side feet standing position, a semi-tandem position and a full-tandem position. Gait speed will be measured over a four m course at the person's usual pace. The faster of two trials will be used for the analysis. For the five-repetition chair-stand test, participants will be asked to perform five repetitions of standing up and sitting down from a chair without using hands and the performance will be timed. Each of the three SPPB subtasks will be categorized into a five-level score, with zero representing the inability to do the test and four corresponding to the highest level of performance.

The six-min walk test will be performed according to the guidelines of the American Thoracic Society [34]. The test will be conducted indoors on a 30-m track. After remaining seated for 15 min, participants will be requested to walk on the track as fast as possible for six min. The test will be interrupted if participants show any sign of chest pain, intolerable dyspnea, leg cramps, staggering, diaphoresis, a pale or ashen appearance, or any other complaint. The distance walked (m) will be used for the analysis.

Disability status will be assessed through the activities of daily living (ADL) and instrumental ADL (IADL) scales [35,36].

2.6. Measurement of Physical Activity Levels and Energy Expenditure

Free-living physical activity levels and energy expenditure will be quantified through a SenseWear® armband (SWA, BodyMedia, Inc., Pittsburgh, PA, USA) over seven consecutive days. The SWA is a body monitor wearable on the back of the arm that enables the continuous monitoring of physical activity at low intensities and during unstructured or intermittent activities. The SWA utilizes a unique combination of sensors to measure the amount of heat being dissipated by the body as well as the skin and near-armband temperature. Measures of galvanic skin response to physical and emotional stimuli are also recorded. A two-axis accelerometer tracks the arm movements and provides information about the body position. Wireless transmission, communication, and wired data download are ensured by a radio and a data port. All information collected is integrated and processed by software using proprietary algorithms according to the participant's characteristics (sex, age, height, and weight) to provide minute-by-minute estimates of energy expenditure during different levels of physical activity [37].

2.7. Collection of Blood Samples

Blood will be drawn in the morning by venipuncture of the median cubital vein. For plasma separation, blood will be collected using tubes containing ethylenediaminetetraacetic acid (EDTA) and will subsequently be centrifuged at 1000× *g* for 15 min at 4 °C. For serum separation, blood samples will be collected in tubes without EDTA and will be kept at room temperature for 30 min to allow clotting. Afterward, samples will be centrifuged as described above. Aliquots will be prepared from the upper clear fraction (plasma or serum) and stored at −80 °C until analysis.

2.8. Determination of Biomarkers of Inflammation and Neurodegeneration

A biomarker panel has been designed based on previous investigations by our group in older populations, including older adults with neurodegenerative conditions [10,38–43]. Markers of systemic inflammation will be assayed as described elsewhere [40]. Briefly, a set of 27 pro- or anti-inflammatory mediators (e.g., cytokines, chemokines, and growth factors) will be measured in serum samples using the Bio-Plex Pro™ Human Cytokine 27-plex Assay kit (#M500KCAF0Y, Bio-Rad, Hercules, CA, USA)

(Table 2). Experiments will be run on a Bio-Plex® System with Luminex xMap Technology (Bio-Rad) and data will be acquired on the Bio-Plex Manager Software 6.1 (Bio-Rad) with instrument default settings. Outliers will be automatically removed by optimization of standard curves across all analytes and results will be obtained as concentration (pg/mL).

Table 2. Serum inflammatory biomolecules assayed by multiplex immunoassay.

Biomarker Class	Components of the Biomarker Panel
Cytokines	IFNγ, IL1β, IL1Ra, IL2, IL4, IL5, IL6, IL7, IL8, IL9, IL10, IL12, IL13, IL15, IL17, TNF-α
Chemokines	CCL5, CCL11, IP-10, MCP-1, MIP-1α, MIP-1β
Growth Factors	FGF-β, G-CSF, GM-CSF, PDGF-BB

Abbreviations: IFN, interferon; IL, interleukin; IL1Ra, interleukin 1 receptor agonist; TNF-α, tumor necrosis factor-alpha; CCL, C-C motif chemokine ligand; IP: interferon-induced protein; MCP-1: monocyte chemoattractant protein 1; MIP: macrophage inflammatory protein; FGF, fibroblast growth factor; G-CSF, granulocyte colony-stimulating factor; GM-CSF, granulocyte macrophage colony-stimulating factor; PDGF-BB, platelet-derived growth factor BB.

Traditional and automated enzyme-linked immunosorbent assays will be run to quantify serum levels of amyloid beta (aa1-42) (#DAB142, R&D Systems, Minneapolis, MN, USA), neurofilament light polypeptide (#SPCKB-PS-002448-000190, R&D Systems), Tau (#NBP2-62749, Novus Biologicals, Littleton, CO, USA), and Tau [p Ser739] (#NBP2-66711, Novus Biologicals) proteins according to the manufacturer's instructions.

2.9. *Measurement of Plasma Fatty Acid Concentrations*

A panel of 14 fatty acids that were previously associated with AD [44] will be measured in plasma by a gas chromatography-electron ionization-mass spectrometry (GC-EI-MS) validated methodology (Eureka One Lab Division Kit, code GC75010; Ancona, Italy) according to the manufacturer's instructions (Table 3).

Table 3. Plasma fatty acids assayed by gas chromatography-electron ionization-mass spectrometry.

Fatty Acid	Fragments
Tetradecanoic acid	74.1; 87.0; 143.0
Hexadecanoic acid	227.0; 270.0
Cis-9-hexadecenoic acid	55.0; 81.0; 96.0; 237.0
Heptadecanoic acid (internal standard)	74.1; 87.0; 284.0
Octadecanoic acid	75.0; 255.0; 298.0
Cis-9-octadecenoic acid	81.1; 96.0; 264.0
Trans-9-octadecenoic acid	81.1; 96.0; 264.0
All cis 9,12-octadecadienoic acid	67.0; 81.1; 95.0
Trans-9,trans-12-octadecadienoic acid	67.0; 81.1; 95.0
All cis 9,12,15 octadecatrienoic acid	67.0; 79.1; 93.0; 95.0
All cis 6,9,12 octadecatrienoic acid	67.0; 79.1; 93.0; 95.0
All cis 8,11,14 eicosatrienoic acid	67.0; 79.1; 93.0
All cis 5,8,11,14 eicosatetraenic acid	67.0; 79.1; 91.0
All cis 5,8,11,14,17 eicosapentaenoic acid	67.0; 79.1; 91.0; 105.0

Following extraction and washing, samples will be treated with a derivatizing solution for 15 min at 100 °C, diluted and directly injected into the GC-MS. GC-MS analyses will be performed on a Trace GC Ultra (Thermo Fisher Scientific, Waltham, MA, USA) equipped with Durabond HP-88 column, 100 m × 0.25 mm × 0.2 μm film thickness (Agilent Technologies, Santa Clara, CA, USA), and connected to an ISQ mass spectrometer (Thermo Fisher Scientific). One μL of the sample will be injected in split mode (1:10 ratio), with injector temperature set at 250 °C. Helium will be used as carrier gas and the flow-rate will be maintained constant at 1 mL/min. The initial oven temperature of 100 °C will be held for 1 min, then raised to 220 °C at 10 °C/min, and maintained for 4 min. Afterward, the oven temperature will be increased up to 240 °C at 10 °C/min and held for 10 min. The mass transfer line will be maintained at 270 °C and the ion source at 200 °C. Analyses will be performed in the selected ion monitoring (SIM) mode. Ions monitored in the analysis are shown in Table 3. The analytical process will be monitored using fatty acid controls (code GC75019, level 1 and level 2) manufactured by Eureka One Lab Division.

2.10. Measurement of Serum Concentrations of Amino Acids and Derivatives

The serum concentration of a panel of 37 amino acids and derivatives will be determined by ultraperformance liquid chromatography/MS (UPLC/MS), as described elsewhere [9]. The panel has been chosen based on previous work by our group in older populations [9,41,45,46] and will include: 1-methylhistidine, 3-methylhistidine, 4-hydroxyproline, α-aminobutyric acid, β-alanine, β-aminobutyric acid, γ-aminobutyric acid, alanine, aminoadipic acid, anserine, arginine, asparagine, aspartic acid, carnosine, citrulline, cystathionine, cystine, ethanolamine, glutamic acid, glycine, histidine, isoleucine, leucine, lysine, methionine, ornithine, phenylalanine, phosphoethanolamine, phosphoserine, proline, sarcosine, serine, taurine, threonine, tryptophan, tyrosine, and valine.

Briefly, 50 μL of the sample will be added to 100 μL 10% (w/v) sulfosalicylic acid containing an internal standard mix (50 μM) (Cambridge Isotope Laboratories, Inc., Tewksbury, MA, USA). The mixture will be centrifuged at 1000× *g* for 15 min. Seventy μL of borate buffer and 20 μL of AccQ Tag reagents (Waters Corporation, Milford, MA, USA) will be added to 10 μL of the obtained supernatant and heated at 55 °C for 10 min. Next, samples will be loaded onto a CORTECS UPLC C18 column 1.6 μm, 2.1 mm × 150 mm (Waters Corporation) for chromatographic separation (ACQUITY H-Class, Waters Corporation). Elution will be accomplished at 500 μL/min flow-rate with a linear gradient (9 min) from 99:1 to 1:99 water 0.1% formic acid/acetonitrile 0.1% formic acid. Finally, analytes will be detected on an ACQUITY QDa single quadrupole mass spectrometer equipped with an electrospray source operating in positive mode (Waters Corporation). Amino acid controls manufactured by the MCA laboratory of the Queen Beatrix Hospital (Winterswijk, The Netherlands) will be used to monitor the analytic process.

2.11. Statistical Analysis

Since no studies have explored the multi-marker profile of MCI chosen in ODINO, no power analysis could be run for sample size calculation. Based on the number of older persons diagnosed with MCI at our center, we estimate that 120 participants will be enrolled over one year. The rate of losses to follow-up over three years is expected to be 20%, leaving a total of 100 cases from whom all variables of interest will be collected. Participants will be censored at the end of follow-up, at the date of conversion to dementia, or at the date of death, as applicable.

Descriptive statistics will be run for all variables. After ascertainment of data distribution, comparisons of variables of interest among MCI subtypes will be performed via analysis of variance (ANOVA) or Kruskal –Wallis H for continuous variables, as appropriate. Categorical variables will be compared through χ^2 statistics. Survival analyses (Kaplan–Meier and Cox regression) will be used to investigate the impact of clinical variables on conversion to dementia. Descriptive statistics will be performed using SAS version 9.2 (SAS Institute Inc., Cary, NC, USA).

In the cross-sectional stage of the study, to characterize MCI subtypes, different classification strategies will be enacted. First, a purely discriminant approach will be adopted. In discriminant classification, data are used to build a predictive model aiming at assigning each individual to one specific group (in this case, any of the MCI subtypes). To this aim, a classification strategy based on partial least squares discriminant analysis (PLS-DA) will be adopted [47]. PLS-DA offers the advantage of processing datasets containing a high number of variables even if they are highly correlated with one another. The analysis of the whole dataset, which encompasses multi-block data, will be performed through the recently developed sequential and orthogonalized covariance selection (SO-CovSel) [48]. The method, which allows a highly parsimonious variable selection, was used by our group for the identification of a gut microbial, inflammatory and metabolic fingerprint in older adults with physical frailty & sarcopenia [41].

To rule out the possibility of chance correlations and to estimate the reliability of predictive models, a thorough validation process by means of double cross-validation (DCV) and permutation tests will be operated. DCV consists of two nested loops of cross-validation; the external loop is used to unbiasedly estimate the predictive ability of the model parameters that are optimized on the basis of the internal loop. The classification ability of the optimal model is then expressed by means of various figures of merit, such as the number of misclassifications (NMC), the area under the receiver operating characteristic (AUROC) curve, and the discriminant Q2 (DQ2), the distributions of which under the null hypothesis are estimated via permutation tests [49]. This allows the establishment of the statistical significance of the observed discrimination. A detailed description of PLS-DA, SO-CovSel, and DCV procedures may be found elsewhere [9,41,43].

Classification will be repeated with a modeling approach based on soft independent modeling of class analogies (SIMCA) [50], in order to assess the degrees of similarity among MCI subtypes. Modeling approaches use data to define which experimental profiles are to be expected from individuals belonging to a particular category (i.e., the so-called model space). Accordingly, a class model allows the prediction of how likely it is for an individual to belong to a class or not, on the basis of the measurements considered. Since each category (MCI subtype) will be modeled independently from the others, the results of this analytical stage will provide information on the effectiveness of the characterization of different MCI subtypes. The analysis will also indicate how likely it is for specific subtypes of participants to be confounded with one another. A SIMCA model was recently built by our group to verify the accuracy of the classification of older adults with Parkinson's disease based on extracellular vesicle cargo [43].

In the longitudinal phase of the study, multilevel ANOVA (MANOVA) coupled with simultaneous component analysis (SCA) or with PLS-DA will be used to characterize the time dynamics of participants over the follow-up [51]. MANOVA may be considered to be equivalent to multivariate ANOVA for repeated measurements. In such a configuration, SCA, which under the constraints of MANOVA is identical to principal component analysis, or PLS-DA allows applying the method to data matrices containing a high number of possibly correlated variables. The statistical significance of MANOVA will be assessed non-parametrically by permutation tests. Multivariate statistics will be conducted using functions written in-house and run under the Matlab environment (The Mathworks, Natick, MA, USA).

3. Discussion

The possibility that MCI, at least in some variants, may be a prodromal stage of dementia has ignited a great deal of research primarily aimed at testing possible strategies to impede MCI progression [52]. However, the clinical heterogeneity of MCI and its multifactorial pathogenesis have been major hurdles in identifying meaningful biomarkers and devising effective interventions.

Different cognitive domains may be affected in MCI, which has allowed the clinical identification of various subtypes. In the aMCI, memory loss is predominant and shows a higher risk for further conversion to AD [53]. Non-amnestic MCI subtypes are characterized by impairments in cognitive domains other than memory and have a greater propensity to convert into other forms of dementia, such as diffuse Lewy body dementia and vascular dementia [53]. Both aMCI and naMCI variants can be further categorized into single- and multi-domain subtypes, depending on the number of cognitive domains affected. The distinction between aMCI and naMCI is not only based on neuropsychological parameters, but is sustained by specific brain structural characteristics, primarily involving the hippocampus, the cortical thickness, and the entorhinal cortex [54–57].

The development of biomarkers able to capture complex phenomena, such as MCI and dementia, are highly sought after. Several investigations have been conducted to evaluate candidate biomarkers for MCI. However, these studies investigated differences between MCI and healthy controls, whereas none of them explored distinct features across MCI subtypes [58,59].

The development of cost-effective omics platforms enabling the simultaneous analysis of a vast repertoire of biological mediators in biofluids has shown great value in providing a comprehensive read-out of the environmental and clinical disturbances affecting cell homeostasis in different settings [11]. Metabolomics analyses of the cerebrospinal fluid (CSF) allowed differentiating MCI from dementia in older people, a task in which traditional biomarkers of AD such as Aβ42 failed [60]. Core CSF biomarkers including total (T-Tau) and phosphorylated Tau (P-Tau) protein, Aβ42, and neurofilament light polypeptide have been strongly associated with AD, such that their clinical implementation for diagnostic purposes has been suggested [61]. However, biomarkers specific for MCI are still missing.

Recent advances in lipidomics suggest that fatty acid dysmetabolism and imbalance in fatty acid lipidome are involved in the initiation and progression of several neurodegenerative diseases, including AD [62]. Indeed, fatty acid β-oxidation and its byproducts impact immune cell function, thereby possibly contributing to neuroinflammation [62]. However, no studies have yet addressed whether specific lipid markers distinguish aMCI and naMCI and predict their conversion into dementia. Furthermore, an investigation involving eight prospective cohorts with over 20,000 participants found an inverse association between serum concentrations of branch-chained amino acids and incident AD [63]. Similar to lipids, it is presently unclear whether specific patterns of circulating amid acids are selectively associated with MCI subtypes.

Inflammatory cytokines, including interleukin (IL) 1β, IL6 and tumor necrosis factor-alpha (TNF-α), are involved in local inflammation triggered by amyloid plaque deposition and may induce neurotoxicity when produced chronically, favoring the generation of Aβ peptides [64]. Therefore, a role for these cytokines in inter- and intracellular signaling in microglia and astrocytes has been hypothesized in AD [65]. Interestingly, peripheral inflammatory markers associated with MCI are distinct from those found in patients with AD [66]. Though, no conclusive data are presently available.

Since markers pertaining to a single domain (e.g., inflammatory rather than metabolic) may fail at capturing the intertwined relationship between local and systemic changes, we decided to combine metabolomics analysis and immunoassays to characterize the metabolic and inflammatory profile of older adults with MCI. A biomarker panel has been designed based on previous investigations by our group in older populations, including older adults with neurodegenerative conditions [10,38–43]. This approach was also guided by the recent advances made in the field of geroscience [67]. To gain insights into the etiology of the MCI subtypes, the panel also includes the analysis of circulating levels of T-Tau and P-Tau protein, Aβ42, and neurofilament light polypeptide [61].

Other lines of evidence indicate that specific impairments in physical performance are associated with MCI. In particular, people with aMCI show significant decreases in gait speed and increases in stride time and time variability when changing from a single- to a dual-task [68,69]. Notably, poor gait performance, particularly under dual-tasking, has been proposed as a motor signature of aMCI [8]. It is noteworthy that the co-occurrence of cognitive complaints and slow gait, a condition referred to as motoric cognitive risk syndrome, identifies individuals at especially high risk to progress to dementia [70,71]. Along similar lines, low muscle strength of upper and lower extremities has shown to increase the risk of MCI progression into AD [72]. The neurophysiologic substratum of reduced physical performance in MCI has its roots in the existence of neuronal networks involved in both cognition and lower extremity function. Indeed, atrophy and amyloid deposition in a network encompassing the dorsolateral prefrontal cortex, cingulate gyrus, parietal association areas, basal ganglia, and medial temporal lobes (particularly the hippocampus) are thought to mediate the relationship between cognitive function and physical performance [73]. Thus, individuals who exhibit both cognitive and motor deficits may have greater underlying brain damage [71]. This implies that the simultaneous analysis of cognitive and physical function may help identify a subset of MCI persons at greater risk of conversion to dementia [71,74]. The observation that low muscle mass is commonly observed in conjunction with cognitive impairment [75,76] suggests that the inclusion of body composition analysis might further refine the identification of people more likely to progress from MCI to dementia. Indeed, muscle loss and cognitive dysfunction share a number of predisposing factors, including inflammation, oxidative stress, insulin resistance, and an inactive lifestyle [76]. With regard to the latter, studies have shown that engagement in regular physical activity is negatively associated with the risk of developing dementia [77–79]. Furthermore, findings from a systematic review and meta-analyses indicated that an increase of 500 kcal or 10 MET-h per week was associated with a 10% decrease in the risk of dementia [79].

4. Conclusions

The ODINO study was conceived as an innovative multi-dimensional investigation aimed at exploring biological and physical performance signatures of MCI subtypes. Measures of inflammatory/metabolic markers and physical performance will be integrated through advanced multivariate statistical analyses to gain insights into the heterogeneity of MCI subtypes and their risk to progress toward dementia. The results obtained from the ODINO study and their comparison with those collected from a thoroughly characterized cohort of non-MCI older adults with similar age and sociodemographic characteristics [9–11] may enable discerning pathways involved in "physiologic" age-related cognitive decline from those implicated in the progression of MCI to early dementia. This knowledge will, therefore, pave the way for the clinical implementation of composite biomarkers of MCI. Our results will also allow the possible identification of therapeutic targets amenable for person-tailored interventions that may hold people with MCI back from the doorstep of dementia.

Author Contributions: Conceptualization, A.P. (Anna Picca), D.F., M.C.C., M.R.L.M., R.C., and R.L.; methodology, A.P. (Anna Picca), A.P. (Aniello Primiano), H.J.C.-J., and J.G.; software, F.M. and R.L.; validation, R.C. and R.L.; development of the data analysis plan, A.B. and F.M.; investigation, C.P., D.R., E.M. (Eleonora Meloni); resources, F.M. and R.B.; writing—original draft preparation, A.P. (Anna Picca), E.M. (Emanuele Marzetti) and R.C.; writing—review and editing, F.M., J.G., and R.L.; supervision, E.M. (Emanuele Marzetti), D.F., M.C.C., M.R.L.M., and R.B.; funding acquisition, R.B. All authors have read and agreed to the published version of the manuscript.

Funding: This work was partially funded by Innovative Medicines Initiative–Joint Undertaking (IMI–JU #115621) and by the nonprofit research foundation "Centro Studi Achille e Linda Lorenzon". The funders had no role in study design, data collection and analysis, preparation of the manuscript, or decision to publish.

Conflicts of Interest: The authors declare no conflict of interest. The funders had no role in the design of the study; in the collection, analyses, or interpretation of data; in the writing of the manuscript, or in the decision to publish the results.

References

1. American Psychiatric Association. *Diagnostic and Statistical Manual of Mental Disorders*, 5th ed.; American Psychiatric Association: Washington, DC, USA, 2013; ISBN 0-89042-555-8.
2. Petersen, R.C.; Smith, G.E.; Waring, S.C.; Ivnik, R.J.; Kokmen, E.; Tangelos, E.G. Aging, Memory, and Mild Cognitive Impairment. *Int. Psychogeriatr.* **1997**, *9*, 65–69. [CrossRef] [PubMed]
3. Sachdev, P.S.; Blacker, D.; Blazer, D.G.; Ganguli, M.; Jeste, D.V.; Paulsen, J.S.; Petersen, R.C. Classifying neurocognitive disorders: The DSM-5 approach. *Nat. Rev. Neurol.* **2014**, *10*, 634–642. [CrossRef] [PubMed]
4. Petersen, R.C. Mild cognitive impairment as a clinical entity. *J. Intern. Med.* **2004**, *256*, 183–194. [CrossRef] [PubMed]
5. Morris, J.C.; Storandt, M.; Miller, J.P.; McKeel, D.W.; Price, J.L.; Rubin, E.H.; Berg, L. Mild cognitive impairment represents early-stage Alzheimer disease. *Arch. Neurol.* **2001**, *58*, 397–405. [CrossRef]
6. Levey, A.; Lah, J.; Goldstein, F.; Steenland, K.; Bliwise, D. Mild cognitive impairment: An opportunity to identify patients at high risk for progression to Alzheimer's disease. *Clin. Ther.* **2006**, *28*, 991–1001. [CrossRef]
7. Chi, G.C.; Fitzpatrick, A.L.; Sharma, M.; Jenny, N.S.; Lopez, O.L.; DeKosky, S.T. Inflammatory Biomarkers Predict Domain-Specific Cognitive Decline in Older Adults. *J. Gerontol. A Biol. Sci. Med. Sci.* **2017**, *72*, 796–803. [CrossRef]

8. Montero-Odasso, M.; Oteng-Amoako, A.; Speechley, M.; Gopaul, K.; Beauchet, O.; Annweiler, C.; Muir-Hunter, S.W. The motor signature of mild cognitive impairment: Results from the gait and brain study. *J. Gerontol. Ser. A Biol. Sci. Med. Sci.* **2014**, *69*, 1415–1421. [CrossRef] [PubMed]
9. Calvani, R.; Picca, A.; Marini, F.; Biancolillo, A.; Gervasoni, J.; Persichilli, S.; Primiano, A.; Coelho-Junior, H.J.; Bossola, M.; Urbani, A.; et al. Distinct Pattern of Circulating Amino Acids Characterizes Older Persons with Physical Frailty and Sarcopenia: Results from the BIOSPHERE Study. *Nutrients* **2018**, *10*, 1691. [CrossRef]
10. Marzetti, E.; Picca, A.; Marini, F.; Biancolillo, A.; Coelho-Junior, H.J.; Gervasoni, J.; Bossola, M.; Cesari, M.; Onder, G.; Landi, F.; et al. Inflammatory signatures in older persons with physical frailty and sarcopenia: The frailty "cytokinome" at its core. *Exp. Gerontol.* **2019**, *122*, 129–138. [CrossRef]
11. Picca, A.; Coelho-Junior, H.J.; Cesari, M.; Marini, F.; Miccheli, A.; Gervasoni, J.; Bossola, M.; Landi, F.; Bernabei, R.; Marzetti, E.; et al. The metabolomics side of frailty: Toward personalized medicine for the aged. *Exp. Gerontol.* **2019**, *126*, 110692. [CrossRef]
12. Seeman, T.E.; McEwen, B.S.; Rowe, J.W.; Singer, B.H. Allostatic load as a marker of cumulative biological risk: MacArthur studies of successful aging. *Proc. Natl. Acad. Sci. USA* **2001**, *98*, 4770–4775. [CrossRef]
13. Calvani, R.; Marini, F.; Cesari, M.; Tosato, M.; Anker, S.D.; von Haehling, S.; Miller, R.R.; Bernabei, R.; Landi, F.; Marzetti, E. SPRINTT consortium Biomarkers for physical frailty and sarcopenia: State of the science and future developments. *J. Cachexia Sarcopenia Muscle* **2015**, *6*, 278–286. [CrossRef] [PubMed]
14. Albert, M.S.; DeKosky, S.T.; Dickson, D.; Dubois, B.; Feldman, H.H.; Fox, N.C.; Gamst, A.; Holtzman, D.M.; Jagust, W.J.; Petersen, R.C.; et al. The diagnosis of mild cognitive impairment due to Alzheimer's disease: Recommendations from the National Institute on Aging-Alzheimer's Association workgroups on diagnostic guidelines for Alzheimer's disease. *Alzheimer's Dement.* **2011**, *7*, 270–279. [CrossRef] [PubMed]
15. Hughes, C.P.; Berg, L.; Danziger, W.L.; Coben, L.A.; Martin, R.L. A new clinical scale for the staging of dementia. *Br. J. Psychiatry* **1982**, *140*, 566–572. [CrossRef] [PubMed]
16. Folstein, M.F.; Folstein, S.E.; McHugh, P.R. "Mini-mental state". A practical method for grading the cognitive state of patients for the clinician. *J. Psychiatr. Res.* **1975**, *12*, 189–198. [CrossRef]
17. Carlesimo, G.A.; Caltagirone, C.; Gainotti, G. The Mental Deterioration Battery: Normative data, diagnostic reliability and qualitative analyses of cognitive impairment. The Group for the Standardization of the Mental Deterioration Battery. *Eur. Neurol.* **1996**, *36*, 378–384. [CrossRef]
18. Wechsler, D. A Standardized Memory Scale for Clinical Use. *J. Psychol. Interdiscip. Appl.* **1945**, *19*, 87–95. [CrossRef]
19. Berch, D.B.; Krikorian, R.; Huha, E.M. The Corsi block-tapping task: Methodological and theoretical considerations. *Brain Cogn.* **1998**, *38*, 317–338. [CrossRef]
20. Busch, R.M.; Farrell, K.; Lisdahl-Medina, K.; Krikorian, R. Corsi Block-Tapping task performance as a function of path configuration. *J. Clin. Exp. Neuropsychol.* **2005**, *27*, 127–134. [CrossRef]
21. Shin, M.-S.; Park, S.-Y.; Park, S.-R.; Seol, S.-H.; Kwon, J.S. Clinical and empirical applications of the Rey-Osterrieth Complex Figure Test. *Nat. Protoc.* **2006**, *1*, 892–899. [CrossRef]
22. Zarino, B.; Crespi, M.; Launi, M.; Casarotti, A. A new standardization of semantic verbal fluency test. *Neurol. Sci.* **2014**, *35*, 1405–1411. [CrossRef] [PubMed]
23. Scarpina, F.; Tagini, S. The Stroop Color and Word Test. *Front. Psychol.* **2017**, *8*, 557. [CrossRef] [PubMed]
24. Marra, C.; Gainotti, G.; Scaricamazza, E.; Piccininni, C.; Ferraccioli, M.; Quaranta, D. The Multiple Features Target Cancellation (MFTC): An attentional visual conjunction search test. Normative values for the Italian population. *Neurol. Sci.* **2013**, *34*, 173–180. [CrossRef] [PubMed]
25. Yesavage, J.A.; Sheikh, J.I. 9/Geriatric Depression Scale (GDS). *Clin. Gerontol.* **1986**, *5*, 165–173. [CrossRef]
26. Landi, F.; Russo, A.; Liperoti, R.; Pahor, M.; Tosato, M.; Capoluongo, E.; Bernabei, R.O.G. Midarm muscle circumference, physical performance and mortality: Results from the aging and longevity study in the Sirente geographic area (ilSIRENTE study). *Clin. Nutr.* **2010**, *29*, 441–447. [CrossRef]
27. Marzetti, E.; Cesari, M.; Calvani, R.; Msihid, J.; Tosato, M.; Rodriguez-Mañas, L.; Lattanzio, F.; Cherubini, A.; Bejuit, R.; Di Bari, M.; et al. The "Sarcopenia and Physical fRailty IN older people: Multi-componenT Treatment strategies" (SPRINTT) randomized controlled trial: Case finding, screening and characteristics of eligible participants. *Exp. Gerontol.* **2018**, *113*, 48–57. [CrossRef]
28. Mathiowetz, V.; Wiemer, D.M.; Federman, S.M. Grip and pinch strength: Norms for 6- to 19-year-olds. *Am. J. Occup. Ther. Off. Publ. Am. Occup. Ther. Assoc.* **1986**, *40*, 705–711. [CrossRef]

29. Landi, F.; Calvani, R.; Tosato, M.; Martone, A.M.; Picca, A.; Ortolani, E.; Savera, G.; Salini, S.; Ramaschi, M.; Bernabei, R.; et al. Animal-Derived Protein Consumption Is Associated with Muscle Mass and Strength in Community-Dwellers: Results from the Milan EXPO Survey. *J. Nutr. Health Aging* **2017**, *21*, 1050–1056. [CrossRef]
30. Marzetti, E.; Lees, H.A.; Manini, T.M.; Buford, T.W.; Aranda, J.M.; Calvani, R.; Capuani, G.; Marsiske, M.; Lott, D.J.; Vandenborne, K.; et al. Skeletal muscle apoptotic signaling predicts thigh muscle volume and gait speed in community-dwelling older persons: An exploratory study. *PLoS ONE* **2012**, *7*, e32829. [CrossRef]
31. Richardson, S. The Timed "Up & Go": A Test of Basic Functional Mobility for Frail Elderly Persons. *J. Am. Geriatr. Soc.* **1991**, *39*, 142–148. [CrossRef]
32. de Melo Borges, S.; Radanovic, M.; Forlenza, O.V. Functional mobility in a divided attention task in older adults with cognitive impairment. *J. Mot. Behav.* **2015**, *47*, 378–385. [CrossRef] [PubMed]
33. Guralnik, J.M.; Simonsick, E.M.; Ferrucci, L.; Glynn, R.J.; Berkman, L.F.; Blazer, D.G.; Scherr, P.A.; Wallace, R.B. A short physical performance battery assessing lower extremity function: Association with self-reported disability and prediction of mortality and nursing home admission. *J. Gerontol.* **1994**, *49*, M85–M94. [CrossRef] [PubMed]
34. Crapo, R.O.; Casaburi, R.; Coates, A.L.; Enright, P.L.; MacIntyre, N.R.; McKay, R.T.; Johnson, D.; Wanger, J.S.; Zeballos, R.J.; Bittner, V.; et al. ATS statement: Guidelines for the six-minute walk test. *Am. J. Respir. Crit. Care Med.* **2002**, *166*, 111–117.
35. Katz, S. Assessing self-maintenance: Activities of daily living, mobility, and instrumental activities of daily living. *J. Am. Geriatr. Soc.* **1983**, *31*, 721–727. [CrossRef]
36. Lawton, M.P.; Brody, E.M. Assessment of older people: Self-maintaining and instrumental activities of daily living. *Gerontologist* **1969**, *9*, 179–186. [CrossRef]
37. Almeida, G.J.M.; Wasko, M.C.M.; Jeong, K.; Moore, C.G.; Piva, S.R. Physical activity measured by the SenseWear Armband in women with rheumatoid arthritis. *Phys. Ther.* **2011**, *91*, 1367–1376. [CrossRef]
38. Ponziani, F.R.; Bhoori, S.; Castelli, C.; Putignani, L.; Rivoltini, L.; Del Chierico, F.; Sanguinetti, M.; Morelli, D.; Paroni Sterbini, F.; Petito, V.; et al. Hepatocellular carcinoma is associated with gut microbiota profile and inflammation in nonalcoholic fatty liver disease. *Hepatology* **2019**, *69*, 107–120. [CrossRef]
39. Picca, A.; Guerra, F.; Calvani, R.; Bucci, C.; Lo Monaco, M.R.; Bentivoglio, A.R.; Landi, F.; Bernabei, R.; Marzetti, E. Mitochondrial-derived vesicles as candidate biomarkers in parkinson's disease: Rationale, Design and Methods of the EXosomes in PArkiNson Disease (EXPAND) Study. *Int. J. Mol. Sci.* **2019**, *20*, 2373. [CrossRef]
40. Ponziani, F.R.; Putignani, L.; Paroni Sterbini, F.; Petito, V.; Picca, A.; Del Chierico, F.; Reddel, S.; Calvani, R.; Marzetti, E.; Sanguinetti, M.; et al. Influence of hepatitis C virus eradication with direct-acting antivirals on the gut microbiota in patients with cirrhosis. *Aliment. Pharmacol. Ther.* **2018**, *48*, 1301–1311. [CrossRef]
41. Picca, A.; Ponziani, F.R.; Calvani, R.; Marini, F.; Biancolillo, A.; Coelho-Junior, H.J.; Gervasoni, J.; Primiano, A.; Putignani, L.; Del Chierico, F.; et al. Gut microbial, inflammatory and metabolic signatures in older people with physical frailty and sarcopenia: Results from the BIOSPHERE Study. *Nutrients* **2019**, *12*, 65. [CrossRef]
42. Addolorato, G.; Ponziani, F.R.; Dionisi, T.; Mosoni, C.; Vassallo, G.A.; Sestito, L.; Petito, V.; Picca, A.; Marzetti, E.; Tarli, C.; et al. Gut microbiota compositional and functional fingerprint in patients with alcohol use disorder and alcohol-associated liver disease. *Liver Int.* **2020**, *40*, 878–888. [CrossRef] [PubMed]
43. Picca, A.; Guerra, F.; Calvani, R.; Marini, F.; Biancolillo, A.; Landi, G.; Beli, R.; Landi, F.; Bernabei, R.; Bentivoglio, A.R.; et al. Mitochondrial Signatures in Circulating Extracellular Vesicles of Older Adults with Parkinson's Disease: Results from the EXosomes in PArkiNson's Disease (EXPAND) Study. *J. Clin. Med.* **2020**, *9*, 504. [CrossRef] [PubMed]
44. Cunnane, S.C.; Schneider, J.A.; Tangney, C.; Tremblay-Mercier, J.; Fortier, M.; Bennett, D.A.; Morris, M.C. Plasma and brain fatty acid profiles in mild cognitive impairment and Alzheimer's disease. *J. Alzheimer's. Dis.* **2012**, *29*, 691–697. [CrossRef] [PubMed]
45. Picca, A.; Calvani, R.; Landi, G.; Marini, F.; Biancolillo, A.; Gervasoni, J.; Persichilli, S.; Primiano, A.; Urbani, A.; Bossola, M.; et al. Circulating amino acid signature in older people with Parkinson's disease: A metabolic complement to the EXosomes in PArkiNson Disease (EXPAND) study. *Exp. Gerontol.* **2019**, *128*, 110766. [CrossRef] [PubMed]

46. Calvani, R.; Rodriguez-Mañas, L.; Picca, A.; Marini, F.; Biancolillo, A.; Laosa, O.; Pedraza, L.; Gervasoni, J.; Primiano, A.; Conta, G.; et al. Identification of a Circulating Amino Acid Signature in Frail Older Persons with Type 2 Diabetes Mellitus: Results from the Metabofrail Study. *Nutrients* **2020**, *12*, 199. [CrossRef] [PubMed]
47. Ståhle, L.; Wold, S. Partial least squares analysis with cross-validation for the two-class problem: A Monte Carlo study. *J. Chemom.* **1987**, *1*, 185–196. [CrossRef]
48. Biancolillo, A.; Marini, F.; Roger, J.M. SO-CovSel: A novel method for variable selection in a multiblock framework. *J. Chemom.* **2020**, *34*, e3120. [CrossRef]
49. Westerhuis, J.A.; Hoefsloot, H.C.J.; Smit, S.; Vis, D.J.; Smilde, A.K.; van Velzen, E.J.J.; van Duijnhoven, J.P.M.; van Dorsten, F.A. Assessment of PLSDA cross validation. *Metabolomics* **2008**, *4*, 81–89. [CrossRef]
50. Wold, S.; Sjöström, M. SIMCA: A Method for Analyzing Chemical Data in Terms of Similarity and Analogy. In *Chemometrics, Theory and Application*; Kowalski, B.R., Ed.; American Chemical Society: Washington, DC, USA, 1977; Volume 52, pp. 243–282. ISBN 9780841203792.
51. Westerhuis, J.A.; van Velzen, E.J.J.; Hoefsloot, H.C.J.; Smilde, A.K. Multivariate paired data analysis: Multilevel PLSDA versus OPLSDA. *Metabolomics* **2010**, *6*, 119–128. [CrossRef]
52. Rosenberg, P.B.; Lyketsos, C. Mild cognitive impairment: Searching for the prodrome of Alzheimer's disease. *World Psychiatry* **2008**, *7*, 72–78. [CrossRef]
53. Grundman, M.; Petersen, R.C.; Ferris, S.H.; Thomas, R.G.; Aisen, P.S.; Bennett, D.A.; Foster, N.L.; Jack, C.R.; Galasko, D.R.; Doody, R.; et al. Alzheimer's Disease Cooperative Study. Mild cognitive impairment can be distinguished from Alzheimer disease and normal aging for clinical trials. *Arch. Neurol.* **2004**, *61*, 59–66. [CrossRef] [PubMed]
54. Csukly, G.; Sirály, E.; Fodor, Z.; Horváth, A.; Salacz, P.; Hidasi, Z.; Csibri, É.; Rudas, G.; Szabó, Á. The Differentiation of Amnestic Type MCI from the Non-Amnestic Types by Structural MRI. *Front. Aging Neurosci.* **2016**, *8*, 52. [CrossRef] [PubMed]
55. Mosconi, L.; Perani, D.; Sorbi, S.; Herholz, K.; Nacmias, B.; Holthoff, V.; Salmon, E.; Baron, J.C.; De Cristofaro, M.T.R.; Padovani, A.; et al. MCI conversion to dementia and the APOE genotype: A prediction study with FDG-PET. *Neurology* **2004**, *63*, 2332–2340. [CrossRef] [PubMed]
56. Jack, C.R.; Petersen, R.C.; Xu, Y.C.; O'Brien, P.C.; Smith, G.E.; Ivnik, R.J.; Boeve, B.F.; Waring, S.C.; Tangalos, E.G.; Kokmen, E. Prediction of AD with MRI-based hippocampal volume in mild cognitive impairment. *Neurology* **1999**, *52*, 1397–1403. [CrossRef]
57. deToledo-Morrell, L.; Stoub, T.R.; Bulgakova, M.; Wilson, R.S.; Bennett, D.A.; Leurgans, S.; Wuu, J.; Turner, D.A. MRI-derived entorhinal volume is a good predictor of conversion from MCI to AD. *Neurobiol. Aging* **2004**, *25*, 1197–1203. [CrossRef]
58. Wang, W.; Cheng, X.; Lu, J.; Wei, J.; Fu, G.; Zhu, F.; Jia, C.; Zhou, L.; Xie, H.; Zheng, S. Mitofusin-2 is a novel direct target of p53. *Biochem. Biophys. Res. Commun.* **2010**, *400*, 587–592. [CrossRef]
59. Serra, L.; Giulietti, G.; Cercignani, M.; Spanò, B.; Torso, M.; Castelli, D.; Perri, R.; Fadda, L.; Marra, C.; Caltagirone, C.; et al. Mild cognitive impairment: Same identity for different entities. *J. Alzheimer's. Dis.* **2013**, *33*, 1157–1165. [CrossRef]
60. Jääskeläinen, O.; Hall, A.; Tiainen, M.; van Gils, M.; Lötjönen, J.; Kangas, A.J.; Helisalmi, S.; Pikkarainen, M.; Hallikainen, M.; Koivisto, A.; et al. Metabolic Profiles Help Discriminate Mild Cognitive Impairment from Dementia Stage in Alzheimer's Disease. *J. Alzheimer's Dis.* **2020**, 1–10. [CrossRef]
61. Olsson, B.; Lautner, R.; Andreasson, U.; Öhrfelt, A.; Portelius, E.; Bjerke, M.; Hölttä, M.; Rosén, C.; Olsson, C.; Strobel, G.; et al. CSF and blood biomarkers for the diagnosis of Alzheimer's disease: A systematic review and meta-analysis. *Lancet Neurol.* **2016**, *15*, 673–684. [CrossRef]
62. Bogie, J.F.J.; Haidar, M.; Kooij, G.; Hendriks, J.J.A. Fatty acid metabolism in the progression and resolution of CNS disorders. *Adv. Drug Deliv. Rev.* **2020**. [CrossRef]
63. Tynkkynen, J.; Chouraki, V.; van der Lee, S.J.; Hernesniemi, J.; Yang, Q.; Li, S.; Beiser, A.; Larson, M.G.; Sääksjärvi, K.; Shipley, M.J.; et al. Association of branched-chain amino acids and other circulating metabolites with risk of incident dementia and Alzheimer's disease: A prospective study in eight cohorts. *Alzheimer's Dement.* **2018**, *14*, 723–733. [CrossRef] [PubMed]
64. Wang, W.Y.; Tan, M.S.; Yu, J.T.; Tan, L. Role of pro-inflammatory cytokines released from microglia in Alzheimer's disease. *Ann. Transl. Med.* **2015**, *3*, 136. [CrossRef] [PubMed]

65. Kim, S.-M.; Song, J.; Kim, S.; Han, C.; Park, M.H.; Koh, Y.; Jo, S.A.; Kim, Y.-Y. Identification of peripheral inflammatory markers between normal control and Alzheimer's disease. *BMC Neurol.* **2011**, *11*, 51. [CrossRef] [PubMed]
66. Bermejo, P.; Martín-Aragón, S.; Benedí, J.; Susín, C.; Felici, E.; Gil, P.; Ribera, J.M.; Villar, A.M. Differences of peripheral inflammatory markers between mild cognitive impairment and Alzheimer's disease. *Immunol. Lett.* **2008**, *117*, 198–202. [CrossRef]
67. Justice, J.N.; Ferrucci, L.; Newman, A.B.; Aroda, V.R.; Bahnson, J.L.; Divers, J.; Espeland, M.A.; Marcovina, S.; Pollak, M.N.; Kritchevsky, S.B.; et al. A framework for selection of blood-based biomarkers for geroscience-guided clinical trials: Report from the TAME Biomarkers Workgroup. *GeroScience* **2018**, *40*, 419–436. [CrossRef]
68. Muir, S.W.; Speechley, M.; Wells, J.; Borrie, M.; Gopaul, K.; Montero-Odasso, M. Gait assessment in mild cognitive impairment and Alzheimer's disease: The effect of dual-task challenges across the cognitive spectrum. *Gait Posture* **2012**, *35*, 96–100. [CrossRef]
69. Pedersen, M.M.; Holt, N.E.; Grande, L.; Kurlinski, L.A.; Beauchamp, M.K.; Kiely, D.K.; Petersen, J.; Leveille, S.; Bean, J.F. Mild cognitive impairment status and mobility performance: An analysis from the Boston RISE study. *J. Gerontol. A. Biol. Sci. Med. Sci.* **2014**, *69*, 1511–1518. [CrossRef]
70. Verghese, J.; Wang, C.; Lipton, R.B.; Holtzer, R. Motoric cognitive risk syndrome and the risk of dementia. *J. Gerontol. A Biol. Sci. Med. Sci.* **2013**, *68*, 412–418. [CrossRef]
71. Semba, R.D.; Tian, Q.; Carlson, M.C.; Xue, Q.L.; Ferrucci, L. Motoric cognitive risk syndrome: Integration of two early harbingers of dementia in older adults. *Ageing Res. Rev.* **2020**, *58*, 101022. [CrossRef]
72. Boyle, P.A.; Buchman, A.S.; Wilson, R.S.; Leurgans, S.E.; Bennett, D.A. Association of muscle strength with the risk of Alzheimer disease and the rate of cognitive decline in community-dwelling older persons. *Arch. Neurol.* **2009**, *66*, 1339–1344. [CrossRef]
73. Del Campo, N.; Payoux, P.; Djilali, A.; Delrieu, J.; Hoogendijk, E.O.; Rolland, Y.; Cesari, M.; Weiner, M.W.; Andrieu, S.; Vellas, B. MAPT/DSA Study Group Relationship of regional brain β-amyloid to gait speed. *Neurology* **2016**, *86*, 36–43. [CrossRef]
74. Tian, Q.; Resnick, S.M.; Mielke, M.M.; Yaffe, K.; Launer, L.J.; Jonsson, P.V.; Grande, G.; Welmer, A.-K.; Laukka, E.J.; Bandinelli, S.; et al. Association of Dual Decline in Memory and Gait Speed With Risk for Dementia Among Adults Older Than 60 Years: A Multicohort Individual-Level Meta-analysis. *JAMA Netw. Open* **2020**, *3*, e1921636. [CrossRef]
75. Nourhashémi, F.; Andrieu, S.; Gillette-Guyonnet, S.; Reynish, E.; Albarède, J.-L.; Grandjean, H.; Vellas, B. Is there a relationship between fat-free soft tissue mass and low cognitive function? Results from a study of 7105 women. *J. Am. Geriatr. Soc.* **2002**, *50*, 1796–1801. [CrossRef] [PubMed]
76. Cabett Cipolli, G.; Sanches Yassuda, M.; Aprahamian, I. Sarcopenia Is Associated with Cognitive Impairment in Older Adults: A Systematic Review and Meta-Analysis. *J. Nutr. Health Aging* **2019**, *23*, 525–531. [CrossRef] [PubMed]
77. Rovio, S.; Kåreholt, I.; Helkala, E.-L.; Viitanen, M.; Winblad, B.; Tuomilehto, J.; Soininen, H.; Nissinen, A.; Kivipelto, M. Leisure-time physical activity at midlife and the risk of dementia and Alzheimer's disease. *Lancet Neurol.* **2005**, *4*, 705–711. [CrossRef]
78. Scarmeas, N.; Luchsinger, J.A.; Schupf, N.; Brickman, A.M.; Cosentino, S.; Tang, M.X.; Stern, Y. Physical activity, diet, and risk of Alzheimer disease. *JAMA* **2009**, *302*, 627–637. [CrossRef]
79. Xu, W.; Wang, H.F.; Wan, Y.; Tan, C.-C.; Yu, J.-T.; Tan, L. Leisure time physical activity and dementia risk: A dose-response meta-analysis of prospective studies. *BMJ Open* **2017**, *7*, e014706. [CrossRef]

Article

Targeted Nutritional Intervention for Patients with Mild Cognitive Impairment: The Cognitive impAiRmEnt Study (CARES) Trial 1

Rebecca Power [1,*], John M. Nolan [1], Alfonso Prado-Cabrero [1], Robert Coen [2], Warren Roche [1], Tommy Power [1], Alan N. Howard [3] and Ríona Mulcahy [4,5,*]

1 Nutrition Research Centre Ireland, School of Health Sciences, Carriganore House, Waterford Institute of Technology West Campus, X91 K0EK Waterford, Ireland; jmnolan@wit.ie (J.M.N.); aprado-cabrero@wit.ie (A.P.-C.); warren.roche@ucdconnect.ie (W.R.); tbpower@wit.ie (T.P.)

2 Mercer's Institute for Research on Ageing, St. James's Hospital, D08 NHY1 Dublin, Ireland; rcoen@stjames.ie

3 Howard Foundation, 7 Marfleet Close, Great Shelford, Cambridge CB22 5LA, UK; alan.howard@howard-foundation.com

4 Age-Related Care Unit, Health Service Executive, University Hospital Waterford, Dunmore Road, X91 ER8E Waterford, Ireland

5 Royal College of Surgeons Ireland, 123 Stephen's Green, Saint Peter's, D02 YN77 Dublin, Ireland

* Correspondence: rpower@wit.ie (R.P.); drriona.mulcahy@hse.ie (R.M.); Tel.: +353-01-845-505 (R.P.); +353-51-842-509 (R.M.)

Received: 31 March 2020; Accepted: 18 May 2020; Published: 25 May 2020

Abstract: Omega-3 fatty acids (ω-3FAs), carotenoids, and vitamin E are important constituents of a healthy diet. While they are present in brain tissue, studies have shown that these key nutrients are depleted in individuals with mild cognitive impairment (MCI) in comparison to cognitively healthy individuals. Therefore, it is likely that these individuals will benefit from targeted nutritional intervention, given that poor nutrition is one of the many modifiable risk factors for MCI. Evidence to date suggests that these nutritional compounds can work independently to optimize the neurocognitive environment, primarily due to their antioxidant and anti-inflammatory properties. To date, however, no interventional studies have examined the potential synergistic effects of a combination of ω-3FAs, carotenoids and vitamin E on the cognitive function of patients with MCI. Individuals with clinically confirmed MCI consumed an ω-3FA plus carotenoid plus vitamin E formulation or placebo for 12 months. Cognitive performance was determined from tasks that assessed global cognition and episodic memory. Ω-3FAs, carotenoids, and vitamin E were measured in blood. Carotenoid concentrations were also measured in tissue (skin and retina). Individuals consuming the active intervention ($n = 6$; median [IQR] age 73.5 [69.5–80.5] years; 50% female) exhibited statistically significant improvements ($p < 0.05$, for all) in tissue carotenoid concentrations, and carotenoid and ω-3FA concentrations in blood. Trends in improvements in episodic memory and global cognition were also observed in this group. In contrast, the placebo group ($n = 7$; median [IQR] 72 (69.5–75.5) years; 89% female) remained unchanged or worsened for all measurements ($p > 0.05$). Despite a small sample size, this exploratory study is the first of its kind to identify trends in improved cognitive performance in individuals with MCI following supplementation with ω-3FAs, carotenoids, and vitamin E.

Keywords: mild cognitive impairment; nutrition; omega-3 fatty acids; antioxidant; carotenoids; vitamin E; cognition; episodic memory; older adults; ageing

1. Introduction

Given the growing social and economic burden of cognitive decline on society, emphasis is being placed on preventative strategies to delay the onset and reduce the risk of developing dementia, with particular focus on Alzheimer's disease (AD) as it is the most common form of dementia. Mild cognitive impairment (MCI) is often a transitional phase between the cognitive changes that one expects as one ages and very early dementia. It is recognized as a deterioration in cognitive function that exceeds what is anticipated for an individual based on their age and education level. Importantly, these changes in cognition are not significant enough to impact an individual's independence or ability to perform activities of daily living [1]. MCI is difficult to diagnose, and its prognosis is notoriously unpredictable. While the mortality rate is higher in MCI patients in comparison to cognitively healthy individuals [2], it is comparable to dementia mortality rates [3]. Although MCI is a risk factor for AD (with MCI to dementia conversion rates estimated at 3%–15% annually [4]), it is important to note that some individuals with the condition remain stable and do not progress while others may improve (i.e., revert to a cognitively intact state) upon follow-up assessment. This reversion phenomenon is an inherent feature of MCI and may be explained by the heterogeneity of the condition. While reverting from MCI to a cognitively intact state seems like a positive outcome, importantly, a number of studies have shown that these individuals are in fact at a greater risk of cognitive decline in the future [5,6]. Thus, due to the increased risk of mortality and progression to AD, MCI is an important public health concern.

Despite its complex and dynamic nature, MCI offers a window of opportunity to examine the potential of preventative strategies for modifying or delaying disease progression and improving cognitive outcomes. Given that many risk factors (e.g., vascular disease, diabetes, smoking, physical inactivity, social isolation [7–9]) for MCI are modifiable, shifting focus towards preventative strategies seems prudent. Specifically, there is a growing body of evidence to suggest that good nutrition is important for cognitive performance [10–14], and is associated with a reduced risk of MCI and AD [15–18]. Omega-3 fatty acids (ω-3FAs), xanthophyll carotenoids (oxygen-containing, plant-based pigments), and vitamin E are important constituents of a healthy diet. While these specific nutrients are present in brain tissue [19–21], studies have shown that they are depleted in individuals with MCI and AD in comparison to cognitively healthy individuals [22–24]. Therefore, it is likely that specific population groups (e.g., individuals with MCI, very early-stage AD or individuals with a low ω-3FA or carotenoid index) will benefit from targeted nutritional intervention. Indeed, observational [20,25–29] and interventional [30–33] evidence to date suggests that these nutritional compounds can work independently to optimize the neurocognitive environment [34], primarily due to their antioxidant and anti-inflammatory properties. Interestingly, previous exploratory work has shown that a combination of ω-3FAs and xanthophyll carotenoids can work synergistically to improve cognition in older women [35], and maintain function and quality of life in AD patients [36]. To date, however, no interventional studies have examined the potential synergistic effects of a combination of ω-3FAs, xanthophyll carotenoids, and vitamin E on the cognitive health and function of patients with MCI.

The present study, the Cognitive impAiRmEnt Study (CARES), was designed to investigate the impact of targeted nutritional intervention with ω-3FAs, xanthophyll carotenoids, and vitamin E on cognitive function among individuals with MCI. CARES was a parallel group, double-blind, placebo-controlled, randomized clinical trial studying two populations of interest. The first arm of the trial (CARES Trial 1) examined the impact of targeted nutritional supplementation on cognitive function in individuals with MCI, while the second arm of the trial (CARES Trial 2) investigated the impact of targeted nutritional supplementation on cognitive function in cognitively heathy older adults ($\geq$65 years). Herein, exploratory work from CARES Trial 1 is presented and discussed.

2. Materials and Methods

2.1. Study Design

CARES Trial 1 investigated the impact of 12-month supplementation with ω-3FAs, xanthophyll carotenoids, and vitamin E on cognitive function in individuals with MCI. Individuals were initially identified as potentially suitable for enrolment based on a medical assessment performed by consultant geriatricians and psychiatrists of old age in the South-East catchment area of Ireland. Both amnestic and non-amnestic MCI were included. MCI sub-type classification was not performed. A diagnosis of MCI was based on published criteria [37,38]. Specific eligibility criteria included: self or family member reported memory loss; fulfilled criteria for minimal cognitive impairment; functionally independent in activities of daily living; ≥65 years of age; no rapidly progressive or fluctuating symptoms of memory loss; no established diagnosis of early dementia (consumption of cognitive enhancement therapy such as cholinesterase inhibitors or N-methyl-D-aspartate receptor antagonists); no stroke disease (clinical stroke or stroke on CTB); no depression (under active review); no psychiatric illness (under active review of psychotropic medications); no glaucoma (acute angle); no consumption of carotenoid or fish/cod liver oil supplements; and no fish allergy.

2.2. MCI Screening

Prior to enrolment, all individuals who expressed an interest in participating in the trial completed a screening assessment to confirm eligibility. This included assessing cognitive function using the Repeatable Battery for the Assessment of Neurological Status (RBANS) Record Form A and the Montreal Cognitive Assessment (MoCA) version 7.1. Level of functional ability was assessed using the Bristol Activities of Daily Living Scale (BADLS) and the Alzheimer's Questionnaire (AQ). A brief description of each of these assessments is provided below. In the event where an informant was not present during the assessment, a family member or carer was contacted via telephone to complete functional ability assessments. In circumstances where no informant was available, the researcher administered the questionnaires to the patient. Individuals who fulfilled the criteria for each cognitive and functional assessment were invited to participate in the clinical trial. Individuals with borderline scores from the screening assessments were referred to a consensus panel (via email or conference call) consisting of one consultant geriatrician, one psychiatrist of old age and one clinical neuropsychologist for assessment of eligibility [39]. Eligible individuals were then invited to enroll into the study (see Figure 1). Prior to enrolment, written informed consent was obtained from all individuals. Ethical approval was granted by the Research Ethics Committees of the Waterford Institute of Technology and University Hospital Waterford in Waterford, Ireland in December 2015. CARES (trial registration number: ISRCTN10431469) adhered to the tenets of the Declaration of Helsinki and followed the full code of ethics with respect to recruitment, testing and general data protection regulations as set out by the European Parliament and Council of the European Union.

Eligible individuals were randomized to either the active intervention (now commercially known as Memory Health) containing 1 g of fish oil (of which 430 mg docosahexaenoic acid [DHA] and 90 mg eicosapentaenoic acid [EPA]), the xanthophyll carotenoids lutein (L) (10 mg), *meso*-zeaxanthin (MZ) (10 mg) and zeaxanthin (Z) (2 mg), and 15 mg of vitamin E (α-tocopherol), or placebo (sunflower oil) intervention group. These doses were provided via two oval-size capsules. Each capsule contained equal quantities of fish oil, carotenoids and vitamin E (see Table A1 in the Appendix A section). Carotenoid and vitamin E concentrations were manufactured by Industrial Orgánica (Monterrey, Mexico), while fish oil concentrations were manufactured by Epax (Ålesund, Norway; product number: EPAX1050TG/N non-tuna). The complete formula composition and the concentration of fatty acids of total lipids are available in the Appendix A section (Tables A1 and A2, respectively in Appendix A). Individuals were instructed to consume two capsules per day with a meal. Frequent phone calls were made to ensure compliance. Tablet counting was also performed at follow-up. Study visits were conducted at baseline and 12 months at a single site (Nutrition Research

Centre Ireland [NRCI]). Intervention randomization was performed by an electronic trial management system (Trial Controller) designed by our research group (NRCI). This administration system was also used to document patient information (name and contact details), support the organization and management of capsules required for the clinical trial and assist with the scheduling of study visits. The primary outcome measure of CARES Trial 1 was change in cognitive function. Secondary outcome measures included change in the following variables: macular pigment optical volume (MPOV); visual function; serum xanthophyll carotenoid concentrations (L, Z and MZ); serum vitamin E concentrations (α-tocopherol); and plasma ω-3FA concentrations (EPA and DHA). Development of AD was also recorded.

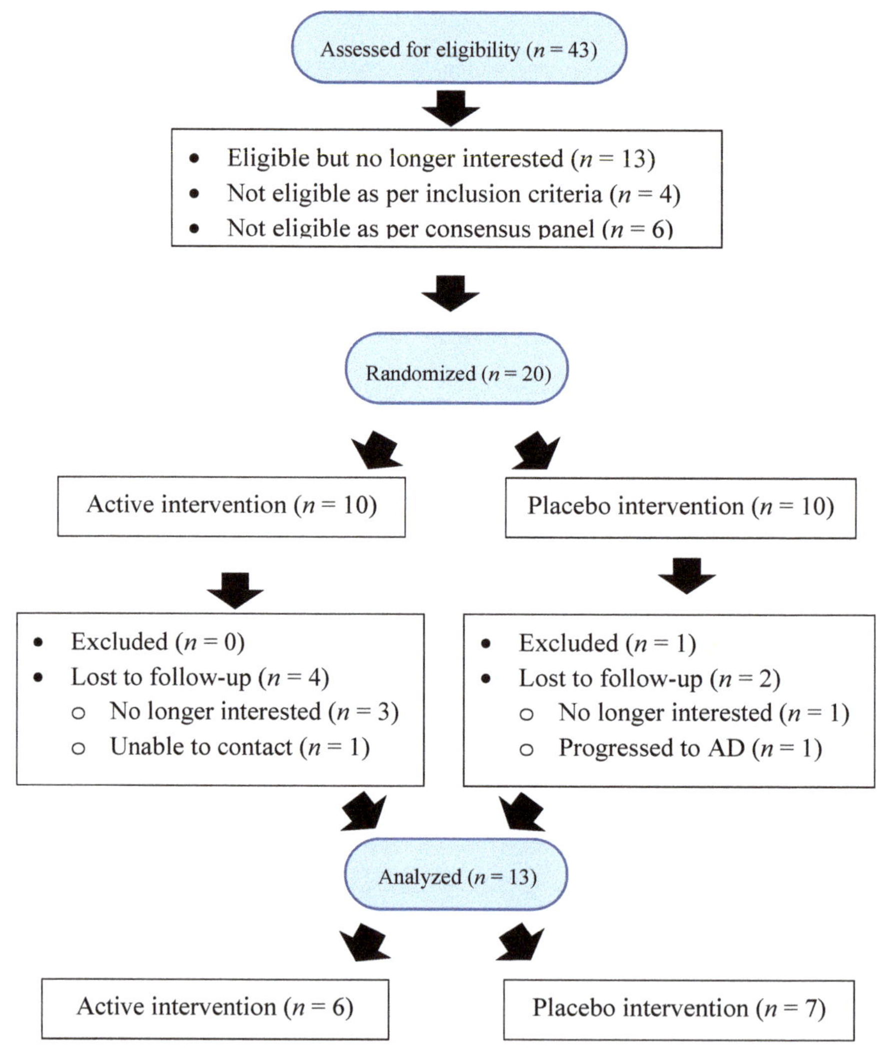

Figure 1. Consolidated Standards of Reporting Trials (CONSORT) flow diagram for the Cognitive impAiRmEnt Study (CARES) Trial 1.

2.3. Assessing Cognitive Function

2.3.1. Global Cognition

The MoCA was used at the screening stage to assess global cognition. It is a short (10-min) cognitive screening tool with high sensitivity and specificity for detecting MCI [40]. Thirty items assess multiple cognitive domains including visuospatial abilities, executive function, phonemic fluency, attention, immediate and delayed recall, language and orientation. The RBANS was used to measure global cognition at screening and at 12-month follow-up visits. Five domains of cognition (immediate memory, visuospatial ability, language, attention and delayed memory) were assessed using 12 sub-tests. A composite or "total index/scale score" was also computed. The RBANS takes approximately 30 min to administer and is a core diagnostic tool for detecting and characterizing dementia [41]. The RBANS yields index standard scores that are based on the raw scores of each subtest. RBANS index scores are metrically scaled, with a mean of 100 and a standard deviation (SD) of 15 for each age group. A score of 100 on any of these measures equates to the average performance of individuals of similar age. Scores of 85 and 115 correspond to 1 SD below and above the mean, respectively, while scores of 70 and 130 are 2 SDs below and above the mean. Approximately 68% of all examinees score between 85 and 115, circa 95% score in the 70 to 130 range, and nearly all examinees obtain scores between 55 and 143 [42]. In the present study, scores < 78 and between 19 and 25 for the RBANS and the MoCA, respectively, were desirable for enrolment.

2.3.2. Specific Cognitive Domains

Additional assessments of specific cognitive domains were performed using the Cambridge neuropsychological test automated battery (CANTAB) Connect Research Software (Cambridge Cognition, Cambridge, UK) [43]. This computerized software program was performed on an iPad and required a finger-operated response. This technology has been previously tested and validated in older adult population groups [44]. The CANTAB protocol [45] was followed in the administration of the test battery and was used to assess comprehension, executive function (working memory), attention (reaction time) and episodic memory at baseline and follow-up visits. Table 1 provides an overview of the CANTAB tests performed.

Spatial memory was also assessed at the screening stage only using the 4 mountains test (4MT) [46]. Using a delayed match-to-sample paradigm, memory for the topographical layout of 4 mountains within a computer-generated landscape is tested. Individuals were asked to recall the spatial configuration of a total of 15 sets of computer-generated landscapes from a shifted viewpoint, which is designed to reflect the role of the hippocampus in spatial cognition. This computerized assessment was performed on an iPad and required a finger operated response. The test takes approximately 20 min to complete and has been used previously among individuals with MCI and AD [47]. A study by Moodley and colleagues [48] suggested that a total 4MT score of ≤8 was associated with 100% sensitivity and 90% specificity for detecting early AD when tested in a UK population, and associated with 100% sensitivity and 50% specificity for detection of MCI and AD when tested in an Italian population group.

Table 1. Tasks performed in CARES to assess cognition using the Cambridge neuropsychological test automated battery (CANTAB).

Cognitive Domain	Task	Description	Outcome Measure	Desirable Score	Performance Ranges
Comprehension	MOT	Individuals must touch the flashing cross shown in different locations on the screen.	Latency (speed of response) Total correct Total errors	Lower Higher Lower	- 0–10 0–10
Executive function (working memory)	SWM	The aim of this test is that, by touching the boxes and using a process of elimination, the individual should find one 'token' in each of the boxes and use them to fill up an empty column on the right-hand side of the screen. The key task instruction is that the computer will never hide a token in the same coloured box, so once a token is found in a box the individual should not return to that box to look for another token.	Between errors Total errors Strategy	Lower Lower Lower	0–90 0–90 2–14
Reaction time	RTI	Individuals must press and hold down a touchscreen button at the bottom of the screen. Circles are presented at the top of the screen (one for simple mode, and five for the five-choice mode). In each case, a yellow spot will appear in one of the circles. Individuals must respond to the spot as quickly as they can by letting go of the button and touching inside the circle where the yellow spot appeared.	Simple reaction time Simple movement time Simple error score Five-choice reaction time Five-choice movement time Five-choice error score	Lower Lower Lower Lower Lower Lower	100–5100 100–5100 0–30 100–5100 100–5100 0–30
Episodic memory	PAL	Boxes are displayed on the screen and open one by one in a randomized order to reveal patterns hidden inside. The patterns are then displayed in the middle of the screen, one at a time, and the individual must touch the box where the pattern was originally located. If the individual makes an error, the patterns are re-presented to remind the individual of their locations.	First attempt memory score No. patterns reached Total errors adjusted	Higher Higher Lower	0–20 2–8 0–70

Performance ranges: the minimum and maximum possible score for each measure; Latency: the speed (milliseconds) of response to the stimulus; Total correct: the number of correct responses; Total errors: the distance between the center of the cross and the location touched; Between errors: the number of times the individual revisits a box in which a token has previously been found; Total errors: the number of times a box is selected that is certain not to contain a token and therefore should not have been visited by the individual; Strategy: for problems with six boxes or more. The number of distinct boxes used by the individual to begin a new search for a token, with the same problem; Simple reaction time: the duration between the onset of the stimulus and the time at which the individual released the button. Calculated for correct trials, where the stimulus could appear in one location only; Simple movement time: the time taken to touch the stimulus after the button has been released. Calculated for correct trials, where the stimulus could appear in one location only; Simple error score: the number of trials where the response status is any error (i.e., an inaccurate response, no response, or a premature response) for the assessment trial where stimuli appear on one location only. The error may be an inaccurate response, no response, or a premature response; Five-choice reaction time: the duration between the onset of the stimulus and the release of the button. Calculated for correct, assessed trials where the stimulus could appear in any of the five locations; Five-choice movement time: the time taken to touch the stimulus after the button has been released. Calculated for correct, assessed trials where the stimulus could appear in any of the five locations; Five-choice error score: the number of trials where the response status is any error (i.e., an inaccurate response, no response, or a premature response) for assessment trials where stimuli appear in any of five locations; First attempt memory score: the number of correct box choices that were made on the first attempt during assessment problems; No. patterns reached: the number of patterns on the last problem reached in the task; Total errors adjusted: the number of times the individual chose the incorrect box for a stimulus on assessment problems, plus an adjustment for the estimated no. errors they would have made on any problems, attempts and recalls they did not reach.

2.4. Assessing Functional Ability

The BADLS is an informant-based, 20-item questionnaire designed to measure the ability of an individual with dementia to carry out activities of daily living such as washing, dressing, preparing food and using transportation [49,50]. It is sensitive to changes in dementia and is regularly used as an outcome measure in clinical trials, where it is world leading as a dementia-specific measure. This outcome is among those recommended by a consensus recommendation of outcome scales for non-drug interventional studies in dementia [51]. A higher score was desirable for this assessment. The AQ is an informant-based screening tool used to detect cognitive impairment. It is regarded as a time efficient and sensitive measure for detecting MCI using structured interview-based questions. The AQ consists of 21 yes/no questions in a weighted format relevant to five different domains: memory, orientation, functional ability, visuospatial and language. The total score is calculated by summing the number of items with a "yes" response. Clinical symptoms known to be highly predictive of AD are given a greater weight in the total score. A score between 5 and 14 points was desirable for this assessment. The AQ has been previously validated and has shown high sensitivity and specificity for detecting MCI [52,53].

2.5. Assessing Nutritional Status

2.5.1. Macular Pigment

The carotenoids L, Z and MZ are preferentially concentrated in the central retina (macula lutea, which is part of the central nervous system) where they are collectively referred to as macular pigment (MP). MP was measured by dual-wavelength autofluorescence (AF) using the Spectralis HRA+OCT MultiColor (Heidelberg Engineering GmbH, Heidelberg, Germany). Pupillary dilation was performed prior to measurement and patient details were entered into the Heidelberg Eye Explorer (HEYEX version 1.7.1.0) software. Dual-wavelength AF in this device uses two excitation wavelengths; one that is well absorbed by MP (486 nm, blue) and one that is not (518 nm, green) [54]. The following acquisition parameters were used: high speed scan resolution, two seconds cyclic buffer size, internal fixation, 30-s movie and manual brightness control. Alignment, focus and illumination were first adjusted in infrared mode. Once the image was evenly illuminated, the laser mode was switched from infrared to blue plus green laser light AF. Using the HEYEX software, the movie images were aligned and averaged, and a MP density map was created. MPOV, calculated as MP average times the area under the curve out to 7° eccentricity [55], is reported here. This system has recently been validated by our research center [56].

2.5.2. Skin Carotenoid Score

Carotenoid concentrations were also measured using the Pharmanex® BioPhotonic Scanner (Salt Lake City, UT, USA). This scanner measures carotenoid levels in human tissue at the skin surface using optical signals (resonant Raman spectroscopy). These signals identify the unique molecular structure of carotenoids, allowing their measurement without interference by other molecular substances. The individual was asked to place a specific point (between the maximal and distal palmar creases, directly below the fifth finger) of their right hand (previously cleaned with hand sanitizer) in front of the scanner's low-energy blue light for 30 s. From this, a skin carotenoid score (SCS) was generated. This provided an indication of the individual's overall antioxidant levels. This was repeated twice more, and an average score was calculated. Based on this result, an individual's score can be classified into three ranges: 0–29,000 = low; 30,000–49,000 = normal; ≥50,000 = high. This technology is safe and has been previously validated [57].

2.6. Biochemical Analysis of Serum Xanthophyll Carotenoids and Vitamin E

2.6.1. Serum Extraction

Non-fasting blood samples were collected at each study visit by standard venipuncture techniques. SST II Advance blood collection tubes (8.5 mL) were inverted at least 5 times to ensure thorough mixing of the silica clot activator. The blood samples were left to clot for 30 min at room temperature and then centrifuged at room temperature at 725 g for 10 min in a Gruppe GC12 centrifuge (Desaga Sarstedt, UK) to separate the serum from the whole blood. Following centrifugation, serum was transferred to light-resistant microtubes and stored at circa −80 °C until extraction. Xanthophyll carotenoids and α-tocopherol were extracted from serum samples as previously described [58] and analyzed by high performance liquid chromatography (HPLC).

2.6.2. Lutein, Zeaxanthin and α-Tocopherol Quantification (Assay 1)

The chromatographic analysis of carotenoids and α-tocopherol was performed on an Agilent 1260 Series HPLC (Agilent Technologies Limited, Santa Clara, CA, USA) equipped with a quaternary pump, autosampler, thermostat column compartment and a photodiode array detector monitoring a wavelength of 450 nm for serum carotenoids and 292 nm for α-tocopherol and the internal standard (IS) α-tocopheryl acetate. The dried samples were reconstituted in 0.2 mL of Methanol:MTBE (9:1, *v/v*), vortexed at the lowest setting for 1 min and pipetted into HPLC vials containing 0.35 mL glass inserts. 0.1 milliliters of each sample was injected in a C30 carotenoid column (250 × 4.6 mm i.d., 3 μm; YMC Europe, Dinslaken, Germany) with a guard column of the same chemistry. HPLC mobile phase A consisted of methanol:MTBE:water (83:15:2, *v/v*), and mobile phase B consisted of methanol:MTBE:water (8:90:2, *v/v*), both with 0.1% BHT. At a flow rate of 1 mL min^{-1}, the gradient initiated at 5% solvent B and increased to 20% in the first 12 min, to 55% over the next 8 min and to 95% over the next 7 min. From 27–30 min, solvent B was held at 95%, and then resumed to initial setting within 3 min. Separations were carried out at 16 °C. Total Z from each sample was automatically collected by the fraction collector in amber eppendorfs.

2.6.3. *Meso*-Zeaxanthin Quantification (Assay 2)

Total Z collected in HPLC system 1 was dried in a vacuum centrifuge and re-suspended in 0.2 mL of hexane:isopropanol (90:10, *v/v*). 0.1 milliliters of the sample was analyzed on another Agilent 1260 Series HPLC system equipped with a Diode Array Detector, binary pump, degasser, thermostatically controlled column compartment and thermostatically controlled high-performance autosampler. The column used for the separation of the stereoisomers of Z was a Daicel Chiralpak IA-3 column, composed of amylose tris (3,5-dimethylphenylcarbamate) bonded to 3 mm silica gel (250 × 4.6 mm i.d.; Chiral Technologies Europe, Cedex, France). The column was protected with a guard column containing a guard cartridge with the same chemistry of the column. Isocratic elution was performed with hexane:isopropanol (90:10, *v/v*) at a flow rate of 0.5 mL min^{-1}. The column temperature was set at 20 °C.

Quantification was performed by constructing a calibration line for each xanthophyll carotenoid analyzed and for α-tocopherol. For each compound of interest, at least five calibration standards were quantified using a UV-Vis spectrophotometer UVmini-1240 (Shimadzu, Kyoto, Japan) with the appropriate molar extinction coefficient (see Appendix A Table A3). These calibrators were analyzed using HPLC system 1 in triplicate, whereas the calibrator of lowest concentration was injected 10 times in order to experimentally calculate the lower limit of quantification (LLOQ) for the compound. The upper limit of quantification (ULOQ) was allocated as the calibrator of highest concentration of each calibration curve analyzed in triplicate (see Appendix A Table A4). Where possible, subject samples that displayed an area in the HPLC chromatogram below the LLOQ or above the ULOQ were re-analyzed in order to obtain an area within the range of the calibration line. If after re-analysis the area of the analyte of interest remained below the LLOQ, this analyte in the subject was marked as 'below LLOQ'.

In order to determine the efficiency and precision of the xanthophyll carotenoid and α-tocopherol quantification methodology, analyte recovery analysis, precision analysis and trueness of sample recovery were performed. Details of this analysis are outlined in the Appendix A section.

2.7. Biochemical Analysis of Plasma Omega-3 Fatty Acids

2.7.1. Plasma Extraction

Lithium heparin blood collection tubes (6 mL) were inverted 8–10 times to ensure thorough mixing and were centrifuged at 4 °C at 3000 rpm for 20 min in a 3–18 K centrifuge (Sigma-Aldrich, St. Louis, MO, USA) to separate red blood cells and plasma. The time of blood collection and time of separation did not exceed 2 h. Following centrifugation, all samples were transferred to light-resistant microtubes and stored at circa −80 °C until the time of analysis. Plasma ω-3FA analysis was performed by gas chromatography (GC). Fatty acid methyl esters (FAME) were prepared as previously described [59]. Briefly, 50 μL of plasma were spiked with 20 μL of 2 mg/mL methyl tricosanoate (Larodan, Solna, Sweden) to assess FAME recovery and saponified with 2 mL of freshly prepared methanolic KOH 0.4 M during 10 min with gentle vortexing at room temperature. The samples were extracted three times with 2 mL of hexane and the combined extracts were dried in a vacuum centrifuge. The pellets were esterified with 2 mL of freshly prepared 5% methanolic sulfuric acid (*v*/*v*) at 80 °C for 30 min in a thermo-block. The FAME produced were extracted three times with 2 mL of hexane and dried in the vacuum centrifuge. The samples were resuspended in 0.4 mL of hexane containing 0.1 mg/mL of methyl heneicosanoate (Larodan) to assess the matrix effect and prepared for GC analysis. Methyl tricosanoate and methyl heneicosanoate 0.1 mg/mL were injected in triplicate to assess recovery and matrix effect, respectively.

2.7.2. DHA and EPA Quantification

FAME were quantified by GC coupled to flame ionization detector (GC-FID) with an Agilent 7890B Gas Chromatographer, using a Thermo 260M142P column (cyanopropylphenyl-based phase, 30 m length, 0.25 mm inner diameter and 0.25 μm film thickness). Nitrogen was used as the carrier gas with a flow rate of 1.5 mL/min and an electronic pressure control at 20.8 psi. Temperature ramp started at 140 °C and was held for 1 min, then followed by an increase of 6 °C min^{-1} until 210 °C, an increase of 2.5 °C min^{-1} until 230 °C and finally an increase of 10 °C min^{-1} until 240 °C, which was maintained for 5 min. Total run time was 26.7 min, with post run temperature at 50 °C and maximum temperature at 250 °C. FAME were identified by comparison with the authentic standard Mixture ME 1220 (Larodan). For FAME quantification, an RF was calculated as follows: a calibration line for methyl docosanoate, methyl undecanoate, methyl heptadecanoate, methyl heneicosanoate, methyl tricosanoate and EPA was prepared with a concentration range of 0.0025–0.5 mg/mL and analyzed in the GC. The resulting calibration lines were forced to pass through the origin of the axes, and the resulting slopes were averaged. The resulting RF was 0.1572 ± 0.0125.

2.8. Additional Biochemical Analysis

Serum and plasma samples were also collected to measure sodium, potassium, chloride, creatinine, total cholesterol, triglycerides, HDL, LDL, folate, vitamin B12, homocysteine, C-reactive protein, thyroid stimulating hormone, and free T4 (see Table A9 in the Appendix A section). One K2EDTA blood collection tube (3 mL) was also used for whole blood analysis. The sample was inverted 8–10 times and refrigerated at 4 °C until sample collection (2–24 h later). This additional analysis was performed by an accredited medical testing service provider (Eurofins Biomnis, Dublin, Ireland).

2.9. Demographic, Health and Lifestyle Data

Demographic, health and lifestyle data, medical history and medication use were recorded via questionnaire. Height and weight measurements were recorded to calculate body mass index (BMI) (kg/m^2). Smoking status was classified into never (smoked < 100 cigarettes in lifetime), past (smoked ≥ 100 cigarettes in lifetime and none in the past year) or current (smoked ≥ 100 cigarettes in lifetime and at least 1 cigarette in the last year) smoker. Alcohol consumption was measured in unit intake per week. One unit of alcohol (10 mL) was the equivalent to one of the following: a single measure of spirits (ABV 37.5%); half a pint of average-strength (4%) lager; two-thirds of a 125 mL glass of average-strength (12%) wine; half a 175 mL glass of average-strength (12%) wine; a third of a 250 mL glass of average-strength (12%) wine. Color fundus photographs were taken to assess the presence of ocular pathology (Zeiss Visucam 200, Carl Zeiss Meditec AG, Jena, Germany).

2.10. Statistical Analysis

The statistical package IBM SPSS version 25 was used and the 5% significance level applied for all analyses. Given that data were not normally distributed, the small sample size and the presence of ranked data, a non-parametric approach was taken. Results were expressed as median (inter quartile range [IQR]) for all variables. Between-group differences (i.e., active versus placebo) were analyzed using Mann–Whitney U, Wilcoxon Signed Rank or Chi-square tests, as appropriate. The Mann–Whitney U test was also used to examine the significance of change in nutrition variables over time between active and placebo intervention groups. Significance values were not computed to examine change in cognition or vision variables over time between both groups due to a lack of statistical power and the small magnitude of change over time observed for these variables. As an alternative, the average percentage change per subject was reported. Of note, percentage change could not be calculated for some variables (e.g., episodic memory) as baseline values were recorded as 0. Thus, the average change per subject was reported.

3. Results

3.1. Baseline Results

Table 2 presents the baseline demographic, health and lifestyle data for active and placebo intervention groups. Table 3 presents the baseline cognitive function and functional ability for active and placebo intervention groups. Baseline variables were statistically comparable between both groups, with the exception of the number of between errors ($p = 0.006$) and total errors ($p = 0.012$) made at stage 8 of the SWM tasks, which were significantly higher in the active group. Additionally, creatinine levels were significantly higher in the active group at baseline ($p = 0.008$), but were within normal ranges (see Table A9 in the Appendix A section). No comprehension or sensorimotor difficulties were observed during the CANTAB assessment (see Materials and Methods section), as the motor screening task (MOT) latency assessment was completed by all individuals at baseline and follow-up. No adverse events were reported by individuals in the active or placebo intervention groups during the trial.

3.2. Observed Change in Nutritional Status

Table 4 summarizes the observed change in nutrition variables for both groups following the 12-month intervention period. Figure 2A–C illustrate the observed changes in MPOV, serum L concentrations and plasma DHA concentrations, respectively. Individuals in the active intervention group exhibited statistically significant improvements in MPOV (62% improvement versus 2% decline for active and placebo groups, respectively; $p = 0.001$) and SCS (79% improvement versus 2% improvement for active and placebo groups, respectively; $p = 0.014$) in comparison to individuals in the placebo group. In terms of biochemical response, individuals in the active intervention group exhibited statistically significant improvements in serum carotenoid concentrations of L and MZ, as well as statistically significant improvements in plasma concentrations of DHA ($p < 0.05$, for all) in comparison to individuals receiving placebo. Serum Z and plasma EPA levels increased in both groups; however, results were not statistically significant ($p > 0.05$, for all). A mixed response to vitamin E supplementation was observed in blood, where levels decreased (−4%) over 12 months in the active intervention group and increased slightly (+1%) in the placebo group.

3.3. Observed Change in Global Cognition

Table 5 shows trends in improvements (ranging from 6% to 18%) in global cognition (as per the RBANS assessment tool) in the active intervention group after 12 months. Specifically, trends in improvements were observed in the immediate memory, attention and delayed memory domains, as well as the total scale score. minor declines were denoted in the visuospatial and language domains (both by 1%). Global cognition results were mixed in the placebo group. Immediate memory, visuospatial and attention domains of the RBANS remained unchanged while language, delayed memory and total scale scores improved after 12 months. Further analysis of the RBANS delayed memory domain in the placebo group suggested that the observed improvement (i.e., a 14% improvement) was driven by one subject. When this subject was removed, an improvement of 4% was denoted. As an example, Figure 3 shows the change in individual scores recorded for the immediate memory domain of the RBANS in both groups.

Table 2. Baseline demographic, health and lifestyle data of active and placebo intervention groups.

Variable	Active (n = 10) Median (IQR)	Placebo (n = 9) Median (IQR)	Sig.
Demographic data			
Age (years)	73.5 (69.5–80.5)	72.0 (69.5–75.5)	0.549
Sex ([n]; [% female])	5 (50.0%)	8 (88.9%)	0.069
Education (years)	17.5 (15.5–21.0)	15.0 (15.0–16.5)	0.095
Health and lifestyle data			
Medications	6.0 (3.0–8.3)	5.0 (2.0–5.5)	0.133
Exercise (min/week)	217.0 (0–326.3)	210.0 (45.0–375.0)	0.720
Smoking ([n]; [%])			0.463
Never	5 (50%)	6 (66.7%)	
Past	5 (50%)	3 (33.3%)	
Current	0	0	
Alcohol consumption ([n]; [%])			0.473
0 units	5 (50.0%)	4 (44.4%)	
1 unit	3 (30.0%)	2 (22.2%)	
2–5 units	0	2 (22.2%)	
6–10 units	2 (20.0%)	1 (11.2%)	
>10 units	0	0	
BMI (kg/m^2)	26.3 (25.8–30.7)	26.7 (25.1–27.8)	0.720
Nutritional status			
MPOV	6987 (2969–9080)	4682 (3740–7311)	0.497
SCS	4.20 (3.75–5.60)	4.65 (4.23–5.40)	0.541
Serum L	1.72 (1.03–2.21)	1.32 (1.04–1.86)	0.815
Serum Z	1.34 (1.06–1.55)	2.03 (1.14–2.12)	0.236
Serum MZ	2.50 (1.85–3.50)	2.15 (1.48–3.78)	0.743
Serum vitamin E	11.0 (8.10–15.55)	8.65 (7.33–11.80)	0.236
Plasma DHA	0.80 (0.55–5.10)	0.45 (0.30–1.10)	0.236
Plasma EPA	1.52 (0.97–2.15)	1.08 (0.95–1.84)	0.541
Folate	12.10 (10.65–14.20)	11.90 (10.88–13.78)	0.888
Vitamin B_{12}	14.10 (12.55–15.15)	12.90 (12.48–13.38)	0.236

Data displayed are median (inter quartile range) for numeric data and actual number and percentages for categorical data; Education: age (years) completed formal education; Medications: the number of prescribed medications consumed; Alcohol consumption: units/week: MPOV: a volume of macular pigment calculated as macular pigment average times the area under the curve out to 7° eccentricity (measured using the Heidelberg Spectralis®). SCS: skin carotenoid score (measured using the Pharmanex BioPhotonic Scanner). Serum lutein, zeaxanthin, *meso*-zeaxanthin and vitamin E concentrations are expressed in μmol/L; Plasma docosahexaenoic acid and eicosapentaenoic acid concentrations expressed in μmol/L; Serum folate concentrations expressed in ng/mL; Serum vitamin B_{12} concentrations expressed in pg/mL. Serum carotenoid and vitamin E data were not available for two individuals in the active intervention and one individual in the placebo group. Plasma DHA and EPA were not available for one individual in the active intervention and one individual in the placebo group.

Table 3. Baseline cognitive function and functional ability data of active and placebo intervention groups.

Variable	Active (n = 10) Median (IQR)	Placebo (n = 9) Median (IQR)	Sig.
Global cognition			
MoCA	21.0 (18.8–24.0)	21.0 (19.0–24.0)	0.842
RBANS immediate memory	78.0 (64.0–82.5)	85.0 (72.5–98.5)	0.156
RBANS visuospatial	101.0 (92.0–109.0)	100.0 (91.5–110.5)	0.842
RBANS language	89.0 (84.5–93.0)	88.0 (82.0–92.0)	0.661
RBANS attention	81.0 (71.0–98.5)	79.0 (77.0–89.5)	0.905
RBANS delayed memory	75.0 (47.0–87.5)	71.0 (58.0–87.0)	<0.999
RBANS total scale	78.0 (75.0–84.0)	82.0 (73.5–85.0)	0.497
4 mountains test	6.5 (6.0–7.8)	7.0 (5.5–7.5)	0.905
Comprehension			
Latency	1165.9 (812.9–1322.2)	1152.8 (991.75–1301.5)	0.842
Total errors	0	0	0.720
Working memory			
Between errors			
Stage 4	1.0 (0.8–2.0)	(1.0–2.5)	0.497
Stage 6	6.0 (4.3–7.5)	6.0 (6.0–7.0)	0.497
Stage 8	16.0 (15.0–17.0)	12.5 (11.0–14.0)	0.006
All stages	21.5 (21.0–26.5)	20.0 (18.0–22.5)	0.182
Total errors			
Stage 4	(0.8–2.0)	2.0 (1.0–2.5)	0.447
Stage 6	6.0 (4.3–7.5)	6.0 (6.0–7.0)	0.497
Stage 8	16.5 (15.0–17.3)	13.0 (11.0–14.8)	0.012
All stages	21.5 (21.0–28.3)	22.0 (18.0–22.5)	0.447
Strategy	10.0 (9.0–11.3)	9.0 (8.5–10.0)	0.315
Reaction time			
Simple reaction time	382.4 (370.7–463.6)	372.9 (354.9–411.5)	0.356
Simple movement time	296.5 (249.6–336.6)	318.9 (277.3–352.3)	0.497
Simple error score	2.0 (0.8–3.5)	1.0 (0–3.0)	0.400
Five-choice reaction time	458.1 (415.7–510.3)	435.1 (406.5–485.5)	0.549
Five-choice movement time	310.3 (286.5–367.0)	313.0 (294.2–335.4)	<0.999
Five-choice error score	0.5 (0–1.3)	1.0 (0–2.0)	0.549
Episodic memory			
First attempt memory score	4.0 (3.5–5.5)	4.0 (2.5–6.5)	0.905
No. patterns reached	6.0 (4.0–6.5)	6.0 (4.0–6.0)	0.549
Total errors adjusted stage 2	0 (0–0.3)	0 (0–0.5)	0.905
Total errors adjusted stage 4	6.0 (2.3–8.8)	6.0 (0.5–9.5)	0.780
Total errors adjusted stage 6	18.5 (12.3–20.0)	20.0 (14.5–20.0)	<0.999
Total errors adjusted stage 8	28.0 (25.3–28.0)	28.0 (28.0–28.0)	0.720
Total errors adjusted all stages	52.5 (42.8–59.0)	48.0 (45.0–58.0)	0.780
Functional ability			
BADLS	20.0 (15.5–20.0)	20.0 (20.0–20.0)	0.243
AQ	8.0 (4.8–13.8)	5.0 (2.5–8.0)	0.113

Data displayed are median (inter quartile range); MoCA: Montreal Cognitive Assessment; RBANS: Repeatable Battery for the Assessment of Neuropsychological Status; BADLS: Bristol Activities of Daily Living Scale; AQ: Alzheimer's Questionnaire.

Table 4. Change in nutritional status over 12 months between active and placebo intervention groups.

Variable	Active Intervention					Placebo Intervention					
	n	Baseline Median (IQR)	12 Months Median (IQR)	%Δ	Outcome	*n*	Baseline Median (IQR)	12 Months (Median (IQR)	%Δ	Outcome	Sig.
Nutrition											
MPOV	6	6987 (2947–9080)	10,363 (5488–12,906)	+62	Improved	7	4682 (3838–7264)	4300 (3827–7277)	−2	Declined	**0.001**
SCS	6	27,250 (18,250–36,705)	37,000 (30,000–60,250)	+79	Improved	7	17,000 (15,000–35,000)	21,000 (14,000–38,000)	+2	Improved	**0.014**
Serum L	5	0.152 (0.107–0.217)	0.562 (0.339–1.388)	+421	Improved	7	0.104 (0.067–0.188)	0.133 (0.067–0.168)	+5	Improved	**0.003**
Serum Z	5	0.059 (0.036–0.078)	0.075 (0.056–0.125)	+58	Improved	7	0.037 (0.035–0.051)	0.042 (0.037–0.049)	+1	Improved	0.247
Serum MZ	5	0	0.068 (0.031–0.234)	-	Improved	7	0	0	0	Unchanged	**0.003**
Serum vit. E	5	23.988 (20.346–30.264)	23.015 (21.010–31.802)	−4	Declined	7	24.426 (22.007–26.274)	24.704 (23.074–26.093)	+1	Improved	>0.999
Plasma DHA	5	235.730 (153.115–258.315)	291.910 (262.170–406.650)	+59	Improved	7	200.480 (165.230–212.230)	234.630 (204.150–267.680)	+17	Improved	**0.048**
Plasma EPA	5	164.500 (90.000–199.070)	178.410 (178.410–205.425)	+6	Improved	7	129.530 (127.550–146.620)	147.020 (108.080–201.850)	+13	Improved	0.639

Data displayed are median (inter quartile range); %Δ: average percentage change per subject; Sig.: Level of significance of the observed change over time between active and placebo intervention groups; MPOV, a volume of macular pigment calculated as macular pigment average times the area under the curve out to 7° eccentricity (measured using the Heidelberg Spectralis®); SCS: skin carotenoid score (obtained using the Pharmanex BioPhotonic Scanner); Serum lutein, zeaxanthin, *meso*-zeaxanthin and vitamin E concentrations are expressed in µmol/L; Plasma docosahexaenoic acid and eicosapentaenoic acid concentrations expressed in µmol/L.

Table 5. Average percentage change in global cognition over 12 months between active and placebo intervention groups.

Variable	Active Intervention					Placebo Intervention				
	n	Baseline Median (IQR)	12 Months Median (IQR)	%Δ	Outcome	*n*	Baseline Median (IQR)	12 Months (Median (IQR)	%Δ	Outcome
Global cognition (RBANS)										
Immediate memory	6	78.0 (73.3–82.5)	91.0 (81.3–100.8)	+18	Improved	7	94.0 (85.0–100.0)	90.0 (81.0–103.0)	0	Unchanged
Visuospatial	6	107.0 (101.5–110.8)	105.0 (101.5–112.0)	−1	Declined	7	96.0 (87.0–109.0)	96.0 (84.0–109.0)	0	Unchanged
Language	6	91.0 (88.0–96.0)	89.0 (88.0–93.0)	−1	Declined	7	88.0 (82.0–92.0)	92.0 (72.0–105.0)	+2	Improved
Attention	6	84.5 (68.0–103.8)	91.0 (79.0–100.0)	+7	Improved	7	79.0 (75.0–85.0)	79.0 (79.0–94.0)	0	Unchanged
Delayed memory	6	85.0 (63.3–96.5)	86.0 (63.0–106.3)	+12	Improved	7	71.0 (60.0–93.0)	90.0 (78.0–98.0)	+14	Improved
Total scale	6	82.0 (78.0–91.5)	87.0 (82.3–99.3)	+6	Improved	7	82.5 (76.3–87.3)	88.5 (78.5–91.3)	+3	Improved

Data displayed are median (inter quartile range); %Δ: average percentage change per subject; RBANS: Repeatable Battery for the Assessment of Neuropsychological Status.

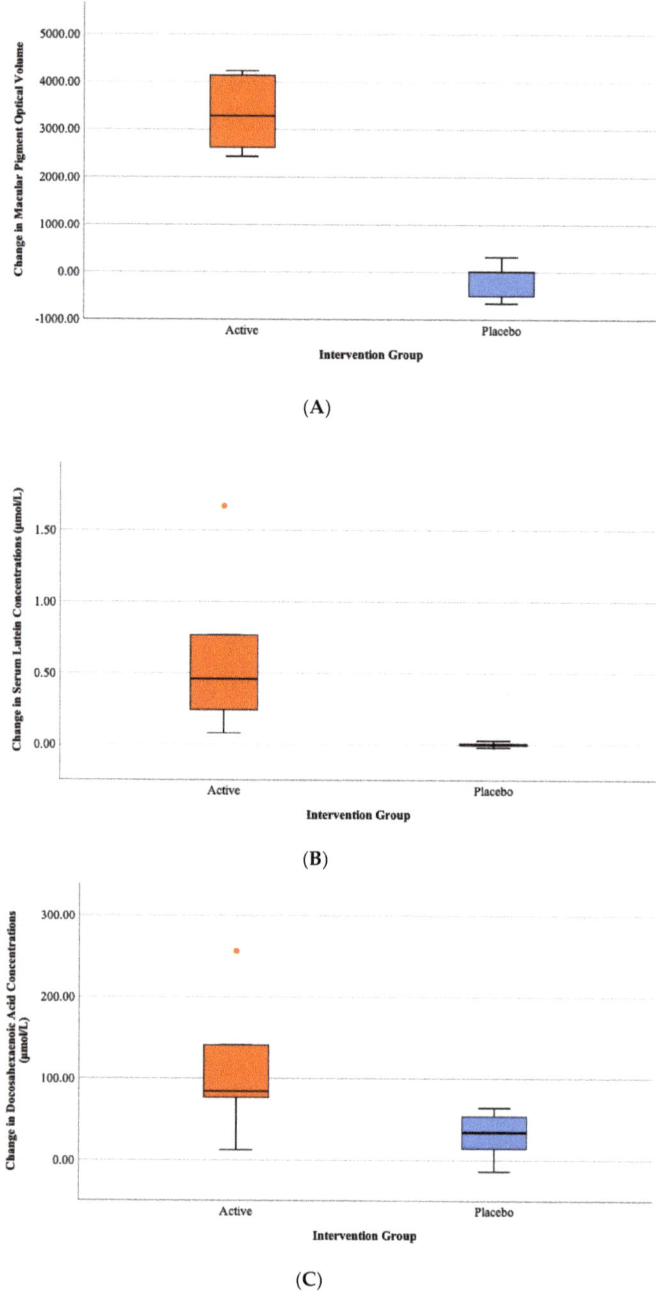

Figure 2. (**A**) Box plots illustrating change in MPOV over 12 months for active and placebo intervention groups. (**B**) Box plot illustrating change in serum lutein over 12 months for active and placebo intervention groups. (**C**) Box plot illustrating change in plasma DHA over 12 months for active and placebo intervention groups.

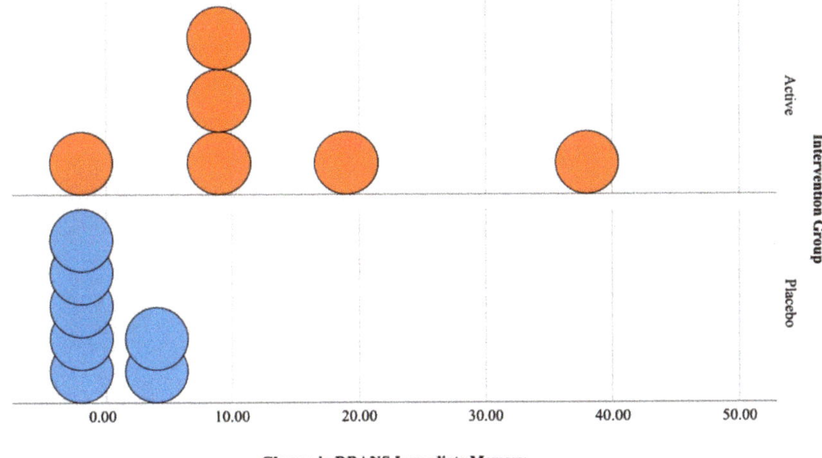

Figure 3. Dot plots illustrating change in individual scores recorded for the immediate memory domain of the RBANS for active and placebo intervention groups.

3.4. Observed Change in Episodic Memory

Table 6 shows trends in improvements in episodic memory in the active intervention group where individuals consuming the nutritional supplement recorded fewer errors at the latter (and more challenging) stages of the paired associated learning (PAL) task in comparison to individuals in the placebo group where scores either remained unchanged or worsened. A minimal improvement (+1%) in the first attempt memory score was also observed among individuals in the active group while individuals in the placebo group declined slightly (by 1%).

3.5. Observed Change in Working Memory

Table 7 summaries the observed changes in working memory following the 12-month intervention period. For both groups, all tasks that assessed working memory either remained unchanged or worsened over the 12-month study period, with the exception of the number of between and total errors made at stage 8 which improved slightly (by two points each) among individuals in the active intervention group and the number of total errors made at stage 4 of the SWM task, which improved slightly (by one point) among individuals in the placebo group. Of note, individuals in the placebo group recorded significantly fewer between and total errors at stage 8 of the SWM tasks at baseline.

3.6. Observed Change in Reaction Time

Table 7 also shows the observed changes in reaction time following the 12-month intervention period. Results for reaction times were also mixed for both groups. Trends in improvements in simple reaction time were observed in both groups (by an average of 2 milliseconds [ms]). In addition, trends in improvements in five-choice reaction time (by an average of 20 ms) were observed among individuals in the active intervention group while declines (by an average of 12 ms) were recorded among individuals receiving placebo. Both simple and five-choice movement times declined in both groups (by an average of 13 ms and 29 ms for active and placebo groups, respectively) with overall trends suggesting a lesser decline among individuals in the active intervention group. Finally, error scores for all reaction time assessments remained unchanged with the exception of the number of errors made during the five-choice reaction time assessment, which declined (by one point) in the placebo group (see Table 7).

Table 6. Average change per subject in episodic memory, working memory and reaction time over 12 months between active and placebo intervention groups.

Variable	Active Intervention					Placebo Intervention			
	n	Baseline Median (IQR)	12 Months Median (IQR)	Δ	Outcome	Baseline Median (IQR)	12 Months (Median (IQR)	Δ	Outcome
Episodic memory (PAL)									
First attempt memory score	6	4.5 (3.3–7.5)	6.5 (2.0–9.8)	+1	Improved	4.0 (3.0–6.0)	3.0 (2.0–6.0)	−1	Declined
Total errors adjusted stage 2	6	0 (0–1.3)	0 (0)	−1	Improved	0 (0)	0 (0–2.0)	−1	Improved
Total errors adjusted stage 4	6	6.0 (2.3–8.0)	7.5 (2.0–11.8)	+2	Declined	6.0 (0–10.0)	8.0 (4.0–10.0)	+2	Declined
Total errors adjusted stage 6	6	15.0 (7.0–18.5)	14.0 (5.8–20.0)	−1	Improved	20.0 (15.0–20.0)	20.0 (15.0–20.0)	0	Unchanged
Total errors adjusted stage 8	6	28.0 (15.3–28.0)	20.5 (8.8–28.0)	−4	Improved	28.0 (28.0–28.0)	28.0 (28.0–28.0)	0	Unchanged
Total errors adjusted all stages	6	54.5 (24.8–54.5)	39.5 (17.8–59.8)	−4	Improved	48.0 (47.0–58.0)	51.0 (48.0–58.0)	+2	Declined

Data displayed are median (inter quartile range); Δ: average change per subject; PAL: paired associated learning; First attempt memory score: the number of correct box choices that were made on the first attempt during assessment problems; Total errors adjusted: the number of times the individual chose the incorrect box for a stimulus on assessment problems, plus an adjustment for the estimated no. errors they would have made on any problems, attempts and recalls they did not reach.

Table 7. Average change per subject in working memory and reaction time over 12 months between active and placebo intervention groups.

Variable	Active Intervention					Placebo Intervention				
	n	Baseline Median (IQR)	12 Months Median (IQR)	Δ	Outcome	n	Baseline Median (IQR)	12 Months (Median (IQR)	Δ	Outcome
Working memory (SWM)										
Between errors stage 4	6	1.0 (0.8–2.0)	1.5 (0.8–2.5)	+1	Declined	7	2.0 (1.0–3.0)	2.0 (0–3.0)	0	Unchanged
Between errors stage 6	6	5.5 (2.0–7.5)	6.0 (5.5–8.8)	+2	Declined	7	6.0 (6.0–7.0)	9.0 (6.0–10.0)	+1	Declined
Between errors stage 8	5	17.0 (15.0–17.0)	16.0 (10.0–18.0)	−2	Improved	5	12.0 (10.5–15.5)	13.0 (12.5–18.0)	+2	Declined
Between errors all stages	5	21.0 (20.0–24.0)	26.0 (15.5–27.0)	−2	Improved	5	20.0 (19.0–23.5)	21.0 (20.0–24.0)	+2	Declined
Total errors stage 4	6	1.0 (0.8–2.0)	1.5 (0.8–2.5)	+1	Declined	7	2.0 (1.0–3.0)	2.0 (0–3.0)	−1	Improved
Total errors stage 6	6	5.5 (2.0–7.5)	6.0 (5.5–9.0)	+2	Declined	7	6.0 (6.0–7.0)	10.0 (6.0–11.0)	+2	Declined
Total errors stage 8	5	17.0 (15.0–18.5)	16.0 (10.5–19.0)	−2	Improved	5	12.0 (10.5–16.5)	15.0 (12.5–18.0)	+2	Declined
Total errors all stages	5	21.0 (20.0–25.5)	26.0 (16.0–28.0)	0	Unchanged	5	22.0 (13.5–22.8)	22.5 (20.5–24.3)	+4	Declined
Strategy	5	10.0 (8.5–12.0)	10.0 (8.5–11.5)	0	Unchanged	5	9.0 (6.8–10.5)	9.0 (9.0–10.5)	+1	Declined
Attention (RTI)										
Simple reaction time	6	381.1 (364.4–463.6)	397.3 (342.8–453.9)	−2	Improved	7	389.7 (340.4–430.6)	369.2 (355.3–448.7)	−2	Improved
Simple movement time	6	302.9 (181.9–432.5)	316.0 (244.7–433.5)	+13	Declined	7	318.9 (274.6–337.7)	344.2 (246.3–449.8)	+29	Declined
Simple error score	6	1.5 (0–2.8)	1.0 (0–2.5)	0	Unchanged	7	1.0 (0–3.0)	1.0 (1.0–3.0)	0	Unchanged
Five-choice reaction time	6	483.7 (420.5–519.5)	445.7 (429.1–477.4)	−20	Improved	7	435.1 (391.2–479.0)	450.7 (393.4–488.3)	+12	Declined
Five-choice movement time	6	316.6 (279.2–443.8)	333.2 (261.8–417.4)	+13	Declined	7	310.5 (289.1–320.0)	332.8 (323.3–416.4)	+29	Declined
Five-choice error score	6	1.0 (0–2.3)	0 (0–3.3)	0	Unchanged	7	1.0 (0–2.0)	1.0 (0–3.0)	+1	Declined

Data displayed are median (inter quartile range); Δ: average change per subject. SWM: spatial working memory; Between errors: the number of times the individual revisits a box in which a token has previously been found; Total errors: the number of times a box is selected that is certain not to contain a token and therefore should not have been visited by the individual; Strategy: for problems with six boxes or more. The number of distinct boxes used by the individual to begin a new search for a token, with the same problem; RTI: reaction time; Simple reaction time: the duration between the onset of the stimulus and the time at which the individual released the button. Calculated for correct trials, where the stimulus could appear in one location only; Simple movement time: the time taken to touch the stimulus after the button has been released. Calculated for correct trials, where the stimulus could appear in one location only; Simple error score: the number of trials where the response status is any error (i.e., an inaccurate response, no response, or a premature response) for the assessment trial where stimuli appear on one location only. The error may be an inaccurate response, no response, or a premature response; Five-choice reaction time: the duration between the onset of the stimulus and the release of the button. Calculated for correct, assessed trials where the stimulus could appear in any of the five locations; Five-choice movement time: the time taken to touch the stimulus after the button has been released. Calculated for correct, assessed trials where the stimulus could appear in any of the five locations; Five-choice error score: the number of trials where the response status is any error (i.e., an inaccurate response, no response, or a premature response) for assessment trials where stimuli appear in any of five locations.

4. Discussion

Given that ω-3FAs, xanthophyll carotenoids, and vitamin E are present in brain tissue, and given their ability to attenuate mechanisms involved in the pathogenesis of AD (namely oxidative stress and neuro-inflammation), it is likely that they play an important neuroprotective role by maintaining and optimizing cognition and reducing the risk of cognitive decline. Importantly, previous studies have shown that cognitively impaired individuals are deficient in these key nutritional compounds in comparison to cognitively healthy individuals. Therefore, it is likely that specific population groups, such as individuals with MCI, will benefit from nutritional intervention. CARES Trial 1 was a parallel group, double-blind, placebo-controlled, randomized clinical trial designed to examine the effect of targeted nutritional intervention on cognitive performance among individuals with MCI. Following 12-month supplementation with a combination of ω-3FAs (DHA and EPA), xanthophyll carotenoids (L, Z and MZ), and vitamin E (α-tocopherol), this exploratory study identified trends in improved performance in episodic memory, immediate memory, attention and delayed memory among individuals with clinically confirmed MCI.

4.1. Significance and Interpretation of Findings

This exploratory study is the first of its kind to examine the impact of a combination of ω-3FAs, xanthophyll carotenoids, and vitamin E on cognition in individuals with MCI. Previous studies have examined the effects of varying combinations of nutritional compounds on the cognitive health of individuals with MCI (see [60] for a review). While many of the studies included in the review performed by Solfrizzi and colleagues [60] observed reductions in brain atrophy in individuals with MCI, no positive effects on cognition were found, with the exception of [61] where improvements in the dementia rating scale were reported following 6-month consumption of a combination of vitamin B_{12}, α-tocopherol, s-adenosylmethionine, N-acetylcysteine and acetyl-L-carnitine. Interestingly, supplementation with ω-3FAs alone have yielded positive results among individuals with MCI. A meta-analysis of 15 interventional trials suggested a benefit of DHA supplementation in terms of improving episodic memory among mildly impaired individuals [62]. Other studies have also reported benefits in memory (episodic, short-term, working, and immediate verbal), processing speed, and attention [63–65] following ω-3FA supplementation. In contrast, supplementation with vitamin E is less promising as, currently, no improvements in cognitive performance have been observed in MCI samples [66] and none have examined the impact of xanthophyll carotenoid supplementation alone on cognition in individuals with MCI. Importantly, to date, none have examined the potential synergistic effect of the ω-3FAs DHA and EPA, the xanthophylls L, Z and MZ, and vitamin E in the α-tocopherol form on cognitive function among individuals with MCI.

In the present exploratory study, individuals with MCI in the active intervention group responded positively to 12-month nutritional supplementation in terms of statistically significant improvements in tissue carotenoid concentrations, as well as statistically significant increases in blood concentrations of serum L, serum MZ and plasma DHA. Of note, individuals in the active intervention group responded poorly to vitamin E (α-tocopherol) supplementation. Reasons underlying the poor vitamin E response following supplementation remain unclear. While in accordance with international recommended dietary allowances (RDAs) [67,68], it is possible that the daily dosage of vitamin E (i.e., 15 mg) used in the present study was too low (in comparison to other interventional studies (e.g., [69,70]) to have any meaningful effect.

Most of the positive outcomes identified in the present exploratory study relate to performance in tasks assessing memory. Memory deficits, which involve difficulties with the encoding, storage and retrieval of information, are commonplace among individuals with MCI. Specifically, impairment in episodic memory has been frequently documented in MCI and is an inherent feature of amnestic MCI [71,72]. Episodic memory refers to the ability to learn, store and retrieve information about experiences that occurred at a particular time and place (e.g., remembering where you parked your car in a multi-story carpark, remembering the details of a family event attended in the past few weeks [73]).

Encoding, retention and retrieval difficulties are likely due to changes in the relevant neurocircuitry including frontal, temporal and medial temporal lobe regions, the hippocampus and adjacent cortical areas [74,75].

4.2. Neurobiological Mechanisms

The brain is a lipid-dense organ containing large amounts of ω-3FAs (and DHA in particular) [76]. Additionally, xanthophyll carotenoids selectively accumulate in brain tissue including frontal and temporal cortices. Ω-3FAs are considered to play an important role in neurological health. It has been suggested that DHA plays an important role in the control and resolution of neuro-inflammation. This role is performed by a number of pathways, including being converted into bioactive lipid metabolites such as endocannabinoid epoxides (molecules that are responsible for antiangiogenic effects, vasodilatory actions, and regulation of platelet aggregation) [77]. It has also been suggested that DHA downregulates the expression of genes involved in the synthesis of pro-inflammatory eicosanoids produced from the ω-6FA arachidonic acid [78]. Despite EPA being stored in the brain in low amounts, it has been demonstrated that this fatty acid is important for neural efficiency. This suggests that EPA may positively influence pathways that regulate high-order cognitive functions [79]. It has also been suggested that EPA can facilitate enzymatic processes required to inhibit neuronal damage from inflammation and oxidative stress [80].

Additionally, xanthophyll carotenoids are premised to be neuroprotective primarily owing to their antioxidant properties. Due to their conjugated double-bond structure, carotenoids are efficient scavengers of reactive oxygen species [81]. The lipid solubility of carotenoids also enables them to reduce the susceptibility of cellular membranes and lipoproteins to oxidative damage through free-radical scavenging [82]. L and Z have been shown to positively impact neural efficiency [83,84] and cellular communication via gap junctions [85]. Carotenoids can also combat inflammation. For example, it has been shown that carotenoids are involved in the modulation of inflammatory cells and pro-inflammatory enzymes, the downregulation of pro-inflammatory molecule production, and the attenuation of inflammatory gene expression [86].

While it cannot be asserted that improvements in specific cognitive domains (as a result of supplementation for example) necessarily negates any pre-existing risk for going on to develop AD, nevertheless it is reasonable to hypothesize that the observed improvements in the encoding and memory retrieval process among individuals consuming the nutritional intervention may reflect favorable changes in the physiological functionality, structural integrity and synaptic activity of brain regions involved in memory, and that these favorable changes may be attributable to the enrichment of the aforementioned nutritional compounds. Moreover, the observed trends in improvements in cognitive outcomes may help to favorably alter the risk profile of these individuals for further cognitive decline in the future by enriching the neurocognitive environment.

4.3. Strengths and Limitations

Strengths of CARES Trial 1 include a comprehensive assessment of MCI using sensitive and validated diagnostic measurement tools at screening, enrolment and follow-up assessments. Furthermore, the use of a consensus panel provided in-depth characterization of all individuals and the implementation of robust inclusion and exclusion criteria ensured a clean dataset. The interpretation, analysis and generalizability of results from CARES Trial 1 were limited due to the lack of statistical power in the trial. In order to ensure sufficient statistical power to test the proposed research hypothesis, CARES Trial 1 aimed to recruit 60 individuals with MCI. In order to achieve this target (and allowing for a 10% attrition rate), it was anticipated that a large number of individuals would have to be screened. Despite increased attempts, the identification and recruitment of individuals with MCI proved extremely challenging. The high rate (30%) of individuals who chose not to participate (despite their eligibility) and high attrition rate (as illustrated in Figure 1) were unforeseen and highlights the challenges of conducting research in the MCI population. A number of attempts were made to address the challenges of identifying and enrolling individuals with MCI. These included widening the recruitment catchment area from one city (i.e., Waterford, Ireland) to the entire South-East region of Ireland and hosting briefing meetings with the relevant consultant geriatricians and psychiatrists of old age in the region. Repeated written communication was also carried out to remind the relevant consultant geriatricians and psychiatrists of old age in the region about the clinical trial (including the project aims and inclusion criteria). Despite increased attempts, MCI baseline numbers remained low and drop-out rates remained high.

The small sample size of CARES Trial 1 also precluded the study from comparing MCI subtypes (i.e., amnestic versus non-amnestic) and examining potential relationships between nutritional status and cognitive outcomes. CARES Trial 1 may also be subject to selection bias, given that individuals were primarily recruited from the clinic setting. Finally, depressive symptoms were not assessed at screening. However, depression under active review was part of the exclusion criteria and may counteract this perceived limitation. Despite these limitations, this exploratory study provides encouraging preliminary data. We have shown that individuals with MCI respond (in tissue and in blood) to targeted nutritional supplementation. Additionally, we have observed trends in improved performance in tasks assessing episodic memory and global cognition (namely immediate memory, delayed memory and attention).

5. Conclusions

In conclusion, the present exploratory study has identified trends in improved performance in episodic memory and global cognition among individuals with clinically confirmed MCI following 12-month targeted nutritional supplementation with a combination of ω-3FAs, xanthophyll carotenoids, and vitamin E. Despite the heterogeneity of MCI, studying individuals with this condition provides a unique opportunity to examine the efficacy of nutrition as a preventative approach in slowing the progression of cognitive impairment and improving cognitive-related outcomes. Given that there has been little clinical success with pharmacological strategies for cognitive decline and AD and given that current thinking surrounding the amyloid hypothesis is being challenged by many in the scientific community [87], shifting focus towards preventative approaches is timely and warranted. The results of the present study are highly promising and highlight the potential of nutrition as a preventative strategy for modifying or delaying MCI progression and improving cognitive outcomes. MCI presents a unique opportunity to examine the potential of nutrition for improving cognitive outcomes in individuals at an early stage of impairment. Despite the small sample size, this exploratory interventional work has not only addressed a gap in the literature but has also shown that individuals with clinically confirmed MCI respond positively to targeted nutritional supplementation. Larger-scaled and appropriately powered interventional trials are clearly warranted to confirm this finding and to explore interactions between nutritional compounds and cognitive status.

Author Contributions: Data curation and formal analysis: R.P., J.M.N., W.R.; investigation, project administration, visualization: R.P., J.M.N.; methodology: R.P., J.M.N., A.P.-C., R.C., T.P., R.M.; resources: R.P., J.M.N., R.M.; writing—original draft: R.P., J.M.N., A.P.-C., R.M.; writing—review & editing: R.P., J.M.N., A.P.-C., R.C., A.N.H., R.M.; supervision: A.P.-C., R.M.; validation: A.P.-C., W.R.; software: W.R.; formula analysis: T.P.; conceptualization: A.N.H., R.M.; funding acquisition: A.N.H. All authors have read and agreed to the published version of the manuscript.

Funding: This research was funded by the Howard Foundation (Registered UK Charity Number: 285822).

Acknowledgments: The authors thank Michael Kirby (University Hospital Waterford, Waterford, Ireland) for his contribution to the consensus panel. The authors also thank Catherine Kelly and Lisa O'Brien (Whitfield Pharmacy, Waterford, Ireland) for the management of the trial supplements and Eurofins Biomnis (Dublin, Ireland) for performing the additional biochemical assessments for the trial.

Conflicts of Interest: Rebecca Power: RP has performed consultancy work for MacuHealth. RP is funded in part by the Howard Foundation. John M. Nolan: JMN does consultancy work as a Director of NOW Science Consultancy Ltd. for companies with an interest in food supplements. Alfonso Prado-Cabrero: APC has performed consultancy work for the Howard Foundation (Cambridge, UK) and MacuHealth LLC™ (Birmingham, MI, USA). These organizations have an interest in commercially available supplements containing the macular carotenoids.

APC has also been involved in a Commercialization Fund Programme from Enterprise Ireland to develop a biotechnological process to produce carotenoids and the fatty acids EPA and DHA. APC is currently supported by VistaMilk SFI Research Centre to develop commercial dairy products enriched in carotenoids. Robert Coen: RC declares no conflict of interest. Warren Roche: WR declares no conflict of interest. Tommy Power: TB declares no conflict of interest. Alan N. Howard: ANH is trustee of Howard Foundation. ANH is also a director of Nutriproducts Ltd. which trades in nutraceuticals on behalf of the Howard Foundation. ANH was involved in design of the study, and the reviewing of the manuscript. Ríona Mulcahy: RM does consultancy work on behalf of the Howard Foundation.

Appendix A

Table A1. Composition of formulation used in CARES Trial 1 (mg/capsule).

Compound Family	Compound	Formulated	Actual
Fatty Acids	Palmitic acid (16:0)		55.44 ± 1.39
	Palmitoleic acid (16:1)		0.84 ± 0.02
	Stearic acid (18:0)		31.60 ± 1.02
	Oleic acid (18:1n9c)		40.24 ± 0.74
	Vaccenic acid (18:1n9t)		6.26 ± 0.06
	Linoleic acid (18:2n6)		28.57 ± 0.78
	α-Linolenic acid (18:3n3)		1.55 ± 0.03
	Eicosenoic acid (20:1n7, n9, n11)		41.16 ± 0.94
	Homo-γ-linolenic acid (20:3)		1.45 ± 0.04
	Arachidonic acid (20:4n6)		14.20 ± 0.20
	Eicosapentaenoic acid (20:5n3)	45	47.89 ± 0.69
	Docosapentaenoic acid (22:5n3)		11.37 ± 0.18
	Docosahexaenoic acid (22:6n3)	215	258.03 ± 2.75
Carotenoids	Lutein	5	5.18 ± 0.06
	Zeaxanthin	1	1.75 ± 0.03
	meso-zeaxanthin	5	6.48 ± 0.24
Vitamin E	α-tocopherol	7.5	6.12 ± 0.04
	Total mg		558.15

Data presented as mean ± SD. Formulations analyzed in triplicate (three capsules analyzed). Average capsule content calculated gravimetric analysis, 0.666 ± 0.001 g ($n = 3$).

Table A2. Concentration of fatty acids (mean ± SD, μmol/L) of plasma total lipids.

	Placebo		Active	
	V1	V2	V1	V2
Miristic acid	158.39 ± 117.06	200.19 ± 98.13	166.02 ± 89.64	147.14 ± 76.60
Palmitic acid	4040.87 ± 689.23	3976.31 ± 504.27	4088.68 ± 855.16	3794.24 ± 832.03
Palmitoleic acid	208.07 ± 80.91	223.47 ± 59.86	231.43 ± 159.19	231.81 ± 153.37
Stearic acid	1144.93 ± 154.75	1150.95 ± 105.75	1017.71 ± 186.56	938.33 ± 132.44
Oleic acid	2368.04 ± 686.01	2693.92 ± 645.78	2383.20 ± 752.13	2405.17 ± 638.84
Vaccenic acid	138.70 ± 21.46	155.97 ± 25.73	148.20 ± 46.87	137.51 ± 45.76
Linoleic acid	2398.38 ± 409.88	2584.15 ± 206.55	2452.42 ± 186.15	2428.33 ± 276.07
γ-linolenic acid	54.37 ± 19.81	51.36 ± 21.40	41.45 ± 20.50	39.65 ± 20.23
α-linolenic acid	88.09 ± 32.29	96.32 ± 46.64	112.40 ± 29.93	101.14 ± 14.93
Eicosenoic acid	493.59 ± 85.53	373.54 ± 46.86	479.94 ± 85.43	369.29 ± 33.35
Homo-γ-linolenic acid	162.94 ± 39.32	166.81 ± 43.94	134.27 ± 36.82	127.83 ± 19.35
Arachidonic acid	789.65 ± 197.72	842.89 ± 271.43	628.37 ± 208.69	588.96 ± 241.35
Eicosapentaenoic acid (EPA)	141.06 ± 43.75	167.51 ± 90.72	148.53 ± 58.71	155.60 ± 60.72
Docosapentaenoic acid (DPA)	55.21 ± 6.88	59.49 ± 16.69	50.74 ± 5.78	49.03 ± 13.64
Docosahexaenoic acid (DHA)	199.96 ± 43.11	231.64 ± 47.20	211.72 ± 56.47	325.91 ± 97.80

Data presented as mean ± SD.

Table A3. Calibration standards used to quantify xanthophyll carotenoids and vitamin E.

	Molecular Weight	Maximum Wavelength	Extinction Coefficient	Solvent	Reference
Lutein	568.88	444	144.8×10^3	Ethanol	[88]
Zeaxanthin	568.88	450	144.2×10^3	Ethanol	[89]
β-cryptoxanthin	552.85	450	135.7×10^3	Hexane	[90]
α-tocopherol	430.71	292	326.5	Ethanol	[91]

Molecular weight (g mol^{-1}); Maximum wavelength (nm); Extinction coefficient (L mol^{-1} cm^{-1}).

Table A4. Regression line, lower and upper limits of quantification for xanthophyll carotenoids and vitamin E.

Compound	Equation	R^2	LLOQ (n = 10)	ULOQ (n = 3)
Lutein	$y = 0.0827x + 0.3409$	0.9997	0.069 ± 0.0007	2.514 ± 0.003
Zeaxanthin	$y = 0.0793x + 1.4721$	0.9998	0.048 ± 0.0003	2.487 ± 0.007
β-cryptoxanthin	$y = 0.0909x + 1.9363$	0.9997	0.071 ± 0.0005	2.460 ± 0.007
α-tocopherol	$y = 0.00254x + 0.04380$	0.9992	4.053 ± 0.0913	69.469 ± 1.011

LLOQ: lower limit of quantification expressed in µmol/L; ULOQ; upper limit of quantification expressed in µmol/L.

Appendix A.1 Analyte Recovery Analysis

L was used as a representative of carotenoid recovery efficiency; recovery efficiency of α-tocopherol was also assessed. To do this, the respective authentic standard was added to a pooled serum sample in a concentration of 85% to 110% ULOQ. The concentration of the analyte was determined by HPLC in this spiked sample, as well as in the pooled serum and in the prepared authentic standard separately. The percentage of recovery of the analyte of interest was determined by adding the area of the analyte in the prepared authentic standard and in the pooled serum, and dividing it by the area of the analyte in the spiked sample. This determination was performed in triplicate for each analyte and on three different days. The efficiency of the recovery of L (Table A5) was 99.4 ± 2.9 % without adjusting with the IS, and 121.6 ± 0.7 adjusting with the IS, at a concentration 92.3 ± 10.4 % of ULOQ (n = 3). The efficiency of recovery of α-tocopherol (Table A6) was 88.2 ± 2.1 % without adjusting with the IS, and 104.1 ± 5.8 % adjusting with the IS, at a concentration of 100.5 ± 11.1 % of ULOQ (n = 3). These results suggested that the method performed to quantify L was exhaustive in terms of L extraction and re-suspension prior to HPLC analysis, but incomplete in this sense for α-tocopherol. Therefore, we did not to use the IS to correct carotenoid concentrations, but used it to correct α-tocopherol concentrations in the samples analysed.

Table A5. Lutein recovery assay.

Assay	% ULOQ	% Recovery No IS	% Recovery with IS
1	81.6	99.4	122.4
2	93.0	102.2	120.9
3	102.4	96.5	121.6
Average	92.3 ± 10.4	99.3 ± 2.9	121.6 ± 0.7

ULOQ; upper limit of quantification; IS: internal standard.

Table A6. α-tocopherol recovery assay.

Assay	%ULOQ	% Recovery No IS	% Recovery with IS
1	88.4	86.1	97.5
2	108.8	90.2	108.6
3	103.5	88.3	106.1
Average	100.2 ± 10.6	88.2 ± 2.1	104.1 ± 5.8

ULOQ; upper limit of quantification; IS: internal standard.

Appendix A.2 HPLC Precision Analysis

The HPLC analysis of the samples were completed in four independent batches, each one performed on a different day. Intra-day and inter-day precision of carotenoid and α-tocopherol analysis were evaluated and quantified in an independently pooled serum sample that was analysed at the beginning, middle and end of each daily analysis. L and α-tocopherol were analysed over runs as witnesses of precision of HPLC analysis. L concentration in the pooled serum averaged 0.151 ± 0.004 μmol/L (Table A7), which is roughly twice the concentration of the LLOQ. Intra-day precision in each day of analysis, expressed as the co-efficient of variation (CV) of the concentration of each analyte was below 5% in L and α-tocopherol, which is below the 15% limit recommended by EMA [92]. Inter-day precision was calculated averaging compound concentration of the pooled samples in each day of analysis. The highest inter-day variability was for α-tocopherol (8.33%, Table A7), which is also below the 15% limit recommended by EMA.

Table A7. Intra-day and inter-day precision of lutein and α-tocopherol in human serum.

μmol/L (*n*; CV)	Intra-Day 1 (*n* = 3)	Intra-Day 2 (*n* = 4)	Intra-Day 3 (*n* = 3)	Intra-Day 4 (*n* = 2)	Inter-Day
Lutein	0.148 ± 0.007 (CV = 4.98%)	0.156 ± 0.003 (CV = 1.89%)	0.153 ± 0.003 (CV = 1.65%)	0.147 ± 0.002 (CV = 1.05%)	0.151 ± 0.004 (CV = 2.71%)
α-tocopherol	26.123 ± 1.095 (CV = 4.19%)	29.031 ± 0.881 (CV = 3.03%)	27.210 ± 0.697 (CV = 2.56%)	23.745 ± 1.057 (CV = 4.45%)	26.527 ± 2.209 (CV = 8.33%)

CV: co-efficient of variation.

Appendix A.3 Trueness of Xanthophyll Carotenoid Quantification

Trueness was assessed with the Certified Standard Material NIST SRM 968f, fat-soluble vitamins in frozen human serum in two different concentrations (level 1 and level 2) using our calibration lines. Only α-tocopherol was below the consensus value for level 1 by 17%; the rest of the compounds were within the limits set by NIST [93] (see Table A8). The IS was not used to calculate carotenoid concentrations, but it was used to calculate α-tocopherol concentration.

Table A8. Trueness of xanthophyll carotenoid quantification.

Compound	NIST Concentration (μmol/L)	
	Level 1 (NIST Consensus Values)	Level 2 (NIST Consensus Values)
Total lutein	0.035 (0.036 ± 0.010)	0.075 (0.087 ± 0.037)
Total lutein + zeaxanthin	0.053 (0.052 ± 0.006)	0.106 (0.115 ± 0.019)
β-cryptoxanthin	0.030 (0.030 ± 0.008)	0.038 (0.044 ± 0.017)
α-tocopherol	4.114 (5.15 ± 0.21)	11.508 (11.85 ± 0.73)

NIST: National Institute of Standards and Technology. Experimental vs. Consensus values (in brackets) equated to the mean of the value of the compound ± expanded uncertainty ($U_{95\%}$).

Table A9. Additional biochemical assessments measured in active and placebo intervention groups.

Variable	Active ($n = 10$) Median (IQR)	Placebo ($n = 9$) Median (IQR)	Sig.
Cholesterol (mmol/L)	4.20 (3.75–5.60)	4.65 (4.23–5.40)	0.541
Triglycerides (mmol/L)	1.72 (1.03–2.21)	1.32 (1.04–1.86)	0.815
HDL (mmol/L)	1.34 (1.06–1.55)	2.03 (1.14–2.12)	0.236
LDL (mmol/L)	2.50 (1.85–3.50)	2.15 (1.48–3.78)	0.743
Homocysteine (μmol/L)	11.0 (8.10–15.55)	8.65 (7.33–11.80)	0.236
C-reactive protein (mg/L)	0.80 (0.55–5.10)	0.45 (0.30–1.10)	0.236
TSH (ulUml)	1.52 (0.97–2.15)	1.08 (0.95–1.84)	0.541
Free T4 (pmol/L)	12.10 (10.85–14.20)	11.90 (10.88–13.78)	0.888
Haemoglobin (g/dL)	14.10 (12.55–15.15)	12.90 (12.48–13.38)	0.236
Sodium (mmol/L)	138.0 (137.50–140.0)	139.0 (138.0–140.0)	0.743
Potassium (mmol/L)	4.15 (3.88–4.83)	4.20 (4.05–4.50)	0.999
Chloride (mmol/L)	101.0 (98.0–104.0)	102.0 (97.0–103.75)	0.999
Creatinine (μmol/L)	74.0 (70.0–86.50)	63.5 (57.5–68.75)	**0.008**

Data displayed are median (inter quartile range): HDL: high-density lipoprotein; LDL: low-density lipoprotein; Free T4: Free thyroxine; TSH: thyroid stimulating hormone. Additional biochemical assessments were not obtained for one individual in the active intervention group and one individual in the placebo group.

References

1. Petersen, R.C.; Roberts, R.O.; Knopman, D.S.; Boeve, B.F.; Geda, Y.E.; Ivnik, R.J.; Smith, G.E.; Jack, C.R., Jr. Mild cognitive impairment: Ten years later. *Arch. Neurol.* **2009**, *66*, 1447–1455. [CrossRef]
2. Bae, J.B.; Han, J.W.; Kwak, K.P.; Kim, B.J.; Kim, S.G.; Kim, J.L.; Kim, T.H.; Ryu, S.H.; Moon, S.W.; Park, J.H.; et al. Impact of Mild Cognitive Impairment on Mortality and Cause of Death in the Elderly. *J. Alzheimers Dis.* **2018**, *64*, 607–616. [CrossRef]
3. Ganguli, M.; Dodge, H.H.; Shen, C.; Pandav, R.S.; DeKosky, S.T. Alzheimer disease and mortality: A 15-year epidemiological study. *Arch. Neurol.* **2005**, *62*, 779–784. [CrossRef]
4. Farias, S.T.; Mungas, D.; Reed, B.R.; Harvey, D.; DeCarli, C. Progression of mild cognitive impairment to dementia in clinic- vs community-based cohorts. *Arch. Neurol.* **2009**, *66*, 1151–1157. [CrossRef]
5. Roberts, R.O.; Knopman, D.S.; Mielke, M.M.; Cha, R.H.; Pankratz, V.S.; Christianson, T.J.; Geda, Y.E.; Boeve, B.F.; Ivnik, R.J.; Tangalos, E.G.; et al. Higher risk of progression to dementia in mild cognitive impairment cases who revert to normal. *Neurology* **2014**, *82*, 317–325. [CrossRef]
6. Lopez, O.L.; Becker, J.T.; Chang, Y.F.; Sweet, R.A.; DeKosky, S.T.; Gach, M.H.; Carmichael, O.T.; McDade, E.; Kuller, L.H. Incidence of mild cognitive impairment in the Pittsburgh Cardiovascular Health Study-Cognition Study. *Neurology* **2012**, *79*, 1599–1606. [CrossRef]
7. Cheng, G.; Huang, C.; Deng, H.; Wang, H. Diabetes as a risk factor for dementia and mild cognitive impairment: A meta-analysis of longitudinal studies. *Intern. Med. J.* **2012**, *42*, 484–491. [CrossRef]
8. Livingston, G.; Sommerlad, A.; Orgeta, V.; Costafreda, S.G.; Huntley, J.; Ames, D.; Ballard, C.; Banerjee, S.; Burns, A.; Cohen-Mansfield, J.; et al. Dementia prevention, intervention, and care. *Lancet* **2017**, *390*, 2673–2734. [CrossRef]
9. DeCarlo, C.A.; MacDonald, S.W.; Vergote, D.; Jhamandas, J.; Westaway, D.; Dixon, R.A. Vascular Health and Genetic Risk Affect Mild Cognitive Impairment Status and 4-Year Stability: Evidence From the Victoria Longitudinal Study. *J. Gerontol. B. Psychol. Sci. Soc. Sci.* **2016**, *71*, 1004–1014. [CrossRef]
10. Valls-Pedret, C.; Sala-Vila, A.; Serra-Mir, M.; Corella, D.; de la Torre, R.; Martinez-Gonzalez, M.A.; Martinez-Lapiscina, E.H.; Fito, M.; Perez-Heras, A.; Salas-Salvado, J.; et al. Mediterranean Diet and Age-Related Cognitive Decline: A Randomized Clinical Trial. *Jama. Intern. Med.* **2015**, *175*, 1094–1103. [CrossRef]
11. Pelletier, A.; Barul, C.; Feart, C.; Helmer, C.; Bernard, C.; Periot, O.; Dilharreguy, B.; Dartigues, J.F.; Allard, M.; Barberger-Gateau, P.; et al. Mediterranean diet and preserved brain structural connectivity in older subjects. *Alzheimers Dement.* **2015**, *11*, 1023–1031. [CrossRef] [PubMed]

12. Zamroziewicz, M.K.; Barbey, A.K. The Mediterranean diet and healthy brain aging: Innovations from nutritional cognitive neuroscience. In *Role of the Mediterranean Diet in the Brain and Neurodegenerative Diseases*; Farooqui, T., Farooqui, A.A., Eds.; Elsevier Publishing Company: New York, NY, USA, 2017.
13. Zwilling, C.E.; Talukdar, T.; Zamroziewicz, M.K.; Barbey, A.K. Nutrient biomarker patterns, cognitive function, and fMRI measures of network efficiency in the aging brain. *Neuroimage* **2019**, *188*, 239–251. [CrossRef] [PubMed]
14. Hardman, R.J.; Kennedy, G.; Macpherson, H.; Scholey, A.B.; Pipingas, A. Adherence to a Mediterranean-Style Diet and Effects on Cognition in Adults: A Qualitative Evaluation and Systematic Review of Longitudinal and Prospective Trials. *Front. Nutr.* **2016**, *3*, 22. [CrossRef] [PubMed]
15. Otaegui-Arrazola, A.; Amiano, P.; Elbusto, A.; Urdaneta, E.; Martinez-Lage, P. Diet, cognition, and Alzheimer's disease: Food for thought. *Eur. J. Nutr.* **2014**, *53*, 1–23. [CrossRef]
16. Lourida, I.; Soni, M.; Thompson-Coon, J.; Purandare, N.; Lang, I.A.; Ukoumunne, O.C.; Llewellyn, D.J. Mediterranean diet, cognitive function, and dementia: A systematic review. *Epidemiology* **2013**, *24*, 479–489. [CrossRef]
17. Roberts, R.O.; Geda, Y.E.; Cerhan, J.R.; Knopman, D.S.; Cha, R.H.; Christianson, T.J.; Pankratz, V.S.; Ivnik, R.J.; Boeve, B.F.; O'Connor, H.M.; et al. Vegetables, unsaturated fats, moderate alcohol intake, and mild cognitive impairment. *Dement. Geriatr. Cogn. Disord.* **2010**, *29*, 413–423. [CrossRef]
18. Singh, B.; Parsaik, A.K.; Mielke, M.M.; Erwin, P.J.; Knopman, D.S.; Petersen, R.C.; Roberts, R.O. Association of mediterranean diet with mild cognitive impairment and Alzheimer's disease: A systematic review and meta-analysis. *J. Alzheimer Dis.* **2014**, *39*, 271–282. [CrossRef]
19. Craft, N.E.; Haitema, T.B.; Garnett, K.M.; Fitch, K.A.; Dorey, C.K. Carotenoid, tocopherol, and retinol concentrations in elderly human brain. *J. Nutr. Health. Aging* **2004**, *8*, 156–162.
20. Johnson, E.J.; Vishwanathan, R.; Johnson, M.A.; Hausman, D.B.; Davey, A.; Scott, T.M.; Green, R.C.; Miller, L.S.; Gearing, M.; Woodard, J. Relationship between serum and brain carotenoids,-tocopherol, and retinol concentrations and cognitive performance in the oldest old from the Georgia Centenarian Study. *J. Aging Res.* **2013**, *2013*, 951786. [CrossRef]
21. Singh, M. Essential fatty acids, DHA and human brain. *Indian. J. Pediatr.* **2005**, *72*, 239–242. [CrossRef]
22. Rinaldi, P.; Polidori, M.C.; Metastasio, A.; Mariani, E.; Mattioli, P.; Cherubini, A.; Catani, M.; Cecchetti, R.; Senin, U.; Mecocci, P. Plasma antioxidants are similarly depleted in mild cognitive impairment and in Alzheimer's disease. *Neurobiol. Aging* **2003**, *24*, 915–919. [CrossRef]
23. Cunnane, S.C.; Schneider, J.A.; Tangney, C.; Tremblay-Mercier, J.; Fortier, M.; Bennett, D.A.; Morris, M.C. Plasma and brain fatty acid profiles in mild cognitive impairment and Alzheimer's disease. *J. Alzheimers Dis.* **2012**, *29*, 691–697. [CrossRef]
24. Nolan, J.; Loskutova, E.; Howard, N.A.; Moran, R.; Mulcahy, R.; Stack, J.; Bolger, M.; Dennison, J.; Akuffo, K.; Owens, N.; et al. Macular Pigment, Visual Function, and Macular Disease among Subjects with Alzheimer's Disease: An Exploratory Study. *J. Alzheimers Dis.* **2014**, *42*, 1191–1202. [CrossRef]
25. Feeney, J.; O'Leary, N.; Moran, R.; O'Halloran, A.M.; Nolan, J.M.; Beatty, S.; Young, I.S.; Kenny, R.A. Plasma Lutein and Zeaxanthin Are Associated With Better Cognitive Function Across Multiple Domains in a Large Population-Based Sample of Older Adults: Findings from The Irish Longitudinal Study on Aging. *J. Gerontol. A. Biol. Sci. Med. Sci.* **2017**, *72*, 1431–1436. [CrossRef]
26. Zamroziewicz, M.K.; Paul, E.J.; Zwilling, C.E.; Johnson, E.J.; Kuchan, M.J.; Cohen, N.J.; Barbey, A.K. Parahippocampal Cortex Mediates the Relationship between Lutein and Crystallized Intelligence in Healthy, Older Adults. *Front. Aging Neurosci.* **2016**, *8*, 297. [CrossRef]
27. Wu, S., Ding, Y.; Wu, F.; Li, R.; Hou, J.; Mao, P. Omega-3 fatty acids intake and risks of dementia and Alzheimer's disease: A meta-analysis. *Neurosci. Biobehav. Rev.* **2015**, *48*, 1–9. [CrossRef]
28. Talukdar, T.; Zamroziewicz, M.K.; Zwilling, C.E.; Barbey, A.K. Nutrient biomarkers shape individual differences in functional brain connectivity: Evidence from omega-3 PUFAs. *Hum. Brain. Mapp.* **2019**, *40*, 1887–1897. [CrossRef]
29. Bowman, G.L.; Dodge, H.H.; Mattek, N.; Barbey, A.K.; Silbert, L.C.; Shinto, L.; Howieson, D.B.; Kaye, J.A.; Quinn, J.F. Plasma omega-3 PUFA and white matter mediated executive decline in older adults. *Front. Aging Neurosci.* **2013**, *5*, 92. [CrossRef]

30. Power, R.; Coen, R.F.; Beatty, S.; Mulcahy, R.; Moran, R.; Stack, J.; Howard, A.N.; Nolan, J.M. Supplemental Retinal Carotenoids Enhance Memory in Healthy Individuals with Low Levels of Macular Pigment in A Randomized, Double-Blind, Placebo-Controlled Clinical Trial. *J. Alzheimers Dis.* **2018**, *61*, 947–961. [CrossRef]
31. Hammond, B.R., Jr.; Miller, L.S.; Bello, M.O.; Lindbergh, C.A.; Mewborn, C.; Renzi-Hammond, L.M. Effects of Lutein/Zeaxanthin Supplementation on the Cognitive Function of Community Dwelling Older Adults: A Randomized, Double-Masked, Placebo-Controlled Trial. *Front. Aging Neurosci.* **2017**, *9*, 254. [CrossRef]
32. Kulzow, N.; Witte, A.V.; Kerti, L.; Grittner, U.; Schuchardt, J.P.; Hahn, A.; Floel, A. Impact of Omega-3 Fatty Acid Supplementation on Memory Functions in Healthy Older Adults. *J. Alzheimers Dis.* **2016**, *51*, 713–725. [CrossRef]
33. Yurko-Mauro, K.; McCarthy, D.; Rom, D.; Nelson, E.B.; Ryan, A.S.; Blackwell, A.; Salem, N., Jr.; Stedman, M. Beneficial effects of docosahexaenoic acid on cognition in age-related cognitive decline. *Alzheimers Dement.* **2010**, *6*, 456–464. [CrossRef]
34. Power, R.; Prado-Cabrero, A.; Mulcahy, R.; Howard, A.; Nolan, J.M. The Role of Nutrition for the Aging Population: Implications for Cognition and Alzheimer's Disease. *Annu. Rev. Food. Sci. Technol.* **2019**, *10*, 619–639. [CrossRef]
35. Johnson, E.J.; McDonald, K.; Caldarella, S.M.; Chung, H.Y.; Troen, A.M.; Snodderly, D.M. Cognitive findings of an exploratory trial of docosahexaenoic acid and lutein supplementation in older women. *Nutr. Neurosci.* **2008**, *11*, 75–83. [CrossRef]
36. Nolan, J.M.; Mulcahy, R.; Power, R.; Moran, R.; Howard, A.N. Nutritional Intervention to Prevent Alzheimer's Disease: Potential Benefits of Xanthophyll Carotenoids and Omega-3 Fatty Acids Combined. *J. Alzheimers Dis.* **2018**, *64*, 367–378. [CrossRef]
37. Petersen, R.C.; Smith, G.E.; Waring, S.C.; Ivnik, R.J.; Tangalos, E.G.; Kokmen, E. Mild cognitive impairment: Clinical characterization and outcome. *Arch. Neurol.* **1999**, *56*, 303–308. [CrossRef]
38. Winblad, B.; Palmer, K.; Kivipelto, M.; Jelic, V.; Fratiglioni, L.; Wahlund, L.O.; Nordberg, A.; Bäckman, L.; Albert, M.; Almkvist, O.; et al. Mild cognitive impairment—Beyond controversies, towards a consensus: Report of the International Working Group on Mild Cognitive Impairment. *J. Intern. Med.* **2004**, *256*, 240–246. [CrossRef]
39. Dubois, B.; Albert, M.L. Amnestic MCI or prodromal Alzheimer's disease? *Lancet Neurol.* **2004**, *3*, 246–248. [CrossRef]
40. Nasreddine, Z.S.; Phillips, N.A.; Bedirian, V.; Charbonneau, S.; Whitehead, V.; Collin, I.; Cummings, J.L.; Chertkow, H. The Montreal Cognitive Assessment, MoCA: A brief screening tool for mild cognitive impairment. *J. Am. Geriatr. Soc.* **2005**, *53*, 695–699. [CrossRef]
41. Randolph, C.; Tierney, M.C.; Mohr, E.; Chase, T.N. The Repeatable Battery for the Assessment of Neuropsychological Status (RBANS): Preliminary clinical validity. *J. Clin. Exp. Neuropsychol.* **1998**, *20*, 310–319. [CrossRef]
42. Randolph, C. *RBANS. Update: Repeatable Battery for the Assessment of Neuropsychological Status Manual*; NCS Pearson; PsychCorp: Bloomington, MN, USA, 2012.
43. Cambridge Cognition. *CANTAB Connect Research: Admin Application User Guide v1.6*; Cambridge Cognition Limited: Cambridge, UK, 2019.
44. Zygouris, S.; Tsolaki, M. Computerized cognitive testing for older adults: A review. *Am. J. Alzheimers Dis. Other. Demen.* **2015**, *30*, 13–28. [CrossRef] [PubMed]
45. Cambridge Cognition. *Product Overview: CANTAB Connect Research v11.10*; Cambridge Cognition Limited: Cambridge, UK, 2019.
46. Chan, D.; Gallaher, L.M.; Moodley, K.; minati, L.; Burgess, N.; Hartley, T. The 4 Mountains Test: A Short Test of Spatial Memory with High Sensitivity for the Diagnosis of Pre-dementia Alzheimer's Disease. *J. Vis. Exp.* **2016**. [CrossRef]
47. Bird, C.M.; Chan, D.; Hartley, T.; Pijnenburg, Y.A.; Rossor, M.N.; Burgess, N. Topographical short-term memory differentiates Alzheimer's disease from frontotemporal lobar degeneration. *Hippocampus* **2010**, *20*, 1154–1169. [CrossRef] [PubMed]
48. Moodley, K.; minati, L.; Contarino, V.; Prioni, S.; Wood, R.; Cooper, R.; D'Incerti, L.; Tagliavini, F.; Chan, D. Diagnostic differentiation of mild cognitive impairment due to Alzheimer's disease using a hippocampus-dependent test of spatial memory. *Hippocampus* **2015**, *25*, 939–951. [CrossRef]

49. Bucks, R.S.; Ashworth, D.L.; Wilcock, G.K.; Siegfried, K. Assessment of activities of daily living in dementia: Development of the Bristol Activities of Daily Living Scale. *Age. Ageing* **1996**, *25*, 113–120. [CrossRef] [PubMed]
50. Bucks, R.S.; Haworth, J. Bristol Activities of Daily Living Scale: A critical evaluation. *Expert. Rev. Neurother.* **2002**, *2*, 669–676. [CrossRef]
51. Moniz-Cook, E.; Vernooij-Dassen, M.; Woods, R.; Verhey, F.; Chattat, R.; De Vugt, M.; Mountain, G.; O'Connell, M.; Harrison, J.; Vasse, E.; et al. A European consensus on outcome measures for pyschosocial intervention research in dementia care. *Aging Ment. Health.* **2008**, *12*, 14–29. [CrossRef]
52. Sabbagh, M.N.; Malek-Ahmadi, M.; Kataria, R.; Belden, C.M.; Connor, D.J.; Pearson, C.; Jacobson, S.; Davis, K.; Yaari, R.; Singh, U. The Alzheimer's questionnaire: A proof of concept study for a new informant-based dementia assessment. *J. Alzheimers Dis.* **2010**, *22*, 1015–1021. [CrossRef]
53. Malek-Ahmadi, M.; Davis, K.; Belden, C.; Laizure, B.; Jacobson, S.; Yaari, R.; Singh, U.; Sabbagh, M.N. Validation and diagnostic accuracy of the Alzheimer's questionnaire. *Age. Ageing.* **2012**, *41*, 396–399. [CrossRef]
54. Trieschmann, M.; Heimes, B.; Hense, H.W.; Pauleikhoff, D. Macular pigment optical density measurement in autofluorescence imaging: Comparison of one- and two-wavelength methods. *Graefes. Arch. Clin. Exp. Ophthalmol.* **2006**, *244*, 1565–1574. [CrossRef]
55. Roche, W.; Green-Gomez, M.; Moran, R.; Nolan, J. The physics of using the Heidelberg Spectralis dual-wavelength autofluorescence method for the measurement of macular pigment volume. *J. Alzheimers Dis.* **2018**, *64*, 1019–1048.
56. Green-Gomez, M.; Bernstein, P.S.; Curcio, C.A.; Moran, R.; Roche, W.; Nolan, J. Standardizing the Assessment of Macular Pigment Using a Dual-Wavelength Autofluorescence Technique. *Transl. Vis. Sci. Technol.* **2019**, *8*, 41. [CrossRef] [PubMed]
57. Zidichouski, J.A.; Mastaloudis, A.; Poole, S.J.; Reading, J.C.; Smidt, C.R. Clinical validation of a noninvasive, Raman spectroscopic method to assess carotenoid nutritional status in humans. *J. Am. Coll. Nutr.* **2009**, *28*, 687–693. [CrossRef] [PubMed]
58. Meagher, K.A.; Thurnham, D.I.; Beatty, S.; Howard, A.N.; Connolly, E.; Cummins, W.; Nolan, J.M. Serum response to supplemental macular carotenoids in subjects with and without age-related macular degeneration. *Br. J. Nutr.* **2013**, *110*, 289–300. [CrossRef] [PubMed]
59. Benner, B.A.; Schantz, M.M.; Powers, C.D.; Schleicher, R.L.; Camara, J.E.; Sharpless, K.E.; Yen, J.H.; Sniegoski, L.T. Standard Reference Material (SRM) 2378 fatty acids in frozen human serum. Certification of a clinical SRM based on endogenous supplementation of polyunsaturated fatty acids. *Anal. Bioanal. Chem.* **2018**, *410*, 2321–2329. [CrossRef]
60. Solfrizzi, V.; Agosti, P.; Lozupone, M.; Custodero, C.; Schilardi, A.; Valiani, V.; Santamato, A.; Sardone, R.; Dibello, V.; Di Lena, L.; et al. Nutritional interventions and cognitive-related outcomes in patients with late-life cognitive disorders: A systematic review. *Neurosci. Biobehav. Rev.* **2018**, *95*, 480–498. [CrossRef]
61. Remington, R.; Lortie, J.J.; Hoffmann, H.; Page, R.; Morrell, C.; Shea, T.B. A Nutritional Formulation for Cognitive Performance in Mild Cognitive Impairment: A Placebo-Controlled Trial with an Open-Label Extension. *J. Alzheimers Dis.* **2015**, *48*, 591–595. [CrossRef]
62. Yurko-Mauro, K.; Alexander, D.D.; Van Elswyk, M.E. Docosahexaenoic acid and adult memory: A systematic review and meta-analysis. *PLoS ONE* **2015**, *10*, e0120391. [CrossRef]
63. Lee, L.K.; Shahar, S.; Chin, A.V.; Yusoff, N.A. Docosahexaenoic acid-concentrated fish oil supplementation in subjects with mild cognitive impairment (MCI): A 12-month randomised, double-blind, placebo-controlled trial. *Psychopharmacology* **2013**, *225*, 605–612. [CrossRef]
64. Mazereeuw, G.; Lanctot, K.L.; Chau, S.A.; Swardfager, W.; Herrmann, N. Effects of omega-3 fatty acids on cognitive performance: A meta-analysis. *Neurobiol. Aging* **2012**, *33*, 1482.e1417–1429. [CrossRef]
65. Bo, Y.; Zhang, X.; Wang, Y.; You, J.; Cui, H.; Zhu, Y.; Pang, W.; Liu, W.; Jiang, Y.; Lu, Q. The n-3 Polyunsaturated Fatty Acids Supplementation Improved the Cognitive Function in the Chinese Elderly with Mild Cognitive Impairment: A Double-Blind Randomized Controlled Trial. *Nutrients* **2017**, *9*. [CrossRef] [PubMed]
66. Farina, N.; Llewellyn, D.; Isaac, M.G.E.K.; Tabet, N. Vitamin E for Alzheimer's dementia and mild cognitive impairment. *Cochrane. Database Syst. Rev.* **2017**, *4*, CD002854. [CrossRef] [PubMed]
67. European Food Safety Authority. Scientific Opinion on Dietary Reference Values for vitamin E as α-tocopherol. *Efsa. J.* **2015**, *13*. [CrossRef]

68. Institute of Medicine. *Dietary Reference Intakes for Vitamin C, Vitamin E, Selenium, and Carotenoids*; The National Academies Press: Washington, DC, USA, 2000.
69. Dysken, M.W.; Sano, M.; Asthana, S.; Vertrees, J.E.; Pallaki, M.; Llorente, M.; Love, S.; Schellenberg, G.D.; McCarten, J.R.; Malphurs, J.; et al. Effect of vitamin E and memantine on functional decline in Alzheimer disease: The TEAM-AD VA cooperative randomized trial. *JAMA* **2014**, *311*, 33–44. [CrossRef]
70. Kryscio, R.J.; Abner, E.L.; Caban-Holt, A.; Lovell, M.; Goodman, P.; Darke, A.K.; Yee, M.; Crowley, J.; Schmitt, F.A. Association of Antioxidant Supplement Use and Dementia in the Prevention of Alzheimer's Disease by Vitamin E and Selenium Trial (PREADViSE). *JAMA Neurol.* **2017**, *74*, 567–573. [CrossRef]
71. Andrés, P.; Vico, H.; Yáñez, A.; Siquier, A.; Ferrer, G.A. Quantifying memory deficits in amnestic mild cognitive impairment. *Alzheimers Dement.* **2019**, *11*, 108–114. [CrossRef]
72. Irish, M.; Lawlor, B.A.; Coen, R.F.; O'Mara, S.M. Everyday episodic memory in amnestic mild cognitive impairment: A preliminary investigation. *BMC Neurosci.* **2011**, *12*, 80. [CrossRef]
73. Hasselmo, M.E. *How We Remember: Brain Mechanisms of Episodic Memory*; MIT Press: Cambridge, MA, USA, 2011.
74. Nyberg, L. Functional brain imaging of episodic memory decline in ageing. *J. Intern. Med.* **2017**, *281*, 65–74. [CrossRef]
75. Nyberg, L.; McIntosh, A.R.; Cabeza, R.; Habib, R.; Houle, S.; Tulving, E. General and specific brain regions involved in encoding and retrieval of events: What, where, and when. *Proc. Natl. Acad. Sci. USA* **1996**, *93*, 11280–11285. [CrossRef]
76. Weiser, M.J.; Butt, C.M.; Mohajeri, M.H. Docosahexaenoic Acid and Cognition throughout the Lifespan. *Nutrients* **2016**, *8*, 99. [CrossRef]
77. McDougle, D.R.; Watson, J.E.; Abdeen, A.A.; Adili, R.; Caputo, M.P.; Krapf, J.E.; Johnson, R.W.; Kilian, K.A.; Holinstat, M.; Das, A. Anti-inflammatory ω-3 endocannabinoid epoxides. *Proc. Natl. Acad. Sci. USA* **2017**, *114*, E6034–E6043. [CrossRef] [PubMed]
78. Smedt-Peyrusse, V.D.; Sargueil, F.; Moranis, A.; Harizi, H.; Mongrand, S.; Layé, S. Docosahexaenoic acid prevents lipopolysaccharide-induced cytokine production in microglial cells by inhibiting lipopolysaccharide receptor presentation but not its membrane subdomain localization. *J. Neurochem.* **2008**, *105*, 296–307. [CrossRef] [PubMed]
79. Bauer, I.; Crewther, S.; Pipingas, A.; Sellick, L.; Crewther, D. Does omega-3 fatty acid supplementation enhance neural efficiency? A review of the literature. *Hum. Psychopharmacol.* **2014**, *29*, 8–18. [CrossRef] [PubMed]
80. Okabe, N.; Nakamura, T.; Toyoshima, T.; Miyamoto, O.; Lu, F.; Itano, T. Eicosapentaenoic acid prevents memory impairment after ischemia by inhibiting inflammatory response and oxidative damage. *J. Stroke. Cereb. Dis.* **2011**, *20*, 188–195. [CrossRef] [PubMed]
81. Sakai, C.; Ishida, M.; Ohba, H.; Yamashita, H.; Uchida, H.; Yoshizumi, M.; Ishida, T. Fish oil omega-3 polyunsaturated fatty acids attenuate oxidative stress-induced DNA damage in vascular endothelial cells. *PLoS ONE* **2017**, *12*, e0187934. [CrossRef] [PubMed]
82. Bouayed, J.; Bohn, T. Dietary Derived Antioxidants: Implications on Health, Nutrition, Well-Being and Health. Available online: https://www.intechopen.com/books/nutrition-well-being-and-health/dietary-derived-antioxidants-implication-on-health (accessed on 31 March 2020).
83. Bovier, E.R.; Hammond, B.R. A randomized placebo-controlled study on the effects of lutein and zeaxanthin on visual processing speed in young healthy subjects. *Arch. Biochem. Biophys.* **2015**, *572*, 54–57. [CrossRef] [PubMed]
84. Lindbergh, C.A.; Mewborn, C.M.; Hammond, B.R.; Renzi-Hammond, L.M.; Curran-Celentano, J.M.; Miller, L.S. Relationship of Lutein and Zeaxanthin Levels to Neurocognitive Functioning: An fMRI Study of Older Adults. *J. Int. Neuropsychol. Soc.* **2016**, 1–12. [CrossRef]
85. Sohl, G.; Maxeiner, S.; Willecke, K. Expression and functions of neuronal gap junctions. *Nat. Rev. Neurosci.* **2005**, *6*, 191–200. [CrossRef]
86. Guest, J.; Grant, R. Carotenoids and Neurobiological Health. *Adv. Neurobiol.* **2016**, *12*, 199–228. [CrossRef]
87. Kametani, F.; Hasegawa, M. Reconsideration of Amyloid Hypothesis and Tau Hypothesis in Alzheimer's Disease. *Front. Neurosci.* **2018**, *12*, 25. [CrossRef] [PubMed]
88. Strain, H.H. *Leaf Xanthophylls*; Carnegie Institution of Washington: Washington, DC, USA, 1938.
89. Zhang, J.; Burgess, J.G. Enhanced eicosapentaenoic acid production by a new deep-sea marine bacterium Shewanella electrodiphila MAR441T. *PLoS ONE* **2017**, *12*, e0188081. [CrossRef] [PubMed]

90. Britton, G. *Carotenoids Volume 1B: Spectroscopy*; Britton, G., Ed.; Birkhauser Verlag: Basel, Switzerland, 1995.
91. Gueguen, S.; Herbeth, B.; Siest, G.; Leroy, P. An Isocratic Liquid Chromatographic Method with Diode-Array Detection for the Simultaneous Determination of alpha-Tocopherol, Retinol, and Five Carotenoids in Human Serum. *J. Chromatogr. Sci.* **2002**, *40*, 69–76. [CrossRef] [PubMed]
92. European Medicines Agency. *Guideline on Bioanalytical Method Validation*; European Medicines Agency: Amsterdam, The Netherlands, 2011.
93. Brown, J.M.; Duewer, D.L.; Burdette, C.Q.; Sniegoski, L.T.; Yen, J.H. *Certification of Standard Reference Material® 968f Fat-Soluble Vitamins in Frozen Human Serum*; Special Publication, Report Number 260-188; National Institute of Standards and Technology: Gaithersburg, MD, USA, 2017.

Article

Semantic Priming in Mild Cognitive Impairment and Healthy Subjects: Effect of Different Time of Presentation of Word-Pairs

Valeria Guglielmi [1], Davide Quaranta [1,*], Ilaria Mega [2], Emanuele Maria Costantini [2], Claudia Carrarini [2], Alice Innocenti [2] and Camillo Marra [2,3]

1 Neurology Unit, Fondazione Policlinico Agostino Gemelli-IRCCS, 00168 Rome, Italy; valeria.guglielmi@policlinicogemelli.it
2 Department of Neuroscience, Catholic University of Sacred Heart, 00168 Rome, Italy; ilaria.mega@gmail.com (I.M.); emanule.m.costantini@gmail.com (E.M.C.); claudia.carrarini@live.it (C.C.); linceinnocenti@gmail.com (A.I.); camillo.marra@policlinicogemelli.it (C.M.)
3 Memory Clinic, Fondazione Policlinico Universitario Agostino Gemelli-IRCCS, 00168 Rome, Italy
* Correspondence: davide.quaranta@policlinicogemelli.it

Received: 24 May 2020; Accepted: 25 June 2020; Published: 29 June 2020

Abstract: Introduction: Semantic memory is impaired in mild cognitive impairment (MCI). Two main hypotheses about this finding are debated and refer to the degradation of stored knowledge versus the impairment of semantic access mechanisms. The aim of our study is to evaluate semantic impairment in MCI versus healthy subjects (HS) by an experiment evaluating semantic priming. **Methods**: We enrolled 27 MCI and 20 HS. MCI group were divided, according to follow up, into converters-MCI and non converters-MCI. The semantic task consisted of 108 pairs of words, 54 of which were semantically associated. Stimuli were presented 250 or 900 ms later the appearance of the target in a randomized manner. Data were analyzed using factorial ANOVA. **Results**: Both HS and MCI answered more quickly for word than for non-word at both stimulus onset asynchrony (SOA) intervals. At 250 ms, both MCI and HS experienced a shorter time of response for related-word than for unrelated words (priming effect), while only the converters-MCI subgroup lost the priming effect. Further, we observed a rather larger Cohen's d effect size in non converters-MCI than in converters-MCI. **Conclusion**: Our data, and in particular the absence of a semantic priming effect in converters-MCI, could reflect the impairment of semantic knowledge rather than the accessibility of semantic stores in MCI individuals that progress to dementia.

Keywords: Alzheimer disease; semantic priming; mild cognitive impairment

1. Introduction

Alzheimer's disease (AD) is a degenerative process that, in its typical evolution, proceeds in the early stages from the medial perirhinal cortex to the entorhinal, and finally, the hippocampus cortex [1]. Memory disorders are consequently the first symptoms of the disease and episodic memory has been always considered a neuropsychological diagnostic marker of AD. Since the medial perirhinal cortex is deeply involved in the organization of item recollection and general knowledge, it is conceivable that impairment of semantic memory could precede that of episodic memory in the early stages of AD [2]. As a matter of fact, subjects with mild AD experienced lower scores on tests of object naming [3–5], and categorical verbal fluency [2,6–8]. Moreover, low performances in different semantic memory tasks have been also reported in mild AD [9–12], with categorical verbal fluency tasks being found impaired in mild cognitive impairment (MCI) [13,14]. Some authors focused on specific deficits of semantic memory emphasizing the presence of an early and specific deficit in naming and knowing of famous people in MCI patients [15]. Other authors explored specific linguistic markers of degradation of the semantic system, such as frequency, age of acquisition (AoA), familiarity and typicality of the words produced in both category fluency tasks [16–18] or procedural speaking in famous writers [19]. Most studies have shown that AD and MCI patients tend to produce, as the disease progresses, more frequent, more typical and early acquired words [20–24].

Nevertheless, the nature of semantic impairment in the early stage of AD is still controversial. Two main hypotheses have been suggested. The first refers the semantic memory deficits to progressive degradation of the stored knowledge; the second suggests an inability of patients to timely recollect the semantic knowledge cause of a deficit of the mechanism that guarantees the access to the information. In this latter case, executive and phonological verbal mechanisms are hypothesized, and worse category fluency was interpreted as a deficit in access to semantic system [25].

The two explanation are not mutually exclusive and probably describe a simple progression of the degradation of the semantic system. Duong et al. studied both automatic and intentional access to the semantic system in a cohort of MCI and AD subjects. They found that the MCI group was only impaired on tasks of intentional access while the AD group was impaired on both types of tasks. This study would suggest that impaired access to the semantic system in MCI precede semantic store degradation observed in AD.

On the other hand, some authors sustain the hypothesis of early semantic degradation from the beginning [9,26–29]. The disproportionate impairment in category fluency relative to letter-based fluency—that is particularly sensitive to executive damage [5,30]—come out on the side of this.

We have previously investigated some aspects of the semantic system, regarding semantic fluency, specifically the typicality of words produced in a semantic verbal fluency task [23]. Our results showed that the MCI group produced more typical words in comparison with the healthy group, while no differences in typicality were observed within the MCI and AD group. These results were interpreted in term of a progressive disruption of semantic system organization, leading the patients to retrieve more typical items of specific categories in respect to less prototypical elements. Our groups also reported that MCI individuals who will eventually progress to dementia produce words that are less related than the ones produced by healthy controls and stable MCI in a category fluency test [31]. This evidence can be interpreted as an effect of reduced strength of conceptual links between items belonging to a given category.

In previous works, semantic priming (SP) has been used to explore the semantic system automatically; SP is an experimental condition of a lexical decision based on the fact that subjects are faster and more accurate in recognizing a target word when it follows a related word as a prime than when it follows an unrelated word [32]. When it happens in a brief interstimulus interval (250 ms or less), the priming effect is automatic and is not influenced by preparatory or intentional strategies [33]. The prime word automatically and diffusely activates the semantic network, allowing a faster recognition of the target word. Several explanations of the priming effect have been proposed. According to the classical spread-activation hypothesis, the retrieval of an item from memory requires the activation of its conceptual representation. The activation can spread to related concepts, that in turn will be more easily retrieved [34]. The compound-cue models claim that a cue to memory contains the target item and other items of the surrounding context. In this model, a relevant effect of familiarity on priming effect is predicted, since cues formed by related words will be more familiar than cues formed by unrelated words, explaining the shorter reaction times in the case of related words [34]. Finally, distributed network models have been developed. According to the so-called proximity models, priming is observed because related primes and targets are close to each other in a high-dimensional semantic space [35]. The semantic proximity models claim that concepts are represented by patterns of activity over interconnected conceptual units, and that related concepts share similar patterns of activity. The other category of distributed models, referred to as learning models [34], attributes SP to incremental learning. In fact, each presentation of a word causes an alteration of the network connections, increasing the probability of producing the same response to the same input, including semantically related targets. According to this class of models, the decay of learning can be very slow, and may cause SP also over long SOAs. The models of SP are embedded in general conceptualizations of semantic memory based on observations obtained from other typologies of investigations. In Alzheimer's disease and MCI, previous studies suggest that spread-activation models or more complex proximity models could be satisfactory to explain the disruption of lexical-semantic system [31,36].

The previous studies on SP in AD have reported contrasting results. Some authors found less-than-normal priming (hypo-priming) in Alzheimer's disease patients compared with controls [37–39], while others reported no differences on priming in AD vs. Healthy subjects [40,41] or even paradoxical increased priming effects (hyper-priming) in Alzheimer's disease patients [42–47]. These opposed results may reflect some differences in the methods used, and also clinical heterogeneity in the samples of individuals studied. The severity of dementia, and therefore of semantic deficits, differed from one study to another, leading to different results in semantic tasks, SP included. Furthermore, the type of paradigm between the prime and the target, that maybe, superordinate, coordinate, or attribute, seems able to influence the priming effects observed. When the semantic relationship is based on the superordinate category, a normal priming effect was observed [41,48,49], while hypo-priming is more frequently reported in experiments in which the target is an attribute of the prime [48,50,51]. Finally, when the prime and the target belong to the same category (coordinate condition) [46,48,49] hypo- or hyper priming is observed.

Few studies have investigated SP in mild cognitive impairment. Duong et al. [25] reported that MCI and healthy individuals both reported SP effect. The same authors, instead, reported that the MCI group was impaired on tasks that required intentional access to the semantic system, suggesting that in early stages of AD, the involvement of the semantic system concerns access to its first.

The aim of the present study is to clarify the subtle semantic impairment in the MCI condition by studying the behavior to an SP paradigm in comparison to a population of healthy subjects (HS). First, we want to verify if a different behavior can be observed in MCI compared with HS and to understand the meaning of the observed behavior. From a theoretical point of view, if in MCI condition patients experienced an early degradation of the semantic system, consequently they do not benefit from SP. On the contrary, if there is difficult access to the semantic system, the semantic facilitation of priming allows a faster time of answer, especially at the brief, pre-attentional interstimulus interval.

2. Materials and Methods

We enrolled 27 individuals with amnestic MCI and 20 age- and education-matched HS. In the patient groups, all subjects underwent a clinical evaluation including medical history, physical and neurological examination, and an extensive neuropsychological evaluation and MRI scans. All the subjects of our sample were native Italian speakers and none of them had a history of traumatic head injury, alcoholism, epilepsy, stroke, nor other relevant neurologic, psychiatric, and general medical diseases. MCI was diagnosed following clinical criteria [52]; the diagnosis of Alzheimer's Disease was based on current clinical criteria [53]. MCI subjects were evaluated after 12 and 24 months. At follow up, patients were assessed according to the clinical dementia rating scale and activity daily living scales. Dementia was diagnosed when CDR >1 and functional impairment were found. Eleven MCI progressed to overt dementia at the 24 months follow-up.

2.1. Neuropsychological Examination

Patients were diagnosed as amnesic MCI after administration of the Mini Mental State examination and a comprehensive battery including learning and long term memory (Rey's Auditory Verbal Learning test- RAVLT) [54]; executive functions (Stroop's test; [55]); visual search and attention (Multiple Features Target Cancellation, MFTC; [56]); working memory (digit span forwards and backwards; [57]); abstract reasoning (Raven's progressive matrices, PM'47; [54]); constructional praxis (copy of figures without and with Landmarks; [54]). Verbal fluency was examined by phonological [54] and categorical verbal fluency tasks [58].

2.2. Lexical and Semantic Priming Procedure

The task consisted of 216 pairs of stimuli (prime-target). Each target stimulus was a word or a non-word preceded by a prime word. Patients were required to decide if the target was a word or not as fast as possible. In total, 108 pairs were a word–non word couple and 108 pairs were a word–word couple. In total, 54 out of the 108 word–word couples were semantically related. The primer–target couples were controlled for word frequency [59]. Stimuli were presented at different intervals of response to the prime and the appearance of the target (SOA) of 250 and 900 ms (Figure 1). The order of appearance of the pairs of word–non word and word–word at different SOA was randomized for each subject. Words were presented in Italian (English translation of the words is reported in the Appendix A). Our task was built with OpenSesame software and was administered by means of a 15″ laptop posed at 0.5 mt from the individual, who had to respond by pressing one out of two buttons marked with a "yes" or "no" label. Time of response for each item was recorded.

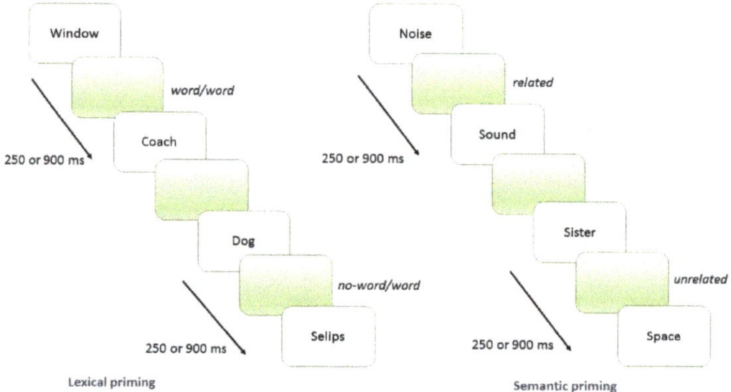

Figure 1. Assessment of the lexical and semantic priming tasks.

We evaluated the 'lexicality effect'; that is, the faster recognition of words compared to non-words. The lexicality effect has been computed as the difference between the mean reaction time between response after that the priming was a word. The 'SP effect' was calculated as the difference in the mean reaction time to word that was preceded by semantically related or semantically unrelated word.

3. Results

MCI and HS were similar in age (MCI: mean age = 72.79; DS = 5.090; HS: mean age = 69.44; DS = 5.515; p = 0.115) and educational level (MCI: mean = 11.61; DS = 2.909; HS: mean = 13.12; DS = 2.711; p = 0.193). At 24 months follow up, 11 subjects (45.8%) were diagnosed as affected by Alzheimer Disease. Then, according to the progression or not to dementia, we divided MCI group in non converters-MCI (n. 13) and converters-MCI (n. 11).

3.1. Lexicality Effect

Both at 250 ms and 900 ms, the factorial ANOVAs comparing MCI vs. HS showed a significant effect of the GROUP factors (F = 458; p <0.001 and F = 441; p<0.001 respectively), WORD factors (word vs. non-word) (F = 1072; $p < 0.001$ and F = 1166; $p < 0.001$ respectively) and of interaction GROUP × WORD (F = 23; $p < 0.001$ and F = 42; $p < 0.001$, respectively).

Table 1 shows response times of the two groups at 250 and 900 ms. Both HS and MCI answered more quickly for word than for non-word, regardless of the presence of a semantic connection between prime and target.

Table 1. Lexicality effect.

	HS		MCI			Non Converters-MCI		Converters-MCI		
SOA 250 ms										
	Mean	**SD**	**Mean**	**SD**	*P*	**Mean**	**SD**	**Mean**	**SD**	*p*
Words	813.37	248.266	1046.15	375.320	<0.001	1052.94	366.878	1036.24	387.545	n.s.
No words	1205.38	483.902	1573.38	471.991	<0.001	1547.61	480.800	1610.85	456.888	n.s.
p	<0.001		<0.001			<0.001		<0.001		
SOA 900 ms										
	Mean	**SD**	**Mean**	**SD**	*P*	**Mean**	**SD**	**Mean**	**SD**	*p*
Words	808.15	247.161	1002.23	331.279	<0.001	996.86	323.493	1010.02	342.552	n.s.
No words	1177.64	460.944	1543.75	462.437	<0.001	1519.25	468.131	1579.34	452.266	n.s.
p	<0.001		<0.001			<0.001		<0.001		

Similarly, patients with MCI were significantly slower than HS both for words ($p < 0.001$) and non-words ($p < 0.001$).

In the comparison between converters-MCI, non converters-MCI and HS at 250 and 900 ms, ANOVA confirmed a significant effect of the GROUP factors (F = 230; $p < 0.001$ at 250 ms and F = 223; $p < 0.001$ at 900 ms), WORD (F = 1144; $p < 0.001$ at 250 ms, and F = 1267; $p < 0.001$ at 900 ms) and of the interaction GROUP x WORD (F = 14; $p < 0.001$ at 250 ms and F = 22; $p < 0.001$ at 900 ms). Both HS and MCI answered more quickly for word than for non-word; non converters-MCI and converters-MCI were significantly slower than HS both for words ($p < 0.001$) that for non-words ($p < 0.001$); the answers time of converters-MCI were not significantly different for the response times of subjects with converters-MCI, both in the presence of words ($p = 0.166$) and in the presence of non-words ($p = 0.9$).

3.1.1. Priming Effect at SOA 250 ms

To evaluate the priming effect, we have considered only reaction times of word–word pairs, comparing semantically related to semantically unrelated pairs of words.

The ANOVA analysis (considering the whole group of MCIs) showed a significant effect of the GROUP factor (F = 131.6; $p < 0.001$) and 'Semantic Relation' Factor (REL) (F = 20.9; $p < 0.001$) but not of the interaction between the two (F = 0.2; $p < 0.678$). On Table 2 are reported the reaction times in the two conditions (pairs of correlated first-target or unrelated strings).

The post-hoc analysis showed that for both related pairs and unrelated words. HS answered more quickly than MCI (in both cases, $p < 0.001$). The response times for related words were lower than the response times for unrelated words in both in HS and MCI (see Table 2). The Cohen's d show only a little effect size in both groups.

In the comparison between non converters-MCI, converters-MCI and HS, ANOVA showed a significant effect of the GROUP factors (F = 69.9; $p < 0.001$) and REL (F = 20.2; $p < 0.001$) but not of the interaction between the two (F = 0.4; $p < 0.650$). On Table 2 is reported the reaction times in the two conditions (pairs of correlated first-target or unrelated strings) in the three groups (non converters-MCI, converters-MCI and HS).

Table 2. Priming Effect at SOAs 250 ms and 900 ms.

	HS		MCI		No Converters-MCI		Converters-MCI	
SOA 250 ms								
	Mean	**SD**	**Mean**	**SD**	**Mean**	**SD**	**Mean**	**SD**
Related words	778.27	187.428	944.01	248.428	969.00	229.787	923.58	261.521
Unrelated words	814.46	212.969	985.38	253.232	1029.33	246.156	949.46	253.915
p	0.015		0.003		0.038		0.321	
Cohen's d	0.18		0.17		**0.25**		0.10	
SOA 900 ms								
	Mean	**SD**	**Mean**	**SD**	**Mean**	**SD**	**Mean**	**SD**
Related words	776.68	214.317	922.60	234.317	930.11	220.048	913.64	250.716
Unrelated words	806.21	204.353	962.37	242.047	982.23	244.109	938.63	238.152
p	0.071		0.003		0.010		0.745	
Cohen's d	0.07		0.17		**0.22**		0.10	

In this condition, we observed a rather larger Cohen's d effect size in non converters-MCI than in converters-MCI. In particular, the effect size in the converters-MCI was similar to the HS group.

The post-hoc analysis showed that for both the related and the unrelated pairs of words, the HS answered more quickly than the subjects non converters-MCI or converters-MCI ($p < 0.001$ in all cases). Response times for related words were shorter than response times for unrelated words in HS and non converters-MCI, but not in converters-MCI (see Table 2). Although both non converters-MCI and HS show a similar priming effect, the Cohen's d is different between these two groups suggesting that the priming in non converters-MCI is preserved only at the cost a higher discrepancy between the two conditions (with related and unrelated words).

3.1.2. Priming Effect at 900 ms

ANOVA analysis showed a significant effect of the GROUP factors ($F = 115.6$; $p < 0.001$) and REL ($F = 17.2$; $p < 0.001$). No interaction between GROUP and REL was found ($F = 0.5$; $p = 0.488$). Table 2 (upper part) reports the mean values of reaction times in the two conditions in HS and MCI.

The post-hoc analysis showed that HS answered more quickly than MCI both for the related pairs and the unrelated words ($p < 0.001$). In HS, the response times for related words did not differ from response time for unrelated words, while MCI answered more quickly for related words that for unrelated words.

Comparing non converters-MCI, converters-MCI and HS, ANOVA showed a significant effect of the GROUP factors ($F = 53.4$; $p < 0.001$) and REL ($F = 16.0$; $p < 0.001$), but no interaction between the two ($F = 0.96$; $p < 0.383$). On Table 2 are reported the means of reaction times in the two conditions in the three groups.

The post-hoc analysis showed that HS had shorter reaction times than non converters-MCI and converters-MCI for both related and unrelated words ($p < 0.001$ in all cases).

Comparing response times for related words vs. unrelated words among the three group, non converters-MCI answered more quickly for related pairs than for unrelated ones, while HS and converters-MCI did not differ in reaction times among related and unrelated words. (see Table 2). The Cohen's d effect size shows that this effect was higher in non converters-MCI than in bot converters-MCI and HS.

4. Discussion

Our data seem to confirm the hypothesis of early degradation of the semantic system in MCI.

A first consideration regards times of reaction: in all tasks, MCI answered significantly slower than HC. These findings are in line with previous studies about priming [42–44,49]. Despite the slower time of reactions, in our sample, lexicality effect was equally represented in HC and MCI, even after splitting MCI in non converters-MCI and converters-MCI. Lexicality effect reflects a physiological learning mechanism: the repeated exposition to verbal stimulus encourages the creation of phonological, syntactic and semantic representations of stimuli. This phenomenon does not happen when stimuli are non words. Our data confirm general findings of the integrity of the phonological, orthographic and syntactic system in early AD. Furthermore, in our study, the presence of lexicality effect in the MCI group demonstrate a good lexical activation and lexical access, even if it happens slower than in HS.

Our data show that at the shorter interval, both HS and MCI had SP effect; nevertheless, when MCI group was divided according to progression to AD, the converters-MCI group lost SP effect, while non converters-MCI had same behavior that HS. At 900 ms, HS lost priming effect, because of the intervention of attentional, strategic and inhibitory mechanisms. Non converters-MCI still showed priming effect, probably because they experienced slower reaction time, and consequently a slower activation of the attentional system. In converters-MCI no priming effect was found, as in the 250 ms interstimulus condition.

In our study MCI group globally considered does not differ from HS, partly because of the intrinsic heterogeneity of MCI population, who belong to subjects with prodromal AD and subjects without progressive memory deficit.

Nevertheless, it is very interesting that the MCI group behaved differently according to disease progression: non converters-MCI had normal priming effect as HS, while converters-MCI lost priming effect. Further, non converters-MCI experienced a larger effect size than converters-MCI; this means that non converters-MCI carry out a greater effort to maintain a SP effect in a compensatory manner. Instead, converters-MCI does not experience any compensatory strategy. This phenomenon happens both at 250 ms and at 900 ms.

These results deserve several considerations: first, the loss of priming effect in converters-MCI—that can be considered as prodromal AD—suggests that semantic system is early impaired regardless of executive functions from the beginning in AD; according to the hypothesis of a semantic store degradation, even in MCI, the priming effect disappears in our converters-MCI because the relationship within semantically close words is lost, preventing the activation of the semantic system or making it ineffective to generate priming effect. This explanation is in line with previous reports about the linguistic features of words produced in the semantic fluency task, by meaning the tendency to generate words with higher typicality and early AoA [20–24]. A possible explanation is that the progressive reduction of the knowledge of the attributes of objects and the relations between similar entities can cause the loss of SP effect that is the counterpart of difficulties observed in the semantic fluency tasks. In general, our findings could be accounted by spread-activation models, or proximity class of diffuse network models, in accordance to previous evidence from studies on semantic fluency [31,36]. On the other hand, learning models are not supported by our findings, since we observed a loss of SP over a long SOA in HS [34].

Some characteristics of the semantic system can help to support our findings. In their study, Mulatti et al. described the cumulative semantic effect in healthy subjects and MCI [51]. The cumulative semantic interference effect refers to a linear increase in the picture naming reaction times which is a function of the already named pictures belonging to the same semantic category to which the named picture belongs. In the author's opinion, this phenomenon is due to the interaction between different cognitive processes involved in picture naming as shared activation, priming and competition: when a representation is activated in the lexicon upon the presentation of a picture, the lexical representations of semantically related items are also activated (shared activation); the activated no-target lexical representations compete with the target lexical representation in a mutually inhibitory way (competition), thus slowing down processing, while any retrieval of a lexical representation facilitates its subsequent retrieval (priming). Analogously, even if in a different experimental condition, in our experiment we observed lexicality effect, but not SP effect in our MCI patients that progress to AD. Our data support the evidence of a loss of cumulative semantic interference effect in MCI converters that could rely on the lack of the shared activation of the semantic features belonging to same categories of knowledge, while the lexical priming still works in the early stages of AD.

The study has several limitations, including the relatively small number of subjects included. MCI patients were diagnosed as MCI due to AD according to clinical and imaging data, but without biomarkers confirmation. However, the rather long follow-up period reduces the possibility of inaccurate identification of MCI individuals both at the baseline and at the follow-up. Furthermore, the number of MCI patients who progressed to dementia is rather low; thus, it is possible that we were not able to detect the SP effect in converters-MCI because of insufficient statistical power. Nevertheless, the number of non-converters MCI was comparable to the number of converters-MCI, yet we were able to detect a significant SP effect, with a Cohen's d higher than the one observed in HS. However, a cautious interpretation of our results is mandatory.

5. Conclusions

In conclusion, our data are of same interest from a theoretical and clinical point of view. As for the theoretical significance, the SP seems a good paradigm to detect subclinical deficit of the semantic system in the early stages of the AD pathology. From a clinical point of view, the different behavior between non converters-MCI and converters-MCI at the SP suggests that this paradigm could be a practical method for evaluating semantic memory in subjects with MCI. If confirmed in larger samples, these results may have significant prognostic and therapeutic implications.

Author Contributions: Concept and design: V.G., D.Q., I.M., C.M. Acquisition, analysis, or interpretation of data: All authors. Drafting of the manuscript: V.G.; D.Q., C.M. Critical revision of the manuscript for important intellectual content: All authors. Statistical analysis: D.Q. Supervision: C.M., D.Q. All authors have read and agreed to the published version of the manuscript.

Funding: This research received no external funding.

Conflicts of Interest: The authors declare no conflict of interest.

Appendix A List of Items

TARGET WORD	RELATED PRIME	UNRELATED PRIME
vinegar	Salad	Hawk
student	Schoolchildren	Peacock
aerial	Cable	Walnut
banana	Gorilla	Die
mouth	Tongue	Species
orange	Peel	Sign
Dog	Peace	Owner
song	Voice	Point
horse	Animal	Example
colour	Red	Need
deluge	Thunder	Ram
dragon	fairy tales	Lip
grass	Lawn	People
summer	Sun	Land
river	Water	Game
giraffe	Height	Accent
Trip	Coach	Gasp
Ant	Bug	Candle
winter	Snow	Number
lake	Fish	Time
snail	Shell	Wax
mum	Sun	News

Sea	Island	Year
pencil	Case	Diamond
mule	Stable	Witch
nose	Face	Day
Ship	Voat	Arm
grandchild	Family	Answer
gandfather	Father	Bottom
track	Trace	Fear
vegetable garden	Vegetable	Loft
package	Courier	Bundle
planet	Heart	Minute
Rain	Weather	Type
spider	Venom	Steam
noise	Sound	Begin
classroom	School	Cold
seed	Grain	Injury
sister	Brother	Space
sauce	Tomato	Sheet
nest	Hibernation	Paint
roof	House	Name
cough	Sickness	Can
trumpet	Brass	Mill
coach	Binary	Steak
wind	Air	Part
worm	Maggot	Elf
volcano	Eruption	Reflex
Paw	Claw	Door
mosquito	Sting	Chest

References

1. Taylor, K.I.; Probst, A. Anatomic localization of the transentorhinal region of the perirhinal cortex. *Neurobiol. Aging* **2008**, *29*, 1591–1596. [CrossRef] [PubMed]
2. Bejanin, A.; Schonhaut, D.R.; La Joie, R.; Kramer, J.H.; Baker, S.L.; Sosa, N.; Ayakta, N.; Cantwell, A.; Janabi, M.; Lauriola, M.; et al. Tau pathology and neurodegeneration contribute to cognitive impairment in Alzheimer's disease. *Brain J. Neurol.* **2017**, *140*, 3286–3300. [CrossRef] [PubMed]
3. Martin, A.; Fedio, P. Word production and comprehension in Alzheimer's disease: The breakdown of semantic knowledge. *Brain Lang.* **1983**, *19*, 124–141. [CrossRef]
4. Huff, F.J.; Collins, C.; Corkin, S.; Rosen, T.J. Equivalent forms of the Boston Naming Test. *J. Clin. Exp. Neuropsychol.* **1986**, *8*, 556–562. [CrossRef]
5. Hodges, J.R.; Salmon, D.P.; Butters, N. Semantic memory impairment in Alzheimer's disease: Failure of access or degraded knowledge? *Neuropsychologia* **1992**, *30*, 301–314. [CrossRef]
6. Ober, B.A.; Dronkers, N.F.; Koss, E.; Delis, D.C.; Friedland, R.P. Retrieval from semantic memory in Alzheimer-type dementia. *J. Clin. Exp. Neuropsychol.* **1986**, *8*, 75–92. [CrossRef]
7. Tröster, A.I.; Salmon, D.P.; McCullough, D.; Butters, N. A comparison of the category fluency deficits associated with Alzheimer's and Huntington's disease. *Brain Lang.* **1989**, *37*, 500–513. [CrossRef]
8. Bayles, K.A.; Tomoeda, C.K.; Trosset, M.W. Naming and categorical knowledge in Alzheimer's disease: The process of semantic memory deterioration. *Brain Lang.* **1990**, *39*, 498–510. [CrossRef]
9. Hodges, J.R.; Patterson, K. Is semantic memory consistently impaired early in the course of Alzheimer's disease? Neuroanatomical and diagnostic implications. *Neuropsychologia* **1995**, *33*, 441–459. [CrossRef]
10. Monsch, A.U.; Bondi, M.W.; Butters, N.; Salmon, D.P.; Katzman, R.; Thal, L.J. Comparisons of verbal fluency tasks in the detection of dementia of the Alzheimer type. *Arch. Neurol.* **1992**, *49*, 1253–1258. [CrossRef] [PubMed]

11. Nutter-Upham, K.E.; Saykin, A.J.; Rabin, L.A.; Roth, R.M.; Wishart, H.A.; Pare, N.; Flashman, L.A. Verbal fluency performance in amnestic MCI and older adults with cognitive complaints. *Arch. Clin. Neuropsychol.* **2008**, *23*, 229–241. [CrossRef]
12. Adlam, A.L.; Bozeat, S.; Arnold, R.; Watson, P.; Hodges, J.R. Semantic knowledge in mild cognitive impairment and mild Alzheimer's disease. *Cortex* **2006**, *42*, 675–684. [CrossRef]
13. Murphy, K.J.; Rich, J.B.; Troyer, A.K. Verbal fluency patterns in amnestic mild cognitive impairment are characteristic of Alzheimer's type dementia. *J. Int. Neuropsychol. Soc.* **2006**, *12*, 570–574. [CrossRef] [PubMed]
14. Amieva, H.; Letenneur, L.; Dartigues, J.F.; Rouch-Leroyer, I.; Sourgen, C.; D'Alchee-Biree, F.; Dib, M.; Barberger-Gateau, P.; Orgogozo, J.M.; Fabrigoule, C. Annual rate and predictors of conversion to dementia in subjects presenting mild cognitive impairment criteria defined according to a population-based study. *Dement. Geriatr. Cogn. Disord.* **2004**, *18*, 87–93. [CrossRef] [PubMed]
15. Joubert, S.; Brambati, S.M.; Ansado, J.; Barbeau, E.J.; Felician, O.; Didic, M.; Lacombe, J.; Goldstein, R.; Chayer, C.; Kergoat, M.J. The cognitive and neural expression of semantic memory impairment in mild cognitive impairment and early Alzheimer's disease. *Neuropsychologia* **2010**, *48*, 978–988. [CrossRef]
16. Schilling, H.E.H.; Rayner, K.; Chumbley, J.I. Comparing naming, lexical decision, and eye fixation times: Word frequency effects and individual differences. *Mem. Cognit.* **1998**, *26*, 1270–1281. [CrossRef]
17. Juhasz, B.J. Age-of-acquisition effects in word and picture identification. *Psychol. Bull.* **2005**, *131*, 684–712. [CrossRef]
18. Holmes, S.J.; Jane Fitch, F.; Ellis, A.W. Age of acquisition affects object recognition and naming in patients with Alzheimer's disease. *J. Clin. Exp. Neuropsychol.* **2006**, *28*, 1010–1022. [CrossRef]
19. Garrard, P.; Maloney, L.M.; Hodges, J.R.; Patterson, K. The effects of very early Alzheimer's disease on the characteristics of writing by a renowned author. *Brain J. Neurol.* **2005**, *128*, 250–260. [CrossRef]
20. Forbes-McKay, K.E.; Ellis, A.W.; Shanks, M.F.; Venneri, A. The age of acquisition of words produced in a semantic fluency task can reliably differentiate normal from pathological age related cognitive decline. *Neuropsychologia* **2005**, *43*, 1625–1632. [CrossRef]
21. Sailor, K.M.; Zimmerman, M.E.; Sanders, A.E. Differential impacts of age of acquisition on letter and semantic fluency in Alzheimer's disease patients and healthy older adults. *Q. J. Exp. Psychol.* **2011**, *64*, 2383–2391. [CrossRef] [PubMed]
22. Venneri, A.; McGeown, W.J.; Hietanen, H.M.; Guerrini, C.; Ellis, A.W.; Shanks, M.F. The anatomical bases of semantic retrieval deficits in early Alzheimer's disease. *Neuropsychologia* **2008**, *46*, 497–510. [CrossRef] [PubMed]
23. Vita, M.G.; Marra, C.; Spinelli, P.; Caprara, A.; Scaricamazza, E.; Castelli, D.; Canulli, S.; Gainotti, G.; Quaranta, D. Typicality of words produced on a semantic fluency task in amnesic mild cognitive impairment: Linguistic analysis and risk of conversion to dementia. *J. Alzheimer's Dis.* **2014**, *42*, 1171–1178. [CrossRef] [PubMed]
24. Venneri, A.; Jahn-Carta, C.; de Marco, M.; Quaranta, D.; Marra, C. Diagnostic and prognostic role of semantic processing in preclinical Alzheimer's disease. *Biomark. Med.* **2018**, *12*, 637–651. [CrossRef] [PubMed]
25. Duong, A.; Whitehead, V.; Hanratty, K.; Chertkow, H. The nature of lexico-semantic processing deficits in mild cognitive impairment. *Neuropsychologia* **2006**, *44*, 1928–1935. [CrossRef]
26. Albert, M.S.; Moss, M.B.; Tanzi, R.; Jones, K. Preclinical prediction of AD using neuropsychological tests. *J. Int. Neuropsychol. Soc.* **2001**, *7*, 631–639. [CrossRef]
27. Chen, P.; Ratcliff, G.; Belle, S.H.; Cauley, J.A.; DeKosky, S.T.; Ganguli, M. Patterns of cognitive decline in presymptomatic Alzheimer disease: A prospective community study. *Arch. Gen. Psychiatry* **2001**, *58*, 853–858. [CrossRef]
28. Hodges, J.R.; Patterson, K.; Graham, N.; Dawson, K. Naming and knowing in dementia of Alzheimer's type. *Brain Lang.* **1996**, *54*, 302–325. [CrossRef] [PubMed]
29. Perry, R.J.; Watson, P.; Hodges, J.R. The nature and staging of attention dysfunction in early (minimal and mild) Alzheimer's disease: Relationship to episodic and semantic memory impairment. *Neuropsychologia* **2000**, *38*, 252–271. [CrossRef]
30. Rosser, A.; Hodges, J.R. Initial letter and semantic category fluency in Alzheimer's disease, Huntington's disease, and progressive supranuclear palsy. *J. Neurol. Neurosurg. Psychiatry* **1994**, *57*, 1389–1394. [CrossRef] [PubMed]

31. Quaranta, D.; Piccininni, C.; Caprara, A.; Malandrino, A.; Gainotti, G.; Marra, C. Semantic Relations in a Categorical Verbal Fluency Test: An Exploratory Investigation in Mild Cognitive Impairment. *Front. Psychol.* **2019**, *10*, 2797. [CrossRef]
32. Collins, A.M.; Loftus, E.F. A spreading-activation theory of semantic processing. *Psychol. Rev.* **1975**, *82*, 407–428. [CrossRef]
33. Neely, J.H. Semantic priming and retrieval from lexical memory: Roles of inhibitionless spreading activation and limited-capacity attention. *J. Exp. Psychol. Gen.* **1977**, *106*, 226–254. [CrossRef]
34. McNamara, P.; Holbrook, J.P. Semantic memory and priming. In *Handbook of Psychology*; Healy, A.F., Proctor, R.W., Eds.; John Wiley & Sons, Inc.: Hoboken, NJ, USA, 2003; Volume 4.
35. Plaut, D.C.; Booth, J.R. Individual and developmental differences in semantic priming: Empirical and computational support for a single-mechanism account of lexical processing. *Psychol. Rev.* **2000**, *107*, 786–823. [CrossRef] [PubMed]
36. Foster, P.S.; Drago, V.; Yung, R.C.; Pearson, J.; Stringer, K.; Giovannetti, T.; Libon, D.; Heilman, K.M. Differential lexical and semantic spreading activation in Alzheimer's disease. *Am. J. Alzheimer's Dis. Other Demen.* **2013**, *28*, 501–507. [CrossRef]
37. Bell, E.E.; Chenery, H.J.; Ingram, J.C. Semantic priming in Alzheimer's dementia: Evidence for dissociation of automatic and attentional processes. *Brain Lang.* **2001**, *76*, 130–144. [CrossRef]
38. Silveri, M.C.; Monteleone, D.; Burani, C.; Tabossi, P. Automatic semantic facilitation in Alzheimer's disease. *J. Clin. Exp. Neuropsychol.* **1996**, *18*, 371–382. [CrossRef] [PubMed]
39. Salmon, D.P.; Butters, N.; Chan, A.S. The deterioration of semantic memory in Alzheimer's disease. *Can. J. Exp. Psychol. Rev. Can. Psychol. Exp.* **1999**, *53*, 108–117. [CrossRef]
40. Perri, R.; Carlesimo, G.A.; Zannino, G.D.; Mauri, M.; Muolo, B.; Pettenati, C.; Caltagirone, C. Intentional and automatic measures of specific-category effect in the semantic impairment of patients with Alzheimer's disease. *Neuropsychologia* **2003**, *41*, 1509–1522. [CrossRef]
41. Ober, B.A.; Shenaut, G.K.; Jagust, W.J.; Stillman, R.C. Automatic semantic priming with various category relations in Alzheimer's disease and normal aging. *Psychol. Aging* **1991**, *6*, 647–660. [CrossRef]
42. Chertkow, H.; Bub, D.; Bergman, H.; Bruemmer, A.; Merling, A.; Rothfleisch, J. Increased semantic priming in patients with dementia of the Alzheimer's type. *J. Clin. Exp. Neuropsychol.* **1994**, *16*, 608–622. [CrossRef] [PubMed]
43. Chertkow, H.; Bub, D.; Seidenberg, M. Priming and semantic memory loss in Alzheimer's disease. *Brain Lang.* **1989**, *36*, 420–446. [CrossRef]
44. Nebes, R.D.; Brady, C.B.; Huff, F.J. Automatic and attentional mechanisms of semantic priming in Alzheimer's disease. *J. Clin. Exp. Neuropsychol.* **1989**, *11*, 219–230. [CrossRef] [PubMed]
45. Balota, D.A.; Watson, J.M.; Duchek, J.M.; Ferraro, F.R. Cross-modal semantic and homograph priming in healthy young, healthy old, and in Alzheimer's disease individuals. *J. Int. Neuropsychol. Soc.* **1999**, *5*, 626–640. [CrossRef]
46. Giffard, B.; Desgranges, B.; Nore-Mary, F.; Lalevée, C.; de la Sayette, V.; Pasquier, F.; Eustache, F. The nature of semantic memory deficits in Alzheimer's disease: New insights from hyperpriming effects. *Brain J. Neurol.* **2001**, *124*, 1522–1532. [CrossRef]
47. Perri, R.; Zannino, G.D.; Caltagirone, C.; Carlesimo, G.A. Semantic priming for coordinate distant concepts in Alzheimer's disease patients. *Neuropsychologia* **2011**, *49*, 839–847. [CrossRef]
48. Rogers, S.L.; Friedman, R.B. The underlying mechanisms of semantic memory loss in Alzheimer's disease and semantic dementia. *Neuropsychologia* **2008**, *46*, 12–21. [CrossRef]
49. Glosser, G.; Friedman, R.B.; Grugan, P.K.; Lee, J.H.; Grossman, M. Lexical semantic and associative priming in Alzheimer's disease. *Neuropsychology* **1998**, *12*, 218–224. [CrossRef]
50. Giffard, B.; Desgranges, B.; Nore-Mary, F.; Lalevée, C.; Beaunieux, H.; de la Sayette, V.; Pasquier, F.; Eustache, F. The dynamic time course of semantic memory impairment in Alzheimer's disease: Clues from hyperpriming and hypopriming effects. *Brain J. Neurol.* **2002**, *125*, 2044–2057. [CrossRef]
51. Mulatti, C.; Calia, C.; De Caro, M.F.; Della Sala, S. The cumulative semantic interference effect in normal and pathological ageing. *Neuropsychologia* **2014**, *65*, 125–130. [CrossRef]

52. Winblad, B.; Palmer, K.; Kivipelto, M.; Jelic, V.; Fratiglioni, L.; Wahlund, L.O.; Nordberg, A.; Backman, L.; Albert, M.; Almkvist, O.; et al. Mild cognitive impairment–beyond controversies, towards a consensus: Report of the International Working Group on Mild Cognitive Impairment. *J. Intern. Med.* **2004**, *256*, 240–246. [CrossRef] [PubMed]
53. McKhann, G.M.; Knopman, D.S.; Chertkow, H.; Hyman, B.T.; Jack, C.R.; Kawas, C.H.; Klunk, W.E.; Koroshetz, W.J.; Manly, J.J.; Mayeux, R.; et al. The diagnosis of dementia due to Alzheimer's disease: Recommendations from the National Institute on Aging-Alzheimer's Association workgroups on diagnostic guidelines for Alzheimer's disease. *Alzheimer's Dement.* **2011**, *7*, 263–269. [CrossRef] [PubMed]
54. Carlesimo, G.A.; Caltagirone, C.; Gainotti, G.; Fadda, L.; Gallassi, R.; Lorusso, S.; Marfia, G.; Marra, C.; Nocentini, U.; Parnetti, L. The mental deterioration battery: Normative data, diagnostic reliability and qualitative analyses of cognitive impairment. *Eur. Neurol.* **1996**, *36*, 378–384. [CrossRef]
55. Caffarra, P.; Vezzadini, G.; Dieci, F.; Zonato, F.; Venneri, A. Una versione abbreviata del test di Stroop: Dati normativi nella popolazione italiana. *Nuova Riv. Neurol.* **2002**, *12*, 111–115.
56. Marra, C.; Gainotti, G.; Scaricamazza, E.; Piccininni, C.; Ferraccioli, M.; Quaranta, D. The Multiple Features Target Cancellation (MFTC): An attentional visual conjunction search test. Normative values for the Italian population. *Neurol. Sci.* **2013**, *34*, 173–180. [CrossRef] [PubMed]
57. Monaco, M.; Costa, A.; Caltagirone, C.; Carlesimo, G.A. Forward and backward span for verbal and visuo-spatial data: Standardization and normative data from an Italian adult population. *Neurol. Sci.* **2013**, *34*, 749–754. [CrossRef]
58. Quaranta, D.; Caprara, A.; Piccininni, C.; Vita, M.G.; Gainotti, G.; Marra, C. Standardization, Clinical Validation, and Typicality Norms of a New Test Assessing Semantic Verbal Fluency. *Arch. Clin. Neuropsychol. Off. J. Natl. Acad. Neuropsychol.* **2016**, *31*, 434–445. [CrossRef]
59. Balota, D.A.; Chumbley, J.I. Are lexical decisions a good measure of lexical access? The role of word frequency in the neglected decision stage. *J. Exp. Psychol. Hum. Percept. Perform.* **1984**, *10*, 340–357. [CrossRef]

Journal of
Personalized
Medicine

Review

Neurophysiological Hallmarks of Neurodegenerative Cognitive Decline: The Study of Brain Connectivity as a Biomarker of Early Dementia

Paolo Maria Rossini [1,*], Francesca Miraglia [1], Francesca Alù [1], Maria Cotelli [2], Florinda Ferreri [3,4], Riccardo Di Iorio [5], Francesco Iodice [1,5] and Fabrizio Vecchio [1]

1. Brain Connectivity Laboratory, Department of Neuroscience & Neurorehabilitation, IRCCS San Raffaele Pisana, 00167 Rome, Italy; fra.miraglia@gmail.com (F.M.); francesca.alu@sanraffaele.it (F.A.); franc.iodice@gmail.com (F.I.); fabrizio.vecchio@uniroma1.it (F.V.)
2. Neuropsychology Unit, IRCCS Istituto Centro San Giovanni di DioFatebenefratelli, 25125 Brescia, Italy; mcotelli@fatebenefratelli.eu
3. Department of Neuroscience, Unit of Neurology and Neurophysiology, University of Padua, 35100 Padua, Italy; fimferreri@yahoo.it
4. Department of Clinical Neurophysiology, Kuopio University Hospital, University of Eastern Finland, 70100 Kuopio, Finland
5. Neurology Unit, IRCCS Polyclinic A. Gemelli Foundation, 00168 Rome, Italy; riccardo.diiorio@policlinicogemelli.it

* Correspondence: paolomaria.rossini@sanraffaele.it; Tel.: +39-06-52253767

Received: 30 March 2020; Accepted: 27 April 2020; Published: 30 April 2020

Abstract: Neurodegenerative processes of various types of dementia start years before symptoms, but the presence of a "neural reserve", which continuously feeds and supports neuroplastic mechanisms, helps the aging brain to preserve most of its functions within the "normality" frame. Mild cognitive impairment (MCI) is an intermediate stage between dementia and normal brain aging. About 50% of MCI subjects are already in a stage that is prodromal-to-dementia and during the following 3 to 5 years will develop clinically evident symptoms, while the other 50% remains at MCI or returns to normal. If the risk factors favoring degenerative mechanisms are modified during early stages (i.e., in the prodromal), the degenerative process and the loss of abilities in daily living activities will be delayed. It is therefore extremely important to have biomarkers able to identify—in association with neuropsychological tests—prodromal-to-dementia MCI subjects as early as possible. MCI is a large (i.e., several million in EU) and substantially healthy population; therefore, biomarkers should be financially affordable, largely available and non-invasive, but still accurate in their diagnostic prediction. Neurodegeneration initially affects synaptic transmission and brain connectivity; methods exploring them would represent a 1st line screening. Neurophysiological techniques able to evaluate mechanisms of synaptic function and brain connectivity are attracting general interest and are described here. Results are quite encouraging and suggest that by the application of artificial intelligence (i.e., learning-machine), neurophysiological techniques represent valid biomarkers for screening campaigns of the MCI population.

Keywords: Alzheimer's disease; mild cognitive impairment; EEG; TMS

1. Introduction

Dementias are of several types; however, the most frequent and diffusely known by the public opinion is Alzheimer's disease (AD), which is characterized by a progressive loss of memory and deterioration of other cognitive functions that significantly interfere with daily life activities [1]. The typical AD clinical phenotype follows a prodromal stage known as mild cognitive impairment

(MCI) that is usually, but not exclusively, characterized by memory loss (amnestic MCI = aMCI). MCI is typically characterized by evidence of an objective impairment of memory and/or of other cognitive domains on neuropsychological testing, but not yet encompassing the standards for dementia diagnosis. It represents an intermediate condition in the elderly between normal cognition and dementia and includes a consistent percentage of subjects (about 50%) in a stage that is prodromal to different types of dementia, including AD (MCI prodromal-to-dementia or prodromal-to-AD). MCI is considered a high-risk population since a significant percentage (from 5 to 20 times higher compared to an age-matched non-MCI population) will develop one type of dementia during a 3- to 5-year follow-up period; the remaining percentage will stay in the MCI condition for the rest of their life or even revert to full normality. MCI prodromal-to-AD (or due-to-AD) cannot be distinguished from those who will not convert on purely clinical grounds. A careful MCI definition requires a comprehensive assessment, including cognitive complaints questionnaires, screening tests (such as Mini-Mental State Examination (MMSE)), an in-depth neuropsychological evaluation (including tests for episodic memory, language, visuo–spatial abilities, and behavioral scales with appropriate normative thresholds [2,3]), functional scales and full neurological examinations. In order to plan optimal and early therapeutic, organizational, and rehabilitative interventions, MCI diagnosis should be combined with the most reliable prognosis on the likelihood and time of eventual progression to dementia. In other words, those MCI subjects who are already in a prodromal-to-dementia condition should be intercepted as early as possible. This goal can nowadays be achieved by combining biomarkers reflecting ongoing neurodegenerative phenomena with the results of neuropsychological tests.

The identification of reliable markers able to intercept those MCI subjects (amnesic, non-amnesic, multi-domain) who are in a prodromal-to-dementia stage represents a goal for all health systems as it would allow early interventions on different risk factors. The risk factors include lifestyle aspects such as obesity, sedentary lifestyle, smoke, low daily cognitive and exercise, and medical conditions such as cardiovascular diseases, diabetes, hypercholesterolemia, and thyroid dysfunction, leading to a significant delay in the daily living autonomies loss even in the absence of a disease-modifying therapy [4–10]. Such a goal would be of paramount importance since—just as an example that might be equally expanded to all countries with an aging population—the costs of dementia in the United States (US) were estimated to be USD 818 billion in 2015, with an increase of 35% compared to 2010. Moreover, MCI prodromal-to-AD subjects are the main targets of many clinical trials with potentially disease-modifying drugs since these drugs have proved ineffective when full symptomatology of AD has already developed, probably because the "neural reserve" has been progressively consumed during the pre-symptomatic and prodromal disease stages. Therefore, early markers predicting with high sensitivity/specificity the evolution from prodromal stages to clinically overt dementia and AD are of pivotal importance in modern public health strategies.

Within this theoretical frame, it seems quite important to have a 1st-level type of low-cost, non-invasive, and widely available biomarker(s) able to screen out from the MCI population those subjects who are non-prodromal-to-dementia, leaving more expensive and highly demanding technologies as a 2nd level approach for a significantly smaller population with a remarkably higher risk of being in a prodromal-to-dementia condition for diagnostic characterization (i.e., AD with amyloid plaques).

2. EEG Biomarkers

Scalp resting state electroencephalographic (EEG) rhythms reflect the summation of oscillatory membrane post-synaptic potentials generated from cortical pyramidal neurons, which play the role of electromagnetic signal sources. These sources were estimated to extend for several square centimeters of the brain cortex [11,12]. These potentials can be considered as the oscillatory output of the resting state cortical system, while inputs include afferents coming from other cortical neural biomasses, thalamo-cortical neurons, and neurons belonging to ascending reticular systems [11].

Practically speaking, EEG data analysis may be divided into a two-step process: first, the signals recorded from all sensors are "de-noised", aiming to improve signal-to-noise ratio by excluding portions of highly noisy data; second, the current density distribution or other parameters of interest are estimated from the cleaned sensor recorded signals. This phase, called preprocessing, is devoted to the extraction of the source under study from the whole population of electromagnetic sources, including also the artefactual ones. Methods improving the signal/noise ratio "separate as much as possible the signal from the noise using information on the specific source under study. In some cases, it is possible to observe neural activity synchronization by supplying to the subject an external stimulus or instructing the subject to perform a specific task. Given the high relevance of analyzing resting-state activity, alternative procedures to enhance the signal to noise ratio were developed, including Blind Source Separation (BSS) methods such as Independent Component Analysis—ICA [13] and semi-BSS methods such as Functional Source Separation—FSS" (see Figure 2) [14,15].

Another important step aims to determine the current density distribution inside the brain, especially in the region of interest. The diverse approaches to solve the so-called inverse-problem (that is the identification of the source(s) within the brain responsible for the distribution of scalp-recorded electromagnetic signals) range from single and multiple dipoles [16] to distributed sources, which include the Multiple Signal Classification—MUSIC [17], the recursively applied and projected MUSIC—RAP-MUSIC [18], the minimum norm estimates—MNE [19], the low-resolution brain electromagnetic tomography—LORETA [20], and the beam-forming and synthetic aperture magnetometry—SAM [21].

The scalp-recorded EEG signals oscillate with rhythms characterized by a spectral content below 50 Hz since the extracerebral layers act as spatial and frequency filters. Two classes of EEG biomarkers for early dementia diagnosis, such as "synchronization" and "connectivity" can be nowadays identified [22]. The term "synchronization" refers to nonlinear oscillatory components of the brain system that reflect a collective oscillatory behavior of cortical neural populations generating EEG rhythms [23]. Synchronization of the cyclic firing of cortical neural populations is the main source of scalp EEG rhythms in both resting state and task-related conditions and produce scalp EEG rhythms: this "synchronization" mechanism must occur at a macroscopic spatial scale of some centimeters. Spectral analysis of EEG rhythms is typically done at fixed frequency bands. Both nonlinear and linear mathematics can estimate the neural current density of EEG cortical sources [24,25]. These procedures model 3D tomographic patterns of EEG cortical generators into a spherical or a magnetic resonance imaging (MRI)-based head model representing electrical properties of the cerebral cortex, skull, and scalp, typically co-registered to Talairach brain atlas [26–30]. Source localization procedures estimate the current intensity of all dipoles (e.g., hundreds to thousands) of the cortical mantle model to explain scalp EEG amplitude/power density.

2.1. EEG Findings in Dementia (including AD)

It is important to clarify that EEG recordings (particularly the routine ones with 19 electrodes) cannot reach a distinct diagnosis of the various types of dementia. In all the studies reported below, the diagnosis of AD was reached with neuropsychological tests eventually combined with other biomarkers dealing with brain metabolism and analysis of beta-amyloid and tau protein metabolites (i.e., fluorodeoxyglucose positron emission tomography (PET–FDG), PET–radioligands, and cerebrospinal fluid (CSF) analysis). Having clarified this important point, one should consider that there is a vast literature on EEG abnormalities in pathological brain aging (for a review, see [31]). Compared to cognitively intact elderly (Nold) subjects, demented (namely, AD) patients contain excessive δ and a significant decrement of posterior α rhythms [32]. Similarly, MCI patients display a significant decrease of α power compared to Nold [33]. Furthermore, a prominent decrease of EEG spectral coherence in the α band in AD has been reported [34,35]. Indeed, the EEG power spectrum in patients with AD compared to age-matched Nold has shown a widespread increase in δ and θ power density and a posterior decrease in α and β power density with lowering of α power density peak in several studies [34–38].

Nonlinear measures of "synchronization" markers pointed to a complexity loss of cerebral dynamics in AD within the same frequency bands [39–45], while the analysis of phase coherence showed differences between AD and Nold [35] and was also able to predict aMCI conversion to AD as demonstrated by neuropsychological follow-up [34]. Cross-validation of EEG source solutions showed that clinical symptoms were positively correlated with abnormalities in β, α, and δ source activities [46,47]. Global cognitive status, as revealed by MMSE scores, correlated negatively with δ/θ source activity and positively with α source activity [22,47–49]. Similar features of EEG sources with some attenuation in amplitude, as seen in AD patients, were also observed in MCI subjects [22,49]. These findings were confirmed by an independent approach based on minimum-norm depth-weighted estimation [50], that showed in AD patients a reduced activity in the precuneus, posterior cingulate, and parietal regions, as well as increased activity in δ or θ sources in inferior parietal cortex, medial temporal cortex, precuneus, and posterior cingulate, compared to aMCI subjects [50].

Occipital, temporal, and parietal α source activities correlated with hippocampal volume, being more evident in aMCI subjects with a greater volume, intermediate in those with a smaller volume, and minimum in AD patients [47]. Moreover, α source activity was statistically linked to the volume of cortical gray matter in aMCI and AD subjects, while a negative correlation was found with δ sources [51,52]. Finally, a negative correlation between EEG α dipolarity (e.g., uniformity of α potential distribution) and p-tau or p-tau/Aβ ratio in cerebrospinal fluid in AD was described [53].

Nonlinearity brain electromagnetic rhythms have attracted substantial attention since the early 1980s [42,54,55], due to the approach based on the chaos theory, aiming at a deterministic characterization of complex time series [56,57] and to the observation that multiple neural processes are governed by nonlinear phenomena which are essential for healthy and adaptive cortical activity, but are also involved in several brain diseases [58]. The early application of nonlinear methods based on the chaos theory to the analysis of spontaneous EEG in AD showed lower correlation dimension (D2) [56] and the largest Lyapunov exponent (L1) [57] compared to Nold, attributable to a reduction of the variables needed to describe the dynamics of the EEG (D2) and to a loss of flexibility in information processing (L1). This is because D2 is a measure of the geometry of the attractor that describes the EEG signals, whereas L1 explains how many similarities diverge over time [54]. Despite their different focus on static and dynamic properties of the ongoing signals, both D2 and L1 parameters paralleled the reduction of complexity seen in the EEG activity of AD patients [45,54,59–61].

Methods of nonlinear EEG analysis can be categorized into three main groups:

- Fractal dimension metrics, including Katz and Higuchi's definitions [62,63].
- Irregularity estimators, including sample entropy [64] and permutation entropy [65].
- Multiscale metrics [66], including multiscale sample entropy and derived approaches such as multiscale dispersion entropy [45].

The concept of fractal dimension refers to a non-integer dimension of a geometric object; this parameter is reduced in AD compared to Nold, especially in temporal–occipital regions [61]. Metrics such as sample entropy (SampEn) can be seen as measures of the production rate of information within a signal (how much information previous samples of the time series provide about the following samples) and its level of predictability [61]. Entropy metrics of spontaneous EEGs showed reduced irregularity in AD. The third major category of nonlinear measures is related to the multiscale behavior of signals and to the concept of complexity ranging between two extremes of fully predictable and deterministic systems and merely random oscillations [67]. Thus, completely ordered (i.e., predictable) or random systems are not physiologically complex [68]. A working measure of complexity (defined as multiscale (sample) entropy (MSE [67])) is based on the measure of entropy (originally SampEn) over multiple temporal scales obtained from "coarse-grained" versions of the signals [60,66], and it has inspired the application of entropy metrics in a multiscale way [66]. MSE has been compared between

AD and Nold [55,68,69]; it has been shown that spontaneous EEG activity in AD is less complex at short temporal scales (associated with higher frequencies), but this tendency reverses at longer temporal scales (related to lower frequencies) [68,69]. This finding is of remarkable interest when arguing about the dependency of the complexity of brain EEG activity on the temporal scales and the frequency range under analysis [70,71]. Arguably, one of the limitations of the nonlinear methods surveyed so far is that they are applicable to single (univariate) signals only. Multivariate versions have become recently available [45,72–74]; however, they should be validated more deeply as a probe for EEG analysis. Finally, we should consider that non-linear analysis of EEG activity has been explored in resting-state and awake conditions; methods also applicable to short time series can now be utilized before, during, and after a task with the aim of increasing sensitivity/specificity to characterize pathological cognitive decline [59,75,76].

Despite a number of limitations, important recent reviews ([44], Rossini et al. (2020)) have summarized the progress in the EEG pattern of demented patients with a neuropsychological profile of AD: generalized slowing of the spectral frequency profile, reduced complexity, and perturbation.

2.2. Brain Connectivity Methods including Graph-Theory

The human brain can be represented as an anatomic-functional matrix (consisting of billions of neurons and their synaptic connections) of network structures at micro–meso–macro-scale levels. Within this matrix of networks, nodes (neuronal assemblies) and links (connecting fibers) cooperate via dynamic aggregations or transient locking/unlocking of their orchestrated firing oscillations [77–80]. Networks continuously re-shape throughout life via plastic modifications mainly governed by long term synaptic potentiation/depression (LTP/LTD) mechanisms ruled by the continuous input bombardment from internal and external environments, including learning/training and maturation/aging processes. Network configuration and excitability are continuously changing even in tens of millisecond time frames, according to the cyclic changes of the cortical state ("cortical uncertainty" of Adrian and Moruzzi, 1939 [81]). Such continuous variability modifies instant-by-instant the efficacy of the brain networks supporting a given skill or task. On this basis, it can be explained why an operating subject can incur cyclic errors during a task even if in apparently stable conditions. Phase synchronization (or coherence), phase-locking, entrainment, cross-frequency (or power synchrony), and phase reset of EEG rhythms measure the degrees of functional and effective connectivity between different brain areas [5,82,83]. As previously said, electromagnetic brain signals are generated by neuronal activities having millisecond time constants and have, therefore, an extremely high temporal discrimination. Because of this, by examining them, one can theoretically follow the dynamics and hierarchies of neuronal assembly connection/disconnection in analogy to the binding/unbinding phenomena of neuronal firing phase coherence, as seen in animal models via microelectrodic recordings. Similar conditions have been recorded in human depth recordings where synchronization mechanisms have been observed to be highly correlated with cognitive performance [84,85].

The stationarity of the resting-state cerebral system (as opposed to non-stationarity) means that the statistical features of scalp electromagnetic brain rhythms are constant during recordings. Stationary conditions can be observed for relatively short periods, usually not longer than tens of seconds [86], during which electromagnetic rhythms can be examined by classical linear frequency analysis [87,88]. Linearity and non-linearity describe the behavior of a neural circuit, in which the output signal strength varies in direct or non-direct proportion to the input signal strength, respectively.

Several tools for EEG analysis exploit graph theory [89], which returns indicators of the balance between the local connectedness and the global integration of a network matrix. Time series of cortical electric neuronal activity can be used for estimating cortical connectivity, based on the following concept: "Two places are functionally connected if their activity time series are similar" in which the 'two places' could be replaced by 'two neuronal assemblies' or 'the neuronal assemblies under two recording electrodes' [90]. However, from a formal point of view, there are many different ways to define the similarity between signals, including those from EEG. Such methods are mainly based on the exact low-resolution electromagnetic tomography (eLORETA) [91], an algorithm representing a linear inverse solution for EEG signals that have no localization error to point sources under ideal (noise-free) conditions [28]. In order to obtain connectivity values, a lagged linear coherence algorithm is applied as a measure of functional connectivity [29,30]. Moving from the scalp-recorded EEG potentials distribution, the cortical 3D mapping of current density (source localization) is carried out via eLORETA, as detailed in previous studies, also providing the proof of its exact zero-error localization property (see [30,91]). Several recent publications from independent groups [91–99] have supported the idea of a correct source localization using eLORETA; such an idea maintains true not only with high-density EEG recordings but also with the standard 20-channel EEG montage (10–20 system).

Human activity from movement to cognitive functions is sustained by time-orchestrated coordination of neuronal aggregates simultaneously firing at multiple brain sites within distributed neuronal networks [85,100–104]. EEG/magnetoencephalographic (MEG) recordings allow for non-invasive measurement of the cyclic firing of neuronal assemblies with high temporal resolution (milliseconds), but with a relatively low spatial resolution (centimeters) and mainly reflect the activity of cortical neurons with little or no contribution from deep brain sources (either in the depth of sulci or in the fronto–orbital and temporo–mesial areas, including the hippocampal formation and insula). Excellent spatial resolution is peculiar of functional MRI (fMRI), reflecting fluctuations of local blood flow and metabolism through the detection of blood-oxygenation-level-dependent (BOLD) changes in the depth of the brain structure. Meanwhile, fMRI has a poor temporal resolution due to the physical properties of hemoglobin relaxation, which is reflected in a remarkable delay between the synchronized and relatively sharp neuronal firing producing the BOLD signal with a smoothing effect on the firing sharpness during the rise/decay phases of the neurovascular coupling. It is also worth mentioning that the BOLD signal is due to transient modifications of energy consumption of neuronal firing, and, therefore, it does not reflect those interneuronal connectivity mechanisms like synchronization/coherence and phase locking-unlocking, which do not require changes of firing frequency/intensity and do not imply energetic fluctuations. Coherence (Coh) [102], partial coherence (pCoh), phase-locking value (PLV) [105], mutual information (MI) [106], and directed transfer function (DTF) [104,107] include mathematical approaches to interneuronal connectivity as probed via EEG/MEG recordings. An adjunctive method is dynamic causal modeling (DCM, [108]), where the modulation of interactions in preselected networks is analyzed [109]. Inverse methods such as BEANFORM in MEG and LORETA in EEG data claim to detect deep sources, but there is the possibility that information from deep structures in the higher frequency rhythms is lost; with such methods, a good source reconstruction can be reached within the framework of their theoretical limitations (Figure 1).

In order to describe properties of large (e.g., whole-brain) networks, the original empirical data can be represented in the form of a graph. Graph theory has been widely applied to MRI tractography (for a review, see [110]), but here is mainly described for applications in EEG/MEG signal analysis. Graphs can be weighted or unweighted, and can be directed or undirected. The first step is to decide what can be considered as a node, and what can be defined as a link [42,89]. Core measures of graph theory can be computed with http://www.brain-connectivity-toolbox.net and adapted by Matlab scripts [97,111,112]. In such scripts, segregation refers to the degree to which network elements form separate clusters and correspond to clustering coefficient (C) [113]:

$$C = \frac{1}{n}\sum_{i \subset N} C_i = \frac{1}{n}\sum_{i \subset N} \frac{2t_i}{k_i(k_i - 1)}$$

While integration refers to the capacity of the network to become interconnected and exchange information [114], it is defined by the characteristic path length (L) coefficient [113]:

$$L = \frac{1}{n}\sum_{i \subset N} L_i = \frac{1}{n}\sum_{i \subset N} \frac{\sum_{j \subset N, j \neq i} d_{ij}}{n - 1}$$

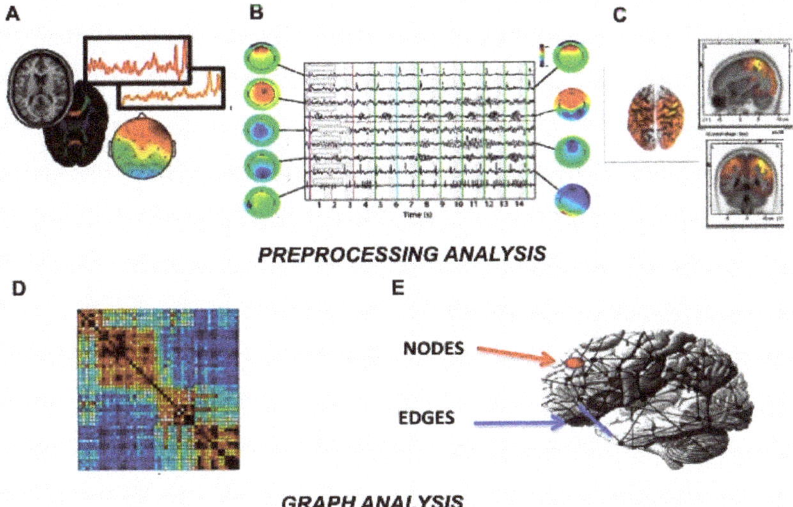

Figure 1. Functional connectivity data analysis: from EEG recordings (**A**), signals are preprocessed with ICA for the artifacts' rejection (**B**), and the eLORETA algorithm is applied to extract EEG sources' localization (**C**). Then, the graph analysis is applied with the construction of an adjacency matrix, square arrays of numbers where rows and columns correspond to nodes, and individual entries correspond to all possible connections (**D**). Nodes can correspond to specific regions, superficial signal recording sites, EEG sources, whereas edges can represent values of functional coupling between nodes (**E**).

The mean clustering coefficient is computed for all nodes of the graph and then averaged. It is a measure for the tendency of network elements to form local clusters [115]. Starting with the definition of L, the weighted characteristic path length Lw represents the shortest weighted path length between two nodes [113,116]. The small-worldness (SW) parameter is defined as the ratio between normalized C and L – Cw and Lw – with respect to the frequency bands. For example, to obtain individual normalized measures, one can divide the characteristic path length and the clustering coefficient by the mean from average values of each parameter in all EEG frequency bands. In this case, it should be stressed that a normalization of the data, with respect to surrogate networks, cannot be done due to the weighted values of the considered networks. The SW coefficient describes the balance between local connectedness and global integration of a network. SW organization is intermediate between that of random networks, in which the short overall path length is associated with a low level of local clustering, and that of regular networks or lattices, with a high level of clustering characterized by a long path length [96]. In this scenario, nodes are linked through relatively few intermediate steps, and most nodes maintain few direct connections. Surrogate analysis plays a pivotal role in testing the significance of functional connections in both bivariate and multi-variate estimators; it also represents a useful approach when applying a data-driven topological filter on statistically significant functional connections [117].

Generally speaking, most of the studies on brain connectivity with various techniques do not report on inter- and/or intra-subject test–retest variability; this is a significant gap for an extensive clinical application. In order to evaluate the within-subject test–retest variability [98], statistical analysis was performed on the normalized characteristic path length of EEG cortical sources for a 10-subject group with two recording sessions at a 2-week interval, introducing the factor Time (First and Second recording sessions). The statistical analyses showed no significant interaction, including time, highlighting the stability of the "Small World" analysis of EEG signals. More recently, findings from 3 recording sessions have been compared from 34 healthy subjects (mean age of 45 years) at a one-week inter-session interval [118] A between-factors analysis of variance (ANOVA) was carried out: Frequency Band (delta, theta, alpha 1, alpha 2, beta 1, beta 2, and gamma) and Time (first, second and third recording) for the Small World parameter. The statistical analysis showed that the interaction, including Time, was not significant ($F(12, 396) = 0.48995$, $p = 0.92057$), highlighting the stability of the proposed parameters at least when carried out in clinically stable subjects. Recently, the importance of reliability studies based on repeat-scan sessions protocol of *connectomics* in any modality has been recognized with publication of a number of freely available papers and datasets [119–121].

Transitivity (Tw) is another graph parameter: it is measured as the fraction of the node's neighbors that are also neighbors of each other [122] and reflects, on average, the prevalence of clustered connectivity around individual nodes, a measure of segregation based on the number of triangles in the network. Tw represents a variant of the clustering coefficient not affected by individual node normalization [123]. More sophisticated methods describing segregation besides the presence of densely interconnected groups of regions also reflect their composition named the network's modular structure (community structure). It reflects the decomposition of networks into groups of nodes, with the maximal content of within-group links (within network connections are dense), and the minimal level of between-group links (between network connections are sparse). The degree to which the network may be subdivided into such clearly delineated and non-overlapping groups is quantified by a single statistic, the modularity (Qw) (Figure 2). Unlike most other network measures, the optimal modular structure for a given network is typically estimated with optimization algorithms. Finally, local efficiency (E_loc^w) is an index of the information transfer efficiency limited to neighboring nodes (i.e., nodes with direct edges to the node of interest) and indicates how mutually interlinked neighboring nodes are [124].

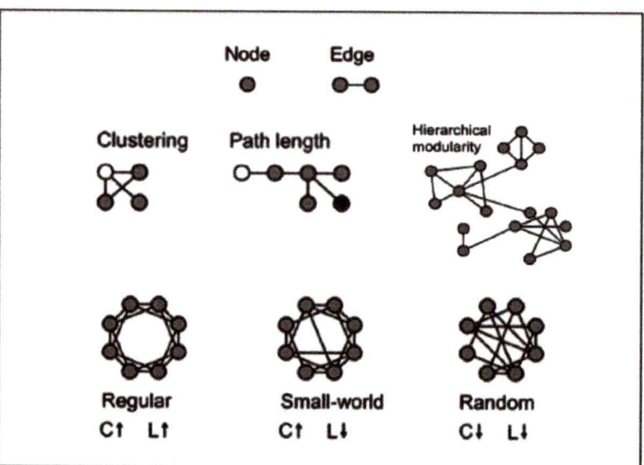

Figure 2. Core measures of graph theory: nodes and edges, clustering coefficient, characteristic path length, Small World index, modularity.

Studies on network hierarchical architecture, as obtained by the analysis of simultaneous EEG oscillations of different frequencies and cross-frequency couplings during a given task performance, have opened new research avenues into cognitive mechanisms [85]. In fact, time modulation of the connectivity pattern of the nodes in a task-related network explains most of the performance variability—i.e., from "excellent" to "poor"—in apparently stable conditions [96,112,125]. In other words, the task–performance level and the task-related choice/behavior contents are largely written in the immediate architecture of the EEG networks' connectivity, preceding the task (by a few seconds, usually).

Each EEG rhythm reflects different mechanisms and a complete view—in time, space, and frequency domains—is needed to obtain a comprehensive analysis of its functional dynamics. It is worth mentioning that, depending upon the frequency content of the examined rhythm, the time discrimination of the activation within the network frame can be as short as few milliseconds (down to 10 msec in the high γ band). Because of this, EEG connectivity analysis facilitates an evaluation of the time hierarchy governing the serial/parallel activation of the nodes and their time/space relationship within a given task-related network (i.e., whether A is active before, after, or in parallel to B).

Aging processes significantly modulate the network configuration of brain connectivity and also affect the time-varying synchronization of rhythmic oscillations in a network organization. Along this line of research, 170 healthy elderly volunteers were submitted to EEG recordings in order to define age-related normative limits [126]. Graph theory functions were applied to eLORETA on cortical sources in order to evaluate the Small World parameter as a representative model of network architecture. The analyses were carried out in the whole brain—as well as for the left and the right hemisphere separately—and in three specific resting=state sub-networks defined as follows: attentional network (AN); frontal network (FN); default mode network (DMN). To evaluate the stability of the investigated graph parameters, a subgroup of 32 subjects underwent three separate EEG recording sessions in identical environmental conditions after a few days interval. Results showed that the whole right/left hemispheric evaluation did not present side differences, but when individual sub-networks were considered, AN and DMN presented in general higher SW in low (delta and/or theta) and high (gamma) frequency bands in the left hemisphere, while for FN the alpha 1 band was lower in the left, with respect to the right hemisphere. It was also evident the test–retest reliability and reproducibility of the present methodology when carried out in clinically stable subjects.

On clinical grounds, it is of interest to the study of conditions that are considered to be prodromal to dementia as in MCI. As previously said, dementias—particularly in their early stages—mainly affect synaptic transmission and therefore represent "disconnection syndromes" [31,44,48,97,127]. A statistically significant difference in the SW organization of those MCI subjects who will progress to AD (Converted MCI, particularly those who can be defined as rapid—i.e., 1–2 years—converters) was recently found [128], the Converted MCI subjects having SW characteristics very similar to those encountered in Alzheimer's patients 1 to 2 years before their conversion (Time 0 of the study). An abnormal increase in graph parameters in Converted, with respect to Stable MCI, for the α rhythm has been observed, along with a decrease for the δ and γ rhythms. Such findings might be interpreted in light of the background physiology of α rhythm, which is usually defined as the "idling rhythms" of the adult brain [129]. Along this vein, it is worth mentioning that, in a population of 145 MCI subjects followed up for 2 years, the receiver operating characteristic (ROC) curve derived from graph-theory EEG analysis showed SW characteristics with a >60% sensitivity (area under the ROC curve (AUC) 0.64, indicating moderate classification accuracy) for classifying the MCI state as a prodromal of AD. These findings are in line with previous studies [97,112,115] in which SW characteristics were decreased in low-frequency bands in patients with AD compared to MCI [128]. That is, the MCI connectivity pattern was less random than that of the AD group. Moreover, significant differences between healthy elderly, MCI subjects and AD patients have been demonstrated by showing that physiological brain aging presents greater specialization (though lower values) of SW characteristics that are higher than normal in low EEG frequencies and lower in α bands. Finally, converted aMCI presented a graph theory pattern practically identical to the AD one. The ROC curves gathered by a combined phenotype and genotype characteristics analysis (obtained at a low cost with widely available apolipoprotein E (ApoE) technology), produced an increase of accuracy up to 91.78% (AUC 0.97, indicating a nearly optimal classification accuracy) for identifying the MCI prodromal-to-AD state [128]. This result is in line with the fact that the $\varepsilon 4$ allele of the APOE gene is the major risk genetic factor for the pathogenesis of late-onset AD [32,130].

This bulk of findings suggests that EEG connectivity analysis, combined with neuropsychological and genetic (i.e., ApoE alleles) evaluation, could be of great help in early MCI prodromal-to-AD identification as a first-line screening method and in intercepting those subjects with a high risk for rapid progression to AD [83,131]. How does the "graph-theoretical" model compete with other types of EEG analysis methods, and how does it contribute to AD diagnosis? Vecchio et al. [128] made a comparative analysis by applying to the same EEG epochs utilized for graph valuation methods of EEG analysis currently used for AD studies, namely, spectral coherence and power spectrum; such methods performed less than graph and showed 51.79% sensitivity, 100% specificity, and 68.86% accuracy.

Several studies converge on the idea that α rhythm is a deterministic chaotic signal involved in several functions, besides others [42], ranging from memory formation to sensorimotor processing and integration [132]. Indeed, there is evidence in the healthy, showing a positive correlation between α frequency and the speed of information processing, as well as cognitive performance [87]. In the adult EEG during resting, awake conditions α rhythms are widely recordable and dominate in the posterior brain areas, while δ rhythms are poorly represented, thus reflecting a condition of likely α-δ "reciprocal inhibition" [31]. Furthermore, it is well known that the anatomical or functional disconnection of lesioned cortical areas generates spontaneous slow oscillations in the δ range in virtually all recorded neurons. In particular, the SW decrease in the δ band represents an increase of functional inhibition. The opposite holds true for the α band.

Gamma rhythms are involved in a variety of cognitive functions, including visual object processing, attention, and memory [133], and are strictly reflecting behavioral performance (accuracy and reaction time) in several memory tasks, including episodic memory, encoding, and retrieval [134]. Gamma oscillations are pivotal in synchronization of the action potentials spike phase, a mechanism that is at the base of EEG connectivity [135]. An SW decrease in the γ band in the MCI-prodromal-to-dementia is in line with previous evidence [98], showing a decrease of SW γ band in AD with respect to MCI and control subjects. The γ band (>30 Hz) mediates information transfer between cortical and hippocampal structures for memory formation [136], particularly through feed-forward mechanisms [137] and coherent phase-coupling between oscillations recorded simultaneously from different neuronal structures [138].

We also explored [131] the EEG functional connectivity in amnesic multidomain-MCI subjects in order to characterize the DMN in converted MCI (cMCI)—those in a prodromal-to-dementia condition who converted to AD during the follow-up—compared to stable MCI (sMCI) subjects. A total of 59 MCI subjects were recruited and divided, after appropriate follow-up, into cMCI or sMCI. They were further divided into MCI with linguistic domain (LD) impairment and in MCI with executive domain (ED) impairment. The Small World (SW) index was computed, restricting to nodes of DMN regions for all frequency bands, and evaluated how they differ between MCI subgroups as assessed through clinical and neuropsychological 4-year follow-ups. Results showed that the SW index significantly decreased in γ band in cMCI compared to sMCI. In cMCI with LD impairment, the SW index significantly decreased in the delta band, while in cMCI with ED impairment, the SW index decreased in delta and γ bands and increased in the alpha1 band. It is argued that the DMN functional alterations in cognitive impairment could reflect an abnormal flow of brain information processing during resting state possibly associated with a status of pre-dementia.

The combination of all the above-mentioned feature extraction techniques results in a wide-ranging collection of features. For this reason, a feature selection process should be preferably carried out in an automated or at least in a semi-automated way. A large number of machine learning algorithms can be used to accomplish this task. A widely used procedure for both feature selection and classification in diagnosing AD applications is the support vector machine (SVM), which achieved up to 98% accuracy in early AD detection [139–141]. One of the major advantages of SVM is that when combined to L1-norm as penalization, it leads to sparse weight vectors and allows feature selection and classification to be accomplished in the same step [142]. An interesting variation of SVM is the Relevance Vector Machine (RVM), which replaces the binary SVM classifier with a soft-decision method based on a probabilistic Bayesian learning framework and outperformed SVM when tested in a fully-automated AD diagnostic system [139].

Recently, we investigated [143] the possibility to automatically classify physiological vs. pathological aging from cortical sources' connectivity based on a support vector machine (SVM) applied to EEG Small World. A total of 295 subjects were recruited: 120 healthy volunteers and 175 AD. Graph theory functions were applied to the undirected and weighted networks obtained by the lagged linear coherence evaluated by eLORETA. A machine-learning classifier (SVM) was then applied. The ROC curve showed an AUC of 0.9 (indicating very high classification accuracy). The resulting classifier showed 83% sensitivity, 100% specificity, and 96% accuracy for the classification of the AD respect to control subjects. Graph theory analysis of brain connectivity from EEG signals provides useful information in distinguishing physiological and pathological age-related brain processes.

In conclusion, EEG connectivity analysis via a combination of source/connectivity biomarkers could represent a promising tool in the identification of AD patient and MCI prodromal-to-dementia subjects. This approach represents a low-cost and non-invasive method reaching high sensitivity/specificity and optimal classification accuracy, which might be combined with other biomarkers with the same characteristics (i.e., ApoE genotyping) for screening large population samples in order to obtain a risk evaluation on an individual basis.

2.3. TMS-EEG Co-Registration for Testing Brain Connectivity

Transcranial magnetic stimulation (TMS) is a non-invasive and painless technique introduced in 1985 by Anthony Barker, that is able to study the excitability, connectivity, and plasticity of the human cerebral cortex. If the coil for the stimulation is precisely localized on the scalp region overlaying the motor cortex, a muscle twitch in the contralateral body segment can be elicited with supra-threshold stimuli. Such responses are called motor-evoked potentials (MEPs); they can be recorded from a target muscle (i.e., the hand) by surface electromyography (EMG) and reflects the activation of corticospinal cells in the primary motor cortex (M1) by single-pulse transcranial magnetic stimulation (spTMS) [144].

The combination of TMS with EEG is considered an important tool to reveal the effective connectivity of brain networks, defined as the influence one neuronal assembly exerts over separate (eventually remote) one(s) through causal or non-causal effects. In fact, the co-registration of the EEG activity—which has a temporal resolution of a few milliseconds and can be simultaneously sampled from a large number of scalp sites—during TMS provides the opportunity of tracking temporal dynamics and inner hierarchies of brain networks that is properly their effective connectivity (for a review see Rossini et al. (2019) [83]).

TMS–EEG has several advantages: (1) Its high temporal resolution conveys precise information about the temporal order of activations of connected cortical areas (either adjacent or remote), defining at the same time the causal interactions (excitatory or inhibitory) between two areas within functional brain networks. (2) Its high temporal resolution allows the identification of critical periods during which the stimulated area and its connections to other brain regions make a critical contribution to the experimental task, thereby enabling to differentiate the connectivity pattern of different cognitive processes related to specific tasks or different brain states and whether or how they are modified by learning and training. Taking into account these points, TMS–EEG co-registration allows the evaluation of the spatio–temporal pattern of neural activity that determines the connections across brain areas, hence providing measures of effective connectivity able to test the predictions of graph theory models [145].

From the first attempt to measure TMS-evoked brain responses made in 1989 by Cracco et al. [146], several efforts have been made to overtake the severe technical limitations related to the coupling of a stimulation artifact (thousands of times higher than the signal of interest) to the recording system. Using a sample-and-hold circuit able to block the acquisition of EEG signal for several milliseconds immediately adjacent to the TMS pulse, TMS-evoked brain EEG responses were successfully measured in 1997, succeeding in tracking TMS-evoked brain activity with a temporal window of a few milliseconds after the stimulus [147,148]. Subsequent studies have begun to describe the scalp topography and to study the possible generator sources of the TMS-evoked EEG potentials (TEPs). Probably, most of the EEG signals record a linear projection of the postsynaptic currents indirectly induced by TMS; then, EEG signals can be used to locate and quantify these synaptic current distributions and evaluate local excitability and functional connectivity in the nervous system. Within the so-called "inductive approach", applying a single TMS pulse on the brain cortex, a network of neuronal connections is triggered and the TMS-induced activation—a summation of post-synaptic potentials—spreads from the stimulation site to other interconnected parts of the brain, producing deflections in scalp EEG signals, starting a few milliseconds after the stimulus and lasting about 300 msec, first in the form of rapid oscillations and then as lower-frequency waves. Increased EEG activity following the magnetic stimulus can be observed in a number of neighboring electrodes, suggesting the spread of TMS-evoked activity to anatomically interconnected cortical areas. Particularly, TMS-evoked activity spreads from the stimulation site ipsilaterally via association fibers, contralaterally via transcallosal fibers, and to subcortical structures and spinal cord via projection fibers.

Therefore, TMS–EEG gives the possibility to study cortico–cortical interactions and how the activity in one area affects the ongoing activity in other areas. It has been suggested that the first part of the TMS-evoked EEG signals reflects the excitability (i.e., the functional state) of the stimulated cortex, whereas the following spatiotemporal distribution over the scalp corresponds to the spread of activation to other cortical areas, i.e., the effective connectivity of the stimulated area (for a review, see Ferreri and Rossini (2013) [148]). The amplitude, latency, and scalp topography of single-pulse TMS-evoked EEG responses have been clearly described [125,149,150]. TMS-evoked EEG averaged responses are generally highly reproducible, provided that the delivery and targeting of TMS (i.e., via neuronavigated stimulation) is well controlled and stable from pulse to pulse and between experiments. Several components of the EEG response to single-pulse TMS applied on the motor cortex have been identified and—benefiting from the knowledge of the anatomical connectivity of the brain as seen by diffusion tensor imaging studies—their spatiotemporal spread has been followed: particularly, single-pulse TMS is able to evoke EEG activity composed at the vertex by a sequence of deflections of negative polarity peaking at approximately 7, 18, 44, 100, and 280 msec, alternating with positive polarity peaks at approximately 13, 30, 60, and 190 msec post-TMS (Figure 3) [150].

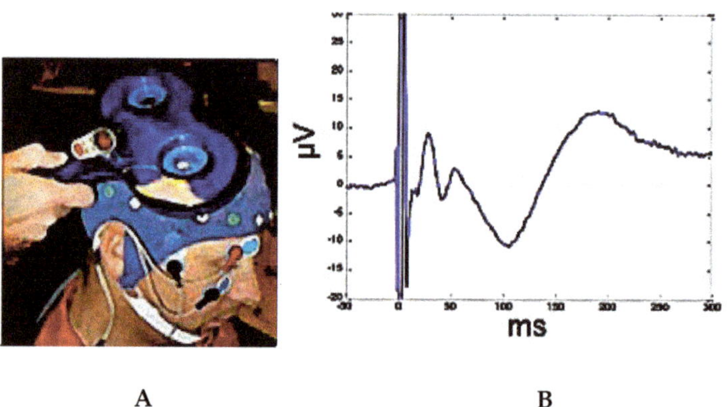

Figure 3. Panel (**A**): Transcranial magnetic stimulation (TMS)- electroencephalographic (EEG) co-registration. Panel (**B**): TMS–EEG evoked responses (TEPs) recorded at vertex during supra-threshold single-pulse stimulation of M1.

However, the previously described pattern of TEPs is not invariable because, in addition to inter-individual differences, it depends on the stimulation intensity, the exact coil location and orientation [149], the local and general state of the cortex [151], the level of vigilance [152], as well as the age of the stimulated brain [153]. Given these unique features, TMS–EEG appears very suitable to test and evaluate the functional brain architecture suggested by graph theory models, both at rest and during cognitive processes. Several TMS–EEG studies on the motor system at rest have demonstrated that the TMS-induced activity spreads from the stimulated node to other nodes of the same motor network: the TMS of the primary motor cortex causes the successive activation of ipsilateral supplementary/premotor areas and contralateral motor region with a short conduction time. The nature of these connections seems to be inhibitory rather than excitatory and depends on the level of cortical activation immediately before and during the stimulation [125,153]. As therefore observed, the restriction of the TMS-activity to the specialized motor network suggests a modular node organization of functional brain architecture at rest. Other studies on the motor cortex at rest have revealed that the stimulation of M1 also generates the late activation of areas outside the "motor" network, including the cingulate gyrus and the temporo–parietal junction. The spread of the later components of TEPs to other areas over the motor network suggests the involvement of further nodes and brain hubs implicated in the transmission of information across the brain. Additional evidence about this bottom-up signal propagation from lower-degree nodes to brain hubs has been provided by studies on the visual system. On the contrary, the existence of mechanisms of top-down modulation has been shown in several studies stimulating multimodal associative areas responsible for high cognitive processes during task performance: it has been demonstrated that the diffusion of TMS-induced activity from these areas across the brain could be divergent depending on the task context, preferring the engagement of one network rather than another. This kind of TMS–EEG approach defines the "interactive approach" and seems to confirm that the targeting of associative areas could correspond to the brain hubs, a subset of high-degree brain sites able to mediate communication between multiple modules or networks according to the cognitive context. These findings, taken together, highlight the potential role of TMS–EEG to test the dynamic changes of cortico–cortical connectivity according to graph theory predictions, identifying both a specialized modular/network segregation during the resting state and a modular/network integration during high cognitive demands with dense connectivity through brain hubs activation.

In addition to standard TEPs, single-pulse TMS or frequency-tuned train of pulses can also trigger or enhance brain oscillatory activity or perturb ongoing rhythms of the targeted cortical area, eliciting event-related phenomena, such as EEG rhythm synchronization or desynchronization. Brain oscillations represent a mechanism through which the communication of neuronal assemblies—by synchronization in specific frequency bands—is rendered more effective, precise, and selectively tuned on the transmission of the relevant information. It has already been demonstrated that different cortical areas, when stimulated, respond at a characteristic frequency, i.e., their natural frequency, and that functionally segregated networks could oscillate at different frequencies at rest [154]. Given these assumptions, some authors speculated that TMS could interact with such oscillatory patterns in the directly stimulated cortical area and in distant areas belonging to the same neural network thus inducing a resonant frequency activity in all "synchronized" areas of the same network by mechanisms of longer-range synchronization (interregional coherence). This frequency-specific "resonant effect" should ensure better information transfer across brain structures and could even determine changes in the behavioral performance [155]. Therefore, the "rhythmic TMS–EEG approach" appears as a promising tool in mapping the natural frequency of different cortical areas and identifying the role of a specific frequency oscillatory activity in distinct brain functions.

With all these premises, despite some technical limitations, it is easy to realize how TMS–EEG can be used to examine normal and altered effective brain connectivity under both physiological and pathological specific conditions, indicating the strengthening or weakening of existing cortico–cortical connections or the recruitment of compensatory networks. Indeed, besides assessment of the general state of the brain, TMS–EEG can be used to track the interactions of brain areas during sensory processing, cognition, or motor control and, moreover, to evaluate such neurological disorders as Alzheimer's disease (AD), characterized by altered connectivity. The few studies regarding this field [156–160], integrating previous observations obtained with the use of the TMS alone (for example, [161,162], already showed that the cortical stimulation in AD patients was associated with significant disruption in TMS-induced activity over several brain areas compared with healthy controls, suggesting a potential role of TMS–EEG as a neurophysiological marker for diagnosis and early identification of mild cognitive impairment (MCI) and AD (Table A1 in Appendix A).

In this context, our research group was able to describe—for the first-time—specific neurophysiological hallmarks of motor cortex functionality in early AD [158]. By using TMS–EEG co-registration, we have demonstrated that in mild AD patients without motor symptoms, the sensorimotor system is strongly hyperexcitable and deeply rearranged with the recruitment of additional neural sources, the activation of reverberant local circuits, and their integration in the distributed network subtending sensorimotor functions. Thus, we have proposed this plastic cortical reorganization would be ensured by the particular organization of the sensorimotor system based on a distributed network with a replicated topographic organization of the same body part and could be interpreted as a compensatory mechanism allowing for the preservation of sensorimotor programming and execution since the preclinical stage trough the MCI stage and over a long period of time in spite of disease progression [158]. Because of such encouraging findings, we are now employing TMS–EEG to investigate hallmarks of sensorimotor cortex functionality in aMCI, assuming they represent the subtending long-term plastic rearrangement induced by the neurodegeneration during the pauci symptomatic prodromal stage and can thus affordably predict the future conversion to AD. TMS–EEG recordings and analysis will be performed both to describe the excitability and effective connectivity of the somatosensory network of the whole aMCI group with respect to a control group, and to investigate baseline differences in these neurophysiological properties between the two groups. Particularly we want to determine (1) whether the sensorimotor networks would show peculiar alterations in aMCI as a whole group, and (2) if there is any hallmark of sensorimotor network disruption able to predict long-term disease progression at the individual level. We are now finalizing a five-year clinical follow-up in a restricted group of aMCI, and, effectively, our preliminary results are promising and indicate that some parameters of the M1 functionality can be used as reliable biomarkers of AD.

3. Conclusions

The time is now right for searching for instrumental biomarkers for early—hopefully, preclinical—diagnosis of dementia in order to contrast as soon as possible all the modifiable risk factors for neurodegeneration, as well as to initiate (as soon as they will become available) disease-modifying drugs. To reach this goal, all the health systems are actually looking for a combination of biomarkers having clear characteristics: high accuracy/specificity/sensitivity, affordable costs, non-invasiveness, and large territorial availability. Neurophysiological techniques have all the required characteristics and are optimal candidates, at least for a 1st level screening.

Author Contributions: P.M.R., F.M., F.V., M.C., F.F., R.D.I.: study concept and design; P.M.R., F.M., F.A., F.V., M.C., F.F., R.D.I., F.I.: drafting the manuscript and figures. All authors have read and agreed to the published version of the manuscript.

Funding: This work was partially supported by the Italian Ministry of Health for Institutional Research (Ricerca Corrente) and by AIFA with the project INTERCEPTOR.

Conflicts of Interest: The authors declare that they have no conflict of interest.

Appendix A

Table A1. Summary of the main studies that analyzed neurophysiological changes associated with the development of dementia. List of abbreviations: electroencephalography (EEG), fluorodeoxyglucose positron emission tomography (FDG–PET), mild cognitive impairment (MCI), Alzheimer's disease (AD), normal elderly (Nold), Small Worldness (SW), transcranial magnetic stimulation (TMS), repetitive transcranial magnetic stimulation (rTMS), motor cortex (M1), Mini Mental State Examination (MMSE), default mode network (DMN).

Author (Year)	Methods	Biomarkers	Subjects	Main Findings	Reference
Babiloni C (2016)	EEG, FDG-PET	spectral analysis (power)	AD	19 AD patients were compared with a group of 40 Nold. The AD group performed FDG-PET. In the AD patients, there was a positive correlation between the Alzheimer's discrimination analysis tool (PALZ) score and the activity of delta sources in the cortical region of interest ($p < 0.05$) suggesting a relationship between resting-state cortical hypometabolism and synchronization of cortical neurons at delta rhythms in AD patients with dementia.	[22]
Rossini PM (2006)	EEG	spectral analysis (linear coupling)	AD, MCI	In 69 MCI, baseline fronto-parietal midline coherence, delta (temporal), theta (parietal, occipital and temporal), and alpha 1 (central, parietal, occipital, temporal, limbic) sources were stronger in MCI Converted than stable subjects ($p < 0.05$). Low midline coherence and weak temporal source were associated with a 10% annual rate AD conversion, while this rate increased up to 40% and 60% when strong temporal delta source and high midline gamma coherence were observed respectively.	[31]
Jelic V (2000)	EEG	spectral analysis (power)	MCI	In 27 MCI patients, progression to AD in a follow up of 21 months was associated with a significantly higher theta relative power and lower beta relative power and mean frequency at the temporal and temporo - occipital derivations	[34]
Adler G (2003)	EEG	spectral analysis (power)	AD	A study with 31 AD compared with 17 Nold. AD patients showed a widespread increase in delta and theta power density and posterior decrease in α and β power density with a lowering of α power density peak.	[35]

Table A1. *Cont.*

Author (Year)	Methods	Biomarkers	Subjects	Main Findings	Reference
Stam CJ (2007)	EEG	spectral analysis (graph theory)	AD	In a study with 15 AD vs. 13 Nold, the characteristic path length L was significantly longer in the AD patients, whereas the cluster coefficient C showed no significant changes. This pattern was still present when L and C were computed as a function of K. A longer path length with a relatively preserved cluster coefficient suggests a loss of complexity and a less optimal organization.	[42]
Jelles B (1999)	EEG	spectral analysis (power)	AD	In a study with 24 probable AD vs. 22 Nold, the correlation dimension (D2) was significantly lower in the Alzheimer patients compared to controls	[43]
Dauwels J (2010)	EEG	spectral analysis (nonlinear coupling)	MCI	Two synchrony measures, Granger casuality, and stochastic event synchrony are able to distinguish MCI patients from age-matched control subjects.	[44]
Azami H (2016)	MEG	entropy	AD	In 36 AD vs. 26 Nold, multiscale dispersion entropy (MDE) values in AD compared with multiscale permutation entropy (MPE) and multiscale entropy (MSE) was significantly lower than their corresponding MSE- and MPE-based values.	[45]
Babiloni C (2009)	EEG, MRI	spectral analysis (power)	AD, MCI	In a study with 35 AD, 80 MCI and 60 Nold, the EEG sources showed a significant linear correlation with hippocampal volume also supported a non-linear correlation with hippocampal volume strongly for the logarithmic one, suggesting that progressive atrophy of hippocampus correlates with decreased cortical alpha power, as estimated by using LORETA source modeling, in the continuum, along MCI and AD conditions.	[47]
Babiloni C (2009)	EEG, MRI	spectral analysis (power)	MCI	Study with 54 MCI subjects with follow-up of 1-year vs. 45 Nold and 50 AD. In MCI, the EEG recordings showed a decreased power of posterior alpha1 and alpha2 sources, suggesting that the resting state EEG alpha sources were sensitive-at least at the group level-to the cognitive decline occurring in the amnesic MCI group over 1 year.	[51]

Table A1. *Cont.*

Author (Year)	Methods	Biomarkers	Subjects	Main Findings	Reference
Jeong J (2004)	EEG	spectral analysis (nonlinear)	AD	EEG in AD showed a lower correlation dimension (D2) and the largest Lyapunov exponent (L1) values than in the healthy. Despite their different focus on static and dynamic properties of the EEGs, the results of both D2 and L1 were associated with a reduction of complexity in EEG activity due to AD	[54]
Smits FM (2016)	EEG	fractal dimension	AD	A comparision between 67 AD vs. 41 Nold showed a reduced fractal dimension in AD compared to healthy especially in temporal-occipital regions	[61]
Escudero J (2006)	EEG	entropy	AD, MCI	In a study with 11 AD and 11 Nold, entropy metrics of spontaneous EEGs in AD and in MCI showed reduced irregularity in AD patients' EEG activity	[69]
Vecchio F (2015)	EEG, MRI/DTI	spectral analysis (graph theory)	AD, MCI	40 subjects, including 9 Nold, 10 MCI, 10 mild AD, 11 moderate AD. Callosal fractional anisotropy (FA) reduction, observed in subjects with Alzheimer's disease (AD) and mild cognitive impairment (MCI), is associated with a loss of brain interhemispheric functional connectivity characterized by increased delta and decreased alpha path length.	[97]
Vecchio F (2014)	EEG	spectral analysis (graph theory)	AD, MCI	Analysis of a database of 378 participants, including AD, MCI, and Nold. Path Length showed a different pattern between normal cognition and dementia as observed in the theta band (MCI subjects are found similar to healthy subjects), while for the normalized Clustering coefficient a significant increment was found for AD group in delta, theta, and alpha 1 bands; the small world parameter presented a significant interaction between AD and MCI groups showing a theta increase in MCI.	[98]

Table A1. *Cont.*

Author (Year)	Methods	Biomarkers	Subjects	Main Findings	Reference
Miraglia F (2016)	EEG	spectral analysis (graph theory)	AD, MCI	30 Nold, 30 aMCI, and 30 AD during eyes closed EC and eyes open EO. In Nold, in EO condition, the brain network is characterized by higher SW in alpha bands and lower SW in beta2 and gamma bands. In aMCI, SW has the same trend, except for delta and theta bands where the network shows less SW. AD shows a similar trend of Nold, but with less fluctuations between EO/EC conditions. aMCI presents SW midway between AD and Nold. In delta and theta bands, in EC, the aMCI group presents network's architecture similar to Nold, while in EO aMCI, SW similar to AD	[112]
de Hann W (2012)	MEG	spectral analysis (graph theory)	AD	In 18 AD vs. 18 Nold, graph spectral analysis confirmed the hub status of the parietal areas and demonstrated a low centrality of the left temporal region in the theta band in AD patients that was strongly related to the MMSE. In AD, impaired network synchronization and a clinically relevant left temporal centrality loss were found	[115]
Vecchio F (2017)	EEG	spectral analysis (graph theory)	AD	In 110 AD and 34 healthy Nold, Alpha band connectivity was negatively correlated, while slow (delta) and fast-frequency (beta, gamma) bands positively correlated with the hippocampal volume of Alzheimer subjects. The larger the hippocampal volume, the lower the alpha, and the higher the delta, beta, and gamma Small World characteristics of connectivity.	[126]
Vecchio F (2018)	EEG, Apo-E allele	spectral analysis (graph theory)	aMCI	145 aMCI classified as Converted to AD (C-MCI, 71) or Stable (S-MCI, 74) according to follow up. Small-World EEG analysis, in combination with an Apo-E allele testing, evaluate on an individual basis with great precision the risk of MCI progression (96.7% sensitivity, 86% specificity and 91.7% accuracy (AUC = 0.97))	[127]

Table A1. *Cont.*

Author (Year)	Methods	Biomarkers	Subjects	Main Findings	Reference
Julkunen P (2011)	TMS-EEG	Cortical Excitability (P30 amplitude)	AD, MCI	In this study with 4 control subjects, 5 MCI and 5 AD, the TMS–EEG response P30 amplitude correlated with cognitive decline and showed good specificity and sensitivity in identifying healthy subjects from those with MCI or AD.	[156]
Casarotto S (2011)	TMS-EEG	Cortical Excitability	AD	In this study with 9 healthy young, 9 healthy elderly, and 9 AD, frontal cortex excitability was not significantly different between healthy young and elderly individuals while was clearly reduced in AD patients.	[157]
Ferreri F (2016)	TMS-EEG	M1 Cortical Excitability and Connectivity	AD	In this study with 12 mild AD patients, the sensorimotor system was found hyperexcitable, and its connectivity disrupted with respect of 12 healthy elderly, despite the lack of clinically evident motor manifestations.	[158]
Bagattini C (2019)	TMS-EEG	Cortical Excitability (P30 amplitude)	AD	In this study with 26 AD patients, the TMS–EEG response P30 amplitude predicted MMSE and face-name memory scores. Particularly higher P30 amplitude predicted poorer cognitive and memory performances.	[159]
Koch G (2018)	rTMS, TEM-EEG	Cortical Excitability and Connectivity	AD	In 14 early AD, a 2-week treatment with rTMS on the precuneus induced a selective improvement in episodic memory. TMS-EEG recording revealed a precuneus enhanced activity and a modification of its functional connectivity within the DMN	[160]
Ferreri F (2003)	TMS	M1 Cortical Excitability (MEP amplitude)	AD	In 16 AD, motor cortex excitability, measured with TMS, was increased, and the center of gravity of motor cortical output, as represented by excitable scalp sites, showed a frontal and medial shift, without correlated changes in the site of maximal excitability (hot-spot).	[161]
Ferreri F (2011)	TMS	M1 Cortical Excitability (MEP amplitude)	AD	In 10 AD patients before and after long-term AchEIs therapy, M1 excitability was found to be unchanged in patients with stabilized cognitive performance during the therapy.	[162]

References

1. McKhann, G.M.; Knopman, D.S.; Chertkow, H.; Hyman, B.T.; Jack, C.R.; Kawas, C.H.; Klunk, W.E.; Koroshetz, W.J.; Manly, J.J.; Mayeux, R.; et al. The diagnosis of dementia due to Alzheimer's disease: Recommendations from the National Institute on Aging-Alzheimer's Association workgroups on diagnostic guidelines for Alzheimer's disease. *Alzheimers Dement.* **2011**, *7*, 263–269. [CrossRef] [PubMed]
2. Costa, A.; Bak, T.; Caffarra, P.; Caltagirone, C.; Ceccaldi, M.; Collette, F.; Crutch, S.; Della Sala, S.; Démonet, J.F.; Dubois, B.; et al. The need for harmonisation and innovation of neuropsychological assessment in neurodegenerative dementias in Europe: Consensus document of the Joint Program for Neurodegenerative Diseases Working Group. *Alzheimers Res. Ther.* **2017**, *9*, 27. [CrossRef] [PubMed]
3. Cerami, C.; Dubois, B.; Boccardi, M.; Monsch, A.U.; Demonet, J.F.; Cappa, S.F. Geneva Task Force for the Roadmap of Alzheimer's Biomarkers. Clinical validity of delayed recall tests as a gateway biomarker for Alzheimer's disease in the context of a structured 5-phase development framework. *Neurobiol. Aging* **2017**, *52*, 153–166. [CrossRef]
4. Geldmacher, D.S. Cost-effectiveness of drug therapies for Alzheimer's disease: A brief review. *Neuropsychiatr. Dis. Treat.* **2008**, *4*, 549–555.
5. D'Amelio, M.; Rossini, P.M. Brain excitability and connectivity of neuronal assemblies in Alzheimer's disease: From animal models to human findings. *Prog. Neurobiol.* **2012**, *99*, 42–60. [CrossRef] [PubMed]
6. Getsios, D.; Blume, S.; Ishak, K.J.; Maclaine, G.; Hernández, L. An economic evaluation of early assessment for Alzheimer's disease in the United Kingdom. *Alzheimers Dement.* **2012**, *8*, 22–30. [CrossRef]
7. Barnett, J.H.; Lewis, L.; Blackwell, A.D.; Taylor, M. Early intervention in Alzheimer's disease: A health economic study of the effects of diagnostic timing. *BMC Neurol.* **2014**, *14*, 101. [CrossRef]
8. Teipel, S.J.; Kurth, J.; Krause, B.; Grothe, M.J.; Alzheimer's Disease Neuroimaging Initiative. The relative importance of imaging markers for the prediction of Alzheimer's disease dementia in mild cognitive impairment—Beyond classical regression. *Neuroimage Clin.* **2015**, *8*, 583–593. [CrossRef]
9. Petersen, R.C. How early can we diagnose Alzheimer disease (and is it sufficient)? The 2017 Wartenberg lecture. *Neurology* **2018**, *91*, 395–402. [CrossRef]
10. Petersen, R.C.; Lopez, O.; Armstrong, M.J.; Getchius, T.S.D.; Ganguli, M.; Gloss, D.; Gronseth, G.S.; Marson, D.; Pringsheim, T.; Day, G.S.; et al. Practice guideline update summary: Mild cognitive impairment: Report of the Guideline Development, Dissemination, and Implementation Subcommittee of the American Academy of Neurology. *Neurology* **2018**, *90*, 126–135. [CrossRef]
11. Nunez, P.L.; Srinivasan, R. A theoretical basis for standing and traveling brain waves measured with human EEG with implications for an integrated consciousness. *Clin. Neurophysiol.* **2006**, *117*, 2424–2435. [CrossRef]
12. Srinivasan, R.; Winter, W.R.; Ding, J.; Nunez, P.L. EEG and MEG coherence: Measures of functional connectivity at distinct spatial scales of neocortical dynamics. *J. Neurosci. Methods* **2007**, *166*, 41–52. [CrossRef]
13. Hyvarinen, A. Blind source separation by nonstationarity of variance: A cumulant-based approach. *IEEE Trans. Neural Netw.* **2001**, *12*, 1471–1474. [CrossRef] [PubMed]
14. Tecchio, F.; Porcaro, C.; Barbati, G.; Zappasodi, F. Functional source separation and hand cortical representation for a brain-computer interface feature extraction. *J. Physiol.* **2007**, *580*, 703–721. [CrossRef] [PubMed]
15. Porcaro, C.; Medaglia, M.T.; Thai, N.J.; Seri, S.; Rotshtein, P.; Tecchio, F. Contradictory reasoning network: An EEG and FMRI study. *PLoS ONE* **2014**, *9*, e92835. [CrossRef] [PubMed]
16. Scherg, M.; Berg, P. Use of prior knowledge in brain electromagnetic source analysis. *Brain Topogr.* **1991**, *4*, 143–150. [CrossRef]
17. Mosher, J.C.; Lewis, P.S.; Leahy, R.M. Multiple dipole modeling and localization from spatio-temporal MEG data. *IEEE Trans. Biomed. Eng.* **1992**, *39*, 541–557. [CrossRef]
18. Mosher, J.C.; Baillet, S.; Leahy, R.M. EEG source localization and imaging using multiple signal classification approaches. *J. Clin. Neurophysiol.* **1999**, *16*, 225–238. [CrossRef]
19. Hämäläinen, M.S.; Ilmoniemi, R.J. Interpreting magnetic fields of the brain: Minimum norm estimates. *Med. Biol. Eng. Comput.* **1994**, *32*, 35–42. [CrossRef]
20. Pascual-Marqui, R.D.; Michel, C.M.; Lehmann, D. Low resolution electromagnetic tomography: A new method for localizing electrical activity in the brain. *Int. J. Psychophysiol.* **1994**, *18*, 49–65. [CrossRef]

21. Vrba, J.; Robinson, S.E. Signal processing in magnetoencephalography. *Methods* **2001**, *25*, 249–271. [CrossRef] [PubMed]
22. Babiloni, C.; Del Percio, C.; Caroli, A.; Salvatore, E.; Nicolai, E.; Marzano, N.; Lizio, R.; Cavedo, E.; Landau, S.; Chen, K.; et al. Cortical sources of resting state EEG rhythms are related to brain hypometabolism in subjects with Alzheimer's disease: An EEG-PET study. *Neurobiol. Aging* **2016**, *48*, 122–134. [CrossRef] [PubMed]
23. Boccaletti, S.; Valladares, D.L.; Pecora, L.M.; Geffert, H.P.; Carroll, T. Reconstructing embedding spaces of coupled dynamical systems from multivariate data. *Phys. Rev. E Stat. Nonlin. Soft Matter Phys.* **2002**, *65*, 035204. [CrossRef] [PubMed]
24. Valdes-Sosa, P.A.; Sanchez-Bornot, J.M.; Sotero, R.C.; Iturria-Medina, Y.; Aleman-Gomez, Y.; Bosch-Bayard, J.; Carbonell, F.; Ozaki, T. Model driven EEG/fMRI fusion of brain oscillations. *Hum. Brain Mapp.* **2009**, *30*, 2701–2721. [CrossRef]
25. Gramfort, A.; Strohmeier, D.; Haueisen, J.; Hämäläinen, M.S.; Kowalski, M. Time-frequency mixed-norm estimates: Sparse M/EEG imaging with non-stationary source activations. *Neuroimage* **2013**, *70*, 410–422. [CrossRef]
26. Talairach, J.; Tournoux, P. *Co-Planar Stereotaxic Atlas of the Human Brain*; Thieme, G., Ed.; George Thieme: Stuttgard, Germany, 1988.
27. Yao, D.; He, B. A self-coherence enhancement algorithm and its application to enhancing three-dimensional source estimation from EEGs. *Ann. Biomed. Eng.* **2001**, *29*, 1019–1027. [CrossRef]
28. Pascual-Marqui, R.D.; Esslen, M.; Kochi, K.; Lehmann, D. Functional imaging with low-resolution brain electromagnetic tomography (LORETA): A review. *Methods Find Exp. Clin. Pharmacol.* **2002**, *24* (Suppl. C), 91–95.
29. Pascual-Marqui, R.D. Discrete, 3D distributed, linear imaging methods of electric neuronal activity. Part 1: Exact, zero error localization. *arXiv* **2007**, arXiv:0710.3341.
30. Pascual-Marqui, R.D. Instantaneous and lagged measurements of linear and nonlinear dependence between groups of multivariate time series: Frequency decomposition. *arXiv* **2007**, arXiv:0711.1455.
31. Rossini, P.M.; Del Percio, C.; Pasqualetti, P.; Cassetta, E.; Binetti, G.; Dal Forno, G.; Ferreri, F.; Frisoni, G.; Chiovenda, P.; Miniussi, C.; et al. Conversion from mild cognitive impairment to Alzheimer's disease is predicted by sources and coherence of brain electroencephalography rhythms. *Neuroscience* **2006**, *143*, 793–803. [CrossRef]
32. Huang, Y.; Mucke, L. Alzheimer mechanisms and therapeutic strategies. *Cell* **2012**, *148*, 1204–1222. [CrossRef] [PubMed]
33. Koenig, T.; Prichep, L.; Dierks, T.; Hubl, D.; Wahlund, L.O.; John, E.R.; Jelic, V. Decreased EEG synchronization in Alzheimer's disease and mild cognitive impairment. *Neurobiol. Aging* **2005**, *26*, 165–171. [CrossRef] [PubMed]
34. Jelic, V.; Johansson, S.E.; Almkvist, O.; Shigeta, M.; Julin, P.; Nordberg, A.; Winblad, B.; Wahlund, L.O. Quantitative electroencephalography in mild cognitive impairment: Longitudinal changes and possible prediction of Alzheimer's disease. *Neurobiol. Aging* **2000**, *21*, 533–540. [CrossRef]
35. Adler, G.; Brassen, S.; Jajcevic, A. EEG coherence in Alzheimer's dementia. *J. Neural Transm.* **2003**, *110*, 1051–1058. [CrossRef] [PubMed]
36. Van der Hiele, K.; Vein, A.A.; Reijntjes, R.H.A.M.; Westendorp, R.G.J.; Bollen, E.L.E.M.; van Buchem, M.A.; van Dijk, J.G.; Middelkoop, H.A.M. EEG correlates in the spectrum of cognitive decline. *Clin. Neurophysiol.* **2007**, *118*, 1931–1939. [CrossRef]
37. Nishida, K.; Yoshimura, M.; Isotani, T.; Yoshida, T.; Kitaura, Y.; Saito, A.; Mii, H.; Kato, M.; Takekita, Y.; Suwa, A.; et al. Differences in quantitative EEG between frontotemporal dementia and Alzheimer's disease as revealed by LORETA. *Clin. Neurophysiol.* **2011**, *122*, 1718–1725. [CrossRef]
38. Scheeringa, R.; Petersson, K.M.; Kleinschmidt, A.; Jensen, O.; Bastiaansen, M.C.M. EEG Alpha Power Modulation of fMRI Resting-State Connectivity. *Brain Connect* **2012**, *2*, 254–264. [CrossRef]
39. Pritchard, W.S.; Duke, D.W.; Krieble, K.K. Dimensional analysis of resting human EEG II: Surrogate-data testing indicates nonlinearity but not low-dimensional chaos. *Psychophysiology* **1995**, *32*, 486–491. [CrossRef]
40. Woyshville, M.J.; Calabrese, J.R. Quantification of occipital EEG changes in Alzheimer's disease utilizing a new metric: The fractal dimension. *Biol. Psychiatry* **1994**, *35*, 381–387. [CrossRef]
41. Stam, C.J.; Jones, B.F.; Nolte, G.; Breakspear, M.; Scheltens, P. Small-world networks and functional connectivity in Alzheimer's disease. *Cereb. Cortex* **2007**, *17*, 92–99. [CrossRef]

42. Stam, C.J. Modern network science of neurological disorders. *Nat. Rev. Neurosci.* **2014**, *15*, 683–695. [CrossRef] [PubMed]
43. Jelles, B.; van Birgelen, J.H.; Slaets, J.P.; Hekster, R.E.; Jonkman, E.J.; Stam, C.J. Decrease of non-linear structure in the EEG of Alzheimer patients compared to healthy controls. *Clin. Neurophysiol.* **1999**, *110*, 1159–1167. [CrossRef]
44. Dauwels, J.; Vialatte, F.; Musha, T.; Cichocki, A. A comparative study of synchrony measures for the early diagnosis of Alzheimer's disease based on EEG. *Neuroimage* **2010**, *49*, 668–693. [CrossRef] [PubMed]
45. Azami, H.; Kinney-Lang, E.; Ebied, A.; Fernandez, A.; Escudero, J. Multiscale dispersion entropy for the regional analysis of resting-state magnetoencephalogram complexity in Alzheimer's disease. *Conf. Proc. IEEE Eng. Med. Biol. Soc.* **2017**, *2017*, 3182–3185. [PubMed]
46. Dierks, T.; Ihl, R.; Frölich, L.; Maurer, K. Dementia of the alzheimer type: Effects on the spontaneous EEG described by dipole sources. *Psychiatry Res. Neuroimaging* **1993**, *50*, 151–162. [CrossRef]
47. Babiloni, C.; Frisoni, G.B.; Pievani, M.; Vecchio, F.; Lizio, R.; Buttiglione, M.; Geroldi, C.; Fracassi, C.; Eusebi, F.; Ferri, R.; et al. Hippocampal volume and cortical sources of EEG alpha rhythms in mild cognitive impairment and Alzheimer disease. *NeuroImage* **2009**, *44*, 123–135. [CrossRef]
48. Babiloni, C.; Vecchio, F.; Lizio, R.; Ferri, R.; Rodriguez, G.; Marzano, N.; Frisoni, G.B.; Rossini, P.M. Resting state cortical rhythms in mild cognitive impairment and Alzheimer's disease: Electroencephalographic evidence. *J. Alzheimers Dis.* **2011**, *26* (Suppl. 3), 201–214. [CrossRef]
49. Canuet, L.; Tellado, I.; Couceiro, V.; Fraile, C.; Fernandez-Novoa, L.; Ishii, R.; Takeda, M.; Cacabelos, R. Resting-state network disruption and APOE genotype in Alzheimer's disease: A lagged functional connectivity study. *PLoS ONE* **2012**, *7*, e46289. [CrossRef]
50. Hsiao, F.-J.; Wang, Y.-J.; Yan, S.-H.; Chen, W.-T.; Lin, Y.-Y. Altered Oscillation and Synchronization of Default-Mode Network Activity in Mild Alzheimer's Disease Compared to Mild Cognitive Impairment: An Electrophysiological Study. *PLoS ONE* **2013**, *8*, e68792. [CrossRef] [PubMed]
51. Babiloni, C.; Del Percio, C.; Lizio, R.; Marzano, N.; Infarinato, F.; Soricelli, A.; Salvatore, E.; Ferri, R.; Bonforte, C.; Tedeschi, G.; et al. Cortical sources of resting state electroencephalographic alpha rhythms deteriorate across time in subjects with amnesic mild cognitive impairment. *Neurobiol. Aging* **2014**, *35*, 130–142. [CrossRef]
52. Babiloni, C.; Lizio, R.; Marzano, N.; Capotosto, P.; Soricelli, A.; Triggiani, A.I.; Cordone, S.; Gesualdo, L.; Del Percio, C. Brain neural synchronization and functional coupling in Alzheimer's disease as revealed by resting state EEG rhythms. *Int. J. Psychophysiol.* **2016**, *103*, 88–102. [CrossRef] [PubMed]
53. Kouzuki, M.; Asaina, F.; Taniguchi, M.; Musha, T.; Urakami, K. The relationship between the diagnosis method of neuronal dysfunction (DIMENSION) and brain pathology in the early stages of Alzheimer's disease. *Psychogeriatrics* **2013**, *13*, 63–70. [CrossRef] [PubMed]
54. Jeong, J. EEG dynamics in patients with Alzheimer's disease. *Clin. Neurophysiol.* **2004**, *115*, 1490–1505. [CrossRef] [PubMed]
55. Hornero, R.; Abásolo, D.; Escudero, J.; Gómez, C. Nonlinear analysis of electroencephalogram and magnetoencephalogram recordings in patients with Alzheimer's disease. *Philos. Trans. A Math. Phys. Eng. Sci.* **2009**, *367*, 317–336. [CrossRef]
56. Grassberger, P.; Procaccia, I. Measuring the strangeness of strange attractors. *Phys. D Nonlinear Phenom.* **1983**, *9*, 189–208. [CrossRef]
57. Wolf, A.; Swift, J.B.; Swinney, H.L.; Vastano, J.A. Determining Lyapunov exponents from a time series. *Phys. D Nonlinear Phenom.* **1985**, *16*, 285–317. [CrossRef]
58. Breakspear, M. Dynamic models of large-scale brain activity. *Nat. Neurosci.* **2017**, *20*, 340–352. [CrossRef]
59. Garn, H.; Waser, M.; Deistler, M.; Benke, T.; Dal-Bianco, P.; Ransmayr, G.; Schmidt, H.; Sanin, G.; Santer, P.; Caravias, G.; et al. Quantitative EEG markers relate to Alzheimer's disease severity in the Prospective Dementia Registry Austria (PRODEM). *Clin. Neurophysiol.* **2015**, *126*, 505–513. [CrossRef]
60. Azami, H.; Escudero, J. Improved multiscale permutation entropy for biomedical signal analysis: Interpretation and application to electroencephalogram recordings. *Biomed. Signal Process. Control* **2016**, *23*, 28–41. [CrossRef]
61. Smits, F.M.; Porcaro, C.; Cottone, C.; Cancelli, A.; Rossini, P.M.; Tecchio, F. Electroencephalographic Fractal Dimension in Healthy Ageing and Alzheimer's Disease. *PLoS ONE* **2016**, *11*, e0149587. [CrossRef]

62. Higuchi, T. Approach to an irregular time series on the basis of the fractal theory. *Phys. D Nonlinear Phenom.* **1988**, *31*, 277–283. [CrossRef]
63. Katz, M.J. Fractals and the analysis of waveforms. *Comput. Biol. Med.* **1988**, *18*, 145–156. [CrossRef]
64. Richman, J.S.; Moorman, J.R. Physiological time-series analysis using approximate entropy and sample entropy. *Am. J. Physiol. Heart Circ. Physiol.* **2000**, *278*, H2039–H2049. [CrossRef] [PubMed]
65. Bandt, C.; Pompe, B. Permutation entropy: A natural complexity measure for time series. *Phys. Rev. Lett.* **2002**, *88*, 174102. [CrossRef] [PubMed]
66. Humeau-Heurtier, A. The Multiscale Entropy Algorithm and Its Variants: A Review. *Entropy* **2015**, *17*, 3110–3123. [CrossRef]
67. Costa, M.; Goldberger, A.L.; Peng, C.-K. Multiscale entropy analysis of biological signals. *Phys. Rev. E Stat. Nonlin. Soft Matter Phys.* **2005**, *71*, 021906. [CrossRef]
68. Yang, A.C.; Tsai, S.-J. Is mental illness complex? From behavior to brain. *Prog. Neuropsychopharmacol. Biol. Psychiatry* **2013**, *45*, 253–257. [CrossRef]
69. Escudero, J.; Abásolo, D.; Hornero, R.; Espino, P.; López, M. Analysis of electroencephalograms in Alzheimer's disease patients with multiscale entropy. *Physiol. Meas.* **2006**, *27*, 1091–1106. [CrossRef]
70. Courtiol, J.; Perdikis, D.; Petkoski, S.; Müller, V.; Huys, R.; Sleimen-Malkoun, R.; Jirsa, V.K. The multiscale entropy: Guidelines for use and interpretation in brain signal analysis. *J. Neurosci. Methods* **2016**, *273*, 175–190. [CrossRef]
71. Azami, H.; Abásolo, D.; Simons, S.; Escudero, J. Univariate and Multivariate Generalized Multiscale Entropy to Characterise EEG Signals in Alzheimer's Disease. *Entropy* **2017**, *19*, 31. [CrossRef]
72. Ahmed, M.U.; Mandic, D.P. Multivariate multiscale entropy: A tool for complexity analysis of multichannel data. *Phys. Rev. E Stat. Nonlin. Soft Matter Phys.* **2011**, *84*, 061918. [CrossRef] [PubMed]
73. Labate, D.; Foresta, F.L.; Morabito, G.; Palamara, I.; Morabito, F.C. Entropic Measures of EEG Complexity in Alzheimer's Disease Through a Multivariate Multiscale Approach. *IEEE Sens. J.* **2013**, *13*, 3284–3292. [CrossRef]
74. Azami, H.; Escudero, J.; Fernández, A. Refined composite multivariate multiscale entropy based on variance for analysis of resting-state magnetoencephalograms in Alzheimer's disease. In Proceedings of the 2016 International Conference for Students on Applied Engineering (ICSAE), Qingdao, China, 29–30 May 2016; pp. 413–418.
75. Morison, G.; Tieges, Z.; Kilborn, K. Analysis of Electroencephalography Activity in Early Stage Alzheimer's Disease Using a Multiscale Statistical Complexity Measure. *Adv. Sci. Lett.* **2013**, *19*, 2414–2418. [CrossRef]
76. Timothy, L.T.; Krishna, B.M.; Nair, U. Classification of mild cognitive impairment EEG using combined recurrence and cross recurrence quantification analysis. *Int. J. Psychophysiol.* **2017**, *120*, 86–95. [CrossRef] [PubMed]
77. Singer, W. The formation of cooperative cell assemblies in the visual cortex. *J. Exp. Biol.* **1990**, *153*, 177–197. [PubMed]
78. Jung, T.P.; Makeig, S.; Westerfield, M.; Townsend, J.; Courchesne, E.; Sejnowski, T.J. Analysis and visualization of single-trial event-related potentials. *Hum. Brain Mapp.* **2001**, *14*, 166–185. [CrossRef]
79. Makeig, S.; Westerfield, M.; Jung, T.P.; Enghoff, S.; Townsend, J.; Courchesne, E.; Sejnowski, T.J. Dynamic brain sources of visual evoked responses. *Science* **2002**, *295*, 690–694. [CrossRef]
80. Fuentemilla, L.; Marco-Pallarés, J.; Grau, C. Modulation of spectral power and of phase resetting of EEG contributes differentially to the generation of auditory event-related potentials. *Neuroimage* **2006**, *30*, 909–916. [CrossRef]
81. Adrian, E.D.; Moruzzi, G. Impulses in the pyramidal tract. *J. Physiol.* **1939**, *97*, 153–199. [CrossRef]
82. Buzsáki, G. Neuroscience: Neurons and navigation. *Nature* **2005**, *436*, 781–782. [CrossRef]
83. Rossini, P.M.; Di Iorio, R.; Bentivoglio, M.; Bertini, G.; Ferreri, F.; Gerloff, C.; Ilmoniemi, R.J.; Miraglia, F.; Nitsche, M.A.; Pestilli, F.; et al. Methods for analysis of brain connectivity: An IFCN-sponsored review. *Clin. Neurophysiol.* **2019**, *130*, 1833–1858. [CrossRef] [PubMed]
84. Uhlhaas, P.J.; Singer, W. Neural synchrony in brain disorders: Relevance for cognitive dysfunctions and pathophysiology. *Neuron* **2006**, *52*, 155–168. [CrossRef]
85. Buzsáki, G.; Schomburg, E.W. What does gamma coherence tell us about inter-regional neural communication? *Nat. Neurosci.* **2015**, *18*, 484–489. [PubMed]

86. Blanco, S.; Quiroga, R.Q.; Rosso, O.A.; Kochen, S. Time-frequency analysis of electroencephalogram series. *Phys. Rev. E Stat. Phys. Plasmas Fluids Relat. Interdiscip. Topics* **1995**, *51*, 2624–2631. [CrossRef] [PubMed]
87. Nunez, P.L. Spatial-Temporal Structures of Human Alpha Rhythms: Theory, Microcurrent Sources, Multiscale Measurements, and Global Binding of Local Networks. *Human Brain Mapping*. 2001. Wiley Online Library. Available online: https://onlinelibrary.wiley.com/doi/full/10.1002/hbm.1030 (accessed on 26 March 2020).
88. Kaplan, A.Y.; Fingelkurts, A.A.; Fingelkurts, A.A.; Borisov, S.V.; Darkhovsky, B.S. Nonstationary nature of the brain activity as revealed by EEG/MEG: Methodological, practical and conceptual challenges. *Signal Process.* **2005**, *85*, 2190–2212. [CrossRef]
89. Miraglia, F.; Vecchio, F.; Rossini, P.M. Searching for signs of aging and dementia in EEG through network analysis. *Behav. Brain Res.* **2017**, *317*, 292–300.
90. Worsley, K.J.; Chen, J.-I.; Lerch, J.; Evans, A.C. Comparing functional connectivity via thresholding correlations and singular value decomposition. *Philos. Trans. R Soc. Lond. B Biol. Sci.* **2005**, *360*, 913–920. [CrossRef]
91. Canuet, L.; Ishii, R.; Pascual-Marqui, R.D.; Iwase, M.; Kurimoto, R.; Aoki, Y.; Ikeda, S.; Takahashi, H.; Nakahachi, T.; Takeda, M. Resting-state EEG source localization and functional connectivity in schizophrenia-like psychosis of epilepsy. *PLoS ONE* **2011**, *6*, e27863. [CrossRef]
92. Barry, R.J.; De Blasio, F.M.; Borchard, J.P. Sequential processing in the equiprobable auditory Go/NoGo task: Children vs. adults. *Clin. Neurophysiol.* **2014**, *125*, 1995–2006. [CrossRef]
93. Aoki, Y.; Ishii, R.; Pascual-Marqui, R.D.; Canuet, L.; Ikeda, S.; Hata, M.; Imajo, K.; Matsuzaki, H.; Musha, T.; Asada, T.; et al. Detection of EEG-resting state independent networks by eLORETA-ICA method. *Front. Hum. Neurosci.* **2015**, *9*, 31. [CrossRef]
94. Ikeda, S.; Mizuno-Matsumoto, Y.; Canuet, L.; Ishii, R.; Aoki, Y.; Hata, M.; Katsimichas, T.; Pascual-Marqui, R.D.; Hayashi, T.; Okamoto, E.; et al. Emotion Regulation of Neuroticism: Emotional Information Processing Related to Psychosomatic State Evaluated by Electroencephalography and Exact Low-Resolution Brain Electromagnetic Tomography. *Neuropsychobiology* **2015**, *71*, 34–41. [CrossRef] [PubMed]
95. Ramyead, A.; Kometer, M.; Studerus, E.; Koranyi, S.; Ittig, S.; Gschwandtner, U.; Fuhr, P.; Riecher-Rössler, A. Aberrant Current Source-Density and Lagged Phase Synchronization of Neural Oscillations as Markers for Emerging Psychosis. *Schizophr. Bull.* **2015**, *41*, 919–929. [CrossRef] [PubMed]
96. Vecchio, F.; Miraglia, F.; Bramanti, P.; Rossini, P.M. Human brain networks in physiological aging: A graph theoretical analysis of cortical connectivity from EEG data. *J. Alzheimers Dis.* **2014**, *41*, 1239–1249. [CrossRef] [PubMed]
97. Vecchio, F.; Miraglia, F.; Curcio, G.; Altavilla, R.; Scrascia, F.; Giambattistelli, F.; Quattrocchi, C.C.; Bramanti, P.; Vernieri, F.; Rossini, P.M. Cortical brain connectivity evaluated by graph theory in dementia: A correlation study between functional and structural data. *J. Alzheimers Dis.* **2015**, *45*, 745–756. [CrossRef] [PubMed]
98. Vecchio, F.; Miraglia, F.; Marra, C.; Quaranta, D.; Vita, M.G.; Bramanti, P.; Rossini, P.M. Human brain networks in cognitive decline: A graph theoretical analysis of cortical connectivity from EEG data. *J. Alzheimers Dis.* **2014**, *41*, 113–127. [CrossRef] [PubMed]
99. Vecchio, F.; Pellicciari, M.C.; Miraglia, F.; Brignani, D.; Miniussi, C.; Rossini, P.M. Effects of transcranial direct current stimulation on the functional coupling of the sensorimotor cortical network. *Neuroimage* **2016**, *140*, 50–56. [CrossRef]
100. Engel, A.K.; Senkowski, D.; Schneider, T.R. Multisensory Integration through Neural Coherence. In *The Neural Bases of Multisensory Processes*; Murray, M.M., Wallace, M.T., Eds.; Frontiers in Neuroscience; CRC Press/Taylor & Francis: Boca Raton, FL, USA, 2012; ISBN 978-1-4398-1217-4.
101. Singer, W.; Gray, C.M. Visual feature integration and the temporal correlation hypothesis. *Annu. Rev. Neurosci.* **1995**, *18*, 555–586. [CrossRef]
102. Gerloff, C.; Richard, J.; Hadley, J.; Schulman, A.E.; Honda, M.; Hallett, M. Functional coupling and regional activation of human cortical motor areas during simple, internally paced and externally paced finger movements. *Brain* **1998**, *121 Pt 8*, 1513–1531. [CrossRef]
103. Singer, W. Neuronal synchrony: A versatile code for the definition of relations? *Neuron* **1999**, *24*, 49–65. [CrossRef]
104. Boenstrup, M.; Feldheim, J.; Heise, K.; Gerloff, C.; Hummel, F.C. The control of complex finger movements by directional information flow between mesial frontocentral areas and the primary motor cortex. *Eur. J. Neurosci.* **2014**, *40*, 2888–2897. [CrossRef] [PubMed]

105. Lachaux, J.P.; Rodriguez, E.; Martinerie, J.; Varela, F.J. Measuring phase synchrony in brain signals. *Hum. Brain Mapp.* **1999**, *8*, 194–208. [CrossRef]
106. Kraskov, A.; Stögbauer, H.; Grassberger, P. Estimating mutual information. *Phys. Rev. E Stat. Nonlin. Soft Matter Phys.* **2004**, *69*, 066138. [CrossRef] [PubMed]
107. Kamiński, M.; Blinowska, K.; Szelenberger, W. Investigation of coherence structure and EEG activity propagation during sleep. *Acta Neurobiol. Exp.* **1995**, *55*, 213–219.
108. Friston, K.J.; Harrison, L.; Penny, W. Dynamic causal modelling. *Neuroimage* **2003**, *19*, 1273–1302. [CrossRef]
109. Kiebel, S.J.; Garrido, M.I.; Moran, R.; Chen, C.-C.; Friston, K.J. Dynamic causal modeling for EEG and MEG. *Hum. Brain Mapp.* **2009**, *30*, 1866–1876. [CrossRef]
110. Crossley, N.A.; Fox, P.T.; Bullmore, E.T. Meta-connectomics: Human brain network and connectivity meta-analyses. *Psychol. Med.* **2016**, *46*, 897–907. [CrossRef]
111. Miraglia, F.; Vecchio, F.; Bramanti, P.; Rossini, P.M. Small-worldness characteristics and its gender relation in specific hemispheric networks. *Neuroscience* **2015**, *310*, 1–11. [CrossRef]
112. Miraglia, F.; Vecchio, F.; Bramanti, P.; Rossini, P.M. EEG characteristics in "eyes-open" versus "eyes-closed" conditions: Small-world network architecture in healthy aging and age-related brain degeneration. *Clin. Neurophysiol.* **2016**, *127*, 1261–1268. [CrossRef]
113. Rubinov, M.; Sporns, O. Complex network measures of brain connectivity: Uses and interpretations. *Neuroimage* **2010**, *52*, 1059–1069. [CrossRef]
114. Sporns, O. The human connectome: Origins and challenges. *Neuroimage* **2013**, *80*, 53–61. [CrossRef]
115. De Haan, W.; van der Flier, W.M.; Wang, H.; Van Mieghem, P.F.A.; Scheltens, P.; Stam, C.J. Disruption of functional brain networks in Alzheimer's disease: What can we learn from graph spectral analysis of resting-state magnetoencephalography? *Brain Connect* **2012**, *2*, 45–55. [CrossRef]
116. Onnela, J.-P.; Saramäki, J.; Kertész, J.; Kaski, K. Intensity and coherence of motifs in weighted complex networks. *Phys. Rev. E Stat. Nonlin. Soft Matter Phys.* **2005**, *71*, 065103. [CrossRef] [PubMed]
117. Moharramipour, A.; Mostame, P.; Hossein-Zadeh, G.-A.; Wheless, J.W.; Babajani-Feremi, A. Comparison of statistical tests in effective connectivity analysis of ECoG data. *J. Neurosci. Methods* **2018**, *308*, 317–329. [CrossRef] [PubMed]
118. Vecchio, F.; Miraglia, F.; Alù, F.; Judica, E.; Cotelli, M.; Pellicciari, M.; Rossini, P. Human brain networks in physiological and pathological aging: Reproducibility of EEG graph theoretical analysis in cortical connectivity. (Unpublished).
119. Zuo, X.-N.; Xing, X.-X. Test-retest reliabilities of resting-state FMRI measurements in human brain functional connectomics: A systems neuroscience perspective. *Neurosci. Biobehav. Rev.* **2014**, *45*, 100–118. [CrossRef] [PubMed]
120. 120Colclough, G.L.; Woolrich, M.W.; Tewarie, P.K.; Brookes, M.J.; Quinn, A.J.; Smith, S.M. How reliable are MEG resting-state connectivity metrics? *Neuroimage* **2016**, *138*, 284–293. [CrossRef]
121. Dimitriadis, S.I.; Drakesmith, M.; Bells, S.; Parker, G.D.; Linden, D.E.; Jones, D.K. Improving the Reliability of Network Metrics in Structural Brain Networks by Integrating Different Network Weighting Strategies into a Single Graph. *Front. Neurosci.* **2017**, *11*, 694. [CrossRef] [PubMed]
122. Watts, D.J.; Strogatz, S.H. Collective dynamics of "small-world" networks. *Nature* **1998**, *393*, 440–442. [CrossRef]
123. Newman, M.E.J. Properties of highly clustered networks. *Phys. Rev. E Stat. Nonlin. Soft Matter Phys.* **2003**, *68*, 026121. [CrossRef]
124. Latora, V.; Marchiori, M. Efficient behavior of small-world networks. *Phys. Rev. Lett.* **2001**, *87*, 198701. [CrossRef]
125. Ferreri, F.; Vecchio, F.; Ponzo, D.; Pasqualetti, P.; Rossini, P.M. Time-varying coupling of EEG oscillations predicts excitability fluctuations in the primary motor cortex as reflected by motor evoked potentials amplitude: An EEG-TMS study. *Hum. Brain Mapp.* **2014**, *35*, 1969–1980. [CrossRef]
126. Vecchio, F.; Miraglia, F.; Judica, E.; Cotelli, M.; Alù, F.; Rossini, P.M. Human brain networks: A graph theoretical analysis of cortical connectivity normative database from EEG data in healthy elderly subjects. *Geroscience* **2020**. [CrossRef] [PubMed]

127. Vecchio, F.; Miraglia, F.; Piludu, F.; Granata, G.; Romanello, R.; Caulo, M.; Onofrj, V.; Bramanti, P.; Colosimo, C.; Rossini, P.M. "Small World" architecture in brain connectivity and hippocampal volume in Alzheimer's disease: A study via graph theory from EEG data. *Brain Imaging Behav.* **2017**, *11*, 473–485. [CrossRef] [PubMed]
128. Vecchio, F.; Miraglia, F.; Iberite, F.; Lacidogna, G.; Guglielmi, V.; Marra, C.; Pasqualetti, P.; Tiziano, F.D.; Rossini, P.M. Sustainable method for Alzheimer dementia prediction in mild cognitive impairment: Electroencephalographic connectivity and graph theory combined with apolipoprotein E. *Ann. Neurol.* **2018**, *84*, 302–314. [CrossRef]
129. Schomer, D.L.; Silva, F.H.L.D. *Niedermeyer's ElectroencephalographyBasic Principles, Clinical Applications, and Related Fields*; Oxford University Press: Oxford, UK, 2017.
130. 130Giri, M.; Zhang, M.; Lü, Y. Genes associated with Alzheimer's disease: An overview and current status. *Clin. Interv. Aging* **2016**, *11*, 665–681.
131. Miraglia, F.; Vecchio, F.; Marra, C.; Quaranta, D.; Alù, F.; Peroni, B.; Granata, G.; Judica, E.; Cotelli, M.; Rossini, P.M. Small World Index in Default Mode Network Predicts Progression from Mild Cognitive Impairment to Dementia. *Int. J. Neural Syst.* **2020**, *30*, 2050004. [CrossRef]
132. Başar, E.; Güntekin, B.; Atagün, I.; Turp Gölbaşı, B.; Tülay, E.; Ozerdem, A. Brain's alpha activity is highly reduced in euthymic bipolar disorder patients. *Cogn. Neurodyn.* **2012**, *6*, 11–20. [CrossRef]
133. Tallon-Baudry, C.; Bertrand, O. Oscillatory gamma activity in humans and its role in object representation. *Trends Cogn. Sci.* **1999**, *3*, 151–162. [CrossRef]
134. Kaiser, J.; Heidegger, T.; Lutzenberger, W. Behavioral relevance of gamma-band activity for short-term memory-based auditory decision-making. *Eur. J. Neurosci.* **2008**, *27*, 3322–3328. [CrossRef]
135. Nikolić, D.; Fries, P.; Singer, W. Gamma oscillations: Precise temporal coordination without a metronome. *Trends Cogn. Sci.* **2013**, *17*, 54–55. [CrossRef] [PubMed]
136. Vinck, M.; Womelsdorf, T.; Buffalo, E.A.; Desimone, R.; Fries, P. Attentional modulation of cell-class-specific gamma-band synchronization in awake monkey area v4. *Neuron* **2013**, *80*, 1077–1089. [CrossRef] [PubMed]
137. Abeles, M. Corticonics: Neural Circuits of the Cerebral Cortex. Available online: /core/books/corticonics/7BF149062695412A32FFC1255C98B410 (accessed on 26 March 2020).
138. Fries, P. A mechanism for cognitive dynamics: Neuronal communication through neuronal coherence. *Trends Cogn. Sci.* **2005**, *9*, 474–480. [CrossRef] [PubMed]
139. Cassani, R.; Falk, T.H.; Fraga, F.J.; Kanda, P.A.M.; Anghinah, R. The effects of automated artifact removal algorithms on electroencephalography-based Alzheimer's disease diagnosis. *Front. Aging Neurosci.* **2014**, *6*, 55. [CrossRef] [PubMed]
140. Fraga, F.J.; Mamani, G.Q.; Johns, E.; Tavares, G.; Falk, T.H.; Phillips, N.A. Early diagnosis of mild cognitive impairment and Alzheimer's with event-related potentials and event-related desynchronization in N-back working memory tasks. *Comput. Methods Programs Biomed.* **2018**, *164*, 1–13. [CrossRef] [PubMed]
141. Trambaiolli, L.R.; Lorena, A.C.; Fraga, F.J.; Kanda, P.A.M.; Anghinah, R.; Nitrini, R. Improving Alzheimer's disease diagnosis with machine learning techniques. *Clin. EEG Neurosci.* **2011**, *42*, 160–165. [CrossRef]
142. Cassani, R.; Estarellas, M.; San-Martin, R.; Fraga, F.J.; Falk, T.H. Systematic Review on Resting-State EEG for Alzheimer's Disease Diagnosis and Progression Assessment. *Dis. Markers* **2018**, *2018*. [CrossRef]
143. Vecchio, F.; Miraglia, F.; Alù, F.; Menna, M.; Judica, E.; Cotelli, M.; Rossini, P. Classification of Alzheimer's Disease respect to physiological aging with innovative EEG Biomarkers in a machine learning implementation. *J. Alzheimer Dis.* **2020**, in press.
144. Barker, A.T.; Jalinous, R.; Freeston, I.L. Non-invasive magnetic stimulation of human motor cortex. *Lancet* **1985**, *1*, 1106–1107. [CrossRef]
145. Ilmoniemi, R.J.; Kicić, D. Methodology for combined TMS and EEG. *Brain Topogr.* **2010**, *22*, 233–248. [CrossRef]
146. Cracco, R.Q.; Amassian, V.E.; Maccabee, P.J.; Cracco, J.B. Comparison of human transcallosal responses evoked by magnetic coil and electrical stimulation. *Electroencephalogr. Clin. Neurophysiol.* **1989**, *74*, 417–424. [CrossRef]
147. Ilmoniemi, R.J.; Virtanen, J.; Ruohonen, J.; Karhu, J.; Aronen, H.J.; Näätänen, R.; Katila, T. Neuronal responses to magnetic stimulation reveal cortical reactivity and connectivity. *Neuroreport* **1997**, *8*, 3537–3540. [CrossRef]
148. Ferreri, F.; Rossini, P.M. TMS and TMS-EEG techniques in the study of the excitability, connectivity, and plasticity of the human motor cortex. *Rev. Neurosci.* **2013**, *24*, 431–442. [CrossRef] [PubMed]

149. Bonato, C.; Miniussi, C.; Rossini, P.M. Transcranial magnetic stimulation and cortical evoked potentials: A TMS/EEG co-registration study. *Clin. Neurophysiol.* **2006**, *117*, 1699–1707. [CrossRef] [PubMed]
150. Ferreri, F.; Pasqualetti, P.; Määttä, S.; Ponzo, D.; Ferrarelli, F.; Tononi, G.; Mervaala, E.; Miniussi, C.; Rossini, P.M. Human brain connectivity during single and paired pulse transcranial magnetic stimulation. *Neuroimage* **2011**, *54*, 90–102. [CrossRef] [PubMed]
151. Ferreri, F.; Ponzo, D.; Hukkanen, T.; Mervaala, E.; Könönen, M.; Pasqualetti, P.; Vecchio, F.; Rossini, P.M.; Määttä, S. Human brain cortical correlates of short-latency afferent inhibition: A combined EEG-TMS study. *J. Neurophysiol.* **2012**, *108*, 314–323. [CrossRef]
152. Massimini, M.; Ferrarelli, F.; Huber, R.; Esser, S.K.; Singh, H.; Tononi, G. Breakdown of cortical effective connectivity during sleep. *Science* **2005**, *309*, 2228–2232. [CrossRef]
153. Ferreri, F.; Guerra, A.; Vollero, L.; Ponzo, D.; Maatta, S.; Mervaala, E.; Iannello, G.; Di Lazzaro, V. Age-related changes of cortical excitability and connectivity in healthy humans: Non-invasive evaluation of sensorimotor network by means of TMS-EEG. *Neuroscience* **2017**, *357*, 255–263. [CrossRef]
154. Rosanova, M.; Casali, A.; Bellina, V.; Resta, F.; Mariotti, M.; Massimini, M. Natural frequencies of human corticothalamic circuits. *J. Neurosci.* **2009**, *29*, 7679–7685. [CrossRef]
155. Assenza, G.; Capone, F.; di Biase, L.; Ferreri, F.; Florio, L.; Guerra, A.; Marano, M.; Paolucci, M.; Ranieri, F.; Salomone, G.; et al. Corrigendum: Oscillatory Activities in Neurological Disorders of Elderly: Biomarkers to Target for Neuromodulation. *Front. Aging Neurosci.* **2017**, *9*, 252. [CrossRef]
156. Julkunen, P.; Jauhiainen, A.M.; Könönen, M.; Pääkkönen, A.; Karhu, J.; Soininen, H. Combining transcranial magnetic stimulation and electroencephalography may contribute to assess the severity of Alzheimer's disease. *Int. J. Alzheimers Dis.* **2011**, *2011*, 654794. [CrossRef]
157. Casarotto, S.; Määttä, S.; Herukka, S.-K.; Pigorini, A.; Napolitani, M.; Gosseries, O.; Niskanen, E.; Könönen, M.; Mervaala, E.; Rosanova, M.; et al. Transcranial magnetic stimulation-evoked EEG/cortical potentials in physiological and pathological aging. *Neuroreport* **2011**, *22*, 592–597. [CrossRef]
158. Ferreri, F.; Vecchio, F.; Vollero, L.; Guerra, A.; Petrichella, S.; Ponzo, D.; Määtta, S.; Mervaala, E.; Könönen, M.; Ursini, F.; et al. Sensorimotor cortex excitability and connectivity in Alzheimer's disease: A TMS-EEG Co-registration study. *Hum. Brain Mapp.* **2016**, *37*, 2083–2096. [CrossRef] [PubMed]
159. Bagattini, C.; Mutann, T.P.; Fracassi, C.; Manenti, R.; Cotelli, M.; Ilmoniemi, R.J.; Miniussi, C.; Bortoletto, M. Predicting Alzheimer's disease severity by means of TMS-EEG coregistration. *Neurobiol. Aging* **2019**, *80*, 38–45. [CrossRef] [PubMed]
160. Koch, G.; Bonnì, S.; Pellicciari, M.C.; Casula, E.P.; Mancini, M.; Esposito, R.; Ponzo, V.; Picazio, S.; Di Lorenzo, F.; Serra, L.; et al. Transcranial magnetic stimulation of the precuneus enhances memory and neural activity in prodromal Alzheimer's disease. *Neuroimage* **2018**, *169*, 302–311. [CrossRef] [PubMed]
161. Ferreri, F.; Pauri, F.; Pasqualetti, P.; Fini, R.; Dal Forno, G.; Rossini, P.M. Motor cortex excitability in Alzheimer's disease: A transcranial magnetic stimulation study. *Ann. Neurol.* **2003**, *53*, 102–108. [CrossRef] [PubMed]
162. Ferreri, F.; Pasqualetti, P.; Määttä, S.; Ponzo, D.; Guerra, A.; Bressi, F.; Chiovenda, P.; Del Duca, M.; Giambattistelli, F.; Ursini, F.; et al. Motor cortex excitability in Alzheimer's disease: A transcranial magnetic stimulation follow-up study. *Neurosci. Lett.* **2011**, *492*, 94–98. [CrossRef] [PubMed]

Journal of *Personalized Medicine*

Review

The Viral Hypothesis in Alzheimer's Disease: Novel Insights and Pathogen-Based Biomarkers

Sean X Naughton [1], Urdhva Raval [1] and Giulio M. Pasinetti [1,2,*]

[1] Department of Neurology, Icahn School of Medicine at Mount Sinai, New York, NY 10029, USA; sean.naughton@mssm.edu (S.X.N.); Urdhva.Raval@mssm.edu (U.R.)

[2] James J. Peters Veterans Affairs Medical Center, Bronx, NY 10468, USA

* Correspondence: giulio.pasinetti@mssm.edu

Received: 18 June 2020; Accepted: 28 July 2020; Published: 29 July 2020

Abstract: Early diagnosis of Alzheimer's disease (AD) and the identification of significant risk factors are necessary to better understand disease progression, and to develop intervention-based therapies prior to significant neurodegeneration. There is thus a critical need to establish biomarkers which can predict the risk of developing AD before the onset of cognitive decline. A number of studies have indicated that exposure to various microbial pathogens can accelerate AD pathology. Additionally, several studies have indicated that amyloid-β possess antimicrobial properties and may act in response to infection as a part of the innate immune system. These findings have led some to speculate that certain types of infections may play a significant role in AD pathogenesis. In this review, we will provide an overview of studies which suggest pathogen involvement in AD. Additionally, we will discuss a number of pathogen-associated biomarkers which may be effective in establishing AD risk. Infections that increase the risk of AD represent a modifiable risk factor which can be treated with therapeutic intervention. Pathogen-based biomarkers may thus be a valuable tool for evaluating and decreasing AD risk across the population.

Keywords: Alzheimer's disease; virus; bacteria; dementia

---

1. Introduction

Alzheimer's disease (AD) is an age-related neurodegenerative disorder. AD results in progressive cognitive decline, and is the most common form of dementia in older adults. This incurable disorder is predicted to affect approximately 100 million people globally by 2050 [1,2]. The characteristic hallmarks of AD pathology are amyloid-beta (Aβ) peptide plaques, tau hyperphosphorylation, and neuroinflammation. Currently, a clinical diagnosis of AD is only possible after disease onset through post-mortem detection of amyloid plaques and neurofibrillary tangles [3]. Cognitive testing can aid in the diagnosis of dementia, but strong cognitive impairments are usually only present at a time point when successful therapeutic intervention is unlikely. Early diagnosis of AD is only possible in rare cases in which the autosomal dominant early onset form of the disease is genetically inherited [4]. Considering the impact and prevalence of AD globally, there is an increasing need to understand and identify biomarkers, in order to detect AD in individuals before the onset of disease and provide mitigating therapeutics. Preclinical detection of biomarkers of Aβ, tau, and other neurodegenerative effects have been extensively studied. Aβ and tau have been detected in cerebrospinal fluid and blood plasma. Neuroimaging via magnetic resonance imaging (MRI) and positron emission tomography (PET), specifically, FDG-PET, amyloid PET and structural MRI can also serve diagnostic roles in AD [1,5]. Interestingly, there is some indication of microbial and viral involvement in AD pathology. Given the relatively high prevalence of certain pathogens, they can serve as biomarkers for preclinical AD. In this review, we explore the role of microbes in AD pathology and the potential of various pathogens as novel biomarkers for AD. This review is based on literature generated from searches conducted between

1 April and 25 July 2020, using standard databases and search engines for scientific literature (PubMed and Google Scholar), using the following keywords: "Alzheimer's Disease", "Pathogen Hypothesis" "Viruses", "Bacteria". Additional references were collected from those discussed in the literature generated through the search.

Amyloid-Beta

The Aβ peptide is integral to AD pathology. Aβ misfolding and the resultant Aβ plaques are thought to be the root cause of cognitive decline in AD. Aβ aggregates are formed from the proteolytic cleavage of a larger type 1 membrane glycoprotein, named amyloid precursor protein (APP). APP is involved in maintaining neuronal homeostasis, neuronal development, signaling, and intracellular transport [6]. APP is cleaved by β-secretases and γ-secretases, to produce an Aβ peptide ranging from 37 to 49 amino acid residues [7]. Aβ aggregates are found in the hippocampus, neocortex, and cerebrovasculature [8]. Aβ exists in different forms, including soluble Aβ, Aβ oligomers, and Aβ plaque forms. These different forms are involved in neurodegeneration at different stages of AD [6]. Aβ plaques induce tau protein hyperphosphorylation and formation neurofibrillary tangles and synaptic dysfunction. These plaques also generate the production of 4-hydroxynonenal, a toxic aldehyde involved in lipid peroxidation, and disruption of cellular homeostasis [9]. Aβ aggregation also leads to DNA damage and the release of inflammatory responses which result in the loss of neuronal synapses and ultimately neuronal death [10]. Intriguingly, Aβ acts as an antimicrobial peptide (AMP) and has been demonstrated to be effective against viruses, bacteria, and fungi. AMPs are a group of defensins, histatins, and cathelicidins that primarily serve to defend the host against a wide variety of pathogens. AMPs can also modulate cytokine release and adaptive immune responses. Aβ has been demonstrated to function like the cathelicidin AMP LL-37. Aβ was shown to be effective against the bacteria Streptococcus pneumoniae and fungus Candida albicans, which are the causative agents of bacterial meningitis and neurocandidiasis, respectively. Aβ also inhibits certain other bacterial species of the genera Pseudomonas, Escherichia, Streptococcus, Staphylococcus, Salmonella, and Enterococcus [11,12]. There is also evidence indicating that Aβ can also inhibit replication of seasonal and pandemic strains of the influenza A and herpes simplex virus 1 (HSV-1) viruses [13]. In fact, Aβ has been shown to be as effective as the antiviral drug Acyclovir at inhibiting HSV-1 neuropathology [14].

2. Pathogens and AD

2.1. Viral Pathogens in Neurodegeneration and AD

The idea that infections may play a role in Alzheimer's disease (AD) pathogenesis dates back nearly 30 years and has been a subject of debate in the field of AD (Figure 1) [15]. Previous studies have suggested that amyloid-β (Aβ) may act as a part of the innate immune system to aggregate around infectious particles. Eimer and colleagues showed that 5XFAD mice infected with herpes simplex virus 1 (HSV-1) showed increased survival rates compared to infected non-transgenic littermates. Moreover, Aβ was found to bind to and entrap HSV1, in a process mediated by fibrillization. Aβ deposition could be triggered by HSV1 infection in young 5XFAD mice, prior to the ordinary development of Aβ deposits [16]. Additionally, brains from Alzheimer's disease patients have been shown to have increased levels of human herpesvirus 6 and human herpesvirus 7 in several key areas [17]. HSV-1 infection has also been shown to drive the development of amyloid fibrillar plaque-like formations in human-induced neural stem cells and 3D human brain-like tissue cultures [18]. While different types of herpesvirus have been associated with AD pathology and detected in the brains of AD patients, there is also evidence to suggest that other viruses (and other types of pathogens) may also play a role in AD. For example, Nimgaonkar and colleagues found that exposure to HSV-2, cytomegalovirus (CMV), or the parasite Toxoplasma gondii (TOX) was associated with cognitive decline in individuals aged 65 and older [19]. Additionally, Ljungan virus (LV) has been detected in the hippocampus of AD brains, but not in age-matched controls [20]. The detection of different viral strains in the brains of different cohorts of patients hints at the idea that viral infection may play a role in AD or that AD may increase susceptibility to neuroinvasion by viruses. It is currently unclear if one or more pathogenic infections might directly stimulate (or accelerate) AD, or if AD creates an environment which facilitates the accumulation of infections in the brain through altered immune function. In this regard, it is interesting to consider the case of HIV, in which immune function is compromised in the presence of a persistent viral infection. In particular, HIV-associated neurocognitive disorders (HAND) have been associated with the presence of β-amyloid, and HAND patients have been shown to have similar cerebrospinal fluid levels of β-amyloid 1-42 when compared to patients with Alzheimer's associated dementia [21,22] Compromised immune function may thus be a critical driver of neurodegeneration by allowing infectious pathogens such as viruses to enter the brain at levels which exceed the capacity of the innate immune system within the brain. A number of different viruses have been shown to enter the brain and cause neurodegeneration, often with accompanying protienopathy. For example, the family of H5N1 avian influenza A viruses responsible for a previous epidemic in Asia have been shown to produce Parkinsonian-like neurodegeneration in mice. H5N1 induced neurodegeneration was accompanied by α-synuclein phosphorylation and neuroinflammation. Interestingly, microgliosis persisted long after the infection resolved and was observable at a time point 90 days from the initial infection [23].

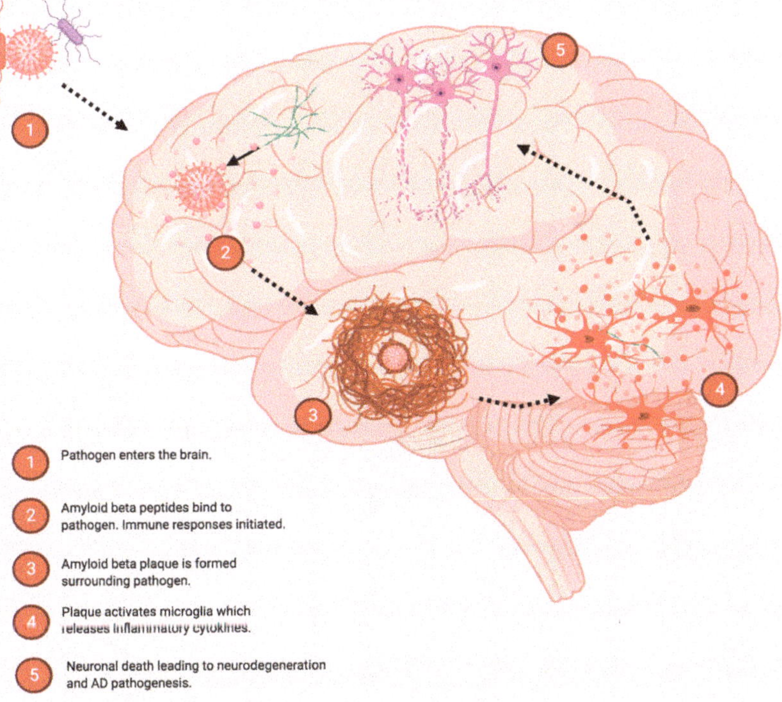

Figure 1. Involvement of Pathogens in Alzheimer's disease (AD). Pathogens such as viruses and bacteria can become entrapped by amyloid-β after entering the brain. Amyloid fibrils form in response to certain pathogens, and infection may play a role in accelerating AD pathology by stimulating inflammation and neurodegeneration.

An intriguing idea which has emerged regarding viral infection and neurodegeneration is the "multi-hit" or "hit and run" hypothesis. This notion is supported by a study from Sadasivan and colleagues, who showed that infection with influenza H1N1 virus 30 days prior to MPTP administration markedly enhanced neurodegeneration in the substantia nigra pars compacta (SNpc) of mice. Furthermore, this enhancement of MPTP induced neurodegeneration could be alleviated by prior vaccination [24]. Thus, it is possible that a viral infection at one time point may later synergize with other factors (i.e., environmental toxins, lifestyle choices, genetic background), to cause or accelerate neurodegeneration. The above-mentioned studies regarding influenza viruses and Parkinsonian like neurodegeneration provide additional insight as to the potential role of viral pathogens in stimulating neurodegeneration independent of β-amyloid. Persistent inflammation stimulated by viral infection may be a critical component in predisposing individuals to AD. Neuroinflammation and neuro-immune interactions have gained attention in recent years as potential driving factors of neurodegeneration [25]. Thus, the pathological effects of increased neuroinflammation paired with its ability to inhibit amyloid clearance may create a bi-directional relationship, whereby viruses (or other pathogens) are able to both increase amyloid activity through direct interactions, while preventing amyloid clearance through stimulation of inflammation.

2.2. Bacterial Pathogens and AD

In addition to the above-mentioned viruses, bacterial pathogens have also been associated with AD. In particular, there is mounting evidence linking periodontal disease and AD. Periodontitis, commonly known as gum disease, is an oral infection resulting in the release of proinflammatory cytokines into the bloodstream and the increase of C-reactive protein. It is caused by the gram-negative anaerobic bacterium Porphyromonas gingivalis [26]. P. gingivalis and its associated toxins, referred to collectively as gingipains, have been identified in 96% of postmortem brain tissue samples of AD patients and are thought to exacerbate AD pathology [27]. Gingipains play a key role in P. gingivalis mediated aggravation of AD. Gingipains are a group of cysteine proteases secreted by P. gingivalis that cause neuronal damage, increased tau production, and increased production of neuro-toxic APOE fragments. Additionally, P. gingivalis induces neuroinflammation, inflammasome activation, and other immune system multiprotein complexes in the brain that result in neurodegeneration and Aβ plaque formation [27,28]. In animal models, P. gingivalis has been demonstrated to travel to the brain following oral inoculation. Interestingly, in mouse models, Aβ1-42 was found to act as an antimicrobial peptide against P. gingivalis. Aβ1-42 inhibited P. gingivalis by disrupting its cell membrane. Notably, P. gingivalis can be detected in the CSF of AD patients and can thus potentially be used as a biomarker for AD [27]. Other bacteria involved in periodontitis include Aggregatibacter actinomycetemcomitans, Prevotella intermedia, Fusobacterium nucleatum, Tannerella forsythensis, Eikenella corrodens, and Treponema denticola. These various bacteria induce inflammation and thus promote neurodegeneration, though there is some evidence alluding to their presence in the brain [26]. The spirochete Borrelia burgdorferi (B. burgdorferi) is the causative agent of Lyme disease, and has also been linked with AD. B. burgdorferi has been detected in the brains of AD patients and is known for its neurodegenerative effects [29,30]. Additionally, Chlamydia pneumoniae (C. pneumoniae) has been detected in the brains of AD patients and may be another factor driving AD pathology [31]. Infection with Helicobacter pylori (H. pylori) has also been associated with increased risk of AD. Infection with H.pylori has been associated with lower cognitive abilities, as well as increased levels of CSF tau and phosphorylated tau among AD patients [32]. Gut microbiome dysbiosis and altered microbiome composition have been implicated in AD and various other neurodegenerative disorders [33]. The gut microbiome produces lipopolysaccharides, neurotoxins, and microbial amyloid. These bacterial products are involved with amyloid plaque formation, neurofibrillary tangles, and neuroinflammation. The gut microbiome composition is also altered in individuals with AD, with the increased relative abundance of bacteria of the genera Verrucomicrobia and Proteobacteria and decreased abundance of Ruminococcus and Butyricicoccus genera [33]. Interestingly, gut microbiota have also been found in the brains of AD patients [34]. The penetrance of gut bacteria into the brain could represent a potential trigger of amyloidosis, that could occur independently of traditional pathogen infection (Figure 2). This could potentially represent a potential mechanism similar to the association between gingivitis and AD, whereby dysregulation of the microbiome could render the host susceptible to amyloidosis and neurodegeneration.

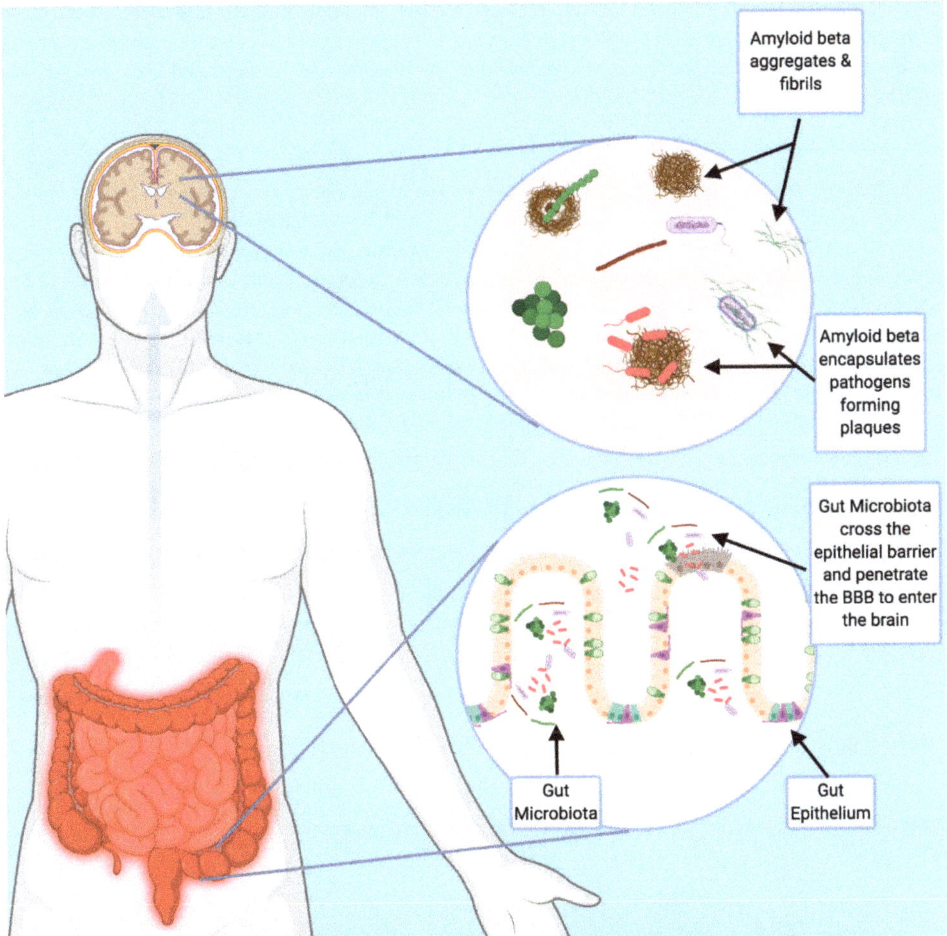

Figure 2. Potential Role of Gut Bacteria in Neurodegeneration. Age-related changes in intestinal permeability and blood brain barrier integrity may allow for penetrance of gut bacteria into the brain and promote the formation of amyloid fibrils. Future studies should focus on the role of gut bacteria as a potential trigger to the amyloid cascade.

2.3. *Other Pathogens and AD*

Fungal infection may also play a role in the pathology of AD. Fungal proteins and DNA have been detected in the brains of AD patients [35]. Additional studies have revealed that fungal proteins and DNA can also be detected peripherally and may make suitable biomarkers (see below for further discussion). The protozoan parasite T. gondii, which is thought to infect up to 50% of the world's population, is known to cause encephalitis and neurological dysfunction. It is thought that T. gondii may be involved with neuroinflammation and olfactory dysfunction in AD pathology [36].

3. Pathogen-Based Biomarkers

Based on the above-mentioned studies linking various pathogens with AD and age-related cognitive decline, it may be rational to use biomarkers based on pathogen exposure to assess the risk for developing AD in elderly individuals. In cases of active infection, intervention with antimicrobial treatments may be a suitable method of reducing AD risk, particularly in patients with advanced age. We discuss below a number of biomarkers based on pathogens which have been linked to AD and cognitive decline (Table 1).

3.1. Antimicrobial Peptides as Biomarkers in AD

Defensins are a family of disulfide knotted antimicrobial peptides that entrap pathogens as a part of the innate immune system [37]. Moreover, α-Defensins 1 and 2 were shown to be elevated in the blood of AD patients and may make a suitable biomarker for detecting AD status [38]. Another anti-microbial protein that might make a suitable biomarker for AD is lactoferrin. Lactoferrin is an antimicrobial peptide that is present in saliva and correlates with AD status. Saliva samples from amnesiac mild cognitive impairment (aMCI) and AD patients showed decreased levels of lactoferrin when compared with controls, and a significant negative correlation was found between lactoferrin and aMCI and AD patients. These results suggest the potential usage of lactoferrin as a non-invasive salivary biomarker for AD [39]. Tears contain a number of antimicrobial proteins that act as part of the innate immune system. Several proteins present in tears have been shown to be differentially expressed in AD patients and may serve as suitable biomarkers. In particular, changes in the expression of the antimicrobial proteins lipocalin-1, dermcidin, lysozyme-C and lacritin have been reported using tear samples of AD patients [40].

Table 1. Pathogen-Based Biomarkers in AD. Numerous pathogens have been associated with AD pathogenesis and the onset of cognitive decline. Biomarkers are listed along with their source and relationship to AD.

	Biomarker	Source	Description	Reference
Antimicrobial Peptides				
	α-Defensin 1	Blood	Increased in blood of AD patients.	[39]
	α-Defensin 2	Blood	Increased in blood of AD patients.	[39]
	Lactoferrin	Saliva	Decreased with AD and aMCI.	[40]
	Lipocalin-1	Tears	Decreased in AD.	[41]
	Dermcidin	Tears	Increased in AD.	[41]
	Lysozyme-C	Tears	Decreased in AD.	[41]
	Lacritin	Tears	Decreased in AD.	[41]
Antibodies				
	IgG against Epstein-Barr Virus	Blood	Correlates with development of aMCI.	[42]
	IgG and IgA against *C. pneumoniae*	Blood	Detectable in patients with vascular dementia.	[43]
	IgG against HSV-2	Blood	Correlates with cognitive decline.	[25]

Table 1. *Cont.*

	Biomarker	Source	Description	Reference
	IgG against CMV	Blood	Correlates with cognitive decline.	[25]
	IgG against *T. gondii*	Blood	Correlates with cognitive decline.	[25]
	IgM against HSV-1	Blood	Associated with increased risk of AD.	[44,45]
	IgG against *H. Pylori*	Blood	Associated with lower MMSE scores.	[35]
Other				
	Fungal Proteins and DNA	CSF, Blood	Detectable in AD patients.	[36,46]
	Gut Microbiome composition	fecal matter	Correlates to gut dysbiosis and cognitive decline.	[6,47]
	Porphyromonas gingivalis	CSF	Identified in 96% of postmortem brain tissue samples of AD patients.	[30]

3.2. Antibodies as Biomarkers for AD

Antibodies against pathogens associated with AD are readily detectable in blood, and may be a reasonable way of establishing AD risk in the elderly. A number of studies have shown correlations between antibodies against various pathogens and cognitive decline. For example, elevated levels of IgG against Epstein–Barr Virus (EBV) have been shown to correlate with the development of aMCI [41]. The presence of C. pneumoniae in AD patients has been well documented (see discussion above), and IgG and IgA antibodies against C. pneumoniae have been detected in patients with vascular dementia [42]. Additionally, IgG antibodies against HSV-2, CMV, and TOX have been shown to correlate with cognitive decline in individuals over the age of 65 and may also serve as rational biomarkers. While the study by [19] did not find any significant correlation between HSV-1 antibodies and cognitive decline, other groups have reported on the possibility of HSV-1 antibodies as potential biomarkers for AD [43,44]. In a study by Roubaud-Baudron et al., the presence of IgG antibodies against H.Pylori was associated with lower scores on the mini-mental state examination (MMSE) and increased CSF tau levels, among AD patients [32]. AD has also been linked with increased T. gondii IgG antibodies. These antibodies can serve as potential AD biomarkers, given the high prevalence of T. gondii infection globally [36].

3.3. Other Potential Biomarkers

Additional pathogens such as fungi may also serve as potential biomarkers in AD. Fungal proteins and DNA have been detected in the CSF for AD patients [45]. Furthermore, fungal polysaccharides, proteins, and DNA have all been detected in blood samples drawn from AD patients [35]. The relationship between the gut microbiome and various neurological diseases has been an area of growing interest in recent years, and may be another option to consider for monitoring the progression of AD. Variation in gut microbiome composition detected from stool samples can also be a preclinical biomarker for AD, given the increased relative abundance of bacteria of the genera Verrucomicrobia and Proteobacteria found in AD. Gut microbiome products such as microbial amyloids and neurotoxin BMAA play a role in neurodegeneration, and can also potentially serve as AD biomarkers [33,46].

4. Conclusions

While the presence of pathogens or antibodies against certain pathogens may not directly indicate a positive diagnosis of AD per se, it is important to note that these biomarkers may be appropriate for determining at-risk cohorts of elderly individuals. The potential to identify at risk individuals and administer prophylactic treatments is of great value to the field of AD research. In particular, the administration of antimicrobial treatments (antiviral, antifungal, antibacterial, anti-parasitic) after positive confirmation of an infection carries little risk, and could be neuroprotective. A recent analysis of data from Taiwan's National Health Insurance Research Database found that treatment with antiherpetic medications was associated with a decreased risk of developing dementia [47]. Furthermore, a clinical trial is currently underway to evaluate the antiviral therapy valacyclovir in the treatment of AD [48].

As it stands, there is currently preclinical evidence to suggest that β-amyloid can directly bind pathogens, and that pathogenic infection can accelerate amyloid pathology in transgenic animal models of AD. Additionally, multiple preclinical studies have demonstrated the neurodegenerative properties of certain pathogens. In terms of clinical data, there are a number of studies providing correlational evidence between the presence of various pathogens and the diagnosis of AD. There is thus a critical missing link between amyloid entrapment of pathogenic microbes, and widespread neurodegeneration. Some insight as to the potential mechanism by which infection may stimulate neurodegeneration can be taken from a recent study examining interferon signaling in response to β-amyloid. Roy and colleagues showed that soluble oligomers interact with nucleic acids (DNA and RNA) or glycosaminoglycans (i.e., heparin), and that these interactions promote the formation of amyloid fibrils. Interestingly, only amyloid fibrils containing nucleic acids promote type 1 interferon response, inflammation, and synaptic loss. Type 1 interferon response was observed across several different transgenic mouse lines, indicating that self-DNA or self-RNA may trigger this response. Additionally, wild type mice that received a hippocampal injection of amyloid fibrils containing RNA showed an inflammatory profile similar to that observed in transgenic models of AD [49]. These findings are very exciting when viewed in the context of other studies, which have shown that amyloid fibrils can form after binding to viral particles such as herpes simplex virus (a double stranded DNA virus). Type 1 interferon response usually occurs as a part of the innate immune response to viral infection; thus, the finding that nucleic acid containing amyloid fibrils stimulates type 1 interferon suggests that Aβ may be an integral component of antiviral defense in the brain. Blocking type 1 interferon response and other immune-related signaling pathways which occur after Aβ entrapment of pathogens may thus be a rational therapeutic strategy in treating AD.

A critical question also becomes whether AD is directly stimulated by one or more pathogens entering the brain, or if AD can occur as a result of dysfunction of Aβ driven anti-viral defenses. Thus, future studies investigating both the function and dysfunction of innate immune responses in the brain will be critical to our understanding and diagnosis of AD. In particular, it is necessary to understand how other factors, such as diet, genetic background, and exposure to environmental toxins may confer susceptibility to neuroinvasion by pathogenic microbes. It is also possible that the reason why so many different pathogens are readily detected in the brains and blood of AD patients is that AD may fundamentally weaken the immune system. Decreased activity or dysfunction of the peripheral immune system may force the innate immune system of the brain to bear a heavy burden when faced with pathogenic infections during AD. Thus, increased reliance on the antimicrobial activity of Aβ during AD may force an already burdened system to a "breaking point", in which severe neurodeneration is facilitated by excessive amyloidosis, microglial activation, and immune dysfunction in the brain.

There is still a great deal of research that needs to be done to establish a direct causal link between AD and pathogenic infection. In particular, several key areas need to be addressed. One critical consideration is that AD may merely increase susceptibility of infection, thus allowing various pathogens to enter brain and making their detection either a secondary occurrence or an artifact. To address this possibility, it will be critical to examine the brains of patients with familial AD to determine if pathogens can be detected that are not present in age-matched controls. This would help one to better understand if AD fundamentally facilitates pathogenic infections in the brain, which would provide insight as to whether infection is a primary or secondary occurrence in sporadic AD. If no pathogens are present in the brains of patients with familial AD, it might suggest that AD is primarily caused by the brain's innate immune system behaving in a dysfunctional manner (i.e., amyloid entrapment of host DNA/RNA as opposed to pathogen DNA/RNA). Another critical area which must be addressed is the high detection rate of different pathogens in various cohorts of AD patients. For example, differing studies have found P. gingivalis, LV, or human herpesvirus (as well as other pathogens) in all or nearly all of the brains from AD patients observed in the respective cohorts of each study [17,20,27]. This would imply that, if we were to extrapolate the findings of each individual cohort to the broader AD population, then all AD patients would be expected to present with multiple pathogenic infections simultaneously. Thus, it is crucial to determine if multiple pathogens can indeed be detected within the same brains of AD patients. If multiple pathogens cannot be detected within a single AD brain, it might imply that geographical differences in exposure to various pathogens might cause a specific pathogen to be overrepresented in one particular cohort. This would also suggest the possibility that multiples pathogens might be independently capable of stimulating the same pathogenic processes within AD. While the underlying mechanisms linking specific infections and AD are not explicitly known, the consistent association of various pathogens in AD cannot be ignored. It is thus critical to evaluate the presence of pathogen related biomarkers in elderly individuals, to aid in the construction of an AD risk profile.

Author Contributions: S.X.N., U.R., and G.M.P. all contributed to the writing—review and editing of this manuscript. All authors have read and agreed to the published version of the manuscript.

Funding: This study was supported by Grant Number P50 AT008661-01 from the NCCIH and the ODS. Pasinetti holds a Senior VA Career Scientist Award. We acknowledge that the contents of this study do not represent the views of the NCCIH, the ODS, the NIH, the U.S. Department of Veterans Affairs, or the United States Government.

Acknowledgments: Figures 1 and 2 were made using Biorender.com.

Conflicts of Interest: The authors declare no conflict of interest. Funding agencies had no role in the writing of this manuscript, or in the decision to publish.

References

1. Nguyen, T.T.; Ta, Q.T.H.; Nguyen, T.K.O.; Nguyen, T.T.D.; Vo, V.G. Role of Body-Fluid Biomarkers in Alzheimer's Disease Diagnosis. *Diagnostics* **2020**, *10*, 326. [CrossRef]
2. Prince, M.J.; Wu, F.; Guo, Y.; Robledo, L.M.G.; O'Donnell, M.; Sullivan, R.; Yusuf, S. The burden of disease in older people and implications for health policy and practice. *Lancet* **2015**, *385*, 549–562. [CrossRef]
3. Perl, D.P. Neuropathology of Alzheimer's Disease. *Mt. Sinai J. Med. J. Transl. Pers. Med.* **2010**, *77*, 32–42. [CrossRef]
4. Giau, V.V.; Bagyinszky, E.; An, S.S.A.; Kim, S. Clinical genetic strategies for early onset neurodegenerative diseases. *Mol. Cell. Toxicol.* **2018**, *14*, 123–142. [CrossRef]
5. Zetterberg, H.; Bendlin, B.B. Biomarkers for Alzheimer's disease-preparing for a new era of disease-modifying therapies. *Mol. Psychiatry* **2020**. [CrossRef]
6. Chen, G.; Xu, T.; Yan, Y.; Zhou, Y.; Jiang, Y.; Melcher, K.; Xu, H.E. Amyloid beta: Structure, biology and structure-based therapeutic development. *Acta Pharmacol. Sin.* **2017**, *38*, 1205–1235. [CrossRef] [PubMed]
7. Nunan, J.; Small, D.H. Regulation of APP cleavage by α-, β- and γ-secretases. *FEBS Lett.* **2000**, *483*, 6–10. [CrossRef]

8. Viswanathan, A.; Greenberg, S.M. Cerebral amyloid angiopathy in the elderly. *Ann. Neurol.* **2011**, *70*, 871–880. [CrossRef] [PubMed]
9. Mattson, M.P. Pathways towards and away from Alzheimer's disease. *Nature* **2004**, *430*, 631–639. [CrossRef] [PubMed]
10. Heneka, M.T.; Golenbock, D.T.; Latz, E. Innate immunity in Alzheimer's disease. *Nat. Immunol.* **2015**, *16*, 229–236. [CrossRef] [PubMed]
11. Soscia, S.J.; Kirby, J.E.; Washicosky, K.J.; Tucker, S.M.; Ingelsson, M.; Hyman, B.; Burton, M.A.; Goldstein, L.E.; Duong, S.; Tanzi, R.E.; et al. The Alzheimer's Disease-Associated Amyloid β-Protein Is an Antimicrobial Peptide. *PLoS ONE* **2010**, *5*, e9505. [CrossRef] [PubMed]
12. Gosztyla, M.L.; Brothers, H.M.; Robinson, S.R. Alzheimer's Amyloid-β is an Antimicrobial Peptide: A Review of the Evidence. *J. Alzheimers Dis.* **2018**, *62*, 1495–1506. [CrossRef] [PubMed]
13. Bourgade, K.; Dupuis, G.; Frost, E.H.; Fülöp, T., Jr. Anti-Viral Properties of Amyloid-β Peptides. *J. Alzheimers Dis.* **2016**, *54*, 859–878. [CrossRef] [PubMed]
14. Lukiw, W.J.; Cui, J.G.; Yuan, L.Y.; Bhattacharjee, P.S.; Corkern, M.; Clement, C.; Kammerman, E.M.; Ball, M.J.; Zhao, Y.; Sullivan, P.M.; et al. Acyclovir or Aβ42 peptides attenuate HSV-1-induced miRNA-146a levels in human primary brain cells. *NeuroReport* **2010**, *21*, 922–927. [CrossRef]
15. Itzhaki, R.F.; Golde, T.E.; Heneka, M.T.; Readhead, B. Do infections have a role in the pathogenesis of Alzheimer disease? *Nat. Rev. Neurol.* **2020**, *16*, 193–197. [CrossRef]
16. Eimer, W.A.; Vijaya Kumar, D.K.; Navalpur Shanmugam, N.K.; Rodriguez, A.S.; Mitchell, T.; Washicosky, K.J.; György, B.; Breakefield, X.O.; Tanzi, R.E.; Moir, R.D. Alzheimer's Disease-Associated β-Amyloid Is Rapidly Seeded by Herpesviridae to Protect against Brain Infection. *Neuron* **2018**, *99*, 56–63.e3. [CrossRef]
17. Readhead, B.; Haure-Mirande, J.-V.; Funk, C.C.; Richards, M.A.; Shannon, P.; Haroutunian, V.; Sano, M.; Liang, W.S.; Beckmann, N.D.; Price, N.D.; et al. Multiscale Analysis of Independent Alzheimer's Cohorts Finds Disruption of Molecular, Genetic, and Clinical Networks by Human Herpesvirus. *Neuron* **2018**, *99*, 64–82.e7. [CrossRef]
18. Cairns, D.M.; Rouleau, N.; Parker, R.N.; Walsh, K.G.; Gehrke, L.; Kaplan, D.L. A 3D human brain–like tissue model of herpes-induced Alzheimer's disease. *Sci. Adv.* **2020**, *6*, eaay8828. [CrossRef]
19. Nimgaonkar, V.L.; Yolken, R.H.; Wang, T.; Chang, C.-C.H.; McClain, L.; McDade, E.; Snitz, B.E.; Ganguli, M. Temporal Cognitive Decline Associated With Exposure to Infectious Agents in a Population-based, Aging Cohort. *Alzheimer Dis. Assoc. Disord.* **2016**, *30*, 216–222. [CrossRef]
20. Niklasson, B.; Lindquist, L.; Klitz, W.; Bank, N.B.; Englund, E. Picornavirus Identified in Alzheimer's Disease Brains: A Pathogenic Path? *J. Alzheimers Dis. Rep.* **2020**, *4*, 141–146. [CrossRef]
21. Canet, G.; Dias, C.; Gabelle, A.; Simonin, Y.; Gosselet, F.; Marchi, N.; Makinson, A.; Tuaillon, E.; Van de Perre, P.; Givalois, L.; et al. HIV Neuroinfection and Alzheimer's Disease: Similarities and Potential Links? *Front. Cell. Neurosci.* **2018**, *12*. [CrossRef] [PubMed]
22. Clifford, D.B.; Fagan, A.M.; Holtzman, D.M.; Morris, J.C.; Teshome, M.; Shah, A.R.; Kauwe, J.S.K. CSF biomarkers of Alzheimer disease in HIV-associated neurologic disease. *Neurology* **2009**, *73*, 1982–1987. [CrossRef] [PubMed]
23. Jang, H.; Boltz, D.; Sturm-Ramirez, K.; Shepherd, K.R.; Jiao, Y.; Webster, R.; Smeyne, R.J. Highly pathogenic H5N1 influenza virus can enter the central nervous system and induce neuroinflammation and neurodegeneration. *Proc. Natl. Acad. Sci. USA* **2009**, *106*, 14063–14068. [CrossRef] [PubMed]
24. Sadasivan, S.; Sharp, B.; Schultz-Cherry, S.; Smeyne, R.J. Synergistic effects of influenza and 1-methyl-4-phenyl-1,2,3,6-tetrahydropyridine (MPTP) can be eliminated by the use of influenza therapeutics: Experimental evidence for the multi-hit hypothesis. *NPJ Park. Dis.* **2017**, *3*, 1–3. [CrossRef]
25. Kinney, J.W.; Bemiller, S.M.; Murtishaw, A.S.; Leisgang, A.M.; Salazar, A.M.; Lamb, B.T. Inflammation as a central mechanism in Alzheimer's disease. *Alzheimers Dement. Transl. Res. Clin. Interv.* **2018**, *4*, 575–590. [CrossRef]
26. Abbayya, K.; Puthanakar, N.Y.; Naduwinmani, S.; Chidambar, Y.S. Association between periodontitis and Alzheimer's disease. *North Am. J. Med. Sci.* **2015**, *7*, 241. [CrossRef]
27. Dominy, S.S.; Lynch, C.; Ermini, F.; Benedyk, M.; Marczyk, A.; Konradi, A.; Nguyen, M.; Haditsch, U.; Raha, D.; Griffin, C.; et al. Porphyromonas gingivalis in Alzheimer's disease brains: Evidence for disease causation and treatment with small-molecule inhibitors. *Sci. Adv.* **2019**, *5*, eaau3333. [CrossRef]

28. Venegas, C.; Kumar, S.; Franklin, B.S.; Dierkes, T.; Brinkschulte, R.; Tejera, D.; Vieira-Saecker, A.; Schwartz, S.; Santarelli, F.; Kummer, M.P.; et al. Microglia-derived ASC specks cross-seed amyloid-β in Alzheimer's disease. *Nature* **2017**, *552*, 355–361. [CrossRef]
29. MacDonald, A.B.; Miranda, J.M. Concurrent neocortical borreliosis and Alzheimer's disease. *Hum. Pathol.* **1987**, *18*, 759–761. [CrossRef]
30. MacDonald, A.B. Alzheimer's neuroborreliosis with trans-synaptic spread of infection and neurofibrillary tangles derived from intraneuronal spirochetes. *Med. Hypotheses* **2007**, *68*, 822–825. [CrossRef]
31. Hammond, C.J.; Hallock, L.R.; Howanski, R.J.; Appelt, D.M.; Little, C.S.; Balin, B.J. Immunohistological detection of Chlamydia pneumoniae in the Alzheimer's disease brain. *BMC Neurosci.* **2010**, *11*, 121. [CrossRef] [PubMed]
32. Roubaud-Baudron, C.; Krolak-Salmon, P.; Quadrio, I.; Mégraud, F.; Salles, N. Impact of chronic Helicobacter pylori infection on Alzheimer's disease: Preliminary results. *Neurobiol. Aging* **2012**, *33*, 1009-e11. [CrossRef]
33. Raval, U.; Harary, J.; Zeng, E.; Pasinetti, G.M. The dichotomous role of the gut microbiome in exacerbating and ameliorating neurodegenerative disorders. *Expert Rev. Neurother.* **2020**. [CrossRef] [PubMed]
34. Westfall, S.; Dinh, D.M.; Pasinetti, G.M. Investigation of Potential Brain Microbiome in Alzheimer's Disease: Implications of Study Bias. *J. Alzheimers Dis. JAD* **2020**, *75*, 559–570. [CrossRef] [PubMed]
35. Alonso, R.; Pisa, D.; Marina, A.I.; Morato, E.; Rábano, A.; Carrasco, L. Fungal Infection in Patients with Alzheimer's Disease. *J. Alzheimers Dis.* **2014**, *41*, 301–311. [CrossRef] [PubMed]
36. Prandota, J. Possible Link Between Toxoplasma Gondii and the Anosmia Associated With Neurodegenerative Diseases. *Am. J. Alzheimers Dis. Dementiasr* **2014**, *29*, 205–214. [CrossRef]
37. Zhao, X.; Wu, H.; Lu, H.; Li, G.; Huang, Q. LAMP: A Database Linking Antimicrobial Peptides. *PLoS ONE* **2013**, *8*, e66557. [CrossRef]
38. Watt, A.D.; Perez, K.A.; Ang, C.-S.; O'Donnell, P.; Rembach, A.; Pertile, K.K.; Rumble, R.L.; Trounson, B.O.; Fowler, C.J.; Faux, N.G.; et al. Peripheral α-Defensins 1 and 2 are Elevated in Alzheimer's Disease. *J. Alzheimers Dis.* **2015**, *44*, 1131–1143. [CrossRef]
39. Carro, E.; Bartolomé, F.; Bermejo-Pareja, F.; Villarejo-Galende, A.; Molina, J.A.; Ortiz, P.; Calero, M.; Rabano, A.; Cantero, J.L.; Orive, G. Early diagnosis of mild cognitive impairment and Alzheimer's disease based on salivary lactoferrin. *Alzheimers Dement. Amst. Neth.* **2017**, *8*, 131–138. [CrossRef]
40. Kalló, G.; Emri, M.; Varga, Z.; Ujhelyi, B.; Tőzsér, J.; Csutak, A.; Csősz, É. Changes in the Chemical Barrier Composition of Tears in Alzheimer's Disease Reveal Potential Tear Diagnostic Biomarkers. *PLoS ONE* **2016**, *11*, e0158000. [CrossRef]
41. Shim, S.-M.; Cheon, H.-S.; Jo, C.; Koh, Y.H.; Song, J.; Jeon, J.-P. Elevated Epstein-Barr Virus Antibody Level is Associated with Cognitive Decline in the Korean Elderly. *J. Alzheimers Dis.* **2017**, *55*, 293–301. [CrossRef] [PubMed]
42. Yamamoto, H.; Watanabe, T.; Miyazaki, A.; Katagiri, T.; Idei, T.; Iguchi, T.; Mimura, M.; Kamijima, K. High Prevalence of Chlamydia Pneumoniae Antibodies and Increased High-Sensitive C-Reactive Protein in Patients with Vascular Dementia. *J. Am. Geriatr. Soc.* **2005**, *53*, 583–589. [CrossRef] [PubMed]
43. Lövheim, H.; Gilthorpe, J.; Johansson, A.; Eriksson, S.; Hallmans, G.; Elgh, F. Herpes simplex infection and the risk of Alzheimer's disease: A nested case-control study. *Alzheimers Dement.* **2015**, *11*, 587–592. [CrossRef] [PubMed]
44. Lövheim, H.; Gilthorpe, J.; Adolfsson, R.; Nilsson, L.-G.; Elgh, F. Reactivated herpes simplex infection increases the risk of Alzheimer's disease. *Alzheimers Dement.* **2015**, *11*, 593–599. [CrossRef] [PubMed]
45. Alonso, R.; Pisa, D.; Rábano, A.; Rodal, I.; Carrasco, L. Cerebrospinal Fluid from Alzheimer's Disease Patients Contains Fungal Proteins and DNA. *J. Alzheimers Dis.* **2015**, *47*, 873–876. [CrossRef]
46. Roy Sarkar, S.; Banerjee, S. Gut microbiota in neurodegenerative disorders. *J. Neuroimmunol.* **2019**, *328*, 98–104. [CrossRef]
47. Tzeng, N.-S.; Chung, C.-H.; Lin, F.-H.; Chiang, C.-P.; Yeh, C.-B.; Huang, S.-Y.; Lu, R.-B.; Chang, H.-A.; Kao, Y.-C.; Yeh, H.-W.; et al. Anti-herpetic Medications and Reduced Risk of Dementia in Patients with Herpes Simplex Virus Infections—A Nationwide, Population-Based Cohort Study in Taiwan. *Neurotherapeutics* **2018**, *15*, 417–429. [CrossRef]

48. Devanand, D.P.; Andrews, H.; Kreisl, W.C.; Razlighi, Q.; Gershon, A.; Stern, Y.; Mintz, A.; Wisniewski, T.; Acosta, E.; Pollina, J.; et al. Antiviral therapy: Valacyclovir Treatment of Alzheimer's Disease (VALAD) Trial: Protocol for a randomised, double-blind, placebo-controlled, treatment trial. *BMJ Open* **2020**, *10*, e032112. [CrossRef]
49. Roy, E.R.; Wang, B.; Wan, Y.; Chiu, G.; Cole, A.; Yin, Z.; Propson, N.E.; Xu, Y.; Jankowsky, J.L.; Liu, Z.; et al. Type I interferon response drives neuroinflammation and synapse loss in Alzheimer disease. *J. Clin. Investig.* **2020**, *130*, 1912–1930. [CrossRef]

MDPI
St. Alban-Anlage 66
4052 Basel
Switzerland
Tel. +41 61 683 77 34
Fax +41 61 302 89 18
www.mdpi.com

Journal of Personalized Medicine Editorial Office
E-mail: jpm@mdpi.com
www.mdpi.com/journal/jpm

www.ingramcontent.com/pod-product-compliance
Lightning Source LLC
LaVergne TN
LVHW071619170726
843515LV00009B/2432

* 9 7 8 3 0 3 9 4 3 9 0 3 4 *